McGraw-Hill Ryerson

Chemistry 12

Author Team

Dr. Frank Mustoe
The University of Toronto Schools
Toronto, Ontario

Michael P. Jansen
Crescent School
Toronto, Ontario

Dr. Michael Webb
Michael J. Webb Consulting Inc.
Toronto, Ontario

Christy Hayhoe
Professional Writer
Toronto, Ontario

Dr. Andrew Cherkas
Stouffville District Secondary School
Stouffville, Ontario

Jim Gaylor
Formerly with St. Michael
Catholic Secondary School
Stratford, Ontario

Contributing Author

Jonathan Bocknek
Professional Writer
Slocan Park, British Columbia

Christa Bedry
Professional Writer
Cochrane, Alberta

Consultants

Greg Wisnicki
Anderson Collegiate and Vocational Institute
Whitby, Ontario

Dr. Audrey Chastko
Springbank Community High School
Calgary, Alberta

Ted Doram
Bowness High School
Calgary, Alberta

Probeware Specialist

Kelly Choy
Minnedosa Collegiate
Minnedosa, Manitoba

Technology Consultants

Alex Annab
Head of Science
Iona Catholic Secondary School
Mississauga, Ontario

Toronto Montréal Boston Burr Ridge, IL Dubuque, IA Madison, WI New York San Francisco St. Louis Bangkok Bogotá Caracas Kuala Lumpur Lisbon London Madrid Mexico City Milan New Delhi Santiago Seoul Singapore Sydney Taipei

McGraw-Hill Ryerson Limited

A Subsidiary of The ***McGraw-Hill*** *Companies*

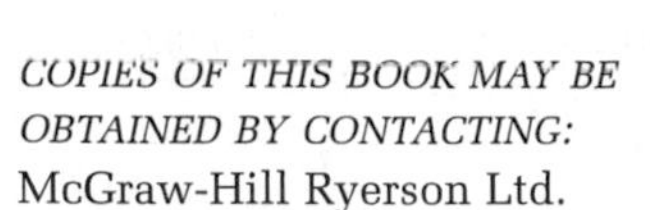

COPIES OF THIS BOOK MAY BE OBTAINED BY CONTACTING:
McGraw-Hill Ryerson Ltd.

E-MAIL:
orders@mcgrawhill.ca

TOLL FREE FAX:
1-800-463-5885

TOLL FREE CALL:
1-800-565-5758

OR BY MAILING YOUR ORDER TO:
McGraw-Hill Ryerson
Order Department,
300 Water Street,
Whitby ON, L1N 9B6

Please quote the ISBN and title when placing your order.

McGraw-Hill Ryerson Chemistry 12

0-07-088686-5

http://www.mcgrawhill.ca

3 4 5 6 7 8 9 0 TRI 0 9 8 7 6 5 4 3 2 1

Printed and bound in Canada

National Library of Canada Cataloguing in Publication Data

Main entry under title:

McGraw-Hill Ryerson chemistry 12

Includes index.

ISBN 0-07-088686-5

1. Chemistry. I. Webb, Michael II. Title. III. Title: Chemistry 12. IV. Title: Chemistry twelve. V. Title: McGraw-Hill Ryerson chemistry twelve.

QD33. M334 2002 540 C2002-900241-9

The Chemistry 12 Team

SCIENCE PUBLISHER: Jane McNulty

PROJECT MANAGER: David Spiegel

SENIOR DEVELOPMENTAL EDITOR: Jonathan Bocknek

DEVELOPMENTAL EDITORS: Christa Bedry, Sara Goodchild, Christy Hayhoe, Winnie Siu

SENIOR SUPERVISING EDITOR: Linda Allison

PROJECT CO-ORDINATOR: Valerie Janicki

PROJECT ASSISTANTS: Melissa Nippard, Janie Reeson

COPY EDITORS: Paula Pettitt-Townsend, May Look

PERMISSIONS EDITOR: Pronk&Associates

SPECIAL FEATURES CO-ORDINATOR: Keith Owen Richards

PRODUCTION CO-ORDINATOR: Jennifer Wilkie

PRODUCTION SUPERVISOR: Yolanda Pigden

DESIGN AND ELECTRONIC PAGE MAKE-UP: Pronk&Associates

SET-UP PHOTOGRAPHY: Ian Crysler

SET-UP PHOTOGRAPHY CO-ORDINATOR: Shannon O'Rourke

TECHNICAL ILLUSTRATION: Theresa Sakno, Jun Park, Pronk&Associates

COVER IMAGE: Ken Edwards/Science Source/Photo Researchers Inc.

Acknowledgements

We extend sincere thanks to the following people: Greg Wisnicki, for his extremely helpful suggestions in his role as consultant and for writing and testing labs during the development of Unit 5; Audrey Chastko, whose astute and detailed comments contributed greatly to the development of this textbook; and Ted Doram, for his excellent insights as a consultant and for testing labs during the development of Unit 1. We are deeply grateful as well to Dr. Frank Mustoe for his help with the set-up photography sessions, and we thank the students of the University of Toronto Schools who participated in these sessions. We also wish to thank the following professional writers who authored the Special Features in *Chemistry 12*: Jess Aldred, Jeremy Boxen, Kirsten Craven, Natasha Marko, Denyse O'Leary, Patrick Rengger, and Erik Spigel. We thank the designers at Pronk&Associates, who collaborated closely with us to bring this book to life. Finally, the *Chemistry 12* development team benefited greatly from the many thoughtful ideas and recommendations provided by our reviewers from across the country, as well as from the comments supplied by our safety reviewer. The authors, publisher, consultants, and editors convey their profound thanks to these talented and dedicated educators. Finally, we acknowledge, with gratitude and respect, Trudy Rising, who initiated McGraw-Hill Ryerson's senior science program, and who worked tirelessly in support of the program, its authors, and its development team.

Pedagogical and Academic Reviewers

Christina Clancy
Loyola Catholic Secondary School
Mississauga, Ontario

Charles Cohen
Community Hebrew
Academy of Toronto
Toronto, Ontario

André Dumais
Hearst High School
Hearst, Ontario

John Eix
Formerly with Upper Canada College
Toronto, Ontario

Christopher Freure
South Lincoln High School
Smithville, Ontario

Theresa H. George
St. Paul Catholic Secondary School
Mississauga, Ontario

Gail Gislason
Crescent Heights High School
Calgary, Alberta

Dana Griffiths
Bishops College
St. John's, Newfoundland

Sarah Houlden
Twin Lakes Secondary School
Orillia, Ontario

Stephen Houlden
Formerly with Toronto
District School Board
Toronto, Ontario

Doug Jones
Sir Winston Churchill Collegiate and
Vocational Institute
Thunder Bay, Ontario

Dorothy Lai
Anderson Collegiate and
Vocational Institute
Whitby, Ontario

Cheryl Madeira
Marshall McLuhan
Catholic Secondary School
Toronto, Ontario

Glen McLeod
Lakehead Public School Board
Thunder Bay, Ontario

Dr. Penny McLeod
Formerly with York Region
District School Board
Aurora, Ontario

Henry Pasma
Cawthra Park Secondary School
Mississauga, Ontario

Cheryl Perkins
Education Consultant
St. John's, Newfoundland

Chris Schramek
John Paul II Catholic
Secondary School
London, Ontario

Brian Schroder
Our Lady of Mount Carmel
Secondary School
Mississauga, Ontario

James Sniatenchuk
Bluevale Collegiate Institute
Waterloo, Ontario

Donna Stack-Durward
St. Mary's High School
Hamilton, Ontario

Laurie Swackhammer
Head of Science
Ancaster High School
Ancaster, Ontario

Frank Villella
St. Thomas More Catholic
Secondary School
Hamilton, Ontario

Accuracy Reviewers

Dr. Michael C. Baird
Queen's University
Kingston, Ontario

Dr. Christopher Flinn
Memorial University
of Newfoundland
St. John's, Newfoundland

Dr. R.J. Gillespie
Department of Chemistry
McMaster University
Hamilton, Ontario

Ian Krouse
Formerly with University of Calgary
Calgary, Alberta
Purdue University
West Lafayette, Indiana

Safety Reviewer

John Henry
Science Teachers Association
of Ontario Safety Committee
Toronto, Ontario

Our cover: The image on our cover shows a computer-generated model of a C_{60} fullerene molecule, containing a molecule of methanol. Fullerenes are spherical molecules of carbon. Since their discovery in 1985, they have fascinated scientists with their perfect geometry and their range of potential applications, including superconductors, rocket fuels, and lubricants. Scientists can manipulate the properties of a fullerene by inserting an atom or a small molecule into it, as shown here, or by bonding a different chemical group to its outside surface. You will examine the connections between structure, bonding, and properties in Unit 2.

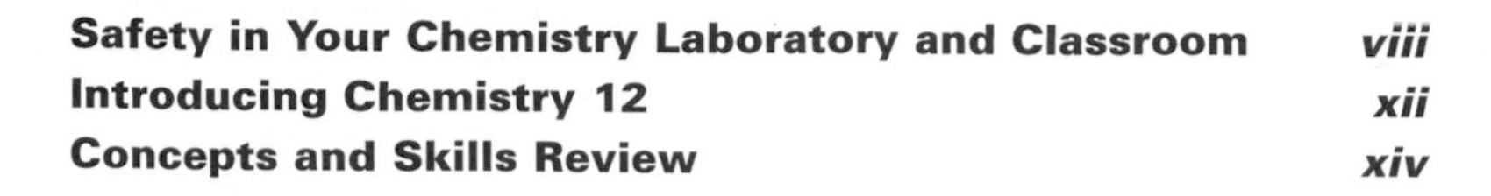

Contents

Safety in Your Chemistry Laboratory and Classroom ***viii***
Introducing Chemistry 12 ***xii***
Concepts and Skills Review ***xiv***

UNIT 1 Organic Chemistry 2

Chapter 1 Classifying Organic Compounds

1.1 Bonding and the Shape of Organic Molecules 5
ExpressLab: Molecular Shapes 6

1.2 Hydrocarbons 12
Careers in Chemistry: The Art and Science of Perfumery 17

1.3 Single-bonded Functional Groups 21
ThoughtLab: Comparing Intermolecular Forces 24

1.4 Functional Groups with the C=O Bond 35
Tools & Techniques: Infrared Spectroscopy 38
Investigation 1-A: Preparing a Carboxylic Acid Derivative 42
Investigation 1-B: Comparing Physical Properties 49
Chapter 1 Review 52

Chapter 2 Reactions of Organic Compounds 56

2.1 The Main Types of Organic Reactions 57

2.2 Reactions of Functional Groups 65
Canadians in Chemistry: Dusanka Filipovic 69
Investigation 2-A: Oxidizing Alcohols 74

2.3 Molecules on a Larger Scale: Polymers and Biomolecules 81
Investigation 2-B: Synthesis of a Polymer 86
Chemistry Bulletin: Degradable Plastics: Garbage That Takes Itself Out 89

2.4 Organic Compounds and Everyday Life 97
ThoughtLab: Risk-Benefit Analyses of Organic Products 100
ThoughtLab: Problem Solving with Organic Compounds 103
Chapter 2 Review 105
Unit 1 Issue: Contemporary Issues Related to Organic Chemistry 110
Unit 1 Review 112

UNIT 2 Structure and Properties 116

Chapter 3 Atoms, Electrons, and Periodic Trends 118

3.1 The Nuclear Atomic Model 119
Investigation 3-A: Atomic Emission Spectra (Teacher Demonstration) 124
Careers in Chemistry: Nuclear Medicine 129

3.2 The Quantum Mechanical Model of the Atom 131

3.3 Electron Configurations and Periodic Trends 139
ThoughtLab: Periodic Connections 151
Chapter 3 Review 159

Chapter 4 Structure and Properties of Substances 162

4.1 Chemical Bonding 163
Investigation 4-A: Properties of Substances 164

4.2 Molecular Shape and Polarity 173
ExpressLab: Using Soap Bubbles to Model Molecular Shape 180
Tools & Techniques: AIM Theory and Electron Density Maps 186

4.3 Intermolecular Forces in Liquids and Solids 190
ThoughtLab: Properties of Liquids 196
Canadians in Chemistry: Dr. R.J. Le Roy 200
Investigation 4-B: Determining the Type of Bonding in Substances 202
Chemistry Bulletin: Ionic Liquids: A Solution to the Problem of Solutions 203
Chapter 4 Review 209
Unit 2 Project: Materials Convention 212
Unit 2 Review 214

UNIT 3 Energy Changes and Rates of Reaction 218

Chapter 5 Energy and Change 220

5.1 The Energy of Physical, Chemical, and Nuclear Processes 221

5.2 Determining Enthalpy of Reaction by Experiment 234
Investigation 5-A: Determining the Enthalpy of a Neutralization Reaction 240

5.3 Hess's Law of Heat Summation 243
Investigation 5-B: Hess's Law and the Enthalpy of Combustion of Magnesium 248

5.4 Energy Sources 256
ThoughtLab: Comparing Energy Sources 258
Careers in Chemistry: Nuclear Safety Supervisor 259
Chemistry Bulletin: Hot Ice 260
Chapter 5 Review 263

Chapter 6 Rates of Chemical Reactions 266

6.1 Expressing and Measuring Reaction Rates 267
ThoughtLab: Average and Instantaneous Reaction Rates 270
Investigation 6-A: Studying Reaction Rates 274

6.2 The Rate Law: Reactant Concentration and Rate 278

6.3 Theories of Reaction Rates 289

6.4 Reaction Mechanisms and Catalysts 297
ThoughtLab: Researching Catalysts 305
Investigation 6-B: Determining the Rate Law Equation for a Catalyzed Reaction 306
Canadians in Chemistry: Dr. Maud L. Menten 308
Chapter 6 Review 311
Unit 3 Project: Developing a Bulletin about Catalysts and Enzymes 314
Unit 3 Review 316

UNIT 4 Chemical Systems and Equilibrium 320

Chapter 7 Reversible Reactions and Chemical Equilibrium 322

7.1 Recognizing Equilibrium 323
ExpressLab: Modelling Equilibrium 325
7.2 Thermodynamics and Equilibrium 328
7.3 The Equilibrium Constant 334
Investigation 7-A: Measuring an Equilibrium Constant 340
7.4 Predicting the Direction of a Reaction 354
Investigation 7-B: Perturbing Equilibrium 358
Chemistry Bulletin: Le Châtelier's Principle: Beyond Chemistry 362
Careers in Chemistry: Anesthesiology: A Career in Pain Management 371
Chapter 7 Review 372

Chapter 8 Acids, Bases, and pH 376

8.1 Explaining the Properties of Acids and Bases 377
ExpressLab: Comparing Acid-Base Reactions 378
8.2 The Equilibrium of Weak Acids and Bases 388
Investigation 8-A: K_a of Acetic Acid 394
8.3 Bases and Buffers 404
8.4 Acid-Base Titration Curves 412
Chapter 8 Review 415

Chapter 9 Aqueous Solutions and Solubility Equilibria 418

9.1 The Acid-Base Properties of Salt Solutions 419
ExpressLab: Testing the pH of Salt Solutions 420
9.2 Solubility Equilibria 430
Investigation 9-A: Determining K_{sp} for Calcium Hydroxide 434
Canadians in Chemistry: Dr. Joseph MacInnis 439
9.3 Predicting the Formation of a Precipitate 443
ThoughtLab: A Qualitative Analysis 450
Chapter 9 Review 452
Unit 4 Issue: Earth in Equilibrium 456
Unit 4 Review 458

UNIT 5 Electrochemistry 462

Chapter 10 Oxidation-Reduction Reactions 464

10.1 Defining Oxidation and Reduction 465
Chemistry Bulletin: Aging: Is Oxidation a Factor? 469
Investigation 10-A: Single Displacement Reactions 470

10.2 Oxidation Numbers 473
ThoughtLab: Finding Rules for Oxidation Numbers 475

10.3 The Half-Reaction Method for Balancing Equations 482
Tools & Techniques: The Breathalyzer Test: A Redox Reaction 491
Investigation 10-B: Redox Reactions and Balanced Equations 492

10.4 The Oxidation Number Method for Balancing Equations 495
Chapter 10 Review 499

Chapter 11 Cells and Batteries 504

11.1 Galvanic Cells 505
Investigation 11-A: Measuring Cell Potentials of Galvanic Cells 510
Careers in Chemistry: Explosives Chemist 514

11.2 Standard Cell Potentials 516
ThoughtLab: Assigning Reference Values 522

11.3 Electrolytic Cells 524
Investigation 11-B: Electrolysis of Aqueous Potassium Iodide 532

11.4 Faraday's Law 538
Investigation 11-C: Electroplating 542

11.5 Issues Involving Electrochemistry 546
Canadians in Chemistry: Dr. Viola Birss 552
Chapter 11 Review 555
Unit 5 Design Your Own Investigation: Electroplating 558
Unit 5 Review 560

Chemistry Course Challenge: The Chemistry of Human Health 564

Appendix A: Answers to Selected and Numerical Chapter and Unit Review Questions 574

Appendix B: Supplementary Practice Problems 579

Appendix C: Alphabetical List of Elements and Periodic Table of the Elements 587

Appendix D: Math and Chemistry 590

Appendix E: Chemistry Data Tables 595

Appendix F: Titration Guidelines 600

Glossary 602

Index 612

Credits 621

Safety in Your Chemistry Laboratory and Classroom

The following *Safety Precautions* symbols appear throughout *Chemistry 12*, whenever an investigation or ExpressLab presents possible hazards.

appears when there is a danger to the eyes, and safety goggles, safety glasses, or a face shield should be worn

appears when substances that could burn or stain clothing are used

appears when objects that are hot or cold must be handled

appears when sharp objects are used, to warn of the danger of cuts and punctures

appears when toxic substances that can cause harm through ingestion, inhalation, or skin absorption are used

appears when corrosive substances, such as acids and bases, that can damage tissue are used

warns of caustic substances that could irritate the skin

appears when chemicals or chemical reactions that could cause dangerous fumes are used and ventilation is required

appears as a reminder to be careful when you are around open flames and when you are using easily flammable or combustible materials

warns of danger of electrical shock or burns from live electrical equipment

Actively engaging in laboratory investigations is essential to gaining a hands-on understanding of chemistry. Following safe laboratory procedures should not be seen as an inconvenience in your investigations. Instead, it should be seen as a positive way to ensure your safety and the safety of others who share a common working environment. Familiarize yourself with the following general safety rules and procedures. It is your responsibility to follow them when completing any of the investigations or ExpressLabs in this textbook, or when performing other laboratory procedures.

General Precautions

- Always wear safety glasses and a lab coat or apron in the laboratory. Wear other protective equipment, such as gloves, as directed by your teacher or by the Safety Precautions at the beginning of each investigation.
- If you wear contact lenses, always wear safety goggles or a face shield in the laboratory. Inform your teacher that you wear contact lenses. Generally, contact lenses should not be worn in the laboratory. If possible, wear eyeglasses instead of contact lenses, but remember that eyeglasses are not a substitute for proper eye protection.
- Know the location and proper use of the nearest fire extinguisher, fire blanket, fire alarm, first aid kit, and eyewash station (if available). Find out from your teacher what type of fire-fighting equipment should be used on particular types of fires. (See "Fire Safety" on page xi.)
- Do not wear loose clothing in the laboratory. Do not wear open-toed shoes or sandals. Accessories may get caught on equipment or present a hazard when working with a Bunsen burner. Ties, scarves, long necklaces, and dangling earrings should be removed before starting an investigation.
- Tie back long hair and any loose clothing before starting an investigation.
- Lighters and matches must not be brought into the laboratory.
- Food, drinks, and gum must not be brought into the laboratory.
- Inform your teacher if you have any allergies, medical conditions, or physical problems (including hearing impairment) that could affect your work in the laboratory.

Before Beginning Laboratory Investigations

- Listen carefully to the instructions that your teacher gives you. Do not begin work until your teacher has finished giving instructions.
- Obtain your teacher's approval before beginning any investigation that you have designed yourself.
- Read through all of the steps in the investigation before beginning. If there are any steps that you do not understand, ask your teacher for help.
- Be sure to read and understand the Safety Precautions at the start of each investigation or Express Lab.

- Always wear appropriate protective clothing and equipment, as directed by your teacher and the Safety Precautions.
- Be sure that you understand all safety labels on materials and equipment. Familiarize yourself with the WHMIS symbols on this page.
- Make sure that your work area is clean and dry.

During Laboratory Investigations

- Make sure that you understand and follow the safety procedures for different types of laboratory equipment. Do not hesitate to ask your teacher for clarification if necessary.
- Never work alone in the laboratory.
- Remember that gestures or movements that may seem harmless could have dangerous consequences in the laboratory. For example, tapping people lightly on the shoulders to get their attention could startle them. If they are holding a beaker that contains an acid, for example, the results could be very serious.
- Make an effort to work slowly and steadily in the laboratory. Be sure to make room for other students.
- Organize materials and equipment neatly and logically. For example, do not place materials that you will need during an investigation on the other side of a Bunsen burner from you. Keep your bags and books off your work surface and out of the way.
- Never taste any substances in the laboratory.
- Never touch a chemical with your bare hands.
- Never draw liquids or any other substances into a pipette or a tube with your mouth.
- If you are asked to smell a substance, do not hold it directly under your nose. Keep the object at least 20 cm away, and waft the fumes toward your nostrils with your hand.
- Label all containers holding chemicals. Do not use chemicals from unlabelled containers.
- Hold containers away from your face when pouring liquids or mixing reactants.
- If any part of your body comes in contact with a potentially dangerous substance, wash the area immediately and thoroughly with water.
- If you get any material in your eyes, do not touch them. Wash your eyes immediately and continuously for 15 min, and make sure that your teacher is informed. A doctor should examine any eye injury. If you wear contact lenses, take them out immediately. Failing to do so may result in material becoming trapped behind the contact lenses. Flush your eyes with water for 15 min, as above.
- Do not touch your face or eyes while in the laboratory unless you have first washed your hands.
- Do not look directly into a test tube, flask, or the barrel of a Bunsen burner.
- If your clothing catches fire, smother it with the fire blanket or with a coat, or get under the safety shower.
- If you see any of your classmates jeopardizing their safety or the safety of others, let your teacher know.

WHMIS (Workplace Hazardous Materials Information System) symbols are used in Canadian schools and workplaces to identify dangerous materials. Familiarize yourself with the symbols below.

Poisonous and Infectious Material Causing Immediate and Serious Toxic Effects

Poisonous and Infectious Material Causing Other Toxic Effects

Flammable and Combustible Material

Compressed Gas

Corrosive Material

Oxidizing Material

Dangerously Reactive Material

Biohazardous Infectious Material

Heat Source Safety

- When heating any item, wear safety glasses, heat-resistant safety gloves, and any other safety equipment that your teacher or the Safety Precautions suggests.
- Always use heat-proof, intact containers. Check that there are no large or small cracks in beakers or flasks.
- Never point the open end of a container that is being heated at yourself or others.
- Do not allow a container to boil dry unless specifically instructed to do so.
- Handle hot objects carefully. Be especially careful with a hot plate that may look as though it has cooled down, or glassware that has recently been heated.
- Before using a Bunsen burner, make sure that you understand how to light and operate it safely. Always pick it up by the base. Never leave a Bunsen burner unattended.
- Before lighting a Bunsen burner, make sure there are no flammable solvents nearby.
- If you do receive a burn, run cold water over the burned area immediately. Make sure that your teacher is notified.
- When you are heating a test tube, always slant it. The mouth of the test tube should point away from you and from others.
- Remember that cold objects can also harm you. Wear appropriate gloves when handling an extremely cold object.

Electrical Equipment Safety

- Ensure that the work area, and the area of the socket, is dry.
- Make sure that your hands are dry when touching electrical cords, plugs, sockets, or equipment.
- When unplugging electrical equipment, do not pull the cord. Grasp the plug firmly at the socket and pull gently.
- Place electrical cords in places where people will not trip over them.
- Use an appropriate length of cord for your needs. Cords that are too short may be stretched in unsafe ways. Cords that are too long may tangle or trip people.
- Never use water to fight an electrical equipment fire. Severe electrical shock may result. Use a carbon dioxide or dry chemical fire extinguisher. (See "Fire Safety" on the next page.)
- Report any damaged equipment or frayed cords to your teacher.

Glassware and Sharp Objects Safety

- Cuts or scratches in the chemistry laboratory should receive immediate medical attention, no matter how minor they seem. Alert your teacher immediately.
- Never use your hands to pick up broken glass. Use a broom and dustpan. Dispose of broken glass as directed by your teacher. Do not put broken glassware into the garbage can.

- Cut away from yourself and others when using a knife or another sharp object.
- Always keep the pointed end of scissors and other sharp objects pointed away from yourself and others when walking.
- Do not use broken or chipped glassware. Report damaged equipment to your teacher.

Fire Safety

- Know the location and proper use of the nearest fire extinguisher, fire blanket, and fire alarm.
- Understand what type of fire extinguisher you have in the laboratory, and what type of fires it can be used on. (See below.) Most fire extinguishers are the ABC type.
- Notify your teacher immediately about any fires or combustible hazards.
- Water should only be used on Class A fires. Class A fires involve ordinary flammable materials, such as paper and clothing. Never use water to fight an electrical fire, a fire that involves flammable liquids (such as gasoline), or a fire that involves burning metals (such as potassium or magnesium).
- Fires that involve a flammable liquid, such as gasoline or alcohol (Class B fires) must be extinguished with a dry chemical or carbon dioxide fire extinguisher.
- Live electrical equipment fires (Class C) must be extinguished with a dry chemical or carbon dioxide fire extinguisher. Fighting electrical equipment fires with water can cause severe electric shock.
- Class D fires involve burning metals, such as potassium and magnesium. A Class D fire should be extinguished by smothering it with sand or salt. Adding water to a metal fire can cause a violent chemical reaction.
- If someone's hair or clothes catch on fire, smother the flames with a fire blanket. Do not discharge a fire extinguisher at someone's head.

Web LINK

CAUTION The Use and Sharing of Computer Files

Be sure to install an up-to-date virus checker on your computer, because even a disk given to you by a close friend could contain a virus. Never open an e-mail file if you do not know the source of the message, since you could, inadvertently, infect your computer with a virus that could destroy files containing many hours of your work. If you discover that you have a virus in your computer, notify your teacher and peers immediately.

Visit the *Chemistry 12* web site at **www.mcgrawhill.ca/links/chemistry12** where you will find links to many useful, reputable educational institutions and organizations—excellent sources of material for research.

Clean-Up and Disposal in the Laboratory

- Clean up all spills immediately. Always inform your teacher about spills.
- If you spill acid or base on your skin or clothing, wash the area immediately with a lot of cool water.
- You can neutralize small spills of acid solutions with sodium hydrogen carbonate (baking soda). You can neutralize small spills of basic solutions with sodium hydrogen sulfate or citric acid.
- Clean equipment before putting it away, as directed by your teacher.
- Dispose of materials as directed by your teacher, in accordance with your local School Board's policies. Do not dispose of materials in a sink or a drain unless your teacher directs you to do so.
- Wash your hands thoroughly after all laboratory investigations.

INTRODUCING *Chemistry 12*

Here is a quick glimpse at the learning that lies before you in this course. Expand your knowledge and skills from your Grade 11 chemistry course, and experience chemistry in action.

Organic compounds are everywhere—in the food you eat, the clothes you wear, the products you use, and even your body. In Unit 1, you will name and classify organic compounds, investigate reactions of organic compounds, and consider risks and benefits of society's use of organic chemistry. At the end of the unit, you will analyze issues related to the development of new products through organic chemistry.

Chemical compounds have an amazing variety of useful properties. For example, superglues have incredible adhesive properties when exposed to moisture in the air or on surfaces. In Unit 2, you will first look at the structure of the atom. You will then use your increased understanding of the atom to explore the relationships between bonding and chemical properties. Your experiences will prepare you for an end-of-unit project about the structure and properties of a material.

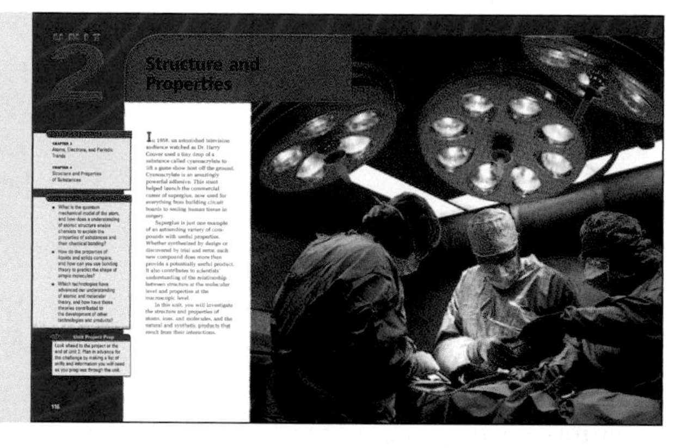

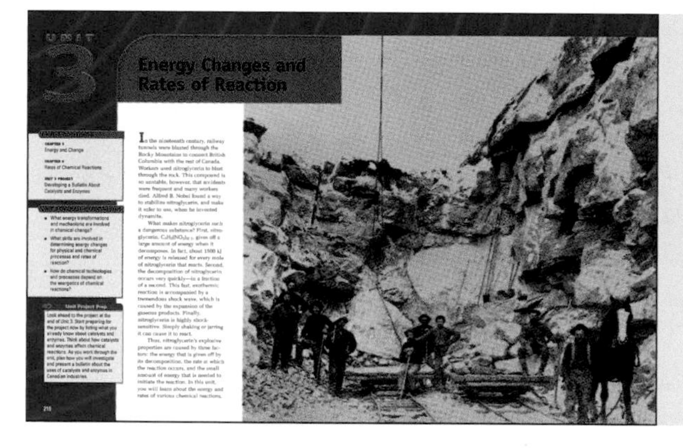

A chemical reaction can have enough energy, and take place quickly enough, to blast through rock. Chemical reactions can also occur with very little energy change or at a very slow rate. In Unit 3, you will learn about the energy and rates of chemical reactions. At the end of Unit 3, you will develop an information bulletin about industrial applications of catalysts and enzymes—substances that affect the rates of chemical reactions.

Hydrogen gas and oxygen gas react to form water. When electricity passes through water, however, it separates into hydrogen and oxygen. Like the formation of water from hydrogen and oxygen, most chemical reactions have the potential to proceed in both directions. They are reversible. In Unit 4, you will examine the equilibrium of reversible reactions. At the end of Unit 4, you will use your understanding of equilibrium to analyze issues surrounding the relationships among Earth's four main biosystems.

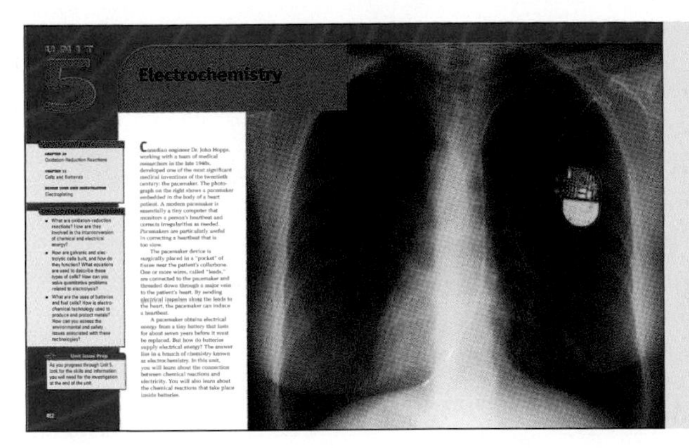

You have probably used many battery-powered devices, such as flashlights and cellular telephones. In Unit 5, you will learn about the connection between chemical reactions and electricity. You will also learn about the chemical reactions that take place inside batteries. Your understanding of electrochemical reactions will prepare you to design an end-of-unit investigation about electroplating.

Following the five units of the textbook is a Chemistry Course Challenge. This is your opportunity to demonstrate your understanding of the concepts covered in the course. You will apply your skills of inquiry to explore the connections between chemistry and advances in medicine and other health sciences. At the end of the challenge, you will communicate your ideas at a conference about the chemistry of human health. By applying your knowledge and skills to real-life situations, you will analyze social issues involving the science and technology of human health.

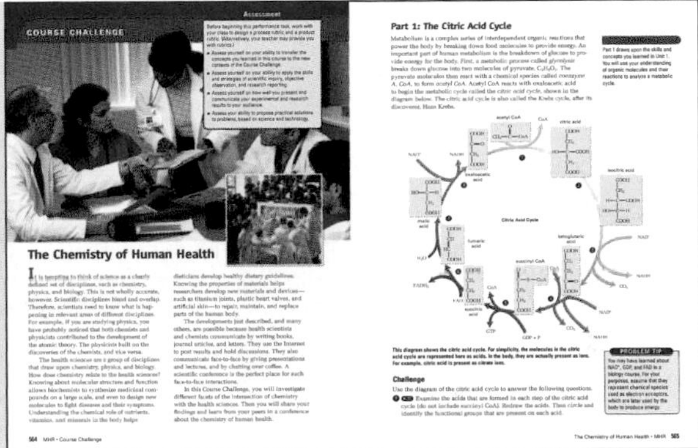

Assessment

You will probably be designing rubrics to assess your end-of-unit project, issue, or investigation. In your rubrics, include criteria that address all the Achievement Chart categories shown below.

COURSE CHALLENGE

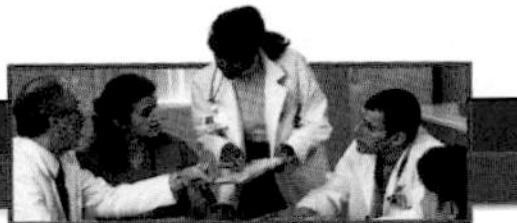

The Chemistry of Human Health

Watch for this feature in text margins throughout the textbook and in the Unit Reviews to help you begin planning for your Course Challenge. Use the cues to trigger your thought processes and point you to a line of research. Keep a file on ideas and preliminary research.

Achievement Chart

During this chemistry course, your skills will be assessed using the Achievement Chart. You will notice that all review questions in *Chemistry 12* are coded according to this chart.

Knowledge and Understanding	Inquiry	Communication	Making Connections
■ understanding of concepts, principles, laws, and theories ■ knowledge of facts and terms ■ transfer of concepts to new contexts ■ understanding of relationships between concepts	■ application of skills and strategies of scientific inquiry ■ application of technical skills and procedures ■ use of tools, equipment, and materials	■ communication of information and ideas ■ use of scientific terminology, symbols, conventions, and standard (SI) units ■ communication for different audiences and purposes ■ use of various forms of communication ■ use of information technology for scientific purposes	■ understanding of connections among science, technology, society, and the environment ■ analysis of social and economic issues involving science and technology ■ assessment of impacts of science and technology on the environment ■ proposal of courses of practical action in relation to science-and-technology-based problems

Review Concepts and Skills Review

This review section summarizes many of the concepts that you encountered in your Grade 11 chemistry course. You may wish to read through this section before continuing with the rest of the textbook, to remind yourself of important terms, equations, and calculations. Sample problems and practice problems are included to help you review your skills. As well, you can refer to the information in this section, as needed, while you work through the textbook. Table R.1 lists the topics that are covered in the Concepts and Skills Review.

Table R.1 Topics Included in Concepts and Skills Review

R.1	Matter	R.9	Types of Chemical Reactions
R.2	Representing Atoms and Ions	R.10	Ionic Equations
R.3	The Periodic Table	R.11	Mole Calculations
R.4	Chemical Bonds	R.12	Concentration Calculations
R.5	Representing Molecules	R.13	Stoichiometric Calculations
R.6	Naming Binary Compounds	R.14	Representing Organic Molecules
R.7	Writing Chemical Formulas	R.15	Isomers of Organic Compounds
R.8	Balancing Chemical Formulas		

R.1 Matter

Chemistry is the study of the properties and changes of matter. **Matter** is defined as anything that has mass and takes up space. All matter can be classified into two groups: pure substances and mixtures.

A **pure substance** has a definite chemical composition. Examples of pure substances are carbon dioxide, CO_2, and nitrogen, N_2. Pure substances can be further classified into elements and compounds.

- An **element** is a pure substance that cannot be separated chemically into any simpler substances. Oxygen gas, $O_{2(g)}$, solid carbon, $C_{(s)}$, and copper metal, $Cu_{(s)}$, are examples of elements.
- A **compound** is a pure substance that results when two or more elements combine chemically to form a different substance. Water, $H_2O_{(\ell)}$, and salt, $NaCl_{(s)}$, are two examples of compounds.

A **mixture** is a physical combination of two or more kinds of matter. Each component in a mixture retains its identity. There are two kinds of mixtures: heterogeneous mixtures and homogeneous mixtures.

- In a **heterogeneous mixture**, the different components are clearly visible. A mixture of sand and table salt is a heterogeneous mixture, with small grains of both sand and salt visible.
- In a **homogeneous mixture** (also called a **solution**) the components are blended together so that the mixture looks like a single substance. A spoonful of table salt in a glass of water results in a homogeneous mixture, since the salt dissolves in the water to produce a salt-water solution.

R.2 Representing Atoms and Ions

An **atom** is the smallest particle of an element that still retains the identity and properties of the element. An atom is composed of one or more protons, neutrons, and electrons. Each atom of an element has the same number of protons in its nucleus.

- The **atomic number** (symbol **Z**) of an element is the number of protons in the nucleus of each atom of the element.
- The **mass number** (symbol **A**) is the total number of protons and neutrons in the nucleus of each atom. If two atoms of an element have the same number of protons, but different numbers of neutrons, they are called **isotopes**. Isotopes have the same atomic number but different mass numbers.
- The **atomic symbol** is different for each element. Some examples of atomic symbols are O for oxygen, Au for gold, and Br for bromine.

Figure R.1 summarizes the notation that is used to express the atomic number, mass number, and atomic symbol of an element.

mass number

A

X

atomic symbol

Z

atomic number

Figure R.1 **The mass number is written at the top left of the atomic symbol. The atomic number is written at the bottom left.**

When an atom gains or loses electrons, it becomes an **ion**. Atoms and ions can be represented by Bohr-Rutherford diagrams or by Lewis structures.

A Bohr-Rutherford diagram shows the number of electrons in each energy level. Figure R.2 shows Bohr-Rutherford diagrams for atoms and ions of magnesium and fluorine.

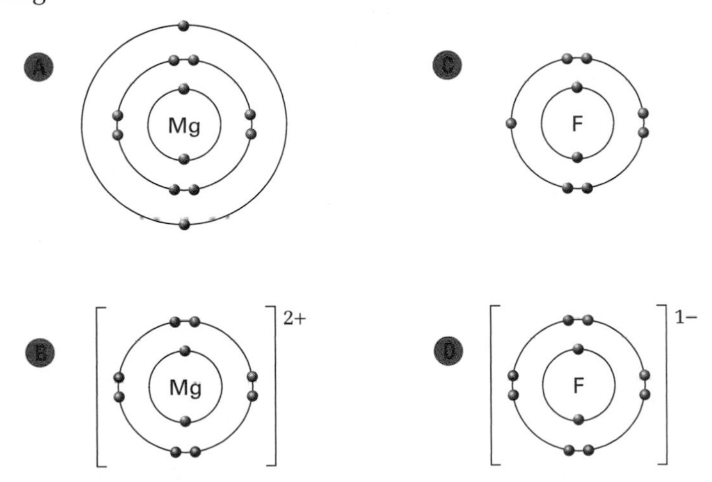

Figure R.2 **Bohr-Rutherford diagrams of (A) a magnesium atom, (B) a magnesium ion, (C) a fluorine atom, and (D) a fluoride ion**

In a Lewis structure of an element, the dots show the number of electrons in the outer, valence shell. The symbol represents the nucleus and the inner energy levels. Figure R.3 shows Lewis structures for atoms and ions of magnesium and fluoride.

Figure R.3 **Lewis structures of (A) a magnesium atom, (B) a magnesium ion, (C) a fluorine atom, and (D) a fluoride ion**

R.3 The Periodic Table

The elements can be organized according to similarities in their properties and atomic structure. The **periodic table** is a system for organizing the elements, by atomic number, into groups (columns) and periods (rows). In the periodic table, elements with similar properties are in the same column. The names that are associated with some sections of the periodic table are shown in Figure R.4.

MAIN GROUP ELEMENTS | TRANSITION ELEMENTS | MAIN-GROUP ELEMENTS

Legend: metals (main group); metals (transition); metals (inner transition); metalloids; nonmetals

Period	1 (IA)	2 (IIA)	3 (IIIB)	4 (IVB)	5 (VB)	6 (VIB)	7 (VIIB)	8	9 (VIIIB)	10	11 (IB)	12 (IIB)	13 (IIIA)	14 (IVA)	15 (VA)	16 (VIA)	17 (VIIA)	18 (VIIIA)
1	1 H 1.01																	2 He 4.003
2	3 Li 6.941	4 Be 9.012											5 B 10.81	6 C 12.01	7 N 14.01	8 O 16.00	9 F 19.00	10 Ne 20.18
3	11 Na 22.99	12 Mg 24.13											13 Al 26.98	14 Si 28.09	15 P 30.97	16 S 32.07	17 Cl 35.45	18 Ar 39.95
4	19 K 39.10	20 Ca 40.08	21 Sc 44.96	22 Ti 47.88	23 V 50.94	24 Cr 52.00	25 Mn 54.94	26 Fe 55.85	27 Co 58.93	28 Ni 58.69	29 Cu 63.55	30 Zn 65.39	31 Ga 69.72	32 Ge 72.61	33 As 74.92	34 Se 78.96	35 Br 79.90	36 Kr 83.80
5	37 Rb 85.47	38 Sr 87.62	39 Y 88.91	40 Zr 91.22	41 Nb 92.91	42 Mo 95.94	43 Tc (98)	44 Ru 101.1	45 Rh 102.9	46 Pd 106.4	47 Ag 107.9	48 Cd 112.4	49 In 114.8	50 Sn 118.7	51 Sb 121.8	52 Te 127.6	53 I 126.9	54 Xe 131.3
6	55 Cs 132.9	56 Ba 137.3	57 La 138.9	72 Hf 178.5	73 Ta 180.9	74 W 183.9	75 Re 186.2	76 Os 190.2	77 Ir 192.2	78 Pt 195.1	79 Au 197.0	80 Hg 200.6	81 Tl 204.4	82 Pb 207.2	83 Bi 209.0	84 Po (209)	85 At (210)	86 Rn (222)
7	87 Fr (223)	88 Ra (226)	89 Ac (227)	104 Rf (261)	105 Db (262)	106 Sg (266)	107 Bh (262)	108 Hs (265)	109 Mt (266)	110 Uun (269)	111 Uuu (272)	112 Uub (277)		114 Uuq (285)		116 Uuh (289)		118 Uuo (293)

INNER TRANSITION ELEMENTS

Period														
6	58 Ce 140.1	59 Pr 140.9	60 Nd 144.2	61 Pm (145)	62 Sm 150.4	63 Eu 152.0	64 Gd 157.3	65 Tb 158.9	66 Dy 162.5	67 Ho 164.9	68 Er 167.3	69 Tm 168.9	70 Yb 173.0	71 Lu 175.0
7	90 Th 232.0	91 Pa (231)	92 U 238.0	93 Np (237)	94 Pu (242)	95 Am (243)	96 Cm (247)	97 Bk (247)	98 Cf (251)	99 Es (252)	100 Fm (257)	101 Md (258)	102 No (259)	103 Lr (260)

- Elements are arranged in seven numbered periods (horizontal rows) and 18 numbered groups (vertical columns).
- Groups are numbered according to two different systems. The current system numbers the group from 1 to 18. An older system numbers the groups from I to VIII, and separates them into two categories labelled A and B.
- The elements in the eight A groups are the main-group elements. They are also called the representative elements. The elements in the ten B groups are known as the transition elements.
- Group 1 (IA) elements are known as *alkali metals*. They react with water to form alkaline, or basic, solutions.
- Group 2 (IIA) elements are known as *alkaline earth metals*. They react with oxygen to form compounds called oxides, which react with water to form alkaline solutions. Early chemists called all metal oxides "earths."
- Group 17 (VIIA) elements are known as *halogens*, from the Greek word *hals*, meaning "salt." Elements in this group combine with other elements to form compounds called salts.
- Group 18 (VIIIA) elements are known as *noble gases*. Noble gases do not combine naturally with any other elements.

Figure R.4 Basic features of the periodic table

The periodic law states that *when the elements are arranged in order of increasing atomic number, a regular repetition of properties is observed.* This statement can be used to predict trends in the properties of the elements. Figure R.5 summarizes the periodic trends of four properties of atoms: atomic size, ionization energy, electron affinity, and electronegativity. These four properties affect the structure of molecules and ions, and they are key to understanding the properties of matter.

Figure R.5 **Periodic trends**

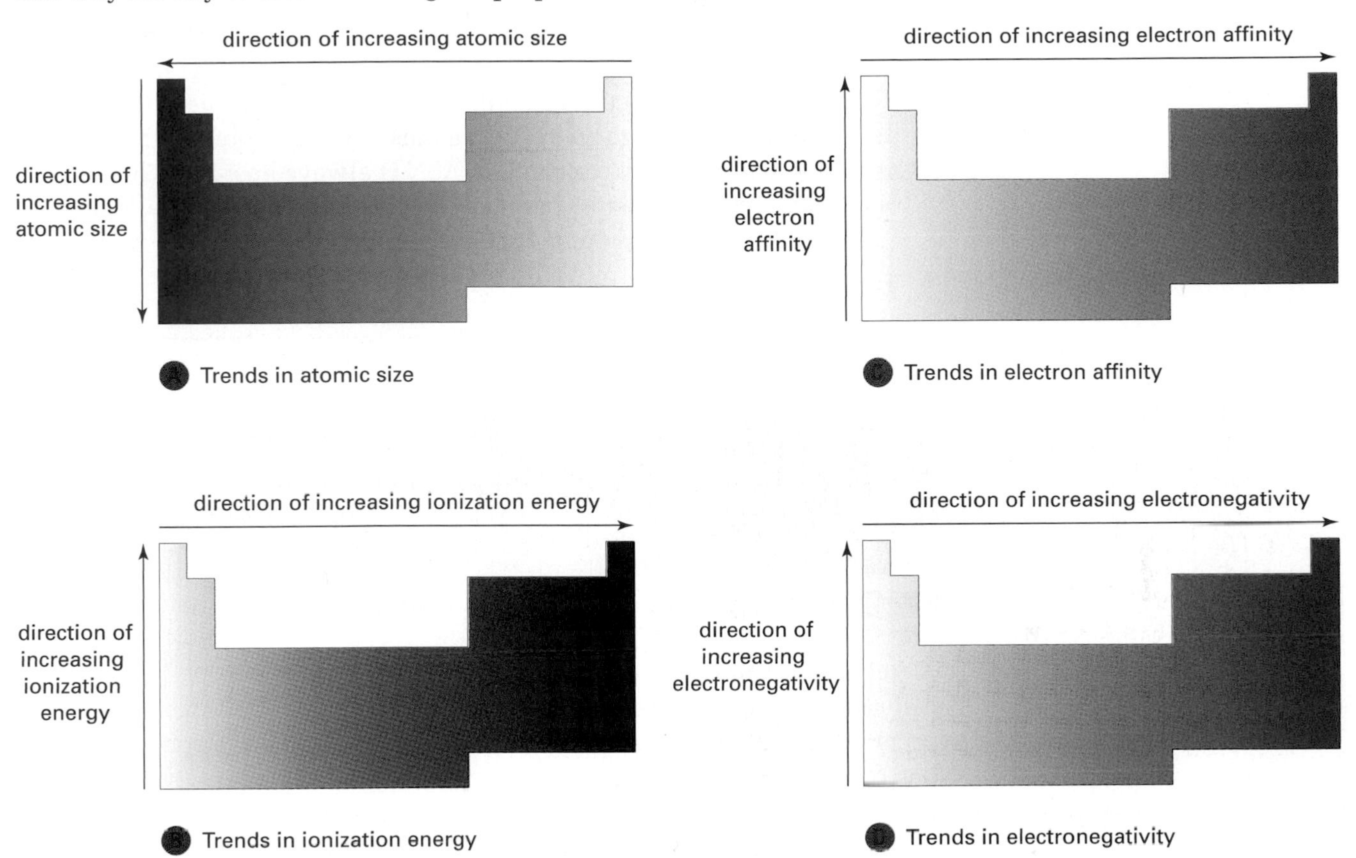

R.4 Chemical Bonds

Bonding involves interactions between the valence electrons of atoms. **Valence electrons** are electrons that occupy the outer energy level of an atom. Bonding usually follows the **octet rule**, which states that atoms attain a more stable electron configuration with eight electrons in their valence shell.

- When electron(s) are transferred between a metal and a non-metal, the electrostatic attraction between the positive metal ion (cation) and the negative non-metal ion (anion) is called an **ionic bond**. Figure R.6 shows the formation of an ionic bond between the metal calcium and the non-metal fluorine.

$$Ca + 2F \rightarrow [Ca]^{2+}[:F:]^{-}[:F:]^{-}$$

Figure R.6 **Lewis structures showing the formation of an ionic bond between calcium and fluorine**

H
H : C : H
H

Figure R.7 Lewis structure showing the covalent bonding in a molecule of CH_4

- When a pair of electrons is shared between two non-metal atoms, the attraction is called a **covalent bond**. A single covalent bond involves one pair of electrons shared between two atoms. A double bond involves two pairs of electrons shared between two atoms, and a triple bond involves three pairs of electrons. Figure R.7 shows the formation of a covalent bond between two non-metals, carbon and hydrogen atoms, in a molecule of methane, CH_4.

The **electronegativity** of an element is a relative measure of the ability of its atoms to attract electrons in a chemical bond. The periodic table in Appendix C gives the electronegativities of the elements. The difference in the electronegativities (ΔEN) of two atoms is used to predict the type of bond that will form. By convention, ΔEN is always positive. Therefore, always subtract the smaller electronegativity from the larger one. Figure R.8 illustrates the range in bond character for different values of ΔEN.

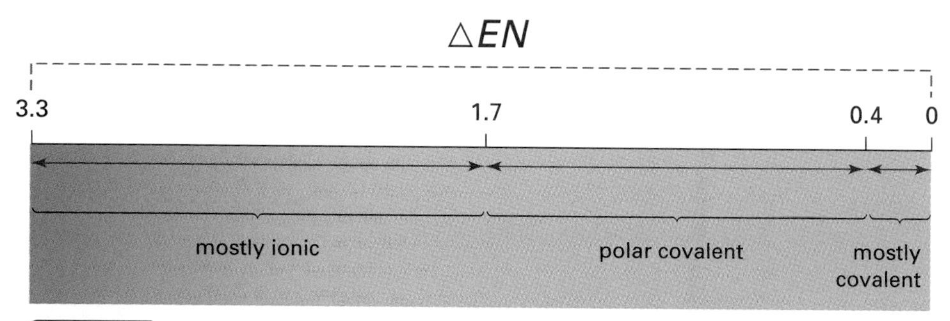

Figure R.8 Chemical bonds range from mostly ionic to mostly covalent.

A **polar covalent bond** is a covalent bond between atoms that have different electronegativities. The electron pair is unevenly shared in a polar covalent bond. Therefore, the atom with the higher electronegativity has a slight negative charge and the atom with the lower electronegativity has a slight positive charge. This separation of charges, or *polarity*, is shown using the Greek letter delta, δ. Figure R.9 gives examples of an ionic bond, a polar covalent bond, and a mostly covalent bond.

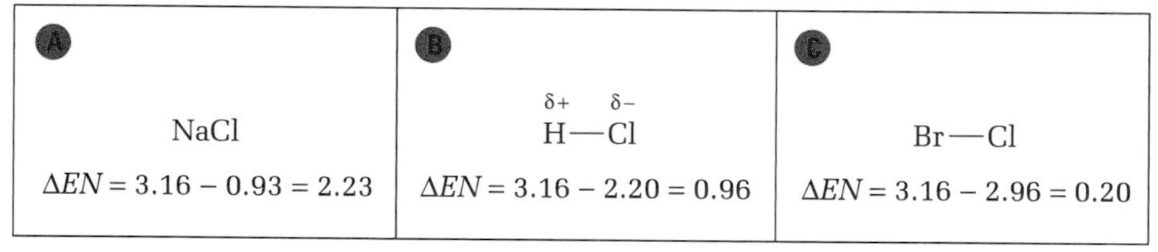

Figure R.9 (A) ionic bond in NaCl, (B) polar covalent bond in HCl, and (C) covalent bond in BrCl

A :O::O:

B :O=O:

Figure R.10 Lewis structures of O_2

R.5 Representing Molecules

The Lewis structure of a molecule shows the electrons that are in the valence shells of all the atoms in the molecule. A common variation of a Lewis structure uses a dash (the symbol of a single bond) to represent one shared pair of electrons. Figure R.10(A) shows a Lewis structure for a molecule of O_2 that contains a double covalent bond. Figure R.10(B) shows a similar structure for the same molecule, in which the electrons of the double bond have been replaced by the symbol of a double bond.

R.6 Naming Binary Compounds

Two non-metals can combine to form a **binary compound**: a compound with only two kinds of atoms. The less electronegative element is usually written on the left, and the more electronegative element is usually written on the right. For example, sulfur and oxygen can combine to form SO_2 and SO_3. Carbon and chlorine can combine to form CCl_4.

Table R.2

Number of atoms	Prefix
1	mono-
2	di-
3	tri-
4	tetra-
5	penta-
6	hexa-
7	hepta-
8	octa-
9	nona-
10	deca-

Naming a Covalent Binary Compound

To name a covalent binary compound, follow these steps.

Step 1 Give the first atom its full name. For example, use sulfur for SO_2 and carbon for CCl_4.

Step 2 Give the second atom its ion name, with the suffix -ide. For example, use oxide for SO_2, and chloride for CCl_4.

Step 3 Use a prefix to indicate the number of each type of atom. Table R.2 lists the prefixes that are commonly used to designate the number of each type of atom in a molecular compound.

Step 4 Put the parts of the name together. Thus, SO_2 is named sulfur dioxide, and CCl_4 is named carbon tetrachloride.

A metal and a non-metal can combine to form an ionic binary compound. In an ionic binary compound, the cation is written on the left and the anion is written on the right. For example, sodium and chlorine can combine to form NaCl. Calcium and chloride can combine to form $CaCl_2$.

Naming an Ionic Binary Compound

To name an ionic binary compound, follow the steps below.

Step 1 Give the metal ion its full name. For example, use sodium for NaCl and calcium for $CaCl_2$.

Step 2 Give the non-metal ion its ion name, with the suffix -ide. For example, use chloride for NaCl and $CaCl_2$.

Step 3 Put the parts of the name together. Thus, NaCl is named sodium chloride, and $CaCl_2$ is named calcium chloride. Note that prefixes are not used for ionic binary compounds. That is, calcium chloride is not named calcium dichloride, even though two chloride ions are present for each calcium ion.

The Classical System and the Stock System

Transition metals may have more than one valence. Two systems are available to name ionic compounds that contain transition metals.

- For the classical system, use the suffix -ous to indicate the metal ion with the lower valence. Use the suffix -ic to indicate the metal ion with the higher valence. The earliest discovered elements are sometimes named using Latin names. For example, FeO is named ferrous oxide, and Fe_2O_3 is named ferric oxide. Other common examples are stannous oxide, SnO, and mercuric nitride, Hg_3N_2.
- The Stock system was devised by the German chemist Alfred Stock. For the Stock system, use Roman numerals to indicate the valence of the metal cation. Place a Roman numeral in brackets after the name of the first element. Examples are copper(II) oxide, CuO, and copper(I) oxide, Cu_2O.

R.7 Writing Chemical Formulas

A chemical formula indicates the type and number of atoms that are present in a compound. You can write the chemical formula of a compound by using the valence of each type of atom. When you write the chemical formula of a neutral compound that contains ions, *the sum of the positive valences plus the negative valences must equal zero.* This statement is known as the **zero sum rule**. To write a chemical formula, follow the steps below.

Writing a Chemical Formula

Step 1 Write the unbalanced formula, placing the element or ion with a positive valence first.

Step 2 Write the valence of each element on top of the appropriate symbol. The names and formulas of commonly used ions are listed in Appendix E.

Step 3 Cross over the numerical value of each valence and write this number as the subscript for the other element or ion in the compound. Do not include negative or positive signs, and do not include the subscript 1 in the formula. Bracket the formula for a polyatomic ion if its subscript is greater than 1.

Step 4 If necessary, divide through by any common factor. (Use your knowledge of chemical bonding to decide when to divide by a common factor. For example, the formula for hydrogen peroxide, H_2O_2, should not be reduced further, since the formula HO cannot represent a neutral molecule.)

The following Sample Problem shows you how to use these steps to write a chemical formula. Note that the steps are listed for only the first two solutions.

Sample Problem

Writing a Chemical Formula From the Name of a Compound

Problem

Write the chemical formula of each compound.

(a) magnesium fluoride
(b) zinc telluride
(c) aluminum carbonate
(d) ammonium phosphite
(e) tin(IV) sulfate
(f) ferrous oxalate

What Is Required?

Use the name of each compound to write a chemical formula for the compound.

What Is Given?

You are given the names of the compounds. From each name, you can identify the types of atoms that are present in the compound.

Plan Your Strategy

Use the periodic table to find the valence of each atom in the name. (Valence numbers usually correspond to the common ion charge. If the compound includes a transition metal with more than one valence, the name of the compound will indicate which valence is used.) Then follow the steps you have just learned to write each chemical formula.

Act on Your Strategy

(a) The valences are +2 for Mg and −1 for F.

Step 1 Magnesium has a positive valence, so write the unbalanced formula as MgF.

Step 2 Write the valences above each element.

$$\overset{+2}{Mg}\ \overset{-1}{F}$$

Step 3 Cross over the numerical value of each valence.

$$MgF_2$$

Step 4 Since there is only one Mg atom, you do not need to divide by a common factor.

(b) The valences are +2 for Zn, and −2 for Te.

Step 1 Zinc has a positive valence, so write the unbalanced formula as ZnTe.

Step 2 Write the valences above each element.

$$\overset{+2}{Zn}\ \overset{-2}{Te}$$

Step 3 Cross over the numerical value of each valence.

$$Zn_2Te_2$$

Step 4 Divide by the common factor, 2, to give the chemical formula ZnTe.

(c) $\overset{+3}{Al}\ \overset{-2}{CO_3}$

$Al_2(CO_3)_3$

(d) $\overset{+1}{NH_4}\overset{-3}{PO_3}$

$(NH_4)_3PO_3$

(e) $\overset{+4}{Sn}\ \overset{-2}{SO_4}$

$Sn_2(SO_4)_4 = Sn(SO_4)_2$

(f) $\overset{+2}{Fe}\ \overset{-2}{C_2O_4}$

$Fe_2(C_2O_4)_2 = FeC_2O_4$

Check Your Solution

If you have time, use each chemical formula to name the compound. Then check that your name and the original name match.

Practice Problems

1. Name each compound.

(a) XeF_4
(b) PCl_5
(c) CO
(d) SF_6
(e) N_2O_4
(f) NaF
(g) $CaCO_3$
(h) BaS_2O_3
(i) NI_3

2. Give both the classical name and the Stock name of each compound.

(a) TiO_2
(b) $CoCl_2$
(c) $NiBr_2$
(d) HgO
(e) $SnCl_2$

Continued ...

Continued ...

3. Write the chemical formula of each compound.
   (a) copper(II) hydroxide
   (b) mercuric nitride
   (c) manganese(II) hydrogen carbonate
   (d) sulfur hexabromide
   (e) potassium chromate
   (f) arsenic trichloride
   (g) iron(III) acetate
   (h) potassium peroxide
   (i) ammonium dichromate
   (j) stannous permanganate
   (k) plumbous nitrate
4. Write the correct chemical formula of each compound.
   (a) zinc perchlorate
   (b) mercury(II) nitride
   (c) stannous fluoride
   (d) ammonium dihydrogen phosphite
   (e) manganese(IV) silicate
   (f) ferric hydroxide
   (g) cobalt(III) nitride
   (h) lead(II) bicarbonate
   (i) silver oxalate
   (j) platinum(IV) oxide
   (k) silicon carbide
   (l) nickel(III) sulfite
   (m) tin(II) carbonate
   (n) aluminum permanganate

R.8 Balancing Chemical Equations

A chemical equation shows the **reactants** (the starting materials) and the **products** (the new materials that form) in a chemical change. Chemical equations are balanced to reflect the fact that atoms are conserved in a chemical reaction. The basic process of balancing an equation involves trial and error—going back and forth between reactants and products to find the correct balance. The systematic approach that is outlined below can be helpful.

Balancing a Chemical Equation

Step 1 List all the atomic species that are involved in the equation to determine which element(s) are not balanced.

Step 2 First balance the most complex substance: the compound that contains the most kinds or the largest number of atoms. Balance the atoms that occur in the largest numbers. Leave hydrogen, oxygen, and elements that occur in smaller numbers until later.

Step 3 Balance any polyatomic ions that occur on both sides of the equation as one unit, rather than as separate atoms.

Step 4 Balance any hydrogen or oxygen atoms that occur in a combined and uncombined state.

Step 5 Balance any other element that occurs in its uncombined state.

Sample Problem

Balancing a Chemical Equation

Problem

Balance the following chemical equation.

$NH_{3(g)} + H_2O_{(\ell)} + Y_2(SO_4)_{3(aq)} \rightarrow (NH_4)_2SO_{4(aq)} + Y(OH)_{3(s)}$

Solution

Step 1 List the atoms on each side of the equation.

1	N	2
2	Y	1
3	SO_4	1
5	H	11
13	O	7

Steps 2 and 3 SO_4^{2-} is a polyatomic ion, and it is present in the most complex substance. It appears on both sides of the equation, so it can be balanced as a unit. Balance SO_4^{2-} by putting a 3 on the right side, in front of $(NH_4)_2SO_{4(aq)}$.

$NH_{3(g)} + H_2O_{(\ell)} + Y_2(SO_4)_{3(aq)} \rightarrow 3(NH_4)_2SO_{4(aq)} + Y(OH)_{3(s)}$

Since there are now 6 N atoms on the right side, put a 6 in front of $NH_{3(g)}$ on the left side. Balance Y by putting a 2 in front of $Y(OH)_{3(s)}$ on the right side.

$6NH_{3(g)} + H_2O_{(\ell)} + Y_2(SO_4)_{3(aq)} \rightarrow 3(NH_4)_2SO_{4(aq)} + 2Y(OH)_{3(s)}$

Step 4 There are now 6 O atoms on the right side and 1 O atom on the left side. (This does not include O atoms in the SO_4^{2-} units, which are already balanced.) Put a 6 in front of H_2O on the left side. This also balances H.

$6NH_{3(g)} + 6H_2O_{(\ell)} + Y_2(SO_4)_{3(aq)} \rightarrow 3(NH_4)_2SO_{4(aq)} + 2Y(OH)_{3(s)}$

Practice Problems

5. Balance each equation.

(a) $H_2O_{(\ell)} + NO_{2(g)} \rightarrow HNO_{3(aq)} + NO_{(g)}$

(b) $NaOH_{(aq)} + Cl_{2(g)} \rightarrow NaCl_{(aq)} + NaClO_{(aq)} + H_2O_{(\ell)}$

(c) $(NH_4)_3PO_{4(aq)} + Ba(OH)_{2(aq)} \rightarrow Ba_3(PO_4)_{2(s)} + NH_4OH_{(aq)}$

(d) $Li_{(s)} + H_2O_{(\ell)} \rightarrow LiOH_{(aq)} + H_{2(g)}$

(e) $NH_{3(g)} + O_{2(g)} \rightarrow N_{2(g)} + H_2O_{(\ell)}$

(f) $Sb_4S_{6(s)} + O_{2(g)} \rightarrow Sb_4O_{6(s)} + SO_{2(g)}$

(g) $Al(NO_3)_{3(aq)} + H_2SO_{4(aq)} \rightarrow Al_2(SO_4)_{3(aq)} + HNO_{3(aq)}$

(h) $Cu(NO_3)_{2(s)} \rightarrow CuO_{(s)} + NO_{2(g)} + O_{2(g)}$

(i) $KI_{(aq)} + MnO_{2(s)} + H_2SO_{4(aq)} \rightarrow K_2SO_{4(aq)} + MnSO_{4(aq)} + H_2O_{(\ell)} + I_{2(s)}$

(j) $C_6H_{6(\ell)} + O_{2(g)} \rightarrow CO_{2(g)} + H_2O_{(g)}$

R.9 Types of Chemical Reactions

Most chemical reactions can be classified as one of four main types of reactions. As you can see in Table R.3, these four types of reactions are classified by counting the number of reactants and products.

- In a **synthesis reaction**, two or more reactants combine to produce a single, different substance.
- In a **decomposition reaction**, a compound breaks down into elements or simpler compounds.
- In a **single displacement reaction**, one element in a compound is replaced by another element.
- In a **double displacement reaction**, the cations of two ionic compounds exchange places, resulting in the formation of two new compounds.

Table R.3 Four Types of Reactions

Name	General form	Example
synthesis reaction	$A + B \rightarrow AB$	$2Cu_{(s)} + S_{(g)} \rightarrow Cu_2S_{(s)}$
decomposition reaction	$AB \rightarrow A + B$	$2HgO_{(g)} \rightarrow 2Hg_{(\ell)} + O_{2(g)}$
single displacement reaction	$A + BC \rightarrow B + AC$	$Zn_{(s)} + CuCl_{2(aq)} \rightarrow Cu_{(s)} + ZnCl_{2(aq)}$
double displacement reaction	$AB + CD \rightarrow CB + AD$	$2NaOH_{(aq)} + CuSO_{4(aq)} \rightarrow Na_2SO_{4(aq)} + Cu(OH)_{2(s)}$

Many combustion reactions do not fit into any of the four categories in Table R.3. Therefore, combustion reactions are usually classified separately. A **combustion reaction** occurs when a compound reacts in the presence of oxygen to form oxides, heat, and light (burning). For example, sulfur reacts with oxygen to produce sulfur dioxide.

$$S_{8(s)} + 8O_{2(g)} \rightarrow 8SO_{2(g)}$$

Combustion reactions are common for compounds that are composed of carbon and hydrogen atoms. These compounds are called *hydrocarbons*.

- **Complete combustion** of a hydrocarbon occurs when the hydrocarbon reacts completely in the presence of sufficient oxygen. The complete combustion of a hydrocarbon produces only water vapour and carbon dioxide gas as products. The complete combustion of propane, C_3H_8, is shown below.

$$C_3H_{8(g)} + 5O_{2(g)} \rightarrow 3CO_{2(g)} + 4H_2O_{(g)}$$

- **Incomplete combustion** of a hydrocarbon occurs when there is not enough oxygen present for the hydrocarbon to react completely. The incomplete combustion of a hydrocarbon produces water, carbon dioxide, carbon monoxide, and solid carbon in varying amounts. More than one balanced equation is possible for the incomplete combustion of a hydrocarbon. One possible equation for the incomplete combustion of propane is shown below.

$$2C_3H_{8(g)} + 7O_{2(g)} \rightarrow 2CO_{2(g)} + 8H_2O_{(g)} + 2CO_{(g)} + 2C_{(s)}$$

R.10 Ionic Equations

When an ionic compound dissolves, the ions break away from their crystal lattice and become mobile in the solution. This process is called **dissociation**. You can summarize the dissociation process using a *dissociation equation*, as illustrated by the following examples.

$$BaSO_{4(s)} \rightarrow Ba^{2+}{}_{(aq)} + SO_4{}^{2-}{}_{(aq)}$$

$$Al(NO_3)_{3(s)} \rightarrow Al^{3+}{}_{(aq)} + 3NO_3{}^{-}{}_{(aq)}$$

Notice that, in addition to being balanced by atoms, these dissociation equations are also balanced by charge. The total net charge is zero on both sides of each equation.

For reactions that occur between ionic compounds in aqueous solution, a **total ionic equation** is used to show all the ions that are present and any ions that combine to form a precipitate. The total ionic equation for the reaction of silver nitrate and potassium chromate is given below.

$$2Ag^{+}{}_{(aq)} + 2NO_3{}^{-}{}_{(aq)} + 2K^{+}{}_{(aq)} + CrO_4{}^{2-}{}_{(aq)} \rightarrow 2K^{+}{}_{(aq)} + 2NO_3{}^{-}{}_{(aq)} + Ag_2CrO_{4(s)}$$

A **net ionic equation** shows only the ions that react and the precipitate that forms. The ions that do not participate in the reaction are called **spectator ions**, and they are omitted from the net ionic equation. In the reaction of silver nitrate and potassium chromate, $K^{+}{}_{(aq)}$ and $NO_3{}^{-}{}_{(aq)}$ are the spectator ions. They are not included in the net ionic equation for the reaction.

$$2Ag^{+}{}_{(aq)} + CrO_4{}^{2-}{}_{(aq)} \rightarrow Ag_2CrO_{4(s)}$$

Practice Problems

6. What type of reaction is represented by each balanced equation?

(a) $2N_2O_{(g)} \rightarrow 2N_{2(g)} + O_{2(g)}$

(b) $NO_{2(g)} + NO_{2(g)} \rightarrow N_2O_{4(g)}$

(c) $8Al_{(s)} + 3Co_3O_{4(s)} \rightarrow 9Co_{(s)} + 4Al_2O_{3(s)}$

7. Write the dissociation equation for each compound.

(a) $MgSO_3$

(b) $Fe(OH)_2$

(c) $Sn(SO_4)_2$

(d) Ag_2S

(e) $Al_2(C_2O_4)_3$

8. Aqueous solutions of several pairs of ionic solids are prepared and then mixed together. For each pair listed below, write

- the dissociation equation that occurs when each solid dissolves in water
- the balanced, total ionic equation
- the net ionic reaction, if a precipitate forms
- the spectator ions

(a) lead nitrate + zinc iodide

(b) potassium chlorate + calcium chloride

(c) ammonium phosphate + copper(II) chlorate

(d) sodium hydroxide + iron(III) nitrate

(e) calcium nitrate + potassium carbonate

R.11 Mole Calculations

The **mole** (symbol **mol**) is the SI base unit for quantity of matter. One mole contains the same number of particles as exactly 12 g of carbon-12. When measured experimentally, one mole is the quantity of matter that contains 6.022×10^{23} items. This number is called the **Avogadro constant (N_A)**. The following concept organizer summarizes the possible conversions between the number of moles (n), mass (m), molar mass (M), number of particles (N), and the Avogadro constant (N_A).

Concept Organizer — Conversions Used in Mole Calculations

Unit analysis

$$\text{mol} = \text{g} \times \frac{\text{mol}}{\text{g}} \qquad \text{mol} = \text{number of particles} \times \frac{\text{mol}}{6.02 \times 10^{23}\ \text{particles}}$$

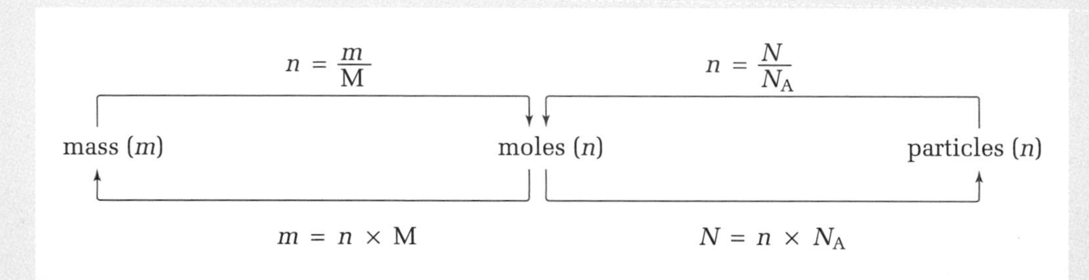

Unit analysis

$$\text{g} = \text{mol} \times \frac{\text{g}}{\text{mol}} \qquad \text{number of particles} = \text{mol} \times \frac{6.02 \times 10^{23}\ \text{particles}}{\text{mol}}$$

Sample Problem

Calculating Mass from Number of Particles

Problem

What is the mass of 2.734×10^{24} formula units of $Cu_3(PO_4)_2$?

What Is Required?

You need to calculate the mass of 2.734×10^{24} formula units of $Cu_3(PO_4)_2$.

What Is Given?

From the periodic table, you can obtain the molar masses of each element in the compound.
You know that one mole contains 6.022×10^{23} particles.

Plan Your Strategy

Step 1 Use the molar masses of the elements to calculate the molar mass of each formula unit.

Step 2 Divide the molar mass of one formula unit by the Avogadro constant to obtain the mass of each formula unit in grams. Multiply this value by the number of formula units to obtain the total mass in grams. These calculations can be performed as a single step.

Act on Your Strategy

Step 1 Molar mass of $Cu_3(PO_4)_2$ = 380.59 g/mol

Step 2 Total mass (g) = 2.734×10^{24} ~~formula units~~

$$\times \frac{1\ \text{mol}}{6.022 \times 10^{23}\ \text{formula units}} \times \frac{380.59\ \text{g}}{\text{mol}}$$

$$= 1.728 \times 10^{3}\ \text{g}$$

Check Your Solution

Make sure that the units cancel out to give the correct unit for the answer.

Practice Problems

9. What is the mass of 3.04×10^{-4} mol of baking soda, $NaHCO_3$?
10. How many moles of iron(III) acetate have a mass of 1.36×10^{3} g?
11. How many moles of chlorine atoms are present in a 9.2×10^{-2} g sample of zirconium(IV) chloride, $ZrCl_4$?
12. What mass of sodium carbonate decahydrate contains 5.47×10^{23} atoms of oxygen?
13. A 5.00 g sample is 88.4% zinc hydroxide. How many atoms of zinc are in this sample?
14. What mass of carbon is found in 8.25×10^{-6} mol of CaC_2?
15. Sample A is 90.2% Fe_3O_4 and has a mass of 4.82 g. Sample B is 100.0% ferric hydroxide and has a mass of 6.0 g. Show, by calculation, which sample contains more atoms of iron.
16. A sample of coal is 2.81% sulfur by mass. How many moles of sulfur are present in 5.00 t of the sample?

R.12 Concentration Calculations

The **concentration** of a solution is an expression of the amount of solute that is present in a given volume of solution. (A **solute** is a substance that is dissolved in a solution.) Table R.4 lists some units of concentration.

Table R.4 Summary of Units of Concentration

Unit of concentration	Mathematical expression	
mass/volume percent (m/v)	$\frac{\text{mass of solute (g)}}{\text{volume of solution}} \times 100\%$	
mass/mass percent (m/m)	$\frac{\text{mass of solute (g)}}{\text{mass of solution (g)}} \times 100\%$	
volume/volume percent (v/v)	$\frac{\text{volume of solute (mL)}}{\text{volume of solution (mL)}} \times 100\%$	
parts per million (ppm)	$\frac{\text{mass of solute (g)}}{\text{mass of solution (g)}} \times 10^{6}$	$\frac{\text{mg}}{\text{L}}$ in H_2O
parts per billion (ppb)	$\frac{\text{mass of solute (g)}}{\text{mass of solution (g)}} \times 10^{9}$	$\frac{\mu\text{g}}{\text{L}}$ in H_2O
molar concentration (mol/L), C	$\frac{\text{mol of solute, } n}{\text{volume of solution, } V}$, also written as $C = \frac{n}{V}$	

Sample Problem

Calculating the Concentration of a Solution

Problem

What volume of 0.0259 mol/L iron(III) nitrate can be prepared from 30.00 g of $Fe(NO_3)_3 \cdot 9H_2O$?

What Is Required?

You need to calculate the volume of 0.0259 mol/L iron(III) nitrate solution that can be made using 30.00 g of $Fe(NO_3)_3 \cdot 9H_2O$.

What Is Given?

The molar concentration of the iron(III) nitrate solution is 0.0259 mol/L.
The mass of $Fe(NO_3)_3 \cdot 9H_2O$ is 30.00 g.

Plan Your Strategy

Step 1 Use the chemical formula $Fe(NO_3)_3 \cdot 9H_2O$ to calculate the molar mass.

Step 2 Use the molar mass to calculate the number of moles of $Fe(NO_3)_3 \cdot 9H_2O$ that are present in 30.00 g.

Step 3 Use the number of moles of $Fe(NO_3)_3 \cdot 9H_2O$ and the molar concentration of the solution of iron(III) nitrate to calculate the volume of the solution. Use the equation $C = \frac{n}{V}$, where C is the molar concentration, n is the number of moles, and V is the volume.

Act on Your Strategy

Step 1 The molar mass of $Fe(NO_3)_3 \cdot 9H_2O$ is 403.97 g/mol.

Step 2 Number of moles of $Fe(NO_3)_3 \cdot 9H_2O$

$$= 30.00 \text{ g } Fe(NO_3)_3 \cdot 9H_2O \times \frac{1 \text{ mol}}{403.97 \text{ g}}$$

$$= 7.426 \times 10^{-2} \text{ mol}$$

Step 3 Volume of solution $(V) = \frac{n}{C}$

$$= \frac{7.426 \times 10^{-2} \text{ mol}}{0.0259 \text{ mol/L}}$$

$$= 2.87 \text{ L}$$

Check Your Solution

The units cancel out to give an answer in litres. The smallest number of significant digits in the question is three, so the answer is also given to three significant digits.

Practice Problems

17. A salt solution has a concentration of 1.00 mol/L. What volume of this solution is needed to prepare 2.00 L of a solution that has a concentration of 0.655 mol/L?

18. A 10.00 g sample of $CaCl_2$ is added to water to make 100.0 mL of solution. Then a 400.0 mL sample of water is added to this solution. Determine the concentration of Cl^- ions in the diluted solution.

19. A 50.0 mL sample of 0.85 mol/L $NaHCO_3$ is diluted to a volume of 250.0 mL. Then a 50.0 mL sample of this dilute solution is evaporated to dryness. What mass of $NaHCO_3$ remains?

20. What volume of 0.502 mol/L KOH solution must be diluted to prepare 1.500 L of 0.100 mol/L KOH?

21. A 500.0 mL sample of a 1.02×10^{-4} mol/L lead(II) acetate solution evaporates to dryness. What mass of $Pb(C_2H_3O_2)_2$ remains?

22. A 13.6 g sample of NaCl and a 7.34 g sample of $CaCl_2$ are dissolved in water to make 200.0 mL of solution. What is the concentration of Cl^- in this solution?

23. A 50.0 g sample of $Al(NO_3)_3$ is dissolved in water to prepare 1500.0 mL of solution. What is the concentration, in mol/L, of NO_3^- ions in the solution?

24. What condition must exist for the concentration of a solution expressed as m/m percent to be the same as its concentration expressed as m/v percent?

25. A sample of lead nitrate, with a mass of 0.00372 g, is completely dissolved in 250.0 mL of water. Assume that no change in volume occurs. Calculate the following concentrations.
    (a) the concentration of the solution, expressed in mol/L
    (b) the concentration of Pb^{2+}, expressed in ppm
    (c) the concentration of the solution, expressed as m/m percent

R.13 Stoichiometric Calculations

Stoichiometry is the study of the relative quantities of reactants and products in a chemical reaction. Calculations in stoichiometry usually involve mole ratios. A **mole ratio** is a ratio that compares the number of moles of different substances in a balanced chemical equation. For example, Table R.5 shows the formation of ammonia, NH_3, from nitrogen, N_2, and hydrogen, H_2. Suppose that you are given the number of moles of nitrogen that react with hydrogen. Using the mole ratio of nitrogen to ammonia, you can predict how many moles of ammonia will be produced.

Table R.5 The Formation of Ammonia

$N_{2(g)}$ +	$3H_{2(g)}$ →	$2NH_{3(g)}$
1 mol	3 mol	2 mol
If you are given 0.5 mol	Then $0.5\text{ mol N}_2 \times \dfrac{3\text{ mol H}_2}{1\text{ mol N}_2} = 1.5\text{ mol H}_2$	And $0.5\text{ mol N}_2 \times \dfrac{3\text{ mol NH}_3}{1\text{ mol N}_2} = 1.0\text{ mol NH}_3$

Stoichiometric problems can usually be solved by following the steps on the next page. Table R.6, on the next page, lists some equations that are useful for stoichiometric calculations.

Solving a Stoichiometric Problem

Step 1 Write a balanced chemical equation.

Step 2 If you are given the mass, number of particles, or volume of a substance, convert this value to the number of moles.

Step 3 Calculate the number of moles of the required substance based on the number of moles of the given substance, using the appropriate mole ratio.

Step 4 Convert the number of moles of the required substance to the appropriate unit, as directed by the question.

CONCEPT CHECK

Table R.6 includes the ideal gas law, $PV = nRT$. Practise manipulating this equation to solve for each of the four variables, and for the gas constant R.

Table R.6 Stoichiometric Calculations

Given	Equation
mass of a reactant or product (m) and molar mass (M)	$m = n \times M$ or $n = \frac{m}{M}$
volume of gas (V) at the known temperature (T) and pressure (P)	$PV = nRT$ or $n = \frac{PV}{RT}$
volume of solution (V) of known molar concentration (C)	$C = \frac{n}{V}$ or $n = C \times V$

Sample Problem

Mass and Particle Stoichiometry

Problem

Passing chlorine gas through molten sulfur produces liquid disulfur dichloride. How many molecules of chlorine react to produce 50.0 g of disulfur dichloride?

What Is Required?

You need to determine the number of molecules of chlorine gas that produce 50.0 g of disulfur dichloride.

What Is Given?

Reactant: chlorine, Cl_2
Reactant: sulfur, S
Product: disulfur dichloride, $S_2Cl_2 \rightarrow$ 50.0 g

Plan Your Strategy

Follow the steps for solving stoichiometric problems.

Act on Your Strategy

Step 1 Write the balanced chemical equation.

$$Cl_{2(g)} + 2S_{(\ell)} \rightarrow S_2Cl_{2(\ell)}$$

Step 2 Convert the number of grams of the product to moles.

$$\frac{50.0 \text{ g } S_2Cl_2}{135 \text{ g/mol}} = 0.370 \text{ mol } S_2Cl_2$$

Step 3 Use the mole ratio to calculate the amount of chlorine.

$$\frac{\text{Amount } Cl_2}{0.370 \text{ mol } S_2Cl_2} = \frac{1 \text{ mol } Cl_2}{1 \text{ mol } S_2Cl_2}$$

$$(0.370 \text{ mol } S_2Cl_2)\frac{\text{Amount } Cl_2}{0.370 \text{ mol } S_2Cl_2} = (0.370 \text{ mol } S_2Cl_2)\frac{1 \text{ mol } Cl_2}{1 \text{ mol } S_2Cl_2}$$

$$\text{Amount } Cl_2 = 0.370 \text{ mol } Cl_2$$

Step 4 Convert the number of moles of chlorine gas to the number of particles.

$0.370 \text{ mol } Cl_2 \times 6.02 \times 10^{23} \text{ molecules/mol}$
$= 2.23 \times 10^{23} \text{ molecules } Cl_2$

Therefore, 2.23×10^{23} molecules of chlorine react to produce 50.0 g of disulfur dichloride.

Check Your Solution

The units are correct. 2.0×10^{23} is about one third of a mole, or 0.33 mol. One third of a mole of disulfur dichloride has a mass of 45 g, which is close to 50 g. The answer is reasonable.

Sample Problem

Calculating the Limiting Reagent and Percentage Yield

Problem

A 18.9 g sample of Cu and a 82.0 mL sample of 16 mol/L HNO_3 are allowed to react. The balanced chemical equation is given below.

$Cu_{(s)} + 4HNO_{3(aq)} \rightarrow Cu(NO_3)_{2(aq)} + 2H_2O_{(\ell)} + 2NO_{2(g)}$

(a) Determine the mass of NO_2 that could be produced.

(b) If 22.6 g of NO_2 is actually produced in this reaction, calculate the percentage yield.

What Is Required?

Predict the mass of NO_2 that should be produced in the reaction. Next, calculate the percentage yield for this reaction, given the mass of NO_2 that is actually produced.

What Is Given?

You know the mass of each reactant: 18.9 g Cu and 82.0 mL of 16 mol/L HNO_3.

Plan Your Strategy

(a) Determine the mass of NO_2. Calculate the number of moles of each reactant that is present. Determine which reactant will be used up first, that is, which is the *limiting reactant*. Use the limiting reactant and the mole ratio to determine the number of moles of NO_2 produced. Convert the number of moles of NO_2 to grams.

(b) Calculate the percentage yield, using the following equation.

$$\text{Percentage yield} = \frac{\text{Actual yield}}{\text{Theoretical yield}} \times 100\,\%$$

Continued ...

Continued ...

Act on Your Strategy

(a) Determine the mass of NO_2.

$$\text{Number of moles of Cu} = 18.9\text{ g Cu} \times \frac{1\text{ mol}}{63.55\text{ g}}$$
$$= 0.297\text{ mol}$$

$$\text{Number of moles of } HNO_3 = C \times V$$
$$= 16.0\ \frac{\text{mol}}{\text{L}} \times 0.0820\text{ L}$$
$$= 1.31\text{ mol}$$

$$0.297\text{ mol Cu} \times \frac{4\text{ mol } HNO_3}{1\text{ mol Cu}} = 1.19\text{ mol } HNO_3$$

If all the Cu reacts, 1.19 mols of HNO_3 will react. More than this amount of HNO_3 is given (1.31 mol). Therefore, HNO_3 is in excess and Cu is the limiting reactant.

$$0.297\text{ mol Cu} \times \frac{2\text{ mol } NO_2}{1\text{ mol Cu}} = 0.594\text{ mol } NO_2$$

$$0.584\text{ mol } NO_2 \times \frac{46.01\text{ g}}{1\text{ mol}} = 27.3\text{ g } NO_2$$

Theoretically, a 27.3 g sample of NO_2 is produced.

(b) Calculate the percentage yield.

$$\text{Percentage yield} = \frac{22.6\text{ g}}{27.3\text{ g}} \times 100\%$$
$$= 82.8\%$$

Check Your Solution

The units are correct. A 27.3 g sample of NO_2 is predicted, but only a 22.6 g sample is produced. Therefore, it makes sense that the percentage yield is 82.8%.

Practice Problems

26. In the smelting process, iron(II) sulfide is converted to iron(III) oxide. The balanced chemical equation is given below.

$4FeS_{(s)} + 7O_{2(g)} \rightarrow 2Fe_2O_{3(s)} + 4SO_{2(g)}$

Calculate the mass of $Fe_2O_{3(s)}$ that is produced when 37.62 g of FeS and 22.56 g of O_2 are allowed to react.

27. 40.00 mL of 0.0256 mol/L gold(III) chloride is treated with 85.00 mL of 0.105 mol/L potassium iodide.

$AuCl_{3(aq)} + 3KI_{(aq)} \rightarrow AuI_{(s)} + 3KCl_{(aq)} + I_{2(aq)}$

What is the theoretical yield of $AuI_{(s)}$ produced?

28. When 300.0 mL of $TiCl_{4(g)}$, at 48.0°C and a pressure of 105.3 kPa, is reacted with 0.4320 g of magnesium, 0.4016 g of titanium is produced.

$TiCl_{4(g)} + 2Mg_{(s)} \rightarrow Ti_{(s)} + 2MgCl_{2(s)}$

Calculate the percentage yield for this reaction.

29. When a sample of solid potassium chlorate is heated strongly, a decomposition reaction occurs. Solid potassium chloride and oxygen gas are produced.

(a) Write the balanced equation for this reaction.

(b) When this reaction was carried out, a mass of 3.78 g of potassium chloride remained after 7.62 g of potassium chlorate decomposed. Calculate the percentage yield of potassium chloride.

30. Sodium carbonate reacts with dilute hydrochloric acid, as shown by the following equation.

$Na_2CO_{3(aq)} + 2HCl_{(aq)} \rightarrow 2NaCl_{(aq)} + H_2O_{(\ell)} + CO_{2(g)}$

(a) A chemist dissolves an impure sample of Na_2CO_3, with a mass of 0.250 g, in water. The chemist determines that 30.4 mL of 0.151 M HCl reacts with the Na_2CO_3 sample. Calculate the percentage purity of the sample.

(b) What volume of CO_2 is produced, at 21.5°C and a pressure of 104.0 kPa, in the reaction described in part (a)?

31. When 15.0 g of copper and 4.83 g of sulfur are heated, a 13.7 g mass of copper(I) sulfide is produced.

$2Cu_{(s)} + S_{(g)} \rightarrow Cu_2S_{(s)}$

What is the percentage yield of Cu_2S?

32. 130.4 mL of 0.459 mol/L $AgNO_3$ and 85.23 mL of 0.251 mol/L $AlCl_3$ are mixed.

$3AgNO_{3(aq)} + AlCl_{3(aq)} \rightarrow Al(NO_3)_{3(aq)} + 3AgCl_{(s)}$

What mass of $AgCl_{(s)}$ precipitates?

33. The following reaction occurs when a lead storage battery in a car is discharging.

$Pb_{(s)} + PbO_{2(s)} + 2H_2SO_{4(aq)} \rightarrow 2PbSO_{4(s)} + 2H_2O_{(\ell)}$

(a) A 3.850 g sample of PbO_2 reacts completely with 2.710 mL of H_2SO_4. Calculate the concentration of H_2SO_4.

(b) What mass of $PbSO_4$ is produced when 30.00 g of H_2SO_4 and 13.6 g of Pb react?

34. A sample of iron(III) oxide, with a mass of 325.0 g reacts with 90.75 L of carbon monoxide at 500.0°C and 1.216 atm.

$Fe_2O_{3(s)} + CO_{(g)} \rightarrow 2Fe_{(s)} + 3CO_{2(g)}$

(a) If a 185.0 g mass of iron is produced, what is the percentage yield for the reaction?

(b) What mass of reactant remains after the reaction stops?

35. The following reaction gives a 45.0% yield of manganese.

$2Al_{(s)} + 3MnO_{(s)} \rightarrow Al_2O_{3(s)} + 3Mn_{(s)}$

What mass of $Mn_{(s)}$ is produced when a 200.0 g sample of $Al_{(s)}$ reacts with 300.0 g of $MnO_{(s)}$?

36. What volume of 0.472 mol/L $AgNO_3$ will precipitate the chloride ion in 40.0 mL of 0.183 mol/L $AlCl_3$?

R.14 Representing Organic Molecules

In organic chemistry, there are many different ways to represent the same molecule. You can represent an organic molecule using a molecular formula, an expanded molecular formula, or a structural diagram.

A **molecular formula** includes the actual number and type of atoms that are present in the molecule, but it gives no information about how the atoms are connected to each other. For example, the molecular formula, C_2H_6O, indicates that there are 2 carbon atoms, 6 hydrogen atoms, and 1 oxygen atom in this molecule.

An **expanded molecular formula** shows the atoms in the order in which they appear in the molecule. You can write the molecular formula C_2H_6O in an expanded form as either CH_3CH_2OH or CH_3OCH_3. Each expanded form shows a different molecule that this molecular formula can represent. When writing an expanded molecular formula, use brackets to indicate groups that are attached to carbon atoms in the main chain. For example, the expanded molecular formula $CH_3C(CH_3)_2CH_2CH_3$ represents a molecule with four carbon atoms attached in a chain, and two additional $—CH_3$ groups attached to the second carbon atom.

A **structural diagram** is a simple drawing of a molecule. Structural diagrams include information about how the atoms are bonded. Figure R.11 illustrates the three types of structural diagrams. Figure R.12 shows structural diagrams for compounds with double and triple bonds.

- A **complete structural diagram** shows all the atoms in a structure and the way they are bonded to one another. Straight lines represent the bonds between the atoms.
- A **condensed structural diagram** is a more compact drawing of the structure. This type of diagram does not show the bonds to hydrogen atoms. Chemists assume that these bonds are present.
- A **line structural diagram** is even simpler than a condensed structural diagram. The end of each line, and the points at which the lines meet, represent carbon atoms. Hydrogen atoms are assumed to be present, but they are not shown. As you can see in Figure R.11(C) and Figure R.12, the lines that represent backbones of single-bonded or double-bonded carbon atoms are usually drawn in a zig-zag pattern. Triple-bonded carbon atoms are drawn in a straight line. (Note that lines are only used for the hydrocarbon portions of a molecule. Other atoms and groups, such as groups containing oxygen, nitrogen, and chlorine atoms, must be written in full.)

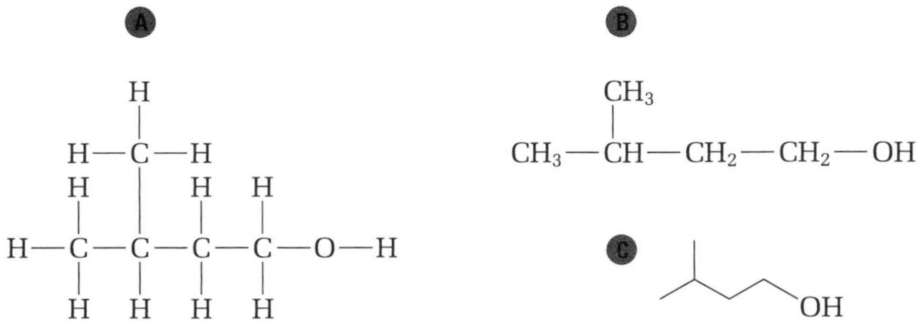

A

```
        H
        |
    H — C — H
    H   |   H   H
    |   |   |   |
H — C — C — C — C — O — H
    |   |   |   |
    H   H   H   H
```

B

$$CH_3—\underset{\displaystyle |}{\overset{CH_3}{}}CH—CH_2—CH_2—OH$$

C

OH

Figure R.11 Complete (A), condensed (B), and line (C) structural diagrams for a compound with the molecular formula $C_5H_{12}O$, and the expanded molecular formula $CH_3CH(CH_3)CH_2CH_2OH$

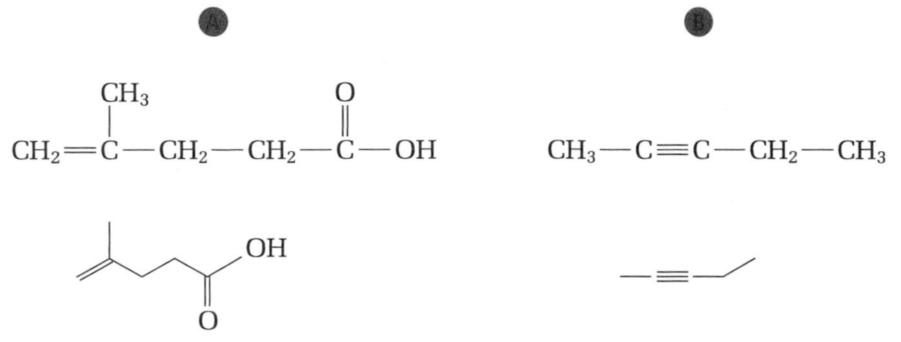

A

$$CH_2{=}C(CH_3){-}CH_2{-}CH_2{-}C(=O){-}OH$$

B

$$CH_3{-}C{\equiv}C{-}CH_2{-}CH_3$$

Figure R.12 (A) a condensed structural diagram and the corresponding line structural diagram for an organic compound with the expanded molecular formula $CH_2{=}C(CH_3)CH_2CH_2COOH$; (B) a condensed structural diagram and the corresponding line structural diagram for a compound with the expanded molecular formula $CH_3C{\equiv}CCH_2CH_3$

Practice Problems

37. What is the molecular formula of a compound with the expanded molecular formula $CH_3CH_2CH(OH)CH_2CH_3$?

38. Draw complete, condensed, and line structural diagrams for each compound.

(a) $CH_3CH_2CH_2CH_2CH_3$

(b) $CH_3CH_2CH(CH_3)CH_2CH_3$

(c) $CH_2{=}CHCH_2CH(CH_3)CH_3$

(d) $CH{\equiv}CCH_2CH_2CH_3$

R.15 Isomers of Organic Compounds

Many molecular formulas represent more than one molecular structure. Compounds that have the same molecular formula, but different structures, are called **structural isomers**. As you saw earlier, the molecular formula C_2H_6O can be represented by the expanded molecular formulas CH_3CH_2OH and CH_3OCH_3. The atoms in these molecules are attached differently to form two different structures. Both molecules, however, have the same molecular formula. Thus, they are isomers of each other. Because isomers have different shapes and bonding, they usually have different physical and chemical properties.

A molecule cannot rotate around a C=C bond. This fact makes a different type of isomer possible. **Cis-trans isomers** (also called **geometric isomers**) are compounds that have the same molecular formula, but different arrangements of atoms around a double carbon-carbon bond.

- In a **cis-isomer**, the two largest groups are on the same side of the double bond.
- In a **trans-isomer**, the two largest groups are on different sides of the double bond.

Figure R.13, in the margin, illustrates a pair of geometric isomers.

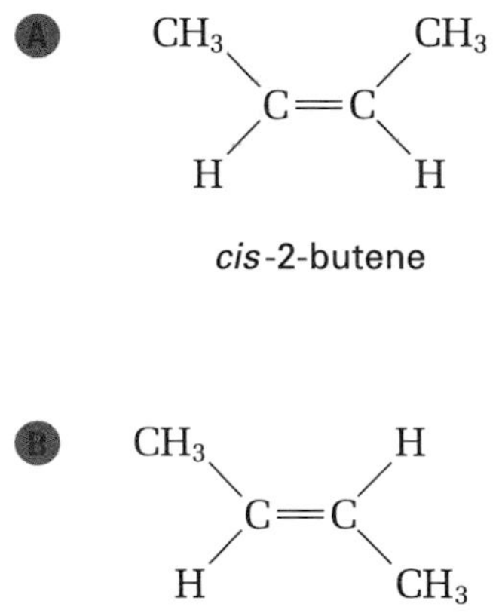

A *cis*-2-butene

B *trans*-2-butene

Figure R.13 Geometric isomers of the compound 2-butene

Practice Problems

39. Identify each diagram as a cis-isomer or a trans-isomer.

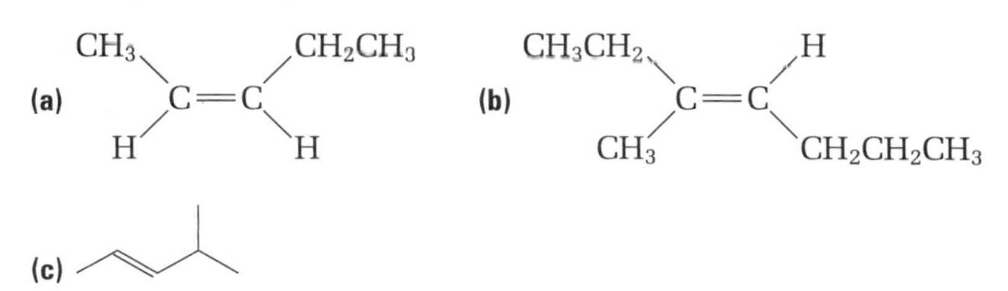

40. (a) Draw condensed structural diagrams for five isomers with the molecular formula C_6H_{12}.

(b) Draw line structural diagrams for five new isomers that also have the molecular formula C_6H_{12}. Include one pair of cis-trans isomers.

Answers to Practice Problems

1.(a) xenon tetrafluoride **(b)** phosphorus pentachloride (phosphorus(V) chloride) **(c)** carbon monoxide **(d)** sulfur hexafluoride **(e)** dinitrogen tetroxide **(f)** sodium fluoride **(g)** calcium carbonate **(h)** barium thiosulfate **(i)** nitrogen triiodide **2.(a)** titanic oxide; titanium(IV) oxide **(b)** cobaltous chloride; cobalt(II) chloride **(c)** nickelous bromide; nickel(II) bromide **(d)** mercuric oxide; mercury(II) oxide **(e)** stannous chloride; tin(II) chloride **3.(a)** $Cu(OH)_2$ **(b)** Hg_3N_2 **(c)** $Mn(HCO_3)_2$ **(d)** SBr_6 **(e)** K_2CrO_4 **(f)** $AsCl_3$ **(g)** $Fe(C_2H_3O_2)_3$ **(h)** K_2O_2 **(i)** $(NH_4)_2Cr_2O_7$ **(j)** $Sn(MnO_4)_2$ **(k)** $Pb(NO_3)_2$ **4.(a)** $Zn(ClO_4)_2$ **(b)** Hg_3N_2 **(c)** SnF_2 **(d)** $NH_4H_2PO_3$ **(e)** $Mn(SiO_4)_2$ **(f)** $Fe(OH)_3$ **(g)** CoN **(h)** $Pb(HCO_3)_2$ **(i)** $Ag_2C_2O_4$ **(j)** PtO_2 **(k)** SiC **(l)** $Ni_2(SO_3)_3$ **(m)** $SnCO_3$ **(n)** $Al(MnO_4)_3$
5.(a) $H_2O_{(\ell)} + 3NO_{2(g)} \rightarrow 2HNO_{3(aq)} + NO_{(g)}$
(b) $2NaOH_{(aq)} + Cl_{2(g)} \rightarrow NaCl_{(aq)} + NaClO_{(aq)} + H_2O_{(\ell)}$
(c) $2(NH_4)_3PO_{4(aq)} + 3Ba(OH)_{2(aq)} \rightarrow Ba_3(PO_4)_{2(s)} + 6NH_4OH_{(aq)}$
(d) $2Li_{(s)} + 2H_2O_{(\ell)} \rightarrow 2LiOH_{(aq)} + H_{2(g)}$ **(e)** $4NH_{3(g)} + 3O_{2(g)} \rightarrow 2N_{2(g)} + 6H_2O_{(\ell)}$
(f) $Sb_4S_{6(s)} + 9O_{2(g)} \rightarrow Sb_4O_{6(s)} + 6SO_{2(g)}$
(g) $2Al(NO_3)_{3(aq)} + 3H_2SO_{4(aq)} \rightarrow Al_2(SO_4)_{3(aq)} + 6HNO_{3(aq)}$
(h) $2Cu(NO_3)_{2(s)} \rightarrow 2CuO_{(s)} + 4NO_{2(g)} + O_{2(g)}$
(i) $2KI_{(aq)} + MnO_{2(s)} + 2H_2SO_{4(aq)} \rightarrow K_2SO_{4(aq)} + MnSO_{4(aq)} + 2H_2O_{(\ell)} + I_{2(s)}$
(j) $2C_6H_{6(\ell)} + 15O_{2(g)} \rightarrow 12CO_{2(g)} + 6H_2O_{(g)}$ **6.(a)** decomposition **(b)** synthesis **(c)** single displacement **7.(a)** $MgSO_{3(s)} \rightarrow Mg^{2+}{}_{(aq)} + SO_3{}^{2-}{}_{(aq)}$
(b) $Fe(OH)_{2(s)} \rightarrow Fe^{2+}{}_{(aq)} + 2OH^-{}_{(aq)}$ **(c)** $Sn(SO_4)_{2(s)} \rightarrow Sn^{4+}{}_{(aq)} + 2SO_4{}^{2-}{}_{(aq)}$
(d) $Ag_2S_{(s)} \rightarrow 2Ag^+{}_{(aq)} + S^{2-}{}_{(aq)}$ **(e)** $Al_2(C_2O_4)_{3(s)} \rightarrow 2Al^{3+}{}_{(aq)} + 3C_2O_4{}^{2-}{}_{(aq)}$
8.(a) $Pb(NO_3)_{2(s)} \rightarrow Pb^{2+}{}_{(aq)} + 2NO_3{}^-{}_{(aq)}$; $ZnI_{2(s)} \rightarrow Zn^{2+}{}_{(aq)} + 2I^-{}_{(aq)}$
$Pb^{2+}{}_{(aq)} + 2NO_3{}^-{}_{(aq)} + Zn^{2+}{}_{(aq)} + 2I^-{}_{(aq)} \rightarrow PbI_{2(s)} + Zn^{2+}{}_{(aq)} + 2NO_3{}^-{}_{(aq)}$
$Pb^{2+}{}_{(aq)} + 2I^-{}_{(aq)} \rightarrow PbI_{2(s)}$; spectator ions: $Zn^{2+}{}_{(aq)}$ and $NO_3{}^-{}_{(aq)}$
(b) $KClO_{3(s)} \rightarrow K^+{}_{(aq)} + ClO_3{}^-{}_{(aq)}$; $CaCl_{2(s)} \rightarrow Ca^{2+}{}_{(aq)} + 2Cl^-{}_{(aq)}$
$K^+{}_{(aq)} + ClO_3{}^-{}_{(aq)} + Ca^{2+}{}_{(aq)} + 2Cl^-{}_{(aq)} \rightarrow$ NR
(c) $(NH_4)_3PO_{4(s)} \rightarrow 3NH_4{}^+{}_{(aq)} + PO_4{}^{3-}{}_{(aq)}$; $Cu(ClO_3)_{2(s)} \rightarrow Cu^{2+}{}_{(aq)} + 2ClO_3{}^-{}_{(aq)}$
$6NH_4{}^+{}_{(aq)} + 2PO_4{}^{3-}{}_{(aq)} + 3Cu^{2+}{}_{(aq)} + 6ClO_3{}^-{}_{(aq)} \rightarrow Cu_3(PO_4)_{2(s)} + 6NH_4{}^+{}_{(aq)} + 6ClO_3{}^-{}_{(aq)}$;
$3Cu^{2+}{}_{(aq)} + 2PO_4{}^{3-}{}_{(aq)} \rightarrow Cu_3(PO_4)_{2(s)}$; spectator ions: $NH_4{}^+{}_{(aq)}$ and $ClO_3{}^-{}_{(aq)}$
(d) $NaOH_{(s)} \rightarrow Na^+{}_{(aq)} + OH^-{}_{(aq)}$; $Fe(NO_3)_{3(s)} \rightarrow Fe^{3+}{}_{(aq)} + 3NO_3{}^-{}_{(aq)}$
$3Na^+{}_{(aq)} + 3OH^-{}_{(aq)} + Fe^{3+}{}_{(aq)} + 3NO_3{}^-{}_{(aq)} \rightarrow Fe(OH)_{3(s)} + 3Na^+{}_{(aq)} + 3NO_3{}^-{}_{(aq)}$
$Fe^{3+}{}_{(aq)} + 3OH^-{}_{(aq)} \rightarrow Fe(OH)_{3(s)}$; spectator ions: $Na^+{}_{(aq)}$ and $NO_3{}^-{}_{(aq)}$
(e) $Ca(NO_3)_{2(s)} \rightarrow Ca^{2+}{}_{(aq)} + 2NO_3{}^-{}_{(aq)}$; $K_2CO_{3(s)} \rightarrow 2K^+{}_{(aq)} + CO_3{}^{2-}{}_{(aq)}$
$Ca^{2+}{}_{(aq)} + 2NO_3{}^-{}_{(aq)} + 2K^+{}_{(aq)} + CO_3{}^{2-}{}_{(aq)} \rightarrow CaCO_{3(s)} + 2K^+{}_{(aq)} + 2NO_3{}^-{}_{(aq)}$
$Ca^{2+}{}_{(aq)} + CO_3{}^{2-}{}_{(aq)} \rightarrow CaCO_{3(s)}$; spectator ions: $K^+{}_{(aq)}$ and $NO_3{}^-{}_{(aq)}$

9. 2.55×10^{-2} g **10.** 5.84 mol **11.** 1.6×10^{-3} mol **12.** 20.0 g **13.** 2.68×10^{22} atoms **14.** 1.98×10^{-4} g **15.** sample A: 3.39×10^{22} atoms; sample B: 3.38×10^{22} atoms **16.** 4.38×10^{3} mol **17.** 1.31 L **18.** 0.360 mol/L **19.** 0.71 g **20.** 0.299 L **21.** 1.66×10^{-2} g **22.** 1.82 mol/L **23.** 0.470 mol/L **24.** For equal masses of solute, volume and mass of the solution are the same. **25.(a)** 4.49×10^{-5} mol/L **(b)** 9.31 ppm Pb^{2+} **(c)** 1.49×10^{-3} m/m percent **26.** 32.17 g **27.** 0.332 g **28.** 94.43% **29.(a)** $2KClO_{3(s)} \rightarrow 2KCl_{(s)} + 3O_{2(g)}$ **(b)** 81.5% **30.(a)** 97.6% **(b)** 54.2 mL **31.** 72.9% **32.** 8.57 g **33.(a)** 11.88 mol/L **(b)** 39.8 g **34.(a)** 95.26% **(b)** 47.30 g of Fe_2O_3 **35.** 105 g **36.** 46.6 mL **37.** $C_5H_{12}O$

38.(a)

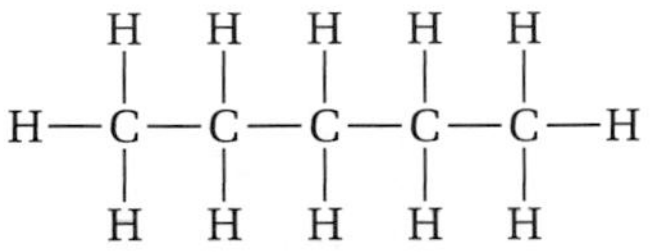

```
   H  H  H  H  H
   |  |  |  |  |
H—C—C—C—C—C—H
   |  |  |  |  |
   H  H  H  H  H
```

$CH_3—CH_2—CH_2—CH_2—CH_3$

(b)

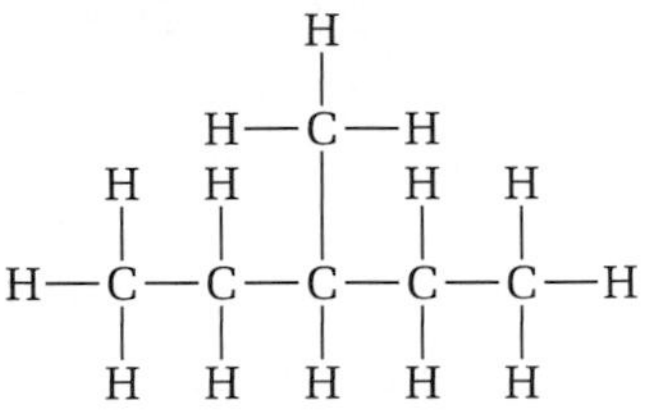

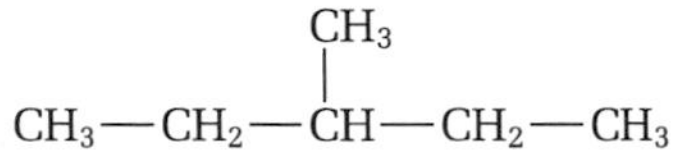

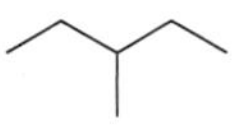

```
         H
         |
       H—C—H
   H  H  |  H  H
   |  |  |  |  |
H—C—C—C—C—C—H
   |  |  |  |  |
   H  H  H  H  H
```

$CH_3—CH_2—CH(CH_3)—CH_2—CH_3$

(c)

```
            H
            |
          H—C—H
   H  H  H  |  H
   |  |  |  |  |
H—C=C—C—C—C—H
         |  |  |
         H  H  H
```

$CH_2{=}CH—CH_2—CH(CH_3)—CH_3$

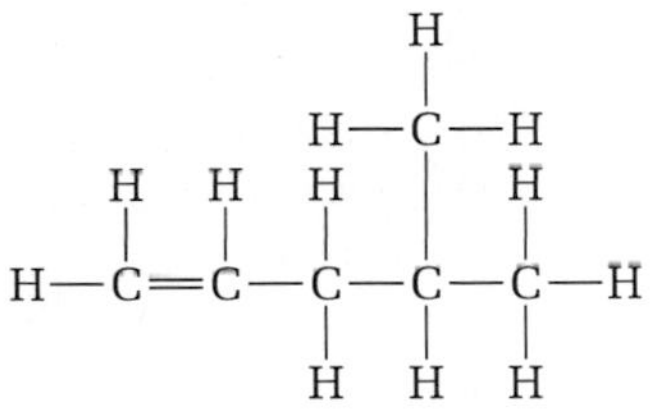

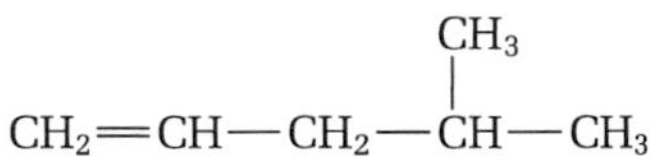

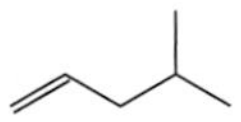

(d)

```
         H  H  H
         |  |  |
H—C≡C—C—C—C—H
         |  |  |
         H  H  H
```

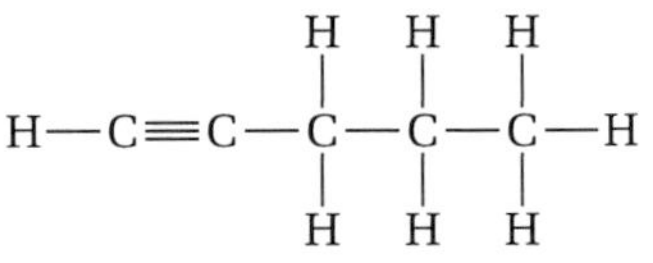

$CH{\equiv}C—CH_2—CH_2—CH_3$

39.(a) cis **(b)** trans **(c)** trans

40.(a) $CH_2{=}CH—CH_2—CH_2—CH_2—CH_3$,

$CH_3—CH{=}CH—CH_2—CH_2—CH_3$,

$CH_3—CH—CH{=}CH—CH_2—CH_3$,

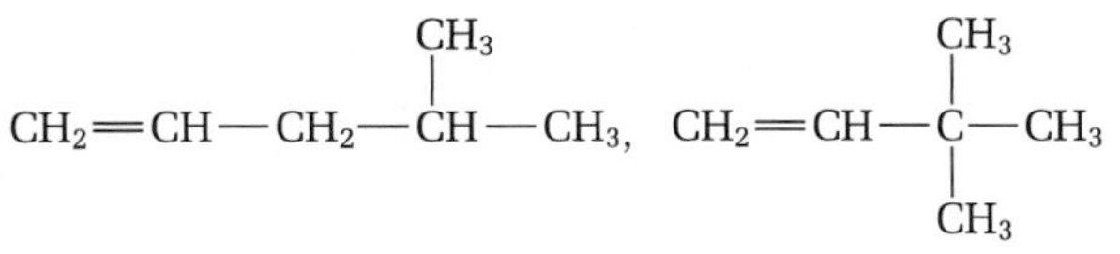

$CH_2{=}CH—CH_2—CH(CH_3)—CH_3$, $CH_2{=}CH—C(CH_3)_2—CH_3$

(b)

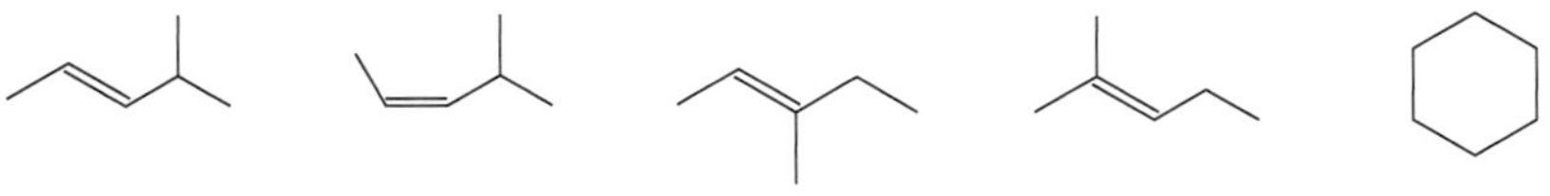

UNIT 1

Organic Chemistry

UNIT 1 CONTENTS

CHAPTER 1
Classifying Organic Compounds

CHAPTER 2
Reactions of Organic Compounds

UNIT 1 ISSUE
Current Issues Related to Organic Chemistry

UNIT 1 OVERALL EXPECTATIONS

- How do the structures of various organic compounds differ? What chemical reactions are typical of these compounds?
- How can you name different organic compounds and represent their structures? What do you need to know in order to predict the products of organic reactions?
- How do organic compounds affect your life? How do they affect the environment?

Unit Issue Prep

Before beginning Unit 1, read pages 110 to 111 to find out about the unit issue. In the unit issue, you will analyze an issue that involves chemistry and society. You can start planning your research as you go through this unit. Which topics interest you the most? How does society influence developments in science and technology?

At this moment, you are walking, sitting, or standing in an "organic" body. Your skin, hair, muscles, heart, and lungs are all made from organic compounds. In fact, the only parts of your body that are *not* mostly organic are your teeth and bones! When you study organic chemistry, you are studying the substances that make up your body and much of the world around you. Medicines, clothing, carpets, curtains, and wood and plastic furniture are all manufactured from organic chemicals. If you look out a window, the grass, trees, squirrels, and insects you may see are also composed of organic compounds.

Are you having a sandwich for lunch? Bread, butter, meat, and lettuce are made from organic compounds. Will you have dessert? Sugar, flour, vanilla, and chocolate are also organic. What about a drink? Milk and juice are solutions of water in which organic compounds are dissolved.

In this unit, you will study a variety of organic compounds. You will learn how to name them and how to draw their structures. You will also learn how these compounds react, and you will use your knowledge to predict the products of organic reactions. In addition, you will discover the amazing variety of organic compounds in your body and in your life.

CHAPTER

1 Classifying Organic Compounds

Chapter Preview

1.1 Bonding and the Shape of Organic Molecules

1.2 Hydrocarbons

1.3 Single-Bonded Functional Groups

1.4 Functional Groups With the C=O bond

Prerequisite Concepts and Skills

Before you begin this chapter, review the following concepts and skills:

- drawing Lewis structures (Concepts and Skills Review)
- writing molecular formulas and expanded molecular formulas (Concepts and Skills Review)
- drawing complete, condensed, and line structural diagrams (Concepts and Skills Review)
- identifying structural isomers (Concepts and Skills Review)

As you wander through the supermarket, some advertising claims catch your eye. "Certified organic" and "all natural" are stamped on the labels of some foods. Other labels claim that the foods are "chemical free." As a chemistry student, you are aware that these labels may be misleading. Are all "chemicals" harmful in food, as some of the current advertising suggests?

Many terms are used inaccurately in everyday life. The word "natural" is often used in a manner suggesting that all natural compounds are safe and healthy. Similarly, the word "chemical" is commonly used to refer to artificial compounds only. The food industry uses "organic" to indicate foods that have been grown without the use of pesticides, herbicides, fertilizers, hormones, and other synthetic chemicals. The original meaning of the word "organic" refers to anything that is or has been alive. In this sense, *all* vegetables are organic, no matter how they are grown.

Organic chemistry is the study of compounds that are based on carbon. Natural gas, rubbing alcohol, aspirin, and the compounds that give fragrance to a rose, are all organic compounds. In this chapter, you will learn how to identify and name molecules from the basic families of organic compounds. You will be introduced to the shape, structure, and properties of different types of organic compounds.

What is the word "organic" intended to mean here? How is this meaning different from the scientific meaning of the word?

Bonding and the Shape of Organic Molecules

1.1

Early scientists defined organic compounds as compounds that originate from living things. In 1828, however, the German chemist Friedrich Wohler (1800–1882) made an organic compound called urea, $CO(NH_2)_2$, out of an inorganic compound called ammonium cyanate, NH_4CN. Urea is found in the urine of mammals. This was the first time in history that a compound normally made only by living things was made from a non-living substance. Since Wohler had discovered that organic compounds can be made without the involvement of a life process, a new definition was required. **Organic compounds** are now defined as compounds that are based on carbon. They usually contain carbon-carbon and carbon-hydrogen bonds.

The Carbon Atom

There are several million organic compounds, but only about a quarter of a million inorganic compounds (compounds that are not based on carbon). Why are there so many organic compounds? The answer lies in the bonding properties of carbon.

As shown in Figure 1.1, *each carbon atom usually forms a total of four covalent bonds.* Thus, a carbon atom can connect to as many as four other atoms. Carbon can bond to many other types of atoms, including hydrogen, oxygen, and nitrogen.

$$\cdot\dot{\underset{\cdot}{C}}\cdot + 4H\cdot \rightarrow H:\overset{H}{\underset{H}{\overset{\cdot\cdot}{\underset{\cdot\cdot}{C}}}}:H$$

Figure 1.1 **This Lewis structure shows methane, the simplest organic compound. The carbon atom has four valence electrons, and it obtains four more electrons by forming four covalent bonds with the four hydrogen atoms.**

In addition, *carbon atoms can form strong single, double, or triple bonds with other carbon atoms.* In a single carbon-carbon bond, one pair of electrons is shared between two carbon atoms. In a double bond, two pairs of electrons are shared between two atoms. In a triple bond, three pairs of electrons are shared between two atoms.

Molecules that contain only single carbon-carbon bonds are *saturated.* In other words, all carbon atoms are bonded to the maximum number of other atoms: four. No more bonding can occur. Molecules that contain double or triple carbon-carbon bonds are *unsaturated.* The carbon atoms on either side of the double or triple bond are bonded to less than four atoms each. There is potential for more atoms to bond to each of these carbon atoms.

Carbon's unique bonding properties allow the formation of a variety of structures, including chains and rings of many shapes and sizes. Figure 1.2 on the next page illustrates some of the many shapes that can be formed from a backbone of carbon atoms. This figure includes examples of three types of structural diagrams that are used to depict organic molecules. (The Concepts and Skills Review contains a further review of these types of structural diagrams.)

Section Preview/ Specific Expectations

In this section, you will

- **discuss** the use of the terms *organic, natural,* and *chemical* in advertising
- **demonstrate** an understanding of the three types of carbon-carbon bonding and the shape of a molecule around each type of bond
- **communicate** your understanding of the following terms: *organic chemistry, organic compounds, tetrahedral, trigonal planar, linear, bent, electronegativity, bond dipole, polar, molecular polarity*

Web LINK

www.mcgrawhill.ca/links/chemistry12

In the chapter opener, you considered how the terms "natural" and "chemical" are used inaccurately. A *natural* substance is a substance that occurs in nature and is not artificial. A *chemical* is any substance that has been made using chemical processes in a laboratory. A chemical can also be defined as any substance that is composed of atoms. This definition covers most things on Earth. Go to the web site above, and click on **Web Links** to find out where to go next. Look up some *natural* poisons, pesticides, and antibiotics that are produced by animals, plants, and bacteria. Then look up some beneficial *chemicals* that have been synthesized by humans. Make a poster to illustrate your findings.

A few carbon compounds are considered to be inorganic. These include carbon dioxide, CO_2, and and carbon compounds containing complex negative ions (for example, CO_3^{2-}, HCO_3^-, and OCN^-).

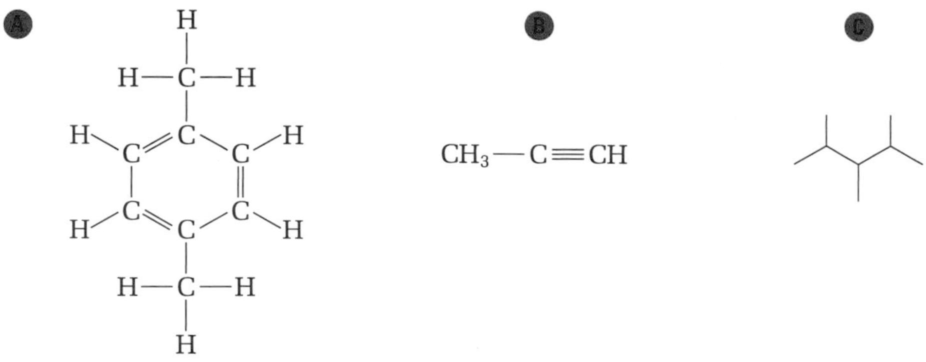

Figure 1.2 (A) This complete structural diagram shows all the bonds in the molecule. (B) This condensed structural diagram shows only carbon-carbon bonds. (C) This line structural diagram uses lines to depict carbon-carbon bonds.

(B) $CH_3—C≡CH$

Carbon compounds in which carbon forms only single bonds have a different shape than compounds in which carbon forms double or triple bonds. In the following ExpressLab, you will see how each type of bond affects the shape of a molecule.

ExpressLab Molecular Shapes

The type of bonding affects the shape and movement of a molecule. In this ExpressLab, you will build several molecules to examine the shape and character of their bonds.

Procedure

1. Build a model for each of the following compounds. Use a molecular model kit or a chemical modelling computer program.

$CH_3—CH_2—CH_2—CH_3$ butane

$H_2C═CH—CH_2—CH_3$ 1–butene

$H_2C═CH—CH═CH_2$ 1,3–butadiene

$H_3C—C≡C—CH_3$ 2–butyne

2. Identify the different types of bonds in each molecule.
3. Try to rotate each molecule. Which bonds allow rotation around the bond? Which bonds prevent rotation?
4. Examine the shape of the molecule around each carbon atom. Draw diagrams to show your observations.

Analysis

1. Which bond or bonds allow rotation to occur? Which bond or bonds are fixed in space?
2. (a) Describe the shape of the molecule around a carbon atom with only single bonds.
   (b) Describe the shape of the molecule around a carbon atom with one double bond and two single bonds.
   (c) Describe the shape of the molecule around a carbon atom with a triple bond and a single bond.
   (d) Predict the shape of a molecule around a carbon atom with two double bonds.
3. Molecular model kits are a good representation of real atomic geometry. Are you able to make a quadruple bond between two atoms with your model kit? What does this tell you about real carbon bonding?

As you observed in the ExpressLab, the shape of a molecule depends on the type of bond. Table 1.1 describes some shapes that you must know for your study of organic chemistry. In Unit 2, you will learn more about why different shapes and angles form around an atom.

Table 1.1 Common Molecular Shapes in Organic Molecules

Central atom	Shape	Diagram
carbon with four single bonds	The shape around this carbon atom is **tetrahedral**. That is, the carbon atom is at the centre of an invisible tetrahedron, with the other four atoms at the vertices of the tetrahedron. This shape results because the electrons in the four bonds repel each other. In the tetrahedral position, the four bonded atoms and the bonding electrons are as far apart from each other as possible.	H, C, H, H, H; 109.5°
carbon with one double bond and two single bonds	The shape around this carbon atom is **trigonal planar**. The molecule lies flat in one plane around the central carbon atom, with the three bonded atoms spread out, as if to touch the corners of a triangle.	H, H, C=C, CH_3, CH_3; 120°, 120° — O=C, H_3C, CH_3; 120°
carbon with two double bonds or one triple bond and one single bond	The shape around this carbon atom is **linear**. The two atoms bonded to the carbon atom are stretched out to either side to form a straight line.	H—C≡C—CH_3; 180°
oxygen with two single bonds	A single-bonded oxygen atom forms two bonds. An oxygen atom also has two pairs of non-bonding electrons, called lone pairs. Since there are a total of four electron pairs around a single-bonded oxygen atom, the shape around this oxygen atom is a variation of the tetrahedral shape. Because there are only two bonds, however, the shape around a single-bonded oxygen atom is usually referred to as **bent**.	lone pairs; O, H, H; 104.5°

Three-Dimensional Structural Diagrams

Two-dimensional structural diagrams of organic compounds, such as condensed structural diagrams and line structural diagrams, work well for flat molecules. As shown in the table above, however, molecules containing single-bonded carbon atoms are not flat.

You can use a three-dimensional structural diagram to draw the tetrahedral shape around a single-bonded carbon atom. In a three-dimensional diagram, wedges are used to give the impression that an atom or group is coming forward, out of the page. Dashed or dotted lines are used to show that an atom or group is receding, or being pushed back into the page. In Figure 1.3, the Cl atom is coming forward, and the Br atom is behind. The two H atoms are flat against the surface of the page.

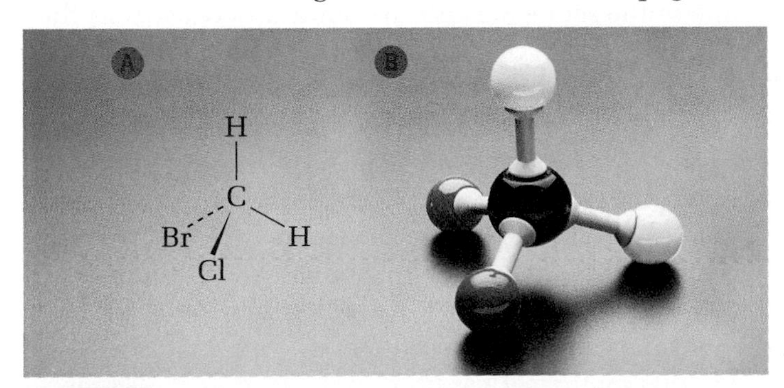

Figure 1.3 **(A) Three-dimensional structural diagram of the bromochloromethane molecule, $BrClCH_2$ (B) Ball-and-stick model**

CONCEPT CHECK

The following diagram shows 1-bromoethanol. (You will learn the rules for naming molecules such as this later in the chapter.) Which atom or group is coming forward, out of the page? Which atom or group is receding back, into the page?

CH_3
C
Br H
HO

Molecular Shape and Polarity

The three-dimensional shape of a molecule is particularly important when the molecule contains polar covalent bonds. As you may recall from your previous chemistry course, a *polar covalent bond* is a covalent bond between two atoms with different electronegativities.

Electronegativity is a measure of how strongly an atom attracts electrons in a chemical bond. The electrons in a polar covalent bond are attracted more strongly to the atom with the higher electronegativity. This atom has a partial negative charge, while the other atom has a partial positive charge. Thus, every polar bond has a **bond dipole**: a partial negative charge and a partial positive charge, separated by the length of the bond. Figure 1.4 illustrates the polarity of a double carbon-oxygen bond. Oxygen has a higher electronegativity than carbon. Therefore, the oxygen atom in a carbon-oxygen bond has a partial negative charge, and the carbon atom has a partial positive charge.

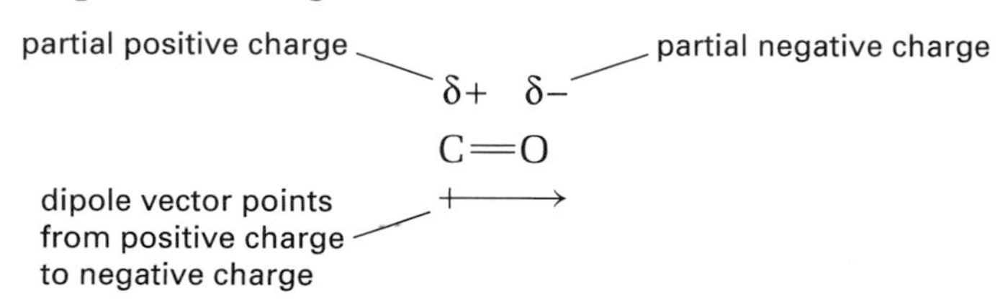

Figure 1.4 **Dipoles are often represented using *vectors*. Vectors are arrows that have direction and location in space.**

CONCEPT CHECK

In this unit, you will encounter the following polar bonds: C—I, C—F, C—O, O—H, N—H, and C—N. Use the electronegativities in the periodic table to discover which atom in each bond has a partial negative charge, and which has a partial positive charge.

Other examples of polar covalent bonds include C—O, O—H, and N—H. Carbon and hydrogen attract electrons to almost the same degree. Therefore, when carbon is bonded to another carbon atom or to a hydrogen atom, the bond is not usually considered to be polar. For example, C—C bonds are considered to be non-polar.

Predicting Molecular Polarity

A molecule is considered to be **polar**, or to have a **molecular polarity**, when the molecule has an overall imbalance of charge. That is, the molecule has a region with a partial positive charge, and a region with a partial negative charge. Surprisingly, not all molecules with polar bonds are polar molecules. For example, a carbon dioxide molecule has two polar C=O bonds, but it is not a polar molecule. On the other hand, a water molecule has two polar O—H bonds, and it *is* a polar molecule. How do you predict whether or not a molecule that contains polar bonds has an overall molecular polarity? To determine molecular polarity, you must consider the shape of the molecule and the bond dipoles within the molecule.

If equal bond dipoles act in opposite directions in three-dimensional space, they counteract each other. A molecule with identical polar bonds that point in opposite directions is not polar. Figure 1.5 shows two examples, carbon dioxide and carbon tetrachloride. Carbon dioxide, CO_2, has two polar C=O bonds acting in opposite directions, so the molecule is non-polar. Carbon tetrachloride, CCl_4, has four polar C—Cl bonds in a tetrahedral shape. You can prove mathematically that four identical dipoles, pointing toward the vertices of a tetrahedron, counteract each other exactly. (Note that this mathematical proof only applies if all four bonds are identical.) Therefore, carbon tetrachloride is also non-polar.

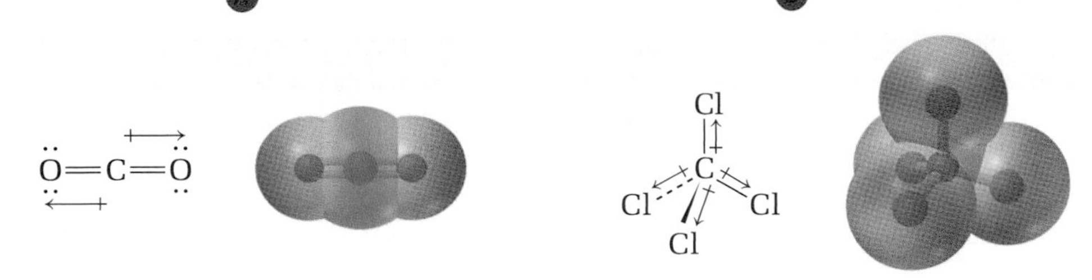

Figure 1.5 The red colour indicates a region of negative charge, and the blue colour indicates a region of positive charge. In non-polar molecules, such as carbon dioxide (A) and carbon tetrachloride (B), the charges are distributed evenly around the molecule.

If the bond dipoles in a molecule do not counteract each other exactly, the molecule is polar. Two examples are water, H_2O, and chloroform, $CHCl_3$, shown in Figure 1.6. Although each molecule has polar bonds, the bond dipoles do not act in exactly opposite directions. The bond dipoles do not counteract each other, so these two molecules are polar.

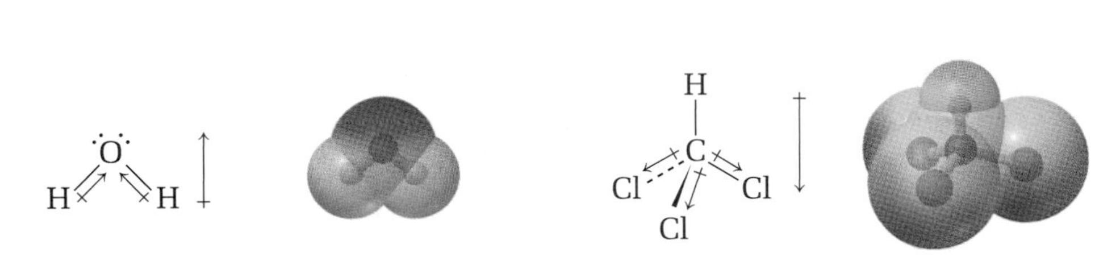

Figure 1.6 In polar molecules, such as water (A) and chloroform (B), the charges are distributed unevenly around the molecule. One part of the molecule has an overall negative charge, and another part has an overall positive charge.

The steps below summarize how to predict whether or not a molecule is polar. The Sample Problem that follows gives three examples.
Note: For the purpose of predicting molecular polarity, you can assume that C—H bonds are non-polar. In fact, they have a very low polarity.

Step 1 Does the molecule have polar bonds? If your answer is no, see below. If your answer is yes, go to step 2.

If a molecule has no polar bonds, it is non-polar.

Examples: $CH_3CH_2CH_3$, $CH_2{=}CH_2$

Step 2 Is there more than one polar bond? If your answer is no, see below. If your answer is yes, go to step 3.

If a molecule contains only one polar bond, it is polar.

Examples: CH_3Cl, $CH_3CH_2CH_2Cl$

Step 3 Do the bond dipoles act in opposite directions and counteract each other? Use your knowledge of three-dimensional molecular shapes to help you answer this question. If in doubt, use a molecular model to help you visualize the shape of the molecule.

If a molecule contains bond dipoles that do not counteract each other, the molecule is polar.

Examples: H_2O, $CHCl_3$

If the molecule contains dipoles that counteract each other, the molecule is non-polar.

Examples: CO_2, CCl_4

CHEM FACT

The polarity of a molecule determines its solubility. Polar molecules attract each other, so polar molecules usually dissolve in polar solvents, such as water. Non-polar molecules do not attract polar molecules enough to compete against the strong attraction between polar molecules. Therefore, non-polar molecules are not usually soluble in water. Instead, they dissolve in non-polar solvents, such as benzene.

Sample Problem

Molecular Polarity

Problem

Use your knowledge of molecular shape and polar bonds to predict whether each molecule has an overall molecular polarity.

(a) $CH_3—CH_3$

(b) $CH_3—CH_2—O—H$

(c) 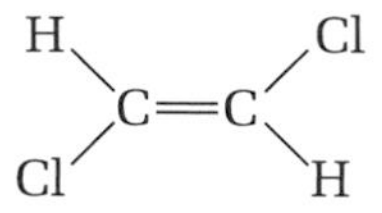 (H)(Cl)C=C(Cl)(H)

Solution

(a) Step 1 Does the molecule have polar bonds? H—C and C—C bonds are usually considered to be non-polar. Thus, this molecule is non-polar.

(b) Steps 1 and 2 Does the molecule have polar bonds? Is there more than one polar bond? The C—O and O—H bonds are polar.

Step 3 Do the bond dipoles counteract each other? The shape around oxygen is bent, and the dipoles are unequal. Therefore, these dipoles do not counteract each other. The molecule has an overall polarity.

(c) Steps 1 and 2 Does the molecule have polar bonds? Is there more than one polar bond? The C—Cl bonds are polar.

Step 3 Do the bond dipoles counteract each other? If you make a model of this molecule, you can see that the C—Cl dipoles act in opposite directions. They counteract each other. Thus, this molecule is non-polar.

Practice Problems

1. Predict and sketch the three-dimensional shape around each single-bonded atom.
   (a) C and O in CH_3OH
   (b) C in CH_4
2. Predict and sketch the three-dimensional shape of each multiple-bonded molecule.
   (a) $HC≡CH$
   (b) $H_2C=O$
3. Identify any polar bonds that are present in each molecule in questions 1 and 2.
4. For each molecule in questions 1 and 2, predict whether the molecule as a whole is polar or non-polar.

Section Summary

In this section, you studied carbon bonding and the three-dimensional shapes of organic molecules. You learned that you can determine the polarity of a molecule by considering its shape and the polarity of its bonds. In Unit 2, you will learn more about molecular shapes and molecular polarity. In the next section, you will review the most basic type of organic compound: hydrocarbons.

Section Review

1 MC How are the following statements misleading? Explain your reasoning.

(a) "You should eat only organic food."

(b) "All-natural ingredients make our product the healthier choice."

(c) "Chemicals are harmful."

2 K/U Classify each bond as polar or non-polar.

(a) C—O **(b)** C—C **(c)** C—N **(d)** C=C **(e)** C=O

3 K/U Describe the shape of the molecule around the carbon atom that is highlighted.

```
      H  H                  H  O  H  H
      |  |                  |  ‖  |  |
(a) H—C—C—H          (b) H—C—C—C—C—H
      |  |                  |     |  |
      H  H                  H     H  H
```

4 K/U Identify each molecule in question 3 as either polar or non-polar. Explain your reasoning.

5 I Identify the errors in the following structural diagrams.

(a) $HC{=}CH{-}CH_2{-}CH_3$ **(b)**

6 C Use your own words to explain why so many organic compounds exist.

1.2 Hydrocarbons

Section Preview/ Specific Expectations

In this section, you will

- **distinguish** among the following classes of organic compounds: alkanes, alkenes, alkynes, and aromatic compounds
- **draw** and **name** hydrocarbons using the IUPAC system
- **communicate** your understanding of the following terms: *hydrocarbons, aliphatic hydrocarbon, aromatic hydrocarbon, alkane, cycloalkane, alkene, functional group, alkyne, alkyl group*

In this section, you will review the structure and names of hydrocarbons. As you may recall from your previous chemistry studies, **hydrocarbons** are the simplest type of organic compound. Hydrocarbons are composed entirely of carbon and hydrogen atoms, and are widely used as fuels. Gasoline, propane, and natural gas are common examples of hydrocarbons. Because they contain only carbon and hydrogen atoms, hydrocarbons are non-polar compounds.

Scientists classify hydrocarbons as either aliphatic or aromatic. An **aliphatic hydrocarbon** contains carbon atoms that are bonded in one or more chains and rings. The carbon atoms have single, double, or triple bonds. Aliphatic hydrocarbons include straight chain and cyclic alkanes, alkenes, and alkynes. An **aromatic hydrocarbon** is a hydrocarbon based on the aromatic benzene group. You will encouter this group later in the section. Benzene is the simplest aromatic compound. Its bonding arrangement results in special molecular stability.

Alkanes, Alkenes, and Alkynes

An **alkane** is a hydrocarbon that has only single bonds. Alkanes that do not contain rings have the formula C_nH_{2n+2}. An alkane in the shape of a ring is called a **cycloalkane.** Cycloalkanes have the formula C_nH_{2n}. An **alkene** is a compound that has at least one double bond. Straight-chain alkenes with one double bond have the same formula as cycloalkanes, C_nH_{2n}.

A double bond involves two pairs of electrons. In a double bond, one pair of electrons forms a single bond and the other pair forms an additional, weaker bond. The electrons in the additional, weaker bond react faster than the electrons in the single bond. Thus, carbon-carbon double bonds are more reactive than carbon-carbon single bonds. When an alkene reacts, the reaction almost always occurs at the site of the double bond.

A **functional group** is a reactive group of bonded atoms that appears in all the members of a chemical family. Each functional group reacts in a characteristic way. Thus, functional groups help to determine the physical and chemical properties of compounds. For example, the reactive double bond is the functional group for an alkene. In this course, you will encounter many different functional groups.

An **alkyne** is a compound that has at least one triple bond. A straight-chain alkyne with one triple bond has the formula C_nH_{2n-2}. Triple bonds are even more reactive than double bonds. The functional group for an alkyne is the triple bond.

Figure 1.7 gives examples of an alkane, a cycloalkane, an alkene, and an alkyne.

CONCEPT CHECK

The molecular formula of benzene is C_6H_6. Remember that each carbon atom must form a total of four bonds. A single bond counts as one bond, a double bond counts as two bonds, and a triple bond counts as three bonds. Hydrogen can form only one bond. Draw a possible structure for benzene.

cyclopentane, C_5H_{10}

$CH_3CH_2CH_2CH_3$

butane, C_4H_{10}

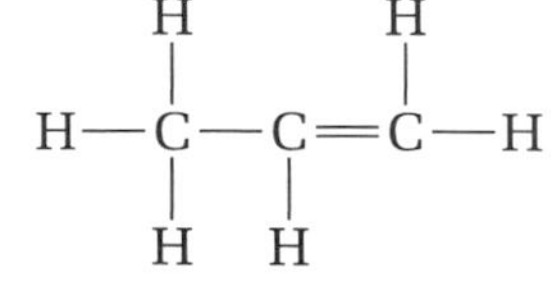

propene, C_3H_6

$CH_3—C≡C—CH_2—CH_2—CH_3$

2-hexyne, C_6H_{10}

Figure 1.7 Identify each compound as an alkane, a cycloalkane, an alkene, or an alkyne.

General Rules for Naming Organic Compounds

The International Union of Pure and Applied Chemistry (IUPAC) has set standard rules for naming organic compounds. The systematic (or IUPAC) names of alkanes and most other organic compounds follow the same pattern, shown below.

prefix + root + suffix

The Root: How Long Is the Main Chain?

The root of a compound's name indicates the number of carbon atoms in the main (parent) chain or ring. Table 1.2 gives the roots for hydrocarbon chains that are up to ten carbons long. To determine which root to use, count the carbons in the main chain, or main ring, of the compound. If the compound is an alkene or alkyne, the main chain or ring must include the multiple bond.

Table 1.2 Root Names

Number of carbon atoms	1	2	3	4	5	6	7	8	9	10
Root	-meth-	-eth-	-prop-	-but-	-pent-	-hex-	-hept-	-oct-	-non-	-dec-

Figure 1.8 shows some hydrocarbons, with the main chain or ring highlighted.

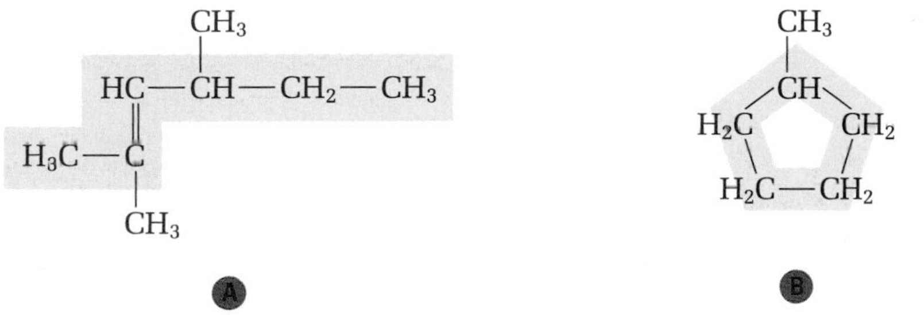

Figure 1.8 (A) There are six carbons in the main chain. The root is -hex-. (B) There are five carbons in the main ring. The root is -pent-.

The Suffix: What Family Does the Compound Belong To?

The suffix indicates the type of compound, according to the functional groups present. (See Table 1.4 on page 22.) As you progress through this chapter, you will learn the suffixes for different chemical families. In your previous chemistry course, you learned the suffixes *-ane* for alkanes, *-ene* for alkenes, and *-yne* for alkynes. Thus, an alkane composed of six carbon atoms in a chain is called *hexane*. An alkene with three carbons is called *propene*.

The Prefix: What Is Attached to the Main Chain?

The prefix indicates the name and location of each branch and functional group on the main carbon chain. Most organic compounds have branches, called alkyl groups, attached to the main chain. An **alkyl group** is obtained by removing one hydrogen atom from an alkane. To name an alkyl group, change the -ane suffix to -yl. For example, $—CH_3$ is the alkyl group that is derived from methane, CH_4. It is called the *methyl* group, taken from the root meth-. Table 1.3 gives the names of the most common alkyl groups.

Table 1.3 Common Alkyl Groups

methyl	ethyl	propyl	isopropyl
$—CH_3$	$—CH_2CH_3$	$—CH_2CH_2CH_3$	CH_3 / $—CH$ \\ CH_3
butyl	**sec-butyl**	**iso-butyl**	**tert-butyl**
$—CH_2CH_2CH_2CH_3$	CH_3 / $—CH$ \\ CH_2CH_3	CH_3 / $—CH_2—CH$ \\ CH_3	CH_3 | $—C—CH_3$ | CH_3

Read the steps below to review how to name hydrocarbons. Then examine the two Sample Problems that follow.

How to Name Hydrocarbons

Step 1 Find the root: Identify the longest chain or ring in the hydrocarbon. If the hydrocarbon is an alkene or an alkyne, make sure that you include any multiple bonds in the main chain. Remember that the chain does not have to be in a straight line. Count the number of carbon atoms in the main chain to obtain the root. If it is a cyclic compound, add the prefix -cyclo- before the root.

Step 2 Find the suffix: If the hydrocarbon is an alkane, use the suffix -ane. Use -ene if the hydrocarbon is an alkene. Use -yne if the hydrocarbon is an alkyne. If more than one double or triple bond is present, use the prefix di- (2) or tri- (3) before the suffix to indicate the number of multiple bonds.

Step 3 Give a position number to every carbon atom in the main chain. Start from the end that gives you the lowest possible position number for the double or triple bond, if there is one. If there is no double or triple bond, number the compound so that the branches have the lowest possible position numbers.

Step 4 Find the prefix: Name each branch as an alkyl group, and give it a position number. If more than one branch is present, write the names of the branches in alphabetical order. Put the position number of any double or triple bonds *after* the position numbers and names of the branches, just before the root. This is the prefix. **Note:** Use the carbon atom with the *lowest* position number to give the location of a double or triple bond.

Step 5 Put the name together: prefix + root + suffix.

Sample Problem

Naming Alkanes

Problem

Name the following alkanes.

(a) 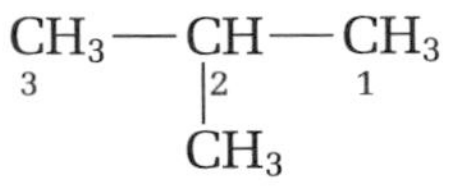 $CH_3—CH—CH_3$ (carbon 2 bears a CH_3 branch; carbons numbered 3, 2, 1 from left)

(b) 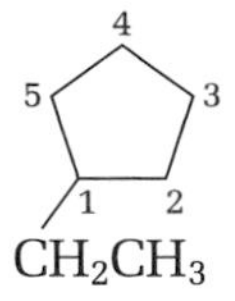 cyclopentane ring numbered 1–5, with CH_2CH_3 attached at carbon 1

Solution

(a) Step 1 Find the root: The longest chain has three carbon atoms, so the root is -prop-.

Step 2 Find the suffix: The suffix is -ane.

Steps 3 and 4 Find the prefix: A methyl group is attached to carbon number 2. The prefix is 2-methyl.

Step 5 The full name is 2-methylpropane.

(b) Steps 1 and 2 Find the root and suffix: The main ring has five carbon atoms, so the root is -pent-. Add the prefix -cyclo-. The suffix is -ane.

Steps 3 and 4 Find the prefix: Start numbering at the ethyl branch. The prefix is 1-ethyl, or just ethyl.

Step 5 The full name is 1-ethylcyclopentane.

PROBLEM TIPS

- Use hyphens to separate words from numbers. Use commas to separate numbers.
- If there is a ring, it is usually taken as the main chain. Follow the same rules to name cyclic compounds that have branches attached. Include the prefix -cyclo- after the names and position numbers of the branches, directly before the root: for example, 2-methyl-1-cyclohexene.

Sample Problem

Naming an Alkene

Problem

Name the following alkene.

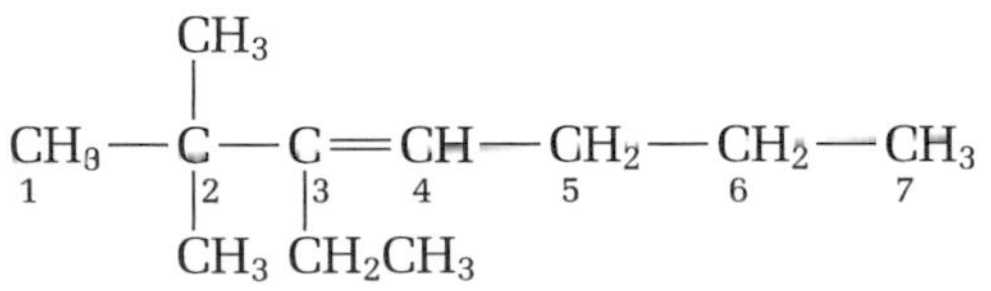

$CH_3—C(CH_3)_2—C(CH_2CH_3)=CH—CH_2—CH_2—CH_3$ (carbons numbered 1 to 7 from left)

Solution

Step 1 Find the root: The longest chain in the molecule has seven carbon atoms. The root is -hept-.

Step 2 Find the suffix: The suffix is -ene. The root and suffix together are -heptene.

Step 3 Numbering the chain from the left, in this case, gives the smallest position number for the double bond.

Step 4 Find the prefix: Two methyl groups are attached to carbon number 2. One ethyl group is attached to carbon number 3. There is a double bond at position 3. The prefix is 3-ethyl-2,2-dimethyl-3-.

Step 5 The full name is 3-ethyl-2,2-dimethyl-3-heptene.

To draw a condensed structural diagram of a hydrocarbon, follow the steps below. Then examine the Sample Problem that follows.

How to Draw Hydrocarbons

Step 1 Draw the carbon atoms of the main chain. Leave space after each carbon atom for bonds and hydrogen atoms to be added later. Number the carbon atoms.

Step 2 Draw any single, double, or triple bonds between the carbon atoms.

Step 3 Add the branches to the appropriate carbon atoms of the main chain.

Step 4 Add hydrogen atoms so that each carbon atom forms a total of 4 bonds. Remember that double bonds count as 2 bonds and triple bonds count as 3 bonds.

Sample Problem

Drawing an Alkane

Problem

Draw a condensed structural diagram for 3-ethyl-2-methylhexane.

Solution

Step 1 The main chain is hexane. Therefore, there are six carbon atoms.

Step 2 This compound is an alkane, so all carbon-carbon bonds are single.

Step 3 The ethyl group is attached to carbon number 3. The methyl group is attached to carbon number 2.

Step 4 Add hydrogen atoms so that each carbon atom forms 4 bonds.

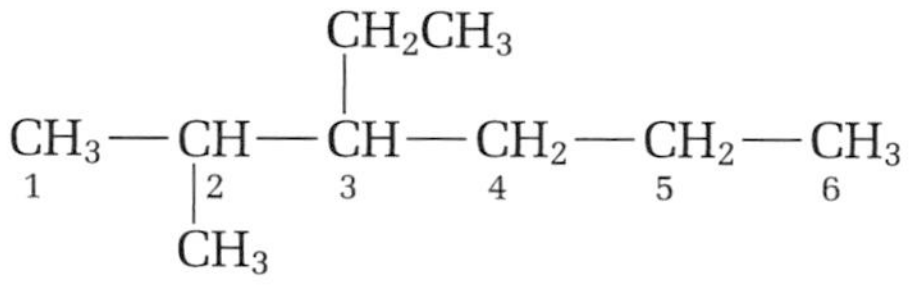

CH_2CH_3 (on carbon 3)

$CH_3 — CH — CH — CH_2 — CH_2 — CH_3$

1 2 3 4 5 6

CH_3 (on carbon 2)

Practice Problems

5. Name each hydrocarbon.

(a) $H_3C — CH_3$

(b)

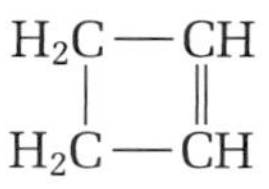

$H_2C — CH$

$H_2C — CH$ (ring, with a double bond between the two CH)

(c) $CH_3 — C{=}CH — CH_2 — CH_3$, with CH_3 on the C

(d)

(e)

(f) CH_3 CH_3

(g)

Electronic Learning Partner

The Chemistry 12 Electronic Learning Partner has a video that compares models of hydrocarbons.

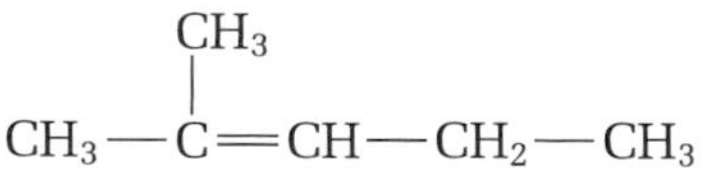

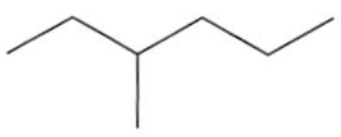

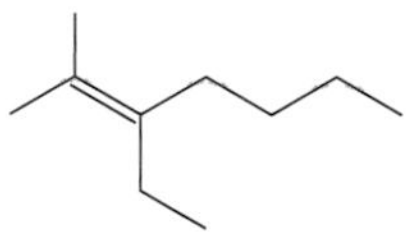

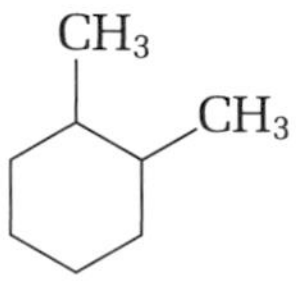

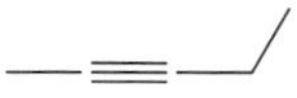

6. Draw a condensed structural diagram for each hydrocarbon.
   (a) propane
   (b) 4-ethyl-3-methylheptane
   (c) 3-methyl-2,4,6-octatriene
7. Identify any errors in the name of each hydrocarbon.
   (a) 2,2,3-dimethylbutane
   (b) 2,4-diethyloctane
   (c) 3-methyl-4,5-diethyl-2-nonyne
8. Correct any errors so that each name matches the structure beside it.
   (a) 4-hexyne $CH_3—CH_2—CH{=}CH—CH_2—CH_3$
   (b) 2,5-hexene $CH_3—C{\equiv}C—C{\equiv}C—CH_3$
9. Use each *incorrect* name to draw the corresponding hydrocarbon. Examine your drawing, and rename the hydrocarbon correctly.
   (a) 3-propyl-2-butene
   (b) 1,3-dimethyl-4-hexene
   (c) 4-methylpentane

Careers in Chemistry

The Art and Science of Perfumery

Since 1932, The Quiggs have manufactured perfume compounds for cosmetics, toiletries, soaps, air fresheners, candles, detergents, and industrial cleaning products.

Jeff Quigg says the mixing of a perfume is "a trial and error process." An experienced perfumer must memorize a vast library of hundreds or even thousands of individual scents and combinations of scents. Perfume ingredients can be divided into *natural essential oils* (derived directly from plants) and *aromatic chemicals* (synthetically produced fragrance components).

Essential oils are organic compounds derived from flowers, seeds, leaves, roots, resins, and citrus fruits. The structures of many fragrant compounds have been studied, and processes for making these valuable compounds in a laboratory have been developed. There are now approximately 5000 synthetically produced chemicals that are available to a perfumer. These chemicals include vanillin, rose oxides, and the *damascones*, or rose ketones.

An aspiring perfumer must have a discriminating sense of smell. As well, a perfumist should obtain at least a bachelor of science degree in chemistry, or a degree in chemical engineering. There are few formal schools for perfumers, so companies usually train perfumers in-house. The training takes five to ten years to complete.

Although inventors are trying to develop electronic and artificial noses to detect odours, they have not yet been able to duplicate the sensitive nose of a skilled, trained, and talented perfumer.

Making Career Connections

1. Perfume schools exist, but admission is very competitive. One of these schools is the Institut Supérieur International du Parfum, de la Cosmétique et de l'Aromatique Alimentaire (ISIPCA, or International High Institute of Perfume, Cosmetic and Food Flavouring). The ISIPCA is located in Versailles, France. You can find out more about the ISIPCA by logging onto **www.mcgrawhill.ca/links/chemistry12** and clicking on **Web Links**. Use the Internet or a library to find out more about perfume schools and training for perfumers.
2. The fragrance industry is closely linked to the flavour industry. Many of the skills required of a perfumer are also required of a flavourist. Find out more about the flavour industry. Contact the chemistry department of a university to find out more about flavour chemistry.

Aromatic Compounds

If you completed the Concept Check activity on page 12, you drew a possible structure for benzene. For many years, scientists could not determine the structure of benzene. From its molecular formula, C_6H_6, scientists reasoned that it should contain two double bonds and one triple bond, or even two triple bonds. Benzene, however, does not undergo the same reactions as other compounds with double or triple bonds.

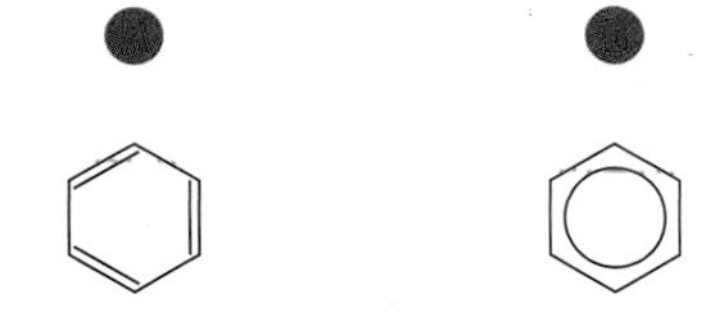

Figure 1.9 Two representations of the benzene molecule.

We know today that benzene is a cyclic compound with the equivalent of three double bonds and three single bonds, as shown in Figure 1.9(A). However, the electrons that form the double bonds in benzene are spread out and shared over the whole molecule. Thus, benzene actually has six identical bonds, each one half-way between a single and a double bond. These bonds are much more stable than ordinary double bonds and do not react in the same way. Figure 1.9(B) shows a more accurate way to represent the bonding in benzene. Molecules with this type of special electron sharing are called *aromatic compounds*. As mentioned earlier, benzene is the simplest aromatic compound.

Figure 1.10 illustrates some common aromatic compounds. To name an aromatic compound, follow the steps below. Figure 1.11 gives an example.

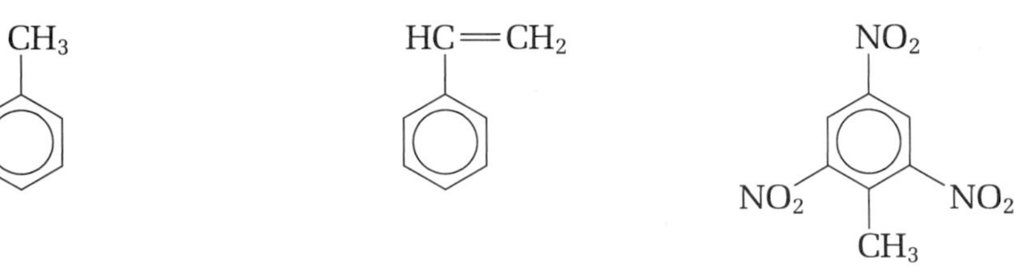

methylbenzene (toluene) | phenylethene (styrene) | 2,4,6-trinitromethylbenzene (trinitrotoluene, TNT)

Figure 1.10 The common name for methylbenzene is toluene. Toluene is used to produce explosives, such as trinitrotoluene (TNT). Phenylethene, with the common name styrene, is an important ingredient in the production of plastics and rubber.

PROBLEM TIP

Often an organic compound has more than one type of branch. When possible, number the main chain or ring of the compound to give the most important branches the lowest possible position numbers. The table below ranks some branches (and other groups) you will encounter in this chapter, from the highest priority to the lowest priority.

The Priority of Branches

Highest priority	$-OH$
	$-NH_2$
	$-F$, $-Cl$, $-Br$, $-I$
	$-CH_2CH_2CH_3$
	$-CH_2CH_3$
Lowest priority	$-CH_3$

Naming an Aromatic Hydrocarbon

Step 1 Number the carbons in the benzene ring. If more than one type of branch is attached to the ring, start numbering at the carbon with the highest priority (or most complex) group. (See the Problem Tip.)

Step 2 Name any branches that are attached to the benzene ring. Give these branches position numbers. If only one branch is attached to a benzene ring, you do not need to include a position number.

Step 3 Place the branch numbers and names as a prefix before the root, benzene.

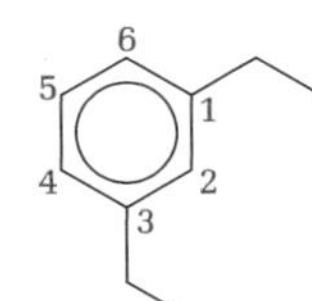

Figure 1.11 Two ethyl groups are present. They have the position numbers 1 and 3. The name of this compound is 1,3-diethylbenzene.

Chemists do not always use position numbers to describe the branches that are attached to a benzene ring. When a benzene ring has only two branches, the prefixes ortho-, meta-, and para- are sometimes used instead of numbers.

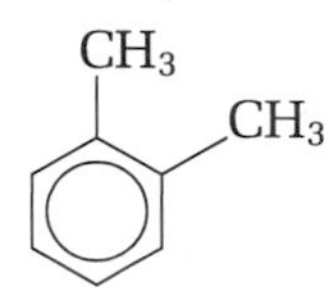

1,2-dimethylbenzene
ortho-dimethylbenzene
(common name:
ortho-xylene)

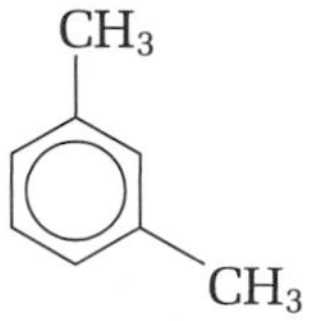

CH_3
CH_3

1,3-dimethylbenzene
meta-dimethylbenzene
(common name:
meta-xylene)

CH_3
CH_3

1,4-dimethylbenzene
para-dimethylbenzene
(common name:
para-xylene)

Practice Problems

10. Name the following aromatic compound.

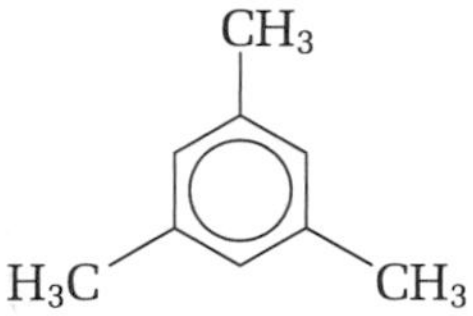

CH_3
H_3C CH_3

11. Draw a structural diagram for each aromatic compound.

(a) 1-ethyl-3-methylbenzene

(b) 2-ethyl-1,4-dimethylbenzene

(c) para-dichlorobenzene (**Hint:** *Chloro* refers to the chlorine atom, Cl.)

12. Give another name for the compound in question 11(a).

13. Draw and name three aromatic isomers with the molecular formula $C_{10}H_{14}$. (*Isomers* are compounds that have the same molecular formula, but different structures. See the Concepts and Skills Review for a review of structural isomers.)

Section Summary

In this section, you reviewed how to name and draw alkanes, alkenes, and alkynes. You also learned how to name aromatic hydrocarbons. The names of all the other organic compounds you will encounter in this unit are based on the names of hydrocarbons. In the next section, you will learn about organic compounds that have single bonds to halogen atoms, oxygen atoms, and nitrogen atoms.

CHEM FACT

Kathleen Yardley Lonsdale (1903–1971) used X-ray crystallography to prove that benzene is a flat molecule. All six carbon atoms lie in one plane, forming a regular hexagonal shape. The bonds are all exactly the same length. The bond angles are all 120°.

Section Review

1 K/U Name each hydrocarbon.

(a) $CH_3—CH_2—CH_2—CH_2—CH_2—CH_2—CH_3$

(b)

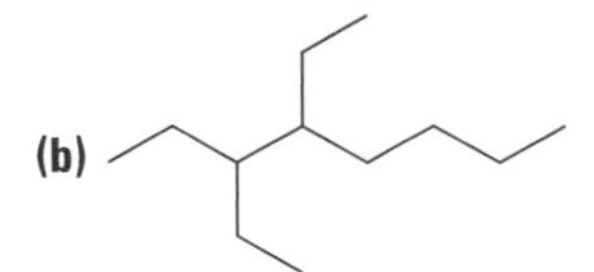

(c) $CH_3—CH(CH_3)—CH_2—CH=CH_2$

(d) 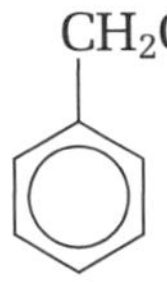 CH_2CH_3 attached to a benzene ring

2 C Draw a condensed structural diagram for each hydrocarbon.

(a) cyclopentane

(b) 2-methyl-2-butene

(c) 1,4-dimethylbenzene (common name: *para*-xylene)

(d) 3-ethyl-2,3,4-trimethylnonane

3 C Draw and name all the isomers that have the molecular formula C_4H_{10}.

4 C Draw a line structural diagram for each hydrocarbon. (See the Concepts and Skills Review for a review of structural diagrams and cis-trans isomers.)

(a) pentane

(b) 2-methylpropane

(c) 1-ethyl-3-methylcyclohexane

(d) *trans*-2,5-dimethyl-3-heptene

5 C Draw and name twelve possible isomers that have the molecular formula C_6H_{10}.

6 I Use a molecular model set to build a model of the benzene ring. Examine your model. Does your model give an accurate representation of benzene's bonding system? Explain your answer.

7 C Draw and name all the isomers that have the molecular formula C_5H_{10}. Include any cis-trans isomers and cyclic compounds.

8 C Draw two different but correct structures for the benzene molecule. Explain why one structure is more accurate than the other.

Single-Bonded Functional Groups

1.3

When you cut yourself, it is often a good idea to swab the cut with rubbing alcohol to disinfect it. Most rubbing alcohols that are sold in drugstores are based on 2-propanol (common name: isopropanol), C_3H_8O. You can also swab a cut with a rubbing alcohol based on ethanol, C_2H_6O. Often it is hard to tell the difference between these two compounds. Both have a sharp smell, and both evaporate quickly. Both are effective at killing bacteria and disinfecting wounds. What is the connection between these compounds? Why is their behaviour so similar?

Functional Groups

Both 2-propanol and ethanol contain the same functional group, an **—OH (hydroxyl) group**, as shown in Figure 1.12. Because ethanol and 2-propanol have the same OH functional group, their behaviour is similar.

$$CH_3—\underset{\underset{OH}{|}}{CH}—CH_3$$

2-propanol

$$CH_3—CH_2—OH$$

ethanol

Figure 1.12 **Ethanol and 2-propanol both belong to the alcohol family.**

The **general formula** for a family of simple organic compounds is *R + functional group*. The letter *R* stands for any alkyl group. (If more than one alkyl group is present, *R′* and *R″* are also used.) For example, the general formula R—OH refers to any of the following compounds:

CH_3OH, CH_3CH_2OH, $CH_3CH_2CH_2OH$, $CH_3CH_2CH_2CH_2OH$, etc.

Organic compounds are named according to their functional group. Generally, the suffix of a compound's name indicates the most important functional group in the molecule. For example, the suffix -ene indicates the presence of a double bond, and the suffix -ol indicates the presence of a hydroxyl group.

Functional groups are a useful way to classify organic compounds, for two reasons:

1. *Compounds with the same functional group often have similar physical properties.* In the next two sections, you will learn to recognize various functional groups. You will use functional groups to help you predict the physical properties of compounds.
2. *Compounds with the same functional group react chemically in very similar ways.* In Chapter 2, you will learn how compounds with each functional group react.

Table 1.4, on the next page, lists some of the most common functional groups.

Section Preview/ Specific Expectations

In this section, you will

- **distinguish** among the following classes of organic compounds: alkyl halides, alcohols, ethers, and amines
- **describe** the effects of intermolecular forces on the physical properties of alcohols, ethers, and amines
- **draw** and **name** alkyl halides, alcohols, ethers, and amines using the IUPAC system
- **identify** the common names of some organic compounds
- **communicate** your understanding of the following terms: *—OH (hydroxyl) group, general formula, intermolecular forces hydrogen bonding, dipole-dipole interactions, dispersion forces, alcohol, parent alkane, alkyl halide (haloalkane), ether, alkoxy group, amine*

CONCEPT CHECK

Each organic family follows a set pattern. You have just seen that you can represent the hydrocarbon part of a functional family by the letter *R*. All the structures below belong to the primary amine family. What is the functional group for this family? Write the general formula for an amine.

$$CH_3—NH_2$$

$$CH_3—CH_2—NH_2$$

$$CH_3—CH_2—CH_2—NH_2$$

Table 1.4 Common Functional Groups

Type of compound	Suffix	Functional group	Example
alkane	-ane	none	propane
alkene	-ene	$>C=C<$	propene
alkyne	-yne	$-C\equiv C-$	propyne
alcohol	-ol	$-\overset{\vert}{\underset{\vert}{C}}-OH$	propanol
amine	-amine	$-\overset{\vert}{\underset{\vert}{C}}-N<$	propanamine
aldehyde	-al	$-\overset{O}{\overset{\Vert}{C}}-H$	propanal
ketone	-one	$-\overset{O}{\overset{\Vert}{C}}-$	propanone
carboxylic acid	-oic acid	$-\overset{O}{\overset{\Vert}{C}}-OH$	propanoic acid
ester	-oate	$-\overset{O}{\overset{\Vert}{C}}-O-$	methyl propanoate
amide	-amide	$-\overset{O}{\overset{\Vert}{C}}-N<$	propanamide

Physical Properties and Forces Between Molecules

Organic compounds that have the same functional group often have similar physical properties, such as boiling points, melting points, and solubilities. Physical properties are largely determined by **intermolecular forces**, the forces of attraction and repulsion between particles. Three types of intermolecular forces are introduced below. You will examine these forces further in Chapter 4.

- **Hydrogen bonding** is a strong intermolecular attraction between the hydrogen atom from an N—H, O—H, or F—H group on one molecule, and a nitrogen, oxygen, or fluorine atom on another molecule.
- The attractive forces between polar molecules are called **dipole-dipole interactions**. These forces cause polar molecules to cling to each other.
- **Dispersion forces** are attractive forces that occur between all covalent molecules. These forces are usually very weak for small molecules, but they strengthen as the size of the molecule increases.

The process that is outlined on the next page will help you to predict the physical properties of organic compounds by examining the intermolecular forces between molecules. As you progress through the chapter, referring back to this process will enable you to understand the reasons behind trends in physical properties.

Intermolecular Forces and Physical Properties

Draw two or three molecules of the same organic compound close together on a page. If you are considering the solubility of one compound in another, sketch the two different molecules close together. Ask the following questions about the intermolecular interactions between the molecules of each compound:

1. **Can the molecules form hydrogen bonds?**

   If the molecules have O—H, N—H, or H—F bonds, they can form hydrogen bonds with themselves and with water. The diagram to the right illustrates hydrogen bonding between water molecules. If the molecules contain O, N, or F atoms that are *not* bonded to hydrogen atoms, they may accept hydrogen bonds from water, even though they cannot form hydrogen bonds with themselves.

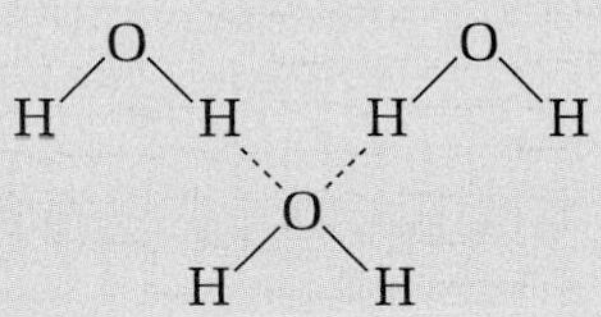

   Molecules that can form hydrogen bonds with themselves have a higher boiling point than similar molecules that cannot form hydrogen bonds with themselves. For example, alcohols can form hydrogen bonds, but alkanes cannot. Therefore, alcohols have higher boiling points than alkanes.

   Molecules that can form hydrogen bonds with water, or can accept hydrogen bonds from water, are usually soluble in water. For example, many alcohols are soluble in water because they can form hydrogen bonds with water.

2. **Are the molecules polar?**

   The molecules are polar if they have polar bonds, and if these bonds do not act in opposite directions and counteract each other. Polar molecules are attracted to each other by dipole-dipole forces.

   Polar molecules usually have a higher boiling point than similar non-polar molecules. Also, polar molecules that can form hydrogen bonds have an even higher boiling point than polar molecules that cannot form hydrogen bonds. For example, ethanol, CH_3CH_2OH, is polar. Its molecules can form hydrogen bonds. Methoxymethane, CH_3OCH_3, is an isomer of ethanol. It is also polar, but its molecules cannot form hydrogen bonds. Thus, ethanol has a higher boiling point than methoxymethane. Both of these compounds have a higher boiling point than the non-polar molecule ethane, CH_3CH_3.

   Polar molecules with a large non-polar hydrocarbon part are less polar than polar molecules with a smaller non-polar hydrocarbon part. For example, octanol, $CH_3CH_2CH_2CH_2CH_2CH_2CH_2CH_2OH$, is less polar than ethanol, CH_3CH_2OH.

   Polar molecules with a large hydrocarbon part are less soluble in water than polar molecules with a smaller hydrocarbon part. For example, octanol, $CH_3CH_2CH_2CH_2CH_2CH_2CH_2CH_2OH$, is less soluble in water than ethanol, CH_3CH_2OH.

continued on the next page

3. How strong are the dispersion forces?

Dispersion forces are weak intermolecular forces. They are stronger, however, when the hydrocarbon part of a molecule is very large. Thus, a large molecule has stronger dispersion interactions than a smaller molecule.

A molecule with a greater number of carbon atoms usually has a higher boiling point than the same type of molecule with fewer carbon atoms. For example, hexane, $CH_3CH_2CH_2CH_2CH_2CH_3$ has a higher boiling point than ethane, CH_3CH_3.

The melting points of organic compounds follow approximately the same trend as their boiling points. There are some anomalies, however, due to more complex forces of bonding in solids.

In the following ThoughtLab you will use the process in the box above to predict and compare the physical properties of some organic compounds.

ThoughtLab  Comparing Intermolecular Forces

Intermolecular forces affect the physical properties of compounds. In this ThoughtLab, you will compare the intermolecular forces of different organic compounds.

Procedure

1. Draw three molecules of each compound below.
   (a) propane, $CH_3CH_2CH_3$
   (b) heptane, $CH_3CH_2CH_2CH_2CH_2CH_2CH_3$
   (c) 1-propanol, $CH_3CH_2CH_2OH$
   (d) 1-heptanol, $CH_3CH_2CH_2CH_2CH_2CH_2CH_2OH$
2. For each compound, consider whether or not hydrogen bonding can occur between its molecules. Use a dashed line to show any hydrogen bonding.
3. For each compound, consider whether or not any polar bonds are present.
   (a) Use a different-coloured pen to identify any polar bonds.
   (b) Which compounds are polar? Which compounds are non-polar? Explain your reasoning.
4. Compare your drawings of propane and heptane.
   (a) Which compound has stronger dispersion forces? Explain your answer.
   (b) Which compound has a higher boiling point? Explain your answer.
5. Compare your drawings of 1-propanol and 1-heptanol.
   (a) Which compound is more polar? Explain your answer.
   (b) Which compound is more soluble in water? Explain your answer.

Analysis

1. Which compound has a higher solubility in water?
   (a) a polar compound or a non-polar compound
   (b) a compound that forms hydrogen bonds with water, or a compound that does not form hydrogen bonds with water
   (c) $CH_3CH_2CH_2OH$ or $CH_3CH_2CH_2CH_2CH_2OH$
2. Which compound has stronger attractions between molecules?
   (a) a polar compound or a non-polar compound
   (b) a compound without O—H or N—H bonds, or
   a compound with O—H or N—H bonds
3. Which compound is likely to have a higher boiling point?
   (a) a polar compound without O—H or N—H bonds, or a polar compound with O—H or N—H bonds
   (b) $CH_3CH_2CH_2OH$ or $CH_3CH_2CH_2CH_2CH_2OH$
4. Compare boiling points and solubilities in water for each pair of compounds. Explain your reasoning.
   (a) ammonia, NH_3, and methane, CH_4
   (b) pentanol, $C_5H_{11}OH$, and pentane, C_5H_{12}

Compounds With Single-Bonded Functional Groups

Alcohols, alkyl halides, ethers, and amines all have functional groups with single bonds. These compounds have many interesting uses in daily life. As you learn how to identify and name these compounds, think about how the intermolecular forces between their molecules affect their properties and uses.

Alcohols

An **alcohol** is an organic compound that contains the —OH functional group. Depending on the position of the hydroxyl group, an alcohol can be *primary*, *secondary*, or *tertiary*. Figure 1.13 gives some examples of alcohols.

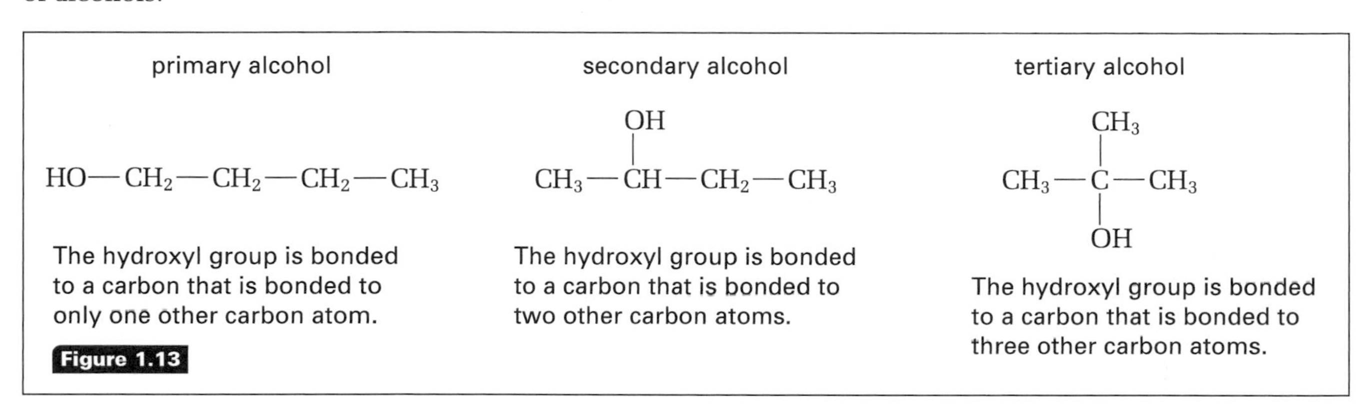

primary alcohol
HO—CH₂—CH₂—CH₂—CH₃
The hydroxyl group is bonded to a carbon that is bonded to only one other carbon atom.
secondary alcohol
OH
CH₃—CH—CH₂—CH₃
The hydroxyl group is bonded to a carbon that is bonded to two other carbon atoms.
tertiary alcohol
CH₃
CH₃—C—CH₃
OH
The hydroxyl group is bonded to a carbon that is bonded to three other carbon atoms.

Figure 1.13

Table 1.5 lists some common alcohols and their uses. Alcohols are very widely used, and can be found in drug stores, hardware stores, liquor stores, and as a component in many manufactured products.

Table 1.5 Common Alcohols and Their Uses

Name	Common name(s)	Structure	Boiling point	Use(s)
methanol	wood alcohol, methyl alcohol	$CH_3—OH$	64.6°C	• solvent in many chemical processes • component of automobile antifreeze
ethanol	grain alcohol, ethyl alcohol	$CH_3—CH_2—OH$	78.2°C	• solvent in many chemical processes • component of alcoholic beverages • antiseptic liquid
2-propanol	isopropanol, isopropyl alcohol, rubbing alcohol	$(CH_3)_2CH—OH$ (two CH_3 groups bonded to CH—OH)	82.4°C	• antiseptic liquid
1,2-ethanediol	ethylene glycol	$HO—CH_2—CH_2—OH$	197.6°C	• main component of automobile antifreeze

Alcohols are named from the **parent alkane**: the alkane with the same basic carbon structure. Follow the steps on the next page to name an alcohol. The Sample Problem that follows gives an example.

Web LINK

www.mcgrawhill.ca/links/chemistry12

Methanol and ethanol are produced industrially from natural, renewable resources. Go to the web site above, and click on **Web Links** to find out where to go next. Research the processes that produce these important chemicals. From where do they obtain their raw materials?

How to Name an Alcohol

Step 1 Locate the longest chain that contains an —OH group attached to one of the carbon atoms. Name the parent alkane.

Step 2 Replace the -e at the end of the name of the parent alkane with -ol.

Step 3 Add a position number before the root of the name to indicate the location of the —OH group. (Remember to number the main chain of the hydrocarbon so that the hydroxyl group has the lowest possible position number.) If there is more than one —OH group, leave the -e in the name of the parent alkane, and put the appropriate prefix (di-, tri-, or tetra-) before the suffix -ol.

Step 4 Name and number any other branches on the main chain. Add the name of these branches to the prefix.

Step 5 Put the name together: prefix + root + suffix.

PROBLEM TIP

If an organic compound is complex, with many side branches, the main chain may not be obvious. Sketch the compound in your notebook or on scrap paper. Circle or highlight the main chain.

Sample Problem

Naming an Alcohol

Problem

Name the following alcohol. Identify it as primary, secondary, or tertiary.

$$\begin{array}{l} \qquad\qquad\qquad\qquad\quad CH_2—CH_2—CH_3 \\ \qquad\qquad\qquad\qquad\quad | \\ HO—CH_2—CH_2—CH—CH_3 \end{array}$$

Solution

Step 1 The main chain has six carbon atoms. The name of the parent alkane is hexane.

Step 2 Replacing -e with -ol gives hexanol.

Step 3 Add a position number for the —OH group, to obtain 1-hexanol.

Step 4 A methyl group is present at the third carbon. The prefix is 3-methyl.

Step 5 The full name is 3-methyl-1-hexanol. This is a primary alcohol.

Practice Problems

14. Name each alcohol. Identify it as primary, secondary, or tertiary.

(a) $CH_3—CH_2—CH_2—OH$

(b) OH

(c)

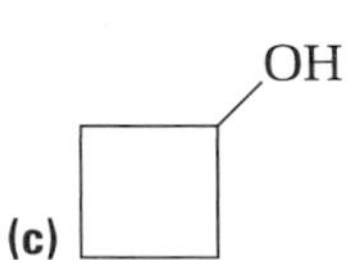

(d)
$$\begin{array}{l} \qquad\qquad\qquad\quad OH \\ \qquad\qquad\qquad\quad | \\ CH_3—CH—CH—CH_2—CH_3 \\ \qquad\quad | \\ \qquad\quad OH \end{array}$$

(e) OH

15. Draw each alcohol.
   (a) methanol
   (b) 2-propanol
   (c) 2,2-butanediol
   (d) 3-ethyl-4-methyl-1-octanol
   (e) 2,4-dimethyl-1-cyclopentanol

16. Identify any errors in each name. Give the correct name for the alcohol.
   (a) 1,3-heptanol

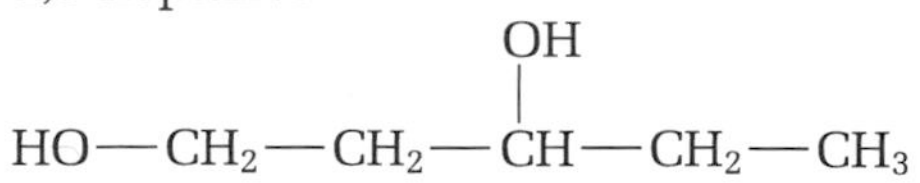

```
                      OH
                      |
HO—CH2—CH2—CH—CH2—CH3
```

   (b) 3-ethyl-4-ethyl-1-decanol

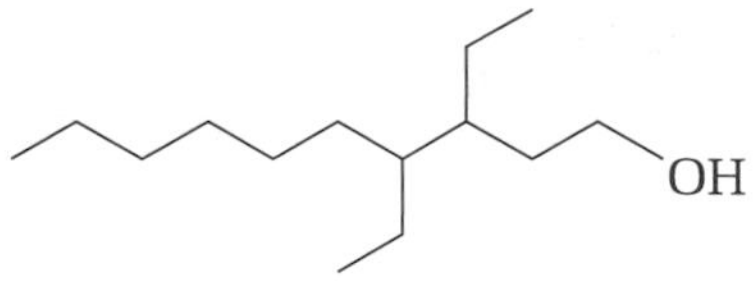

OH

   (c) 1,2-dimethyl-3-butanol

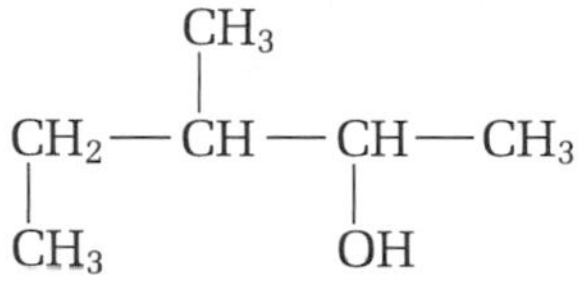

```
       CH3
       |
CH2—CH—CH—CH3
|           |
CH3         OH
```

17. Sketch a three-dimensional diagram of methanol. **Hint:** Recall that the shape around an oxygen atom is *bent*.

Table 1.6 lists some common physical properties of alcohols. As you learned earlier in this chapter, alcohols are polar molecules that experience hydrogen bonding. The physical properties of alcohols depend on these characteristics.

Table 1.6 Physical Properties of Alcohols

Polarity of functional group	The O—H bond is very polar. As the number of carbon atoms in an alcohol becomes larger, the alkyl group's non-polar nature becomes more important than the polar O—H bond. Therefore small alcohols are more polar than alcohols with large hydrocarbon portions.
Hydrogen bonding	Alcohols experience hydrogen bonding with other alcohol molecules and with water.
Solubility in water	The capacity of alcohols for hydrogen bonding makes them extremely soluble in water. Methanol and ethanol are *miscible* (infinitely soluble) with water. The solubility of an alcohol decreases as the number of carbon atoms increases.
Melting and boiling points	Due to the strength of the hydrogen bonding, most alcohols have higher melting and boiling points than alkanes with the same number of carbon atoms. Most alcohols are liquids at room temperature.

Additional Characteristics of Alcohols

- Alcohols are extremely flammable, and should be treated with caution.
- Most alcohols are poisonous. Methanol can cause blindness or death when consumed. Ethanol is consumed widely in moderate quantities, but it causes impairment and/or death when consumed in excess.

Alkyl Halides

An **alkyl halide** (also known as a **haloalkane**) is an alkane in which one or more hydrogen atoms have been replaced with halogen atoms, such as F, Cl, Br, or I. The functional group of alkyl halides is R—X, where X represents a halogen atom. Alkyl halides are similar in structure, polarity, and reactivity to alcohols. To name an alkyl halide, first name the parent hydrocarbon. Then use the prefix fluoro-, chloro-, bromo-, or iodo-, with a position number, to indicate the presence of a fluorine atom, chlorine atom, bromine atom, or iodine atom. The following Sample Problem shows how to name an alkyl halide.

Sample Problem

Naming an Alkyl Halide

Problem

Name the following compound.

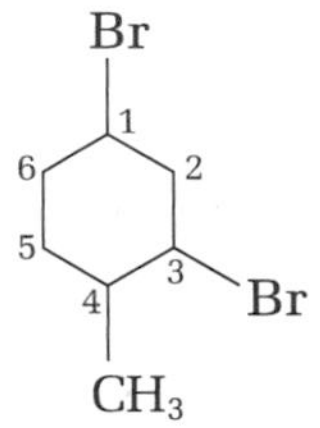

Br
1 2 3 4 5 6
Br
CH_3

Solution

The parent hydrocarbon of this compound is cyclohexane. There are two bromine atoms attached at position numbers 1 and 3. Therefore, part of the prefix is 1,3-dibromo-. There is also a methyl group at position number 4. Because the groups are put in alphabetical order, the full prefix is 1,3-dibromo-4-methyl-. (The ring is numbered so that the two bromine atoms have the lowest possible position numbers. See the Problem Tip on page 18.) The full name of the compound is 1,3-dibromo-4-methylcyclohexane.

Practice Problems

18. Draw a condensed structural diagram for each alkyl halide.

(a) bromoethane

(b) 2,3,4-triiodo-3-methylheptane

19. Name the alkyl halide at the right. Then draw a condensed structural diagram to represent it.

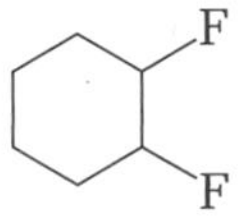

F
F

20. Draw and name an alkyl halide that has three carbon atoms and one iodine atom.

21. Draw and name a second, different alkyl halide that matches the description in the previous question.

Ethers

Suppose that you removed the H atom from the —OH group of an alcohol. This would leave space for another alkyl group to attach to the oxygen atom.

$$CH_3CH_2{-}OH \rightarrow CH_3CH_2{-}O{-} \rightarrow CH_3CH_2{-}O{-}CH_3$$

The compound you have just made is called an ether. An **ether** is an organic compound that has two alkyl groups joined by an oxygen atom. The general formula of an ether is R—O—R. You can think of alcohols and ethers as derivatives of the water molecule, as shown in Figure 1.14. Figure 1.15 gives two examples of ethers.

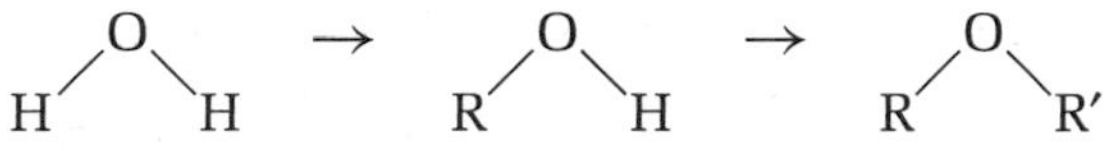

$$H{-}O{-}H \rightarrow R{-}O{-}H \rightarrow R{-}O{-}R'$$

water alcohol ether

Figure 1.14 **An alcohol is equivalent to a water molecule with one hydrogen atom replaced by an alkyl group. Similarly, an ether is equivalent to a water molecule with both hydrogen atoms replaced by alkyl groups.**

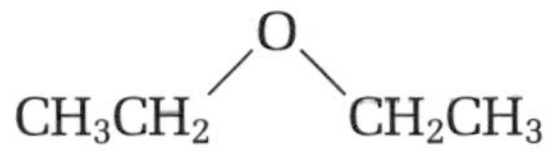

$$CH_3CH_2{-}O{-}CH_2CH_3$$

ethoxyethane
(common name:
diethyl ether)

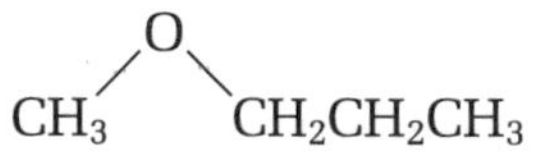

$$CH_3{-}O{-}CH_2CH_2CH_3$$

1-methoxypropane
(common name:
methyl propyl ether)

Figure 1.15 **Until fairly recently, ethoxyethane was widely used as an anaesthetic. It had side effects, such as nausea, however. Compounds such as 1-methoxypropane are now used instead.**

To name an ether, follow the steps below. The Sample Problem then shows how to use these steps to give an ether its IUPAC name and its common name.

How to Name an Ether

IUPAC Name

Step 1 Choose the longest alkyl group as the parent alkane. Give it an alkane name.

Step 2 Treat the second alkyl group, along with the oxygen atom, as an **alkoxy group** attached to the parent alkane. Name it by replacing the -yl ending of the corresponding alkyl group's name with -oxy. Give it a position number.

Step 3 Put the prefix and suffix together: alkoxy group + parent alkane.

Common Name

Step 1 List the alkyl groups that are attached to the oxygen atom, in alphabetical order.

Step 2 Place the suffix -ether at the end of the name.

Sample Problem

Naming an Ether

Problem

Give the IUPAC name and the common name of the following ether.

$CH_3—CH_2—O—CH_2—CH_2—CH_3$

Solution

IUPAC Name

Step 1 The longest alkyl group is based on propane.

$CH_3—CH_2—O—CH_2—CH_2—CH_3$

Step 2 The alkoxy group is based on ethane (the ethyl group). It is located at the first carbon atom of the propane part. Therefore, the prefix is 1-ethoxy-.

Step 3 The full name is 1-ethoxypropane.

Common Name

Step 1 The two alkyl groups are ethyl and propyl.

Step 2 The full name is ethyl propyl ether.

Practice Problems

22. Use the IUPAC system to name each ether.

(a) $H_3C—O—CH_3$

(b) $H_3C—O—CH(CH_3)_2$ (CH bonded to two CH_3 groups)

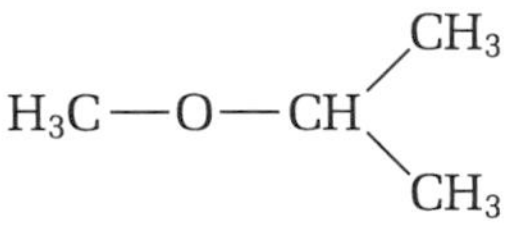

(c) $CH_3—CH_2—CH_2—CH_2—O—CH_3$

23. Give the common name for each ether.

(a) $H_3C—O—CH_2CH_3$

(b) $H_3C—O—CH_3$

24. Draw each ether.

(a) 1-methoxypropane

(b) 3-ethoxy-4-methylheptane

(c) tert-butyl methyl ether

25. Sketch diagrams of an ether and an alcohol with the same number of carbon atoms. Generally speaking, would you expect an ether or an alcohol to be more soluble in water? Explain your reasoning.

Table 1.7 describes some physical properties of ethers. Like alcohols, ethers are polar molecules. Ethers, however, cannot form hydrogen bonds with themselves. The physical properties of ethers depend on these characteristics.

Table 1.7 Physical Properties of Ethers

Polarity of functional group	The bent shape around the oxygen atom in an ether means that the two C—O dipoles do not counteract each other. Because a C—O bond is less polar than an O—H bond, an ether is less polar than an alcohol.
Hydrogen bonding	Because there is no O—H bond in an ether, hydrogen bonding does not occur between ether molecules. Ethers can accept hydrogen bonding from water molecules.
Solubility in water	Ethers are usually soluble in water. The solubility of an ether decreases as the size of the alkyl groups increases.
Melting and boiling points	The boiling points of ethers are much lower than the boiling points of alcohols with the same number of carbon atoms.

Additional Characteristics of Ethers

- Like alcohols, ethers are extremely flammable and should be used with caution.

Amines

An organic compound with the functional group $—NH_2$, —NHR, or $—NR_2$ is called an **amine**. The letter N refers to the nitrogen atom. The letter R refers to an alkyl group attached to the nitrogen. The general formula of an amine is $R—NR'_2$. Amines can be thought of as derivatives of the ammonia molecule, NH_3. They are classified as *primary*, *secondary*, or *tertiary*, depending on how many alkyl groups are attached to the nitrogen atom. Note that the meanings of "primary," "seconday," and "tertiary" are slightly different from their meanings for alcohols. Figure 1.16 gives some examples of amines.

primary amine

$CH_3—CH_2—NH_2$

A primary amine has one alkyl group and two hydrogen atoms attached to the nitrogen.

secondary amine

$CH_3—CH_2—NH(CH_3)$ (with CH_3 attached to NH)

A secondary amine has two alkyl groups and one hydrogen atom attached to the nitrogen.

tertiary amine

$CH_3—CH_2—N(CH_3)—CH_2—CH_3$ (with CH_3 attached to N)

A tertiary amine has three alkyl groups attached to the nitrogen atom.

Figure 1.16

To name an amine, follow the steps below. The Sample Problem illustrates how to use these steps to name a secondary amine.

How to Name an Amine

Step 1 Identify the largest hydrocarbon group attached to the nitrogen atom as the parent alkane.

Step 2 Replace the -e at the end of the name of the parent alkane with the new ending -amine. Include a position number, if necessary, to show the location of the functional group on the hydrocarbon chain.

Step 3 Name the other alkyl group(s) attached to the nitrogen atom. Instead of position numbers, use the letter N- to locate the group(s). (If two identical alkyl groups are attached to the nitrogen atom, use N,N-.) This is the prefix.

Step 4 Put the name together: prefix + root + suffix.

Sample Problem

Naming a Secondary Amine

Problem

Name the following secondary amine.

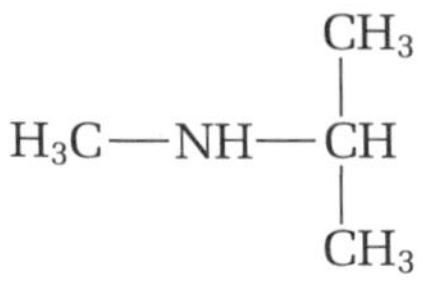

$$H_3C-NH-\underset{\displaystyle CH_3}{\overset{\displaystyle CH_3}{\overset{|}{\underset{|}{CH}}}}$$

Solution

Step 1 The propyl group is the largest of the two hydrocarbon groups attached to the nitrogen atom. Therefore, the parent alkane is propane.

Step 2 Replacing the -e with -amine gives propanamine. The position number of the functional group in the propane chain is 2.

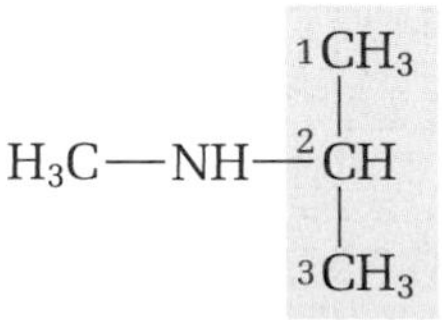

$$H_3C-NH-{}^{2}\underset{\displaystyle {}^{3}CH_3}{\overset{\displaystyle {}^{1}CH_3}{\overset{|}{\underset{|}{CH}}}}$$

Step 3 A methyl group is also attached to the nitrogen atom. The corresponding prefix is N-methyl-.

Step 4 The full name is N-methyl-2-propanamine.

CHEM FACT

It is also common and correct to name amines with each alkyl branch listed as an attachment before the suffix -amine. In this system of nomenclature, the molecules in Figure 1.16 are ethylamine, methyl ethyl amine, and methyl diethyl amine. Several other methods of naming amines exist, but they will not be covered in this course.

Practice Problems

26. Name each amine.

(a) CH_3-NH_2

(b) $C(CH_3)_3CH_2-\underset{\displaystyle H}{\underset{|}{N}}-CH_2CH_3$

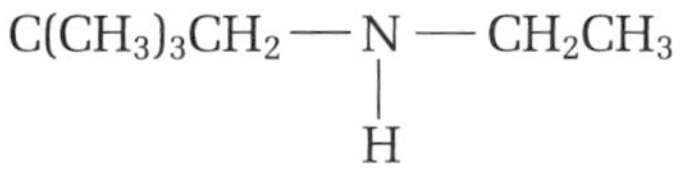

(c) $CH_3-CH_2-\underset{\displaystyle NH_2}{\underset{|}{CH}}-CH_3$

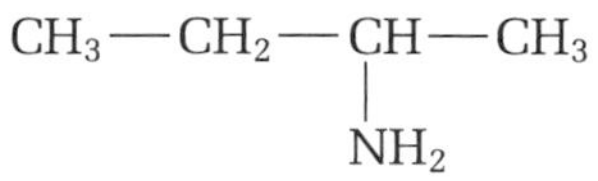

(d) (cyclopentyl)$-\overset{\displaystyle CH_3}{\overset{|}{N}}-CH_3$

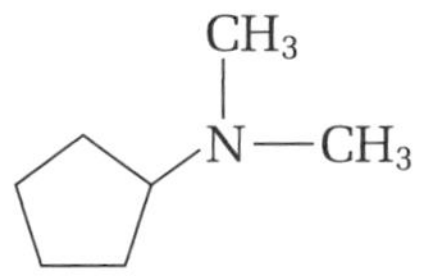

27. Draw a condensed structural diagram for each amine.

(a) 2-pentanamine

(b) cyclohexanamine

(c) N-methyl-1-butanamine

(d) N,N-diethyl-3-heptanamine

28. Classify each amine in the previous question as primary, secondary, or tertiary.

29. Draw and name all the isomers with the molecular formula $C_4H_{11}N$.

Amines are polar compounds. Primary and secondary amines can form hydrogen bonds, but tertiary amines cannot. Table 1.8 lists some common physical properties of amines.

Table 1.8 Physical Properties of Amines

Polarity of functional group	C—N and N—H bonds are polar. Thus, amines are usually polar.
Hydrogen bonding	The presence of one or more N—H bonds allows hydrogen bonding to take place.
Solubility in water	Because of hydrogen bonding, amines with low molecular masses (four or less carbon atoms) are completely miscible with water. The solubility of an amine decreases as the number of carbon atoms increases.
Melting and boiling points	The boiling points of primary and secondary amines (which contain N—H bonds) are higher than the boiling points of tertiary amines (which do not contain an N—H bond). The higher boiling points are due to hydrogen bonding between amine molecules.

Additional Characteristics of Amines

- Amines are found widely in nature. They are often toxic. Many amines that are produced by plants have medicinal properties. (See Figure 1.17.)
- Amines with low molecular masses have a distinctive fishy smell. Also, many offensive odours of decay and decomposition are caused by amines. For example, cadavarine, $H_2NCH_2CH_2CH_2CH_2CH_2NH_2$, contributes to the odour of decaying flesh. This compound gets its common name from the word "cadaver," meaning "dead body."
- Like ammonia, amines act as weak bases. Since amines are bases, adding an acid to an amine produces a salt. This explains why vinegar and lemon juice (both acids) can be used to neutralize the fishy smell of seafood, which is caused by basic amines.

Figure 1.17 **(A) Aniline is an aromatic amine that is useful for preparing dyes. (B) Adrenaline is a hormone that is produced by the human body when under stress. (C) Quinine is an effective drug against malarial fever.**

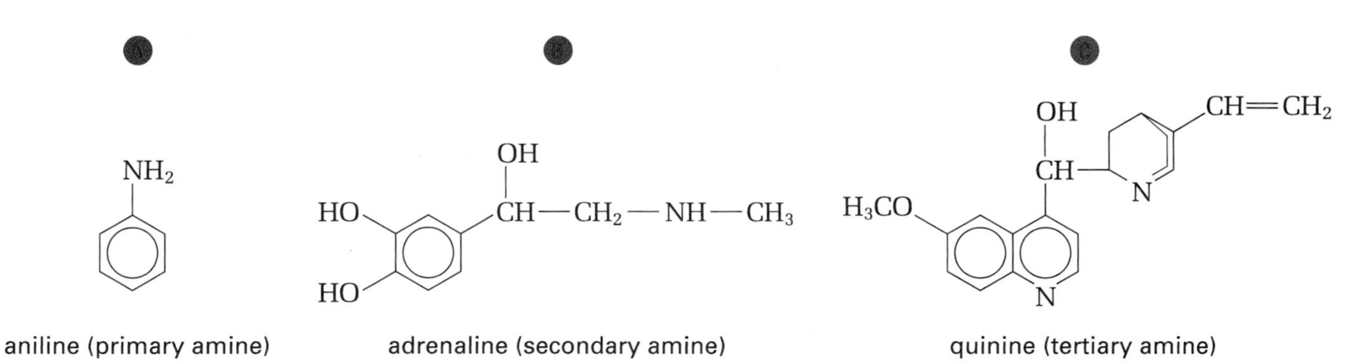

Section Summary

In this section, you learned how to recognize, name, and draw members of the alcohol, alkyl halide, ether, and amine families. You also learned how to recognize some of the physical properties of these compounds. In the next section, you will learn about families of organic compounds with functional groups that contain the C=O bond.

Section Review

1 K/U Name each compound.

(a) 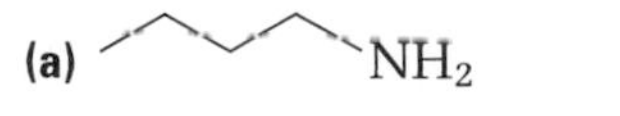

(c) 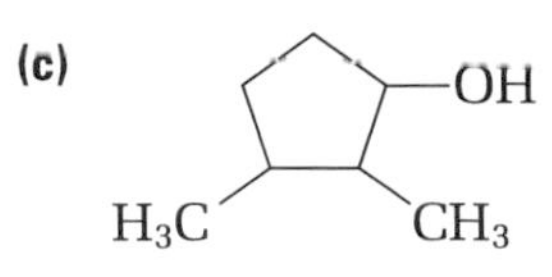

(b) $CH_3CH_2—O—CH(CH_3)CH_2CH_3$

2 K/U Write the IUPAC name for each compound.

(a) $CH_3CH_2CH_2CH_2CH_2CH_2CH_2OH$

(b) $CH_3CH(OH)CH_2CH_3$

(c) $CH_3CH_2NH_2$

(d) $(CH_3)_2NH$

(e) $CH_3CH_2OCH_3$

(f) $CH_3CH(Cl)CH_3$

3 C Draw a condensed structural diagram for each compound.

(a) 3-heptanol

(b) N-ethyl-2-hexanamine

(c) 3-methoxypentane

4 C Draw and name three isomers that have the molecular formula $C_5H_{12}O$.

5 I Name the following compounds. Then rank them, from highest to lowest boiling point. Explain your reasoning.

(a) $H_3C—CH_3$ $H_3C—O—CH_3$ $CH_3—CH_2—OH$

(b) $CH_3—CH_2—CH_3$ $CH_3—CH_2—CH_2—NH_2$ $CH_3—N(CH_3)—CH_3$

6 C Draw cyclohexanol and cyclohexane. Which compound do you expect to be more soluble in water? Explain your reasoning.

7 I Name the following compounds. Which compound do you expect to be more soluble in benzene? Explain your reasoning.

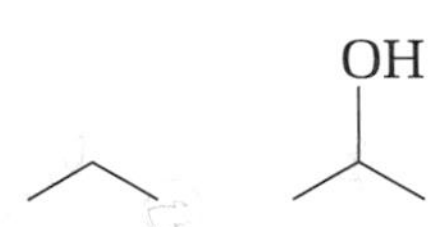

8 I Name these compounds. Then rank them, from highest to lowest molecular polarity. Explain your reasoning.

$CH_3—CH_2—CH_2—CH_3$

$HO—CH_2—CH_2—CH_2—CH_3$

$CH_3CH_2—O—CH_2CH_3$

Unit Issue Prep

Organic compounds are used in a wide variety of applications all around you. If you want to prepare for your Unit 1 Issue, research the use of organic compounds as fuel, medicines, and food additives.

Functional Groups With the C═O Bond

1.4

Some of the most interesting and useful organic compounds belong to families you are about to encounter. For example, the sweet taste of vanilla and the spicy scent of cinnamon have something in common: a carbonyl group. A **carbonyl group** is composed of a carbon atom double-bonded to an oxygen atom. In this section, you will study the structures and properties of organic compounds that have the C═O group.

Aldehydes and Ketones

Aldehydes and ketones both have the carbonyl functional group. An **aldehyde** is an organic compound that has a double-bonded oxygen on the last carbon of a carbon chain. The functional group for an aldehyde is

$$-\overset{\overset{\large O}{\|}}{C}H$$

The general formula for an aldehyde is R—CHO, where R is any alkyl group. Figure 1.18 shows the first two aldehydes.

$$H\overset{\overset{\large O}{\|}}{C}H \qquad\qquad H_3C-\overset{\overset{\large O}{\|}}{C}H$$

methanal (common name: formaldehyde)

ethanal (common name: acetaldehyde)

Figure 1.18 Methanal is made from methanol. It is used to preserve animal tissues and to manufacture plastics. Ethanal is also used as a preservative, and in the manufacture of resins and dyes.

When the carbonyl group occurs within a hydrocarbon chain, the compound is a ketone. A **ketone** is an organic compound that has a double-bonded oxygen on any carbon within the carbon chain. The functional group of a ketone is

$$-\overset{\overset{\large O}{\|}}{C}-$$

The general formula for a ketone is RCOR′, where R and R′ are alkyl groups. Figure 1.19 shows the simplest ketone, propanone.

Like the other organic compounds you have encountered, the names of aldehydes and ketones are based on the names of the parent alkanes. To name an aldehyde, follow the steps below.

How to Name an Aldehyde

Step 1 Name the parent alkane. Always give the carbon atom of the carbonyl group the position number 1.

Step 2 Replace the -e at the end of the name of the parent alkane with -al. The carbonyl group is always given position number 1. Therefore, you do not need to include a position number for it.

To name a ketone, follow the steps on the next page. The Sample Problem that follows gives examples for naming both aldehydes and ketones.

Section Preview/ Specific Expectations

In this section, you will

- **distinguish** among the following classes of organic compounds: aldehydes, ketones, carboxylic acids, esters, and amides
- **describe** physical properties of aldehydes, ketones, carboxylic acids, esters, and amides
- **draw** and **name** aldehydes, ketones, carboxylic acids, esters, and amides using the IUPAC system
- **identify** the common names of some organic compounds
- **build** molecular models of organic compounds
- **communicate** your understanding of the following terms: *carbonyl group, aldehyde, ketone, carboxylic acid, carboxyl group, derivative, ester, amide*

$$CH_3-\overset{\overset{\large O}{\|}}{C}-CH_3$$

propanone (common name: acetone)

Figure 1.19 Propanone is the main component of most nail polish removers. It is used as a solvent, and in the manufacture of plastics.

How to Name a Ketone

Step 1 Name the parent alkane. Remember that the main chain must contain the C=O group.

Step 2 If there is one ketone group, replace the -e at the end of the name of the parent alkane with -one. If there is more than one ketone group, keep the -e suffix and add a suffix such as -dione or -trione.

Step 3 For carbon chains that have more than four carbons, a position number is needed for the carbonyl group. Number the carbon chain so that the carbonyl group has the lowest possible number.

Sample Problem

Drawing and Naming Aldehydes and Ketones

Problem

Draw and name seven isomers with the molecular formula $C_5H_{10}O$.

Solution

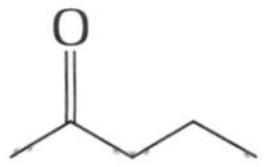

2-pentanone

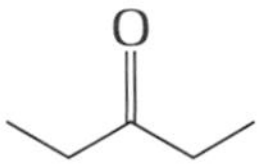

3-pentanone

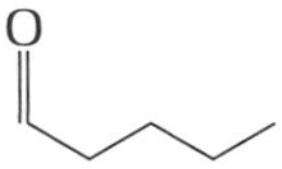

pentanal

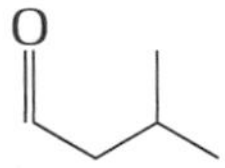

3-methylbutanal

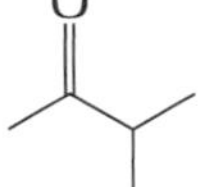

3-methyl-2-butanone

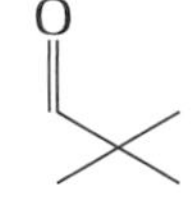

2,2-dimethylpropanal

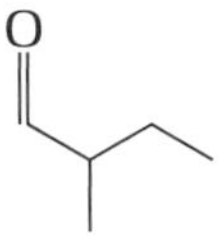

2-methylbutanal

Practice Problems

30. Name each aldehyde or ketone.

(a) 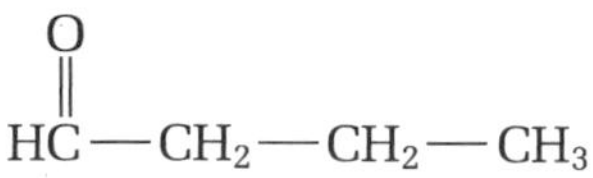 $HC(=O)-CH_2-CH_2-CH_3$

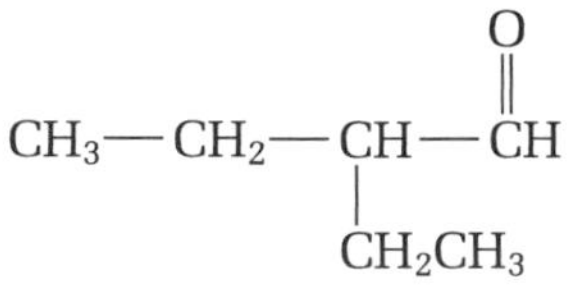

(b)

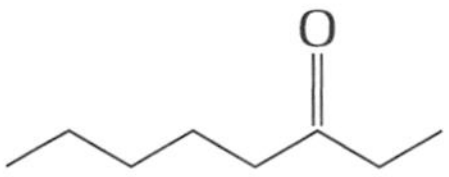

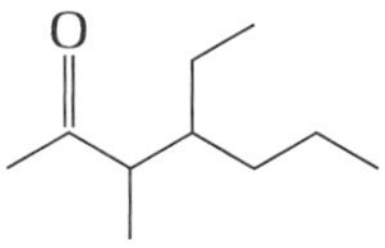

(c) 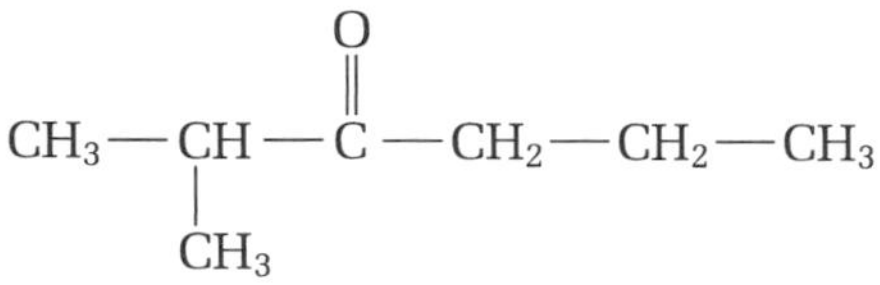 $CH_3-CH(CH_3)-C(=O)-CH_2-CH_2-CH_3$

(d) $CH_3-CH_2-CH(CH_2CH_3)-CH(=O)$

(e)

31. Draw a condensed structural diagram for each aldehyde or ketone.

(a) 2-propylpentanal

(b) cyclohexanone

(c) 4-ethyl-3,5-dimethyloctanal

32. Is a compound with a C=O bond and the molecular formula C_2H_4O an aldehyde or a ketone? Explain.

33. Draw and name five ketones and aldehydes with the molecular formula $C_6H_{12}O$.

Table 1.9 Physical Properties of Aldehydes and Ketones

Polarity of functional group	The C=O bond is polar, so aldehydes and ketones are usually polar.
Hydrogen bonding	Hydrogen bonding cannot occur between molecules of these compounds, since there is no O—H bond. The oxygen atom, however, can accept hydrogen bonds from water, as shown here.
Solubility in water	Aldehydes and ketones with low molecular masses are very soluble in water. Aldehydes and ketones with a large non-polar hydrocarbon part are less soluble in water.
Melting and boiling points	The boiling points of aldehydes and ketones are lower than the boiling points of the corresponding alcohols. They are higher than the boiling points of the corresponding alkanes.

Additional Characteristics of Aldehydes and Ketones

- In general, aldehydes have a strong pungent smell, while ketones smell sweet. Aldehydes with higher molecular masses have a pleasant smell. For example, cinnamaldehyde gives cinnamon its spicy smell. (See Figure 1.20.) Aldehydes and ketones are often used to make perfumes. The rose ketones (shown in Figure 1.21) provide up to 90% of the characteristic rose odour. Perfumers mix organic compounds, such as the rose ketones, to obtain distinctive and attractive scents.
- Since aldehydes and ketones are polar, they can act as polar solvents. Because of the non-polar hydrocarbon part of their molecules, aldehydes and ketones can also act as solvents for non-polar compounds. For example, 2-propanone (common name: acetone) is an important organic solvent in the chemical industry.
- Table 1.10 compares the boiling points of an alkane, an alcohol, and an aldehyde with the same number of carbon atoms. You can see that the boiling point of an alcohol is much greater than the boiling point of an alkane or an aldehyde.

Table 1.10 Comparing Boiling Points

Compound	Structure	Boiling point
propane	$CH_3CH_2CH_3$	42.1°C
1-propanol	$CH_3CH_2CH_2OH$	97.4°C
propanal	CH_3CH_2CHO	48.8°C

Figure 1.20 Cinnemaldehyde gives cinnamon its spicy smell.

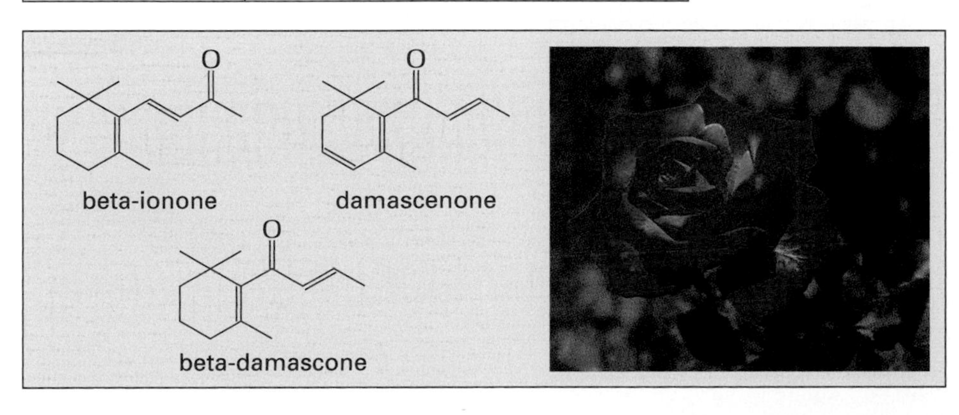

Figure 1.21 The rose ketones provide the sweet smell of a rose.

Tools & Techniques

Infrared Spectroscopy

How do researchers know when they have produced or discovered a new compound? How do they analyze the structure of a molecule?

A technique called *infrared (IR) spectroscopy* is valuable in the study of organic compounds. This technique allows researchers to determine the kinds of bonds and functional groups that are present in a molecule. Using a more advanced analysis, researchers are even able to determine other groups and bonds that are nearby. This information, paired with the molecular formula of a compound, helps researchers puzzle out the precise structure of an unknown molecule.

An infrared spectrometer works by shining infrared light through a sample of a compound. Organic molecules absorb light at certain frequencies in the range of 450 nm to 4000 nm. A sensor on the other side of the sample detects the amount of light that is absorbed by the sample at each wavelength.

When a molecule absorbs light at a certain frequency, it means that a specific bond is stretching, bending, or vibrating. The frequency where each bond absorbs light energy to stretch, bend, or vibrate is very specific. The absorption results in a decrease in intensity of the light transmitted, as measured by the sensor. This is how an infrared spectrum is formed for a molecule. The spectrum is as specific to a certain molecule as your fingerprint is to you. The table below shows some typical locations for peaks on an infrared spectrum.

Table 1.11

Bond or functional group	Wavelength (μm)
O—H	2.8–3.1
N—H	2.9–3.0
C—H	3.4–3.5
carboxylic acid group	2.8–4.0 and 5.8–5.9
aldehyde	3.4–3.7 and 5.7–5.8
ketone	5.8–5.9
ester	5.7–5.8 and 7.7–8.5
alkene	5.9–6.2
alkyne	4.5–4.8
haloalkane	13.4–20

The shape of the peak for each functional group is affected by other groups and atoms in the vicinity. Examine the spectrum for 2-pentanone, shown below. This compound has many C—H bonds, and one C═O bond. The jagged peak in the range of 3.4 μm to 3.5 μm is caused by the various C—H bonds in the molecule absorbing light and stretching. The peak between 5.8 μm and 5.9 μm is caused by the C═O bond stretching. The peaks between 6.7 μm and 7.7 μm represent the C—H bonds bending.

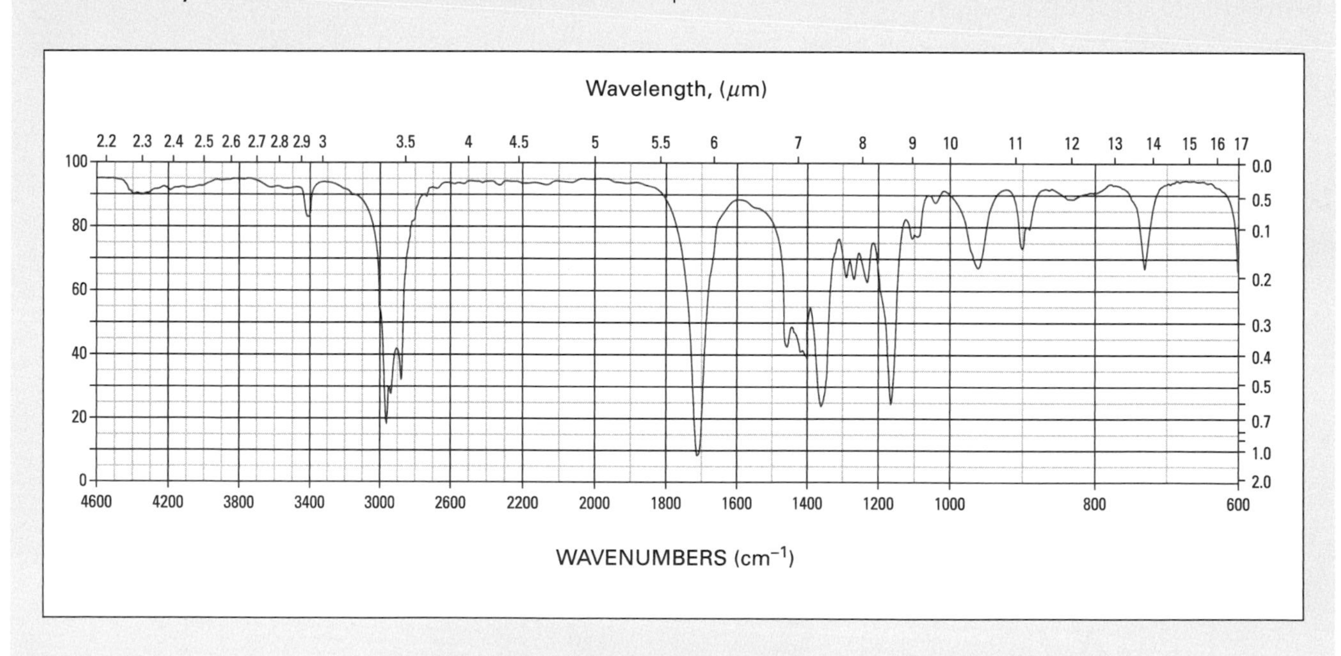

Carboxylic Acids

You are already familiar with one carboxylic acid. In fact, you may sprinkle it over your French fries or your salad, as shown in Figure 1.22. Vinegar is a 5% solution of acetic acid in water. The IUPAC name for acetic acid is ethanoic acid, CH_3COOH.

A **carboxylic acid** is an organic compound with the following functional group:

$$-\overset{\displaystyle O}{\overset{\|}{C}}-OH$$

This —COOH group is called the **carboxyl group**. The general formula for a carboxylic acid is R—COOH. Figure 1.23 shows some common carboxylic acids.

Figure 1.22 Many salad dressings contain vinegar.

$$H\overset{\displaystyle O}{\overset{\|}{C}}-OH$$

methanoic acid (common name: formic acid)

$$CH_3-\overset{\displaystyle O}{\overset{\|}{C}}-OH$$

ethanoic acid (common name: acetic acid)

$$C_6H_5-\overset{\displaystyle O}{\overset{\|}{C}}-OH$$

benzoic acid

$$HO-\overset{\displaystyle O}{\overset{\|}{C}}-CH_2-\underset{\displaystyle OH}{\overset{\displaystyle COOH}{C}}-CH_2-\overset{\displaystyle O}{\overset{\|}{C}}-OH$$

citric acid

Figure 1.23 Common carboxylic acids

To name a simple carboxylic acid, follow the steps below. Figure 1.24 gives some examples of carboxylic acid names.

How to Name a Carboxylic Acid

Step 1 Name the parent alkane.

Step 2 Replace the -e at the end of the name of the parent alkane with -oic acid.

Step 3 The carbon atom of the carboxyl group is always given position number 1. Name and number the branches that are attached to the compound.

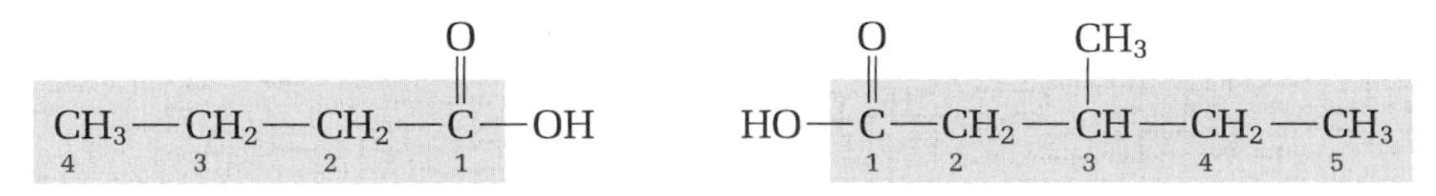

butanoic acid

3-methylpentanoic acid

Figure 1.24 Examples of carboxylic acid names

COURSE CHALLENGE

In the Course Challenge at the end of this book, you will practise recognizing organic compounds. Prepare a list of tips to help you recognize each type of organic compound you have encountered.

CONCEPT CHECK

Organic compounds that have one or two carbon atoms are usually known by their common names. These common names are based on the Latin words *formica* (ant) and *acetum* (vinegar). Give the IUPAC names for formaldehyde, formic acid, acetaldehyde, and acetic acid.

Practice Problems

34. Name each carboxylic acid.

(a) 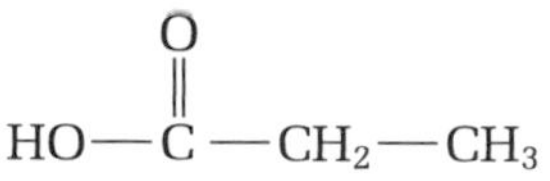 $HO{-}\overset{\overset{\displaystyle O}{\|}}{C}{-}CH_2{-}CH_3$

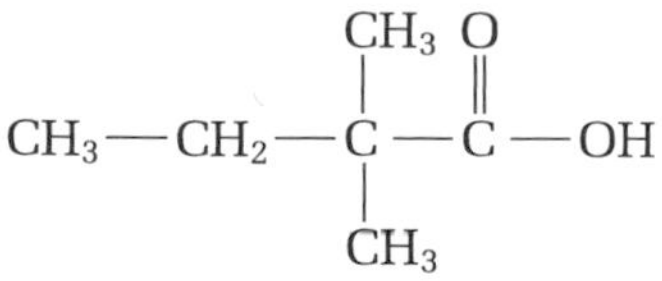

(b) $CH_3{-}CH_2{-}\overset{\overset{\displaystyle CH_3}{|}}{\underset{\underset{\displaystyle CH_3}{|}}{C}}{-}\overset{\overset{\displaystyle O}{\|}}{C}{-}OH$

(c)

O

OH

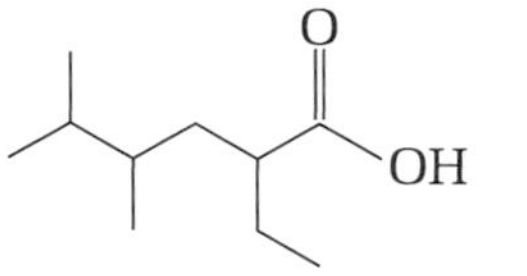

35. Draw a condensed structural diagram for each carboxylic acid.

(a) hexanoic acid

(b) 3-propyloctanoic acid

(c) 3,4-diethyl-2,3,5-trimethylheptanoic acid

36. Draw a line structural diagram for each compound in question 35.

37. Draw and name two carboxylic acids with the molecular formula $C_4H_8O_2$.

Table 1.12 lists some of the physical properties of carboxylic acids. Notice that carboxylic acids have even stronger hydrogen bonding than alcohols.

Table 1.12 Physical Properties of Carboxylic Acids

Polarity of functional group	Due to the presence of the polar O–H and C=O bonds, carboxylic acids are polar compounds.
Hydrogen bonding	The hydrogen bonding between carboxylic acid molecules is strong, as shown here: 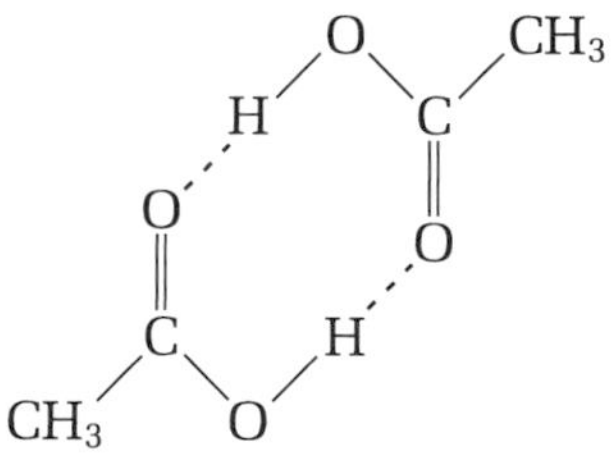 (O, CH_3, H, C, O, O, C, H, CH_3, O) Hydrogen bonding also occurs between carboxylic acid molecules and water molecules.
Solubility in water	Carboxylic acids with low molecular masses are very soluble in water. The first four simple carboxylic acids (methanoic acid, ethanoic acid, propanoic acid, and butanoic acid) are miscible with water. Like alcohols, ketones, and aldehydes, the solubility of carboxylic acids in water decreases as the number of carbon atoms increases.
Melting and boiling points	Because of the strong hydrogen bonds between molecules, the melting and boiling points of carboxylic acids are very high.

Additional Characteristics of Carboxylic Acids

- Carboxylic acids often have unpleasant odours. For example, butanoic acid has the odour of stale sweat.
- The —OH group in a carboxylic acid does *not* behave like the basic hydroxide ion, OH^-. Oxygen has a high electronegativity (attraction to electrons) and there are two oxygen atoms in the carboxylic acid functional group. These electronegative oxygen atoms help to carry the extra negative charge that is caused when a positive hydrogen atom dissociates. This is why the hydrogen atom in a carboxylic acid is able to dissociate, and the carboxylic acid behaves like an acid.

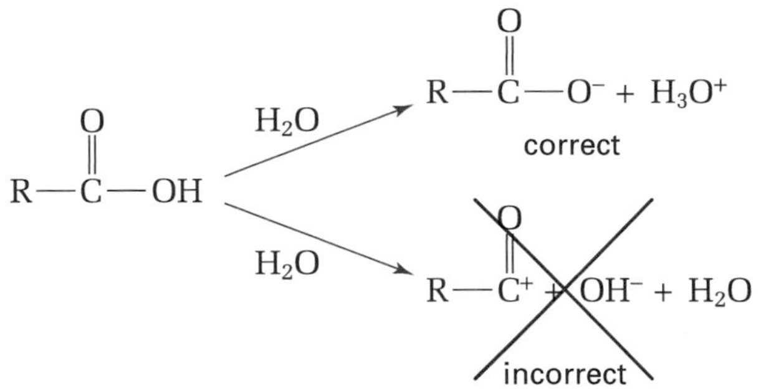

$$R{-}C(=O){-}OH \xrightarrow{H_2O} R{-}C(=O){-}O^- + H_3O^+$$

correct

$$R{-}C(=O){-}OH \xrightarrow{H_2O} R{-}C^+(=O) + OH^- + H_2O$$

incorrect

- Figure 1.25 compares the melting and boiling points of a carboxylic acid with the melting and boiling points of other organic compounds. As you can see, the melting and boiling points of the carboxylic acid are much higher than the melting and boiling points of the other compounds. This is due to the exceptionally strong hydrogen bonding between carboxylic acid molecules.

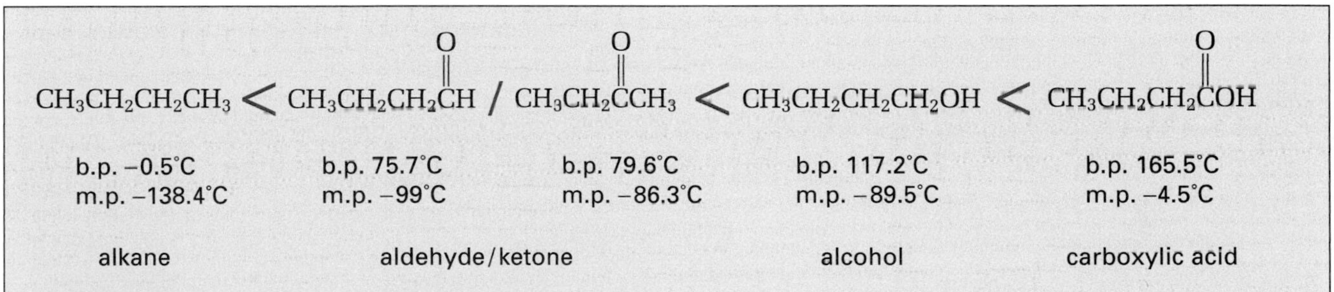

$$CH_3CH_2CH_2CH_3 < CH_3CH_2CH_2CH(=O) \,/\, CH_3CH_2C(=O)CH_3 < CH_3CH_2CH_2CH_2OH < CH_3CH_2CH_2C(=O)OH$$

alkane	aldehyde/ketone		alcohol	carboxylic acid
b.p. −0.5°C	b.p. 75.7°C	b.p. 79.6°C	b.p. 117.2°C	b.p. 165.5°C
m.p. −138.4°C	m.p. −99°C	m.p. −86.3°C	m.p. −89.5°C	m.p. −4.5°C

Figure 1.25

Derivatives of Carboxylic Acids

Exchanging the hydroxyl group of a carboxylic acid with a different group produces a *carboxylic acid derivative*. In the following investigation, you will react two carboxylic acids with primary alcohols to produce different organic compounds.

Investigation 1-A

MICROSCALE

SKILL FOCUS

Predicting

Performing and recording

Analyzing and interpreting

Preparing a Carboxylic Acid Derivative

The reaction between a carboxylic acid and an alcohol produces an organic compound that you have not yet seen. In this investigation, you will prepare this new type of compound and observe one of its properties.

Questions

What kind of compound will form in the reaction of a carboxylic acid with an alcohol? What observable properties will the new compounds have?

Safety Precautions

- The products of these reactions are very flammable. Before starting, make sure that there is *no* open flame in the laboratory. Use a hot plate, *not* a Bunsen burner.
- Carry out all procedures in a well-ventilated area. Use a fume hood for steps involving butanoic acid.
- Concentrated sulfuric acid, ethanoic acid, and butanoic acid are all extremely corrosive. Wear goggles, gloves, and an apron while performing this investigation. Treat the acids with extreme care. If you spill any acid on your skin, wash it with plenty of cold water and notify your teacher.

Materials

retort stand with two clamps
thermometer
hot plate
2 small beakers (50 or 100 mL)
100 mL beaker
2 large beakers (250 mL)
4 plastic micropipettes
2 graduated cylinders (10 mL)
medicine dropper
stopper or paper towel
distilled water
6 mol/L sulfuric acid
ethanoic acid (glacial acetic acid)
butanoic acid
ethanol
1-propanol
ice

Procedure

1. Before starting the investigation, read through the Procedure. Then prepare your equipment as follows.
   (a) Make sure that all the glassware is very clean and dry.
   (b) Prepare a hot-water bath. Heat about 125 mL of tap water in a 250 mL beaker on the hot plate to 60°C. Adjust the hot plate so the temperature remains between 50°C and 60°C.
   (c) Prepare a cold-water bath. Place about 125 mL of a mixture of water and ice chips in the second 250 mL beaker. As long as there are ice chips in the water, the temperature of the cold-water bath will remain around 0°C.
   (d) Place about 5 mL of distilled water in a small beaker (or graduated cylinder). You will use this in step 8.
   (e) Set up the retort stand beside the hot-water bath. Use one clamp to hold the thermometer in the hot-water bath. Make sure that the thermometer is not touching the walls or the bottom of the beaker. Use a stopper or wrap a piece of paper towel around the thermometer to hold the thermometer in place carefully but firmly. You will use the other clamp to steady the micropipettes after you fill them with reaction mixture and place them in the hot-water bath.

(f) Label three pipettes according to each trial:
- ethanoic acid + ethanol
- ethanoic acid + 1-propanol
- butanoic acid + ethanol

(g) Cut off the bulb of one of the micropipettes, halfway along the wide part of the bulb (see the diagram below.) You will use the bulb as a cap to prevent vapours from escaping during the reactions.

2. Use the graduated cylinder to measure 1.0 mL of ethanol into the 100 mL beaker. As you do so, you will get a whiff of the odour of the alcohol. (CAUTION Do not inhale the alcohol vapour directly.) Record your observations.

3. Use the graduated cylinder to add 1.0 mL of ethanoic acid to the ethanol. As you do so, you will get a whiff of the odour of the acid. (CAUTION Do not inhale the acid directly.) Record your observations.

4. Your teacher will *carefully* add 4 drops of sulfuric acid to the alcohol/acid mixture.

5. Suction the mixture up into the appropriately labelled micropipette. Invert the micropipette. Place it, bulb down, in the hot-water bath. (See the diagram below.) Place the cap over the tip of the pipette. Use a clamp to hold the pipette in place.

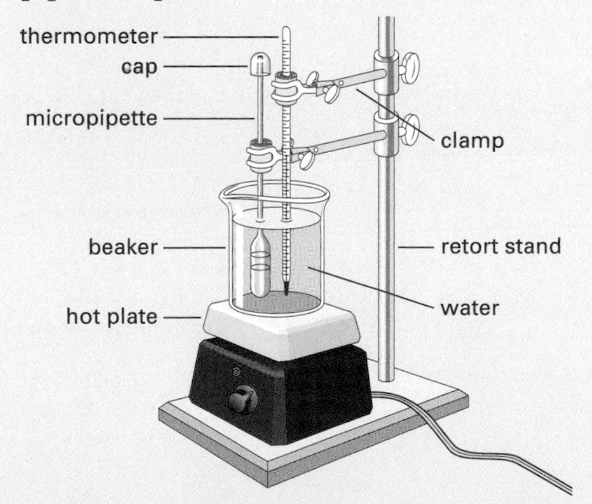

6. Leave the pipette in the hot water for about 10 min to 15 min. Use the thermometer to monitor the temperature of the hot water. The temperature should stay between 50°C and 60°C.

7. After 10 to 15 min in the hot-water bath, place the pipette in the cold-water bath. Allow it to cool for about 5 min.

8. Carefully squeeze a few drops of the product onto a watch glass. Mix it with a few drops of distilled water. To smell the odour of the compound, take a deep breath. Use your hand to wave the aroma toward your nose as you breathe out. Record your observations of the aroma.

9. Repeat the procedure for the other two trials. (CAUTION Butanoic acid has a strong odour. Do not attempt to smell it.)
Note: While the first reaction mixture is being heated, you may wish to prepare the other mixtures and put them in the hot-water bath. Working carefully, you should be able to stabilize more than one micropipette with the clamp. If you choose to do this, make sure that your materials are clearly labelled. Also remember to keep a record of the time at which each micropipette was introduced to the hot-water bath.

10. Dispose of all materials as your teacher directs. Clean all glassware thoroughly with soap and water.

Analysis

1. Make up a table to organize your data. What physical property did you observe?
2. How do you know that a new product was formed in each reaction? Explain.

Conclusion

3. Describe the odour of each product that was formed. Compare the odours to familiar odours (such as the odours of plants, flowers, fruits, and animals) to help you describe them.

Application

4. Research the organic compounds that are responsible for the smell and taste of oranges, pineapples, pears, oil of wintergreen, and apples. Find and record the chemical structure of each compound.

Carboxylic Acid Derivatives

The strong-smelling compounds you prepared in Investigation 1-A do not fit into any of the organic families you have studied so far. According to their molecular formulas, however, they are isomers of carboxylic acids. They are esters. Because an ester is obtained by replacing the —OH group of a carboxylic acid with a different group, it is called a **derivative** of a carboxylic acid. Carboxylic acids have several important derivatives. In this section, you will study two of these derivatives: esters and amides.

If you have access to probeware, do Probeware Investigation 1-A, or a similar investigation from a probeware company.

Esters

An **ester** is an organic compound that has the following functional group:

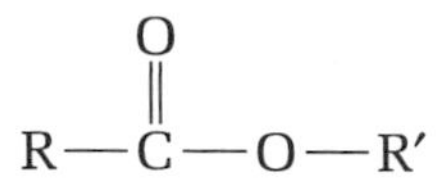

The general formula for an ester is RCOOR′, where R is a hydrogen atom or a hydrocarbon, and R′ is a hydrocarbon. You can think of an ester as the product of a reaction between a carboxylic acid and an alcohol, as shown in Figure 1.26.

$$CH_3CH_2CH_2C(=O)—OH + HO—CH_2CH_3 \rightarrow CH_3CH_2CH_2C(=O)—O—CH_2CH_3 + H_2O$$

carboxylic acid | alcohol | ester | water

Figure 1.26 When heated, butanoic acid reacts with ethanol to produce an ester called ethyl butanoate.

To name an ester, you must recognize that an ester can be thought of as having two distinct parts. The main part of the ester contains the —COO group. This part comes from the parent acid. When numbering the main chain of a carboxylic acid, the carbon atom in the carboxyl group is always given position number 1. The second part of an ester is the alkyl group.

To name an ester, follow the steps below.

How to Name an Ester

Step 1 Identify the main part of the ester, which contains the C=O group. This part comes from the parent acid. Begin by naming the parent acid.

Step 2 Replace the *-oic acid* ending of the name of the parent acid with -oate.

Step 3 The second part of an ester is the alkyl group that is attached to the oxygen atom. Name this as you would name any other alkyl group.

Step 4 Put the two names together. Note that esters are named as two words. (See Figure 1.27.)

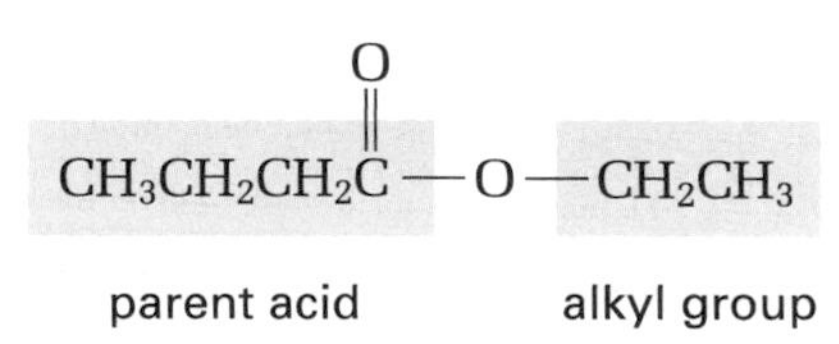

Figure 1.27 The main part of this molecule is based on butanoic acid. The other part of the ester is an ethyl group. The full name is ethyl- + -butanoate → ethyl butanoate

Sample Problem

Naming and Drawing Esters

Problem

Name and draw three esters that have the molecular formula $C_4H_8O_2$.

Solution

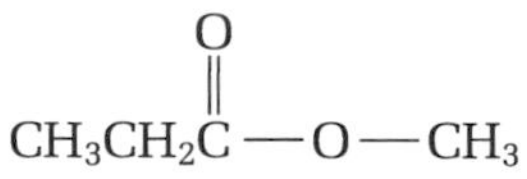

$$CH_3CH_2\overset{\overset{\displaystyle O}{\|}}{C}-O-CH_3$$

methyl propanoate

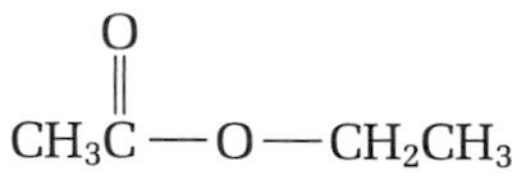

$$CH_3\overset{\overset{\displaystyle O}{\|}}{C}-O-CH_2CH_3$$

ethyl ethanoate

$$H-\overset{\overset{\displaystyle O}{\|}}{C}-O-CH_2CH_2CH_3$$

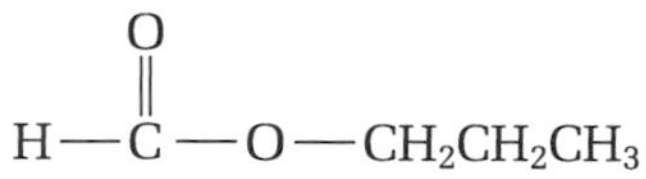

propyl methanoate

Practice Problems

38. Name each ester.

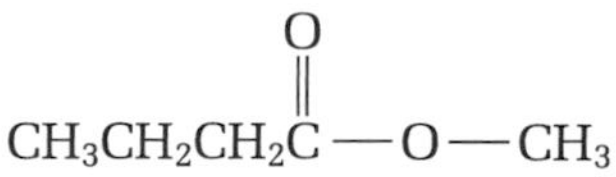

(a) 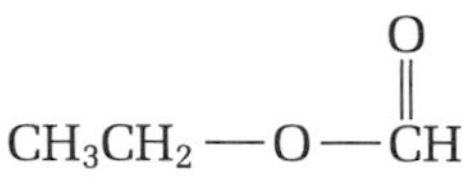 $CH_3CH_2-O-\overset{\overset{\displaystyle O}{\|}}{C}H$

(b) $CH_3CH_2CH_2\overset{\overset{\displaystyle O}{\|}}{C}-O-CH_3$

(c) $CH_3CH_2CH_2CH_2\overset{\overset{\displaystyle O}{\|}}{C}-O-CH_2CH_2CH_2CH_2CH_3$

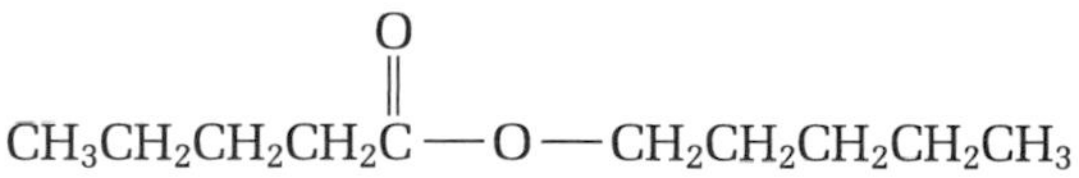

39. For each ester in the previous question, name the carboxylic acid and the alcohol that are needed to synthesize it.

40. Draw each ester.

(a) methyl pentanoate

(b) heptyl methanoate

(c) butyl ethanoate

(d) propyl octanoate

(e) ethyl 3,3-dimethylbutanoate

41. Write the molecular formula of each ester in the previous question. Which esters are isomers of each other?

42. Draw and name five ester isomers that have the molecular formula $C_5H_{10}O_2$.

Table 1.13, on the next page, describes some of the physical properties of esters. As you will see, esters have different physical properties than carboxylic acids, even though esters and carboxylic acids are isomers of each other.

Table 1.13 Physical Properties of Esters

Polarity of functional group	Like carboxylic acids, esters are usually polar molecules.
Hydrogen bonding	Esters do not have an O–H bond. Therefore, they cannot form hydrogen bonds with other ester molecules.
Solubility in water	Esters can accept hydrogen bonds from water. Therefore, esters with very low molecular masses are soluble in water. Esters with carbon chains that are longer than three or four carbons are not soluble in water.
Melting and boiling points	Because esters cannot form hydrogen bonds, they have low boiling points. They are usually volatile liquids at room temperature.

Additional Characteristics of Esters

- Esters often have pleasant odours and tastes, so they are used to produce perfumes and artificial flavours. In fact, the characteristic tastes and smells of many fruits come from esters. (See Figure 1.28.)

Amides

Figure 1.28 Octyl ethanoate is found in oranges.

An ester can be thought of as the combination of a carboxylic acid and an alcohol. Similarly, you can think of an amide as the combination of a carboxylic acid and ammonia or an amine. An **amide** is an organic compound that has a carbon atom double-bonded to an oxygen atom and single-bonded to a nitrogen atom.

Amides have the functional group below:

O
‖
—C—N—
|

The general formula for an amide is R—CO—NR_2. R can stand for a hydrogen atom or an alkyl group. Figure 1.29 gives some examples of amides.

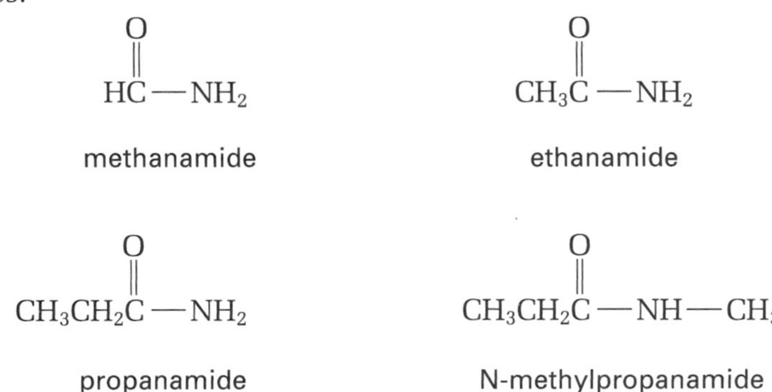

Figure 1.29 Examples of amides

To name an amide, follow the steps on the next page. The Sample Problem that follows illustrates how to use these steps. Later, Table 1.14 describes some physical properties of amides.

How to Name an Amide

Step 1 Locate the part of the amide that contains the C=O group. Name the parent carboxylic acid that this part derives from. **Note:** The carbon in the C=O group is always given position number 1.

Step 2 Replace the *-oic acid* ending of the name of the parent acid with the suffix -amide.

Step 3 Decide whether the compound is a primary, secondary, or tertiary amide:

- If there are two hydrogen atoms (and no alkyl groups) attached to the nitrogen atom, the compound is a *primary amide* and needs no other prefixes.
- If there is one alkyl group attached to the nitrogen atom, the compound is a *secondary amide*. Name the alkyl group, and give it location letter N- to indicate that it is bonded to the nitrogen atom.
- If there are two alkyl groups, the compound is a *tertiary amide*. Place the alkyl groups in alphabetical order. Use location letter N- before each group to indicate that it is bonded to the nitrogen atom. If the two groups are identical, use N,N-.

Step 4 Put the name together: prefix + root + suffix.

Sample Problem

Naming a Secondary Amide

Problem

Name the following compound.

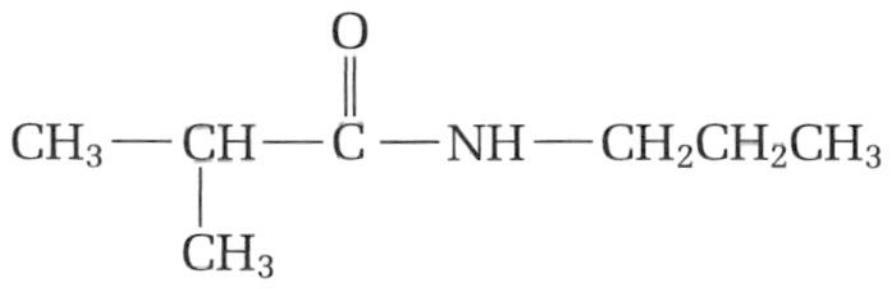

$$CH_3-\underset{\substack{|\\CH_3}}{CH}-\overset{\substack{O\\\|}}{C}-NH-CH_2CH_2CH_3$$

Solution

Step 1 The parent acid is 2-methylpropanoic acid.

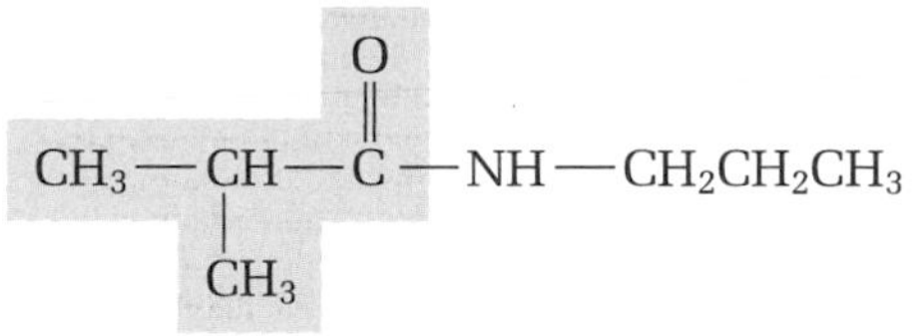

$$CH_3-\underset{\substack{|\\CH_3}}{CH}-\overset{\substack{O\\\|}}{C}-NH-CH_2CH_2CH_3$$

Step 2 The base name for this compound is 2-methylpropanamide. (The root is -propan-, and the suffix is -amide.)

Step 3 A propyl group is attached to the nitrogen atom. Thus, the compound's prefix is N-propyl-.

Step 4 The full name of the compound is N-propyl-2-methylpropanamide.

Practice Problems

43. Name each amide.

(a) $CH_3CH_2CH_2\overset{\overset{\large O}{\|}}{C}—NH_2$

(b) $H_3C—NH—\overset{\overset{\large O}{\|}}{C}CH_2CH_2CH_2CH_2CH_3$

(c) N, O

44. Draw each amide.

(a) nonanamide

(b) N-methyloctanamide

(c) N-ethyl-N-propylpropanamide

(d) N-ethyl-2,4,6-trimethyldecanamide

45. Name each amide.

(a) CH_3CONH_2

(b) $CH_3CH_2CH_2CH_2CH_2CH_2CONHCH_3$

(c) $(CH_3)_2CHCON(CH_3)_2$

46. Draw a line structural diagram for each amide in the previous question.

Table 1.14 Physical Properties of Amides

Polarity of functional group	Because the nitrogen atom attracts electrons more strongly than carbon or hydrogen atoms, the C–N and N–H bonds are polar. As a result, the physical properties of amides are similar to the physical properties of carboxylic acids.
Hydrogen bonding	Because primary amides have two N–H bonds, they have even stronger hydrogen bonds than carboxylic acids. Secondary amides also experience hydrogen bonding.
Solubility in water	Amides are soluble in water. Their solubility decreases as the non-polar hydrocarbon part of the molecule increases in size.
Melting and boiling points	Primary amides have much higher melting and boiling points than carboxylic acids. Many simple amides are solid at room temperature.

Additional Characteristics of Amides

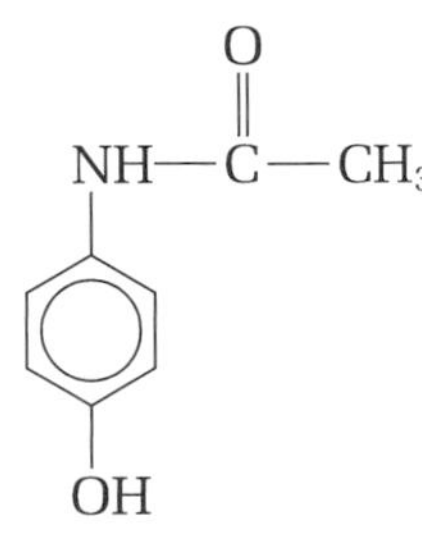

acetaminophen

- An amide called acetaminophen is a main component of many painkillers.
- Urea, another common example of an amide, is made from the reaction between carbon dioxide gas, CO_2, and ammonia, NH_3. Urea was the first organic compound to be synthesized in a laboratory. It is found in the urine of many mammals, including humans, and it is used as a fertilizer.

O
H_2N NH_2

urea

Comparing Physical Properties

In this chapter, you have learned how to recognize many different types of organic compounds. In the first section, you learned how to use polar bonds and the shape of a molecule to determine its molecular polarity. The following investigation allows you to apply what you have learned to predict and compare the physical properties of various organic compounds.

SKILL FOCUS

Predicting

Performing and recording

Investigation 1-B

Comparing Physical Properties

In this investigation, you will examine the differences between molecules that contain different functional groups. As you have learned, the polarity and hydrogen bonding abilities of each functional group affect how these molecules interact among themselves and with other molecules. You will examine the shape of each molecule and the effects of intermolecular forces in detail to make predictions about properties.

Prediction

How can you predict the physical properties of an organic compound by examining its structure?

Materials

molecular model kit

Procedure

1. Choose a parent alkane that has four to ten carbon atoms (butane to decane).
2. By adding functional groups to your parent alkane, build a model of a molecule from each class of organic compounds:
   - aromatic hydrocarbons (section 1.2)
   - alcohols, ethers, amines (section 1.3)
   - aldehydes, ketones, carboxylic acids, esters, amides (section 1.4)
3. Draw a condensed structural diagram and a line structural diagram of each compound. Include the IUPAC name of the compound.

Analysis

1. Look at your aromatic hydrocarbon, your alcohol, your ether, and your carboxylic acid. Rank these compounds from
   - least polar to most polar
   - lowest to highest boiling point

   Record your rankings, and explain them.
2. Compare the alcohol you made with an alcohol that has a different number of carbon atoms, made by a classmate. Which alcohol is more soluble in water, yours or your classmate's? Which alcohol is more polar? Explain your reasoning.
3. Choose two of your compounds at random. Which compound do you think has the higher melting point? Explain your reasoning.
4. Consider all the compounds you made. Which compound do you think is the most soluble in water? Explain your reasoning.
5. Which compound do you think is the most soluble in benzene? Explain your reasoning.
6. Where possible, predict the odour of each organic compound you made.

Conclusions

7. Look up the boiling point, melting point, and solubility of each compound in a reference book, such as *The CRC Handbook of Chemistry and Physics*. Compare your findings with the predictions you made in the Analysis.
8. Challenge a classmate to name each compound you made and identify its functional group.

Application

9. Use reference books and the Internet to discover any useful applications of the compounds you made. Prepare a short report for your class, explaining how you think each compound's physical properties may affect its usefulness in real life.

Section Summary

In this chapter, you learned how to recognize, name, and predict the physical properties of organic compounds that belong to the alcohol, ether, amine, aldehyde, ketone, carboxylic acid, ester, and amide families. You discovered many important uses for organic compounds. You know that 2-propanol (isopropyl alcohol) is used as an antiseptic. Acetone (a ketone) is the main component of nail polish remover. Esters give many fruits and processed foods their distinctive odours and tastes. In the next chapter, you will learn about different chemical reactions that are typical of each functional family. You will take a detailed look at some common organic compounds in your life, and learn about some of the benefits and risks of using organic compounds.

The Chemistry 12 Electronic Learning Partner has a review activity to practise naming functional groups.

Section Review

1 K/U Write the IUPAC name for each compound.

(a) CH_3OH

(b) CH_3CH_2OH

(c) 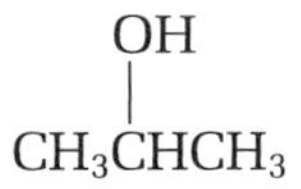 $CH_3\overset{\overset{OH}{|}}{C}HCH_3$

(d) 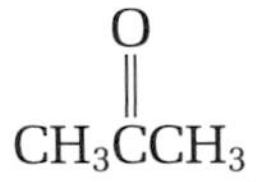 $CH_3\overset{\overset{O}{\|}}{C}CH_3$

(e)  $H\overset{\overset{O}{\|}}{C}H$

(f) 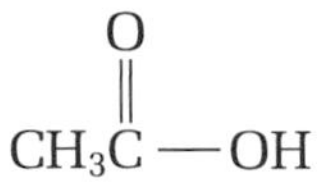 $CH_3\overset{\overset{O}{\|}}{C}-OH$

(g) 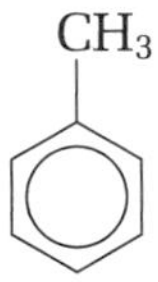 CH_3

2 K/U Write the common name for each compound in question 1.

3 K/U Name each compound.

(a) 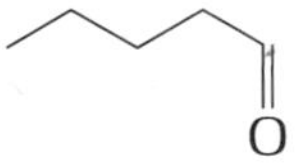 O

(b) 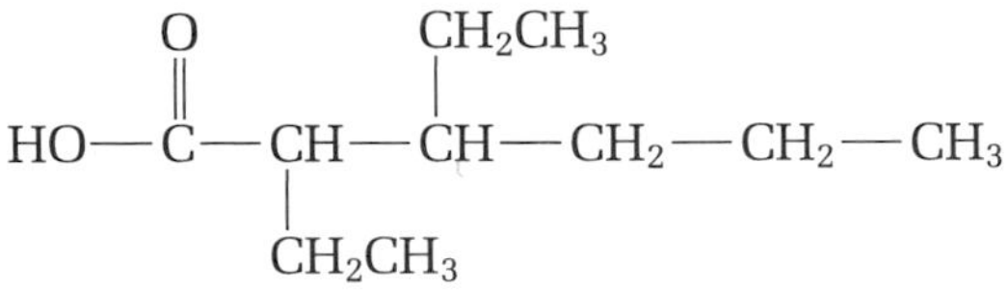 $HO-\overset{\overset{O}{\|}}{C}-\underset{\underset{CH_2CH_3}{|}}{CH}-\overset{\overset{CH_2CH_3}{|}}{CH}-CH_2-CH_2-CH_3$

(c) $CH_3CH_2\overset{\overset{O}{\|}}{C}-NH-CH_2CH_3$

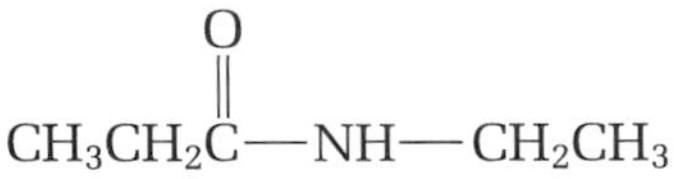

(d) 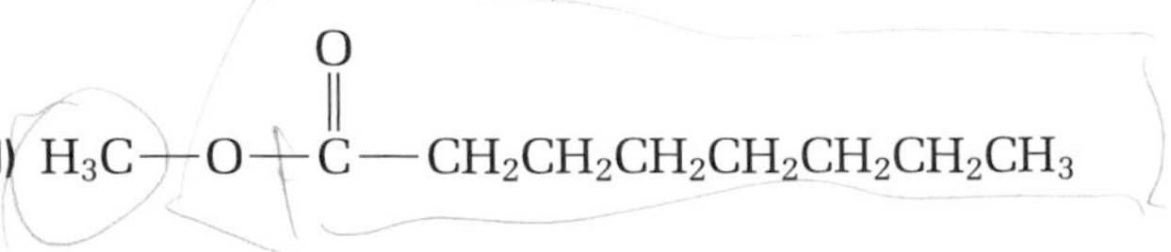 $H_3C-O-\overset{\overset{O}{\|}}{C}-CH_2CH_2CH_2CH_2CH_2CH_2CH_3$

4 **I** Identify the family that each organic compound belongs to.

(a) $CH_3CH_2CONH_2$

(b) $CH_3CH_2CH(CH_3)CH_2CH_2COOH$

(c) $CH_3CH_2C(CH_3)_2CH_2CHO$

(d) CH_3COOCH_3

5 **K/U** Name each organic compound in question 4.

6 **C** Draw and name one carboxylic acid and one ester with the molecular formula $C_6H_{12}O_2$.

7 **C** Draw and name one primary amide, one secondary amide, and one tertiary amide with the molecular formula $C_6H_{13}ON$.

8 **I** Identify the functional groups in each molecule.

(a) vanillin, a flavouring agent

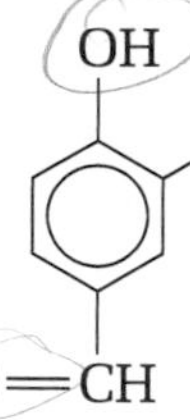

(b) DEET, an insect repellant

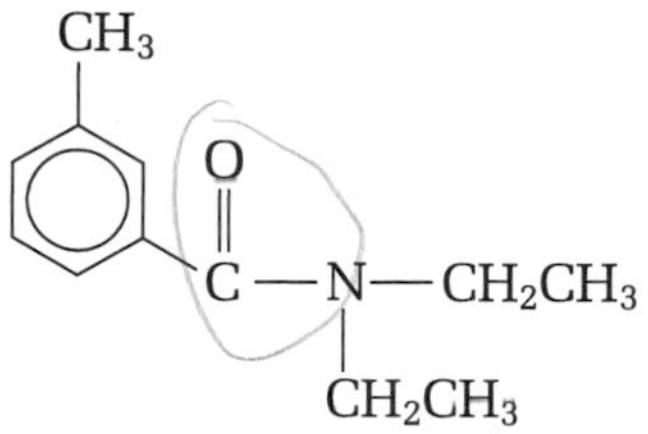

(c) penicillin V, an antibiotic (This compound also contains a functional group that is unfamiliar to you. Identify the atom(s) in the unfamiliar functional group as part of your answer.)

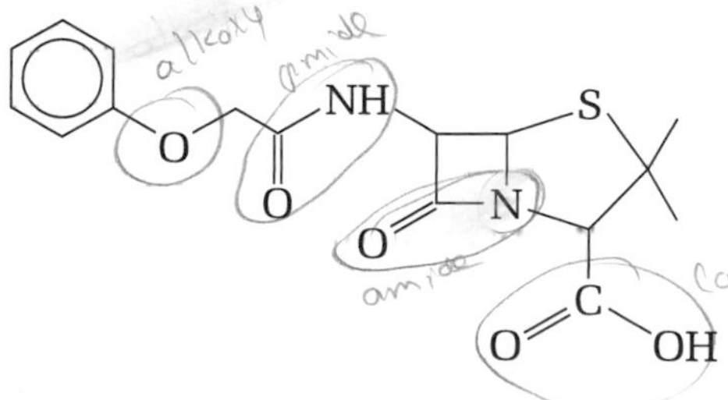

9 **K/U** Consider the compounds CH_3CH_2COH, CH_3CH_2COOH, and CH_3COOCH_3. Which compound has the highest boiling point? Use diagrams to explain your reasoning.

10 **K/U** How could you use your sense of smell to distinguish an amine from a ketone?

CHAPTER 1 Review

Reflecting on Chapter 1

Summarize this chapter in the format of your choice. Here are a few ideas to use as guidelines:

- Explain the appropriate use of the terms "natural," "organic," and "chemical."
- Explain how the polarity of a molecule depends on its shape and on any polar bonds that are present.
- Describe the shape and structure of a benzene molecule.
- Draw and name an example of an alcohol, an alkyl halide, an ether, and an amine.
- Draw and name an example of an aldehyde, a ketone, a carboxylic acid, an ester, and an amide.

Reviewing Key Terms

For each of the following terms, write a sentence that shows your understanding of its meaning.

organic chemistry	organic compounds
tetrahedral	trigonal planar
linear	bent
bond dipole	polar
molecular polarity	hydrocarbons
aliphatic hydrocarbon	aromatic hydrocarbon
alkane	cycloalkane
alkene	functional group
alkyne	alkyl group
—OH (hydroxyl) group	general formula
intermolecular forces	hydrogen bonding
dipole-dipole interactions	dispersion forces
alcohol	parent alkane
alkyl halide (haloalkane)	ether
alkoxy group	amine
carbonyl group	aldehyde
ketone	carboxylic acid
carboxyl group	derivative
ester	amide

Knowledge/Understanding

1. What is an organic compound?
2. Examine the following molecules. Then answer the questions below.

CH_4 $\quad CH_3—\overset{O}{\overset{\|}{C}}—CH_3 \quad CH_3—CH{=}CH_2$

(a) Predict the shape around the central carbon atom in each molecule.
(b) Identify any polar bonds that are present in each molecule.
(c) Describe each molecule as polar or non-polar.

3. Classify each molecule as polar or non-polar.
(a) CH_3OH (b) CH_3CH_3 (c) CH_3NH_2 (d) $H_2C{=}CH_2$ (e) $H_2C{=}O$
4. Describe, in words, the functional group for each type of organic compound.
(a) alkene (b) alcohol (c) amine (d) ketone (e) carboxylic acid
5. Give the general formula for each type of organic compound.
(a) alcohol (b) amine (c) ester (d) amide
6. What is the difference between each two organic compounds?
(a) alcohol and ether
(b) amine and amide
(c) carboxylic acid and ester
(d) cyclic alkene and benzene
7. What is the difference between a saturated compound and an unsaturated compound?
8. Use the suffix to name the functional family of each type of organic compound.
(a) ethanamine
(b) 2,3-dimethyloctane
(c) cyclohexanol
(d) methyl propanoate
(e) N-methylpentanamide
(f) 3-ethyl heptanoic acid
9. Identify the functional group(s) in each molecule.

(a) $CH_3CH_2\overset{OH}{\overset{|}{C}}HCH_2CH_3$ (b) 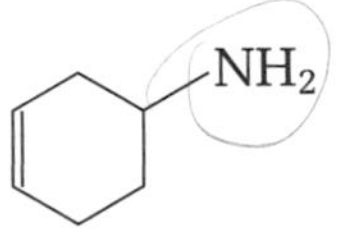 [cyclohexane ring]—NH_2

(c) 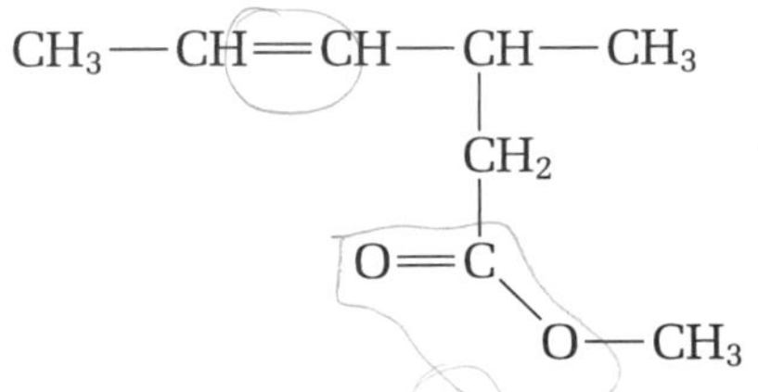 $CH_3—CH{=}CH—CH(—CH_2—C({=}O)—O—CH_3)—CH_3$

(d) $HC{\equiv}C—CH_2—\overset{O}{\overset{\|}{C}}—CH_2—CH_2—O—CH_3$

(e) 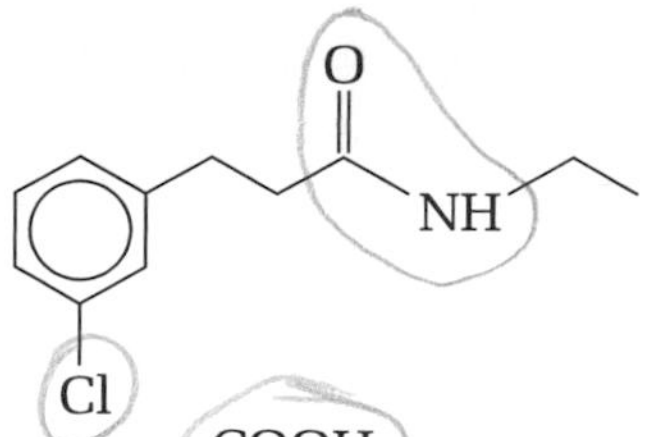 [structure: 3-chlorophenyl–CH_2CH_2–C(=O)–NH–CH_2CH_3; labels: O, NH, Cl]

(f) $CH_3CH_2CH(COOH)CH_2CH_2CHO$

10. Classify each type of organic compound.

(a) $CH_3—NH_2$

(b) $CH_3CH_2CH(CH_2CH_3)CH_3$

(c) $CH_3—CH_2—\overset{O}{\overset{\|}{C}}H$

(d) $CH_3—O—\overset{O}{\overset{\|}{C}}—CH_2—CH(CH_3)—CH_3$

(e) $HO—\overset{O}{\overset{\|}{C}}—CH(CH_2—CH_3)—CH_2—CH_3$

(f) $CH_3CH(CH_3)CH_2\overset{O}{\overset{\|}{C}}CH_3$

11. Identify each alcohol as primary, secondary, or tertiary.

(a) $CH_3—CH(OH)—CH_3$

(b) $CH_3—CH_2—C(CH_3)(CH_2CH_3)—OH$

(c)  [line structure: propan-1-ol, OH]

(d) 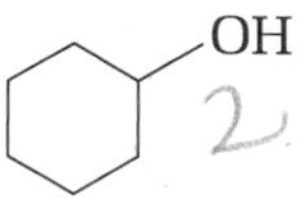 [line structure: cyclohexyl–OH]

12. Identify each amine as primary, secondary, or tertiary.

(a) 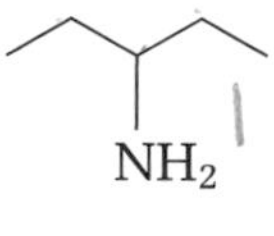 [line structure: butan-2-amine, NH_2]

(b) $CH_3—NH—CH_3$

(c) $CH_3—CH_2—CH(N(CH_3)_2)—CH_3$

(d) 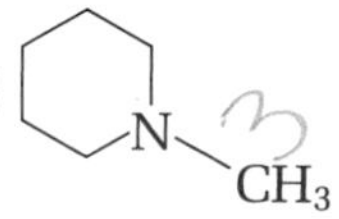 [line structure: N-methylpiperidine, N, CH_3]

13. (a) Suppose that you have an alcohol and a ketone with the same molecular mass. Which compound has a higher boiling point?

(b) Use your understanding of hydrogen bonding to explain your answer to part (a).

14. (a) Benzene is a non-polar solvent. Is an alcohol or an alkane more soluble in benzene?

(b) Use molecular polarity to explain your answer to part (a).

15. Give the IUPAC name for each compound.

(a) $CH_3—CH(Br)—CH_3$

(b) 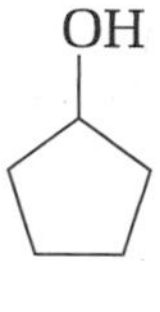 [line structure: cyclopentane with OH]

(c) $HO—\overset{O}{\overset{\|}{C}}—CH_2—CH_2—CH_2—CH_3$

(d) $CH_3—CH_2—\overset{O}{\overset{\|}{C}}—O—CH_2—CH_3$

(e) $CH_3—CH(CH_3)—C(CH_3)(CH_2—CH_3)—CH_2—\overset{O}{\overset{\|}{C}}—CH_2—CH_3$

16. Name each compound. Then identify the family of organic compounds that it belongs to.

(a) 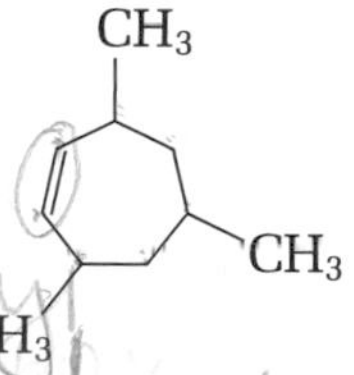 [ring structure: cycloheptene with three CH_3 groups]

(b) 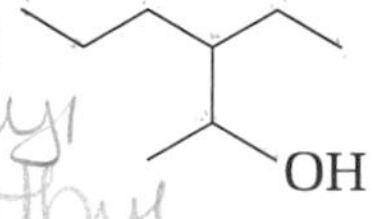 [line structure with OH]

(c) 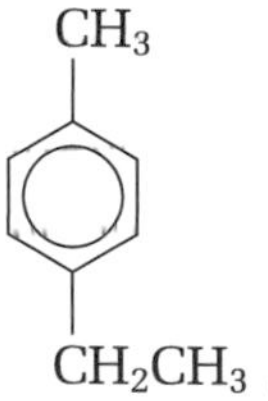 [benzene ring with CH_3 and CH_2CH_3 substituents]

(d) 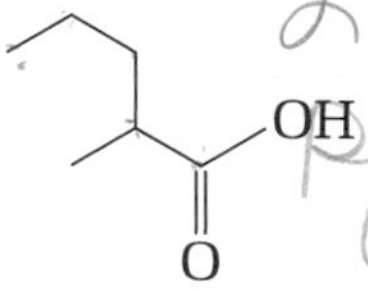 [line structure with C(=O)OH; labels: OH, O]

(e) $CH_3—CH_2—CH_2—CH_2—N(CH_3)—CH_2—CH_3$

(f) $CH_3CH_2\overset{O}{\overset{\|}{C}}—NH—CH_2CH_2CH_2CH_3$

17. Give the common name for each compound.

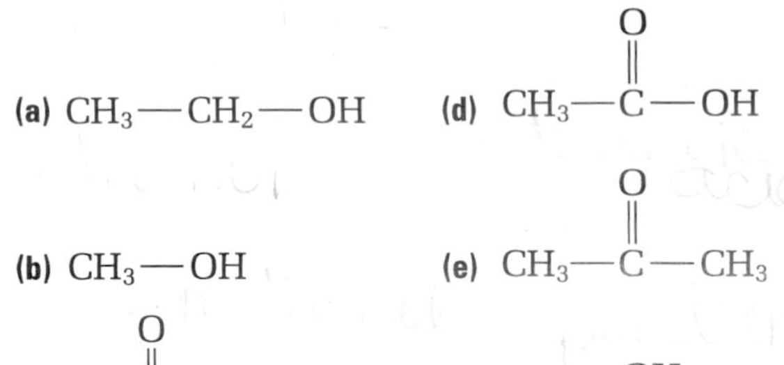

Inquiry

18. Suppose that you are working with five unknown compounds in a chemistry laboratory. Your teacher tells you that these compounds are ethane, ethanol, methoxymethane, ethanamine, and ethanoic acid.
   (a) Use the following table of physical properties to identify each unknown compound.

Compound	Solubility in water	Hydrogen bonding	Boiling point	Odour	Molecular polarity
A	infinitely soluble	strong	17°C	fishy	polar
B	not soluble	none	−89°C	odourless	non-polar
C	soluble	accepts hydrogen bonds from water, but cannot form hydrogen bonds between its molecules	−25°C	sweet	polar
D	infinitely soluble	very strong	78°C	sharp, antiseptic smell	very polar
E	infinitely soluble	extremely strong	118°C	sharp, vinegar smell	very polar

   (b) Draw a complete structural diagram for each compound.

Communication

19. Draw a condensed structural diagram for each compound.
   (a) 1-propanamine
   (b) 3-ethylpentane
   (c) 4-heptanol
   (d) propanoic acid
   (e) cyclobutanol
   (f) methoxyethane
   (g) 1,1-dibromobutane
   (h) 2-methyl-3-octanone
   (i) hexanal
   (j) N-ethylpropanamide
   (k) methyl butanoate

20. Draw a structural diagram for each compound.
   (a) 3,4-dimethylheptanoic acid
   (b) 3-ethyl-3-methyl-1-pentyne
   (c) N-ethyl-2,2-dimethyl-3-octanamine
   (d) N-ethyl-N-methylhexanamide
   (e) 1,3-dibromo-5-chlorobenzene

21. Draw a structural diagram for each compound. Then identify the family of organic compounds that it belongs to.
   (a) 4-ethylnonane
   (b) 4-propylheptanal
   (c) 3,3-dimethyl-2-hexanamine
   (d) 2-methoxypentane
   (e) para-dimethylbenzene

22. Identify the error in each name. Then correct the name.
   (a) 2-pentanal
   (b) 1,3-dimethylpropane
   (c) 2,2-dimethylbenzene
   (d) N,N-diethyl-N-methylpentanamide
   (e) 1-methylpropanoic acid

23. Draw and name all the isomers that have the molecular formula $C_4H_{10}O$.

24. Draw and name six isomers that have the molecular formula $C_4H_8O_2$.

Making Connections

25. List at least three common organic compounds. Describe how these compounds are used in everyday life.

26. The terms "organic," "chemical," and "natural" have slightly different meanings in science than they do in everyday life.
   (a) Compare the use of the term "organic" in the phrases "organic vegetables" and "organic compound." Where might each phrase be used?
   (b) Are chemicals harmful when added to food? In your answer, consider the true meaning of the word "chemical."
   (c) Are natural compounds always safer and more healthy than artificial compounds? Give reasons for your answer.

Answers to Practice Problems and Short Answers to Section Review Questions:

Practice Problems: **3.** C—O, O—H, C=O **4.** polar, non-polar, non-polar, polar **5.(a)** ethane **(b)** cyclobutene **(c)** 2-methyl-2-pentene **(d)** 3-methylhexane **(e)** 3-ethyl-2-methyl-2-heptene **(f)** 1,2-dimethylcyclohexane **(g)** 2-pentyne **7.(a)** 2,2,3-trimethylbutane **(b)** 5-ethyl-3-methylnonane **(c)** 3-methyl-4,5-diethyl-2-nonene **8.(a)** 3-hexene **(b)** 2,4-hexadiyne **9.(a)** 3-methyl-2-hexene **(b)** 4-methyl-2-heptene **(c)** 2-methylpentane **10.** 1,3,5-trimethylbenzene **12.(a)** meta-ethylmethylbenzene **13.** some isomers include ortho-, meta-, and para-diethylbenzene; 1,2,3,4-tetramethylbenzene; and 1-methyl-2-propylbenzene **14.(a)** 1-propanol; primary **(b)** 2-butanol; secondary **(c)** cyclobutanol; secondary **(d)** 2,3-pentanediol; secondary **(e)** 2,4-dimethyl-1-heptanol; primary **16.(a)** 1,3-pentanediol **(b)** 3,4-diethyl-1-decanol, **(c)** 3-methyl-2-pentanol **19.** 1,2-difluorocyclohexane **20.** 1-iodopropane **21.** 2-iodopropane **22.(a)** methoxymethane **(b)** 2-methoxypropane **(c)** 1-methoxybutane **23.(a)** ethyl methyl ether **(b)** dimethyl ether **25.** alcohol **26.(a)** methanamine **(b)** N-ethyl-2, 2-dimethyl-1-propanamine **(c)** 2-butanamine **(d)** N,N-dimethylcyclopentanamine **28.(a)** primary **(b)** primary **(c)** secondary **(d)** tertiary **29.** 1-butanamine, 2-butanamine, 2-methyl-1-propanamine, N-methyl-1-propanamine, N-methyl-2-propanamine, N-ethylethanamine, N,N-dimethylethanamine, 2-methyl-2-propanamine **30.(a)** butanal **(b)** 3-octanone **(c)** 2-methyl-3-hexanone **(d)** 2-ethylbutanal **(e)** 4-ethyl-3-methyl-2-heptanone **32.** aldehyde **33.** hexanal, 2-hexanone, 3-hexanone, 3-methyl-2-pentanone, 3,3-dimethylbutanal; many other isomers are possible **34.(a)** propanoic acid **(b)** 2,2-dimethylbutanoic acid **(c)** 2-ethyl-4,5-dimethylhexanoic acid **37.** butanoic acid, 2-methylpropanoic acid **38.(a)** ethyl methanoate **(b)** methyl butanoate **(c)** pentyl pentanoate **39.(a)** methanoic acid, ethanol **(b)** butanoic acid, methanol **(c)** petanoic acid, petanol **41.(a)** and **(c)**; **(b)** and **(e)** **42.** methyl butanoate, ethyl propanoate, propyl ethanoate, butyl methanoate, methyl 2-methylpropanoate; other isomers exist **43.(a)** butanamide **(b)** N-methylhexanamide **(c)** N,N-diethyl-3-methylheptanamide **45.(a)** ethanamide **(b)** N-methyheptanamide **(c)** N,N-dimethyl-2-methylpropanamide

Section Review: **1.1:** **2.(a)** polar **(b)** non-polar **(c)** polar **(d)** non-polar **(e)** polar **3.(a)** tetrahedral **(b)** trigonal planar **4.(a)** non-polar **(b)** polar **1.2:** **1.(a)** heptane **(b)** 3,4-diethyloctane **(c)** 4-methyl-1-pentene, **(d)** ethylbenzene **3.** butane, 2-methylpropane **5.** 1-hexyne, 2-hexyne, 3-hexyne, cyclohexene, 1-methylcyclopentene, 3-methylcyclopentene, 4-methylcyclopentene, 1,2-dimethylcyclobutene, 1,3-dimethylcyclobutene, 2,3-dimethycyclobutene, 1-ethylcyclobutene, 3-ethylcyclobutene **7.** 1-pentene, cyclopentane, trans-2-pentene, cis-2-pentene, 2-methyl-2-butene, 2-methyl-1-butene, 3-methyl-1-butene, methylcyclobutane, ethylcyclopropane, 1,1-dimethylcyclopropane, 1,2-dimethylcyclopropane **1.3:** **1.(a)** 1-butanamine **(b)** 2-ethoxybutane **(c)** 2,3-dimethylcyclopentanol **2.(a)** 1-heptanol **(b)** 2-butanol **(c)** ethanamine **(d)** N-methylmethanamine **(e)** methoxyethane **(f)** 2-chloropropane **4.** 1-butanol, 2-butanol, 2-methyl-2-propanol, methoxypropane, ethoxyethane **5.(a)** ethanol > methoxymethane > ethane **(b)** 1-propanamine > N,N-dimethylmethanamine > propane **6.** cyclohexanol **7.** propane, 2-propanol; propane **8.** 1-butanol > ethoxyethane > butane **1.4:** **1.(a)** methanol **(b)** ethanol **(c)** 2-propanol **(d)** 2-propanone **(e)** methanal **(f)** ethanoic acid **(g)** methylbenzene **2.(a)** wood alcohol **(b)** grain alcohol **(c)** rubbing alcohol **(d)** acetone **(e)** formaldehyde **(f)** acetic acid **(g)** toluene **3.(a)** pentanal **(b)** 2,3-diethylhexanoic acid **(c)** N-ethylpropanamide **(d)** methyl octanoate **4.(a)** amide **(b)** carboxylic acid **(c)** aldehyde **(d)** ester **5.(a)** propanamide **(b)** 4-methylhexanoic acid **(c)** 3,3-dimethylpentanal **(d)** methyl ethanoate **6.** hexanoic acid, propyl propanoate **7.** hexanamide, N-methylpentanamide, N,N-dimethylbutanamide **8.(a)** benzene ring, methoxy group, aldehyde group, hydroxyl group **(b)** benzene ring, amide group **(c)** benzene ring, ether linkage, two amide groups, acid group, sulfur atom **9.** CH_3CH_2COOH **10.** An amine has a distinctive “fishy” smell. A ketone smells sweet.

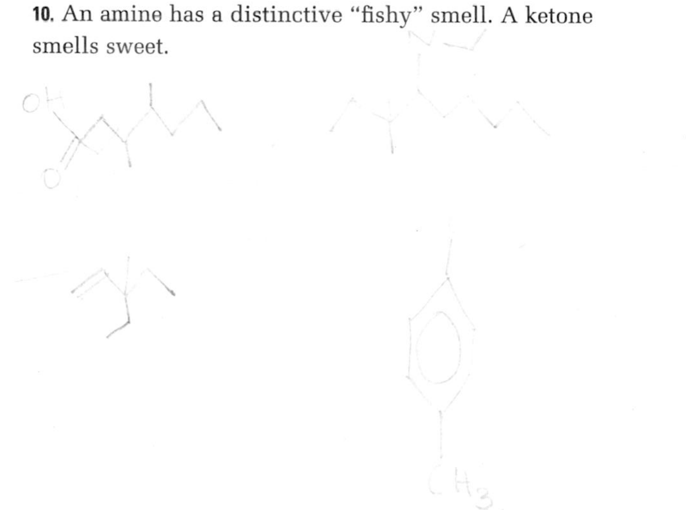

CHAPTER

2 Reactions of Organic Compounds

Chapter Preview

2.1 The Main Types of Organic Reactions

2.2 Reactions of Functional Groups

2.3 Molecules on a Larger Scale: Polymers and Biomolecules

2.4 Organic Compounds and Everyday Life

Prerequisite Concepts and Skills

Before you begin this chapter, review the following concepts and skills:

- identifying functional groups (Chapter 1, sections 1.3 and 1.4)
- naming and drawing organic compounds (Chapter 1, sections 1.3 and 1.4)

Cancer is one of the leading causes of death in Canada. It is a disease in which cells mutate and grow at uncontrolled rates, disrupting the body's normal functions. TAXOL™, shown at the bottom of the page, is an organic compound that is found in the bark of the Pacific yew tree. After many tests and reviews, TAXOL™ was approved for use in treating ovarian cancer.

The bark of a large yew tree yields only enough TAXOL™ for a single treatment. However, cancer patients require repeated treatments over a long period of time. Working in a laboratory, chemists found the first solution to this problem. They developed several different methods to make, or *synthesize*, TAXOL™, from simple, widely available chemicals. Unfortunately, these methods are expensive and time-consuming.

When studying the Pacific yew's close relative, the European yew, chemists made an exciting discovery that led to the second solution. The needles and twigs of the European yew contain 10-deacetylbaccatin III, shown at the bottom of the page. A few reaction steps can transform 10-deacetylbaccatin III into TAXOL™. Instead of destroying an entire Pacific yew tree for a single treatment of TAXOL™, scientists can now use the needles from a European yew. The parent tree is not harmed.

As you have already learned, the study of organic chemistry is really the study of functional groups. Each functional group undergoes specific, predictable reactions. Chemists used their knowledge of functional groups to design the reactions that convert 10-deacetylbaccatin III into TAXOL™. In this chapter, you will learn how to predict the products of various organic reactions, such as addition, substitution, and elimination reactions.

The Main Types of Organic Reactions

2.1

Figure 2.1(A) shows raw fruit. The crisp, sharp-tasting fruit becomes soft and sweet when it is cooked. Figure 2.1(B) shows a chemist accelerating the tranformation of ethanol into ethanoic acid, by adding potassium dichromate and sulfuric acid.

What do these reactions have in common? They are both examples of organic reactions. In this section, you will take a quick look at the main types of organic reactions. You will concentrate on simply recognizing these types of organic reactions. In the next section, you will examine the reactions of specific functional groups and learn how to predict the products of organic reactions.

Figure 2.1 Organic reactions take place all around you, not only in a science lab.

Addition, Substitution, and Elimination Reactions

Addition reactions, substitution reactions, and elimination reactions are the three main types of organic reactions. Most organic reactions can be classified as one of these three types.

Addition Reactions

In an **addition reaction**, atoms are added to a double or triple bond. One bond of the multiple bond breaks so that two new bonds can form. To recognize an addition reaction, remember that *two* compounds usually react to form *one* major product. (Sometimes two isomers are formed.) The product has more atoms bonded to carbon atoms than the organic reactant did. A general example of addition to an alkene is given below.

$$-CH{=}CH- \; + \; XY \rightarrow -\underset{}{\overset{X}{\overset{|}{C}}}H-\overset{Y}{\overset{|}{C}}H-$$

Addition reactions are common for alkenes and alkynes. Addition reactions can also occur at a C=O bond. Some examples of addition reactions are shown on the next page.

Section Preview/ Specific Expectations

In this section, you will

- **describe** different types of organic reactions, including addition, substitution, elimination, oxidation, and reduction
- **predict** the products of organic reactions, including addition, substitution, elimination, oxidation, and reduction
- **communicate** your understanding of the following terms: *addition reaction, substitution reaction, elimination reaction, oxidation, reduction, condensation reaction, hydrolysis reaction*

Example 1

$$\begin{array}{c} H \quad\quad H \\ \diagdown \quad \diagup \\ C{=}C \\ \diagup \quad \diagdown \\ H \quad\quad H \end{array} \quad + \quad HBr \quad \rightarrow \quad H{-}\overset{\overset{Br}{|}}{\underset{\underset{H}{|}}{C}}{-}\overset{\overset{H}{|}}{\underset{\underset{H}{|}}{C}}{-}H$$

ethene — hydrobromic acid — bromoethane

Example 2

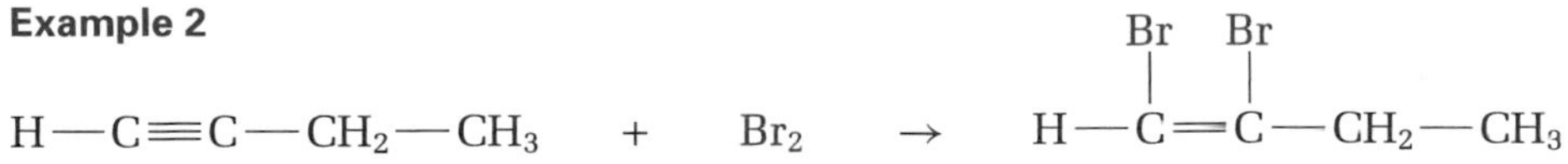

$$H{-}C{\equiv}C{-}CH_2{-}CH_3 \quad + \quad Br_2 \quad \rightarrow \quad H{-}\overset{\overset{Br}{|}}{C}{=}\overset{\overset{Br}{|}}{C}{-}CH_2{-}CH_3$$

butyne — bromine — 1,2-dibromobutene

Substitution Reactions

In a **substitution reaction**, a hydrogen atom or a functional group is replaced by a different functional group. To help you recognize this type of reaction, remember that *two* compounds usually react to form *two* different products. The organic reactant(s) and the organic product(s) have the same number of atoms bonded to carbon.

CONCEPT CHECK

You can express an addition reaction algebraically, using the equation $a + b \rightarrow ab$. Come up with similar equations for substitution and elimination reactions.

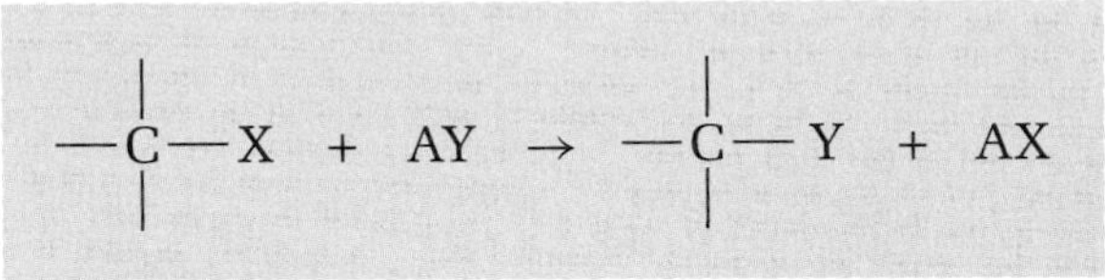

$$-\overset{|}{\underset{|}{C}}{-}X \;+\; AY \;\rightarrow\; -\overset{|}{\underset{|}{C}}{-}Y \;+\; AX$$

Alcohols, alkyl halides, and aromatic compounds commonly undergo substitution reactions, as shown in these examples.

Example 1

$$CH_3{-}CH_2{-}OH \quad + \quad HI \quad \rightarrow \quad CH_3{-}CH_2{-}I \quad + \quad HOH$$

ethanol — hydroiodic acid — iodoethane — water

Example 2

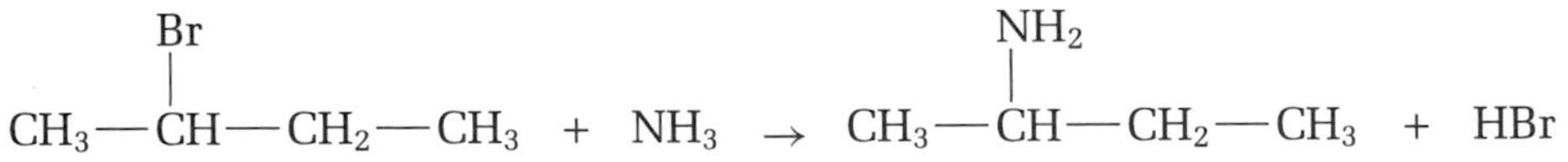

$$CH_3{-}\overset{\overset{Br}{|}}{CH}{-}CH_2{-}CH_3 \;+\; NH_3 \;\rightarrow\; CH_3{-}\overset{\overset{NH_2}{|}}{CH}{-}CH_2{-}CH_3 \;+\; HBr$$

2-bromobutane — ammonia — 2-butanamine — hydrobromic acid

Example 3

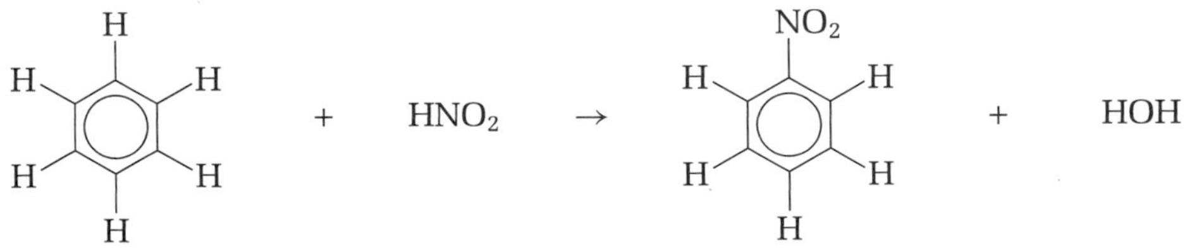

$$C_6H_6 \quad + \quad HNO_2 \quad \rightarrow \quad C_6H_5NO_2 \quad + \quad HOH$$

benzene — nitrous acid — nitrobenzene — water

Elimination Reactions

In an **elimination reaction**, atoms are removed from a molecule to form a double bond. This type of reaction is the reverse of an addition reaction. *One* reactant usually breaks up to give *two* products. The organic product typically has fewer atoms bonded to carbon atoms than the organic reactant did.

$$-\overset{X}{\underset{|}{\overset{|}{C}}}-\overset{Y}{\underset{|}{\overset{|}{C}}}- \rightarrow \;\; >C=C< \;\; + \;\; XY$$

Alcohols often undergo elimination reactions when they are heated in the presence of strong acids, such as sulfuric acid, H_2SO_4, which acts as a catalyst. (See the first example below.) Alkyl halides also undergo elimination reactions to produce alkenes. (See the second example.)

Example 1

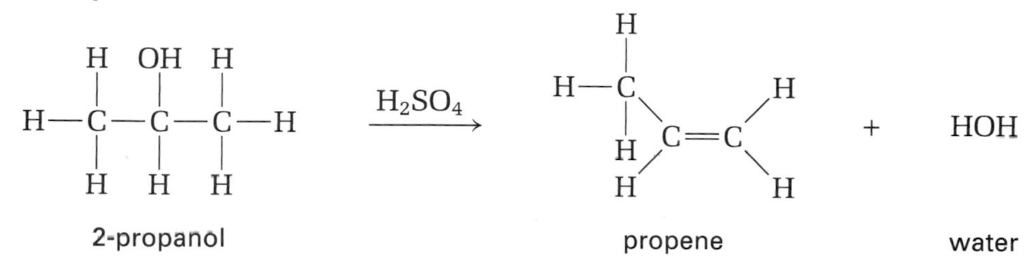

CONCEPT CHECK

In the first example of an elimination reaction, the strong acid, H_2SO_4, does not count as a reactant. It is not directly involved in the reaction. It is a *catalyst*: a compound that speeds up a reaction but is not consumed by it.

Example 2

$$H-\overset{H}{\underset{H}{C}}-\overset{Br}{\underset{H}{C}}-H \rightarrow \;\; H_2C=CH_2 \;\; + \;\; HBr$$

bromoethane → ethene + hydrobromic acid

Oxidation and Reduction

An important type of organic reaction occurs when there is a change in the number of hydrogen or oxygen atoms that are bonded to carbon. In Unit 5, you will take a close look at oxidation-reduction reactions in terms of the transfer of electrons. As you will learn, oxidation and reduction always occur together. One reactant is oxidized while the other reactant is reduced. In this unit, however, you will focus on the organic reactant only. Therefore, you will deal with oxidation and reduction separately, as they apply to organic compounds. In organic chemistry, oxidation and reduction are defined by the changes of the bonds to carbon atoms in the organic reactant.

Oxidation

In organic chemistry, **oxidation** is defined as a reaction in which a carbon atom forms more bonds to oxygen, O, or less bonds to hydrogen, H. An oxidation that involves the formation of double C=O bonds may also be classified as an elimination reaction.

For example, alcohols can be oxidized to produce aldehydes and ketones. Oxidation occurs when an organic compound reacts with an oxidizing agent. Common oxidizing agents include acidified potassium permanganate, $KMnO_4$, acidified potassium dichromate, $K_2Cr_2O_7$, and ozone, O_3. The symbol [O] is used to symbolize an oxidizing agent, as shown below. Note that equations for the oxidation of organic compounds are often left unbalanced. The purpose of the equation is to show the changes in the organic reactant only.

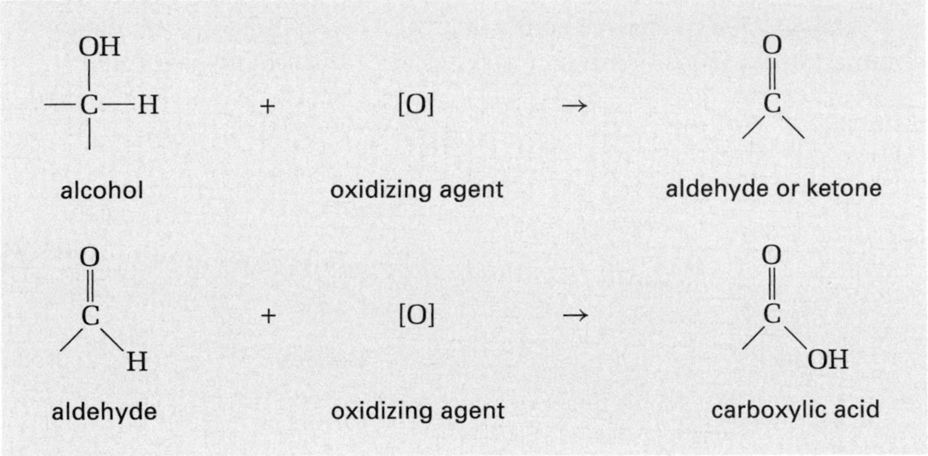

$$-\overset{OH}{\underset{|}{\overset{|}{C}}}-H \quad + \quad [O] \quad \rightarrow \quad \overset{O}{\overset{\|}{C}}$$

alcohol + oxidizing agent → aldehyde or ketone

$$\overset{O}{\overset{\|}{C}}-H \quad + \quad [O] \quad \rightarrow \quad \overset{O}{\overset{\|}{C}}-OH$$

aldehyde + oxidizing agent → carboxylic acid

To identify an oxidation, count and compare the number of C—H and C—O bonds in both the reactant and product. Try it for the following example.

$$H-\overset{H}{\underset{H}{C}}-\overset{OH}{\underset{H}{C}}-H \quad + \quad [O] \quad \rightarrow \quad H-\overset{H}{\underset{H}{C}}-\overset{O}{\overset{\|}{C}}-H$$

ethanol → ethanal

Reduction

In organic chemistry, **reduction** is defined as a reaction in which a carbon atom forms fewer bonds to oxygen, O, or more bonds to hydrogen, H. Often, a C=O bond or C=C bond is reduced to a single bond by reduction. A reduction that transforms double C=C or C=O bonds to single bonds may also be classified as an addition reaction. Aldehydes, ketones, and carboxylic acids can be reduced to become alcohols. Alkenes and alkynes can be reduced by the addition of H_2 to become alkanes.

Reduction occurs when an organic compound reacts with a reducing agent. Common reducing agents are lithium aluminum hydride, $LiAlH_4$, and hydrogen gas over a platinum catalyst, H_2/Pt. The symbol [H] is used to symbolize a reducing agent. As is the case for oxidation, equations showing the reduction of organic compounds are often left unbalanced.

$$\underset{\text{aldehyde or ketone}}{\overset{O}{\overset{\|}{C}}} + \underset{\text{reducing agent}}{[H]} \rightarrow \underset{\text{alcohol}}{-\overset{OH}{\overset{|}{C}}-H}$$

$$\underset{\text{alkene}}{>C=C<} + \underset{\text{reducing agent}}{[H]} \rightarrow \underset{\text{alkane}}{-\overset{H}{\overset{|}{C}}-\overset{H}{\overset{|}{C}}-}$$

To identify a reduction, count and compare the number of C—H and C—O bonds in both the reactant and the product. Try it for the following example.

$$\underset{\text{propanone}}{H_3C-\overset{O}{\overset{\|}{C}}-CH_3} + [H] \rightarrow \underset{\text{2-propanol}}{H_3C-\underset{H}{\underset{|}{\overset{OH}{\overset{|}{C}}}}-CH_3}$$

Other Important Organic Reactions

In this chapter, you will also encounter the following classes of organic reactions: condensation reactions and hydrolysis reactions. Condensation and hydrolysis reactions are both types of substitution reactions.

CHEM FACT

When you see two arrows that point in opposite directions in a chemical equation, the reaction can proceed in both directions. This type of reaction is called an *equilibrium reaction*. You will learn more about equilibrium reactions in Unit 4.

Condensation Reactions

In a **condensation reaction**, two organic molecules combine to form a single organic molecule. A small molecule, usually water, is produced during the reaction. For example, a carboxylic acid and an alcohol can condense to form an ester.

$$\underset{\text{carboxylic acid}}{H_3C-\overset{O}{\overset{\|}{C}}-OH} + \underset{\text{alcohol}}{HO-CH_2-CH_2-\overset{CH_3}{\overset{|}{CH}}-CH_3} \xrightleftharpoons{H_2SO_4} \underset{\text{ester}}{H_3C-\overset{O}{\overset{\|}{C}}-O-CH_2-CH_2-\overset{CH_3}{\overset{|}{CH}}-CH_3} + \underset{\text{water}}{H_2O}$$

Hydrolysis Reactions

In a **hydrolysis reaction**, water adds to a bond, splitting it in two. This reaction is the reverse of a condensation reaction. For example, water can add to an ester or amide bond. A carboxylic acid and an alcohol are produced if an ester bond is hydrolyzed, as shown in the example below. A carboxylic acid and an amine are produced if an amide bond is hydrolyzed.

$$\underset{\text{ester}}{H_3C-\overset{O}{\overset{\|}{C}}-O-CH_2-CH_2-\overset{CH_3}{\overset{|}{CH}}-CH_3} + \underset{\text{water}}{H_2O} \xrightleftharpoons{H_2SO_4} \underset{\text{carboxylic acid}}{H_3C-\overset{O}{\overset{\|}{C}}-OH} + \underset{\text{alcohol}}{HO-CH_2-CH_2-\overset{CH_3}{\overset{|}{CH}}-CH_3}$$

The following Sample Problems show how to identify different types of organic reactions.

Sample Problem

Identifying Addition, Substitution, and Elimination Reactions

Problem

Identify each reaction as an addition, substitution, or elimination reaction.

(a) $HO—CH_2CH_2CH_3 \rightarrow CH_2{=}CHCH_3 + H_2O$

(b) $H_2C{=}CHCH_2CH_3 + H_2 \rightarrow CH_3CH_2CH_2CH_3$

(c) $CH_3CH(CH_3)CH_2CH_2Br + NaOH \rightarrow CH_3CH(CH_3)CH_2CH_2OH + NaBr$

Solution

(a) A double bond is formed. One reactant becomes two products. The organic product has fewer atoms bonded to carbon. Thus, this reaction is an elimination reaction.

(b) A double bond becomes a single bond. Two reactants become one product. The organic product has more atoms bonded to carbon. Thus, this reaction is an addition reaction.

(c) No double bond is broken or formed. Two reactants form two products. An atom in the organic reactant (Br) is replaced by a different group of atoms (OH). Thus, this is a substitution reaction.

Sample Problem

Identifying Oxidation and Reduction

Problem

Identify each reaction as an oxidation or a reduction. (The oxidizing and reducing agents are not shown.)

(a) 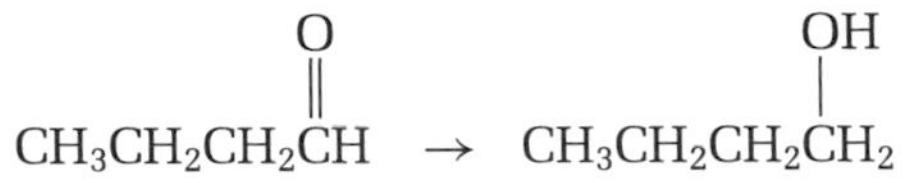 $CH_3CH_2CH_2CH{=}O \rightarrow CH_3CH_2CH_2CH_2OH$

(b) 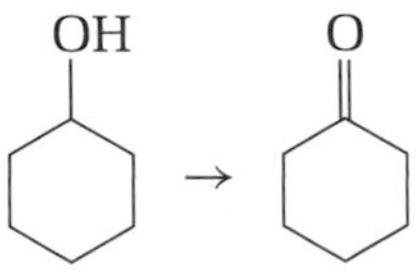 cyclohexanol (cyclohexane ring with OH) $\rightarrow$ cyclohexanone (cyclohexane ring with =O)

Solution

(a) Count the number of C—H bonds and C—O bonds in the reactant and in the product.

Reactant: 8 C—H bonds, 2 C—O bonds (or 1 C=O bond)
Product: 9 C—H bonds, 1 C—O bond

The product has *gained* a C—H bond and *lost* a C—O bond. Thus, this is a reduction.

(b) Count the number of C—H bonds and C—O bonds in the reactant and in the product. Although the C—H bonds are not shown, you know that a carbon atom forms a total of four bonds.

Reactant: 11 C—H bonds, 1 C—O bond
Product: 10 C—H bonds, 2 C—O bonds (or 1 C=O bond)

The product has *lost* a C—H bond and *gained* a C—O bond. Thus, this is an oxidation.

Practice Problems

1. Identify each reaction as an addition, substitution, or elimination reaction.

(a) $CH_3{-}CH_2{-}CH(OH){-}CH_3 + HBr \rightarrow CH_3{-}CH_2{-}CH(Br){-}CH_3 + HOH$

(b) $CH_3{-}CH{=}CH{-}CH(CH_3){-}CH_3 + Cl_2 \rightarrow CH_3{-}CH(Cl){-}CH(Cl){-}CH(CH_3){-}CH_3$

(c) chlorocyclopentane $\rightarrow$ cyclopentene + HCl

2. Identify each reaction as an oxidation or a reduction. The oxidizing and reducing agents are not shown.

(a) $CH_3{-}CH(CH_3){-}CH(CH_2CH_3){-}CH_2{-}\overset{O}{\overset{\|}{C}}H \rightarrow CH_3{-}CH(CH_3){-}CH(CH_2CH_3){-}CH_2{-}\overset{O}{\overset{\|}{C}}{-}OH$

(b) $(CH_3CH_2)(H)C{=}C(CH_3)_2 \rightarrow CH_3CH_2CH_2CH(CH_3)CH_3$

(c) cyclohexane-1,4-dione $\rightarrow$ cyclohexane-1,4-diol

3. Classify each reaction in two different ways: for example, as oxidation *and* as an elimination reaction.

(a) $(CH_3)(H)C{=}C(H)(CH_2CH_3) + H_2 \rightarrow CH_3CH_2CH_2CH_2CH_3$

(b) $CH_3{-}\overset{O}{\overset{\|}{C}}{-}OH + HO{-}CH_3 \rightleftharpoons CH_3{-}\overset{O}{\overset{\|}{C}}{-}O{-}CH_3 + HOH$

4. Identify the type of reaction.

(a) $CH_3{-}CH{=}CH_2 + HOH \rightarrow CH_3{-}CH(OH){-}CH_3$

(b) $CH_3{-}CH_2{-}CH(OH){-}CH_2{-}CH_3 \rightarrow CH_3{-}CH_2{-}\overset{O}{\overset{\|}{C}}{-}CH_2{-}CH_3$

Section Summary

In this section, you were introduced to some of the main types of organic reactions: addition, substitution, and elimination reactions; oxidation and reduction; and condensation and hydrolysis reactions. In the next section, you will take a close look at each type of reaction. You will find out how organic compounds, such as alcohols and carboxylic acids, can react in several different ways.

Section Review

1 K/U Identify each reaction as an addition, substitution, or elimination reaction.

(a) $H_2C{=}CH_2 + Br_2 \rightarrow H_2\overset{\overset{Br}{|}}{C}{-}\overset{\overset{Br}{|}}{C}H_2$

(b) $CH_3CH_2\overset{\overset{OH}{|}}{C}HCH_3 \rightarrow CH_3CH{=}CHCH_3 + H_2O$

(c) 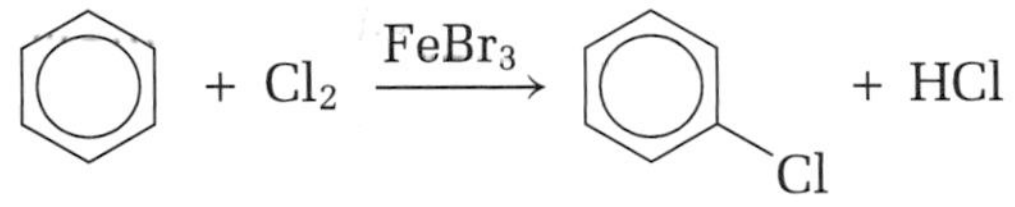 $C_6H_6 + Cl_2 \xrightarrow{FeBr_3} C_6H_5Cl + HCl$

(d) $CH_3CH_2CH_2Br + H_2NCH_2CH_3 \rightarrow CH_3CH_2CH_2NHCH_2CH_3 + HBr$

2 K/U Identify each reaction as an oxidation or a reduction. (Oxidizing and reducing agents are not shown.)

(a) $CH_3CH_2CH_2OH \rightarrow CH_3CH_2\overset{\overset{O}{\|}}{C}H$

(b) $CH_3CH_2\overset{\overset{O}{\|}}{C}CH_3 \rightarrow CH_3CH_2\overset{\overset{OH}{|}}{C}HCH_3$

(c) $H_2C{=}CHCH_3 \rightarrow CH_3CH_2CH_3$

(d) 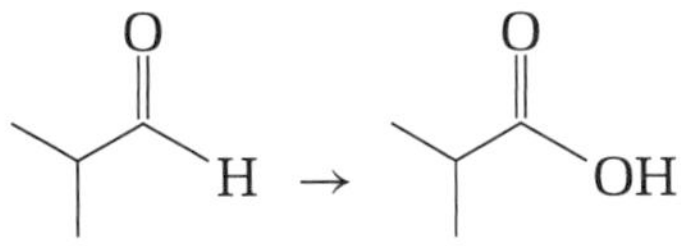 $(CH_3)_2CH\overset{\overset{O}{\|}}{C}H \rightarrow (CH_3)_2CH\overset{\overset{O}{\|}}{C}OH$

(e) $CH_3C{\equiv}CCH_3 \rightarrow CH_3CH_2CH_2CH_3$

3 **(a)** K/U Draw a complete structural diagram for each compound that is involved in the following reaction.

N-ethylpentanamide + water → pentanoic acid + ethanamine

(b) K/U Is this reaction a condensation reaction or a hydrolysis reaction? Explain your reasoning.

4 C In your own words, describe each type of organic reaction. Include an example for each type.

(a) addition **(b)** substitution **(c)** elimination

5 C The following reaction can be classified as an addition reaction. It can also be classified as a reduction. Explain why this reaction fits into both classes.

$CH_3CH{=}CHCH_3 + H_2 \rightarrow CH_3CH_2CH_2CH_3$

Reactions of Functional Groups

2.2

In the past, mirrors were rare and valuable, made from sheets of polished copper or silver. Although they were expensive, they usually produced only warped and dim images. Modern mirrors are everywhere. They are cheaply made of glass, with a very thin layer of silver over the back of the glass. Because the surface of the glass and the layer of silver are so smooth, the image is almost perfect. One method for depositing a thin layer of silver over glass is shown in Figure 2.2.

Figure 2.2 **The "silver mirror test" is used to distinguish an aldehyde from a ketone. Tollen's reagent, $Ag(NH_3)_2OH$, acts as an oxidizing agent. When it is mixed with an aldehyde, the aldehyde oxidizes to the salt of a carboxylic acid. The silver ions in Tollen's reagent are reduced to silver atoms, and coat the glass of the reaction container with solid silver metal.**

In this section, you will learn how to predict the reactions of different functional groups. You studied the most common reaction of alkanes, combustion, in your previous chemistry course. For this reason, the reactions of alkanes will not be considered here. The reactions of amines and ethers will be left for a later chemistry course.

Reactions of Alkenes and Alkynes

Because alkenes and alkynes have reactive double or triple bonds, they undergo addition reactions. Some common atoms and groups of atoms that can be added to a double or triple bond include

- H and OH (from H_2O)
- H and X (from HX), where X = Cl, Br, or I
- X and X (from X_2), where X = Cl, Br, or I
- H and H (from H_2)

Depending on which atoms are added to the multiple bond, the product may be an alcohol, alkyl halide, alkane, or alkene (if atoms are added to a triple bond).

In this section, you will

- **describe** various addition, substitution, elimination, oxidation, reduction, condensation, and hydrolysis reactions
- **predict**, **draw**, and **name** the products of various organic reactions
- **investigate** the oxidation of alcohols
- **communicate** your understanding of the following terms: *Markovnikov's rule*, *esterification reaction*

Addition to Alkenes

The product of an addition reaction depends on the symmetry of the reactants. A *symmetrical alkene* has identical groups on either side of the double bond. Ethene, $CH_2{=}CH_2$, is an example of a symmetrical alkene. An alkene that has different groups on either side of the double bond is called an *asymmetrical alkene*. Propene, $CH_3CH{=}CH_2$, is an example of an asymmetrical alkene.

The molecules that are added to a multiple bond can also be classified as symmetrical or asymmetrical. For example, chlorine, Cl_2, breaks into two identical parts when it adds to a multiple bond. Therefore, it is a symmetrical reactant. Water, HOH, breaks into two different groups (H and OH) when it adds to a multiple bond, so it is an asymmetrical reactant.

In Figures 2.3 and 2.4, at least one of the reactants is symmetrical. When one or more reactants in an addition reaction are symmetrical, only one product is possible.

$$\underset{\text{2-butene (symmetrical)}}{CH_3CH{=}CHCH_3} + \underset{\text{water (asymmetrical)}}{HOH} \rightarrow \underset{\text{2-butanol}}{CH_3CH(OH){-}CH(H)CH_3}$$

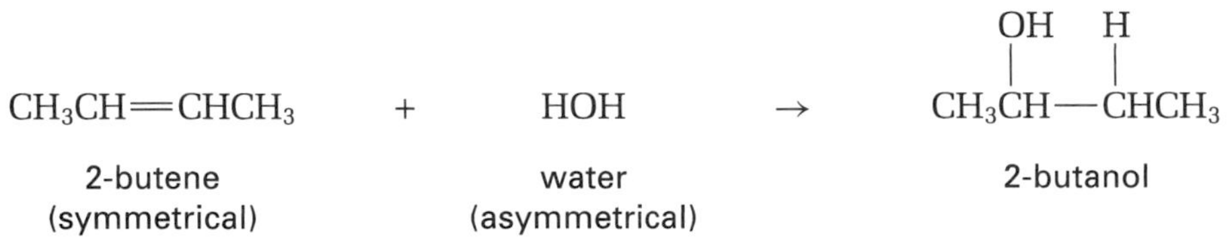

Figure 2.3 The addition of water to 2-butene

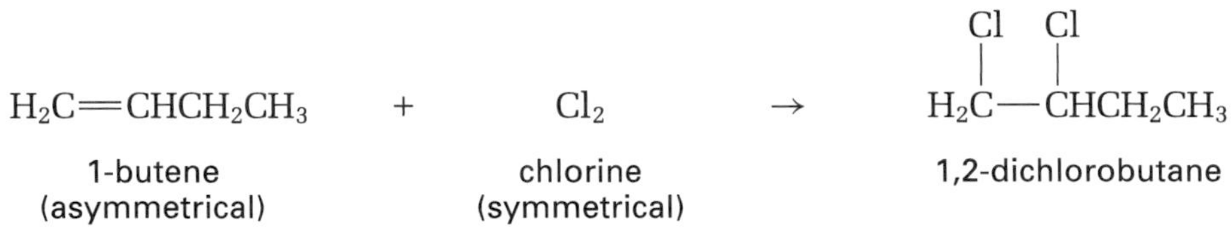

$$\underset{\text{1-butene (asymmetrical)}}{H_2C{=}CHCH_2CH_3} + \underset{\text{chlorine (symmetrical)}}{Cl_2} \rightarrow \underset{\text{1,2-dichlorobutane}}{H_2C(Cl){-}CH(Cl)CH_2CH_3}$$

Figure 2.4 The addition of chlorine to 1-butene

When both reactants are asymmetrical, however, more than one product is possible. This is shown in Figure 2.5. The two possible products are isomers of each other.

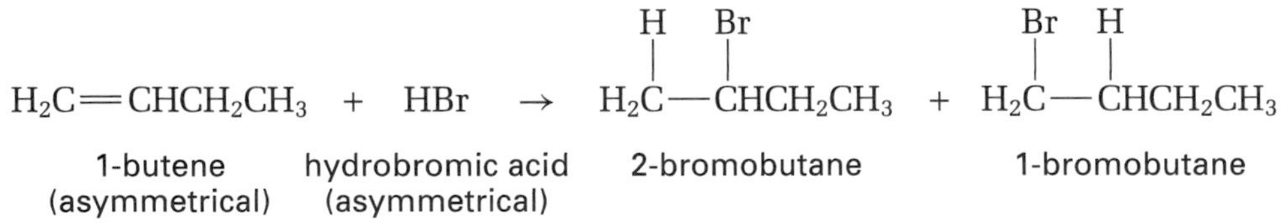

$$\underset{\text{1-butene (asymmetrical)}}{H_2C{=}CHCH_2CH_3} + \underset{\text{hydrobromic acid (asymmetrical)}}{HBr} \rightarrow \underset{\text{2-bromobutane}}{H_2C(H){-}CH(Br)CH_2CH_3} + \underset{\text{1-bromobutane}}{H_2C(Br){-}CH(H)CH_2CH_3}$$

Figure 2.5 The addition of hydrobromic acid to 1-butene

Although both products are possible, 2-bromobutane is the observed product. This observation is explained by Markovnikov's rule. **Markovnikov's rule** states that *the halogen atom or OH group in an addition reaction is usually added to the more substituted carbon atom—the carbon atom that is bonded to the largest number of other carbon atoms.* Think of the phrase "the rich get richer." The carbon with the most hydrogen atoms receives even more hydrogen atoms in an addition reaction. According to Markovnikov's rule, the addition of two asymmetrical reactants forms primarily one product. Only a small amount of the other isomer is formed. The following Sample Problem shows how to use Markovnikov's rule to predict the products of addition reactions.

Sample Problem

Using Markovnikov's Rule

Problem

Draw the reactants and products of the following incomplete reaction.

2-methyl-2-pentene + hydrochloric acid →?

Use Markovnikov's rule to predict which of the two isomeric products will form in a greater amount.

Solution

According to Markovnikov's rule, *the hydrogen atom will go to the double-bonded carbon with the larger number of hydrogen atoms.* Thus, the chlorine atom will go to the other carbon, which has the larger number of C—C bonds. The product 2-chloro-2-methylpentane will form in the greater amount. Small amounts of 3-chloro-2-methylpentane will also form.

$$CH_3{-}\overset{\overset{\displaystyle CH_3}{|}}{C}{=}CH{-}CH_2{-}CH_3 \quad + \quad HCl \quad \rightarrow$$

2-methyl-2-pentene — hydrochloric acid

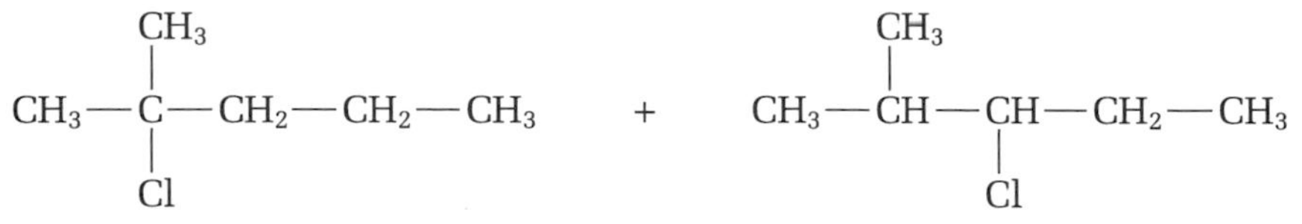

$$CH_3{-}\underset{\underset{\displaystyle Cl}{|}}{\overset{\overset{\displaystyle CH_3}{|}}{C}}{-}CH_2{-}CH_2{-}CH_3 \quad + \quad CH_3{-}\overset{\overset{\displaystyle CH_3}{|}}{CH}{-}\underset{\underset{\displaystyle Cl}{|}}{CH}{-}CH_2{-}CH_3$$

2-chloro-2-methylpentane (major product) — 3-chloro-2-methylpentane (minor product)

Practice Problems

5. Draw the reactants and products of the following reaction.

3-ethyl-2-heptene + HOH → 3-ethyl-3-heptanol + 3-ethyl-2-heptanol

Use Markovnikov's rule to predict which of the two products will form in the greater amount.

6. Name the reactants and products of each reaction. Use Markovnikov's rule to predict which of the two products will form in the greater amount.

(a) $CH_2{=}CHCH_2CH_2CH_2CH_3 + HBr \rightarrow$

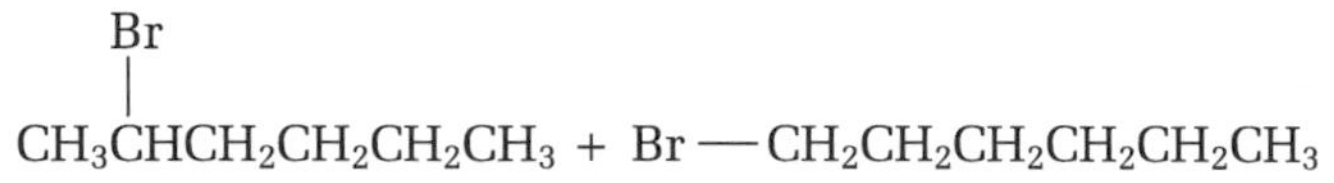

$$CH_3\overset{\overset{\displaystyle Br}{|}}{C}HCH_2CH_2CH_2CH_3 + Br{-}CH_2CH_2CH_2CH_2CH_2CH_3$$

Continued ...

Addition reactions form the basis for tests that distinguish alkenes and alkynes from alkanes. Bromine, Br_2, has a deep reddish-brown colour. When bromine is added to an alkene or alkyne, an addition reaction takes place. As the bromine is used up, the brown colour of the bromine disappears. Since alkanes cannot undergo addition reactions, no reaction takes place when bromine is added to an alkane.

Go to the Chemistry 12 Electronic Learning Partner for more information about the bonding in ethyne.

Continued ...

(b) $CH_3-\overset{\overset{CH_3}{|}}{C}=CH-CH_2-CH_3 + HOH \rightarrow$

$$CH_3-\underset{\underset{OH}{|}}{\overset{\overset{CH_3}{|}}{C}}-CH_2-CH_2-CH_3 + CH_3-\overset{\overset{CH_3}{|}}{CH}-\underset{\underset{OH}{|}}{CH}-CH_2-CH_3$$

7. Draw the major product of each reaction.

(a) $CH_3CH{=}CH_2 + Br_2 \rightarrow$

(b) $CH_2{=}CH_2 + HOH \rightarrow$

(c) $CH_2{=}CHCH_2CH_3 + HBr \rightarrow$

(d) $(CH_3)_2C{=}CHCH_2CH_2CH_3 + HCl \rightarrow$

8. For each reaction, name and draw the reactants that are needed to produce the given product.

(a) $? + ? \rightarrow CH_3CH(Cl)CH_3$

(b) $? + ? \rightarrow Br-CH_2CH_2-Br$

(c) $? + HOH \rightarrow CH_3CH_2\underset{\underset{CH_3}{|}}{\overset{\overset{OH}{|}}{C}}CH_2CH_3$

(d) $CH_2{=}CHCH_3 + ? \rightarrow CH_3CH_2CH_3$

Addition to Alkynes

Since alkynes have triple bonds, two addition reactions can take place in a row. If one mole of a reactant, such as HCl, Br_2, or H_2O, is added to one mole of an alkyne, the result is a substituted alkene.

$$\underset{\text{propyne (1 mol)}}{H-C{\equiv}C-CH_3} + \underset{\text{bromine (1 mol)}}{Br_2} \rightarrow \underset{\text{1,2-dibromopropene}}{H-\overset{\overset{Br}{|}}{C}=\underset{\underset{Br}{|}}{C}-CH_3}$$

If two moles of the reactant are added to one mole of an alkyne, the reaction continues one step further. A second addition reaction takes place, producing an alkane.

$$\underset{\text{propyne (1 mol)}}{H-C{\equiv}C-CH_3} + \underset{\text{bromine (2 mol)}}{2Br_2} \rightarrow \underset{\text{1,1,2,2-tetrabromopropane}}{H-\underset{\underset{Br}{|}}{\overset{\overset{Br}{|}}{C}}-\underset{\underset{Br}{|}}{\overset{\overset{Br}{|}}{C}}-CH_3}$$

Like alkenes, asymmetrical alkynes follow Markovnikov's rule when an asymmetrical molecule, such as H_2O or HBr, is added to the triple bond. An example is given below.

$$\underset{\text{1-butyne (1 mol)}}{H-C{\equiv}C-CH_2CH_3} + \underset{\text{hydrobromic acid (1 mol)}}{HBr} \rightarrow \underset{\text{2-bromo-1-butene (major product)}}{H_2C=\overset{\overset{Br}{|}}{C}-CH_2CH_3} + \underset{\text{1-bromo-1-butene (minor product)}}{\overset{\overset{Br}{|}}{HC}=CH-CH_2CH_3}$$

Canadians in Chemistry

Dusanka Filipovic

When Dusanka Filipovic was a teenager, she liked hiking, playing volleyball, and working as a camp counsellor. Now she is a successful chemical engineer who develops and markets technology that helps the environment. What do her early pursuits have to do with her engineering career? "Both require creativity, persistence, and an ability to work under pressure," says Filipovic.

Filipovic was born and grew up in Belgrade, in the former Yugoslavia. She attended the University of Belgrade, and graduated with a degree in chemical engineering. Attracted by Canada's technical advances, she came here after winning a scholarship to study at McMaster University in Hamilton, Ontario. Canada has become her home.

Filipovic has worked hard to break down barriers for women in the fields of engineering and business. In 1974, she became the first female professional engineer employed by a major chemical producing company. The National Museum of Science and Technology in Ottawa, Ontario, has featured her work as part of an exhibit on women inventors.

Since the 1980s, Filipovic has concentrated on developing environmentally friendly technologies. She is the co-inventor of a patented process known as Blue Bottle™ technology. This process is used to recover and recycle halogenated hydrocarbons, such as chlorofluorocarbons (CFCs) and other ozone-depleting substances (ODS), from damaged or unused residential refrigerators and automotive air conditioners.

Using *adsorption*, the binding of molecules or particles to a surface, Blue Bottle™ technology acts as a selective molecular sieve to capture refrigerant gases so that they can be safely stored and re-used. A non-pressurized Blue Bottle™ cylinder, packed with an adsorbent, synthetic zeolite called Halozite™, is connected to the back of a refrigeration or air-conditioning unit. Zeolites are porous aluminosilicate minerals that commonly contain sodium and calcium as major cations (positively charged ions) and are capable of ion exchange. (Zeolites contain water molecules that allow reversible dehydration, and they are often used as water softeners.) The Halozite™ adsorbs the refrigerants that are released from the unit at ambient temperatures, under atmospheric pressure.

Once the adsorbent is saturated, the Blue Bottle™ cylinder is sent to a central reclamation facility, where the refrigerants are reclaimed and stored. The Blue Bottle™ cylinders can be re-used after the refrigerants have been collected. Reclaimed halogenated hydrocarbons can be used as refrigerants, solvents, cleaners, fumigants, and fire retardants.

In 1991, Filipovic formed her own company to commercialize Blue Bottle™ technology. In 1999, she founded a new company, which uses a process similar to Blue Bottle™ technology to capture and convert greenhouse gas emissions from hospital operating rooms. These emissions had previously been discharged into the atmosphere.

Filipovic has this advice for aspiring engineers and scientists: "Once you recognize what it is you want to pursue, make sure you take advantage of every training opportunity offered to you. I became an engineer because I wanted to be able to create something new, and monitor and enjoy the results of my work . . . I am working to make a significant contribution that will live on after me."

Reactions of Aromatic Compounds

As you learned in Chapter 1, aromatic compounds do not react in the same way that compounds with double or triple bonds do. Benzene's stable ring does not usually accept the addition of other atoms. Instead, aromatic compounds undergo substitution reactions. A hydrogen atom or a functional group that is attached to the benzene ring may be replaced by a different functional group. Figure 2.6 shows two possible reactions for benzene. Notice that iron(III) bromide, $FeBr_3$, is used as a catalyst in the substitution reaction. An addition reaction does *not* occur because the product of this reaction would be less stable than benzene.

Electronic Learning Partner

Go to the Chemistry 12 Electronic Learning Partner for more information about aspects of material covered in this section of the chapter.

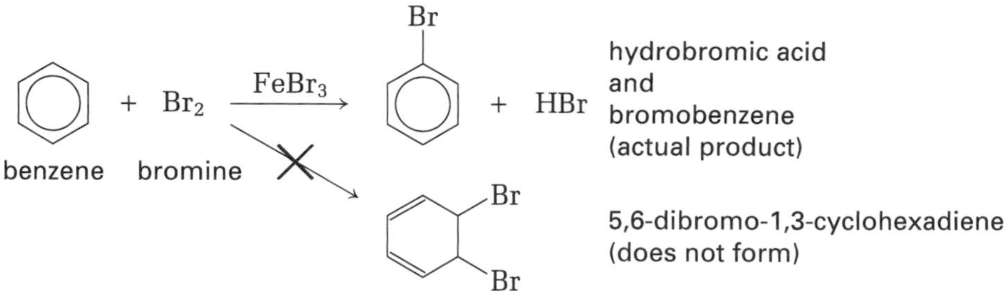

benzene + bromine (Br_2) $\xrightarrow{FeBr_3}$ bromobenzene (actual product) + hydrobromic acid and (HBr)

benzene + bromine ✗→ 5,6-dibromo-1,3-cyclohexadiene (does not form)

Figure 2.6 **The bromine does not add to benzene in an addition reaction. Instead, one of the H atoms on the benzene ring is replaced with a Br atom in a substitution reaction.**

Reactions of Alcohols

Alcohols can react in several ways, depending on the reactants and on the conditions of the reaction. For example, alcohols can undergo substitution with halogen acids, elimination to form alkenes, and oxidation to form aldehydes, ketones, or carboxylic acids.

Substitution Reactions of Alcohols

When a halogen acid, such as HCl, HBr, or HI, reacts with an alcohol, the halogen atom is substituted for the OH group of the alcohol. This is shown in Figure 2.7(A). The reverse reaction takes place when an alkyl halide reacts with OH^- in a basic solution. See Figure 2.7(B).

(A) $$CH_3—CH_2—OH + HCl \rightarrow CH_3—CH_2—Cl + H_2O$$

(B) $$CH_3—CH_2—Cl + OH^- \rightarrow CH_3—CH_2—OH + Cl^-$$

Figure 2.7 **(A) Ethanol reacts with hydrochloric acid to produce chloroethane. (B) In a basic solution, the reverse reaction takes place. A hydroxide ion reacts with chloroethane to produce ethanol.**

Elimination Reactions of Alcohols

When an alcohol is heated in the presence of the strong acid and dehydrating agent, H_2SO_4, an elimination reaction takes place. This type of reaction is shown in Figure 2.8, on the next page. The OH group and one H atom leave the molecule, and water is produced. As a result, the molecule forms a double bond. Because water is produced, this type of reaction is also called a *dehydration* (meaning "loss of water") reaction.

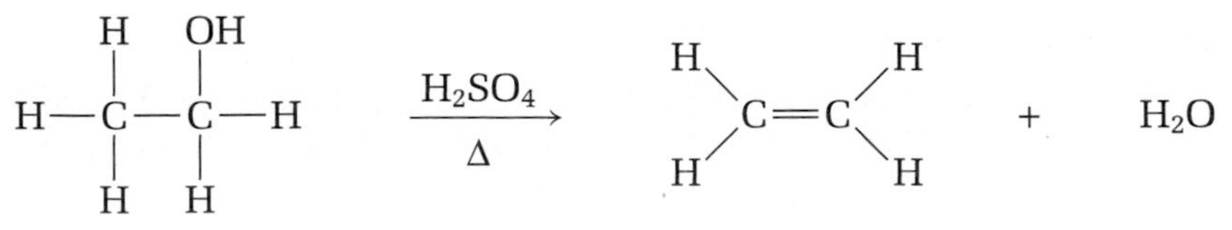

$$\mathrm{H{-}C(H)(H){-}C(OH)(H){-}H} \xrightarrow[\Delta]{H_2SO_4} \mathrm{(H)_2C{=}C(H)_2} + H_2O$$

ethanol ethene water

Figure 2.8 The Δ symbol is used in chemistry to represent heat added to a reaction.

Oxidation of Alcohols

In the presence of an oxidizing agent, an alcohol is oxidized to form an aldehyde or a ketone.

- A primary alcohol is oxidized to an aldehyde. If the aldehyde is oxidized further, it becomes a carboxylic acid.

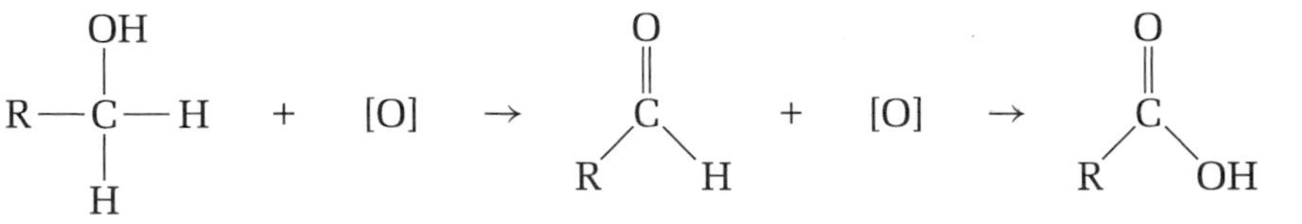

$$\mathrm{R{-}C(OH)(H){-}H} + [O] \rightarrow \mathrm{R{-}C({=}O){-}H} + [O] \rightarrow \mathrm{R{-}C({=}O){-}OH}$$

primary alcohol aldehyde carboxylic acid

- A secondary alcohol is oxidized to a ketone. All the carbon bonding sites are now occupied with bonds to carbon and oxygen, so no further oxidation is possible.

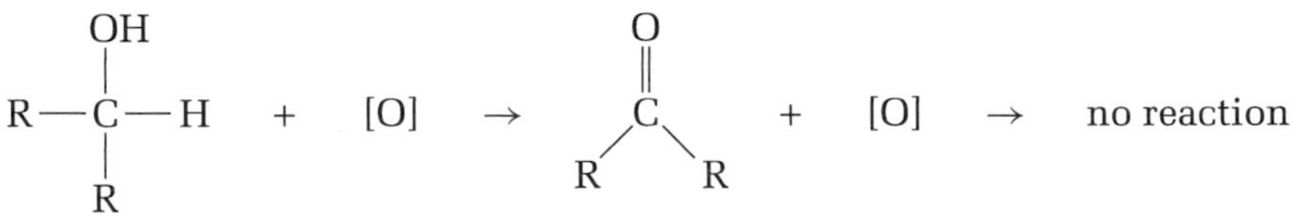

$$\mathrm{R{-}C(OH)(R){-}H} + [O] \rightarrow \mathrm{R{-}C({=}O){-}R} + [O] \rightarrow \text{no reaction}$$

secondary alcohol ketone

- A tertiary alcohol cannot be oxidized. Carbon can form up to four bonds, but all possible bonding sites are already occupied. C—H bonds can be broken by an oxidizing agent. On the carbon atom of a tertiary alcohol, however, there is no hydrogen atom available to be removed. C—C bonds are too strong to be broken by an oxidizing agent.

$$\mathrm{R{-}C(OH)(R){-}R} + [O] \rightarrow \text{no reaction}$$

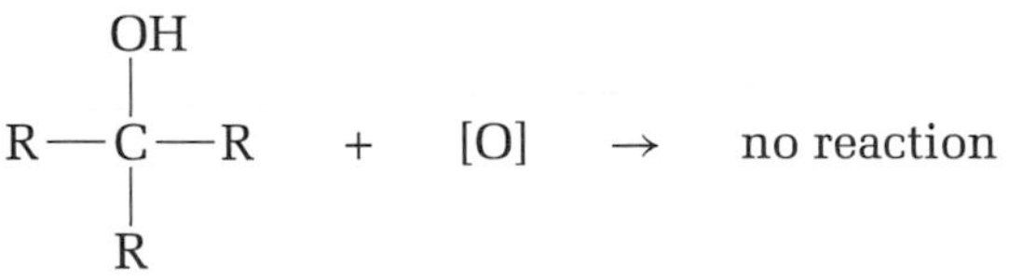

tertiary alcohol

The following Sample Problem shows how to predict the products of reactions of alcohols.

Sample Problem

Predicting the Reaction of an Alcohol

Problem

Name each type of reaction. Then predict and name the products.

(a) 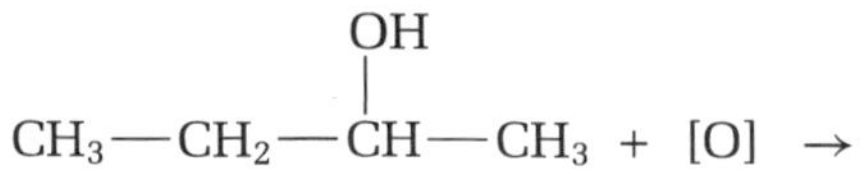 $CH_3—CH_2—CH(OH)—CH_3 + [O] \rightarrow$

(b) 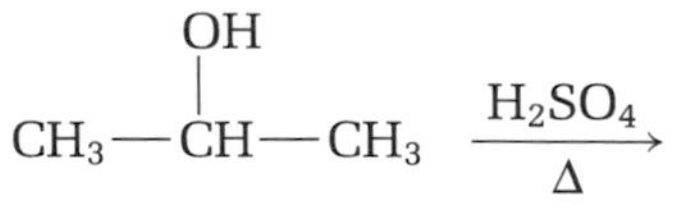 $CH_3—CH(OH)—CH_3 \xrightarrow[\Delta]{H_2SO_4}$

(c) 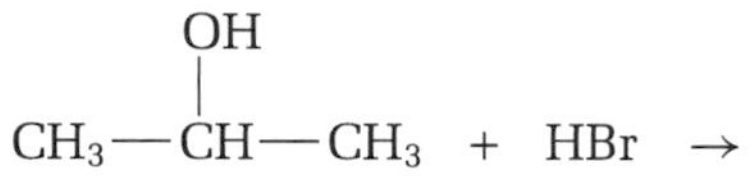 $CH_3—CH(OH)—CH_3 + HBr \rightarrow$

Solution

(a) The reactant is a secondary alcohol, and an oxidizing agent is present. Therefore, this must be oxidation. You know that a secondary alcohol is oxidized to a ketone. The ketone must have the same carbon-carbon bonds, and the same number of carbon atoms, as the reacting alcohol. Therefore, the ketone must be 2-butanone.

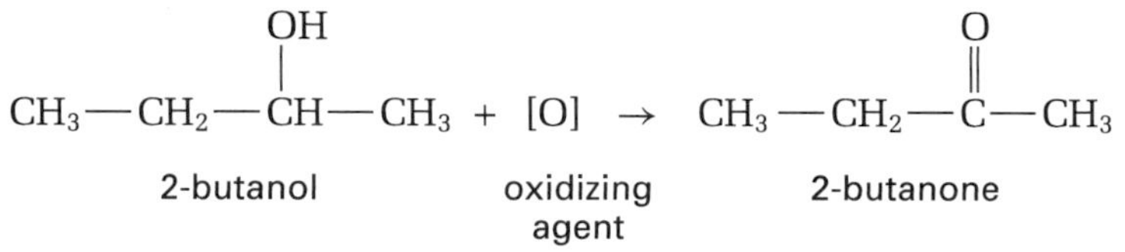

$$\underset{\text{2-butanol}}{CH_3—CH_2—CH(OH)—CH_3} + \underset{\text{oxidizing agent}}{[O]} \rightarrow \underset{\text{2-butanone}}{CH_3—CH_2—C(=O)—CH_3}$$

(b) This reaction takes place in the presence of heat and sulfuric acid, H_2SO_4. It is an elimination reaction. The product is an alkene, with the same number of carbon atoms as the reacting alcohol. Since this reaction is an elimination reaction, a small molecule (in this case, water) must be eliminated as the second product. The organic product is propene.

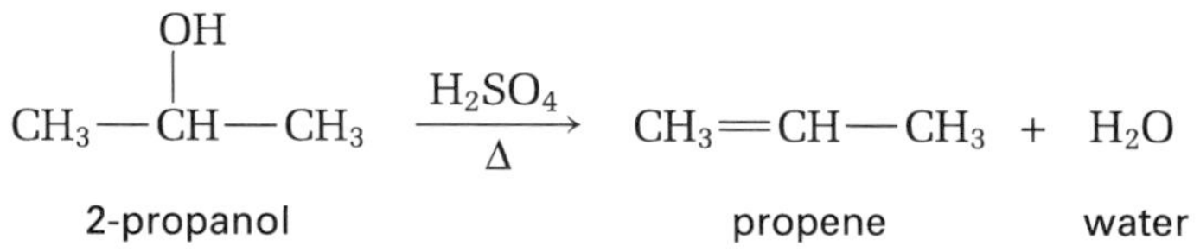

$$\underset{\text{2-propanol}}{CH_3—CH(OH)—CH_3} \xrightarrow[\Delta]{H_2SO_4} \underset{\text{propene}}{CH_3=CH—CH_3} + \underset{\text{water}}{H_2O}$$

(c) In this reaction, an alcohol reacts with hydrobromic acid, HBr. This is a substitution reaction. The product is an alkyl halide, with the same carbon-carbon bonds, and the same number of carbon atoms, as the reacting alcohol. The alkyl halide is 2-bromopropane. The second product is water.

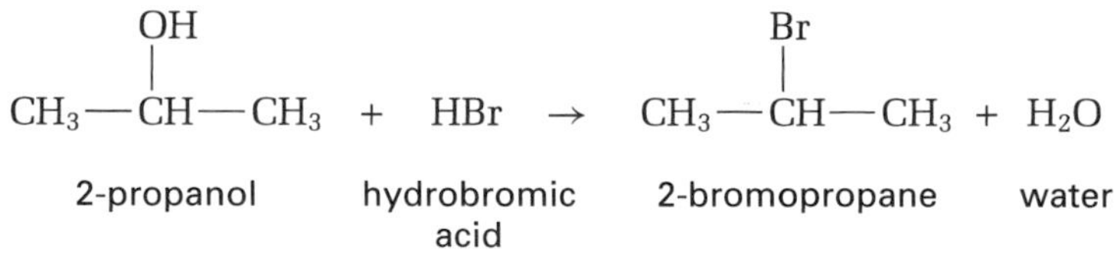

$$\underset{\text{2-propanol}}{CH_3—CH(OH)—CH_3} + \underset{\text{hydrobromic acid}}{HBr} \rightarrow \underset{\text{2-bromopropane}}{CH_3—CH(Br)—CH_3} + \underset{\text{water}}{H_2O}$$

PROBLEM TIP

In many elimination reactions, water is produced. The formation of water is a strong driving force for many reactions.

Practice Problems

9. Name each type of reaction.

(a) 1-propanol + HCl → 1-chloropropane + H_2O

(b) 1-butanol + [O] → butanal

(c) $CH_3CH_2CH_2Cl + NaOH \rightarrow CH_3CH_2CH_2OH + NaCl$

(d) 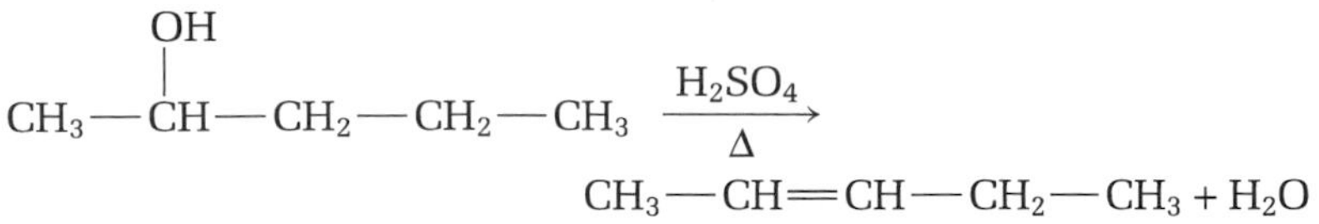 $CH_3—\underset{}{\overset{OH}{\overset{|}{C}H}}—CH_2—CH_2—CH_3 \xrightarrow[\Delta]{H_2SO_4} CH_3—CH{=}CH—CH_2—CH_3 + H_2O$

10. Draw the structures of the reactants and products in parts (a) and (b) of question 9.

11. Name each type of reaction.

(a) $CH_3—CH_2—CH_2—CH_2—OH \xrightarrow[\Delta]{H_2SO_4}$

(b) $CH_3—CH_2—CH_2—CH_2—CH_2—OH + [O] \rightarrow$ (i) + [O] → (ii)

(c) 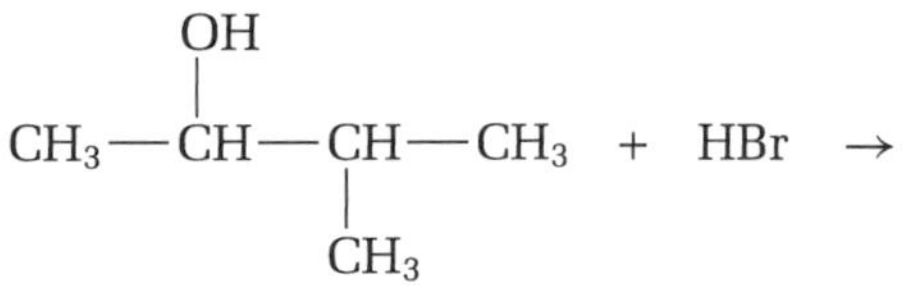 $CH_3—\overset{OH}{\overset{|}{C}H}—\underset{\underset{CH_3}{|}}{CH}—CH_3 + HBr \rightarrow$

(d) 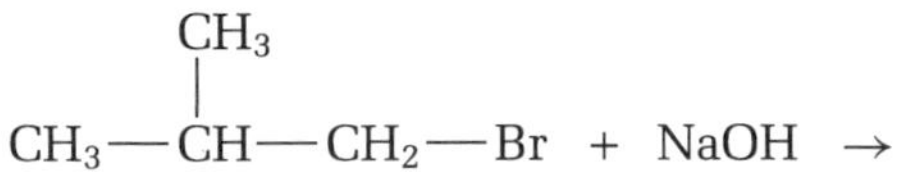 $CH_3—\overset{CH_3}{\overset{|}{C}H}—CH_2—Br + NaOH \rightarrow$

(e) 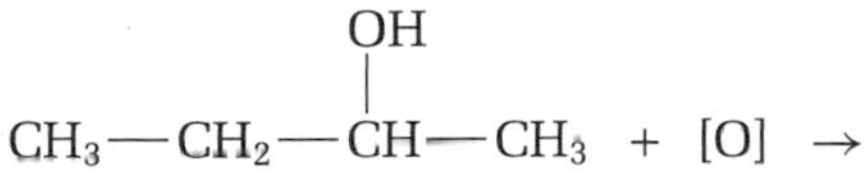 $CH_3—CH_2—\overset{OH}{\overset{|}{C}H}—CH_3 + [O] \rightarrow$

(f) 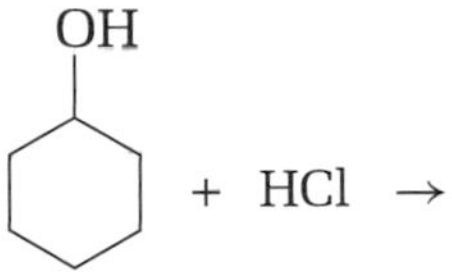 (cyclohexane ring with OH) + HCl →

12. Draw and name the products of each reaction in question 11.

13. Is the following reaction possible? Why or why not?

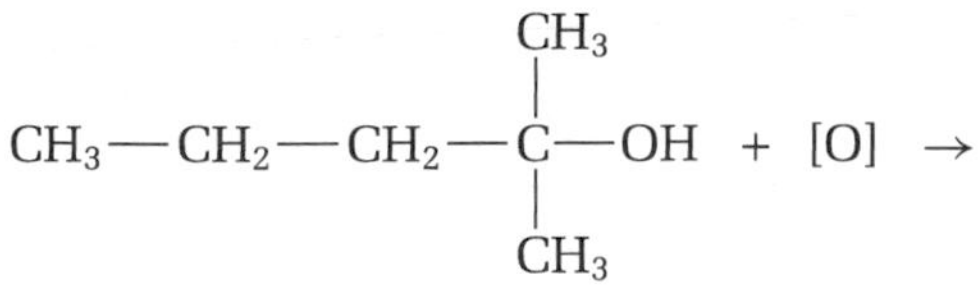

$$CH_3—CH_2—CH_2—\underset{\underset{CH_3}{|}}{\overset{\overset{CH_3}{|}}{C}}—OH + [O] \rightarrow$$

In the following investigation, you will carry out the oxidation of different alcohols.

Investigation 2-A

SKILL FOCUS

Predicting

Performing and recording

Analyzing and interpreting

Oxidizing Alcohols

Acidified potassium permanganate solution, $KMnO_{4(aq)}$, acts as an oxidizing agent when it comes in contact with alcohols. In this investigation, you will discover how potassium permanganate reacts with a primary, a secondary, and a tertiary alcohol.

Questions

How do primary, secondary, and tertiary alcohols react with an oxidizing agent?

Prediction

Predict which of the following alcohols will react with an oxidizing agent. Draw and name the product you expect for each reaction.

- 2-propanol
- 1-butanol
- 2-methyl-2-propanol

Safety Precautions

Handle the sulfuric acid solution with care. It is corrosive. Wipe up any spills with copious amounts of water, and inform your teacher.

Materials

2-propanol, 1-butanol, 2-methyl-2-propanol
0.01 mol/L $KMnO_4$ solution
1 mol/L H_2SO_4
distilled water
5 test tubes with stoppers
dropper
stopwatch or clock

Procedure

1. Place 3 mL $KMnO_4$, 3 mL water, and 2 mL acid in a test tube. This mixture should not oxidize quickly. You will use it as a standard to compare the colour of the other test tubes.
2. Prepare each of the following mixtures, one at a time.
   - Test tube 1: 3 mL $KMnO_4$, 3 mL 2-propanol, 2 mL acid
   - Test tube 2: 3 mL $KMnO_4$, 3 mL 1-butanol, 2 mL acid
   - Test tube 3: 3 mL $KMnO_4$, 3 mL 2-methyl-2-propanol, 2 mL acid
3. Place a stopper in each test tube, and shake the test tube gently to mix the contents. Record the reaction time for each mixture. Use the change in the colour to determine if a reaction has taken place.
4. Dispose of your mixtures as directed by your teacher.

Analysis

1. Draw the structure of each alcohol you used. Label each alcohol as primary, secondary, or tertiary.
2. (a) Which alcohol(s) reacted with the potassium permanganate? How did you know?
   (b) Which alcohol(s) did not react with the potassium permanganate? Explain your observations.
3. Which alcohol(s) reacted the most quickly? Which reacted slowly? Make a general statement that summarizes your observations.
4. Write a chemical equation for each reaction.

Conclusions

5. Which alcohol reacted the fastest: primary, secondary, or tertiary?
6. Were there any sources of experimental error? What could you improve if you did this investigation again?

Reactions of Aldehydes and Ketones

As you observed in the investigation, different alcohols react differently with an oxidizing agent. In the same way, aldehydes and ketones react differently with oxidizing and reducing agents, even though aldehydes and ketones have similar structures.

Oxidation of Aldehydes

In the presence of an oxidizing agent, an aldehyde is oxidized to produce a carboxylic acid. The hydrogen atom that is bonded to the carbon atom of the C=O bond is replaced with an OH group.

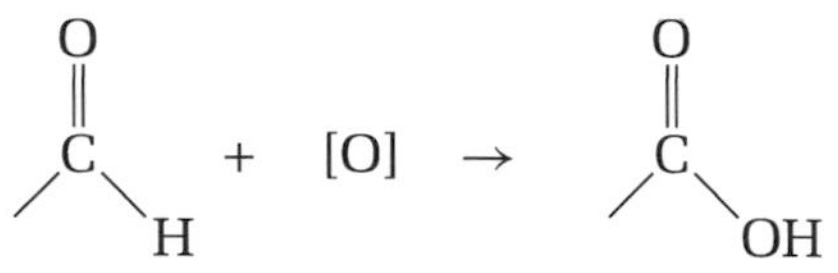

$$\text{—C(=O)H} + [O] \rightarrow \text{—C(=O)OH}$$

aldehyde   carboxylic acid

Ketones do not usually undergo oxidation. Like tertiary alcohols, ketones do not have a hydrogen atom that is available to be removed. Carbon-carbon bonds are too strong to be broken by an oxidizing agent.

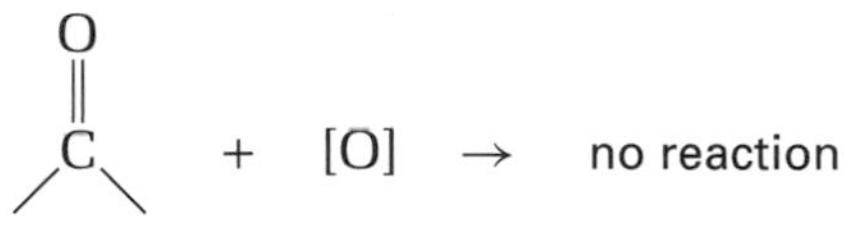

$$\text{—C(=O)—} + [O] \rightarrow \text{no reaction}$$

ketone

Reduction of Aldehydes and Ketones

You learned earlier that primary alcohols are oxidized to aldehydes, and secondary alcohols are oxidized to ketones. You can think of the reduction of aldehydes and ketones as the reverse of these reactions. Aldehydes can be reduced to produce primary alcohols. Ketones can be reduced to produce secondary alcohols.

$$\text{RC(=O)H} + [H] \rightarrow \text{R—CH(OH)—H}$$

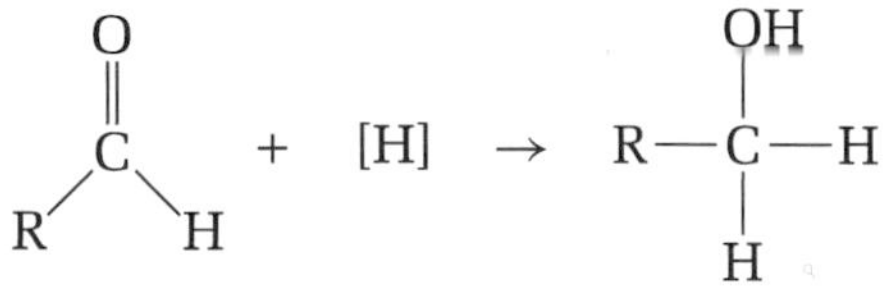

aldehyde   primary alcohol

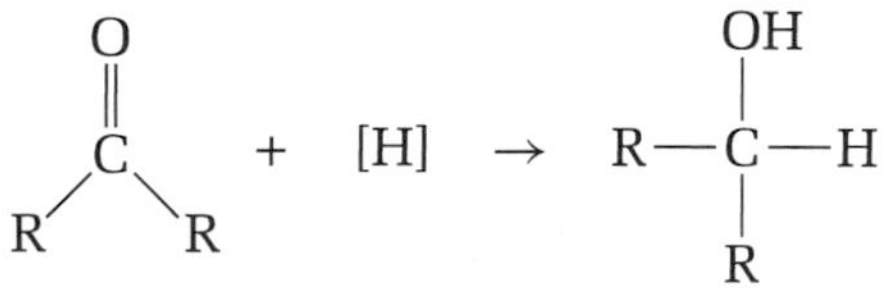

$$\text{RC(=O)R} + [H] \rightarrow \text{R—C(OH)(R)—H}$$

ketone   secondary alcohol

CONCEPT CHECK

Explain why the transformation of an aldehyde into an alcohol is a reduction not an oxidation.

Reactions of Carboxylic Acids

Like other acids, a carboxylic acid reacts with a base to produce a salt and water.

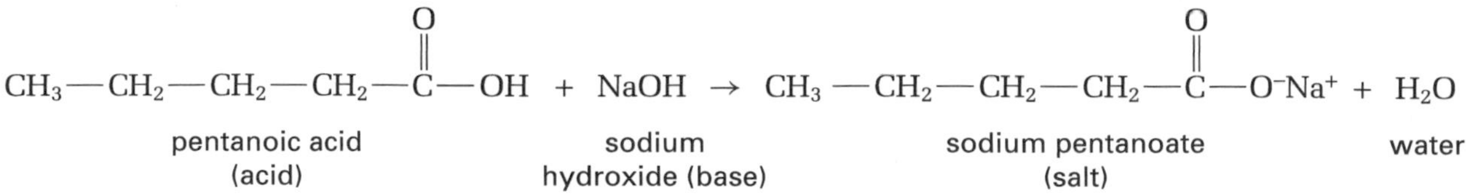

$$CH_3{-}CH_2{-}CH_2{-}CH_2{-}\overset{\overset{\displaystyle O}{\|}}{C}{-}OH + NaOH \rightarrow CH_3{-}CH_2{-}CH_2{-}CH_2{-}\overset{\overset{\displaystyle O}{\|}}{C}{-}O^{-}Na^{+} + H_2O$$

pentanoic acid (acid) | sodium hydroxide (base) | sodium pentanoate (salt) | water

As you learned earlier, a carboxylic acid reacts with an alcohol to produce an ester. Water is the second product of this reaction. A strong acid, such as H_2SO_4, is used to catalyze (speed up) the reaction. The reverse reaction can also occur. (See "Reactions of Esters and Amides," below.)

$$R{-}\overset{\overset{\displaystyle O}{\|}}{C}{-}OH + HO{-}R' \xrightleftharpoons{H_2SO_4} R{-}\overset{\overset{\displaystyle O}{\|}}{C}{-}O{-}R' + H_2O$$

carboxylic acid | alcohol | ester | water

The reaction of a carboxylic acid with an alcohol to form an ester is called an **esterification reaction**. An esterification reaction is one type of condensation reaction. In a similar type of condensation reaction, an amide can be formed from a carboxylic acid and an amine, but this process is slightly longer and will not be discussed here.

Carboxylic acids can be reduced to form aldehydes and alcohols.

Reactions of Esters and Amides

Both esters and amides undergo hydrolysis reactions. In a hydrolysis reaction, the ester or amide bond is *cleaved*, or split in two, to form two products. As mentioned earlier, the hydrolysis of an ester produces a carboxylic acid and an alcohol. The hydrolysis of an amide produces a carboxylic acid and an amine. There are two methods of hydrolysis: acidic hydrolysis and basic hydrolysis. Both methods are shown in Figure 2.9. Hydrolysis usually requires heat. In acidic hydrolysis, the ester or amide reacts with water in the presence of an acid, such as H_2SO_4. In basic hydrolysis, the ester or amide reacts with the OH^- ion, from NaOH or water, in the presence of a base. Soap is made by the basic hydrolysis of ester bonds in vegetable oils or animal fats.

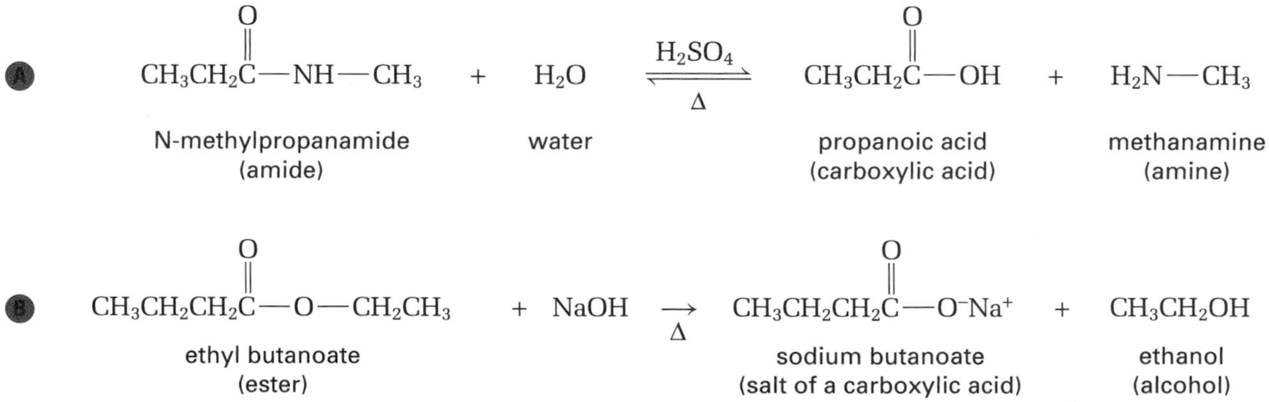

A

$$CH_3CH_2\overset{\overset{\displaystyle O}{\|}}{C}{-}NH{-}CH_3 + H_2O \underset{\Delta}{\xrightleftharpoons{H_2SO_4}} CH_3CH_2\overset{\overset{\displaystyle O}{\|}}{C}{-}OH + H_2N{-}CH_3$$

N-methylpropanamide (amide) | water | propanoic acid (carboxylic acid) | methanamine (amine)

B

$$CH_3CH_2CH_2\overset{\overset{\displaystyle O}{\|}}{C}{-}O{-}CH_2CH_3 + NaOH \xrightarrow[\Delta]{} CH_3CH_2CH_2\overset{\overset{\displaystyle O}{\|}}{C}{-}O^{-}Na^{+} + CH_3CH_2OH$$

ethyl butanoate (ester) | sodium butanoate (salt of a carboxylic acid) | ethanol (alcohol)

Figure 2.9 (A) The acid hydrolysis of an amide produces a carboxylic acid and an amine. (B) The basic hydrolysis of an ester produces the salt of a carboxylic acid and an alcohol.

Sample Problem

Predicting the Products of More Organic Reactions

Problem

Identify each type of reaction. Then predict and name the product(s).

(a) 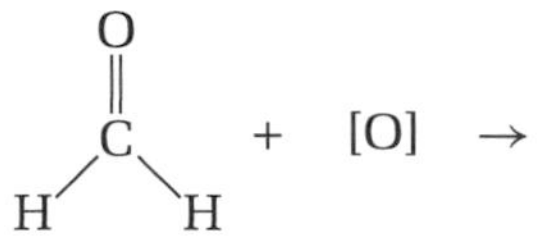 $HCHO$ (H–C(=O)–H) $+ [O] \rightarrow$

(b) $H_3C–C(=O)–CH_3 + [H] \rightarrow$

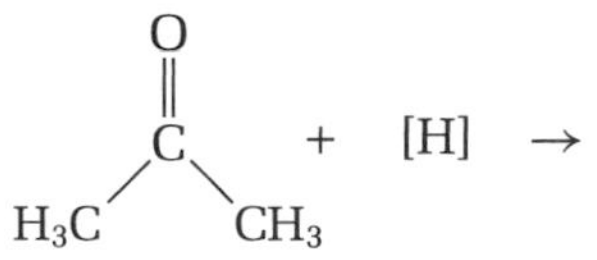

(c) $CH_3COOH + CH_3CH_2OH \rightleftharpoons$ (structure $CH_3\overset{O}{\overset{\|}{C}}OH$)

Solution

(a) This reaction involves an aldehyde and an oxidizing agent. Thus, it is an oxidation. (If the reactant were a ketone, no reaction would occur.) The product of this reaction is a carboxylic acid called methanoic acid.

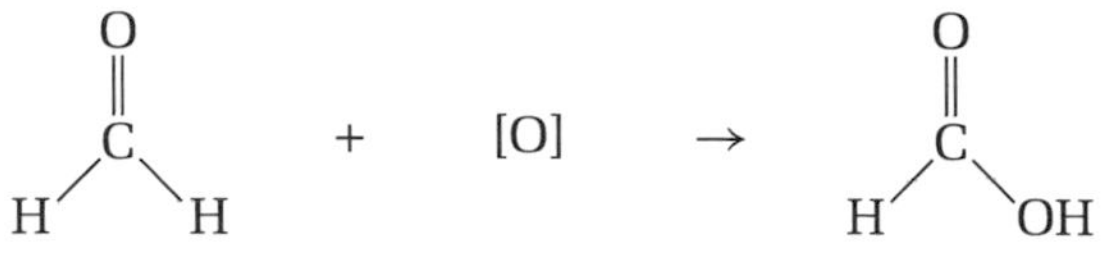

$$H–C(=O)–H + [O] \rightarrow H–C(=O)–OH$$

methanal — oxidizing agent — methanoic acid

(b) This reaction involves a ketone and a reducing agent. Thus, it is a reduction. The product of this reaction is an alcohol called 2-propanol.

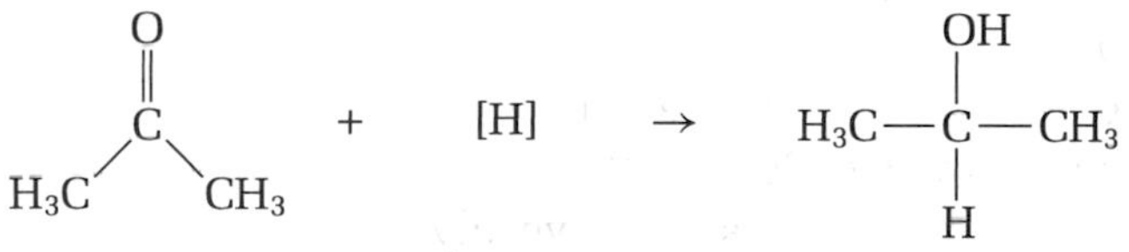

$$H_3C–C(=O)–CH_3 + [H] \rightarrow H_3C–CH(OH)–CH_3$$

propanone — reducing agent — 2-propanol

(c) This reaction occurs between a carboxylic acid and an alcohol. It is an esterification, or condensation, reaction. The product of this reaction is an ester called ethyl ethanoate.

$$CH_3\overset{O}{\overset{\|}{C}}OH + CH_3CH_2OH \rightleftharpoons CH_3\overset{O}{\overset{\|}{C}}OCH_2CH_3$$

ethanoic acid — ethanol — ethyl ethanoate

Practice Problems

14. Identify each type of reaction. (Oxidizing and reducing agents are not shown.)

(a) 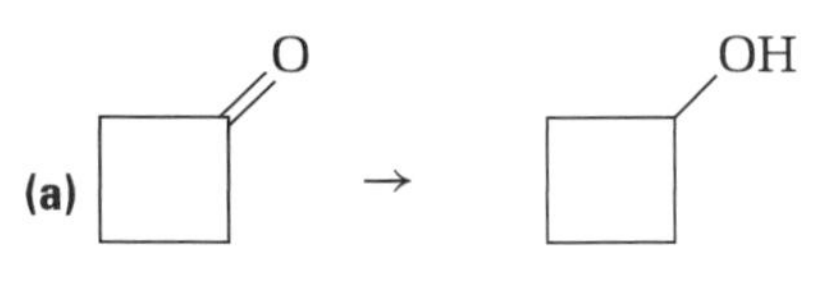

(b) $(CH_3)_3C-C(=O)-C(CH_3)_3 \rightarrow (CH_3)_3C-CH(OH)-C(CH_3)_3$

(c) 2-pentanone → 2-pentanol

(d) hexanoic acid + ethanol ⇌ ethyl hexanoate + water

(e) butyl methanoate + water $\xrightleftharpoons[\Delta]{H_2SO_4}$ methanoic acid + 1-butanol

15. (a) Name the reactants and products of the first two reactions in question 14.

(b) Draw the reactants and products of the last three reactions in question 14.

16. Name and draw the product(s) of each reaction.

(a) hexanal + [O] →

(b) octanal + [H] →

(c) propanoic acid + methanol ⇌

(d) propyl ethanoate + water $\xrightleftharpoons[\Delta]{H_2SO_4}$

(e) 3-hexanone + [H] →

(f) 2-propanol + 3-methylpentanoic acid ⇌

17. Name and draw the reactant(s) in each reaction.

(a) ? + [H] → $CH_3CH_2CH(CH_3)CH_2OH$

(b) ? + [O] → ? + [O] → $CH_3CH_2CH_2CH_2COOH$

(c) ? + ? → [structure: ester with O and C=O]

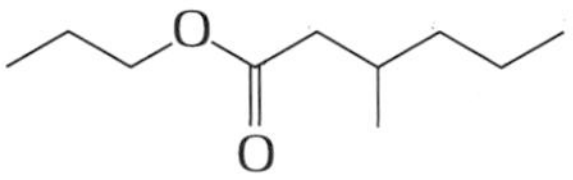

Concept Organizer — Organic Reactions

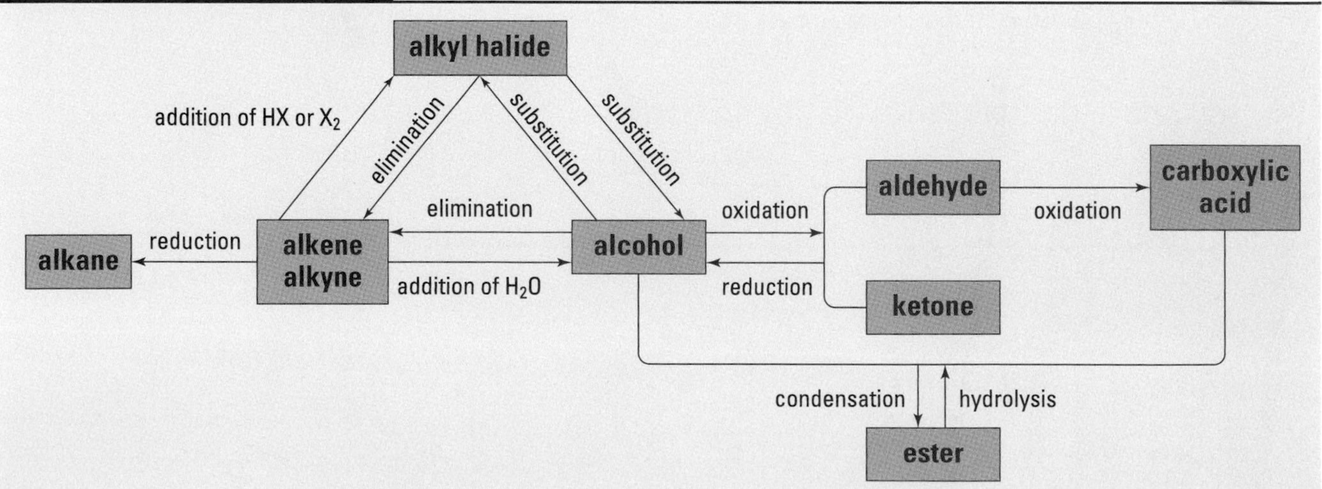

Section Summary

In this section, you learned about the reactions of alkenes, alkynes, aromatic compounds, alcohols, aldehydes, ketones, carboxylic acids, amides, and esters. You learned how to use Markovnikov's rule to predict the major product of an elimination reaction. You also learned how to predict the products of other types of reactions. In the next section, you will encounter a special branch of organic chemistry, which deals with much larger molecules.

Section Review

1 K/U Identify each reaction as one of the following types of reactions: addition, substitution, elimination, oxidation, reduction, condensation, or hydrolysis.

(a) 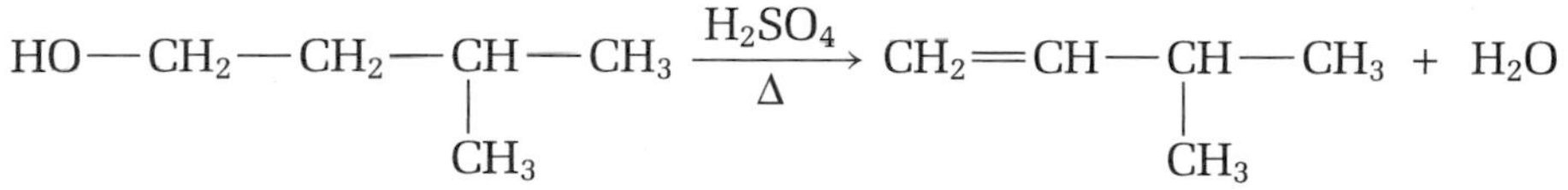 $HO-CH_2-CH_2-\underset{\underset{CH_3}{|}}{CH}-CH_3 \xrightarrow[\Delta]{H_2SO_4} CH_2=CH-\underset{\underset{CH_3}{|}}{CH}-CH_3 + H_2O$

(b) 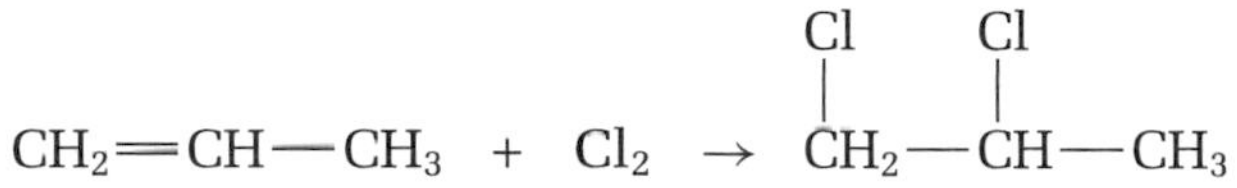 $CH_2=CH-CH_3 + Cl_2 \rightarrow \overset{\overset{Cl}{|}}{CH_2}-\overset{\overset{Cl}{|}}{CH}-CH_3$

(c) 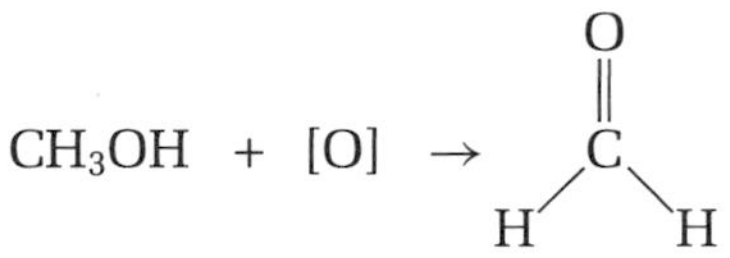 $CH_3OH + [O] \rightarrow H-\overset{\overset{O}{\|}}{C}-H$

(d) 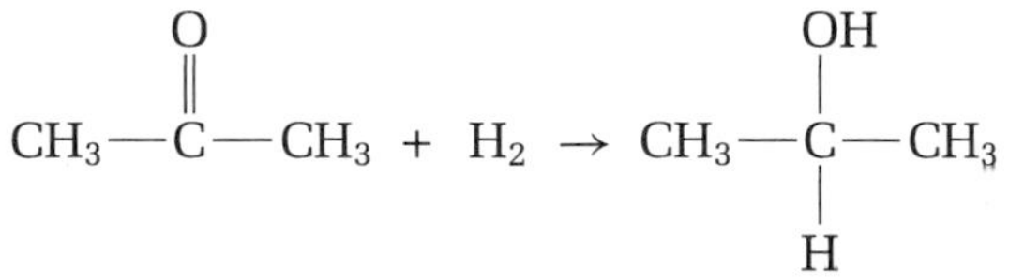 $CH_3-\overset{\overset{O}{\|}}{C}-CH_3 + H_2 \rightarrow CH_3-\underset{\underset{H}{|}}{\overset{\overset{OH}{|}}{C}}-CH_3$

(e) 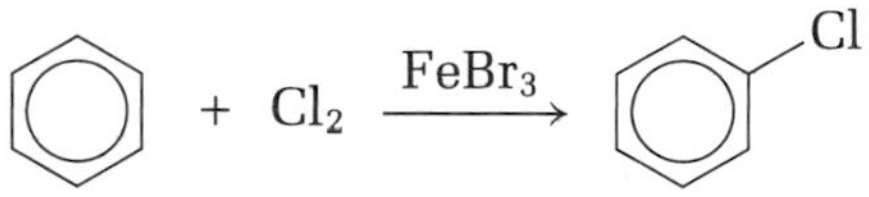 $C_6H_6 + Cl_2 \xrightarrow{FeBr_3} C_6H_5Cl$

(f) 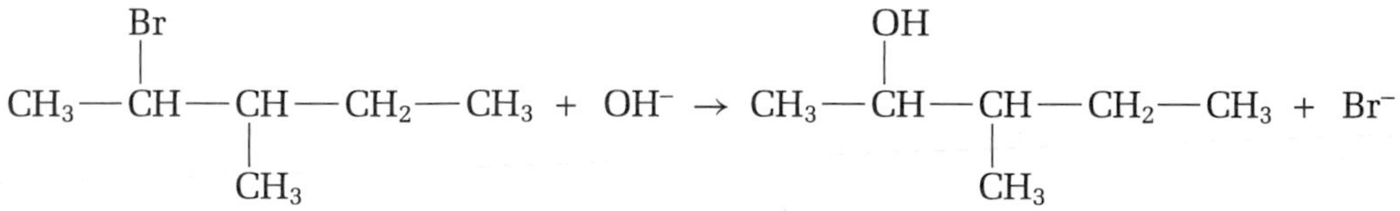 $CH_3-\overset{\overset{Br}{|}}{CH}-\underset{\underset{CH_3}{|}}{CH}-CH_2-CH_3 + OH^- \rightarrow CH_3-\overset{\overset{OH}{|}}{CH}-\underset{\underset{CH_3}{|}}{CH}-CH_2-CH_3 + Br^-$

(g)  $CH_3-CH_2-CH_2-\overset{\overset{O}{\|}}{C}-OH + CH_3-OH \xrightarrow{H_2SO_4} CH_3-CH_2-CH_2-\overset{\overset{O}{\|}}{C}-O-CH_3 + H_2O$

2 I Draw and name the product(s) of each addition or elimination reaction.

(a) $CH_3CH=CHCH_3 + Br_2 \rightarrow$

(b) $CH_2=CHCH_2CH_3 + HCl \rightarrow$

(c) $CH_3CH_2CH_2CH_2CH_2CH_2OH \xrightarrow[\Delta]{H_2SO_4}$

(d) 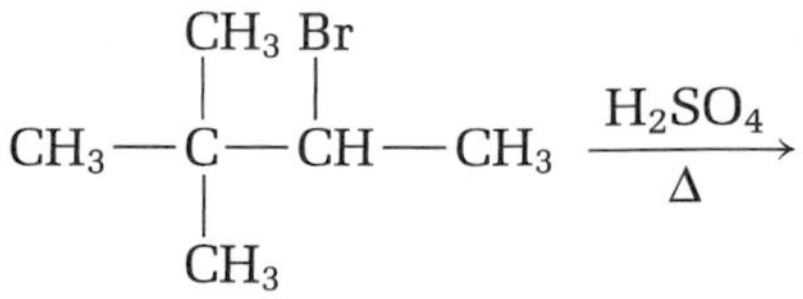 $CH_3-\underset{\underset{CH_3}{|}}{\overset{\overset{CH_3}{|}}{C}}-\overset{\overset{Br}{|}}{CH}-CH_3 \xrightarrow[\Delta]{H_2SO_4}$

3 **I** Draw and name the products of each substitution reaction.

(a) benzene + bromine $\xrightarrow{FeBr_3}$

(b) $CH_3CH_2CH_2CH_2CH_2OH + HBr \rightarrow$

4 **I** Draw and name the product of each oxidation or reduction.

(a) 2-pentanol + [O] →

(b) propanal + [O] →

(c) 2,3-dimethyl-1-butanol + [O] →

(d) $CH_2{=}CH_2 + [H] \rightarrow$

(e) 3-pentanone + [H] →

(f) $CH_3CH_2C{\equiv}CH + [H] \rightarrow (i) + [H] \rightarrow (ii)$

(g) 2-ethyl-3,4-dimethyloctanal + [H] →

5 **I** Name and draw the reactant(s) in each reaction.

(a) $? + H_2O \rightarrow (CH_3)_2CHOH + CH_3CH_2COOH$

(b) ? + [H] → (cyclopentane)

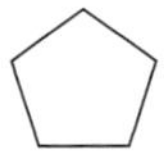

(c) ? + ? → 2-butyl methanoate

(d) ? + [O] → (cyclohexanone)

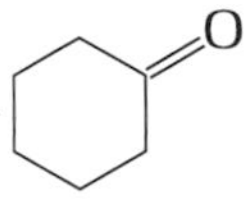

(e) $? + H_2O \rightarrow$ ethanamine + heptanoic acid

6 **(a)** **K/U** State Markovnikov's rule. Under what circumstances is this rule important?

(b) **K/U** Which product of the following reaction is formed in the greater amount?

$$CH_2{=}CHCH_2CH(CH_3)CH_3 + HBr \rightarrow CH_2(Br)CH_2CH_2CH(CH_3)CH_3 + CH_3CH(Br)CH_2CH(CH_3)CH_3$$

7 **I** At least one of the following reactions is not possible. Identify the impossible reaction(s), and explain your reasoning.

(a) $CH_3CH_2C(=O)CH_3 + [O] \rightarrow$

(b) $CH_3{-}CH_2{-}OH + [O] \rightarrow CH_3{-}C(=O){-}OH$

(c)  (benzene) $+ Cl_2 \rightarrow$ (5,6-dichloro-1,3-cyclohexadiene: ring with Cl, Cl)

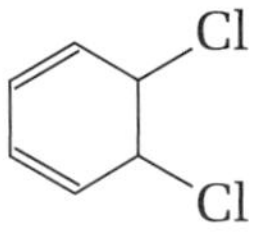

(d) $CH_3{-}C(OH)(CH_3){-}CH_2{-}CH_3 + [O] \rightarrow$

8 **C** Copy the Concept Organizer into your notebook. Then make it more complete by giving an example of each type of reaction.

Molecules on a Larger Scale: Polymers and Biomolecules

2.3

Try to imagine a day without plastics. From simple plastic bags to furniture and computer equipment, plastics are an intergral part of our homes, schools, and work places. So far in this unit, you have encountered fairly small organic molecules. Many of the organic molecules that are used industrially, such as plastics, and many of the essential molecules in your body, are much larger in size.

A **polymer** is a very long molecule that is made by linking together many smaller molecules called **monomers**. To picture a polymer, imagine taking a handful of paper clips and joining them into a long chain. Each paper clip represents a monomer. The long chain of repeating paper clips represents a polymer. Many polymers are made of just one type of monomer. Other polymers are made from a combination of two or more different monomers. Figure 2.10 shows an example of joined monomers in a polymer structure.

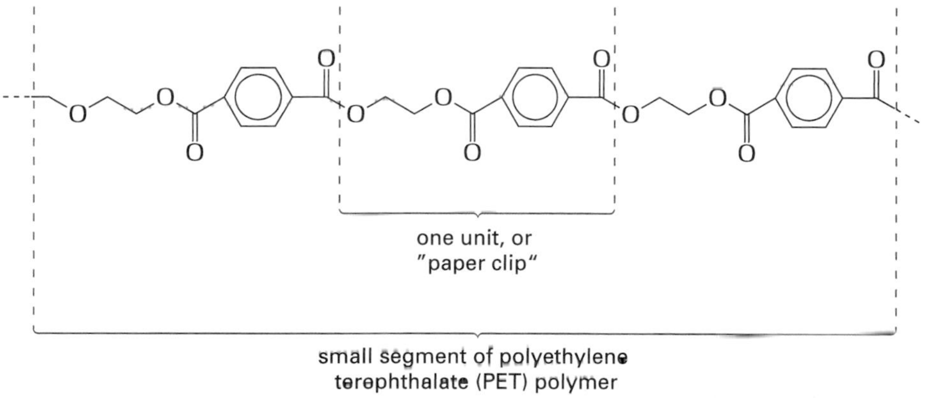

one unit, or "paper clip"

small segment of polyethylene terephthalate (PET) polymer

Figure 2.10 **Polyethylene terephthalate (PET) is the plastic that is used to make soft drink bottles.**

Polymers that can be heated and moulded into specific shapes and forms are commonly known as **plastics**. All plastics are synthetic (artificially made) polymers. For example, polyethene is a common synthetic polymer that is used to make plastic bags. Ethene, $CH_2=CH_2$, is the monomer for polyethene. Adhesives, rubber, chewing gum, and Styrofoam™ are other important materials that are made from synthetic polymers.

Natural polymers are found in living things. For example, glucose, $C_6H_{12}O_6$, is the monomer for the natural polymer starch. You will learn more about natural polymers later in this section.

Some polymers, both synthetic and natural, can be spun into long, thin fibres. These fibres are woven into natural fabrics (such as cotton, linen, and wool) or synthetic fabrics (such as rayon, nylon, and polyester). Figure 2.11, on the next page, shows some polymer products.

Section Preview/ Specific Expectations

In this section, you will

- **communicate** your understanding of addition and condensation polymerization
- **synthesize** a polymer
- **describe** some organic compounds in living organisms, and **explain** their function
- **describe** the variety and importance of polymers and other plastics in your life
- **analyze** some risks and benefits of the production and use of polymers
- **communicate** your understanding of the following terms: *polymer, monomers, plastics, addition polymerization, condensation polymerization, nylons, polyamides, polyesters, biochemistry, protein, amino acids, carbohydrate, saccharide, monosaccharide, disaccharide, polysaccharide, starch, DNA (2-deoxyribonucleic acid), nucleotides, RNA (ribonucleic acid), lipids, fats, oils, waxes*

Web LINK

www.mcgrawhill.ca/links/chemistry12

There are two main types of plastics: *thermoplastics* and *thermosets*. What is the difference between them? What is each type of plastic used for?

Polyethylene terephthalate (PET) can be recycled successfully into new and interesting materials. Research the trade name Ecospun™ on the Internet. What is this material, what is it made from, and what is it used for?

To answer these questions, go to the web site above and click on **Web Links**.

CONCEPT CHECK

Why does "polyethene" end with the suffix -ene, if it does not contain any double bonds?

Figure 2.11 Both synthetic and natural polymers are used to make clothing.

Making Synthetic Polymers: Addition and Condensation Polymerization

Synthetic polymers are extremely useful and valuable. Many polymers and their manufacturing processes have been patented as corporate technology. Polymers form by two of the reactions you have already learned: addition reactions and condensation reactions.

The name of a polymer is usually written with the prefix poly- (meaning "many") before the name of the monomer. Often the common name of the monomer is used, rather than the IUPAC name. For example, the common name of ethene is ethylene. Polyethene, the polymer that is made from ethene, is often called polyethylene. Similarly, the polymer that is made from chloroethene (common name: vinyl chloride) is named polyvinylchloride (PVC). The polymer that is made from propene monomers (common name: propylene) is commonly called polypropylene, instead of polypropene.

Addition polymerization is a reaction in which monomers with double bonds are joined together through multiple addition reactions to form a polymer. Figure 2.12 illustrates the addition polymerization reaction of ethene to form polyethene. Table 2.1, on the next page, gives the names, structures, and uses of some common polymers that are formed by addition polymerization.

$$H_2C{=}CH_2 + H_2C{=}CH_2 \rightarrow -\underset{\displaystyle H}{\overset{\displaystyle H}{C}}-\underset{\displaystyle H}{\overset{\displaystyle H}{C}}-\underset{\displaystyle H}{\overset{\displaystyle H}{C}}-\underset{\displaystyle H}{\overset{\displaystyle H}{C}}- \xrightarrow{H_2C{=}CH_2}$$

$$-\underset{\displaystyle H}{\overset{\displaystyle H}{C}}-\underset{\displaystyle H}{\overset{\displaystyle H}{C}}-\underset{\displaystyle H}{\overset{\displaystyle H}{C}}-\underset{\displaystyle H}{\overset{\displaystyle H}{C}}-\underset{\displaystyle H}{\overset{\displaystyle H}{C}}-\underset{\displaystyle H}{\overset{\displaystyle H}{C}}- \xrightarrow{H_2C{=}CH_2} \text{etc.}$$

Figure 2.12 The formation of polyethene from ethene

Table 2.1 Examples of Addition Polymers

Name	Structure of monomer	Structure of polymer	Uses
polyethene	$H_2C{=}CH_2$ ethene	$\cdots-\overset{H}{\underset{H}{C}}-\overset{H}{\underset{H}{C}}-\overset{H}{\underset{H}{C}}-\overset{H}{\underset{H}{C}}-\overset{H}{\underset{H}{C}}-\overset{H}{\underset{H}{C}}-\cdots$ polyethene	• plastic bags • plastic milk, juice, and water bottles • toys
polystyrene	$H_2C{=}CH-C_6H_5$ (benzene ring on CH) styrene	$\cdots-CH_2-CH(C_6H_5)-CH_2-CH(C_6H_5)-\cdots$ (benzene rings on CH) polystyrene	• styrene and Styrofoam™ cups • insulation • packaging
polyvinylchloride (PVC, vinyl)	$H_2C{=}CH-Cl$ vinyl chloride	$\cdots-CH_2-\underset{Cl}{CH}-CH_2-\underset{Cl}{CH}-\cdots$ polyvinylchloride (PVC)	• building and construction materials • sewage pipes • medical equipment
polyacrylonitrile	$H_2C{=}CH-CN$ acrylonitrile	$\cdots-\overset{H}{\underset{H}{C}}-\overset{H}{\underset{CN}{C}}-\overset{H}{\underset{H}{C}}-\overset{H}{\underset{CN}{C}}-\cdots$ polyacrylonitrile	• paints • yarns, knit fabrics, carpets, and wigs

Condensation polymerization is a reaction in which monomers are joined together by the formation of ester or amide bonds. A second smaller product, usually water, is produced by this reaction. For condensation polymerization to occur, each monomer must have two functional groups (usually one at each end of the molecule).

Nylon-66 is made by the condensation polymerization of the dicarboxylic acid adipic acid, and 1,6-diaminohexane, an amine. (The number 66 comes from the fact that each of the two reactants contains six carbon atoms.) This reaction results in the formation of amide bonds between monomers, as shown in Figure 2.13. Condensation polymers that contain amide bonds are called **nylons** or **polyamides**. Condensation polymers that contain ester bonds are called **polyesters**. Polyesters result from the esterification of diacids and dialcohols.

Go to the Chemistry 12 Electronic Learning Partner for more information about Nylon-66.

$$\underset{\text{1,6-diaminohexane}}{H-\overset{H}{N}-(CH_2)_6-\overset{H}{N}-H} \quad \underset{\text{adipic acid}}{HO-\overset{O}{\overset{\|}{C}}-(CH_2)_4-\overset{O}{\overset{\|}{C}}-OH} \quad \underset{\text{1,6-diaminohexane}}{H-\overset{H}{N}-(CH_2)_6-\overset{H}{N}-H}$$

$$\downarrow$$

$$\cdots-NH-(CH_2)_6-\underset{\text{amide bond}}{NH-\overset{O}{\overset{\|}{C}}}-(CH_2)_4-\underset{\text{amide bond}}{\overset{O}{\overset{\|}{C}}-NH}-(CH_2)_6-NH-\cdots + \underset{\text{water}}{H_2O}$$

Figure 2.13 **Nylon-66 is made from adipic acid and 1,6-diaminohexane.**

Table 2.2 shows two common polymers that are formed by condensation. Notice that Dacron™, a polyester, contains ester linkages between monomers. Nylon-6, a polyamide, contains amide linkages between monomers.

Table 2.2 Examples of Condensation Polymers

Name	Structure	Uses
Dacron™ (a polyester)	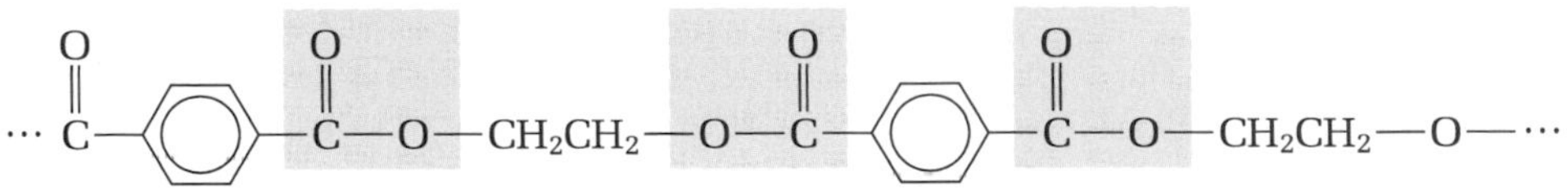 $\cdots\text{C(=O)}-\text{C}_6\text{H}_4-\text{C(=O)}-\text{O}-\text{CH}_2\text{CH}_2-\text{O}-\text{C(=O)}-\text{C}_6\text{H}_4-\text{C(=O)}-\text{O}-\text{CH}_2\text{CH}_2-\text{O}-\cdots$	• synthetic fibres used to make fabric for clothing and surgery
Nylon-6 (a polyamide)	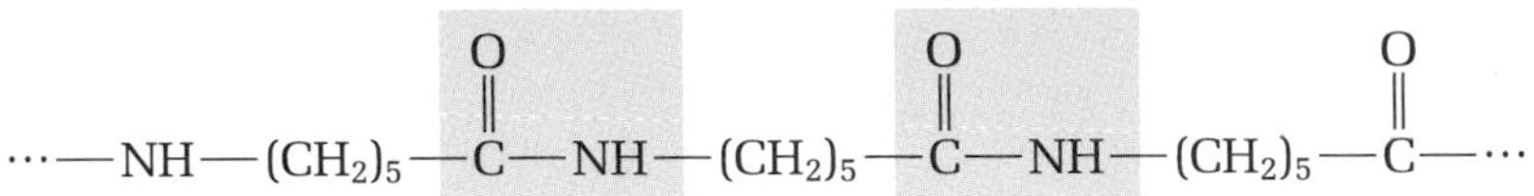 $\cdots-\text{NH}-(\text{CH}_2)_5-\text{C(=O)}-\text{NH}-(\text{CH}_2)_5-\text{C(=O)}-\text{NH}-(\text{CH}_2)_5-\text{C(=O)}-\cdots$	• tires • synthetic fibres used to make rope and articles of clothing, such as stockings

The following Sample Problem shows how to classify a polymerization reaction.

Sample Problem

Classifying a Polymerization Reaction

Problem

Tetrafluoroethene polymerizes to form the slippery polymer that is commonly known as Teflon™. Teflon™ is used as a non-stick coating in frying pans, among other uses. Classify the following polymerization reaction, and name the product. (The letter *n* indicates that many monomers are involved in the reaction.)

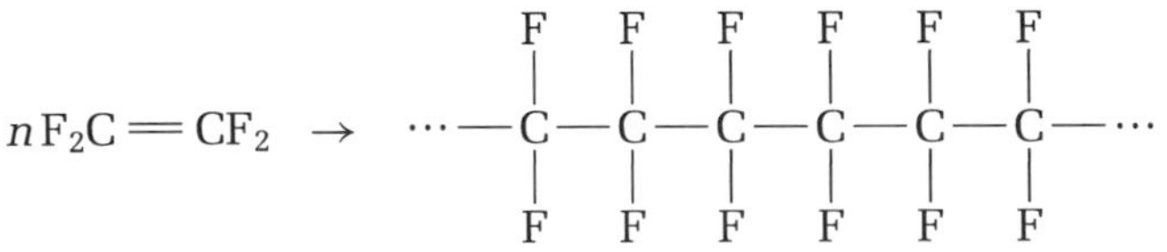

$$n\text{F}_2\text{C}=\text{CF}_2 \rightarrow \cdots-\text{CF}_2-\text{CF}_2-\text{CF}_2-\text{CF}_2-\text{CF}_2-\text{CF}_2-\cdots$$

Solution

The monomer reactant of this polymerization reaction contains a double bond. The product polymer has no double bond, so an addition reaction must have occurred. Thus, this reaction is an addition polymerization reaction. Since the monomer's name is tetrafluoroethene, the product's name is polytetrafluoroethene.

Practice Problems

18. A monomer called methylmethacrylate polymerizes to form an addition polymer that is used to make bowling balls. What is the name of this polymer?

19. Classify each polymerization reaction as an addition or condensation polymerization reaction.

(a)

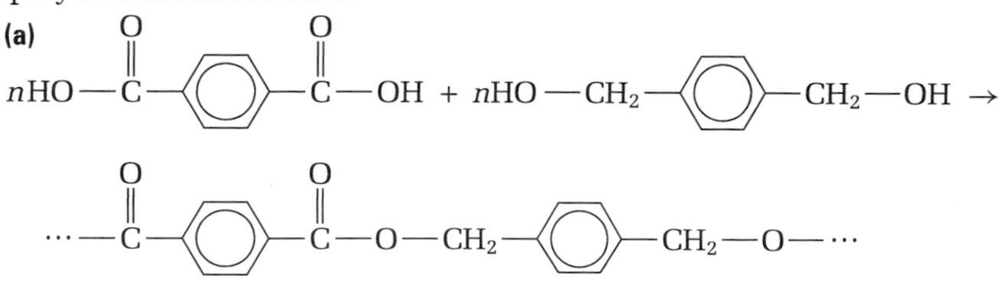

(b) $n\text{CH}_2\!=\!\text{CH(CN)} \rightarrow \cdots\!-\!\text{CH}_2\!-\!\text{CH(CN)}\!-\!\text{CH}_2\!-\!\text{CH(CN)}\!-\!\cdots$

(c) $n\text{HO}\!-\!\text{CH}_2\!-\!\text{C(=O)}\!-\!\text{OH} \rightarrow \cdots\!-\!\text{O}\!-\!\text{CH}_2\!-\!\text{C(=O)}\!-\!\text{O}\!-\!\text{CH}_2\!-\!\text{C(=O)}\!-\!\cdots$

20. Draw the product of each polymerization reaction. Include at least two linkages for each product.

(a) $n\text{HO}\!-\!\text{CH}_2\text{CH}_2\text{CH}_2\!-\!\text{OH} + n\text{HO}\!-\!\text{C(=O)}\!-\!\text{CH}_2\!-\!\text{C(=O)}\!-\!\text{OH} \rightarrow$

(b) $n\text{H}_2\text{C}\!=\!\text{CH(CH}_3) \rightarrow$

(c) $n\text{H}_2\text{NCH}_2\!-\!\text{C}_6\text{H}_4\!-\!\text{CH}_2\text{NH}_2 + n\text{HO}\!-\!\text{C(=O)(CH}_2)_6\text{C(=O)}\!-\!\text{OH} \rightarrow$

21. Classify each polymer as an *addition polymer* (formed by addition polymerization) or a *condensation polymer* (formed by condensation polymerization). Then classify each condensation polymer as either a polyester or a nylon (polyamide).

(a) $\cdots\!-\!\text{CH}_2\!-\!\text{CH(Br)}\!-\!\text{CH}_2\!-\!\text{CH(Br)}\!-\!\cdots$

(b) $\cdots\!-\!\text{NH}\!-\!\text{CH}_2\!-\!\text{NH}\!-\!\text{C(=O)}\!-\!\text{CH}_2\text{CH}_2\!-\!\text{C(=O)}\!-\!\text{NH}\!-\!\text{CH}_2\!-\!\text{NH}\!-\!\cdots$

(c) $\cdots\!-\!\text{O}\!-\!\text{CH}_2\text{CH}_2\!-\!\text{C(=O)}\!-\!\text{O}\!-\!\text{CH}_2\text{CH}_2\!-\!\text{C(=O)}\!-\!\text{O}\!-\!\text{CH}_2\text{CH}_2\!-\!\text{C(=O)}\!-\!\cdots$

(d) $\cdots\!-\!\text{O}\!-\!\text{C}_6\text{H}_4\!-\!\text{O}\!-\!\text{C(=O)CH}_2\text{C(=O)}\!-\!\text{O}\!-\!\text{C}_6\text{H}_4\!-\!\text{O}\!-\!\text{C(=O)CH}_2\text{C(=O)}\!-\!\cdots$

22. Draw the structure of the repeating unit for each polymer in question 21. Then draw the structure of the monomer(s) used to prepare each polymer.

Electronic Learning Partner

Go to the Chemistry 12 Electronic Learning Partner for more information about aspects of material covered in this section of the chapter.

In the following investigation, you will examine three polymers and synthesize a cross-linked addition polymer.

Investigation 2-B

SKILL FOCUS

Performing and recording

Analyzing and interpreting

Synthesis of a Polymer

In this investigation, you will examine three different polymers. First, you will examine the addition polymer sodium polyacrylate. This polymer contains sodium ions trapped inside the three-dimensional structure of the polymer. When placed in distilled water, the concentration of sodium ions inside the polymer is much greater than the concentration of sodium ions outside the polymer. The concentration imbalance causes water molecules to move by diffusion into the polymer. As a result, the polymer absorbs many times its own mass in distilled water.

$$\cdots-CH_2-\underset{\underset{\underset{O^-Na^+}{|}}{\underset{|}{C=O}}}{CH}-CH_2-\underset{\underset{\underset{O^-Na^+}{|}}{\underset{|}{C=O}}}{CH}-\cdots$$

sodium polyacrylate

You will also examine polyvinyl alcohol, another addition polymer. Polyvinyl alcohol is made by the polymerization of hydroxyethene, also known as vinyl alcohol.

$$\cdots-CH_2-\underset{\underset{OH}{|}}{CH}-CH_2-\underset{\underset{OH}{|}}{CH}-\cdots$$

polyvinyl alcohol

Using an aqueous solution of polyvinyl alcohol, you will prepare a cross-linked polymer, commonly known as "Slime."

Questions

How much water can sodium polyacrylate absorb? What properties do polymers have in common? How do polymers differ?

Prediction

How can you tell if polymerization has occurred? List one or more changes you would expect to see in a liquid solution as polymerization takes place.

Materials

Part 1

about 0.5 g sodium polyacrylate powder
distilled water
500 mL beaker
100 mL graduated cylinder
sodium chloride
balance

Part 2

10 mL graduated cylinder
50 mL beaker
20 cm × 20 cm piece of a polyvinyl alcohol bag (Alternative: Your teacher will prepare a solution of 40 g polyvinyl alcohol powder in 1 L of water. You will be given 20 mL of this solution.)
25 mL hot tap water
stirring rod
food colouring
5 mL of 4% borax solution (sodium tetraborate decahydrate)

Safety Precautions

- Wear an apron, safety glasses, and gloves while completing this investigation.
- Wash your hands thoroughly after this investigation.
- Sodium polyacrylate is irritating to eyes and nasal membranes. Use only a small amount, and avoid inhaling the powder.

Procedure

Part 1 Examination of a Water-Absorbent Polymer

1. Examine the sodium polyacrylate powder. Record your observations.
2. Record the mass of a dry, empty 500 mL beaker. Place about 0.5 g of the powder in the bottom of the beaker. Record the mass of the beaker again, to find the exact mass of the powder.
3. Use a graduated cylinder to measure 100 mL of distilled water. Very slowly, add the water to the beaker. Stop frequently to observe what happens. Keep track of how much water you add, and stop when no more water can be absorbed. (You may need to add less or more than 100 mL.) Record the volume of water at which the powder becomes saturated and will no longer absorb water.
4. Add a small amount of sodium chloride to the beaker. Record your observations. Continue adding sodium chloride until no more changes occur. Record your observations.
5. Dispose of excess powder and the gelled material as directed by your teacher. Do not put the powder or gel down the sink!

Part 2 Synthesis of a Cross-Linked Addition Polymer

1. Before starting, examine the polyvinyl alcohol bag. Record your observations.
2. Place 25 mL of hot tap water into a small beaker.
3. Add the piece of polyvinyl alcohol bag to the hot water. Stir the mixture until the polymer has dissolved.
4. Add a few drops of food colouring to the mixture, and stir again.
5. Add 5 mL of the borax solution, and stir.
6. Examine the "Slime" you have produced. Record your observations.

Analysis

1. Calculate the mass of water that was absorbed by the sodium polyacrylate polymer. (**Note:** 1 mL of water = 1 g)
2. Use the mass of the sodium polyacrylate polymer and the mass of the water it absorbed to calculate the mass/mass ratio of the polymer to water.
3. What practical applications might the sodium polyacrylate polymer be used for?
4. What happened when you added the salt? Come up with an explanation for what you observed. Search for a discussion of sodium polyacrylate on the Internet to check your explanation.

Conclusions

5. How do you think a plastic bag made of polyvinyl alcohol might be useful?
6. What change(s) alerted you to the fact that polymerization was occurring when you added borax to the dissolved solution of polyvinyl alcohol? Explain your observations.
7. Compare the properties of the three polymers you observed in this investigation. Were there any similarities? How were they different?

Applications

8. Sodium polyacrylate absorbs less tap water than distilled water. Design and carry out an investigation to compare the mass of tap water absorbed to the mass of distilled water absorbed by the same mass of powder.
9. Research the use of polyvinyl alcohol (PVA) bags in hospitals.

Risks of the Polymer Industry

In the 1970s, workers at an American plastics manufacturing plant began to experience serious illnesses. Several workers died of liver cancer before the problem was traced to its source: prolonged exposure to vinyl chloride. Vinyl chloride (or chloroethene, C_2H_3Cl) is used to make polyvinylchloride (PVC). Unfortunately, vinyl chloride is a powerful carcinogen. Government regulations have now restricted workers' exposure to vinyl chloride. Trace amounts of this dangerous chemical are still present, however, as pollution in the environment.

The manufacture and disposal of PVC creates another serious problem. Dioxin, shown in Figure 2.14, is a highly toxic chemical. It is produced during the manufacture and burning of PVC plastics.

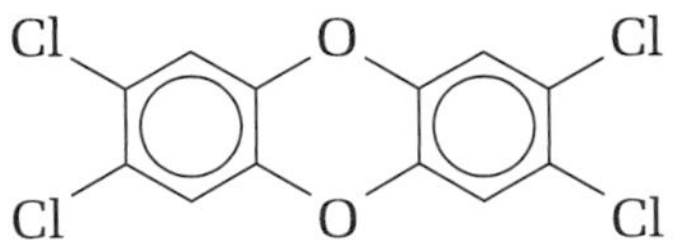

Figure 2.14 Dioxin has been shown to be extremely carcinogenic to some animals. As well, it is suspected of disrupting hormone activity in the body, leading to reduced fertility and birth defects.

Many synthetic polymers, including most plastics, do not degrade in the environment. What can be done with plastic and other polymer waste? One solution may be the development and use of *degradable plastics*. These are polymers that break down over time when exposed to environmental conditions, such as light and bacteria. The Chemistry Bulletin on the next page gives more information about degradable plastics.

Natural Polymers

Much of our technology has been developed by observing and imitating the natural world. Synthetic polymers, such as those you just encountered, were developed by imitating natural polymers. For example, the natural polymer *cellulose* provides most of the structure of plants. Wood, paper, cotton, and flax, are all composed of cellulose fibres. Figure 2.15 shows part of a cellulose polymer.

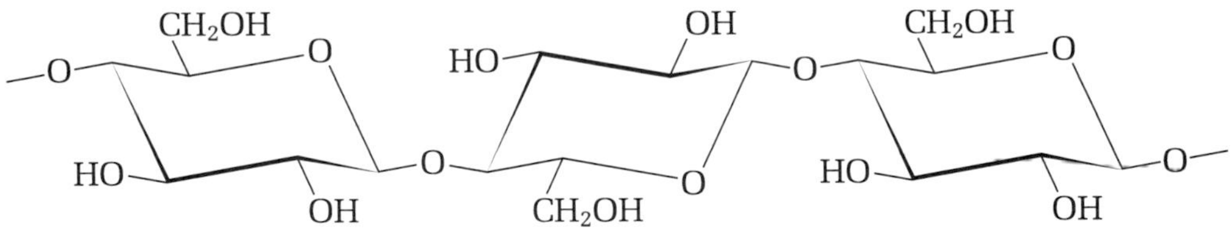

Figure 2.15 Like other polymers, cellulose is made of repeating units. It is the main structural fibre in plants, and it makes up the fibre in your diet.

Biochemistry is the study of the organic compounds and reactions that occur in living things. Proteins, starches, and nucleic acids are important biological polymers. Some biological monomers, like amino acids and sugars, are biologically important all on their own. You will also encounter lipids, a class of large biological molecules that are not polymers.

Chemistry Bulletin

Science | Technology | Society | Environment

Degradable Plastics: Garbage That Takes Itself Out

Did you know that the first plastics were considered too valuable to be thrown out? Today, many plastics are considered disposable. Plastics now take up nearly one third of all landfill space, and society's use of plastics is on the rise. Recycling initiatives are helping to reduce plastic waste. Another solution to this problem may involve the technology of degradable plastics.

All plastics are polymers, long molecules made of repeating units. Because polymers are extremely stable, they resist reaction. Most conventional plastics require hundreds or even thousands of years to break down.

Degradable plastics are an important step forward in technology. These new plastics can break down in a relatively short time—within six months to a year. There are two main kinds of degradable plastics: biodegradable plastics and photodegradable plastics. *Biodegradable plastics* are susceptible to decay processes that are similar to the decay processes of natural objects, such as plants. *Photodegradable plastics* disintegrate when they are exposed to certain wavelengths of light.

The ability of a degradable plastic to decay depends on the structure of its polymer chain. Biodegradable plastics are often manufactured from natural polymers, such as cornstarch and wheat gluten. Micro-organisms in the soil can break down these natural polymers. Ideally, a biodegradable plastic would break down completely into carbon dioxide, water, and biomass within six months, just like a natural material.

Photodegradable plastic is a Canadian contribution to the environmental health of the planet. It was developed in Canada more than 25 years ago by Dr. James Guillet, of the University of Toronto. This plastic incorporates light-absorbing, or *chromomorphic*, molecular groups as part of its backbone chain. The chromomorphic groups change their structure when they are exposed to particular frequencies of light (usually ultraviolet). The structural changes in these groups cause the backbone to break apart. Although the photodegradable process is not very effective at the bottom of a landfill, it is ideally suited to the marine environment. Plastics that are discarded in oceans and lakes usually float at the surface, and thus are exposed to light. The rings that are used to package canned drinks can strangle marine wildlife. These rings are now made with photodegradable technology in many countries.

Several nations are conducting further research into other marine uses for this technology.

Making Connections

1. When a degradable plastic decays, the starch or cellulose backbone may break down completely into carbon dioxide, water, and organic matter. What do you think happens to F, Cl, or other heteroatom groups on the polymer chain as the backbone decays? Why might this be a problem?
2. Several notable scientists have cautioned against the use of biodegradable and photodegradable plastics in landfills. The bottom of a landfill has no light, or oxygen for the metabolism of bacteria. Why is this a problem? What changes must be made to landfill design, if biodegradable and photodegradable plastics are going to be included in landfills?
3. Do degradable plastics disappear as soon as they are discarded? What are the dangers of public confidence in the degradability of plastics?

Amino Acids and Proteins

A **protein** is a natural polymer that is composed of monomers called **amino acids**. Proteins are found in meat, milk, eggs, and legumes. Wool, leather, and silk are natural materials that are made of proteins. Your fingernails, hair, and skin are composed of different proteins. Proteins carry out many important functions in your body, such as speeding up chemical reactions (enzymes), transporting oxygen in your blood (hemoglobin), and regulating your body responses (hormones).

There are 20 common amino acids in the human body. All amino acids contain a carboxylic acid group and an amino group. Amino acids can be linked by amide bonds to form proteins. Each amino acid has a different side chain, which is attached to the centre carbon atom. Figure 2.16, below, shows the structure of an amino acid. The letter G represents the side chain of the amino acid. Examples of side chains include $—CH_3$ (for the amino acid alanine), $—CH_2CH_2CONH_2$ (for glutamine), and $—CH_2OH$ (for serine).

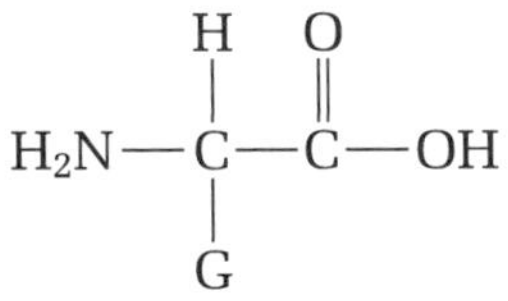

Figure 2.16 The structure of an amino acid

Biology LINK

Studies have shown that honey has the same quick absorption and long-lasting energy-burst effect that many sports drinks do. This is because honey is high in fructose, a simple fruit sugar that has been found to be the best sugar for athletes. Unpasteurized honey also contains hydrogen peroxide and enzymes, which aid in digestion.

The shape and biological function of a protein depends on its sequence of amino acids. A typical protein, such as insulin, contains at least 50 amino acid groups. Depending on the sequence of these amino acids, an infinite number of different proteins are possible. Each species of animal has its own distinct proteins. The DNA in your body contains the blueprints for making specific proteins for your body's structure and function.

Carbohydrates

A **carbohydrate** (also called a **saccharide**) is a biological molecule that contains either an aldehyde or ketone group, along with two or more hydroxyl groups. The C=O group at one end of a simple saccharide reacts with an OH group at the other end to produce the most common formation, a ring. Figure 2.17 shows the ring forms of some of the simplest carbohydrates.

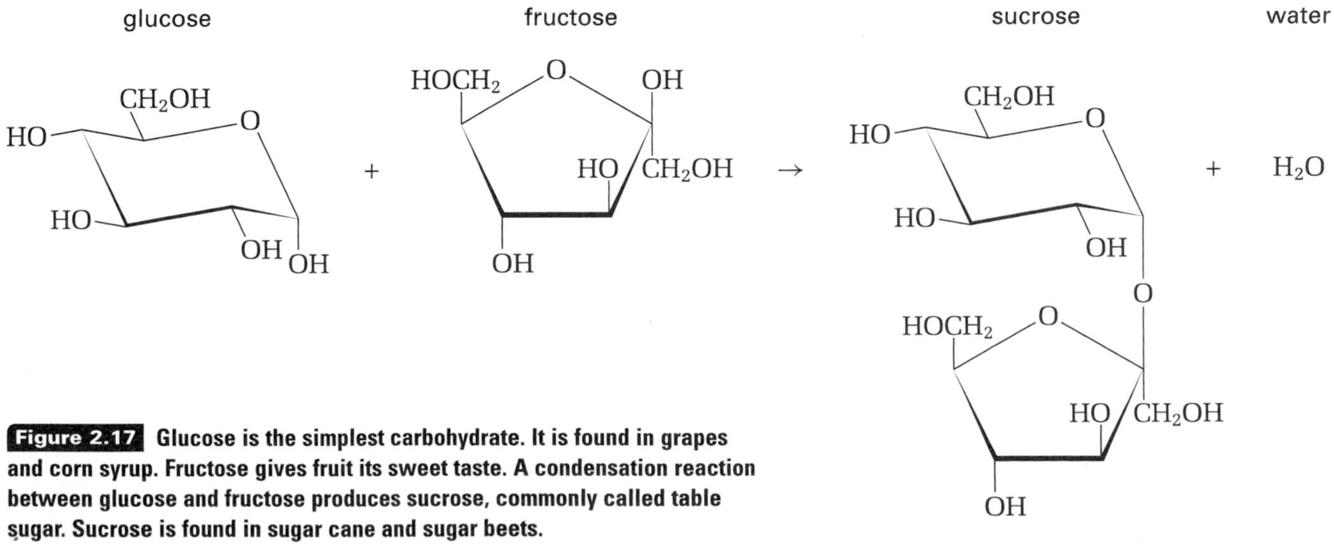

Figure 2.17 Glucose is the simplest carbohydrate. It is found in grapes and corn syrup. Fructose gives fruit its sweet taste. A condensation reaction between glucose and fructose produces sucrose, commonly called table sugar. Sucrose is found in sugar cane and sugar beets.

Carbohydrates are found in foods such as bread, pasta, potatoes, and fruits. They are the primary source of energy for the body. In a process called *cellular respiration*, carbohydrates combine with inhaled oxygen and are oxidized to produce carbon dioxide and water, plus energy. As shown in Figure 2.18, carbon dioxide and water vapour from cellular respiration are expelled in your breath.

Figure 2.18 **On a cold day, the water vapour in your breath condenses on contact with the air, forming a visible cloud.**

A **monosaccharide** is composed of one saccharide unit. Thus, it is the smallest molecule possible for a carbohydrate. A monosaccharide is often called a simple carbohydrate, or simple sugar. Glucose, seen in Figure 2.17, on the previous page, is one example of a monosaccharide. Other monosaccharides include fructose, galactose, and mannose.

Sucrose (shown in Figure 2.17) is a **disaccharide**, containing two saccharide monomers. To produce a disaccharide, an —OH group on one monomer reacts with an —OH group on another monomer to form an ether —O— linkage (also shown in Figure 2.17). This reaction is a condensation, because two molecules combine and produce a molecule of water. During digestion, carbohydrates undergo hydrolysis reactions, which break the ether bonds between the monomers.

Polysaccharides are polymers that contain many saccharide units. The most important polysaccharides are those in which glucose is the monomer. Plants use the glucose polysaccharide *cellulose* as a structural material. (See Figure 2.15 for the structure of cellulose.) They store energy in the starches *amylose* and *amylopectin*. A **starch** is a glucose polysaccharide that is used by plants to store energy. Humans can digest starches. Animals store energy in the glucose polysaccharide *glycogen*, illustrated in Figure 2.19

COURSE CHALLENGE

How can you use your knowledge of organic chemistry to identify reactions between biological molecules? You will apply your learning later on, in the Chemistry Course Challenge.

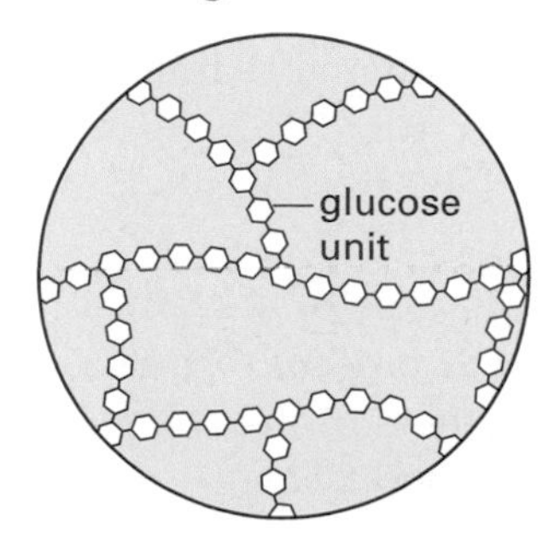

Figure 2.19 **Glycogen is the major storage polysaccharide in animals. This tiny portion of a tangled glycogen granule shows the highly branched polymer, made of glucose units.**

Nucleotides and Nucleic Acids

Humans and cats both begin life as fertilized eggs, which look fairly similar. Why does a fertilized cat egg develop into a kitten, while a fertilized human egg develops into a child (Figure 2.20)? A very special type of polymer, located in the cell nucleus, directs the formation of the proteins from which an organism develops.

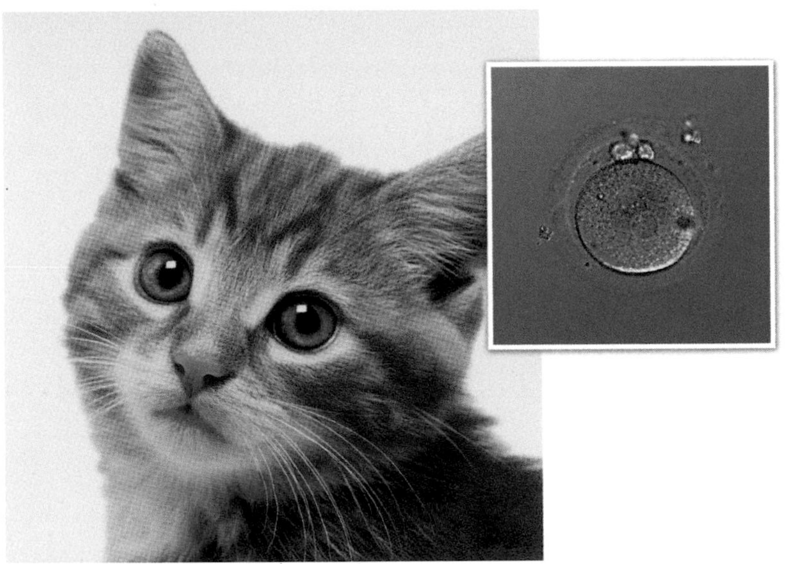

Figure 2.20 DNA in the cells of your body direct the production of proteins. Since your bones, muscles, and organs are built from proteins that are specific to humans, your body is different from the body of a cat.

DNA (short for **2-deoxyribonucleic acid**) is the biological molecule that determines the shape and structure of all organisms. It is found mostly in the nuclei of cells. Each strand of DNA is a polymer that is composed of repeating units, called **nucleotides**. A single strand of DNA may have more than one million nucleotides. Each nucleotide consists of three parts: a sugar, a phosphate group, and a base.

DNA is one type of nucleic acid. The other type of nucleic acid is called **RNA** (short for **ribonucleic acid**). RNA is present throughout a cell. It works closely with DNA to produce the proteins in the body. Table 2.3, on the next page, shows the structures of RNA and DNA.

Table 2.3 Structures of DNA and RNA

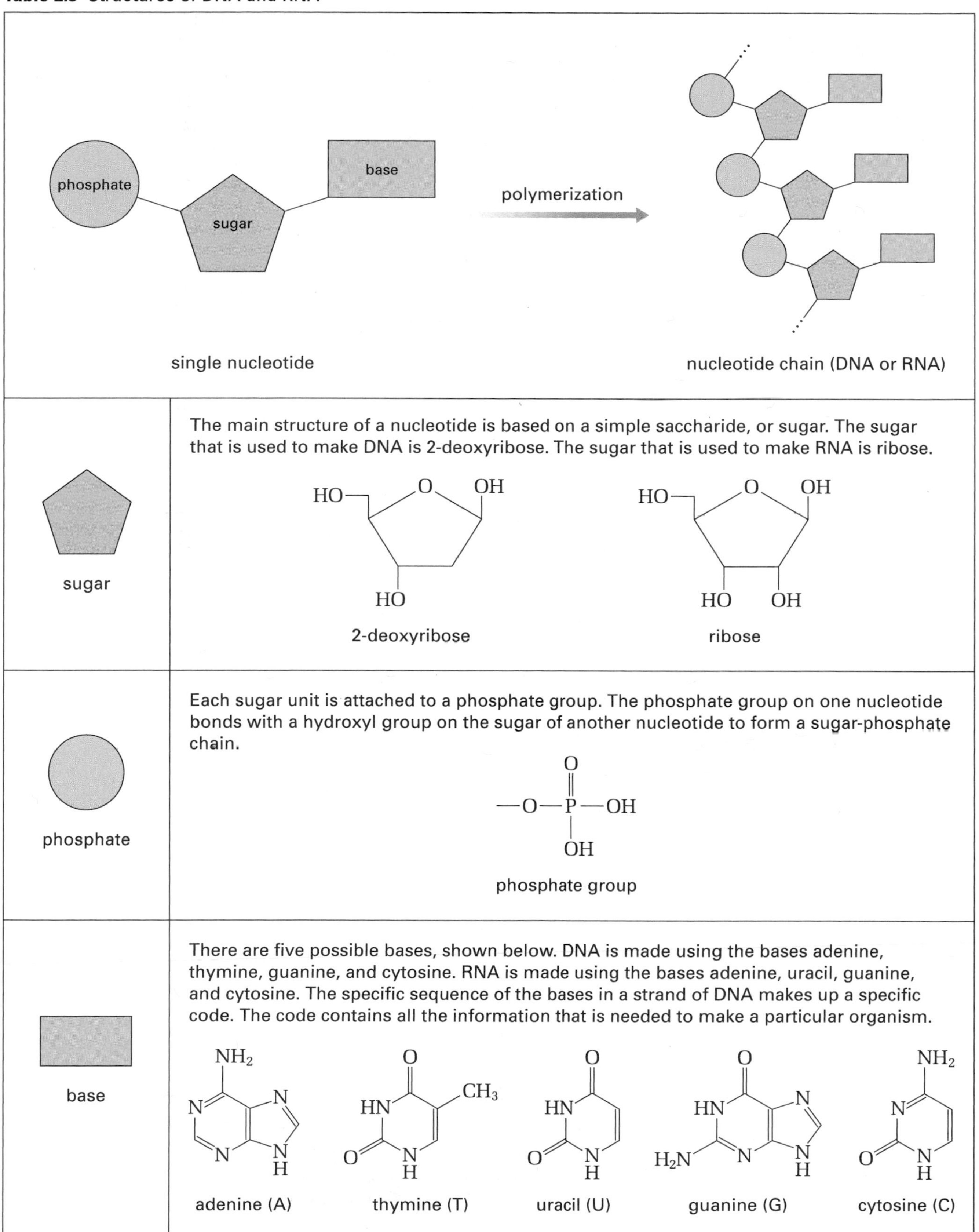

phosphate / sugar / base — single nucleotide	polymerization → nucleotide chain (DNA or RNA)
sugar	The main structure of a nucleotide is based on a simple saccharide, or sugar. The sugar that is used to make DNA is 2-deoxyribose. The sugar that is used to make RNA is ribose. 2-deoxyribose ribose
phosphate	Each sugar unit is attached to a phosphate group. The phosphate group on one nucleotide bonds with a hydroxyl group on the sugar of another nucleotide to form a sugar-phosphate chain. phosphate group
base	There are five possible bases, shown below. DNA is made using the bases adenine, thymine, guanine, and cytosine. RNA is made using the bases adenine, uracil, guanine, and cytosine. The specific sequence of the bases in a strand of DNA makes up a specific code. The code contains all the information that is needed to make a particular organism. adenine (A) thymine (T) uracil (U) guanine (G) cytosine (C)

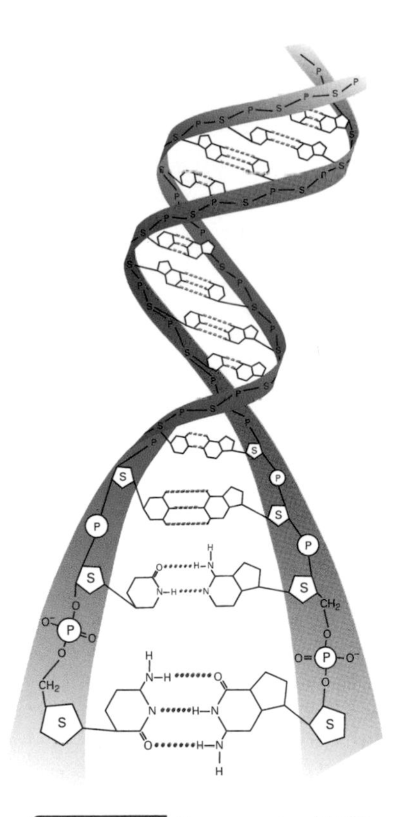

Figure 2.21 Two strands of DNA are held together by hydrogen bonds to form a double-helix molecule.

A complete molecule of DNA consists of two complementary strands. These strands are held together by hydrogen bonds between the bases on the nucleotide units, as shown in Figure 2.21.

DNA has two main functions in the body:

- It replicates itself exactly whenever a cell divides. Thus, every cell in an organism contains the same information.
- It contains codes that allow each type of protein that is needed by the body to be made by the body's cells. RNA assists in this process.

Lipids

The last biological molecules that you will examine in this course are not polymers. They are, however, very large in size. **Lipids** are defined as biological molecules that are not soluble in water, but are soluble in non-polar solvents, such as benzene and hexanes.

From your previous studies of solubility, you might hypothesize that lipid molecules must have large hydrocarbon parts, since they are soluble in hydrocarbon solvents. Your hypothesis would be correct. Lipids contain large hydrocarbon chains, or other large hydrocarbon structures.

Several types of lipids may be familiar to you. Fats, oils, and some waxes are examples of lipids. **Fats** are lipids that contain a glycerol molecule, bonded by ester linkages to three long-chain carboxylic acids. **Oils** have the same structure as fats, but are classified as being liquid at room temperature, while fats are solids at room temperature. **Waxes** are esters of long-chain alcohols and long-chain carboxylic acids.

Other lipids include steroids, such as cholesterol, and terpenes, which are plant oils such as oil of turpentine, or oil of cedar. Figure 2.22 shows the lipid *glyceryl trioleate*, which is present in olive oil.

$$
\begin{array}{l}
\mathrm{H_2C{-}O{-}C(=O){-}(CH_2)_7{-}CH{=}CH{-}(CH_2)_7{-}CH_3} \\
\mathrm{\;\;|} \\
\mathrm{HC{-}O{-}C(=O){-}(CH_2)_7{-}CH{=}CH{-}(CH_2)_7{-}CH_3} \\
\mathrm{\;\;|} \\
\mathrm{H_2C{-}O{-}C(=O){-}(CH_2)_7{-}CH{=}CH{-}(CH_2)_7{-}CH_3}
\end{array}
$$

glycerol　　3 fatty acids

glyceryl trioleate

Figure 2.22 A fat or an oil is produced when three long-chain carboxylic acids, called *fatty acids*, bond with a glycerol unit to form three ester bonds. (Note that the three fatty acids are not always identical.) A hydrolysis reaction can split these three ester bonds to produce glycerol and a combination of the salts of fatty acids, better known as soap.

Lipids carry out many important functions in the body. Their most important function is the long-term storage of energy. One gram of fat contains two-and-a-quarter times more energy than a gram of carbohydrate or protein. When an animal consumes more carbohydrates than are needed at the time, the body converts the excess carbohydrates to fat. Later, when insufficient carbohydrates are available for energy, the body may break down some of these fats to use for energy. Carbohydrates are still needed in your diet, however, for the proper functioning of your cells.

In addition to storing energy, lipids have other vital functions:

- Cell membranes are made of a double layer of *phospholipids*. These are lipids that are polar at one end and non-polar at the other end.
- Lipids such as cholesterol and testosterone act as hormones to regulate body functions. (Not all hormones are lipids. Some hormones are proteins. Other hormones are neither lipids nor proteins.)
- Several vitamins (such as vitamins A, D, and E) are lipids. These vitamins are known as *fat-soluble*.
- The fatty tissues in your body are made from lipids. They act like protective "packaging" around fragile organs, such as the heart. They insulate the body from excessive heat or cold, and they act as padding to reduce impact from collisions.

Section Summary

In this section, you learned how to recognize addition and condensation polymerization reactions. You examined the structures and functions of several important biological molecules, such as proteins, amino acids, carbohydrates, DNA, and lipids. In the next section, you will examine the risks and benefits of manufacturing and using organic compounds.

Section Review

1 K/U In this section, you learned about biological polymers.

(a) Name three types of biological polymers.

(b) Give an example of each type of biological polymer.

(c) Name the monomer unit for each type of biological polymer.

2 K/U Classify each polymerization reaction as an addition reaction or a condensation reaction.

(a) $nH_2C{=}CH(F) \rightarrow \cdots{-}CH_2{-}CH(F){-}CH_2{-}CH(F){-}\cdots$

(b) $nHO{-}C_6H_4{-}OH + nHO{-}C(=O){-}(CH_2)_4{-}C(=O){-}OH \rightarrow$

$\cdots{-}O{-}C_6H_4{-}O{-}C(=O){-}(CH_2)_4{-}C(=O){-}O{-}C_6H_4{-}O{-}\cdots$

3 I Draw the product of each polymerization reaction, and classify the reaction. Then circle and identify any amide or ester bonds in the product.

(a) $nH_2C{=}CH_2 \rightarrow$

(b) $nHO{-}CH_2{-}OH + nHO{-}C(=O){-}C_6H_4{-}C(=O){-}OH \rightarrow$

(c) $nHO{-}C(=O){-}(CH_2)_3{-}C(=O){-}OH + nH_2N{-}CH(CH_3){-}NH_2 \rightarrow$

Unit Issue Prep

Before you research your issue on organic chemistry, think about the importance of organic compounds in everyday life. How are polymers important in medicine? How are they important in industry?

4 C What function does each type of biological molecule perform in living organisms?

(a) carbohydrate
(b) protein
(c) DNA
(d) lipid

5 I Identify each type of biological molecule. Copy the structure into your notebook, and circle the functional group(s).

(a) $H_2N—CH(CH_3)—C(=O)—OH$

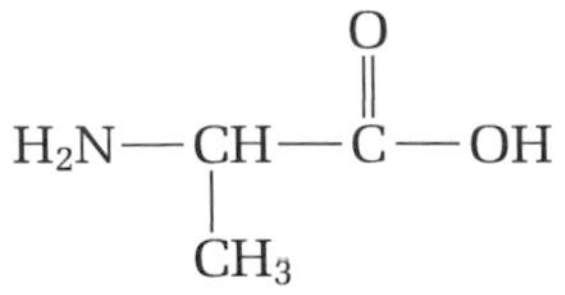

(b) HO, O, OH, OH, CH_2OH, OH

(c) $H_2N—CH(CH_2C_6H_5)—C(=O)—OH$

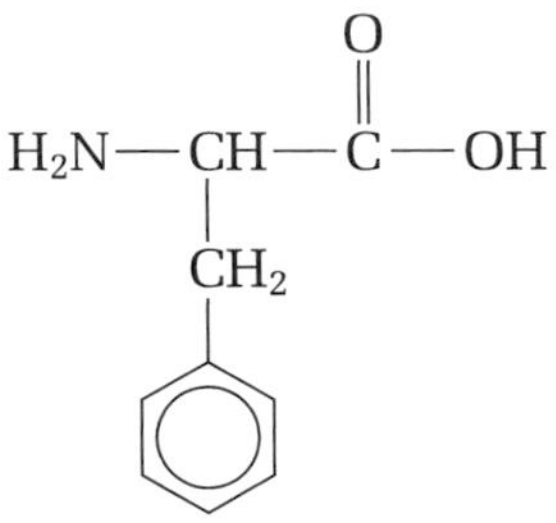

(d) $H_2C—O—C(=O)—(CH_2)_{14}—CH_3$
$HC—O—C(=O)—(CH_2)_{12}—CH_3$
$H_2C—O—C(=O)—(CH_2)_{16}—CH_3$

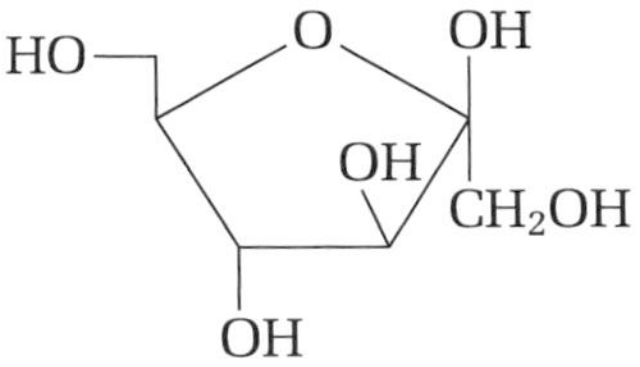

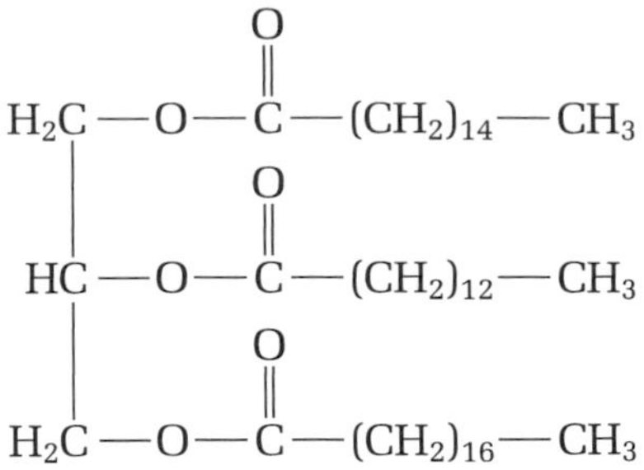

(e) CH_2OH, O, HO, HO, OH, HO, O, O, OH, CH_2OH, OH

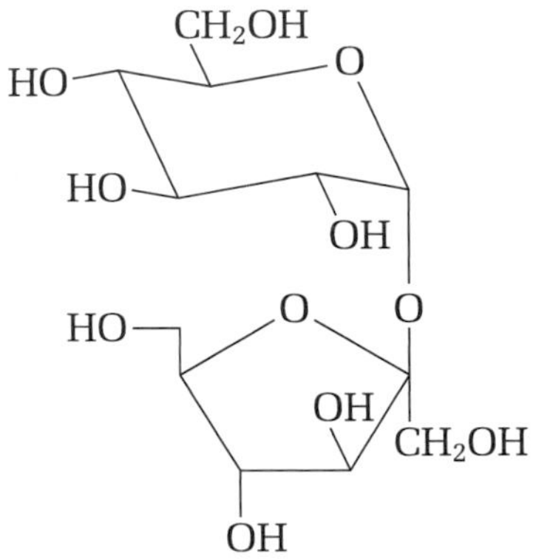

6 MC Gore-tex™ is a popular fabric that provides protection from rain while allowing perspiration to escape. What polymer coating is used to make Gore-tex™ fabric? What special properties does this polymer have? What else is this polymer used for? Research the answers to these questions.

Organic Compounds and Everyday Life

2.4

Section Preview/ Specific Expectations

In this section, you will

- **discuss** the variety and importance of organic compounds in your life
- **analyze** the risks and benefits of the production and use of organic compounds
- **explain** how organic chemistry has helped to solve problems related to human health and the environment

You are already aware that organic compounds are an inescapable part of your daily life. As you know, much of your body is composed of organic compounds. The food you eat is made from organic compounds, as are the clothes you wear. The fuel that heats your home and powers buses and cars is also organic. When you are sick or injured, you may use organic compounds to fight infection or to reduce pain and swelling. Modern processed foods include many flavourings and colourings—another use of organic compounds. (See Figure 2.23.)

Figure 2.23 In Chapter 1, you learned about esters that are used as flavourings. Not all flavourings belong to the ester family. Menthol is an organic compound that is found in the mint plant. It is used as a flavouring in many substances, including toothpaste, chewing gum, and cough syrups. What functional group is present in menthol? How might you guess this from the name?

In this section, you will examine the uses of some common organic compounds. You will also carry out an analysis of some of the risks and benefits that are involved in making and using organic compounds.

Risk-Benefit Analysis

How can you make the most informed decision possible when analyzing the risks or benefits of a product or issue? The following steps are one way to carry out a risk-benefit analysis for the manufacture and use of an organic compound.

Step 1 Identify possible risks and benefits of the organic product. Ask questions such as these:

- What are the direct uses of this product? (For example, is it used as a pharmaceutical drug for a specific disease or condition? Is it an insecticide for a specific insect "pest"?)
- What are the indirect uses of this product? (For example, is it used in the manufacture of a different organic or inorganic product?)
- What are the direct risks of making or using this product? (For example, is the product an environmental pollutant, or does it contain compounds that may be harmful to the environment? Can the workers who manufacture this product be harmed by exposure to it?)

- What are the indirect risks of making or using this product? (For example, are harmful by-products produced in the manufacture of this product? How much energy does the manufacture of this product require, and where does the energy come from? What, if any, side effects does this product cause?)

Step 2 Research the answers to your questions from step 1. Ensure that your information comes from reliable sources. If there is controversy on the risks and benefits of the product, find information that covers both sides of the issue.

Step 3 Decide on your own point of view. In your opinion, are the benefits greater than the risks? Do the risks outweigh the benefits?

Step 4 Consider possible alternatives that may lessen the risks. Ask questions such as the following:
- Does a similar, less harmful, product exist?
- Is there a better way to manufacture or dispose of this product?
- What safety precautions can be taken to reduce the risks?

Risks and Benefits of Organic Compounds

The next four tables outline common uses of various organic compounds, as well as possible complications resulting from each use. The ThoughtLab that follows provides an opportunity for you to work with and extend the information in the tables. Of course, the compounds listed are only a few of the many useful organic compounds that are available to society.

Table 2.4 Uses and Possible Complications of Selected Pharmaceutical Drugs

Organic compound	Source	Common uses	Side effects and/or concerns
salbutalmol (Ventolin®)	• developed from ephedrine, a compound that occurs naturally in the ma huang (ephedra) plant	• used as asthma medication	• possible side effects include tremors, nausea, heart palpitations, headaches, and nervousness
lidocaine	• developed from cocaine, a compound that occurs naturally in the coca leaf	• used as a local anaesthetic, in throat sprays and sunburn sprays	• possible side effects include skin rash or swelling
morphine	• occurs naturally in the opium poppy plant, which is still the simplest and cheapest source of morphine	• one of the strongest pain-killers known	• highly addictive • possible side effects include dizziness, drowsiness, and breathing difficulty
acetylsalicylic acid (A.S.A., Aspirin®)	• developed from salicin, found in the bark of the willow tree	• reduces fever, inflammation, and pain • used to prevent heart attacks by thinning blood	• possible side effects include stomach upset • more rare side effects include stomach bleeding and Reye's syndrome
acetaminophen (Tylenol™)	• developed from acetyl salicylic acid	• reduces fever and pain	• possible side effects include stomach upset • can cause liver damage in high doses

Table 2.5 Uses and Possible Complications of Selected Pesticides

Pesticide	Common uses	Side effects and/or concerns
parathion	• insecticide • kills insects by disrupting their nerve impulses	• one of the most toxic organophosphate pesticides • highly toxic to humans, birds, aquatic invertebrates, and honeybees
carbaryl	• insecticide • similar to parathion, but safer for humans	• more expensive than parathion • generally low toxicity to humans and most other mammals • highly toxic to fish, aquatic invertebrates, and honey bees
endrin	• insecticide • also controls rodents, such as mice	• persists in soil up to 12 years • highly toxic to fish • banned in some countries and restricted use in many others
pyrethrin	• natural pesticide • breaks down quickly in the environment	• irritating to skin, eyes, respiratory system • highly toxic • can be dangerous to humans and pets within range of the pesticide application • very toxic to aquatic life
resmethrin	• synthetic pyrethroid pesticide • 75 times less toxic to humans and pets than pyrethrin	• low toxicity to humans and other mammals (causing skin, eye, and respiratory system irritation) • irritating to skin, eyes, and respiratory system • highly toxic to fish and honey bees

Table 2.6 Uses and Possible Complications of Selected Food Additives

Organic Compound	Common uses	Side effects and/or concerns
aspartame	• artificial sweetener	• significant controversy over possible health risks
menthol	• flavouring in many substances, including toothpaste, chewing gum, and cough syrups	• may cause allergic reaction
MSG (monosodium-glutamate)	• flavour enhancer for foods	• may cause allergic reaction known as MSG symptom complex, with symptoms such as nausea and headache • may worsen already severe asthma
Red #40	• food colouring	• may cause allergic reaction

Table 2.7 Uses and Possible Complications of Selected Common Organic Compounds

Organic compound	Common uses	Side effects and/or concerns
tetrachloroethene (Perc™)	• solvent used to dry-clean clothing	• toxic • may damage central nervous system, kidneys, and liver
ethylene glycol	• automobile antifreeze	• toxic, especially to small pets and wildlife • may damage central nervous system, kidneys, and heart
acetone	• nail polish remover • industrial solvent	• highly flammable • irritating to throat, nose, and eyes

ThoughtLab  Risk-Benefit Analyses of Organic Products

As you know, organic products provide many of the necessities and comforts of your life. They are also responsible for a great deal of damage, including environmental pollution and harm to human health. In this Thought Lab, you will perform risk-benefit analyses on three organic compounds.

Procedure

1. Choose three of the organic compounds listed in Tables 2.4 through 2.7. Research and record the chemical structure of each compound.
2. Follow the four steps for a risk-benefit analysis, given at the beginning of this section. You will need to carry out additional research on the compounds of your choice.

Analysis

For each compound, answer the following questions.

1. What are the practical purposes of this organic compound?
2. **(a)** What alternative compounds (if any) can be used for the same purpose?
   (b) What are the benefits of using your particular compound, instead of using the alternatives in part (a)?
3. **(a)** What are the direct risks of using this compound?
   (b) What are the indirect risks of using this compound?
4. In your opinion, should this compound continue to be used? Give reasons for your answer.
5. What safety precautions could help to reduce possible harm—to people as well as to the environment—resulting from this compound?

Problem Solving With Organic Compounds

Many chemicals and products have been developed as solutions to health, safety, and environmental problems. Sometimes, however, a solution to one problem can introduce a different problem. Read the three articles on the next few pages. Then complete the ThoughtLab that follows.

Replacing CFCs—At What Cost?

At the beginning of the twentieth century, refrigeration was a relatively new technology. Early refrigerators depended on the use of toxic gases, such as ammonia and methyl chloride. Unfortunately, these gases sometimes leaked from refrigerators, leading to fatal accidents. In 1928, a new, "miracle" compound was developed to replace these toxic gases. Dichlorodifluoromethane, commonly known as Freon®, was a safe, non-toxic alternative. Freon® and other chlorofluorocarbon compounds, commonly referred to as CFCs, were also used for numerous other products and applications. They were largely responsible for the development of many conveniences, such as air-conditioning, that we now take for granted.

Today we know that CFCs break up when they reach the ozone layer, releasing chlorine atoms. The chlorine atoms destroy ozone molecules faster than the ozone can regenerate from oxygen gas. Studies in the past ten years have shown dramatic drops in ozone concentration at specific locations. Since ozone protects Earth from the Sun's ultraviolet radiation, this decrease in ozone has led to increases in skin cancer, as well as damage to plants and animals. In addition, CFCs are potent greenhouse gases and contribute to global warming. Through the Montréal Protocol, and later "Earth Summit" gatherings, many countries—including Canada—have banned CFC production.

Substitutes for CFCs are available, but none provide a completely satisfactory alternative. Hydrofluorocarbons (HFCs) are organic compounds that behave like CFCs, but do not harm the ozone layer. For example, 1,1,1,2-tetrafluoroethane and 1,1-difluoroethane are HFCs that can be used to replace CFCs in refrigerators and air conditioners. Unfortunately, HFCs are also greenhouse gases.

Simple hydrocarbons can also be used as CFC substitutes. Hydrocarbons such as propane, 2-methylpropane (common name: isobutane), and butane are efficient aerosol propellants. These hydrocarbons are stable and inexpensive, but they are extremely flammable.

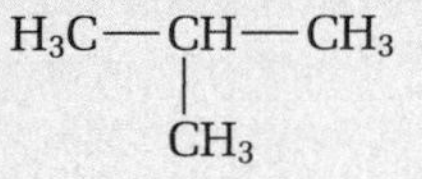

$H_3C—CH(CH_3)—CH_3$

isobutane

$CH_3CH_2CH_3$

propane

$CH_3CH_2CH_2CH_3$

butane

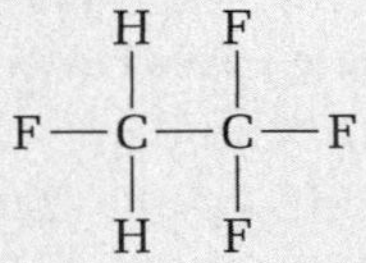

$CHF_2—CH_2F$ (H, F / F—C—C—F / H, F)

1,1,1,2-tetrafluoroethane

$CHF_2—CH_3$ (F, H / F—C—C—H / H, H)

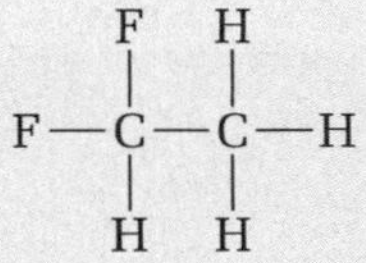

1,1-difluoroethane

Revisiting DDT: Why Did it Happen?

Dichlorodiphenyltrichloroethane, better known as DDT, is a well-known pesticide that has caused significant environmental damage. It is easy to point fingers, blaming scientists and manufacturers for having "unleashed" this organic compound on an unsuspecting world. It is more difficult, however, to understand why this environmental disaster took place.

Insects consume more than one third of the world's crops each year. In addition, insects such as mosquitoes spread life-threatening diseases, including malaria and encephalitis. Weeds reduce crop yields by taking over space, using up nutrients, and blocking sunlight. Some weeds even poison the animals that the crops are intended to feed. Crop damage and low crop yield are significant problems for countries undergoing food shortages and famines.

DDT was one of the first pesticides that was developed. For many years, it was used successfully to protect crops and fight disease epidemics. Paul Mueller, the scientist who discovered DDT's use as a pesticide, was awarded the Nobel Prize in 1948.

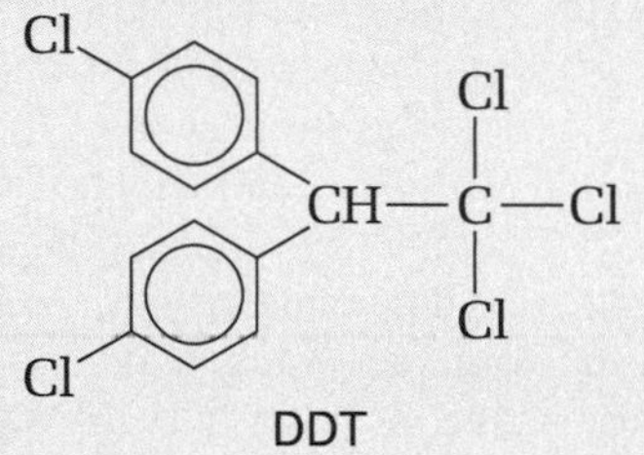

DDT

Later, however, tests revealed that DDT does not readily decompose in the environment. Instead, DDT remains present in the soil for decades after use. The hydrocarbon part of this molecule makes it soluble in the fatty tissues of animals. As it passes through the food chain, DDT accumulates. Animals that are higher in the food chain, such as large birds and fish, contain dangerous concentrations of this chemical in their tissues. A high DDT concentration in birds causes them to lay eggs with very thin shells, which are easily destroyed. Today DDT is no longer produced or used in Canada. It is still used, however, in many developing countries.

Knocking on the Car Door

Automobile fuels are graded using *octane numbers*, which measure the combustibility of a fuel. A high octane number means that a fuel requires a higher temperature and/or higher pressure to ignite. Racing cars with high-compression engines usually run on pure methanol, which has an octane number of 120.

Gasoline with too low an octane number can cause "knocking" in the engine of a car, when the fuel ignites too easily and burns in an uncontrolled manner. Knocking lowers fuel efficiency, and it can damage the engine.

As early as 1925, two of the first automobile engineers became aware of the need to improve the octane number of fuels. Charles Kettering advocated the use of a newly developed compound called tetra-ethyl lead, $Pb(C_2H_5)_4$. This compound acts as a catalyst to increase the efficiency of the hydrocarbon combustion reaction. Henry Ford believed that ethanol, another catalyst, should be used instead of tetra-ethyl lead. Ethanol could be produced easily from locally grown crops. As we now know, ethanol is also much better for the environment.

Tetra-ethyl lead became the chosen fuel additive. Over many decades, lead emissions from car exhausts accumulated in urban ponds and water systems. Many waterfowl that live in urban areas experience lead poisoning. Lead is also dangerous to human health.

Leaded fuels are now banned across Canada. In unleaded gasoline, simple organic compounds are added instead of lead compounds. These octane-enhancing compounds include methyl-t-butyl ether, t-butyl alcohol, methanol, and ethanol. Like lead catalysts, these compounds help to reduce engine knocking. In addition, burning ethanol and methanol produces fewer pollutants than burning hydrocarbon fuels, which contain contaminants. Since they can be made from crops, these alcohols are a renewable resource.

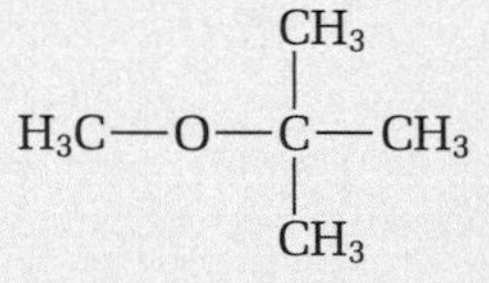

2-methoxy-2-methyl propane (methyl tert-butyl ether)

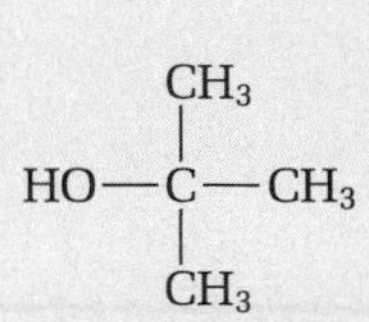

1,1-dimethyl ethanol (tert-butyl alcohol)

ThoughtLab  Problem Solving with Organic Compounds

In this ThoughtLab, you will consider several situations in which organic compounds were used to help solve a health, safety, or environmental problem.

Procedure

1. Choose two of the situations that are discussed in the three newspaper articles. Come up with a third situation, involving organic compounds, that you have heard or read about recently. For the three situations you have chosen, answer the following questions.
   (a) What was the original problem?
   (b) How was this problem resolved? What part did organic compounds play in the solution?
   (c) What further problems (if any) were introduced by the solution in part (b)?
   (d) Have these additional problems been resolved? If so, how have they been resolved?
   (e) Do you foresee any new problems arising from the solutions in part (d)? Explain why or why not.

Analysis

1. Use your knowledge of organic chemistry to describe how organic chemistry has helped to provide a solution to
   (a) a human health problem
   (b) a safety problem
   (c) an environmental problem
2. In your opinion, is it worth coming up with solutions to problems, if the solutions carry the possibility of more problems? Explain your answer.

Section Summary

In this section, you encountered some of the ways that organic compounds make our lives easier. As well, you learned about some of the risks involved in using organic compounds. You carried out risk-benefit analyses on several organic compounds. Finally, you examined some situations in which organic chemistry has helped to solve problems related to human health, safety, and the environment.

Section Review

1 MC Think about the risks and benefits of organic compounds.
(a) Describe five specific benefits you obtain from organic compounds.
(b) Describe two risks from organic compounds, and explain how these risks may affect you.

2 C Prepare a brochure, booklet, or web page explaining the practical uses of organic compounds to a younger age group. Your booklet should include the following information:
- a simple explanation of what an organic compound is
- a description of some benefits of organic compounds
- specific examples of at least three different organic compounds that you studied in this chapter, plus their uses
- specific examples of at least three different organic compounds that you did *not* study in this chapter, plus their uses

Unit Issue Prep

Before researching your issue at the end of Unit 1, decide how you will use the information and skills you learned in this section. What are some possible direct and indirect risks and benefits involved in the issue you have chosen?

3. C Research persistent organic pollutants (POPs) on the Internet. Use your research to prepare a poster about POPs for your community. Include the following information:
   - examples of three POPs, with descriptions of their negative effects on the environment and/or human health
   - a description of what the Canadian government has done to address the problem of POPs
   - suggestions for ways that your community can avoid or reduce harm from organic pollutants
4. MC In section 2.3, you learned about some of the risks and benefits of polymers. Choose a synthetic polymer from section 2.3. Carry out a risk-benefit analysis of this polymer.
5. C Scientists and horticulturalists who work with and sell the synthetic pesticide *pyrethroid* are concerned about public perception. Many people buy the natural pesticide *pyrethrin*, even though it is more toxic than pyrethroid. Why do you think this happens? Do you think this happens for other products that have natural and synthetic alternatives? Write a brief editorial outlining your opinions and advice to consumers, to help them make informed product choices.

CHAPTER 2 Review

Reflecting on Chapter 2

Summarize this chapter in the format of your choice. Here are a few ideas to use as guidelines:

- Describe the different types of organic reactions, and give an example of each type.
- Compare addition polymerization reactions with condensation polymerization reactions.
- List the biomolecules you learned about in this chapter, and give an example of each.
- List some benefits of the use of organic compounds.
- List some risks from the use of organic compounds.

Reviewing Key Terms

For each of the following terms, write a sentence that shows your understanding of its meaning.

addition reaction	substitution reaction
elimination reaction	oxidation
reduction	condensation reaction
hydrolysis reaction	esterification reaction
Markovnikov's rule	monomers
polymer	addition polymerization
plastics	nylons
condensation polymerization	polyamides
	biochemistry
polyesters	amino acids
protein	saccharide
carbohydrate	disaccharide
monosaccharide	starch
polysaccharide	nucleotides
DNA (2-deoxyribonucleic acid)	lipids
	oils
RNA (ribonucleic acid)	fats
waxes	

Knowledge/Understanding

1. Describe each type of organic reaction, and give an example.
   - **(a)** addition
   - **(b)** substitution
   - **(c)** elimination
   - **(d)** oxidation
   - **(e)** reduction
   - **(f)** condensation
   - **(g)** hydrolysis
2. What is the connection between a condensation reaction and an esterification reaction?
3. How is an addition reaction related to an elimination reaction?
4. **(a)** In your own words, define an esterification reaction and a hydrolysis reaction. Give two examples of each type of reaction.
   (b) Explain the connection between these two types of reactions.
5. Identify each reaction as an addition reaction, a substitution reaction, or an elimination reaction.
   - **(a)** $CH_2{=}CH_2 + H_2 \rightarrow CH_3CH_3$
   - **(b)** benzene $+ \ I_2 \xrightarrow{FeBr_3}$ iodobenzene (C_6H_5–I) $+ \ HI$
   - **(c)** $CH_3CH_2CH_2OH \rightarrow CH_3CH{=}CH_2 + H_2O$
   - **(d)** $CH_3CH_2CH_2OH + HCl \rightarrow CH_3CH_2CH_2Cl + H_2O$
   - **(e)** $CH_3CH(CH_3)CH{=}CH_2 + HBr \rightarrow CH_3CH(CH_3)CH(Br)CH_3$
   - **(f)** $CH_3CH_2CH{=}CHCH_2CH_3 + H_2O \rightarrow CH_3CH_2CH(OH)CH_2CH_2CH_3$
   - **(g)** $CH_3CH_2Br + NH_3 \rightarrow CH_3CH_2NH_2 + HBr$
6. Identify each reaction as an oxidation or a reduction. (Oxidizing and reducing agents are not shown.)
   - **(a)** $CH_3CH_2CH(CH_3)CH{=}O \rightarrow CH_3CH_2CH(CH_3)C({=}O)OH$
   - **(b)** (skeletal structures) ketone with C=O $\rightarrow$ alcohol with OH
   - **(c)** $HO{-}CH_2{-}CH_2{-}CH(CH_2OH){-}CH_3 \rightarrow HC({=}O){-}CH_2{-}CH(HC{=}O){-}CH_3$

7. Describe each type of polymerization, and give an example.
   (a) addition polymerization
   (b) condensation polymerization
8. What is the difference between a nylon and a polyester? How are they similar?
9. How can the following reaction be an example of both an addition reaction *and* a reduction?
   $CH_2{=}CH_2 + H_2 \rightarrow CH_3CH_3$
10. (a) What is Markovnikov's rule? Why does it apply to the following reaction?
    $CH_3CH{=}CH_2 + HBr \rightarrow ?$
    (b) Name and draw the two isomeric products of this reaction.
    (c) Which product is formed in the greater amount?
11. Classify the following oxidation as an addition, substitution, or elimination reaction. Explain your reasoning.
    $CH_3CH_2OH + [O] \rightarrow CH_3CHO$
12. Consider the reaction of an aldehyde with lithium aluminum hydride, $LiAlH_4$, to produce an alcohol. Why is this reaction a reduction, not an oxidation?
13. A short section of the polymer *polypropene* is shown below.
    $\cdots{-}CH_2{-}CH(CH_3){-}CH_2{-}CH(CH_3){-}\cdots$
    (a) Draw and name the monomer that is used to make this polymer.
    (b) What type of polymer is polypropene? Explain your reasoning.
    (c) What is the common name of this polymer?
14. (a) What is the difference between a protein and an amino acid?
    (b) How are proteins important to living organisms?
15. (a) Define the term "lipid."
    (b) Give three different examples of lipids.
    (c) List four foods you eat that contain lipids.
    (d) How are lipids important to your body?
16. (a) Distinguish between monosaccharides, disaccharides, and polysaccharides.
    (b) Draw and name an example of each.
    (c) How are carbohydrates important to living organisms?
17. (a) What is the generic monomer of a polysaccharide polymer?
    (b) What is the monomer of cellulose?

Inquiry

18. Draw and name the product(s) of each incomplete reaction. (**Hint:** Do not forget to include any second products, such as H_2O or HBr.)
    (a) $CH_3CH{=}CHCH_3 + Br_2 \rightarrow$
    (b) $HO{-}CH_2CH_2CH_2CH_2CH_3 + HBr \rightarrow$
    (c) $CH_3CH_2\overset{\overset{\displaystyle O}{\|}}{C}H + [O] \rightarrow$
    (d) $HO{-}CH_2CH_2CH_3 \xrightarrow[\Delta]{H_2SO_4}$
    (e) 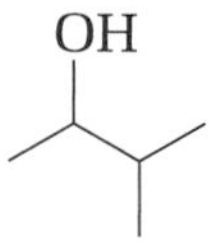 [skeletal structure: 3-methyl-2-pentanol-type alcohol with OH] $+ [O] \rightarrow$
    (f) $CH_3CH_2\overset{\overset{\displaystyle O}{\|}}{C}OH + HOCH_3 \xrightarrow{H_2SO_4}$
    (g) $CH_3CH{=}CHCH_2CH_3 + H_2 \rightarrow$
    (h) $CH_3{-}\overset{\overset{\displaystyle O}{\|}}{C}{-}O{-}CH_2CH_3 + H_2O \xrightarrow[\Delta]{H_2SO_4}$
19. Draw the product(s) of each incomplete reaction. **Hint:** Do not forget to include the second product, such as H_2O or HBr, for a substitution reaction.
    (a) $CH_3CH_2{-}\overset{\overset{\displaystyle O}{\|}}{C}{-}OH + CH_3CH_2CH_2OH \rightarrow$
    (b) $CH_3CH_2C{\equiv}CH + Cl_2 \rightarrow \text{(i)} + Cl_2 \rightarrow \text{(ii)}$
    (c) 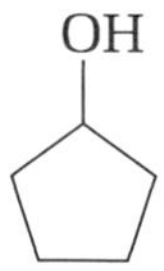 [cyclopentanol: cyclopentane ring with OH] $\xrightarrow[\Delta]{H_2SO_4}$
    (d) 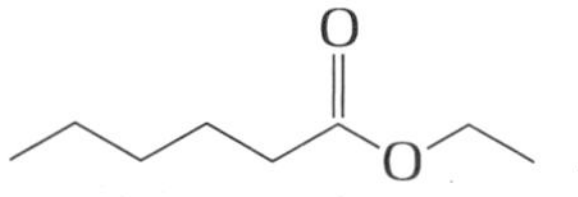 [skeletal structure: ethyl ester with C=O and O] $+ H_2O \xrightarrow[\Delta]{H_2SO_4}$
    (e) $CH_3{-}\overset{\overset{\displaystyle O}{\|}}{C}{-}\underset{\underset{\displaystyle CH_3}{|}}{CH}{-}CH_3 + [H] \rightarrow$

(f) $H_2C{=}CHCH_2CH(CH_3)_2 + HOH \rightarrow$ (i) (major product) + (ii) (minor product)

(g) $CH_3CH_2CH_2CH_2CH_2{-}OH + [O] \rightarrow$ (i) + [O] $\rightarrow$ (ii)

(h) 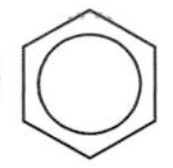 (benzene) $+ \; Cl_2 \xrightarrow{FeBr_3}$

(i) $CH_3CH_2CH(Br)CH_3 + OH^- \rightarrow$

(j) $HO{-}CH_2CH_2CH_2CH_3 \; + \; HC({=}O){-}OH \rightarrow$

(k) $nCH_2{=}CH(OH) \xrightarrow{\text{polymerization}}$

(l) $nHO(CH_2)_7OH + nHOC({=}O)CH_2C({=}O)OH \xrightarrow{\text{polymerization}}$

20. Draw and name the reactant(s) in each reaction.

(a) $? \; + \; Cl_2 \rightarrow H{-}C(Cl)(H){-}C(Cl)(H){-}H$

(b) $? \; + \; [O] \; \rightarrow \; HC({=}O)CH_2CH_3$

(c) $HC{\equiv}C{-}CH_3 \; + \; ? \rightarrow$ $C(Br)(H){=}C(Br)(CH_3)$

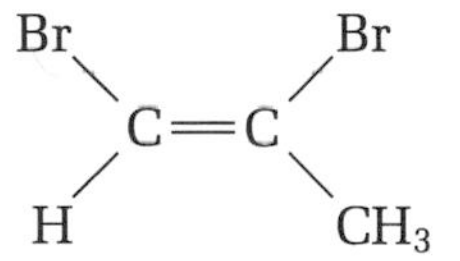

(d) $? \; + \; [H] \; \rightarrow CH_3{-}CH(CH_3){-}CH(OH){-}CH_3$

(e) $? \; + \; ? \rightarrow CH_3CH_2C({=}O)OCH_2CH_2CH_2CH_3$

(f) $? \; + \; [H] \; \rightarrow$ (skeletal structure) OH

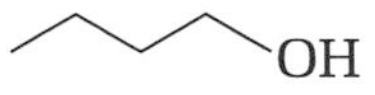

(g) $? \; + \; [O] \; \rightarrow \; ? \; + \; [O] \; \rightarrow$ (skeletal structure) C(=O)OH

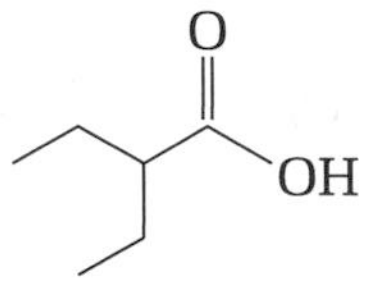

(h) $? \xrightarrow[\Delta]{H_2SO_4} \square + H_2O$

(i) $? \xrightarrow{\text{polymerization}} \cdots{-}CH_2{-}CH(CH_2CH_3){-}CH_2{-}CH(CH_2CH_3){-}\cdots$

21. Identify each type of reaction in questions 18, 19, and 20.

22. **(a)** Write a brief procedure that you could use to carry out the following reactions in a laboratory.

- oxidation of 1-pentanol
- oxidation of 2-pentanol
- oxidation of 2-methyl-2-butanol

(b) Describe the results you would expect.

(c) Write a chemical equation for each reaction. Show the products, if any.

23. In Investigation 1-A in Chapter 1, you carried out a condensation reaction to produce an ester called ethyl ethanoate. In this chapter, you learned about the reverse reaction: hydrolysis.

(a) Write a balanced chemical equation, showing the reactants and the products of the hydrolysis of ethyl ethanoate.

(b) Write a step-by-step procedure to carry out the acid hydrolysis of ethyl ethanoate.

(c) List the materials and equipment you would need to carry out this reaction.

24. At the beginning of section 2.2, you learned about Tollen's reagent. Tollen's reagent oxidizes aldehydes to produce the salt of a carboxylic acid. How does this reagent work as a test to distinguish between aldehydes and ketones?

25. One type of condensation reaction (which you did not study in this chapter) takes place when an acid, usually H_2SO_4, is added to an alcohol at temperatures that are lower than the temperatures for an elimination reaction. In this type of condensation reaction, two alcohol molecules combine to form an ether. A water molecule is eliminated as a second product of the reaction.

(a) Write a chemical equation to show the condensation reaction of ethanol to produce ethoxyethane.

(b) Classify this condensation reaction as an addition reaction, a substitution reaction, or an elimination reaction. Explain your answer.
(c) A different reaction occurs if the acid HCl is used instead of H_2SO_4. Name this type of reaction, and write the equation for it.

Communication

26. Use a molecular model kit to build an example of a DNA nucleotide monomer. Join your nucleotide to your classmates' nucleotides to build a short strand of DNA.
27. Make a list of five materials or substances you use every day that are made from organic compounds. Write a paragraph that describes how your life would be different without these five materials or substances.
28. How has organic chemistry helped to solve the problems caused by the use of the following substances?
(a) leaded gasoline (b) CFCs

Making Connections

29. Suppose that the Canadian government proposed a $3 million grant to scientists who are working on the development of synthetic pesticides. Would you support this grant? Write a brief paragraph that explains your point of view.
30. Styrofoam™ "clamshells" were used for packaging fast food until a few years ago. Today, large fast food corporations use paper packaging instead. Why has Styrofoam™ use been reduced? What problems does this product cause? Research Styrofoam™ (made from a polymer called polystyrene) to answer these questions.
31. Do you think Canada should cut back on the manufacture and use of synthetic polymers? Write a brief paragraph that explains your point of view. Include at least three good reasons to back up your point of view. If necessary, research the information you need.
32. Steroids are one type of lipid. Research answers to the following questions.
(a) What is the structure of a steroid?
(b) What are some common examples of steroids? How are steroids used in medicine?
(c) How are steroids misused in sports?
33. Muscle-building athletes sometimes drink beverages that contain amino acids.
(a) Why might this type of drink help to build muscles?
(b) Research the benefits and risks of this type of drink.
34. Early settlers in Canada often made their own soap using animal fats and lye.
(a) What type of chemical would you expect lye to be? Explain your reasoning.
(b) Use a reference book or dictionary to find the chemical name for lye.
(c) Look through books about early pioneer life to find an early recipe for making soap.
(d) Use the recipe you found to write a full laboratory procedure for making soap like the early settlers did. Include a list of materials and a diagram, if necessary.

Answers to Practice Problems and Short Answers to Section Review Questions

Practice Problems: **1.(a)** substitution **(b)** addition **(c)** elimination **2.(a)** oxidation **(b)** reduction **(c)** reduction **3.(a)** addition/reduction **(b)** condensation/substitution **4.(a)** addition **(b)** oxidation **5.** 3-ethyl-3-heptanol **6.(a)** 1-bromohexane (minor), 2-bromohexane (major) **(b)** 2-methyl-2-pentanol (major), 2-methyl-3-pentanol (minor) **8.(a)** propene + hydrochloric acid **(b)** ethene + bromine **(c)** 3-methyl-2-pentene, 2-ethyl-1-butene **(d)** hydrogen gas **9.(a)** substitution **(b)** oxidation **(c)** substitution **(d)** elimination **11.(a)** elimination **(b)** oxidation **(c)** substitution **(d)** substitution **(e)** oxidation **(f)** substitution **12.(a)** 1-butene + water **(b)** pentanal, pentanoic acid **(c)** 2-bromo-3-methylbutane + water **(d)** 2-methyl-1-propanol + NaBr **(e)** butanone **(f)** chlorocyclohexane + water **13.** no, because tertiary alcohols cannot be oxidized. **14.(a)** reduction **(b)** reduction **(c)** reduction **(d)** esterification **(e)** hydrolysis **15.(a)** cyclobutanone, cyclobutanol; 2,2,4,4-tetramethyl-3-pentanone, 2,2,4,4-tetramethyl-3-pentanol **16.(a)** hexanoic acid **(b)** 1-octanol **(c)** methyl propanoate + water **(d)** 1-propanol + ethanoic acid **(e)** 3-hexanol **(f)** 2-propyl (or isopropyl) 3-methylpentanoate **17.(a)** 2-methylbutanal **(b)** 1-pentanol, pentanal **(c)** 1-propanol, 3-methylhexanoic acid **18.** polymethylmethacrylate (PMMA) **19.(a)** condensation **(b)** addition **(c)** condensation **21.(a)** addition **(b)** condensation; nylon **(c)** condensation; polyester **(d)** condensation; polyester

Section Review: 2.1: 1.(a) addition **(b)** elimination **(c)** substitution **(d)** substitution **2.(a)** oxidation **(b)** reduction **(c)** reduction **(d)** oxidation **(e)** reduction **3.(b)** hydrolysis **2.2: 1.(a)** elimination **(b)** addition **(c)** oxidation **(d)** reduction or addition **(e)** substitution **(f)** substitution **(g)** condensation **2.(a)** 2,3-dibromobutane **(b)** 1-chlorobutane (minor) + 2-chlorobutane (major) **(c)** 1-hexene + water **(d)** 3,3-dimethyl-1-butene + HBr **3.(a)** bromobenzene **(b)** 1-bromopentane + water **4.(a)** 2-pentanone **(b)** propanoic acid **(c)** 2,3-dimethylbutanal **(d)** ethane **(e)** 3-pentanol **(f)** 1-butene, butane **(g)** 2-ethyl-3,4-dimethyl-1-octanol **5.(a)** 2-propyl propanoate **(b)** cyclopentene **(c)** methanoic acid, 2-butanol **(d)** cyclohexanol **(e)** N-ethyl heptanamide **6.(b)** 2-bromo-4-methylpentane **7.(a)**, **(c)**, and **(d)** **2.3: 1.(a)** protein, starch, nucleic acid **(b)** insulin, amylose, DNA **(c)** amino acid, monosaccharide (glucose), nucleotide **2.(a)** addition **(b)** condensation **3.(a)** addition polymerization **(b)** condensation polymerization **(c)** condensation polymerization **5.(a)** amino acid **(b)** monosaccharide **(c)** amino acid **(d)** lipid **(e)** disaccharide

Current Issues Related to Organic Chemistry

Background

Have you ever heard the saying "necessity is the mother of invention"? Almost all new products are developed to meet a need. As you have seen in this unit, however, new products carry risks as well as benefits. Often, the development of a new product is opposed by groups of people who are concerned with possible risks.

There may be several points of view about any scientific development. These points of view are influenced by various factors such as the economy, society, and politics.

In this unit issue, you will explore the development of new products through organic chemistry. You will choose one issue from the list below, and research the different points of view. Then you will present your findings, and your own point of view, to your class.

- *Can biofuel be used to replace petroleum-based fuels?* Biofuel can be made from biological materials, such as plants and animal fats. Biodiesel and ethanol are the two most common biological fuels. As part of your research, find out what biofuel is used for. Think about factors that may be holding back the sale of biofuel on the Canadian market.
- *Are Canadian processes for approving new pharmaceutical drugs adequate, too extensive, or not extensive enough?* Some Canadians complain about the long waiting period for a new foreign drug to be approved in Canada. Why is there a long waiting period? What testing processes are needed for approval of a new drug in Canada?
- *Why are some drugs so expensive?* The Canadian government recently increased its drug patents from 17 years to 20 years. This should help drug companies make sufficient profits to cover the enormous costs of developing and testing new drugs. Find out about the costs and procedures that are involved in developing and testing new drugs. Who finances drug research? Is it fair for drug companies to hold patents? What happens when these patents are not honoured?

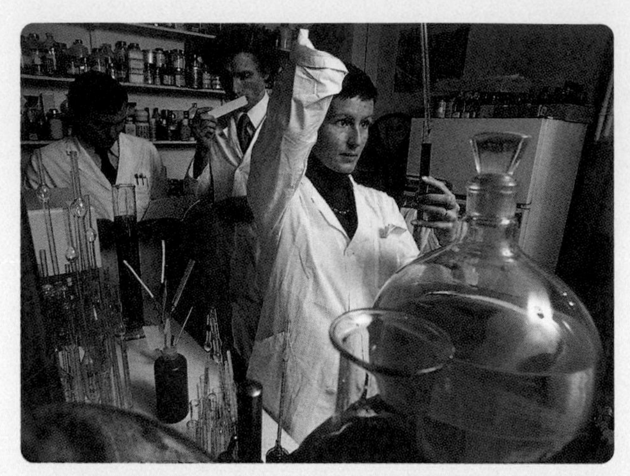

- *Should all plastics that are made in the future be biodegradable?* Biopolymers, such as proteins and starches, are used to produce one type of biodegradable plastic. Find out more about biodegradable plastics. How are they made, and how do they break down? What are the risks and benefits?
- *Are food additives necessary? Should they be regulated more closely?* Almost all processed foods contain food additives. Choose three or more food additives, not including any compounds that you may have researched in Chapter 2. Research the uses, structures, and purposes of the food additives you have chosen.
- *Decide on a different current issue to research and analyze.* If you want to research a different issue, have your teacher approve your plan before you begin.

Plan and Present

1. Choose an issue to research.
2. Who is involved in the issue? Research and report on the different points of view of action groups, corporations, and government groups. Ensure that your research covers a wide range of sources and resources. Some possible contacts are given below:
   - environmental groups, which often provide one strong point of view on an issue
   - research scientists at universities, who may present information on the development of a technology

- Environment Canada and other Canadian government agencies, which can usually provide scientifically sound information on a range of topics

3. Find out general public opinion on the issue by taking a survey in your community.
4. Decide on your own point of view, and provide information to back up your opinion.
5. When researching, consider the following question: Does everyone who is involved in the issue have a good understanding of the related science? If your answer is no, provide specific proof to back up your opinion. For example, include quotations from your sources, where applicable. Could the issue be resolved by providing factual information? Explain.
6. Prepare a presentation for your class. Your presentation can take the form of a pamphlet, an illustrated research report, a poster, an oral or multimedia presentation, a video, a web page, or another creative method that has been approved by your teacher. Include the following information:
   - a description of the factors or needs that led to the development of the process, technology, or product you are analyzing
   - structures and equations for any specific organic compounds and reactions that relate to the issue
   - a list of the benefits and risks that are involved in the issue
   - a description of the various points of view on the issue, along with your own point of view
   - a list of groups or people who might benefit from the process, technology, or product
   - a list of groups or people who might benefit if the process, technology, or product was stopped or changed
   - a list of government and/or industry influences (Who financed the research? Have any laws or bills been passed to regulate the development of the process, technology, or product?)

Evaluate the Results

1. Take notes on the presentations of other members of your class. If you have a class debate, take notes on the various points of view.
2. What did you learn? Has your point of view on any issue changed? If so, how?

Assessment

After you have explored this issue, answer the following questions to assess your work.

- Was your analysis scientifically accurate? Was it complete? Was it objective?
- Was your explanation of the related science clear and concise?
- Did you use at least three primary research sources? Did you reference these sources in your presentation or report?
- Did you include interesting or unusual information?
- Did you consider all points of view in your presentation?
- Did your research influence the opinions of others in your class?

Web LINK

www.mcgrawhill.ca/links/chemistry12

Go to the web site above, and click on **Web Links** to find appropriate sites related to this unit issue.

UNIT 1 Review

Knowledge/Understanding

Multiple Choice

In your notebook, write the letter for the best answer to each question.

1. Carbon atoms can bond with each other to form
   (a) single bonds only
   (b) double bonds only
   (c) single and double bonds
   (d) single, double, and triple bonds
   (e) single, double, triple, and quadruple bonds
2. The three-dimensional shape around a single-bonded carbon atom is
   (a) linear
   (b) trigonal planar
   (c) tetrahedral
   (d) pyramidal
   (e) bent
3. The length of a carbon-carbon bond in a benzene molecule is
   (a) halfway between a single bond and a double bond
   (b) halfway between a double bond and a triple bond
   (c) the same as a single bond
   (d) the same as a double bond
   (e) the same as a triple bond
4. The functional group of an alcohol is
   (a) $-OH$
   (b) $-C{=}C-$
   (c) $-NR_2$
   (d) $-C(O)NR_2$
   (e) $-OR$
5. The functional group of an amide is
   (a) $-OH$
   (b) $-C{=}C-$
   (c) $-NR_2$
   (d) $-C(O)NR_2$
   (e) $-OR$
6. The compound $NH_2CH_2CH_2CH_3$ has the IUPAC name
   (a) methanamine
   (b) ethanamine
   (c) propanamine
   (d) butanamine
   (e) butanamide
7. The compound $CH_3CH(CH_3)CH_2CH_2OH$ has the IUPAC name
   (a) 1-pentanol
   (b) 1-butanol
   (c) 2-methyl-4-butanol
   (d) 3-methyl-1-butanol
   (e) 3-methyl-1-pentanol
8. The compound $CH_3CH_2CH_2C(O)CH_2CH_3$ has the IUPAC name
   (a) ethoxypentane
   (b) ethoxypropane
   (c) ethyl butanoate
   (d) 3-pentanone
   (e) 3-hexanone
9. $CH_3CH_2C(CH_3)_2CH_2CH_2CH_2COOH$ has the IUPAC name
   (a) 5,5-dimethylheptanoic acid
   (b) 3,3-dimethylheptanoic acid
   (c) 4,4-dimethylheptanoic acid
   (d) 2,2-dimethylheptanoic acid
   (e) 3,3-dimethyl heptanoate
10. The oxidation of pentanal produces
    (a) pentanol
    (b) pentanal
    (c) pentanoic acid
    (d) pentanamine
    (e) pentanamide
11. The esterification of methanol and butanoic acid produces
    (a) 1-methyl butanoic acid
    (b) 4-methyl butanoic acid
    (c) butyl methanoate
    (d) methyl butanoate
    (e) butanol and methanoic acid
12. The polymerization of propene, $CH_3CH{=}CH_2$, can be classified as
    (a) a hydrolysis reaction
    (b) an elimination reaction
    (c) a substitution reaction
    (d) a condensation reaction
    (e) an addition reaction
13. Ketones are reduced to produce
    (a) primary alcohols
    (b) secondary alcohols
    (c) tertiary alcohols
    (d) aldehydes
    (e) carboxylic acids

14. Choose the most correct definition.
   (a) Organic compounds are based on carbon, and they usually contain carbon-nitrogen and carbon-silicon bonds.
   (b) Organic compounds are based on nitrogen, and they usually contain carbon-nitrogen and carbon-hydrogen bonds.
   (c) Organic compounds are based on carbon, and they usually contain carbon-hydrogen and carbon-carbon bonds.
   (d) Organic compounds are based on hydrogen, and they usually contain carbon-hydrogen and carbon-oxygen bonds.
   (e) Organic compounds are based on oxygen, and they usually contain carbon-oxygen and carbon-carbon bonds.

15. Which statement is *not* correct?
   (a) A C–O bond is polar.
   (b) An N–H bond is polar.
   (c) An O–H bond is polar.
   (d) A C–C bond is polar.
   (e) A C–Cl bond is polar.

16. Which description of an addition reaction is accurate?
   (a) Two molecules are combined, and a small molecule, such as water, is produced as a second product.
   (b) A hydrogen atom or functional group is replaced with a different functional group.
   (c) Atoms are added to a double or triple carbon-carbon bond.
   (d) A double carbon-carbon bond is formed when atoms are removed from a molecule.
   (e) Carbon atoms form more bonds to oxygen or less bonds to hydrogen.

Short Answer

In your notebook, write a sentence or a short paragraph to answer each question.

17. Is it accurate to say that most healthy foods are chemical-free? Explain your answer.

18. What is the difference between an amide and an amine? Use examples to explain your answer.

19. Do hydrocarbons have functional groups? Use examples to explain your answer.

20. Are the boiling points of carboxylic acids higher or lower than the boiling points of alcohols? Explain your answer.

21. A primary amine can form intermolecular hydrogen bonds, but a tertiary amine cannot. Explain this statement.

22. What happens when an oxidizing agent, such as potassium permanganate, is added to a tertiary alcohol? Explain.

23. What happens when an oxidizing agent is added to a secondary alcohol? Explain.

24. What is the common name for ethanol?

25. What is the IUPAC name for acetic acid?

26. The aldehyde CH_3CH_2CHO is named propanal, rather than 1-propanal. Why is the position number unnecessary?

27. Name and draw the product of the oxidation of 3-ethyl-4-methyl-1-hexanol. If this product is oxidized further, what second product will be formed?

Inquiry

28. Write the IUPAC name for each compound.
   (a) $CH_3CH_2CH_2COOH$
   (b) CH_3OH
   (c) $CH_3CH_2CH_2CH_2COOCH_3$
   (d) $CH_3CH_2C(O)CH_3$
   (e) $CH_3CH_2NHCH_3$
   (f) $CH_3CH_2CH(CH_3)CH(CH_3)CH_2CH_2CHO$

29. Give a common name for each compound.
   (a) methanol
   (b) 2-propanol
   (c) ethoxyethane

30. Write the IUPAC name for each compound.

   (a) 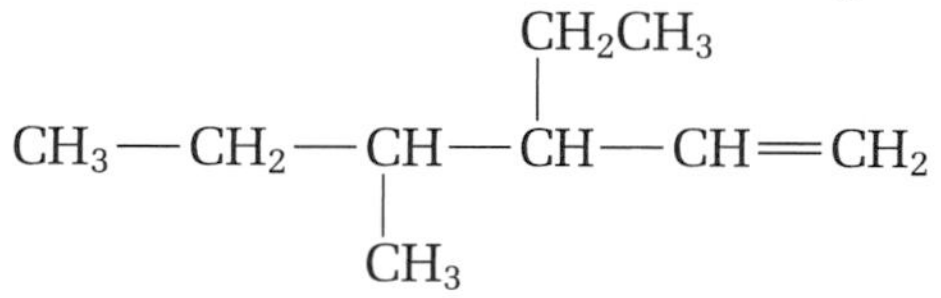 $CH_3—CH_2—CH(CH_3)—CH(CH_2CH_3)—CH{=}CH_2$

   (b) $CH_3—O—CH_2—CH_2—CH_3$

   (c) 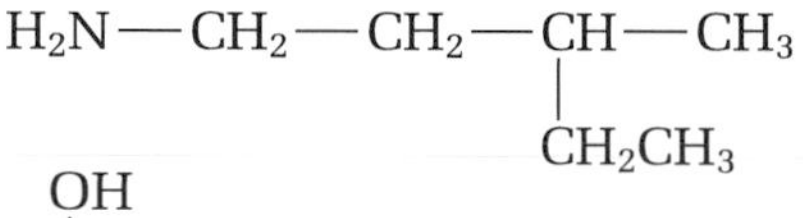 $H_2N—CH_2—CH_2—CH(CH_2CH_3)—CH_3$

   (d) 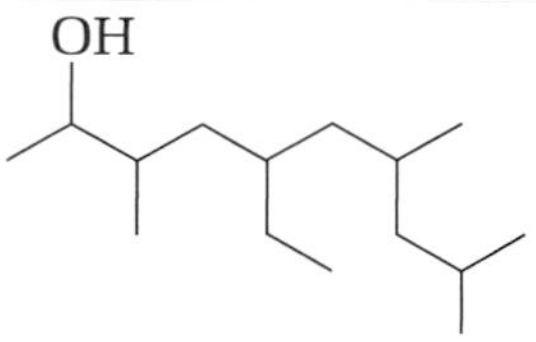 OH

(e) $CH_3—CH_2—\overset{\overset{\displaystyle O}{\|}}{C}—O—CH_2—CH_2—CH_2—CH_3$

31. Name each compound.

(a) O OH

(b) $CH_3—CH_2—NH—CH_2—CH_2—CH_2—CH_3$

(c)

(d) $CH_3—CH_2—CH_2—\overset{\overset{\displaystyle O}{\|}}{C}—NH_2$

(e) O

(f) OH CH_2CH_3 CH_3

(g) $H—\overset{\overset{\displaystyle O}{\|}}{C}—CH_2—CH_2—\underset{\underset{\displaystyle CH_3}{|}}{\overset{\overset{\displaystyle CH_3}{|}}{C}}—CH_2—\underset{\underset{\displaystyle CH_3}{|}}{\overset{\overset{\displaystyle CH_3}{|}}{C}}—CH_3$

(h) CH_3 CH_3CH_2 CH_3

(i) $CH_3—\underset{\underset{\displaystyle CH_2CH_3}{|}}{N}—\overset{\overset{\displaystyle O}{\|}}{C}—CH_3$

(j) CH_3 CH_3 CH_3 O

32. What is wrong with the following names? Use each incorrect name to draw a structure. Then correctly rename the structure you have drawn.
(a) 2-propyl-4-pentanone
(b) 3,3-trimethylhexane
(c) 2,5-octanol

33. Imagine that you dip one finger into a beaker of rubbing alcohol and another finger into a beaker of water. When you take your fingers out of the beakers, you observe that one finger feels colder than the other. Which finger feels colder, and why?

34. You are given two beakers. One contains 1-pentanol, and the other contains pentanoic acid. Describe the materials you would need to distinguish between the two liquids, in a laboratory investigation.

35. You are given three test tubes that contain colourless liquids. One test tube contains benzene, another contains ethanol, and the third contains 2,4-hexadiene. Design a procedure that will tell you the contents of each test tube. Describe your expected observations. **CAUTION** Do not try your procedure in a lab. Benzene is carcinogenic.

36. Predict the product(s) of each reaction.
(a) $CH_3CH_2CH{=}CHCH_3 + I_2 \rightarrow$
(b) $CH_3CH_2CH_2OH + HCl \rightarrow$
(c) $CH_3CH(CH_3)CHO + [O] \rightarrow$
(d) 1-propanol + methanoic acid →
(e) 1-heptanol $\xrightarrow[\Delta]{H_2SO_4}$

37. Predict the product(s) of each reaction.
(a) $CH_2{=}CHCH_3 + H_2O \rightarrow$
(b) propyl decanoate + water →
(c) $CH_3CH(CH_3)CH_2CH_2Br + H_2O \rightarrow$
(d) $CH_3CH(OH)CH_3 + KMnO_4 \rightarrow$
(e) $CH_2{=}CHCH_2CH_2OH + 2HCI \rightarrow$

38. Draw and name the missing reactant(s) in each reaction.
(a) ? + HBr → $CH_2{=}CHBr$
(b) ? + H_2O → $CH_3CH_2CH_2CH(CH_3)CH_2OH$ + HCl
(c) ? + ? → butyl heptanoate + H_2O
(d) ? + [H] → 2-methylpropanal
(e) ? + [O] → 3-ethyl-4,4-dimethyl-2-pentanone

Communication

39. Draw a structural diagram for each compound.
    (a) methanamine
    (b) 2-octanol
    (c) 1,2-dichlorobutane
    (d) pentanamide
    (e) methoxynonane
40. Draw a structural diagram for each compound.
    (a) propyl propanoate
    (b) 2,2,4,5-tetramethylheptanal
    (c) 2-methylpropanoic acid
    (d) N-ethyl-N-methyl butanamide
    (e) 4-ethyl-3,3-dimethyl-2-heptanone
41. Prepare a poster that describes condensation polymerization to students who do not take science. Use the polymerization of Nylon-6 as an example. Include information on the practical uses of Nylon-6. What natural product was nylon designed to replace? When nylon was first invented in the 1930s, what consumer products were made from it? What part did nylon play in World War II?
42. Draw a concept web that summarizes what you have learned about organic chemistry. Include the following topics:
    - functional groups
    - physical properties of organic compounds
    - reactivity of organic compounds
    - polymers, both natural and synthetic
    - biological molecules and their functions
    - practical applications of organic chemistry
    - risks and benefits of organic chemistry
43. In your opinion, are the risks of manufacturing and using pesticides greater or less than the benefits? Write a short (half-page) essay that explains your point of view. Include specific examples to back up your argument. (You may want to carry out further research on pesticides, and the effects of pesticides, before writing your essay.)
44. How are organic compounds important to your health and lifestyle? Write a short paragraph that describes any benefits you obtain from organic compounds.
45. Choose one of the biological molecules described in Chapter 2. Research this molecule until you discover three interesting facts about it, which you did not know before. Share your findings with your class.

Making Connections

46. When was plastic cling wrap first invented? What problem was it created to solve? Use the Internet or other resources to find the answers to these questions.
47. Organic pollutants can build up in living things by a process called bioaccumulation.
    (a) Use the Internet to research bioaccumulation. Where do most organic pollutants accumulate in the body of an animal?
    (b) Explain your answer to the previous question, using your knowledge of the solubility of organic compounds.
48. Most dyes are organic compounds. Before artificial dyes were invented, people prepared natural dyes from roots, berries, leaves, and insects.
    (a) Research some dyes used in the past and in the present. Find the chemical structure of each dye. Identify the functional group(s) in each dye.
    (b) How do artificial dyes differ from natural dyes? How have artificial dyes affected the fashion industry?

COURSE CHALLENGE

The Chemistry of Human Health

Think about these questions as you plan for your Chemistry Course Challenge.

- How can you identify the family to which an organic compound belongs?
- What reactions do each functional group commonly undergo?

UNIT 2

Structure and Properties

UNIT 2 CONTENTS

CHAPTER 3
Atoms, Electrons, and Periodic Trends

CHAPTER 4
Structure and Properties of Substances

UNIT 2 OVERALL EXPECTATIONS

- What is the quantum mechanical model of the atom, and how does a understanding of atomic structure enable chemists to explain the properties of substances and their chemical bonding?
- How do the properties of liquids and solids compare, and how can you use bonding theory to predict the shape of simple molecules?
- Which technologies have advanced our understanding of atomic and molecular theory, and how have those theories contributed to the development of other technologies and products?

Unit Project Prep

Look ahead to the project at the end of Unit 2. Plan in advance for the challenge by making a list of skills and information you will need as you progress through the unit.

In 1958, an astonished television audience watched as Dr. Harry Coover used a tiny drop of a substance called cyanoacrylate to lift a game show host off the ground. Cyanoacrylate is an amazingly powerful adhesive. This stunt helped launch the commercial career of superglue, now used for everything from building circuit boards to sealing human tissue in surgery.

Superglue is just one example of an astounding variety of compounds with useful properties. Whether synthesized by design or discovered by trial and error, each new compound does more than provide a potentially useful product. It also contributes to scientists' understanding of the relationship between structure at the molecular level and properties at the macroscopic level.

In this unit, you will investigate the structure and properties of atoms, ions, and molecules, and the natural and synthetic products that result from their interactions.

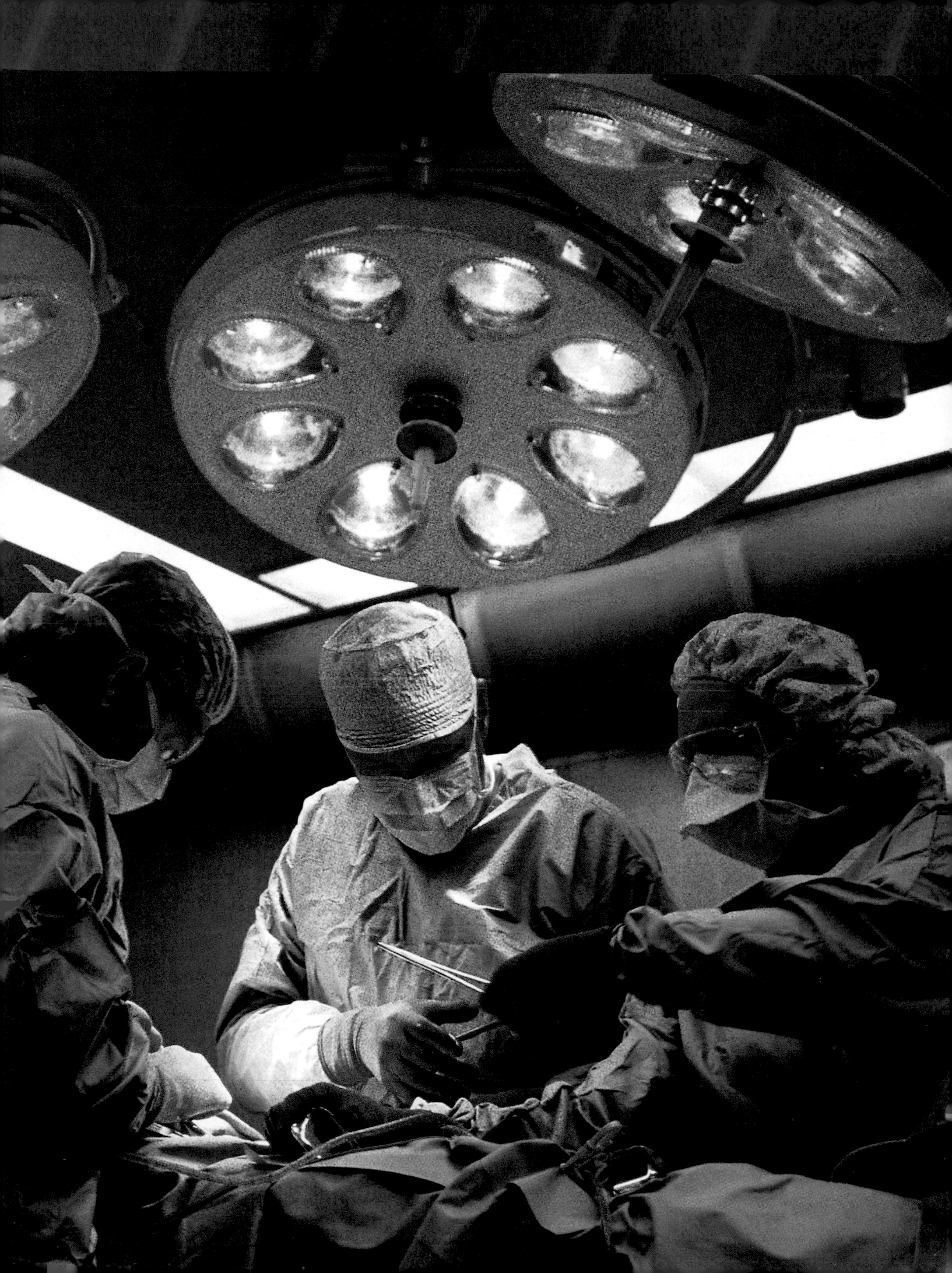

CHAPTER

3 Atoms, Electrons, and Periodic Trends

Chapter Preview

3.1 The Nuclear Atomic Model

3.2 The Quantum Mechanical Model of the Atom

3.3 Electron Configurations and Periodic Trends

Prerequisite Concepts and Skills

Before you begin this chapter, review the following concepts and skills:

- identifying subatomic particles and their properties (from previous studies)
- describing the structure and organization of the periodic table (from previous studies)
- explaining periodic trends for properties such as atomic radius, first ionization energy, and electron affinity (from previous studies)

The northern lights, also called the *aurora borealis*, are a fleeting feature of the night sky in northern parts of the world. For centuries, people who saw the shimmering, dancing sheets of coloured lights were awed by the phenomenon, and tried to explain it. Inuit that lived near Hudson's Bay told that the lights were caused by torches held by spirits to lead souls of the dead to the afterlife. In Greenland and Labrador, people thought the lights could help predict the weather. In Europe in the Middle Ages, their appearance was an omen of impending war or plague.

Today, scientists explain the northern lights as resulting from streams of protons and electrons—plasma—emanating from the Sun. When the electrons from the plasma interact with gaseous atoms in Earth's upper atmosphere, the atoms emit coloured light.

Nineteenth-century atomic theory could not explain why the aurora exhibits a limited range of specific wavelengths, rather than the full visible spectrum. Early in the twentieth century, however, scientists developed a revolutionary new model of the atom. This model, and the theory that supports it, helped to account for many puzzling phenomena that the existing atomic theory had failed to explain. Among these phenomena are the characteristic colours of the northern lights.

In this chapter, you will learn about the developments that led to the modern model of the atom. You will also learn about the model itself and how it relates to periodic trends and the periodic table.

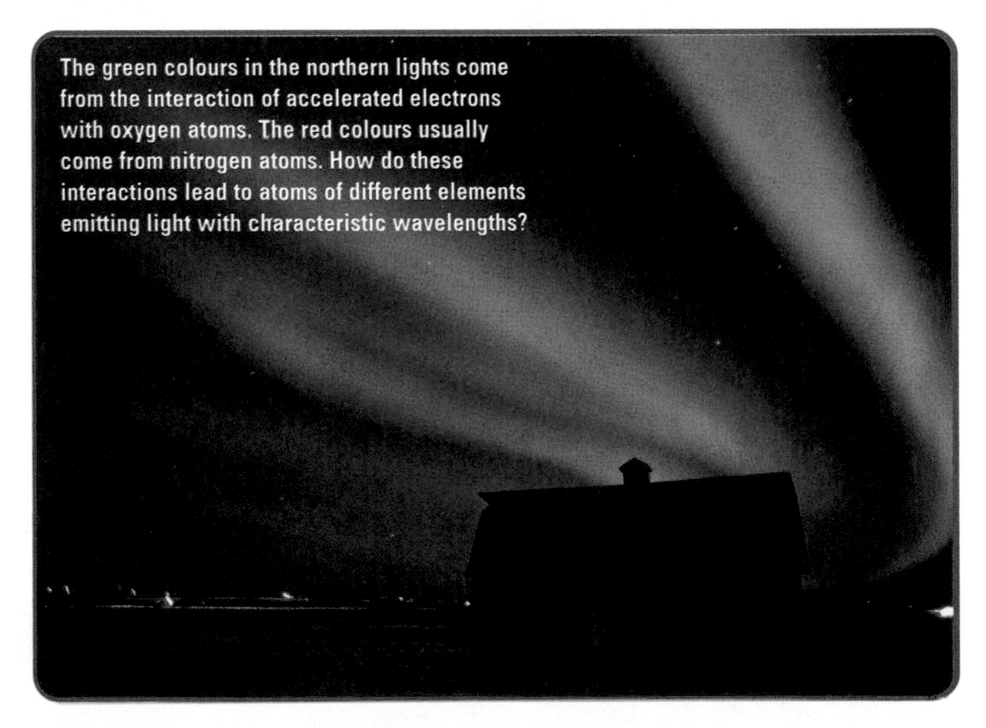
The green colours in the northern lights come from the interaction of accelerated electrons with oxygen atoms. The red colours usually come from nitrogen atoms. How do these interactions lead to atoms of different elements emitting light with characteristic wavelengths?

The Nuclear Atomic Model

3.1

Toward the close of the nineteenth century, chemists had two invaluable conceptual tools to aid them in their understanding of matter. The first was John Dalton's atomic theory, which you have studied intensively in previous chemistry courses. Dalton's atomic theory, first published in 1809, provided chemists with a framework for describing and explaining the behaviour of matter during chemical reactions. As you can see in Figure 3.1, the model of the atom that resulted from this theory was very simple.

The second conceptual tool was Dmitri Mendeleev's periodic table, which listed the known elements in order of increasing atomic mass. The resulting organizational chart arranged elements so that those with similar chemical properties were grouped together in the same column.

As you know, Dalton's atomic theory no longer applies in its original form, and Mendeleev's periodic table has undergone many changes. For example, scientists later discovered that atoms are not the most basic unit of matter because they *are* divisible. As well, the modern periodic table lists the elements in order of their atomic number, not their atomic mass. Of course, it also includes elements that had not been discovered in Mendeleev's time. Even so, in modified form, both of these inventions are still studied and used today in every chemistry course around the world.

When Mendeleev invented the periodic table, he was well-acquainted with Dalton's atomic theory. He knew nothing, however, about subatomic particles, and especially the electron, which is the foundation for the modern periodic table's distinctive shape. Because the original periodic table developed out of experimental observations, chemists did not need an understanding of atomic structure to develop it. (As you will see in section 3.3, however, the periodic table easily accommodates details about atomic structure. In fact, you will learn that the modern periodic table's distinctive design is a natural consequence of atomic structure.)

The First Step Toward the Modern Atomic Model

Chemists needed Dalton's atomic theory to advance their understanding of matter and its behaviour during chemical reactions. His atomic model, however, was inadequate for explaining the behaviour of substances. For example, Dalton designed a system of symbols to show how atoms combine to form other substances. Figure 3.2 on the next page shows several of these symbols. As you will no doubt notice, Dalton correctly predicted the formulas for carbon dioxide and sulfur trioxide, but ran into serious trouble with water, ammonia, and methane. Dalton's attempt at molecular modelling highlights a crucial limitation with his atomic model. Chemists could not use it to explain *why* atoms of elements combine in the ratios in which they do. This inability did not prevent chemists from pursuing their studies. It did, however, suggest the need for a more comprehensive atomic model.

Section Preview/ Specific Expectations

In this section, you will

- **describe** and **explain** the experimental observations that led to Rutherford's nuclear model of the atom
- **describe** and **explain** the modifications that Niels Bohr made to the nuclear model of the atom
- **distinguish** between a quantized model of the atom and the model proposed by Rutherford
- **communicate** your understanding of the following terms: *nuclear model, emission spectrum, absorption spectrum, quantum, photons*

CONCEPT CHECK

In your notebook, list the main ideas in Dalton's atomic theory. Explain how this theory enabled chemists to explain the three mass laws: the law of conservation of mass, the law of definite proportions, and the law of multiple proportions.

Figure 3.1 The model of the atom in 1809. The atom, as Dalton pictured it, was a tiny, solid, indestructible sphere.

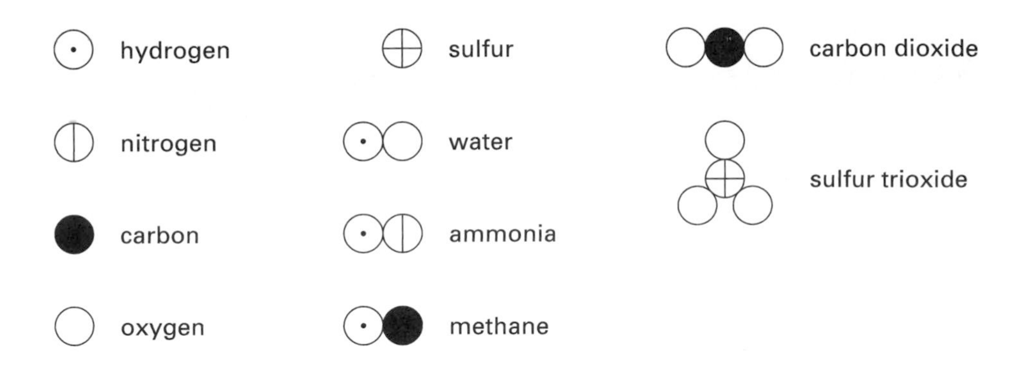

Figure 3.2 Examples of Dalton's system of symbols for atoms and molecules

The Discovery of the Electron Requires a New Atomic Model

In 1897, Dalton's idea of an indivisible atom was shattered with a startling announcement. A British scientist, Joseph John Thomson, had discovered the existence of a negatively charged particle with mass less than $\frac{1}{1000}$ that of a hydrogen atom. This particle was, of course, the electron. (Later calculations showed that the mass of an electron is $\frac{1}{1837}$ that of a hydrogen atom.) It took several years for chemists to consider the consequences of this discovery. They realized that if atoms contain electrons, atoms must also contain a positive charge of some kind to balance the negative charge. The atomic model that Thomson eventually proposed is shown in Figure 3.3. Keep in mind that scientists had not yet discovered the proton. Therefore, in this model, the *entire sphere* carries a uniform, positive charge.

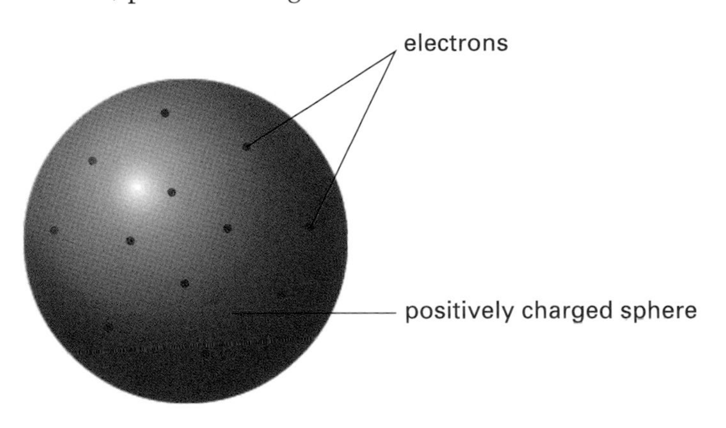

Figure 3.3 The atomic model in 1903. Thomson viewed the atom as a positively charged sphere embedded with sufficient numbers of electrons to balance (neutralize) the total charge.

Scientists initially described radioactivity solely in terms of radiation. The idea of radioactive *particles* first appeared around the turn of the twentieth century. In 1909, Ernest Rutherford reported confidently that the alpha particle was, in fact, a helium nucleus, $^{4}_{2}He$, with a 2+ charge. Scientists still had not discovered the proton by this time, so the nature of the helium nucleus (or any other atomic nucleus) was still unknown. In 1919, the existence of the proton was confirmed experimentally by, appropriately enough, Rutherford himself.

Rutherford's Nuclear Model of the Atom

In the last four years of the nineteenth century, scientists in France—notably Henri Becquerel and Marie and Pierre Curie—discovered that certain elements are radioactive. That is, their atoms naturally emit positively charged particles (alpha particles), negatively charged particles (beta particles), and energy (gamma radiation).

Working in his laboratory at Montreal's McGill University, Ernest Rutherford studied the chemistry of radioactive elements intensively from 1898 to 1907. His efforts would lead to a Nobel Prize in Chemistry in 1908. Greater achievements lay ahead, however.

In 1909, two of Rutherford's students reported observations that cast doubts on Thomson's atomic model. As part of ongoing investigations into the nature and properties of radioactive emissions, they aimed alpha (α) particles at extremely thin metal foils. A small number of the alpha particles, about one in every 8000, were deflected significantly by the atoms that made up the metal foils. These observations were inconsistent with Thomson's atomic model. The researchers expected the alpha particles to pass through the metal atoms with, at most, deflections averaging $\frac{1}{200}$ of a degree. Therefore, deflections of 90° *and more* strained the credibility of Thomson's model. Rutherford encouraged further investigation. Either the observations and data were flawed, or Thomson's atomic model was.

As you may recall from earlier studies, the flaw lay in Thomson's model. In 1911, Rutherford published the results of the now-famous gold-foil experiment, shown in Figure 3.4. On the basis of this experiment, Rutherford suggested that the deflections he and his students observed were caused by an encounter between an alpha particle and an intense electric field at the centre of the atom.

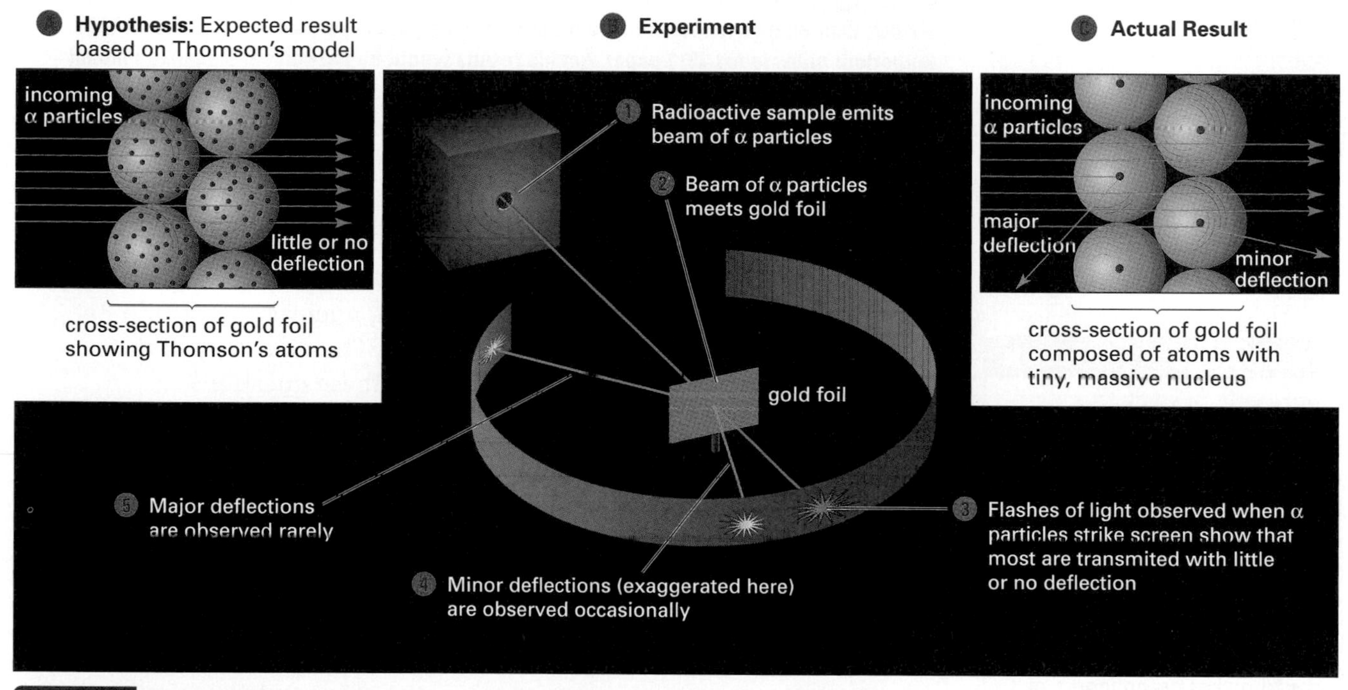

Figure 3.4 **The hypothesis, experiment, and results of Rutherford's gold foil experiment. The experimental hypothesis and design owed much to the contributions of Rutherford's students, Hans Geiger (of Geiger-counter fame) and Ernest Marsden.**

Rutherford performed several calculations that led him to an inescapable conclusion: the atom is made up mainly of empty space, with a small, massive region of concentrated charge at the centre. Soon afterward, the charge on this central region was determined to be positive, and was named the atomic nucleus. Because Rutherford's atomic model, shown in Figure 3.5 on the next page, pictures electrons in motion around an atomic nucleus, chemists often call this the **nuclear model** of the atom. You may also see it referred to as a planetary model because the electrons resemble the planets in motion around a central body.

Contrary to common belief, Rutherford's 1911 paper on the discovery of the nucleus does *not* associate the nucleus with a positive charge. In fact, Rutherford acknowledged that the results of his experiment were independent of the sign of the charge. He suggested that the charge of the nucleus, as well as other questions raised by his paper, were subjects for later investigation. Sometime between 1911 and 1913, someone identified the charge on the nucleus as positive. Whether or not this person was Rutherford is not clear.

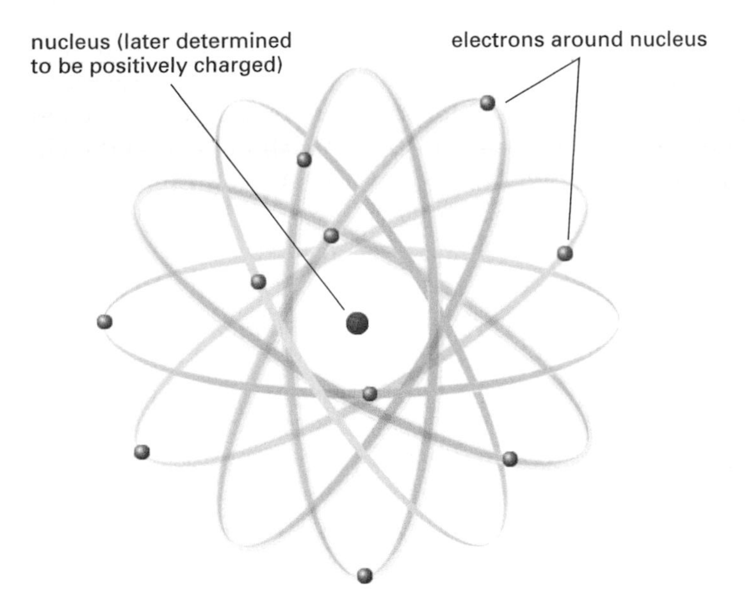

Figure 3.5 The atomic model in 1911. A Japanese scientist, Hantaro Nagaoka, proposed a similar, disk-shaped model with electrons orbiting a positively charged nucleus, in 1904. Rutherford notes in his 1911 paper that his results would be the same if Nagaoka's model were correct.

Prelude to a New Atomic Model

Rutherford's atomic model solved problems inherent in Thomson's atomic model, but it also raised others. For example, an atomic nucleus composed entirely of positive charges should fly apart due to electrostatic forces of repulsion. Furthermore, Rutherford's nuclear atom could not adequately explain the total mass of an atom. The discovery of the neutron, in 1932, eventually helped to settle these questions.

There was a more significant problem, however. Rutherford's atomic model seemed to contradict the laws of nineteenth-century physics. According to these assumptions, an electron in motion around a central body *must* continuously give off radiation. Consequently, one should be able to observe a continuous spectrum (a "rainbow") of light energy as the electron gives off its radiation.

Because the electron should also lose energy as a result of this radiation, the radius of its orbit should continuously decrease in size until it spirals into the nucleus. This predicted annihilation of the atom would occur in a fraction of a second. However, atoms were not seen to destabilize as predicted by this model.

Nineteenth-century physics could not explain why Rutherford's model corresponded to a stable atom. The solution to this problem marked a turning point in the history of chemistry and physics. To appreciate the scope of the problem, and the impact of its eventual solution, it is necessary to remain a short while longer in the nineteenth century. In the following pages, you will examine a phenomenon that scientists of the time could not explain.

Technology LINK

Chemists used spectral analysis during the nineteenth century to analyze substances and, sometimes, to discover new elements. Another common technique for analyzing substances, often used in conjunction with spectral analysis is a flame test. One of the foremost practitioners of this technique was a chemist named Robert Bunsen. Find out why he invented his famous burner. Carry on your research to investigate other ways that chemists use spectral analysis to examine the composition of substances. Select an appropriate medium to report your findings.

The Problem of Atomic Spectra

You probably recall that visible light is part of a broader continuum of energy called the electromagnetic spectrum. Figure 3.6 reviews the key properties of the electromagnetic spectrum of energy.

Despite differences in wavelengths and frequency, all electromagnetic radiation travels at the same speed in a vacuum: 3.00×10^{-8} m/s. This value, the speed of light, is a constant represented by the symbol *c*.

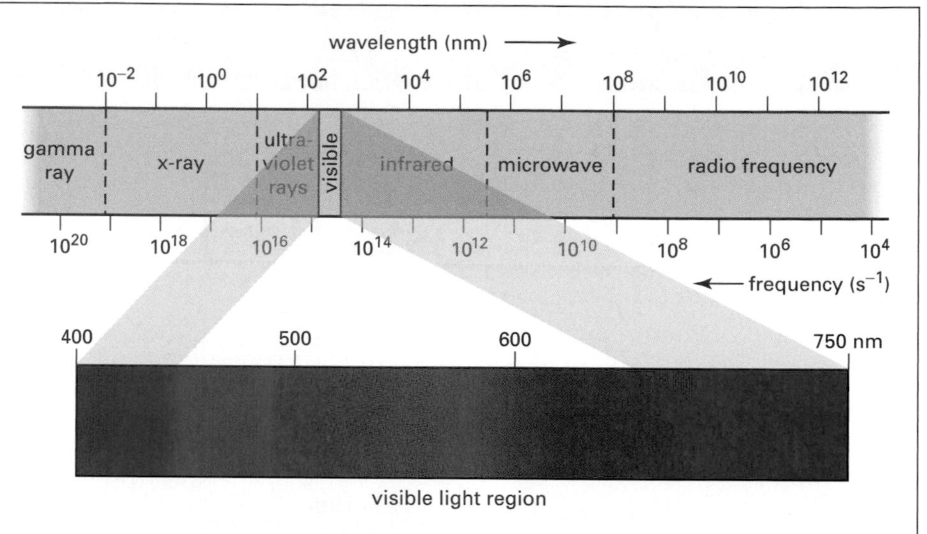

The relationship between frequency (symbol ν, measured in Hz) and wavelength (symbol λ, usually measured in nm or m, depending on the energy). As wavelength increases, frequency decreases. As wavelength decreases, frequency increases. (Note that 1 Hz = 1 s^{-1}. Note also that the frequency symbol, ν, is the Greek letter nu, not the letter v.)

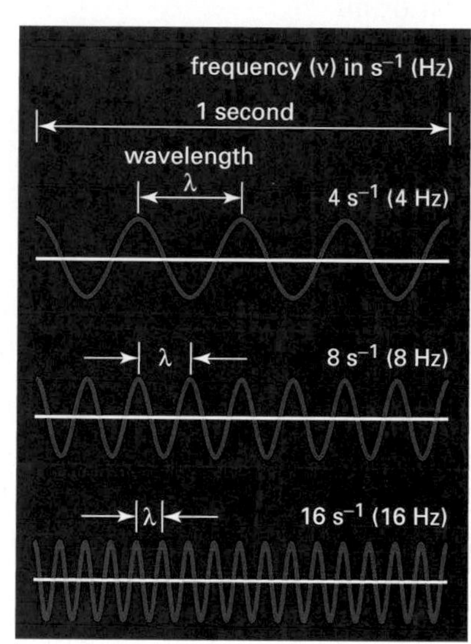

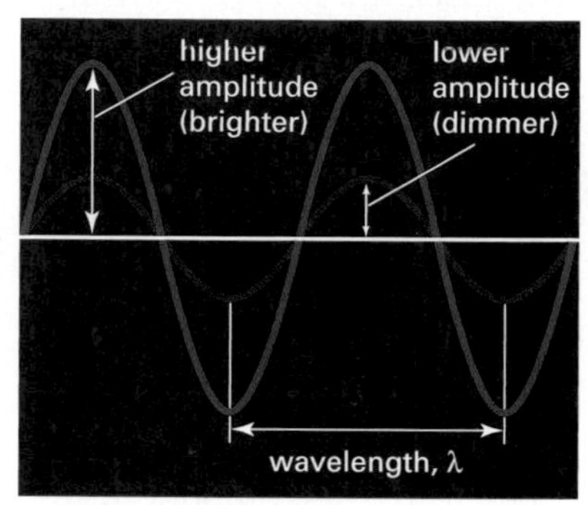

The amplitude (height) of a wave represents its intensity. A wave with a higher amplitude has a greater intensity (is brighter) than a wave with a lower amplitude.

Figure 3.6 **The electromagnetic spectrum and its properties**

The visible portion of the electromagnetic spectrum is called a continuous spectrum, because the component colours are indistinct. They appear "smeared" together into a continuum of colour. According to nineteenth-century physics, part of the energy emitted by electrons *should* be observable as a continuous spectrum. This is not, however, the case. Instead, when atoms absorb energy (for example, when they are exposed to an electric current), you observe a pattern of discrete (distinct), coloured lines separated by spaces of varying length. See Figure 3.7. You can also observe this *line spectrum* for hydrogen, and for other atoms, in Investigation 3-A.

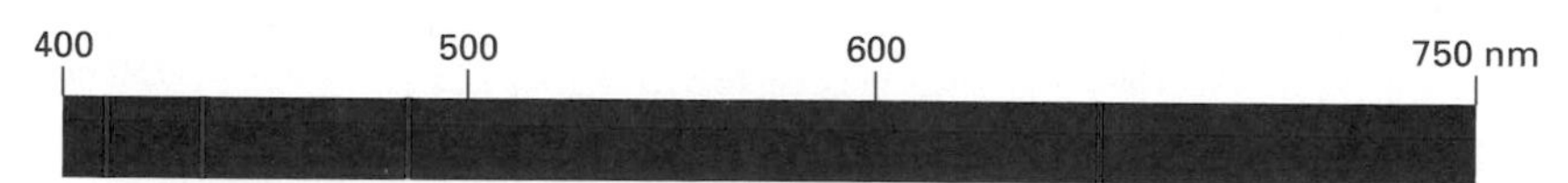

Figure 3.7 **The discrete, coloured lines of this spectrum are characteristic of hydrogen atoms. No other atoms display this pattern of coloured lines.**

SKILL FOCUS

Analyzing and interpreting

Communicating results

Investigation 3-A

Atomic Emission Spectra (Teacher Demonstration)

When a high voltage current is passed through a glass tube that contains hydrogen gas at low pressure, the gas glows with a pinkish-purple colour. If you observe this gas discharge tube through a spectroscope or a diffraction grating, you can see distinct, coloured lines. This type of spectrum is called a line spectrum or an emission spectrum. Chemists first observed spectra like this in the mid-1800s.

Question

Can the Rutherford model of the atom explain the emission spectra of elements?

Predictions

The Rutherford model of the atom described electrons as tiny masses in motion around a nucleus. According to this model, excited atoms should emit a continuous spectrum of light energy as their electrons give off radiation. In fact, according to this model, atoms should continually radiate energy whether they are excited or not. Read through the procedure and predict whether your observations will support or contradict the Rutherford model. Give reasons for your answer.

Safety Precautions

- A very high voltage is required to operate the gas discharge tubes. Do not come into contact with the source while viewing the tubes.
- Do not work with the gas discharge tubes yourself. Your teacher will demonstrate how it works.

Materials

gas discharge tube apparatus (teacher use only)
low pressure hydrogen gas tube and tubes for other available elements
spectroscope or diffraction grating

Procedure

1. Practise using the spectroscope or diffraction grating by observing an incandescent light bulb. Point the slit of the spectroscope toward the bulb and move the spectroscope until you can clearly see the spectrum.
2. Record the appearance of the spectrum from the incandescent bulb.
3. Your teacher will set up the gas discharge tube apparatus to demonstrate the emission spectra of hydrogen and other elements.
4. Observe the hydrogen discharge tube, and note its colour when high-voltage current is applied to it.
5. With the lights dimmed, examine the hydrogen emission spectrum with the aid of a spectroscope or diffraction grating. Make a careful sketch to record your observations. Note in particular the colours you see. Estimate their wavelengths. (Use the figure below to help you estimate the wavelengths.)
6. Follow steps 4 and 5 for the emission spectra produced by other gas discharge tubes that your teacher has available.

400 500 600 750 nm
H

Analysis

1. According to Rutherford's model of the atom and nineteenth-century physics, electrons could emit or absorb any quantity of energy.
   (a) Compare the spectrum you observed for the incandescent bulb to the spectrum that Rutherford's model predicts for hydrogen.
   (b) Compare the spectrum you observed for hydrogen to the spectrum you observed for the incandescent bulb. Do your observations seem to support or contradict Rutherford's model? Explain your answer.
   (c) How do you think Rutherford's model would have to be modified in order to account for your observations?
2. How do you think the lines you observed in the hydrogen emission spectrum relate to the energy of the electrons in a hydrogen atom? Include a diagram in your answer.
3. Compare the hydrogen emission spectrum to the other spectra you observed. Why do you think emission spectra are different for different elements?
4. Why do you think that scientists initially focused their attention on the hydrogen emission spectrum rather than the emission spectrum of another element?
5. Each element has a characteristic emission spectrum. Why is it only possible to observe the emission spectra of selected elements in a high school laboratory?

Conclusion

6. Reread your prediction and state whether it was accurate or not. Explain your answer in detail.

Applications

7. The emission spectrum of each element is characteristic of that element. In other words, it is a "fingerprint" that can be used to identify the element.
   (a) Suggest reasons to explain why each element has a distinct and characteristic emission spectrum.
   (b) Helium was discovered in the Sun before it was discovered on Earth. How do you think this was possible? Use print resources or the Internet to help you answer the question.
8. Examine the spectrum below. It is the *absorption spectrum* for sodium. Absorption spectra are produced when atoms absorb light of specific wavelengths.
   (a) If you examined the emission spectrum for sodium in the laboratory, compare it to the absorption spectrum for sodium. (If you did not, look for a picture of the emission spectrum for sodium in a book or on the Internet.)
   (b) In general, how do you think emission spectra compare to absorption spectra?
   (c) Instruments based on absorption spectra are used extensively in analytical chemistry. One such instrument is an atomic absorption spectrophotometer. Use print and electronic resources to find out how the instrument works, and for what applications it is used.

Na

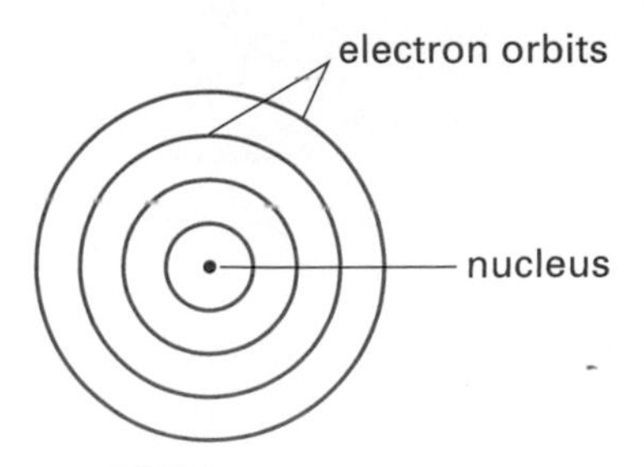

Figure 3.8 The model of the atom in 1913. Although Bohr pictured the atom in three dimensions, it is presented here, in two dimensions, to simplify some of Bohr's revolutionary concepts.

The Bohr Model of the Atom

Scientists of the nineteenth century lacked the concepts necessary to explain line spectra. Even in the first decade of the twentieth century, a suitable explanation proved elusive. This changed in 1913 when Niels Bohr, a Danish physicist and student of Rutherford, proposed a new model for the hydrogen atom. This model retained some of the features of Rutherford's model. More importantly, it was able to explain the line spectrum for hydrogen because it incorporated several new ideas about energy. As you can see in Figure 3.8, Bohr's atomic model pictures electrons in orbit around a central nucleus. Unlike Rutherford's model, however, in which electrons may move anywhere within the volume of space around the nucleus, Bohr's model imposes certain restrictions.

The Bohr Atomic Model

1. The atom has only specific, allowable energy levels, called stationary states. Each stationary state corresponds to the atom's electrons occupying fixed, circular orbits around the nucleus.
2. While in one of its stationary states, atoms do *not* emit energy.
3. An atom changes stationary states by emitting or absorbing a specific quantity of energy that is exactly equal to the difference in energy between the two stationary states.

On what basis could Bohr make these claims? In 1900, a physicist named Max Planck proposed an idea that was so revolutionary that he himself was unwilling to accept its implications. Planck suggested that *matter, at the atomic level, can absorb or emit only discrete quantities of energy.* Each of these specific quantities is called a **quantum** of energy. In other words, Planck said that the energy of an atom is *quantized*. Something that is quantized can exist only in certain discrete amounts. It is not continuous. In this sense, the rungs of a ladder are quantized, while the smooth slope of a ramp is not. Unlike ladders and ramps, however, a quantum is an extremely small "packet" of energy.

Although Planck said that the energy of matter is quantized, he continued to describe energy as travelling in the form of waves. He was, in fact, unwilling to consider that energy might have particle-like properties. In 1905, however, Albert Einstein was prepared to make just such an assertion. Light, according to Einstein, is also quantized. It occurs as quanta of electromagnetic energy that have particle-like properties. These particle-like "packets" of energy were later called **photons**. In Einstein's view, light (and, by extension, all electromagnetic energy) travels in the form of photons of energy. Light is emitted as photons of energy, and light is absorbed as photons of energy.

Web LINK

www.mcgrawhill.ca/links/chemistry12

The concept of quantization enabled physicists to solve problems that nineteenth-century physics could not. One of these involved the thermal properties of solids when they are heated to incandescence. The other involved the induction of electrical current in metals when they are exposed to only specific frequencies of electromagnetic radiation. Use the Internet to investigate the phenomena known as blackbody radiation and the photoelectric effect. Write a brief report outlining the nature of these phenomena, why nineteenth-century physics could not explain them, and how the new idea of quantization could. Go to the web site above. Click on **Web Links** to find out where to go next.

How Bohr's Atomic Model Explains the Spectrum for Atomic Hydrogen

If you re-read the key points of Bohr's atomic model above, you can see that he is applying the new ideas of quanta in his model. Bohr proposed that the energy that is emitted and absorbed by an atom must have specific values. The change in energy when an electron moves to higher or lower energy levels is not continuous. It is, rather, quantized.

Look again at the line spectrum for hydrogen shown in Figure 3.7. The energy that is associated with the coloured lines in this spectrum corresponds to the change in energy of an electron as it moves to higher or lower energy levels. For example, when a hydrogen atom is exposed to electrical current, or to another form of electromagnetic energy, its electron absorbs photons of energy. The atom is now said to be in an *excited* state. Another common way for atoms to become excited is through atomic collisions. For example, when two hydrogen atoms collide, some of the kinetic energy from one atom is transferred to the other atom. The electron of the second hydrogen atom absorbs this energy and is excited to a higher energy level.

When the electron of a hydrogen atom that has been excited to the third energy level falls to the second energy level, it emits light of certain energy. Specifically, an electron that makes a transition from the third energy level to the second energy level emits a photon of red light with a wavelength of 656 nm. Because line spectra result when atoms in an excited state *emit* photons as they fall to a lower energy level, these spectra are also called **emission spectra**.

Figure 3.9 shows the energy transitions that are responsible for the coloured lines in hydrogen's emission spectrum. Notice the use of the symbol n to designate the allowed energy levels for the hydrogen atom: $n = 1$, $n = 2$, and so on. This symbol, n, represents a positive integer (such as 1, 2, or 3), and is called a *quantum number*. You will learn more about the significance of quantum numbers in section 3.2.

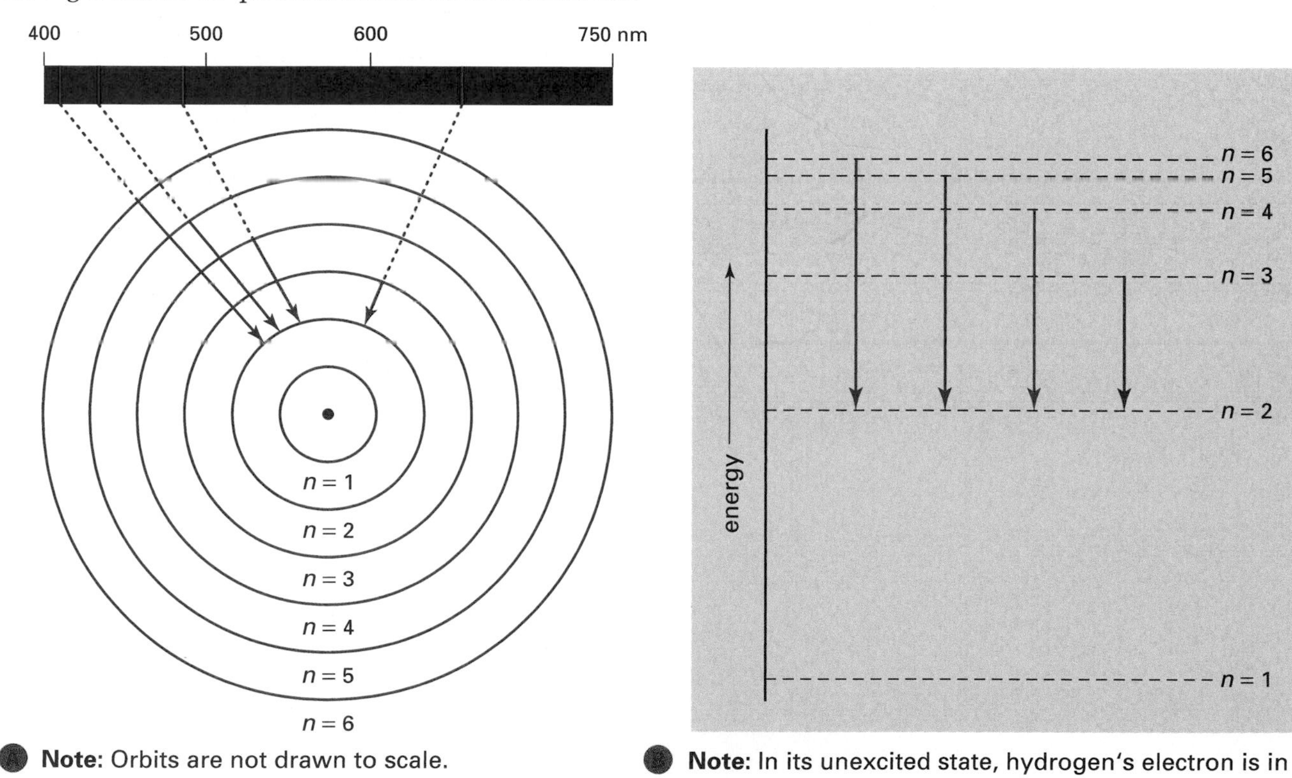

A **Note:** Orbits are not drawn to scale.

B **Note:** In its unexcited state, hydrogen's electron is in the energy level closest to the nucleus: $n = 1$. This is the lowest-possible energy level, representing a state of greatest stability for the hydrogen atom.

Figure 3.9 **How the Bohr model explains the coloured lines in hydrogen's emission spectrum. When an excited electron falls from a higher energy level to a lower energy level (shown by the downward-pointing arrows), it emits a photon with a specific wavelength that corresponds to one of the coloured lines in the spectrum.**

Another kind of spectrum that is related to an atom's emission spectrum is an **absorption spectrum**. This kind of spectrum results when electrons of atoms absorb photons of certain wavelengths, and so are excited from lower energy levels to higher energy levels. Figuro 3.10 shows the absorption spectrum for hydrogen. As you can see, discrete, dark lines appear in the precise locations (wavelengths) as those of the coloured lines in hydrogen's emission spectrum.

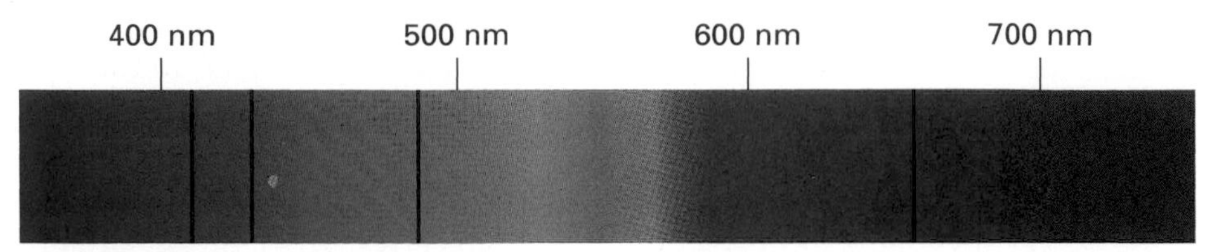

Figure 3.10 **When light passes through gaseous hydrogen atoms, the electrons in the hydrogen atoms absorb photons of red, green, blue, and purple light at specific wavelengths. When the electrons return to lower energy levels, they re-emit photons with these same wavelengths. However, most of the photons are emitted in a different direction from the one in which the electrons absorbed the light. As a result, dark lines appear in the spectrum where photons of those particular wavelengths are absent.**

The Successes and Limitations of the Bohr Atomic Model

An atom has an infinite number of energy levels. In other words, n ranges from 1 to ∞ (infinity). What is the change in energy when a hydrogen atom is excited from $n = 1$ to $n = \infty$? Use the equation below to determine the value. Then consult the periodic table in Appendix C and examine hydrogen's atomic properties. To what property of hydrogen does this value correspond? Is this finding surprising to you? Explain why.

$$\Delta E = E_{\text{final}} - E_{\text{initial}} = -2.18 \times 10^{-18}\text{J} \left(\frac{1}{n^2_{\text{final}}} - \frac{1}{n^2_{\text{initial}}}\right)$$

Bohr's realization that the atom's energy is quantized—that electrons are restricted to specific energy levels (orbits)—was an astounding achievement. As you have seen, this model successfully predicted the coloured lines in the visible-light portion of hydrogen's emission spectrum. It also successfully predicted other lines, shown in Figure 3.11, that earlier chemists had discovered in the ultraviolet and infrared portions of hydrogen's emission spectrum.

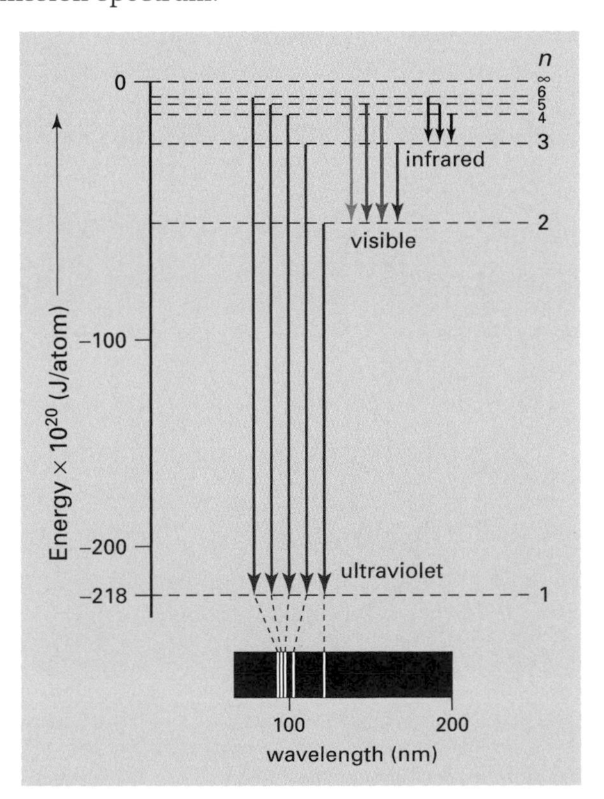

Figure 3.11 **The ultraviolet series of lines appear as a result of electrons falling from higher energy levels to $n = 1$. The infrared lines appear as a result of electrons falling from higher energy levels to $n = 3$.**

There was a problem with Bohr's model, however. It successfully explained *only* one-electron systems. That is, it worked fine for hydrogen and for ions with only one electron, such as He^{+}, Li^{2+}, and Be^{3+}. Bohr's model was unable, however, to explain the emission spectra produced by atoms with two or more electrons. Either Bohr's model was a coincidence, or it was an oversimplification in need of modification. Further investigation was in order.

Careers in Chemistry

Nuclear Medicine

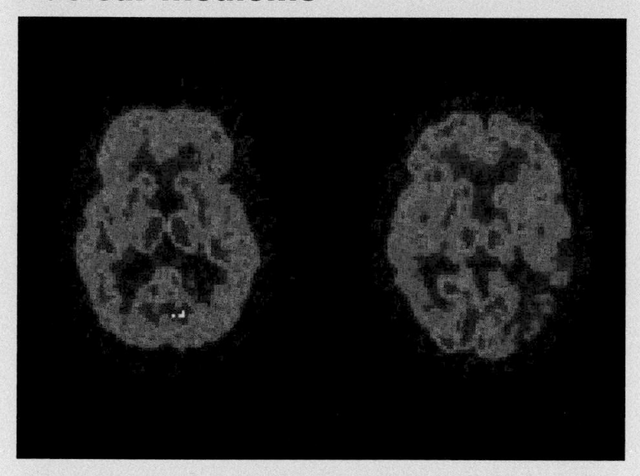

This scan, a product of a nuclear medical technology called PET, shows changes in metabolic activity in different parts of the brain. The red colour indicates greater activity.

The discovery of the atomic nucleus affected all sciences, including biology and its applications in the field of medicine. During her graduate studies at the University of Western Ontario, Dr. Karen Goulenchyn excelled in theoretical mathematics, but switched to medicine because she was drawn to practical applications of science in people's lives. An interest in computers led her to nuclear medicine, which she now practices at the Civic Hospital in Ottawa.

Nuclear medicine is used chiefly in medical diagnosis. A radiopharmaceutical—a relatively harmless compound with a low dose of radiation—is swallowed or injected into the patient and tracked through the bloodstream by instruments such as a PET (positron emission tomography) camera. The nuclear physician can use the results to create a three-dimensional computer image that evaluates the function as well as the structure of an organ. This procedure enables the physician to diagnose cancers and other tissue irregularities without the need for more invasive techniques such as exploratory surgery.

The dose of radiation that patients receive is similar to that of a diagnostic x-ray. However, because many patients are alarmed by the word "nuclear," Dr. Goulenchyn says that her team must explain the procedure tactfully "so that they don't run away." Some physicians ease fears by calling the discipline "molecular medicine."

Dr. Goulenchyn sees an increase of 6 to12 percent of patients annually, primarily for suspected cancer or heart disease. Her greatest frustration is the time she must spend lobbying for new equipment, particularly for PET, a technique for measuring the concentrations of positron-emitting radioisotopes within the tissues. Positrons are a form of radiation identical to a beam of electrons, except that the charge of a positron is positive. PET scans can assess biochemical changes in the body, especially abnormal ones. These scans provide greater accuracy in determining whether a current or proposed therapy is effective.

A career in nuclear medicine, Dr. Goulenchyn notes, requires excellence in computers, science, and people skills, with a strong background in physiology. It also offers an unusual benefit for a physician: regular working hours. She appreciates the time this provides for her family, as well as for her work with the Canadian Association of Nuclear Medicine, for which she served as president from 1996 to 1998.

Section Summary

You have seen how scientists in the late nineteenth and early twentieth century developed and modified the atomic model. Changes in this model resulted from both experimental evidence and new ideas about the nature of matter and energy. By 1913, chemists and physicists had a working model that pointed tantalizingly in a promising direction. During the third decade of the twentieth century, the promise was fulfilled. In the next section, you will learn how physicists extended the ideas of Planck, Einstein, and Bohr to develop an entirely new branch of physics, and a new model of the atom.

Section Review

1 (a) K/U In what ways did Rutherford's model of the atom differ from Thomson's?

(b) On the basis of what evidence did Rutherford propose his model?

2 K/U Why did Rutherford's atomic model require modification? Be as specific as possible in your answer.

3 (a) C Imagine that you are Niels Bohr. Explain your ideas about the atom to scientists who are familiar with the work of Planck and Einstein.

(b) Imagine that you are a nineteenth-century scientist who has never heard of Planck and Einstein's ideas. You have just reviewed Bohr's proposed model for the atom. Write a paragraph describing your response to Bohr's model.

4 C List and arrange the types of electromagnetic radiation

(a) in order from longest wavelength to shortest wavelength

(b) in order from highest frequency to lowest frequency

5 C Compare visible light radiation to ultraviolet radiation. How are they different? How are they the same?

6 I Explain the significance, in terms of absorption or emission of photons, of the following statements.

(a) An electron moves from $n = 1$ to $n = 6$.

(b) An electron moves from $n = 5$ to $n = 2$.

(c) An electron moves from $n = 4$ to $n = 3$.

7 C Arrange the following in order from lowest energy to highest energy, and justify your sequence: $n = 7$, $n = 2$, $n = 5$, $n = 4$, $n = 1$.

The Quantum Mechanical Model of the Atom

3.2

You have seen how Bohr's model of the atom explains the emission spectrum of hydrogen. The emission spectra of other atoms, however, posed a problem. A mercury atom, for example, has many more electrons than a hydrogen atom. As you can see in Figure 3.12, mercury has more spectral lines than hydrogen does. The same is true for other many-electron atoms. Observations like these forced Bohr and other scientists to reconsider the nature of energy levels. The large spaces between the individual colours suggested that there are energy differences between individual energy levels, as stated in Bohr's model. The smaller spaces between coloured lines, however, suggested that there were smaller energy differences *within* energy levels. In other words, scientists hypothesized that there are *sublevels* within each energy level. Each of these sublevels has its own slightly different energy.

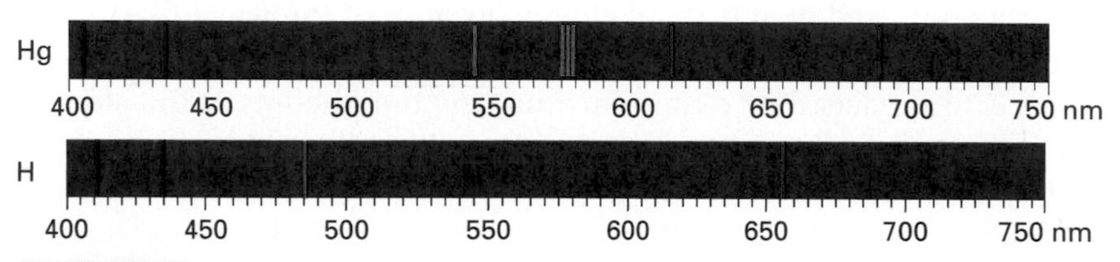

Figure 3.12 The emission spectrum for mercury shows that it has more spectral lines than the emission spectrum for hydrogen.

It was fairly straightforward to modify Bohr's model to include the idea of energy sublevels for the hydrogen spectrum and for atoms or ions with only one electron. There was a more fundamental problem, however. The model still could not explain the spectra produced by many-electron atoms. Therefore, a simple modification of Bohr's atomic model was not enough. The many-electron problem called for a new model to explain spectra of all types of atoms. However, this was not possible until another important property of matter was discovered.

The Discovery of Matter Waves

By the early 1920s, it was standard knowledge that energy had matter-like properties. In 1924, a young physics student named Louis de Broglie stated a hypothesis that followed from this idea. What if, de Broglie wondered, matter has wave-like properties?

He developed an equation that enabled him to calculate the wavelength associated with any object — large, small, or microscopic. For example, a baseball with a mass of 142 g and moving with a speed of 25.0 m/s has a wavelength of 2×10^{-34} m. Objects that you can see and interact with, such as a baseball, have wavelengths so small that they do not have any significant observable effect on the object's motion. However, for microscopic objects, such as electrons, the effect of wavelength on motion becomes very significant. For example, an electron moving at a speed of 5.9×10^{6} m/s has a wavelength of 1×10^{-10} m. The size of this wavelength is greater than the size of the hydrogen atom to which it belongs. (The calculated atomic radius of hydrogen is 5.3×10^{-11} m.)

Section Preview/ Specific Expectations

In this section, you will

- **describe** the quantum mechanical model of the atom and its historical development
- **state** the meaning and significance of the first three quantum numbers
- **communicate** your understanding of the following terms: *quantum mechanical model of the atom, orbitals, ground state, principal quantum number (n), orbital-shape quantum number (l), magnetic quantum number (m_l)*

De Broglie's hypothesis of matter waves received experimental support in 1927. Researchers observed that streams of moving electrons produced diffraction patterns similar to those that are produced by waves of electromagnetic radiation. Since diffraction involves the transmission of waves through a material, the observation seemed to support the idea that electrons had wave-like properties.

The Quantum Mechanical Model of the Atom

Figure 3.13 The model of the atom in 1927. The fuzzy, spherical region that surrounds the nucleus represents the volume in which electrons are most likely to be found.

In 1926, an Austrian physicist, Erwin Schrödinger, used mathematics and statistics to combine de Broglie's idea of matter waves and Einstein's idea of quantized energy particles (photons). Schrödinger's mathematical equations and their interpretations, together with another idea called Heisenberg's uncertainty principle (discussed below), resulted in the birth of the field of *quantum mechanics*. This is a branch of physics that uses mathematical equations to describe the wave properties of sub-microscopic particles such as electrons, atoms, and molecules. Schrödinger used concepts from quantum mechanics to propose a new atomic model: the **quantum mechanical model of the atom**. This model describes atoms as having certain allowed quantities of energy because of the wave-like properties of their electrons.

Figure 3.13 depicts the volume surrounding the nucleus of the atom as being indistinct or cloud-like because of a scientific principle called the *uncertainty principle*. The German physicist Werner Heisenberg proposed the uncertainty principle in 1927. Using mathematics, Heisenberg showed that it is impossible to know both the position and the momentum of an object beyond a certain measure of precision. (An object's momentum is a property given by its mass multiplied by its velocity.) According to this principle, if you can know an electron's precise position and path around the nucleus, as you would by defining its orbit, you cannot know with certainty its velocity. Similarly, if you know its precise velocity, you cannot know with certainty its position. Based on the uncertainty principle, Bohr's atomic model is flawed because you cannot assign fixed paths (orbits) to the motion of electrons.

Clearly, however, electrons exist. And they must exist *somewhere*. To describe where that "somewhere" is, scientists used an idea from a branch of mathematics called statistics. Although you cannot talk about electrons in terms of certainties, you *can* talk about them in terms of probabilities. Schrödinger used a type of equation called a wave equation to define the probability of finding an atom's electrons at a particular point within the atom. There are many solutions to this wave equation, and each solution represents a particular wave function. Each wave function gives information about an electron's energy and location within an atom. Chemists call these wave functions **orbitals**.

In further studies of chemistry and physics, you will learn that the wave functions that are solutions to the Schrödinger equation have no direct, physical meaning. They are mathematical ideas. However, the square of a wave function does have a physical meaning. It is a quantity that describes the probability that an electron is at a particular point within the atom at a particular time. The square of each wave function (orbital) can be used to plot three-dimensional probability distribution graphs for that orbital. These plots help chemists visualize the space in which electrons are most likely to be found around atoms. These plots are also referred to as electron probability density graphs. Note: Although orbitals are wave functions without associated physical characteristics like shape and size, chemists often use the term orbitals when they *mean* three-dimensional probability distribution graphs. To simplify discussion, this textbook will discuss the size and shapes of orbitals. However, what is really meant is their associated probability distribution graph, which is calculated using the square of the wave function.

Each orbital has its own associated energy, and each represents information about where, inside the atom, the electrons would spend most of their time. Scientists cannot determine the actual paths of the moving electrons. However, orbitals indicate where there is a high probability of finding electrons.

Figure 3.14A represents the probability of finding an electron at any point in space when the electron is at the lowest energy level ($n = 1$) of a hydrogen atom. Where the density of the dots is greater, there is a higher probability of finding the electron. This graph is "fuzzy-looking" because the probability of finding the electron anywhere in the $n = 1$ energy level of a hydrogen atom is never zero. Farther from the nucleus, the probability becomes very small, but it will still never reach zero. Therefore, because the shape of the $n = 1$ orbital for hydrogen represents the level of probability of finding an electron, and since the probability never reaches zero, you have to select a "cut-off" level of probability. A level of probability is usually expressed as a percentage. Therefore, the contour line in Figure 3.14B defines an area that represents 95 percent of the probability graph. This two-dimensional shape is given three dimensions in Figure 3.14C. What this means is that, at any time, there is a 95 percent chance of finding the electron within the volume defined by the spherical contour.

CONCEPT CHECK

Distinguish clearly between an electron orbit, as depicted in Bohr's atomic model, and an electron orbital, as depicted in the quantum mechanical model of the atom.

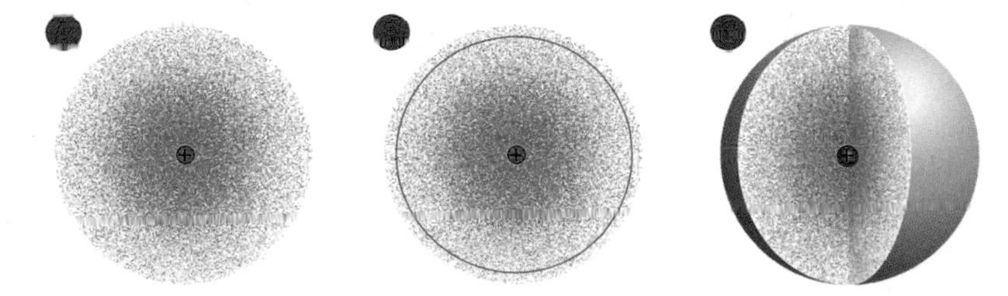

Figure 3.14 **Electron density probability graphs for the lowest energy level in the hydrogen atom. These diagrams represent the probability of finding an electron at any point in this energy level.**

Quantum Numbers and Orbitals

Figure 3.14 showed electron-density probabilities for the lowest energy level of the hydrogen atom. This is the most stable energy state for hydrogen, and is called the **ground state**. The quantum number, n, for a hydrogen atom in its ground state is 1. When $n = 1$ in the hydrogen atom, its electron is associated with an orbital that has a characteristic energy and shape. In an excited state, the electron is associated with a different orbital with its own characteristic energy and shape. This makes sense, because the electron has absorbed energy in its excited state, so its total energy increases and its motion changes. Figure 3.15 on the next page compares the sizes of hydrogen's atomic orbitals when the atom is in its ground state and when it is in an excited state.

Figure 3.15 The relationship between orbital size and quantum number for the hydrogen atom. As n increases, the electron's energy increases and orbital size increases.

Orbitals have a variety of different possible shapes. Therefore, scientists use three quantum numbers to describe an atomic orbital. One quantum number, n, describes an orbital's energy level and size. A second quantum number, l, describes an orbital's shape. A third quantum number, m_l, describes an orbital's orientation in space. These three quantum numbers are described further below. The Concept Organizer that follows afterward summarizes this information. (In section 3.3, you will learn about a fourth quantum number, m_s, which is used to describe the electron inside an orbital.)

The First Quantum Number: Describing Orbital Energy Level and Size

The **principal quantum number (n)** is a positive whole number that specifies the energy level of an atomic orbital and its relative size. The value of n, therefore, may be 1, 2, 3, and so on. A higher value for n indicates a higher energy level. A higher n value also means that the size of the energy level is larger, with a higher probability of finding an electron farther from the nucleus. The greatest number of electrons that is possible in any energy level is $2n^2$.

The Second Quantum Number: Describing Orbital Shape

The second quantum number describes an orbital's shape, and is a positive integer that ranges in value from 0 to $(n - 1)$. Chemists use a variety of names for the second quantum number. For example, you may see it referred to as the angular momentum quantum number, the azimuthal quantum number, the secondary quantum number, or the orbital-shape quantum number.

Regardless of its name, the second quantum number refers to the energy sublevels within each principal energy level. The name that this book uses for the second quantum number is **orbital-shape quantum number (l)**, to help you remember that the value of l determines orbital shape. (You will see examples of orbital shapes near the end of this section.)

The value of n places precise limits on the value of l. Recall that l has a maximum value of $(n - 1)$. So, if $n = 1$, $l = 0$ (that is, $1 - 1$). If $n = 2$, l may be either 0 or 1. If $n = 3$, l may be either 0, 1, or 2. Notice that the number of possible values for l in a given energy level is the same as the value of n. In other words, if $n = 2$, then there are only two possible sublevels (two types of orbital shapes) at this energy level.
Each value for l is given a letter: s, p, d, or f.

- The $l = 0$ orbital has the letter s.
- The $l = 1$ orbital has the letter p.
- The $l = 2$ orbital has the letter d.
- The $l = 3$ orbital has the letter f.

To identify an energy sublevel (type of orbital), you combine the value of n with the letter of the orbital shape. For example, the sublevel with $n = 3$ and $l = 0$ is called the $3s$ sublevel. The sublevel with $n = 2$ and $l = 1$ is the $2p$ sublevel.

There are, in fact, additional sublevels beyond $l = 3$. However, for chemical systems known at this time, only the *s*, *p*, *d*, and *f* sublevels are required.

The Third Quantum Number: Describing Orbital Orientation

The **magnetic quantum number (m_l)** is an integer with values ranging from $-l$ to $+l$, including 0. This quantum number indicates the orientation of the orbital in the space around the nucleus. The value of m_l is limited by the value of l. If $l = 0$, m_l can be only 0. In other words, for a given value of n, there is only one orbital, of s type ($l = 0$). If $l = 1$, m_l may have one of three values: −1, 0, or +1. In other words, for a given value of n, there are three orbitals of *p* type ($l = 1$). Each of these *p* orbitals has the same shape and energy, but a different orientation around the nucleus. Notice that for any given value of l, there are $(2l + 1)$ values for m_l.

The total number of orbitals for any energy level n is given by n^2. For example, if $n = 2$, it has a total of 4 (that is, 2^2) orbitals (an *s* orbital and three *p* orbitals). The Sample Problem below shows further use of this calculation.

The letters used to represent energy sublevels are abbreviations of names that nineteenth-century chemists used to describe the coloured lines in emission spectra. These names are *s*harp, *p*rincipal, *d*iffuse, and *f*undamental.

Concept Organizer **The Relationship Among the First Three Quantum Numbers**

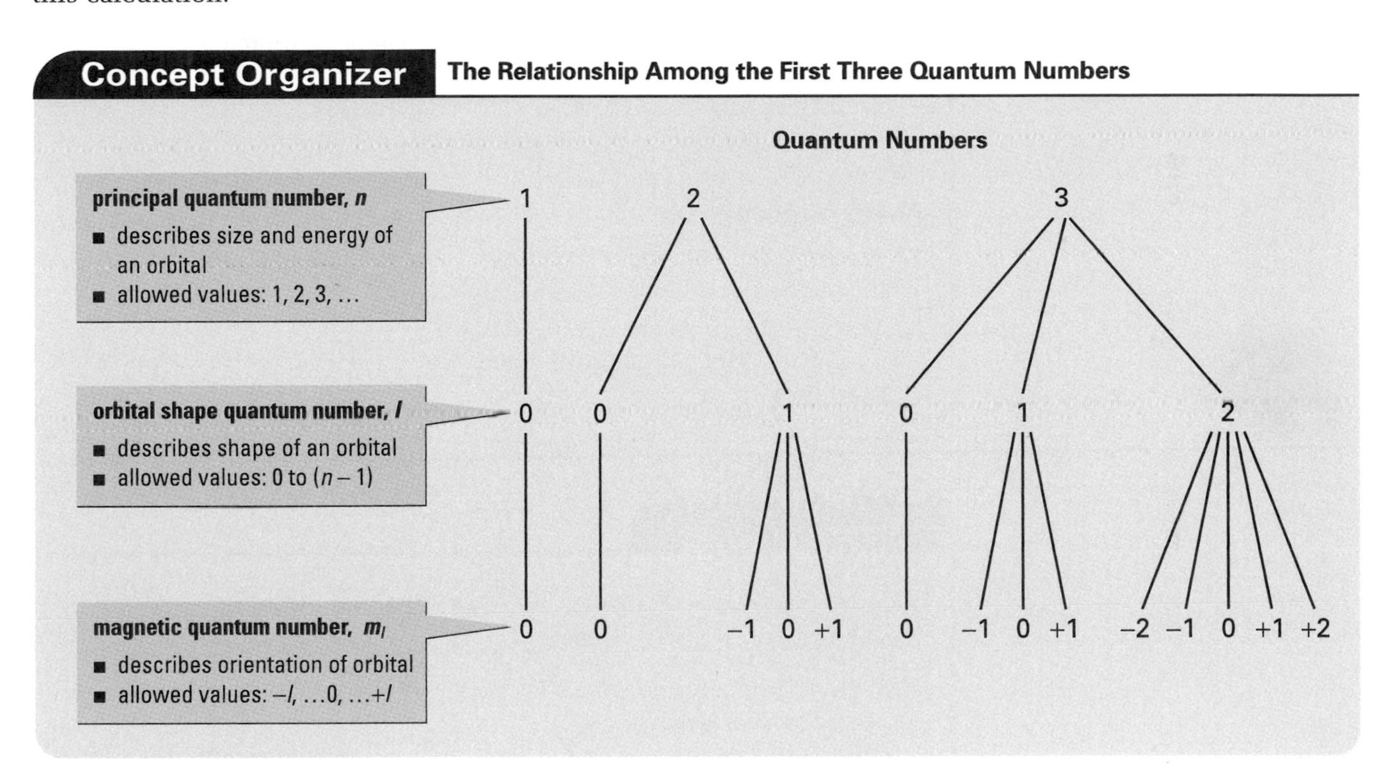

Sample Problem

Determining Quantum Numbers

Problem

(a) If $n = 3$, what are the allowed values for l and m_l, and what is the total number of orbitals in this energy level?

(b) What are the possible values for m_l if $n = 5$ and $l = 1$? What kind of orbital is described by these quantum numbers? How many orbitals can be described by these quantum numbers?

Continued ...

Continued ...

Solution

(a) The allowed values for l are integers ranging from 0 to $(n - 1)$. The allowed values for m_l are integers ranging from $-l$ to $+l$ including 0. Since each orbital has a single m_l value, the total number of values for m_l gives the number of orbitals.
To find l from n:
If $n = 3$, l may be either 0, 1, or 2.
To find m_l from l:
If $l = 0$, $m_l = 0$
If $l = 1$, m_l may be $-1, 0, +1$
If $l = 2$, m_l may be $-2, -1, 0, +1, +2$
Since there are a total of 9 possible values for m_l, there are 9 orbitals when $n = 3$.

(b) You determine the type of orbital by combining the value for n with the letter used to identify l. You can find possible values for m_l from l, and the total of the m_l values gives the number of orbitals.
To name the type of orbital:
$l = 1$, which describes a p orbital
Since $n = 5$, the quantum numbers represent a $5p$ orbital.
To find m_l from l:
If $l = 1$, m_l may be $-1, 0, +1$
Therefore, there are 3 possible $5p$ orbitals.

Check Your Solution

(a) Since the total number of orbitals for any given n is n^2, when $n = 3$, the number of orbitals must be 9 (that is, 3^2).

(b) The number m_l values is equivalent to $2l + 1 : 2(1) + 1 = 3$. Since the number of orbitals equals the number of m_l values, the answer of 3 must be correct.

Practice Problems

1. What are the allowed values for l in each of the following cases?
   (a) $n = 5$ **(b)** $n = 1$
2. What are the allowed values for m_l, for an electron with the following quantum numbers:
   (a) $l = 4$ **(b)** $l = 0$
3. What are the names, m_l values, and total number of orbitals described by the following quantum numbers?
   (a) $n = 2, l = 0$ **(b)** $n = 4, l = 3$
4. Determine the n, l, and possible m_l values for an electron in the $2p$ orbital.
5. Which of the following are allowable sets of quantum numbers for an atomic orbital? Explain your answer in each case.
   (a) $n = 4, l = 4, m_l = 0$ **(c)** $n = 2, l = 0, m_l = 0$
   (b) $n = 3, l = 2, m_l = 1$ **(d)** $n = 5, l = 3, m_l = -4$

Shapes of Orbitals

An orbital is associated with a size, a three-dimensional shape, and an orientation around the nucleus. Together, the size, shape, and position of an orbital represent the probability of finding a specific electron around the nucleus of an atom. Figure 3.16 shows the probability shapes associated with the *s*, *p*, and *d* orbitals. (The *f* orbitals have been omitted due to their complexity. You may study these orbitals in post-secondary school chemistry courses.)

Notice that the *overall shape* of an atom is a combination of all its orbitals. Thus, the overall shape of an atom is spherical. Be careful, however, to distinguish for yourself between the overall spherical shape of the atom, and the spherical shape that is characteristic of *only* the *s* orbitals.

Finally, it is important to be clear about what orbitals are when you view diagrams such as those in Figure 3.16. Orbitals, remember, are solutions to mathematical equations. Those solutions, when manipulated, describe the motion and position of the electron in terms of probabilities. Contour diagrams, such as those shown here and in numerous print and electronic resources, *appear solid*. It therefore becomes easy to begin thinking about orbitals as physical "containers" that are "occupied" by electrons. In some ways, this is unavoidable. Try to remind yourself, now and then, of the following:

- Electrons have physical substance. They have a mass that can be measured, and trajectories that can be photographed. They exist in the physical universe.
- Orbitals are mathematical descriptions of electrons. They do not have measurable physical properties such as mass or temperature. They exist in the imagination.

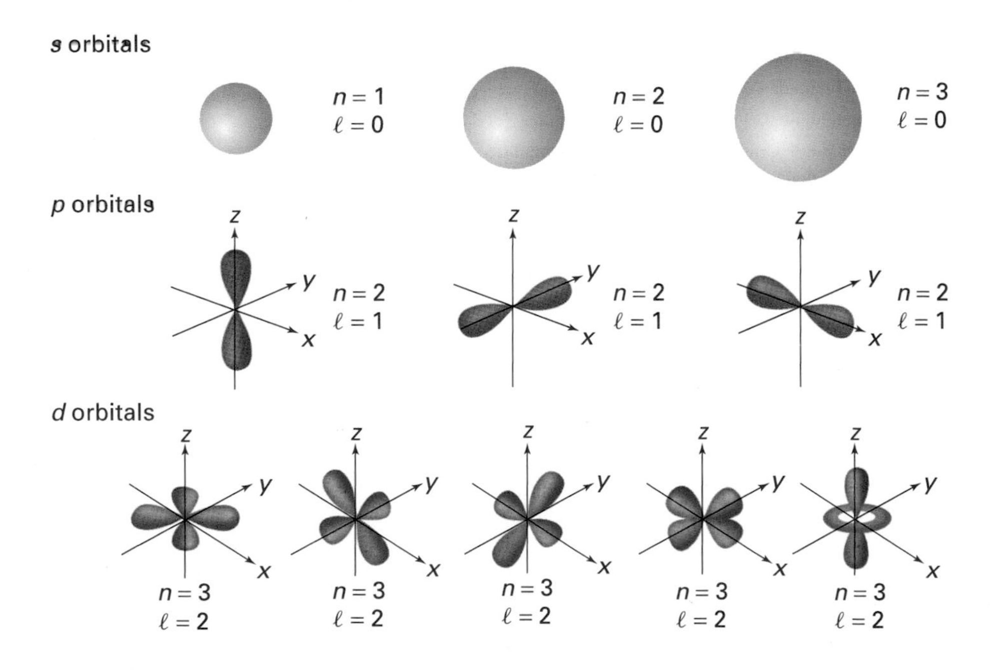

Figure 3.16 **Shapes of the *s*, *p*, and *d* orbitals. Orbitals in the *p* and *d* sublevels are oriented along or between perpendicular *x*, *y*, and *z* axes.**

Section Summary

In this section, you saw how the ideas of quantum mechanics led to a new, revolutionary atomic model—the quantum mechanical model of the atom. According to this model, electrons have both matter-like and wave-like properties. Their position and momentum cannot both be determined with certainty, so they must be described in terms of probabilities. An orbital represents a mathematical description of the volume of space in which an electron has a high probability of being found. You learned the first three quantum numbers that describe the size, energy, shape, and orientation of an orbital. In the next section, you will use quantum numbers to describe the total number of electrons in an atom and the energy levels in which they are most likely to be found in their ground state. You will also discover how the ideas of quantum mechanics explain the structure and organization of the periodic table.

Section Review

1. K/U Explain how the quantum mechanical model of the atom differs from the atomic model that Bohr proposed.
2. K/U List the first three quantum numbers, give their symbols, and identify the property of orbitals that each describes.
3. C Design a chart that shows all the possible values of l and m_l for an electron with $n = 4$.
4. C Agree or disagree with the following statement: The meaning of the quantum number n in Bohr's atomic model is identical to the meaning of the principal quantum number n in the quantum mechanical atomic model. Justify your opinion.
5. I Identify any values that are incorrect in the following sets of quantum numbers.
   - **(a)** $n = 1, l = 1, m_l = 0$; name: $1p$
   - **(b)** $n = 4, l = 3, m_l = +1$; name: $4d$
   - **(c)** $n = 3, l = 1, m_l = -2$; name: $3p$
6. I Fill in the missing values in the following sets of quantum numbers.
   - **(a)** $n =?, l =?, m_l = 0$; name: $4p$
   - **(b)** $n = 2, l = 1, m_l = 0$; name: ?
   - **(c)** $n = 3, l = 2, m_l = -2$; name: ?
   - **(d)** $n =?, l =?, m_l =?$; name: $2s$

Electron Configurations and Periodic Trends

3.3

For the hydrogen atom (and other single-electron systems), all orbitals that have the same value for n have the same energy. You can see this in the energy level diagram shown in Figure 3.17. Hydrogen's single electron is in its lowest (most stable) energy state when it is in the 1s orbital. In other words, the 1s orbital is the ground state for hydrogen. When a hydrogen atom is in an excited state, its electron may be found in any of its other orbitals, depending on the amount of energy absorbed.

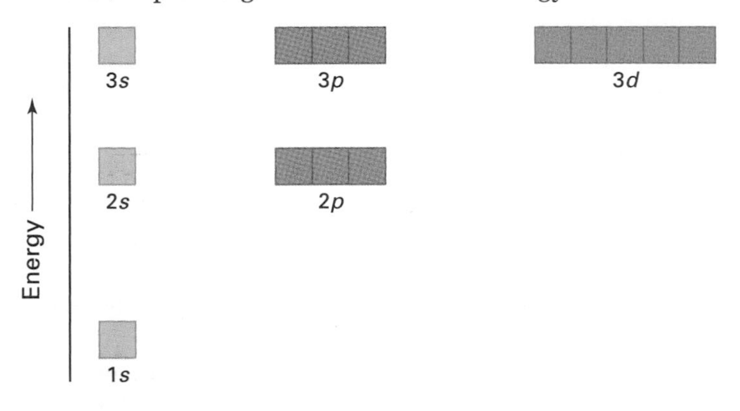

Figure 3.17 **For the hydrogen atom, orbital energy depends only on the value of *n*. For example, all four $n = 2$ orbitals have the same energy. All nine $n = 3$ orbitals have the same energy. What must the value of *l* be for each of these orbitals?**

Atoms with two or more electrons have orbitals with shapes that are similar to those of hydrogen. However, the interactions among the additional electrons result in orbitals with the same value of n, but different *energies*. For example, Figure 3.18 shows an energy level diagram for the lithium atom, which has three electrons. You can see that the 2s orbital has a lower energy than the 2p orbitals. Similarly, the 3p orbitals are lower in energy than the 3d orbitals. Notice, however, that all the orbitals within a sublevel have the same energy. For example, the three p orbitals in the 3p sublevel have the same energy.

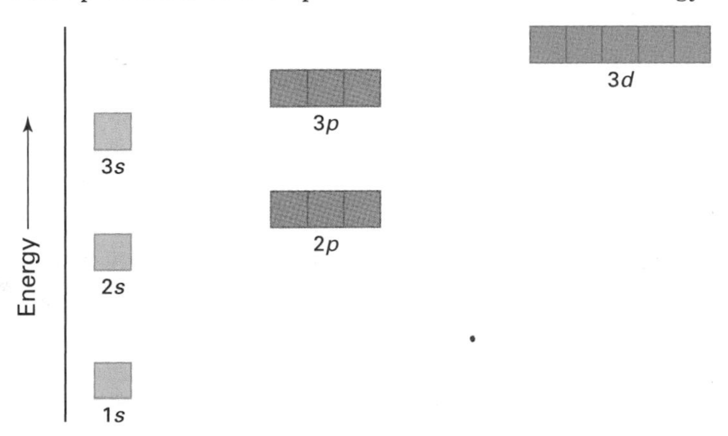

Figure 3.18 **In the lithium atom, and for all other multi-electron atoms, orbitals in different energy sublevels differ in energy.**

Section Preview/ Specific Expectations

In this section, you will

- **write** electron configurations and **draw** orbital diagrams for atoms of elements in the periodic table
- **explain** the significance of the exclusion principle and Hund's rule in writing electron configurations
- **list** characteristics of the *s*, *p*, *d*, and *f* blocks of elements in the periodic table
- **explain** the relationship between electron configurations and the position and properties of elements in the periodic table
- **communicate** your understanding of the following terms: *spin quantum number (m_s), Pauli exclusion principle, electron configuration, aufbau principle, orbital diagram, Hund's rule, atomic radius, ionization energy, electron affinity*

Since low energy usually means stability, it is reasonable that the most stable energy state (ground state) for any atom is one in which its electrons are in the lowest possible energy level. As you know, this is the 1*s* orbital ($n = 1$). However, the electrons of most atoms are not "packed" into this orbital. In fact, experimental evidence shows that only two atoms have all their electrons in the 1*s* orbital: hydrogen and helium. To explain how and why this is the case, you must consider another property of the electron.

The Fourth Quantum Number: A Property of the Electron

As you learned from the previous section, three quantum numbers—n, l, and m_l—describe the energy, size, shape, and spatial orientation of an orbital. A fourth quantum number describes a property of the electron that results from its particle-like nature. Experimental evidence suggests that electrons spin about their axes as they move throughout the volume of their atoms. Like a tiny top, an electron can spin in one of two directions, each direction generating a magnetic field. The **spin quantum number (m_s)** specifies the direction in which the electron is spinning. This quantum number has only two possible values: $+\frac{1}{2}$ or $-\frac{1}{2}$.

In 1925, an Austrian physicist, Wolfgang Pauli, proposed that *only two electrons of opposite spin could occupy an orbital.* This proposal became known as the **Pauli exclusion principle**. What the exclusion principle does is place a limit on the total number of electrons that may occupy any orbital. That is, an orbital may have a maximum of two electrons only, each of which must have the opposite spin direction of the other. It may also have only one electron of either spin direction. An orbital may also have no electrons at all.

Summarizing the Four Quantum Numbers for Electrons in Atoms

Quantum Number Name	Symbol	Allowed Values	Property
principal	n	positive integers (1, 2, 3, etc.)	orbital size and energy
orbital-shape	l	integers from 0 to $(n - 1)$	orbital shape
magnetic	m_l	integers from $-l$ to $+l$	orbital orientation
spin	m_s	$+\frac{1}{2}$ or $-\frac{1}{2}$	electron spin direction

The Pauli Exclusion Principle and Quantum Numbers

Another way of stating the exclusion principle is that *no two electrons in an atom have the same four quantum numbers.* This important idea means that each electron in an atom has its own unique set of four quantum numbers. For example, compare the quantum numbers that distinguish a ground state hydrogen atom from a helium atom. (Recall that a helium atom has two electrons. Note also that m_s quantum number is given as $+\frac{1}{2}$. It could just as easily have a value of $-\frac{1}{2}$. By convention, chemists usually use the positive value first.)

Atom	Electron	Quantum numbers
Hydrogen	lone	$n = 1, l = 0, m_l = 0, m_s = +\frac{1}{2}$
Helium	first	$n = 1, l = 0, m_l = 0, m_s = +\frac{1}{2}$
	second	$n = 1, l = 0, m_l = 0, m_s = -\frac{1}{2}$

Each of a lithium atom's three electrons also has its own unique set of quantum numbers, as you can see below. (Assume these quantum numbers represent a ground state lithium atom.)

Atom	Electron	Quantum numbers
Lithium	first	$n = 1, l = 0, m_l = 0, m_s = +\frac{1}{2}$
	second	$n = 1, l = 0, m_l = 0, m_s = -\frac{1}{2}$
	third	$n = 2, l = 0, m_l = 0, m_s = +\frac{1}{2}$

Consider the significance of lithium's quantum numbers. The first two electrons occupy the 1*s* orbital, and have the same sets of quantum numbers as helium's two electrons. According to the Pauli exclusion principle, the 1*s* orbital ($n = 1$) is now "full" because it contains the maximum number of electrons: two. For lithium's third electron, n cannot equal 1. Therefore, the next principal energy level for the electron is $n = 2$. If $n = 2$, l may have a value of 0 or 1. Remember that an atom's ground state represents its lowest energy state. Because an orbital with $l = 0$ (an *s* orbital) has a lower energy than an orbital with $l = 1$ (a *p* orbital), you would expect a high probability of finding lithium's third electron in the *s* orbital given by $n = 2, l = 0$. In fact, experimental evidence supports this expectation. You know that if $l = 0$, m_l has only one possible value: 0. Finally, by convention, m_s is $+\frac{1}{2}$.

It is possible to distinguish atoms by writing sets of quantum numbers for each of their electrons. However, writing quantum numbers for an atom such as uranium, which has 92 electrons, would be mind-bogglingly tedious. Fortunately, chemists have developed a "shortcut" to represent the number and orbital arrangements of electrons in each atom. As you will see shortly, these electron configurations, as they are called, are intimately connected to the structure and logic of the periodic table. In learning how to write electron configurations, you will discover this connection. You will also find out why orbital energies differ (for example, why a 2*s* orbital is lower in energy than a 2*p* orbital).

Note: Before proceeding, it is important to keep the following in mind. According to quantum mechanics:

- Electrons do not "occupy" orbitals. Nor do orbitals "contain" electrons.
- Electrons do not "fill" orbitals one at a time. Nor do electrons have properties that designate individual electrons as first, second, third, fourth, etc.
- Orbitals do not really have substance or shapes. Orbitals, recall, are mathematical equations that can be used to calculate probability densities for electrons. The shapes, and the impression of substance, result from the contour lines used to set a limit on the extent of those probabilities.

Nevertheless, it takes time and experience for most people to think about and talk about atoms in completely quantum mechanical terms. It is much simpler to think about orbitals as if they have substance. Therefore, in the text that follows, you will read about electrons "occupying" and "filling" orbitals. Strictly speaking, this terminology is not scientifically correct. However, it will help you understand some challenging concepts more easily.

CONCEPT CHECK

Write a set of quantum numbers for each electron of the following atoms: beryllium (Be), boron (B), and carbon (C). Look closely for evidence of a pattern. Write a short paragraph describing this pattern and explaining why you would have expected it.

An Introduction to Electron Configurations

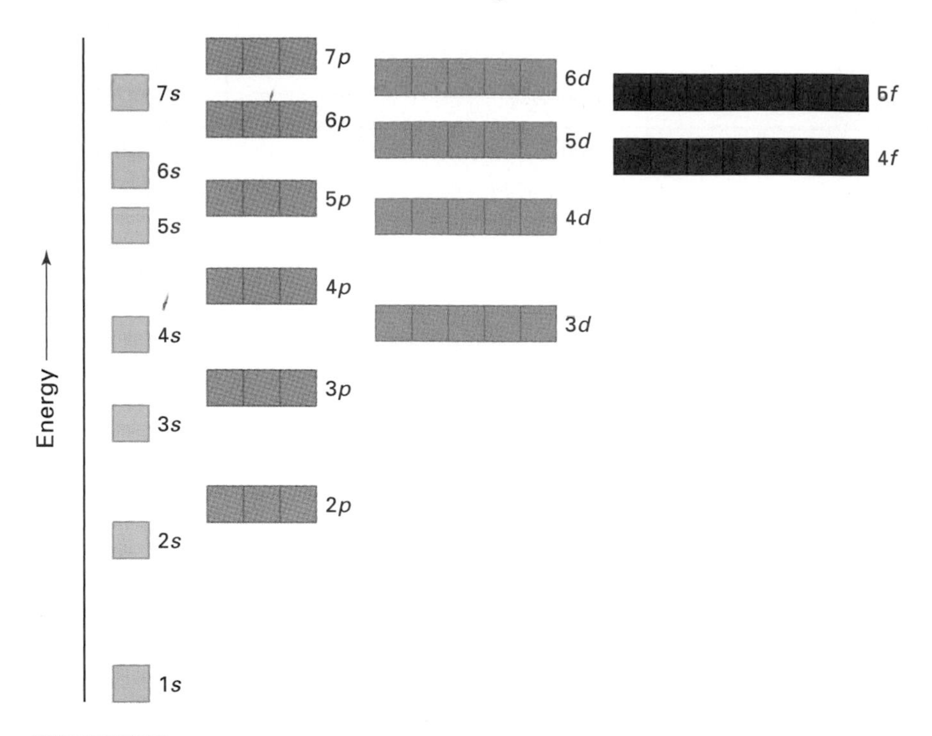

Figure 3.19 Atomic orbitals and their relative energies

Figure 3.19 shows the total number of orbitals in which electrons are likely to be found for all naturally occurring atoms and most synthetic atoms. The orbitals, represented by boxes, are listed in order of increasing energy. You will refer to this diagram often as you learn to write electron configurations.

An atom's **electron configuration** is a shorthand notation that shows the number and arrangement of electrons in its orbitals. Because the value of n ranges to infinity, there are an infinite number of electron configurations that are possible for each atom. For each atom, all but one of these represent the atom in an excited state. However, *an atom's chemical properties are mainly associated with its ground state electron configuration*. Therefore, unless stated otherwise, you can assume that any given electron configuration in this book represents an atom in its ground state. For example, the ground state electron configuration for a hydrogen atom is $1s^1$ (pronounced "one ess one"). The superscript 1 indicates that only one electron is in the s orbital. The ground state electron configuration for a helium atom is $1s^2$ (pronounced "one ess two"), where the superscript 2 indicates that there are two electrons in the s orbital. For a lithium atom, it is $1s^2 2s^1$. Again, the superscript numbers indicate the number of electrons in each of the orbitals.

CONCEPT CHECK

Refer to the sets of quantum numbers for hydrogen and helium that you saw earlier. Then use the quantum numbers for lithium to infer why a lithium atom has the ground state electron configuration that it does.

When you are learning to write electron configurations, it is helpful to start with the first element in the periodic table (hydrogen), and "build up" the electronic structure of its atom by adding an electron to its lowest available energy level. Then you move on to the next element (helium) by adding a proton (and several neutrons) to the nucleus, and by adding an electron to the appropriate orbital. This imaginary process of building up the ground state electronic structure for each atom, in order of atomic number, is called the **aufbau principle**. (*Aufbau* comes from a German word that means "to build up.")

Writing Electron Configurations

Electron configurations, as they are used in this book, provide information about the first two quantum numbers, n and l. (Electron configurations may also reflect the third quantum number, m_l, but this notation goes beyond the scope of this chemistry course.) The electron configuration below represents a boron atom in its ground state.

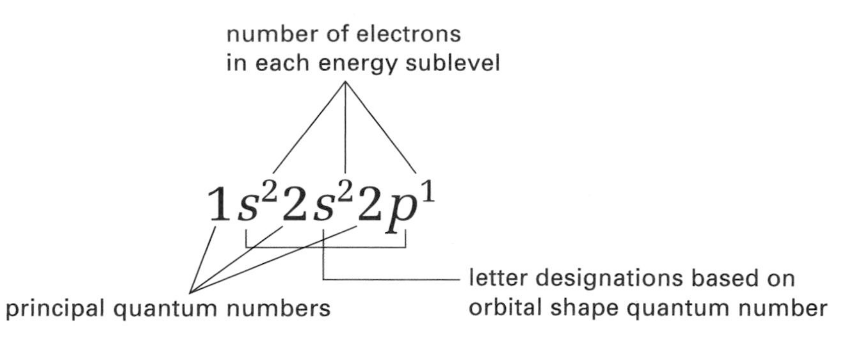

From the few examples of electron configurations that you have seen so far, you may have noticed that they do not include information about the spin direction of the electron. Once you have considered the electron configurations for a variety of atoms, you will see that you can safely infer where electrons with opposite spins have been paired together.

There is, however, a system that chemists use alongside electron configurations to help them plot and keep track of electrons in their orbitals. An **orbital diagram** uses a box for each orbital in any given principal energy level. (Some chemists use a circle or a line instead of a box.) An empty box [ ] represents an orbital in which there are no electrons (an "unoccupied" orbital). A box that has an upward-pointing arrow [↑] represents an orbital with an electron that spins in one direction. A box with a downward-pointing arrow [↓] represents an orbital with an electron that spins in the opposite direction. You can probably surmise that a box with oppositely pointing arrows [↑↓] represents a filled orbital.

You are nearly ready to begin "building up" the electronic structures of the atoms, using both electron configurations and orbital diagrams. Below you will find some guidelines to help you. Refer to these guidelines and to Figure 3.19 as you work your way through the information that follows.

Guidelines for "Filling" Orbitals

1. Place electrons into the orbitals in order of increasing energy level.
2. Each set of orbitals of the same energy level must be completely filled before proceeding to the next orbital or series of orbitals.
3. Whenever electrons are added to orbitals of the same energy sub-level, each orbital receives one electron before any pairing occurs.
4. When electrons are added singly to separate orbitals of the same energy, the electrons must all have the same spin.

Guidelines 3 and 4 together comprise what is known as **Hund's rule**. If you "obey" Hund's rule, no two electrons can have the same set of quantum numbers. Thus, this rule follows from Pauli's exclusion principle.

Writing Electron Configurations for Periods 1 and 2

Below are the electron configurations for atoms of the first ten elements, along with their orbital diagrams.

Figure 3.20 presents the same basic information about the first ten elements, arranged in the form of the periodic table. Using these electron configurations, orbital diagrams, and Figure 3.20, you can probably follow the logic that leads to the electronic structure of each atom. As you examine and compare this information, note the following.

- The energy of each orbital (or each group of orbitals) increases as you move from left to right.
- Boron's fifth electron must go into the 2*p* energy sublevel. Since $l = 1$, m_l may be −1, 0, or +1. The fifth electron can go into any of these orbitals, because they all have the same energy. When you draw orbital diagrams, it is customary to place the electron in the first available box, from left to right.
- With carbon, you must apply Hund's rule. In other words, carbon's sixth electron must go into the next unoccupied 2*p* orbital. (Experimental evidence confirms that the spin of this sixth electron is, in fact, the same direction as the fifth electron.)
- With oxygen, as with helium's 1s orbital and beryllium's 2s orbital, the last-added (eighth) electron is paired with a 2p electron of opposite spin. In other words, the Pauli exclusion principle applies.

		1*s*	2*s*	2*p*	
H	($Z = 1$)	[↑]			$1s^1$
He	($Z = 2$)	[↑↓]			$1s^2$
Li	($Z = 3$)	[↑↓]	[↑]	[ ][ ][ ]	$1s^2 2s^1$
Be	($Z = 4$)	[↑↓]	[↑↓]	[ ][ ][ ]	$1s^2 2s^2$
B	($Z = 5$)	[↑↓]	[↑↓]	[↑][ ][ ]	$1s^2 2s^2 2p^1$
C	($Z = 6$)	[↑↓]	[↑↓]	[↑][↑][ ]	$1s^2 2s^2 2p^2$
N	($Z = 7$)	[↑↓]	[↑↓]	[↑][↑][↑]	$1s^2 2s^2 2p^3$
O	($Z = 8$)	[↑↓]	[↑↓]	[↑↓][↑][↑]	$1s^2 2s^2 2p^4$
F	($Z = 9$)	[↑↓]	[↑↓]	[↑↓][↑↓][↑]	$1s^2 2s^2 2p^5$
Ne	($Z = 10$)	[↑↓]	[↑↓]	[↑↓][↑↓][↑↓]	$1s^2 2s^2 2p^6$

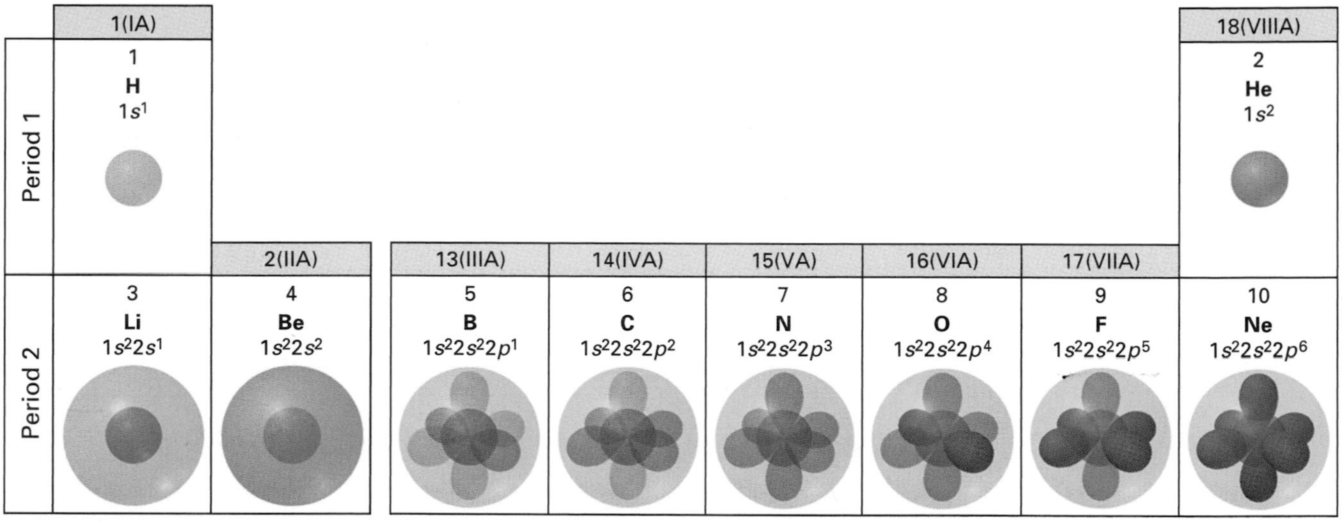

	1(IA)	2(IIA)	13(IIIA)	14(IVA)	15(VA)	16(VIA)	17(VIIA)	18(VIIIA)
Period 1	1 H $1s^1$							2 He $1s^2$
Period 2	3 Li $1s^22s^1$	4 Be $1s^22s^2$	5 B $1s^22s^22p^1$	6 C $1s^22s^22p^2$	7 N $1s^22s^22p^3$	8 O $1s^22s^22p^4$	9 F $1s^22s^22p^5$	10 Ne $1s^22s^22p^6$

Figure 3.20 How orbitals are filled for atoms of the first 10 elements of the periodic table

Electron Configurations and Orbital Diagrams for Period 3

To write the electron configurations and draw the orbital diagrams for atoms of Period 3 elements, you follow the same process as for Period 2. The practice problems below give you the chance to write and draw these representations. In doing so, you will observe a pattern that enables you to take a "shortcut" in writing electron configurations. As the atomic number increases, electron configurations become longer and longer. You know that chemical reactivity depends mainly on an atom's valence electrons: the electrons in the outermost-occupied principle energy level. To highlight these electrons, you can use a simplified notation called a *condensed electron configuration*. This notation places the electron configuration of the noble gas of the previous period in square brackets, using its atomic symbol only. Then you continue with the configuration of the next energy level that is being filled.

The condensed electron configuration for a nitrogen atom, for example, is [He]$2s^22p^3$. The notation [He] is used to represent $1s^2$. For a sodium atom (Z = 11), the condensed electron configuration is [Ne]$3s^1$. Here, [Ne] represents, $1s^22s^22p^6$. Be aware that condensed electron configurations are simply convenient short forms. Thus, [Ne]$3s^1$ does *not* mean that a sodium atom is the same as a neon atom plus one electron. Sodium and neon are different elements because the nuclei of their atoms are completely different.

Practice Problems

6. Use the aufbau principle to write complete electron configurations and complete orbital diagrams for atoms of the following elements: sodium, magnesium, aluminum, silicon, phosphorus, sulfur, chlorine, and argon (atomic numbers 11 through 18).
7. Write condensed electron configurations for atoms of these same elements.

Continued ...

Continued ...

8. Make a rough sketch of the periodic table for elements 1 through 18, including the following information: group number, period number, atomic number, atomic symbol, and condensed electron configuration.
9. A general electron configuration for atoms belonging to any element of group 1 (IA) is ns^1, where n is the quantum number for the outermost occupied energy level. Based on the patterns you can observe so far for elements 1 to 18, predict the general electron configuration for the outermost occupied energy levels of groups 2 (IIA), 13 (IIIA), 14 (IVA), 15 (VA), 16 (VIA), 17 (VIIA), and 18 (VIIIA).

Electron Configurations and Orbital Diagrams for Period 4

You may have noticed that period 3 ended with electrons filling the 3*p* energy sublevel. However, when $n = 3$, l may equal 0, 1, *and* 2. Perhaps you wondered what happened to the 3*d* orbitals ($l = 2$).

Electrons do not start occupying 3*d* orbitals until the 4*s* orbital is filled. Therefore, the 3*d* orbitals are filled starting in period 4. The reason for this change in the expected orbital filling order is that the 4*s* orbital has a lower energy than the 3*d* orbitals. Remember that usually you fill orbitals in order of increasing energy. (If necessary, refer to the Guidelines for "Filling" Orbitals and Figure 3.19 to refresh your memory.)

The chart below shows electron configurations and partial orbital diagrams for the 18 elements of period 4. You would expect the filling pattern shown for potassium ($Z = 19$) through vanadium ($Z = 23$). However, an unexpected deviation from the pattern occurs with chromium ($Z = 24$). The same thing happens with copper ($Z = 29$). All other configurations for period 4 conform to the aufbau principle.

Why do Cr and Cu have electron configurations that are different from what you would predict? The guidelines that you have been using state that you fill orbitals according to increasing energy. In most cases, this results in the most stable ground state configuration for the atom. Experimental evidence indicates, however, that some atoms achieve greater stability with electron configurations that do not conform to predicted patterns. For chromium, the greatest ground state stability results from a configuration in which its 4*s* and 3*d* orbitals are half-filled. For copper, a half-filled 4*s* orbital and a completely filled set of 3*d* orbitals gives the most stable configuration. Similar situations arise for a number of atoms in the remaining periods.

Electron Configurations and Partial Orbital Diagrams for Atoms of Period 4 Elements

Element	Atomic Number	Orbital Diagram (4*s*, 3*d*, and 4*p* Orbitals Only)	Complete Electron Configuration	Condensed Electron Configuration
		4*s* / 3*d* / 4*p*		
K	$Z = 19$	4s: [↑] 3d: [ ][ ][ ][ ][ ] 4p: [ ][ ][ ]	$[1s^2 2s^2 2p^6 3s^2 3p^6]4s^1$	$[Ar]4s^1$
Ca	$Z = 20$	4s: [↑↓] 3d: [ ][ ][ ][ ][ ] 4p: [ ][ ][ ]	$[1s^2 2s^2 2p^6 3s^2 3p^6]4s^2$	$[Ar]4s^2$
Sc	$Z = 21$	4s: [↑↓] 3d: [↑][ ][ ][ ][ ] 4p: [ ][ ][ ]	$[1s^2 2s^2 2p^6 3s^2 3p^6]4s^2 3d^1$	$[Ar]4s^2 3d^1$
Ti	$Z = 22$	4s: [↑↓] 3d: [↑][↑][ ][ ][ ] 4p: [ ][ ][ ]	$[1s^2 2s^2 2p^6 3s^2 3p^6]4s^2 3d^2$	$[Ar]4s^2 3d^2$

		4s	3d	4p		
V	$Z = 23$	[↑↓]	[↑][↑][↑][ ][ ]	[ ][ ][ ]	$[1s^22s^22p^63s^23p^6]4s^23d^3$	$[Ar]4s^23d^3$
Cr	$Z = 24$	[↑]	[↑][↑][↑][↑][↑]	[ ][ ][ ]	$[1s^22s^22p^63s^23p^6]4s^13d^5$	$[Ar]4s^13d^5$
Mn	$Z = 25$	[↑↓]	[↑][↑][↑][↑][↑]	[ ][ ][ ]	$[1s^22s^22p^63s^23p^6]4s^23d^5$	$[Ar]4s^23d^5$
Fe	$Z = 26$	[↑↓]	[↑↓][↑][↑][↑][↑]	[ ][ ][ ]	$[1s^22s^22p^63s^23p^6]4s^23d^6$	$[Ar]4s^23d^6$
Co	$Z = 27$	[↑↓]	[↑↓][↑↓][↑][↑][↑]	[ ][ ][ ]	$[1s^22s^22p^63s^23p^6]4s^23d^7$	$[Ar]4s^23d^7$
Ni	$Z = 28$	[↑↓]	[↑↓][↑↓][↑↓][↑][↑]	[ ][ ][ ]	$[1s^22s^22p^63s^23p^6]4s^23d^8$	$[Ar]4s^23d^8$
Cu	$Z = 29$	[↑]	[↑↓][↑↓][↑↓][↑↓][↑↓]	[ ][ ][ ]	$[1s^22s^22p^63s^23p^6]4s^13d^{10}$	$[Ar]4s^13d^{10}$
Zn	$Z = 30$	[↑↓]	[↑↓][↑↓][↑↓][↑↓][↑↓]	[ ][ ][ ]	$[1s^22s^22p^63s^23p^6]4s^23d^{10}$	$[Ar]4s^23d^{10}$
Ga	$Z = 31$	[↑↓]	[↑↓][↑↓][↑↓][↑↓][↑↓]	[↑][ ][ ]	$[1s^22s^22p^63s^23p^6]4s^23d^{10}4p^1$	$[Ar]4s^23d^{10}4p^1$
Ge	$Z = 32$	[↑↓]	[↑↓][↑↓][↑↓][↑↓][↑↓]	[↑][↑][ ]	$[1s^22s^22p^63s^23p^6]4s^23d^{10}4p^2$	$[Ar]4s^23d^{10}4p^2$
As	$Z = 33$	[↑↓]	[↑↓][↑↓][↑↓][↑↓][↑↓]	[↑][↑][↑]	$[1s^22s^22p^63s^23p^6]4s^23d^{10}4p^3$	$[Ar]4s^23d^{10}4p^3$
Se	$Z = 34$	[↑↓]	[↑↓][↑↓][↑↓][↑↓][↑↓]	[↑↓][↑][↑]	$[1s^22s^22p^63s^23p^6]4s^23d^{10}4p^4$	$[Ar]4s^23d^{10}4p^4$
Br	$Z = 35$	[↑↓]	[↑↓][↑↓][↑↓][↑↓][↑↓]	[↑↓][↑↓][↑]	$[1s^22s^22p^63s^23p^6]4s^23d^{10}4p^5$	$[Ar]4s^23d^{10}4p^5$
Kr	$Z = 36$	[↑↓]	[↑↓][↑↓][↑↓][↑↓][↑↓]	[↑↓][↑↓][↑↓]	$[1s^22s^22p^63s^23p^6]4s^23d^{10}4p^6$	$[Ar]4s^23d^{10}4p^6$

Electron Configurations and the Periodic Table

So far, you have "built up" atoms of the first 36 elements of the periodic table. There are more than 80 elements still remaining. You do not, however, have to build up atoms of these elements to write their electron configurations. You can do it simply by consulting the periodic table! All you need is to recognize the significance of several patterns. You have already observed many of these.

In the process of examining the patterns outlined below, you will learn the filling order for atoms of elements in periods 5, 6, and 7. You will also see why the shape and organization of the periodic table is a direct consequence of the electronic structure of the atoms.

Arranging Elements by Electron Configurations

Figure 3.21 on the next page shows the entire periodic table, with certain segments colour-coded and labelled according to the type of orbital that is being filled. If you built up an atom of every element, filling orbitals in order of increasing energy, you would construct a chart with *precisely* this shape and organization. Figure 3.22 highlights the filling order, and therefore the energy order, of the orbitals when you read the periodic table from left to right.

Elements that appear in the s block and the *p* block are called either the *main group elements* or the *representative elements*. Chemists give them these names because, collectively, these elements are representative of a wide range of physical and chemical properties. Among the main group elements, for example, you will find metals, non-metals, metalloids, highly reactive elements, moderately reactive elements, and unreactive elements. While most main group elements are solids at room temperature, roughly one quarter of them are gases, and one is a liquid.

Elements that appear in the *d* block are called the *transition elements*. They mark the transition from the *p* orbital filling order to the *d* orbital filling order. By the same reasoning, the *f* block elements are called the *inner transition elements*, because they mark a transition from the *d* orbital filling order to the *f* orbital filling order.

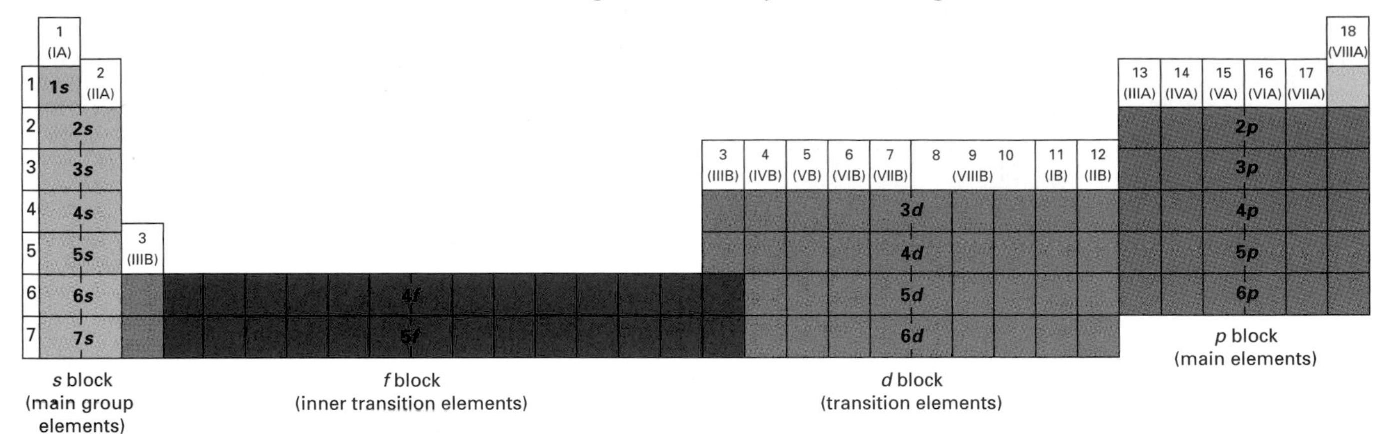

Figure 3.21 **The long form of the periodic table, with the four energy sublevel blocks identified.**

Patterns Involving Group Numbers and Period Numbers

Elements in a group have similar chemical properties because they have similar outer electron configurations. That is, they have the same number of valence electrons. This observation gives rise to three patterns that you can deduce from the periodic table.

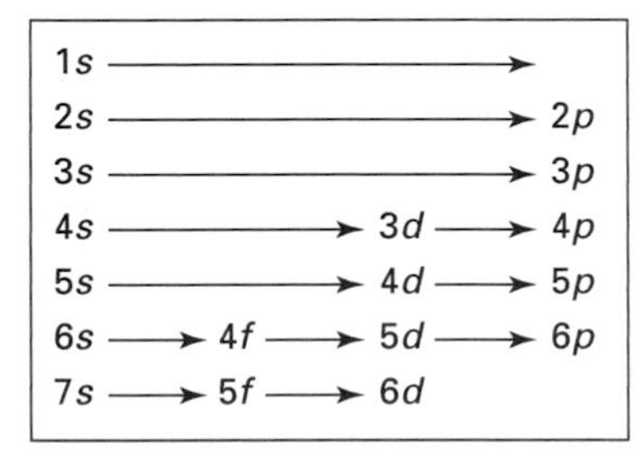

Figure 3.22 **Reading the periodic table from left to right, starting at the top and finishing at the bottom, displays the filling order and the energy order of the atomic orbitals.**

- For main group elements, the last numeral of the group number is the same as the number of valence electrons. For example, phosphorus in group 15 has 5 valence electrons. Strontium in group 2 has 2 valence electrons. (If you use the roman-numeral system for numbering groups, the group number is the number of valence electrons.) The only exception to this pattern is helium (group 18), which has only two valence electrons. Nevertheless, helium's outermost occupied energy level is complete with two electrons.
- The n value of the highest occupied energy level is the period number. For example, atoms of elements that have electrons with $n = 3$ appear in period 3. Atoms of elements that have electrons with $n = 6$ appear in period 6.
- The square of the n value (n^2) equals the total number of orbitals in that energy level. Furthermore, since each orbital may have a maximum of two electrons, the maximum number of electrons in any principle energy level is $2n^2$. For example, with $n = 2$, there are four orbitals: one 2*s* orbital and three 2*p* orbitals. Squaring the n value gives $2^2 = 4$. The total number of electrons in this energy level is *eight*, given by $2n^2$. Notice that there are *eight* elements in period 2. This is no coincidence. Compare this result with the next period. Does the pattern hold? For period 3 elements, $n = 3$. The number of orbitals is $n^2 = 9$. (Verify that for yourself.) The number of electrons is $2n^2 = 18$. However, *there are only eight elements in period 3*. Why? The 4*s* orbital has a lower energy than the 3*p* orbitals, and thus fills first. The 3*d* orbitals fill after the 4*s* orbitals. Therefore, there are 18 elements in period 4. Recall that the 4*f* orbitals do not fill until after the 6*s* orbital fills.

Summarizing Characteristics of *s*, *p*, *d*, and *f* Block Elements

The *s* block includes hydrogen, helium, and the elements of groups 1 (IA) and 2 (IIA). Valence electrons occupy only the *ns* orbitals in this block. Hydrogen atoms and atoms of group 1 have partially filled *s* orbitals, represented by the general electron configuration ns^1. Helium and atoms of group 2 elements have *s* orbitals filled with the maximum number of valance electrons, represented by the general notation ns^2. Because *s* orbitals can hold a total of two electrons, the *s* block elements span two groups.

The *p* block includes elements of groups 13 (IIIA) through 18 (VIIIA). Electron configurations of the *p* block atoms take the general form ns^2np^a, where *a* represents a value ranging from 1 to 6. Because the three *p* orbitals can hold a maximum of six electrons, the *p* block elements span six groups.

Unique among the *p* block elements are the atoms of group 18 (VIIIA) elements: the noble gases, renowned for their chemical stability. Their electron configuration takes the general form ns^2np^6, representing a ground state outer energy level with fully occupied *s* and *p* orbitals. The stability of this configuration is a key factor that drives the formation of chemical bonds.

The *d* block includes all the transition elements. In general, atoms of *d* block elements have filled *ns* orbitals, as well as filled or partially filled *d* orbitals. Generally, the *ns* orbitals fill before the $(n-1)d$ orbitals. However, there are exceptions (such as chromium and copper) because these two sublevels are very close in energy, especially at higher values of *n*. Because the five *d* orbitals can hold a maximum of ten electrons, the *d* block spans ten groups.

The *f* block includes all the inner transition elements. Atoms of *f* block elements have filled *s* orbitals in the outer energy levels, as well as filled or partially filled 4*f* and 5*f* orbitals. In general, the notation for the orbital filling sequence is *ns*, followed by $(n-2)f$, followed by $(n-1)d$, followed by (for period 6 elements) *np*. However, there are many exceptions that make it difficult to predict electron configurations. Because there are seven *f* orbitals, with a maximum of fourteen electrons, the *f* block spans fourteen groups.

CONCEPT CHECK

A teacher asks the class to write electron configurations for magnesium, iron, and tin. Lois does not have a periodic table available. Edward's advice is to use a memory aid that is shown below. Write a short paragraph to explain how to use this memory aid.

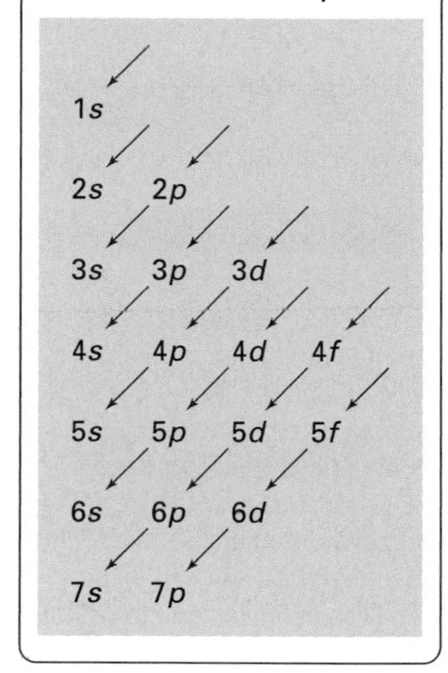

Sample Problem

Determining Quantum Numbers

Problem

You do not have a periodic table. You are told that the condensed electron configuration for strontium is $[Kr]5s^2$. Identify the group number, period number, and orbital block in which strontium appears on the periodic table. Show your reasoning.

Solution

From the electron configuration, you can deduce the energy level of the valence electrons, which tells you the period number for strontium. The number of valence electrons and their sublevel enable you to deduce the group number and the block. Therefore:

Continued ...

Continued ...

The configuration for the valence electrons, s^2, indicates that strontium is in group 2 (IIA). The value of 5 in $5s^2$ tells you that strontium is in period 5. The notation s^2 means that strontium's valence electrons fill the s orbital. Thus, strontium must be in the s block.

Check Your Solution

All elements in group 2 (IIA) have electron configurations that end with the notation s^2. The link between n value and period number is correctly applied. Strontium has an electron configuration with the general notation ns^2, which is characteristic of s block elements.

CHEM FACT

The *d* block elements known to nineteenth century chemists posed a puzzle. Many of these metals were known to react with neutral molecules. In particular, chemists observed that nickel could bond with carbon monoxide. The resulting compound, nickel tetracarbonyl, $Ni(CO)_4$, is a gas. At the time, the nature of the bonding in this compound was unexplainable. Even so, chemists could easily produce and decompose $Ni(CO)_4$. The chemical reactions involved were used to purify nickel, and led to the establishment of a company that later became International Nickel Company, INCO.

Practice Problems

10. Without looking at a periodic table, identify the group number, period number, and block of an atom that has the following electron configurations.

(a) [Ne]$3s^1$ **(b)** [He]$2s^2$ **(c)** [Kr]$5s^2 4d^{10} 5p^5$

11. Use the aufbau principle to write complete electron configurations for the atom of the element that fits the following descriptions.

(a) group 2 (IIA) element in period 4

(b) noble gas in period 6

(c) group 12 (IIB) element in period 4

(d) group 16 (VIA) element in period 2

12. Identify all the possible elements that have the following valence electron configurations.

(a) s^2p^1 **(b)** s^2p^3 **(c)** s^2p^6

13. For each of the elements below, use the aufbau principle to write the full and condensed electron configurations and draw partial orbital diagrams for the valence electrons of their atoms. You may consult the periodic table in Appendix C, or any other periodic table that omits electron configurations.

(a) potassium **(b)** nickel **(c)** lead

Electron Configurations, Atomic Properties, and Periodic Trends

Electron configurations—representations of the electronic structure of atoms—help to determine the atomic and chemical properties of the elements. Properties such as atomic radius, ionization energy, metallic character, and electron affinity are periodic. That is, they follow recurring trends in the periodic table. In a previous chemistry course, you studied these properties and noted their general trends. With your newly developed understanding of the periodic table, you are in a position to re-examine these properties and consider the exceptions to the general trends. Before doing so, carry out the following ThoughtLab, which applies and synthesizes your understanding of the quantum mechanical model of the atom, quantum numbers, electron configurations, and the periodic arrangement of the elements.

ThoughtLab: Periodic Connections

The arrangement of elements in the periodic table is a direct consequence of the allowed values for the four quantum numbers. If the laws of physics allowed these numbers to have different values, the appearance of the periodic table would change. In this activity, you will examine the effect on the periodic arrangement of the elements when one of the quantum numbers has a different set of allowed values.

Part 1

Procedure

1. Suppose, in a different universe, that the quantum number m_l has only one allowed value, $m_l = 1$. All other allowed values for the remaining quantum numbers are unchanged. Therefore, the set of allowed values for all quantum numbers is:
   $n = 1, 2, 3, \ldots$
   $l = 0, 1, 2, \ldots (n - 1)$
   $m_l = 1$
   $m_s = +\frac{1}{2}, -\frac{1}{2}$
   Write a memory aid for the relative order of orbital energies for the first six energy levels.
2. Assume that the exclusion principle and Hund's rule are followed. Demonstrate that the periodic table for the first 30 elements would be as shown below:

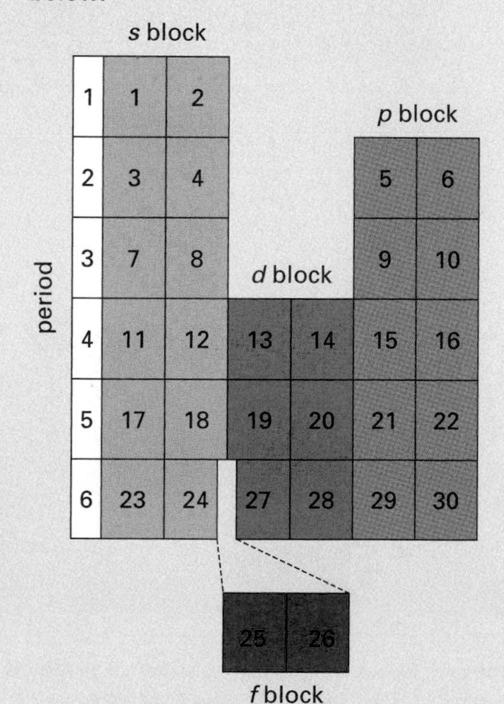

Analysis

1. Explain why there is only one orbital of a given type for each principal quantum number, *n*.
2. What is the principal quantum number of the first *d* block elements, 13 and 14?
3. What are the atomic numbers of the elements that are the noble gases in *our* universe?
4. Which elements would you expect to form cations with a charge of 2+?
5. Which elements would you expect to form anions with a charge of 1–?
6. What is the principal quantum number of the *f* block elements, 25 and 26?
7. What are the atomic numbers of the elements in this alternative universe that would correspond to the noble gases of *our* universe?

Part 2

Procedure

1. In another different universe, the quantum number m_l has the allowed values of $m_l = 0, 1, 2 \ldots + l$. That is, the values of m_l are positive integers, to a maximum value of $+l$. All other allowed values are unchanged, so the set of allowed values is:
   $n = 1, 2, 3, \ldots$
   $l = 0, 1, 2, \ldots (n - 1)$
   $m_l = 0, 1, 2 \ldots + l$
   $m_s = +\frac{1}{2}, -\frac{1}{2}$
   Design the periodic table for the first 30 elements.
2. Label the *s*, *p*, and *d* block elements.
3. Shade or colour the noble gas-like elements.

Analysis

1. Explain why the number of *p* orbitals in this imaginary periodic table is less than the number in the real periodic table.
2. What are the atomic numbers of the elements in this new universe that are likely to have properties similar to the alkali metals of our universe?
3. How many of the first 30 elements are probably non-metals?
4. What is the electron configuration of atomic number 20?

Periodic Trends in Atomic Radius

As you know, atoms are not solid spheres. Their volumes (the extent of their orbitals) are described in terms of probabilities. Nevertheless, the size of an atom — its **atomic radius** — is a measurable property. Chemists can determine it by measuring the distance between the nuclei of bonded, neighbouring atoms. For example, for metals, atomic radius is half the distance between neighbouring nuclei in a crystal of the metal element. For elements that occur as molecules, which is the case for many non-metals, atomic radius is half the distance between nuclei of identical atoms that bonded together with a single covalent bond. In Figure 3.23, the radii of metallic elements represent the radius of an atom in a metallic crystal. The radii of all other elements represent the radius of an atom of the element participating in a single covalent bond with one additional, like atom.

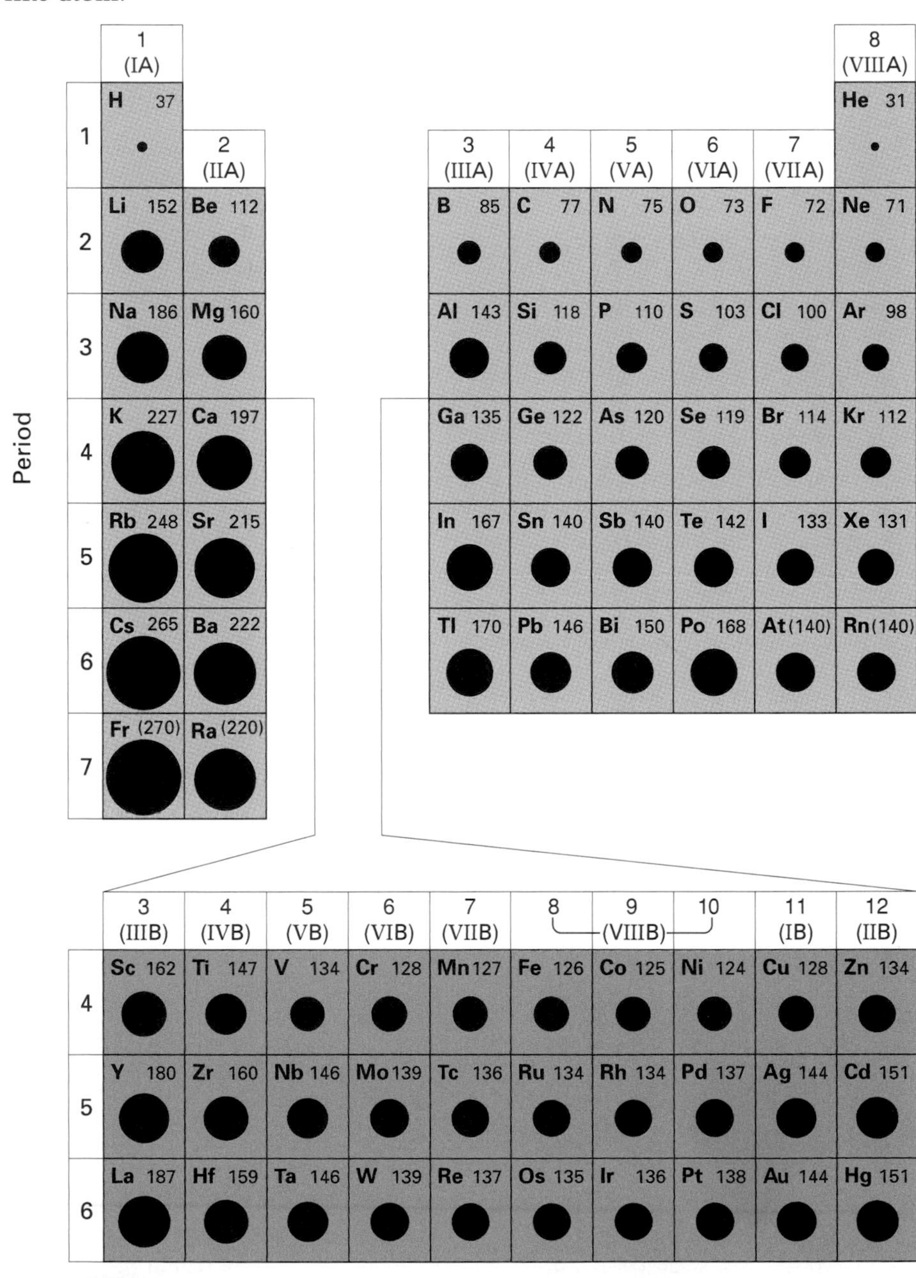

Figure 3.23 Representations of atomic radii for main group and transition elements. (Values for atomic radii are given in picometres. Those in parentheses have only two significant digits.)

Figure 3.23 shows the periodic trends associated with the atomic radius. You can see that atomic radii generally decrease across a period. Furthermore, atomic radii generally increase down a group. Two factors affect differences in atomic radii.

One factor affecting atomic radii is changing n. As n increases, there is a higher probability of finding electrons farther from their nucleus. Therefore, the atomic volume is larger. In other words, atomic radius tends to increase with increasing n, and decrease with decreasing n.

The other factor that affects atomic radii is changing nuclear charge — specifically, the *effective* nuclear charge. Effective nuclear charge is the *net* force of attraction between electrons and the nucleus they surround. Only hydrogen's single electron experiences the full positive charge of its nucleus. For hydrogen, the nuclear charge experienced by the electron is Z, its atomic number. For all other atoms, the nuclear charge that any given electron experiences is offset to some degree by other electrons. Depending on the electron density between the nucleus and any particular electron, the net force of attraction—the effective nuclear charge, Z_{eff} — may be large or small.

As Z_{eff} increases, electrons are attracted more strongly to the nucleus, and the size of the atom decreases. As Z_{eff} decreases, electrons experience a reduced force of attraction, and the size of the atom increases. Valence electrons, especially, experience a smaller Z_{eff} than inner electrons because the inner electrons shield or screen them from the attractive force of the nucleus.

For the main group elements, the combined influences of n, Z_{eff}, and this shielding effect have the following outcomes.

- n governs the trend of increasing atomic radius down a group. Down a group, atoms of each subsequent element have one more level of inner electrons, increasing the shielding effect. Because of the additional shielding, Z_{eff} for the valence electrons changes only slightly. The increase in atomic size, therefore, results from the increasing value of n.
- Z_{eff} governs the trend of decreasing atomic radius across a period. Across a period, atoms of each element have one more electron added to the same outer energy level — n does not change. The shielding effect changes only slightly. However, Z_{eff} changes considerably. The decrease in atomic size across a period, therefore, results from increasing Z_{eff}.

The atoms of transition elements do not display the same general trend as the main group elements. A key reason for this is that electrons are added to inner energy levels — the d orbitals — rather than to the outer energy levels. As a result, Z_{eff} changes relatively little, so atomic size remains fairly constant. In later chemistry courses, you will learn a more complete explanation for the atomic radii of transition-element atoms.

Periodic Trends in Ionization Energy

For the hydrogen atom, the difference in energy between $n = 1$ and $n = \infty$ represents the energy change associated with the complete removal of its electron. (By convention, when $n = \infty$, an electron is considered to be completely removed from its atom.) The energy needed to completely remove one electron from a ground state gaseous atom is called the **ionization energy**. Energy must be added to the atom to remove an electron in order to overcome the force of attraction exerted on the electron by the nucleus. Since multi-electron atoms have two or more electrons, they also have more than one ionization energy.

A gaseous atom's first ionization energy is the least energy required to remove an electron from the outermost occupied energy level. The second ionization energy is always greater than the first ionization energy, because the electron must be removed from a positively charged ion. The same reasoning applies for successive ionization energies.

First ionization energy (IE_1) is closely linked to an atom's chemical reactivity. Atoms with a low IE_1 tend to form cations during chemical reactions. Notice in Figure 3.24 that the atoms with the lowest IE_1 are those belonging to group 1 elements. These elements, also known as the alkali metals, are among the most reactive elements in the periodic table. Atoms of elements with high IE_1 tend to form anions: negatively charged ions. (A notable exception is the noble gases, which do not form ions naturally.)

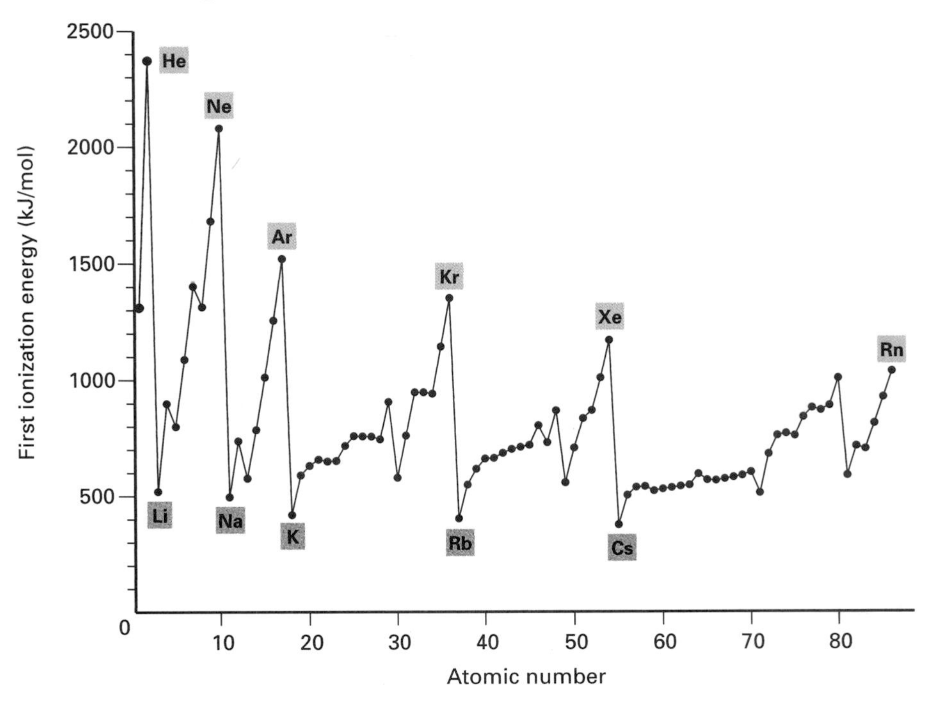

Figure 3.24 Periodic trends involving first ionization energy

Figure 3.24 illustrates several trends associated with ionization energy.

- Ionization energy generally decreases down a group. Notice that this trend is the inverse of the trend for atomic radius. The two trends are, in fact, linked. As atomic radius increases, the distance of valence electrons from the nucleus also increases. There is a decrease, therefore, in the force of attraction exerted by the nucleus on the valence electrons. Thus, less energy is needed to remove one such electron.
- Ionization energy generally increases across a period. Again, this trend is linked to the atomic radius. Across a period, the atomic radius decreases because Z_{eff} increases. The force of attraction between the nucleus and valence electrons is subsequently increased. Therefore, more energy is needed to remove one such electron.

Several minor variations in these trends occur. The variations involve boron and aluminum in group 13 (IIIA) and oxygen and sulfur in group 16 (VIA). The drop in IE_1 with group 13 occurs because electrons start to fill the *np* orbitals. These orbitals are higher in energy than the *ns* orbital, so the additional electron is more easily removed. Group 16 marks the first pairing of the *p* orbital electrons. The np^3 configuration of nitrogen is more stable than the np^4 configuration of oxygen. Electron repulsions increase the orbital energy in oxygen. Therefore, less energy is required to remove the fourth *p* sublevel electron in oxygen.

Periodic trends in ionization energy are linked to trends involving the reactivity of metals. In general, the chemical reactivity of metals increases down a group and decreases across a period. These trends, as well as a further trend from metallic to non-metallic properties across a period, and increasing metallic properties down a group, are shown in Table 3.1.

Table 3.1 Periodic Trends in Reactivity of Metals

Reaction	With oxygen in the air	With water	Reaction with dilute acids
Group 1	React rapidly to form oxides with the general formula M_2O. These compounds are strong bases.	React vigorously with cold water, generating $H_{2(g)}$ and forming a strong base with the general formula MOH.	The reaction is dangerously violent, igniting $H_{2(g)}$ generated.
Group 2	React moderately to form basic oxides with the general formula MO.	React slowly in cold water, generating $H_{2(g)}$ and forming a strong base, $M(OH)_2$.	React quickly generating $H_{2(g)}$.
Group 13	React slowly. Aluminum forms an oxide coating that protects the metal; the compound Al_2O_3 can act as a base or an acid.	Al displaces $H_{2(g)}$ from steam.	Al reacts readily if the protective oxide coating is removed, to generate $H_{2(g)}$.
Group 14	C and Si are non-metals, and form acidic oxides. Ge is a metalloid, and forms an acidic oxide. SnO_2 can act as an acid or a base. PbO_2 is unreactive.	The most metallic elements, Sn and Pb, do not react with water.	Sn and Pb react slowly to form $H_{2(g)}$.
Transition Metals (for example, Fe, Co, Ni, Cu, Zn, Ag, Au)	Zn is the most reactive transition metal, and forms ZnO when the metal is burned in air. Ag and Au do not react.	Zn and Fe displace $H_{2(g)}$ from steam. Fe rusts slowly at room temperature.	Zn reacts readily to generate $H_{2(g)}$, Cu, Ag, and Au do not react.

Magnesium ribbon reacting with oxygen in air

Potassium reacting with water

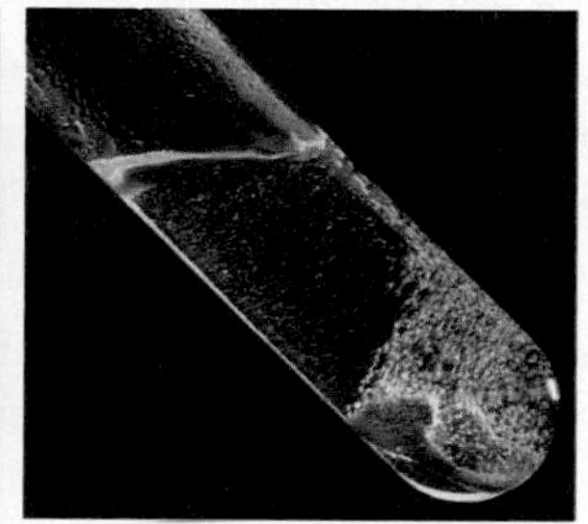
Nickel reacting with dilute acid

Periodic Trends in Electron Affinity

Electron affinity is the change in energy that occurs when an electron is added to a gaseous atom. As was the case with ionization energy, there is more than one value for electron affinity. The first electron affinity results in the formation of an anion with a charge of 1–. For example, when a neutral fluorine atom acquires an electron, as in a bonding situation, the resulting electron configuration of the fluoride ion is the same as that of the noble gas, neon:

$$\text{F}\ (1s^2 2s^2 2p^5) + \text{e}^- \rightarrow \text{F}^-\ (1s^2 2s^2 2p^6)\ \Delta E = \text{EA}_1$$

Fluorine is a highly reactive element. The relative ease with which it gains an electron when it forms bonds is reflected in its high electron affinity.

1 (IA)	2 (IIA)	13 (IIIA)	14 (IVA)	15 (VA)	16 (VIA)	17 (VIIA)	18 (VIIIA)
H −72.8							He (+21)
Li −59.6	Be (+241)	B −26.7	C −122	N 0	O −141	F −328	Ne (+29)
Na −52.9	Mg (+230)	Al −42.5	Si −134	P −72.0	S −200	Cl −349	Ar (+34)
K −48.4	Ca (+156)	Ga −28.9	Ge −119	As −78.2	Se −195	Br −325	Kr (+39)
Rb −46.9	Sr (+167)	In −28.9	Sn −107	Sb −103	Te −190	I −295	Xe (+40)
Cs −45.5	Ba (+52)	Tl −19.3	Pb −35.1	Bi −91.3	Po −183	At −270	Rn (+41)

Figure 3.25 **Electron affinities for the main group elements. Negative values mean that anions form with the release of energy. Positive values mean that energy is absorbed in order to form the anion. The positive values are estimated, because the anions formed for atoms of groups 2 and 18 are unstable.**

Figure 3.25 shows electron affinities for atoms of the main group elements. Positive values mean that energy is required to add an electron. Negative values mean that energy is given off when an electron is added. High negative numbers mean a high electron affinity. Low negative numbers and positive numbers mean a low electron affinity.

Trends for electron affinity are more irregular than those for atomic radius and ionization energy, because factors other than atomic size and Z_{eff} are involved. In future chemistry courses, you will learn about these factors and how they explain the irregularities. However, the property of electron affinity is still significant when you consider it in combination with ionization energy. The trends that result from this combination are important for chemical bonding.

- Atoms of elements in group 17 (VIIA), and to a lesser degree group 16 (VIA), have high ionization energies and high electron affinities. As a result, it takes a great deal of energy to remove electrons from these atoms. However, these atoms attract electrons strongly, and form negative ions in ionic compounds.

- Atoms of elements in group 1 (IA) and group 2 (IIA) have low ionization energies and low electron affinities. Atoms of these elements give up electrons easily, but attract them poorly. Therefore, they form positive ions in ionic compounds.
- Atoms of elements in group 18 (VIIIA) have very high ionization energies and very low electron affinities. Therefore, in nature, they do not gain, give up or share electrons at all. (Under laboratory conditions, only the larger group 18 atoms can be made to form compounds.)

Section Summary

In this section, you have seen how a theoretical idea, the quantum mechanical model of the atom, explains the experimentally determined structure of the periodic table, and the properties of its elements. Your understanding of the four quantum numbers enabled you to write electron configurations and draw orbital diagrams for atoms of the elements. You also learned how to "read" the periodic table to deduce the electron configuration of any element.

Since the start of high school science courses, you have used the periodic table to help you investigate the composition and behaviour of the elements. Your early experiences with the periodic table were limited largely to the first 20 elements, because you could explain their electron structure without the concepts of orbitals and electron configurations. The modern, quantum mechanical model of the atom has broadened your understanding of the elements, the composition of their atoms, and their chemical and physical behaviour in the world around you. You will draw upon these concepts in the next chapter, as you expand your understanding of the forces that are responsible for the millions of kinds of matter in the universe: chemical bonds.

Section Review

1 **(a)** K/U What is the fourth quantum number, and in what way is it different from the other three quantum numbers?

(b) Explain why the fourth quantum number is an example of a quantized value.

2 C Compare the orbitals of a hydrogen atom with the orbitals of all other atoms. In what ways are they similar? In what ways are they different?

3 C Identify the group number, period number, and orbital block of the periodic table for elements whose atoms have the following electron configurations.

(a) $[Kr]5s^2 4d^1$

(b) $[Ar]4s^2 3d^{10} 4p^3$

(c) $[He]2s^2 2p^6$

(d) $[Ne]3s^2 3p^1$

4 C Which requires more energy: removing a valence electron from its atom or removing an electron from an inner energy level? Explain why.

5 K/U On which side of the periodic table would you find an element whose atom is likely to form a cation? Which atomic property is related to this question?

6 K/U In what ways are each of the following related to or affected by one another?

(a) the aufbau principle, the exclusion principle, and Hund's rule

(b) electron configurations and orbital diagrams

(c) ionization energy and atomic radius

(d) ionization energy and electron affinity

(e) effective nuclear charge and valence electrons

(f) the periodic table and chemical properties of the elements

(g) electron configurations and the periodic table

(h) electron configurations and quantum numbers

(i) electron configurations, period number, and group number

(j) the quantum mechanical model of the atom and the periodic table

7 C Use the aufbau principle to write complete and condensed electron configurations for the most common ions for the elements listed below, and explain the significance of any patterns you observe in their electronic structures.

(a) sodium

(b) calcium

(c) chlorine

(d) sulfur

8 I The following data lists the ionization energies for a given atom: IE_1 = 738 kJ/mol; IE_2 = 1451 kJ/mol; IE_3 = 7733 kJ/mol. Predict the valence electron configuration for this atom, and explain your reasoning.

CHAPTER 3 Review

Reflecting on Chapter 4

Summarize this chapter in the format of your choice. Here are a few ideas to use as guidelines:

- Compare the Rutherford, Bohr, and quantum mechanical models of the atom.
- Distinguish between the view of matter and energy in the macroscopic world (the world of everyday experience) and the view of matter and energy in the quantum mechanical world.
- Describe, using examples, the relationship between the four quantum numbers and the electron configurations of atoms.
- Explain the relationship between the electronic structure of atoms and the arrangement of elements in the periodic table.
- Illustrate the trends associated with several atomic and chemical properties of the elements.

Reviewing Key Terms

For each of the following terms, write a sentence that shows your understanding of its meaning.

nuclear model	emission spectrum
absorption spectrum	quantum
photons	quantum mechanical model of the atom
orbitals	ground state
principal quantum number (n)	orbital-shape quantum number (l)
magnetic quantum number (m_l)	spin quantum number (m_s)
Pauli exclusion principle	electron configuration
aufbau principle	orbital diagram
Hund's rule	atomic radius
ionization energy	electron affinity

Knowledge/Understanding

1. Explain the experimental observations and inferences that led Rutherford to propose the nuclear model of the atom.
2. Explain how the Bohr atomic model differs from the Rutherford atomic model, and explain the observations and inferences that led Bohr to propose his model.
3. Both the Rutherford and Bohr atomic models have been described as planetary models. In what ways is this comparison appropriate? In what ways is this comparison misleading?
4. Briefly describe the contributions made by the following physicists to the development of the quantum mechanical model of the atom.
   (a) Planck
   (b) de Broglie
   (c) Einstein
   (d) Heisenberg
   (e) Schrödinger
5. List characteristics of the *s*, *p*, *d*, and *f* blocks to clearly distinguish among the atoms of their elements.
6. Explain how Pauli's exclusion principle and Hund's rule assist you in writing electron configurations.
7. Give the energy level and type of orbital occupied by the electron with the following set of quantum numbers: $n = 3$, $l = 1$, $m_l = 0$, $m_s = +\frac{1}{2}$.
8. Differentiate between a ground state sulfur atom and an excited state sulfur atom. Use electron configurations to illustrate this difference.
9. Which of the following pairs of atoms would you expect to have a larger atomic radius? Explain your choice.
   (a) Na, Mg
   (b) Cl, K
   (c) Ca, Sr
   (d) Al, Ga
   (e) O, F
   (f) Cl, Br
10. Locate each of the following elements on a periodic table, and state the orbital block in which each element is found: U, Zr, Se, Rb, Re, Sr, Dy, Kr.
11. The chemical formulas for the oxides of lithium, beryllium, and boron are: Li_2O, BeO, and B_2O_3. Write the formulas for the oxides of sodium, potassium, magnesium, calcium, aluminum, and gallium. Explain how you determined their formulas.
12. **(a)** How many electron-containing orbitals does an arsenic atom have in the ground state?
    (b) How many of the orbitals are completely filled?
    (c) How many of the orbitals are associated with the atom's fourth principal energy level?
    (d) How many possible orbitals could electrons occupy if the atom were excited?

Inquiry

13. Which of the following is the correct orbital diagram for the third and fourth principal energy levels of a vanadium atom ($Z = 23$)? Justify your answer.

(a) [↑] [↑↓ | ↑ | ↑ | ↑ |] [| |]

(b) [↑] [↑ | ↑ | ↑ | ↑ | ↑] [| |]

(c) [↑↓] [↑ | ↑ | ↑ | |] [| |]

(d) [↑↓] [↑ | ↑ | ↑ | ↑ |] [| |]

14. Each of the following orbital diagrams is incorrect. Identify the errors, explain how you recognized them, and use the aufbau principle to write electron configurations using the corrected orbital diagrams.

(a) carbon: [↑↓] [↑↓] [↑↓ | |]

(b) iron: [↑↓] [↑ | ↑ | ↑ | ↑ | ↑] [↑ | |]

(c) bromine: [↑↓] [↑↓ | ↑↓ | ↑↓ | ↑↓ | ↑↓] [↑ | ↑ | ↑]

15. The electron configurations below represent atoms in excited states. Identify each atom, and write its ground state electron configuration.

(a) $1s^22s^2sp^63s^13p^1$

(b) $1s^22s^22p^63s^23p^44s^1$

(c) $1s^22s^22p^63s^23p^64s^23d^44p^1$

(d) $1s^22s^22p^63s^23p^64s^23d^{10}4p^65s^14d^2$

16. Sketch a blank periodic table into your notebook to fill, roughly, half a page. Write the letters of the questions below in their appropriate place in the periodic table.

(a) smallest atomic radius in period 3

(b) lowest IE_1 in period 5

(c) highest EA in period 17 (VIIA)

(d) largest atomic radius in period 6

(e) most metallic character in period 14 (IVA)

(f) group 3 (IIIA) atom that forms the most basic oxide

(g) forms a 2+ ion with the electron configuration $[Ar]4s^2$

(h) has a condensed ground state electron configuration of $[Ne]3s^23p^2$

(i) period 4 element whose atom, in the ground state, has an outermost energy level that is completely full

(j) a transition metal for which the aufbau principle predicts an electron configuration that would be incorrect (**Note:** There are several possible answers; you are required to provide only one.)

17. On your copy of the periodic table from question 16, colour-code and label the four orbital blocks.

18. Imagine that scientists have successfully synthesized element X, with atomic number 126. Predict the values of n and l for the outermost electron in an atom of this element. State the number of orbitals there would be in this energy sublevel.

Communication

19. In terms of the general periodic trends for groups and periods, explain why ranking the following pairs might prove tricky.

(a) K and Sr, according to atomic size

(b) Mn and Fe, according to first ionization energy

(c) Na and Ca, according to metallic character

20. Explain why it is not possible to measure the size of an atom directly.

21. Identify elements whose atoms have the following valence electron configurations:

(a) $5s^1$

(b) $4s^23d^2$

(c) $3s^2$

(d) $4s^23d^{10}4p^3$

22. Why are there no p block elements in period 1 of the periodic table?

23. At one time, scandium ($Z = 21$) was placed in the same group of the periodic table as aluminum ($Z = 13$).

(a) Use the aufbau principle to write the ground state electron configurations for atoms of these two elements.

(b) What ionic charge would each have in common? How might this have led chemists to place them originally in the same group?

(c) Identify and explain all the evidence you can think of to support their current locations in the periodic table.

24. The values for ionization energy in the periodic table in Appendix C are first ionization energies. Construct a bar graph to show the relative sizes of IE_1 values for the main group elements. If available, use spreadsheet software to plot and render your graph.

Making Connections

25. Scientists have succeeded in synthesizing about 25 elements that do not exist in nature. All except one of these (technetium, $Z = 43$) are members of period 7. All are very dense and radioactive. Many, such as Rutherfordium ($Z = 104$) and Bohrium ($Z = 107$), are quite short-lived, with no known applications outside of the laboratory. An exception is the element plutonium ($Z = 94$). Research several applications of this element, and the properties that make it well-suited for these uses. What risks are associated with plutonium? Do the risks outweigh the benefits? Justify your opinions.
26. Synthesizing elements requires the use of equipment called particle accelerators. These machines, usually the size of buildings, are enormously expensive to build and operate. Do you think this money could be better spent elsewhere? Conduct research to help you examine this question thoroughly and thoughtfully. Present your findings in the form of an editorial that either supports or opposes theoretical scientific research of this kind.

Answers to Practice Problems and Short Answers to Section Review Questions

Practice Problems: **1.(a)** 0, 1, 2, 3, 4 **(b)** 0 **2.(a)** ±4, ±3, ±2, ±1, 0 **(b)** 0 **3.(a)** 2*s*; 0; one **(b)** 4*f*; ±3, ±2, ±1, 0; seven **4.** $n = 2$, $l = 1$, $1\ m_l = 0, \pm1$ **5. (b), (c)**, allowed **6.** Na: $1s^22s^22p^63s^1$; Mg:$1s^22s^22p^63s^2$; Al: $1s^22s^22p^63s^23p^1$; Si: $1s^22s^22p^63s^23p^2$; P: $1s^22s^22p^63s^23p^3$; S: $1s^22s^22p^63s^23p^4$; Cl: $1s^22s^22p^63s^23p^5$; Ar: $1s^22s^22p^63s^23p^6$ **7.** Na:[Ne]$3s^1$; Mg:[Ne]$3s^2$; Al:[Ne]$3s^23p^1$; Si:[Ne]$3s^23p^2$; P:[Ne]$3s^23p^3$; S:[Ne]$3s^23p^4$; Cl:[Ne]$3s^23p^5$; Ar:[Ne]$3s^23p^6$ **9.** 2 (IIA): ns^2; 13 (IIIA): ns^2np^1; 14 (IVA): ns^2np^2; 15 (VA): ns^2np^3; 16 (VIA): ns^2np^4; 17 (VIIA): ns^2np^5; 18 (VIIIA): ns^2np^6 **10.(a)** group1 (IA), period 3, *s* block **(b)** group 2 (IIA), period 2, *s* block **(c)** group17 (VIIA), period 5, *p* block
11.(a) Ca:$1s^22s^22p^63s^23p^64s^2$
(b) Rn:$1s^22s^22p^63s^23p^64s^23d^{10}4p^65s^24d^{10}5p^66s^24f^{14}5d^{10}6p^6$
(c) Zn:$1s^22s^22p^63s^23p^64s^23d^{10}$ **(d)** O:$1s^22s^22p^4$
12.(a) B, Al, Ga, In, Tl **(b)** N, P, As, Sb, Bi **(c)** Ne, Ar, Kr, Xe, Rn
13.(a) K:$1s^22s^22p^63s^23p^64s^1$; [Ar]$4s^1$;
(b) Ni:$1s^22s^22p^63s^23p^64s^23d^8$; [Ar]$4s^23d^8$;
(c) Pb:$1s^22s^22p^63s^23p^64s^23d^{10}4p^65s^24d^{10}5p^66s^24f^{14}5d^{10}6p^2$; [Xe]$6s^24f^{14}5d^{10}6p^2$

Section Review 3.1: **1.(a)** mostly empty space; contains a central, massive, concentration of charge. **(b)** scattering of alpha particles that pass through gold foil **4.(a)** radio waves, microwaves, infrared, visible light, ultraviolet, x-rays, gamma rays **(b)** opposite of (a) **5.** UV has higher energy and frequency, lower wavelength; both travel at the same speed and have wave and particle-like properties **6.(a)** photons absorbed to excite electron to higher energy level **(b)** photons emitted as electron falls to lower energy level **(c)** photons emitted as electrons falls to lower energy level **7.** $n = 1$, $n = 2$, $n = 4$, $n = 5$, $n = 7$ **3.2: 2.** Principal quantum number, *n*, size and energy; orbital-shape quantum number, *l*, shape; magnetic quantum number, m_l, orientation **5.(a)** $l = 0$, 1*s* **(b)** 4*f* **(c)** $m_l \neq -2$, but could be 0, ±1 **6.(a)** $n = 4$, $l = 1$ **(b)** 2*p* **(c)** 3*d* **(d)** $n = 2$, $l = 0$ **3.3: 1.** magnetic spin quantum number, m_s; not dependent on value of any other quantum number, always ±1/2 for any m_l value, not derived from quantum mechanics (describes electron, not orbital) **3.(a)** group 3 (IIIB), period 5, *d* block **(b)** group 15 (VA), period 4, *p* block **(c)** group 18 (VIIIA), period 2, *p* block **(d)** group 13 (IIIA), period 3, *p* block **4.** removal of inner electron, being closer to nucleus, Z_{eff} is greater **5.** left, ionization energy **7.(a)** $1s^22s^22p^6$, [Ne] **(b)** $1s^22s^22p^63s^23p^6$ [Ar] **(c)** $1s^22s^22p^63s^23p^6$ [Ar] **(d)** $1s^22s^22p^63s^23p^6$; stable ions have a noble gas electron configuration **8.** ns^2, large jump between IE_2 and IE_3 indicates that a noble gas electron configuration was attained after 2 electrons are removed

CHAPTER 4

Structures and Properties of Substances

Chapter Preview

4.1 Chemical Bonding
4.2 Molecular Shape and Polarity
4.3 Intermolecular Forces in Liquids and Solids

Prerequisite Concepts and Skills

Before you begin this chapter, review the following concepts and skills:

- identifying the number of valence electrons in an atom (from previous studies; Chapter 3, section 3.2)
- writing the ground state electron configurations of atoms (Chapter 3, section 3.2)
- predicting bond type using electronegativity (from previous studies)

A water molecule has a bent shape, while a carbon dioxide molecule is linear. An ammonia molecule looks like a pyramid, and sulfur hexafluoride (one of the most dense and most stable gaseous compounds) is shaped like an octahedron. In fact, all molecules in nature have a specific shape, which is important to their chemistry.

Why does the shape of a molecule matter? To answer this question, consider that, right now, each nerve cell in your brain is communicating with adjacent nerve cells by releasing molecules called neurotransmitters from one cell to the next. Enzymes are assisting in the chemical breakdown of food in your digestive system. The aroma of cologne, desk wood, or cleansers that you may be smelling is a result of odorous molecules migrating from their sources to specific sites in your nasal passages.

Each of these situations depends on the ability of one molecule with a specific shape to "fit" into a precise location with a corresponding shape. The properties of substances derive from the ways in which particles bond together, the forces that act within and among the compounds they form, and the shapes that result from these interactions.

In this chapter, you will review and extend your understanding of chemical bonding. You will discover how and why each molecule has a characteristic shape, and how molecular shape is linked to the properties of substances. You will also consider the importance of molecular shape to the development of materials with specific applications in the world around you.

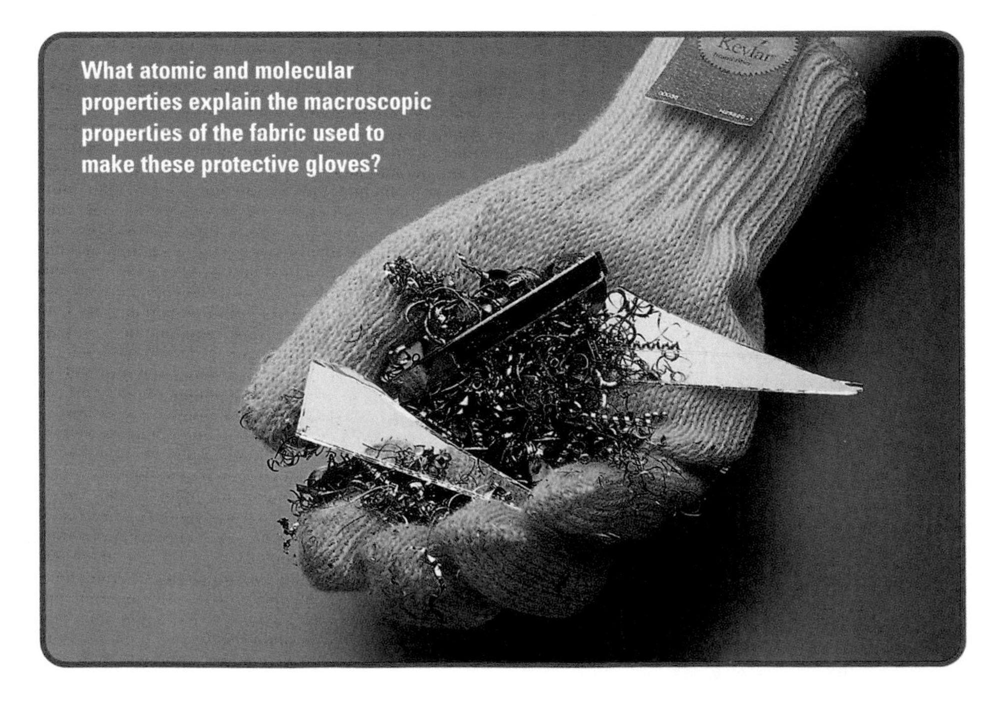

What atomic and molecular properties explain the macroscopic properties of the fabric used to make these protective gloves?

Chemical Bonding

4.1

Of the more than one hundred elements that occur in nature or that have been produced synthetically, only the noble gases exist naturally as single, uncombined atoms. The atoms of all other elements occur in some combined form, bonded together. **Chemical bonds** are electrostatic forces that hold atoms together in compounds.

Why, however, do atoms form bonds at all? The answer to this question involves energy. In nature, systems of lower energy tend to be favoured over systems of higher energy. In other words, *lower-energy systems tend to have greater stability than higher-energy systems.* Bonded atoms, therefore, tend to have lower energy than single, uncombined atoms.

Using Lewis Structures to Represent Atoms in Chemical Bonding

Chemical bonding involves the interaction of valence electrons—the electrons that occupy the outermost principal energy level of an atom. You have used Lewis structures in previous studies to indicate the valence electrons of atoms. Recall that to draw the Lewis structure of an atom, replace its nucleus and inner electrons with its atomic symbol. Then add dots around the atomic symbol to symbolize the atom's valence electrons. Many chemists place the dots starting at the top and continue adding dots clockwise, at the right, then bottom, then left. After you have drawn the first four dots, you begin again at the top, as shown below.

Na· Mg: Al: ·Si: ·P: ·S: ·Cl: :Ar:

In this chapter, you will use Lewis structures often to represent molecules and the simplest formula unit of an ionic solid. Drawing a Lewis structure for a molecule lets you see exactly how many electrons are involved in each bond, and helps you to keep track of the number of valence electrons. In the example below, notice that there are two ways to show the bonding pairs of electrons. Some chemists use dots only. Other chemists show the bonding pairs as lines between atoms. Dots are reserved for representing a *lone pair* (a non-bonding pair) of electrons. You will see the second example, with lines for bonding pairs, used more often in this textbook.

:Ö::C::Ö: or :Ö=C=Ö: (four lone pairs)

Bonding and the Properties of Substances

Chemists classify substances according to their bonds and the forces of attraction that exist between their particles. In the following investigation, you will observe and record data about the properties of five solids. Each solid represents a particular type of bonding. Experimental evidence from this investigation will support some of the explanations you will encounter later in the chapter. Keep a record of your observations and ideas for future reference.

Section Preview/ Specific Expectations

In this section, you will

- **compare** ionic, covalent, and metallic bonding
- **predict** the polarity of various covalent bonds, using electronegativity values
- **observe** and **analyze** the properties of substances, and **infer** the type of bonding
- **communicate** your understanding of the following terms: *chemical bond, ionic bond, covalent bond, lattice energy, bond energy, polar covalent bond, free-electron model, metallic bond*

SKILL FOCUS

Performing and recording

Analyzing and interpreting

Investigation 4-A

Properties of Substances

In this investigation, you will study the properties of five different types of solids: non-polar covalent, polar covalent, ionic, network, and metallic. You will be asked to identify each substance as one of the five types. In some cases, this will involve making inferences and drawing on past knowledge and experience. In others, this may involve process-of-elimination. The emphasis is on the skills and understandings you use to make your decisions. Later, you will be able to assess the validity of your decisions.

Question

What are some of the properties of the different types of solids?

Safety Precautions

- Tie back long hair and any loose clothing.
- Before you light the candle or Bunsen burner, check that there are no flammable solvents nearby.
- Use only a pinch of SiO_2 in steps 6 and 7.

Materials

samples of paraffin wax, $C_{20}H_{42}$ (shavings and a block)
samples of sucrose, $C_{12}H_{22}O_{11}$ (granular and a large piece, such as a candy)
samples of sodium chloride, NaCl (granular and rock salt)
samples of silicon dioxide, SiO_2 (sand and quartz crystals)
samples of tin, Sn (granular and foil)
distilled water
100 mL beaker and stirrer
metal plate (iron or aluminum)
conductivity tester
candle
Bunsen burner
retort stand with ring clamp
timer

Procedure

1. Read the entire Procedure before you begin. Design a table to record your observations.
2. Rub one solid onto the surface of another. Rank the relative hardness of each solid on a scale of 1 to 5. A solid that receives no scratch marks when rubbed by the other four is the hardest (5). A solid that can be scratched by the other four is the softest (1).
3. One at a time, place a small quantity (about 0.05 g) of each solid into 25 mL of distilled water in a 100 mL beaker. Observe and record the solubility of each solid.
4. For solids that dissolved in step 3, test the solution for electrical conductivity.
5. Test each of the five large solid samples for electrical conductivity.
6. One at a time, put a small sample of each solid on the metal plate. Place the plate on the ring stand and heat it with the burning candle. The plate should be just above the flame. Observe how soon the solid melts.
7. Repeat step 6 with a Bunsen flame. Observe how soon the solid melts.

Analysis

1. Based on what you know about bonding, classify each solid as non-polar covalent, polar covalent, ionic, network, or metallic. Give reasons to support your decision.
2. Determine the electronegativity difference between the elements in each of the substances. How does this difference relate to the properties you observed?

Conclusion

3. Based on the properties you observed, write a working definition of each type of solid.

Ionic Bonding

The force of attraction between oppositely charged ions (cations and anions) constitutes an ionic bond. Ionic bonding occurs between atoms of elements that have large differences in electronegativity — usually a metal with a very low electronegativity and a non-metal with a very high electronegativity. The units of ionic compounds such as sodium chloride and magnesium fluoride cannot be separated easily by direct heating of the crystal salts. The ions that make up the ionic solid are arranged in a specific array of repeating units. In solid sodium chloride, for example, the ions are arranged in a rigid lattice structure. In such systems, the cations and anions are arranged so that the system has the minimum possible energy.

Because of the large differences in electronegativity, the atoms in an ionic compound usually come from the *s* block metals and the *p* block non-metals. For example, magnesium in Group 2 (IIA) and fluorine in Group 17 (VIIA) combine to form the ionic compound magnesium fluoride, MgF_2. Figure 4.1 shows a repeating unit in the crystal model of magnesium fluoride. The process that results in the formation of ions can be illustrated with an orbital diagram or with Lewis structures, as shown in Figures 4.2 and 4.3. Use them as a guide for the Practice Problems below. Through bonding, the atoms of each element obtain a valence electron configuration like that of the nearest noble gas. In this case, the nearest noble gas for both ions is neon. This observation reflects the octet rule.

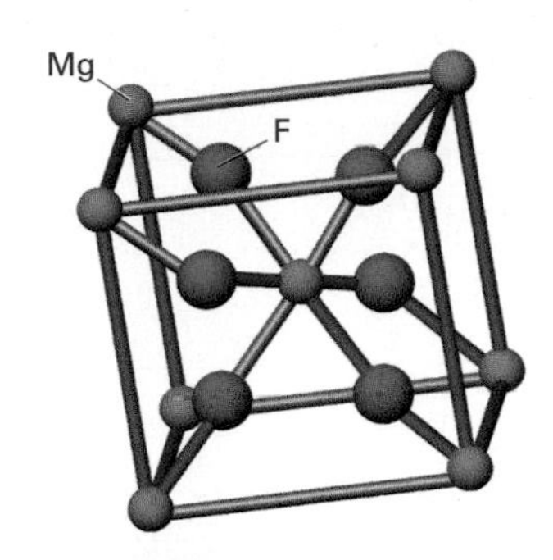

Figure 4.1 A repeating unit in a magnesium fluoride crystal

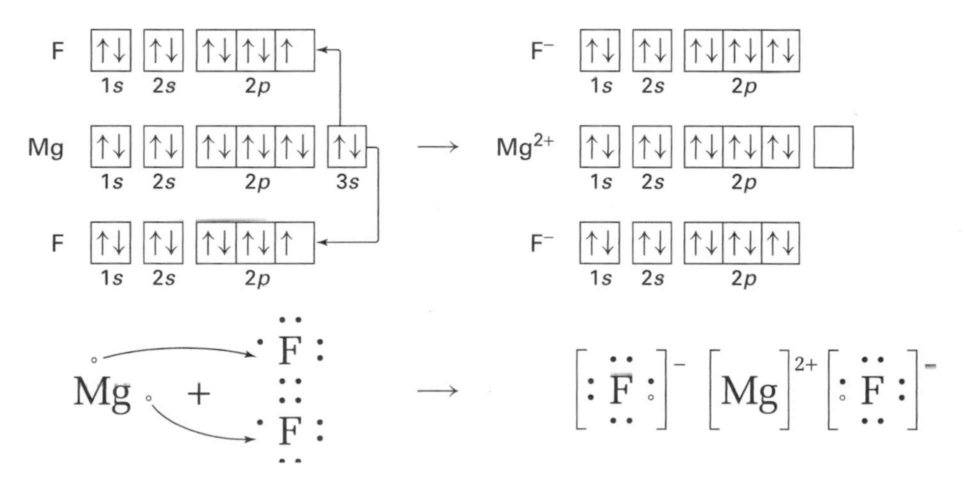

Figure 4.2 An orbital diagram for MgF_2

Figure 4.3 A Lewis structure for MgF_2

Practice Problems

1. Write electron configurations for the following:
   (a) Li^+ (b) Ca^{2+}
   (c) Br^- (d) O^{2-}
2. Draw Lewis structures for the chemical species in question 1.
3. Draw orbital diagrams and Lewis structures to show how the following pairs of elements can combine. In each case, write the chemical formula for the product.
   (a) Li and S (b) Ca and Cl
   (c) K and Cl (d) Na and N

Continued ...

Continued ...

4. To which main group on the periodic table does X belong?
   (a) MgX (b) X_2SO_4
   (c) X_2O_3 (d) XCO_3

Properties of Ionic Solids

Magnesium fluoride has many properties that are characteristic of ionic solids. In general, ionic solids have the following properties:

- crystalline with smooth, shiny surfaces
- hard but brittle
- non-conductors of electricity and heat
- high melting points

Many ionic solids are also soluble in water. Magnesium fluoride, however, is an exception. It has a very low solubility in water.

Lattice Energy and Ionic Bonding

The reaction between gaseous magnesium ions and gaseous fluorine ions is highly exothermic. Chemists have determined that 2957 kJ of energy is released when 1 mol of $Mg^{2+}{}_{(g)}$ and 2 mol of $F^{-}{}_{(g)}$ react to form 1 mol of the stable ionic solid, MgF_2.

The amount of energy given off when an ionic crystal forms from the gaseous ions of its elements is called the **lattice energy**. Thus, the lattice energy of MgF_2 is 2957 kJ/mol. The same amount of energy must be added to break the ionic crystal back into its gaseous ions. What is the relationship between lattice energy and the formation of an ionic bond? For example, when solid magnesium and gaseous fluorine react to form magnesium fluoride, the energy change is 1123 kJ/mol. How does this energy change relate to magnesium fluoride's lattice energy of 2957 kJ/mol?

Remember that, in general, the lower the total energy of a system is, the higher is its stability. Any process that requires the addition of energy usually decreases the stability of the ionic arrangement. Any process that gives off energy usually increases that stability. Ionic crystals such as magnesium fluoride have very high melting points. This means that a large amount of energy is required to break the bonds that hold the ions together in their rigid lattice structure. You might wonder if this energy comes from the process of ionization. However, this is not the case. The ionization energy of the metal is always larger than the electron affinity of the non-metal. In other words, the transfer of electrons between gaseous ions always requires energy. In fact, it is the lattice energy given off (2957 kJ/mol for MgF_2) that favours the formation of ionic bonds.

For magnesium fluoride, the overall decrease in energy of 1123 kJ/mol is accounted for mainly by its large lattice energy. The lattice energies of the magnesium halides decrease from fluoride to iodide. The melting points of these halides reveal a trend in the same direction, with magnesium fluoride having the highest melting point and magnesium iodide having the lowest melting point. In other words, the compound with the largest lattice energy has the highest melting point—it requires the most energy to disrupt the lattice structure.

Math LINK

The electrostatic force of attraction, F, between two charges, q_1 and q_2, separated by a distance, d, is given by the formula: $F = \frac{kq_1q_2}{d^2}$, where k is a constant. How would the charge (q) on two oppositely charged ions, and the internuclear distance between the ions (d) in the crystal lattice, affect the force of attraction between the ions?

Covalent Bonding

Covalent bonding involves a balance between the forces of attraction and repulsion that act between the nuclei and electrons of two or more atoms. This idea is represented in Figure 4.4, with a molecule of hydrogen, H_2. There is an optimum separation for two hydrogen atoms at which their nucleus-electron attractions, nucleus-nucleus repulsions, and electron-electron repulsions achieve this balance. This optimum separation favours a minimum energy for the system, and constitutes the covalent bond between the two hydrogen atoms. Unlike ionic bonding, in which electrons behave as if they are transferred from one atom to another, covalent bonding involves the sharing of pairs of electrons.

In Chapter 3 you learned how the quantum mechanical model applies to the atom. The model can also be extended to explain bonding. A covalent bond may form when two half-filled atomic orbitals from two atoms overlap to share the same region of space. A covalent bond involves the formation of a new orbital, caused by the overlapping of atomic orbitals. The new orbital has energy levels that are lower than those of the original atomic orbitals. Since electrons tend to occupy the lowest available energy level, the new orbitals provide a more energetically favourable configuration than the two atoms had before they interacted.

electron
nucleus
attraction
repulsion
optimum separation

Figure 4.4 Covalent bonding involves forces of attraction and repulsion that occur simultaneously.

Characteristics of Covalent Bonding

In many cases, electron-sharing enables each atom in a covalent bond to acquire a noble gas configuration. For a hydrogen molecule, each atom acquires a filled valence level like that of helium by treating the shared pair of electrons as if it is part of its own composition. As you can see in Figure 4.5, a single *shared pair* of electrons—a bonding pair—fills the valence level of both hydrogen atoms at the same time.

The period 2 non-metals from carbon to fluorine must fill their 2*s* and their three 2*p* orbitals to acquire a noble gas configuration like that of neon. Covalent bonding that involves these elements obeys the octet rule. In the formation of the diatomic fluorine molecule, F_2, for example, the bonding (shared) pair of electrons gives each fluorine atom a complete valence level.

H:H

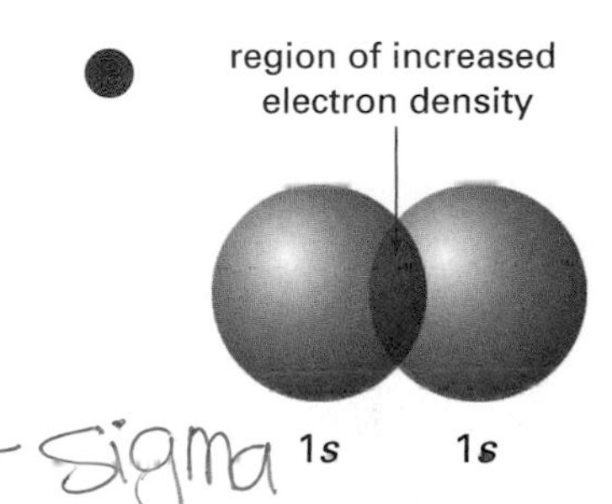

Figure 4.5 Representations of a covalent bond in a molecule of hydrogen (A) Lewis structure (B) Overlap of two 1*s* orbitals

bonding pair — lone pairs

:F:F: or   :F—F:

Each fluorine atom also has three unshared pairs of electrons. These pairs of electrons, called *lone pairs*, are not involved in bonding.

The covalent bond that holds molecules of hydrogen, fluorine, and hydrogen fluoride together is a single bond. It involves a single bonding pair of electrons. Some molecules are bonded together with two shared pairs of electrons. These are called double bonds. Carbon dioxide is an example of a covalent molecule that consists of double bonds.

O::C::O   or   O=C=O

Molecules that are bonded with three shared pairs of electrons have triple bonds. Nitrogen, N_2, another diatomic molecule, is a triple-bonded molecule.

:N⋮⋮N: or   :N≡N:

CONCEPT CHECK

Individual atoms of hydrogen and fluorine are highly reactive, and readily bond together to form molecules of hydrogen fluoride. Draw a Lewis structure for hydrogen fluoride. Label the bonding and lone pairs, and explain why this molecule is stable.

Bond energy is the energy required to break the force of attraction between two atoms in a bond and to separate them. Thus, bond energy is a measure of the strength of a bond. You might expect that the bond energy would increase if more electrons were shared between two atoms because there would be an increase in charge density between the nuclei of the bonded atoms. In other words, you might predict that double bonds are stronger than single bonds, and that triple bonds are stronger than double bonds. The data in Table 4.1 support this prediction.

Table 4.1 Average Bond Energies of Bonds Between Carbon Atoms and Between Nitrogen Atoms

Bond	Bond energy (kJ/mol)	Bond	Bond energy (kJ/mol)
C—C	347	N—N	160
C═C	607	N═N	418
C≡C	839	N≡N	945

CONCEPT CHECK

Consider the differences in properties between covalent compounds and ionic compounds. Do these differences imply that covalent bonds are weaker than ionic bonds? Give evidence to justify your answer.

Predicting Ionic and Covalent Bonds

There are several methods you can use to predict the type of bond in an unknown substance. For example, you can consider the substance's physical properties. In contrast to ionic solids, covalent (molecular) compounds typically have the following properties:

- exist as a soft solid, a liquid, or a gas at room temperature
- have low melting points and boiling points
- are poor conductors of electricity, even in solution
- may not be soluble in water (acetone, however, is one of many exceptions; you will find out why shortly)

What if you know the formula of a compound but don't know how to classify it? In this case, you can use the electronegativity difference between the bonding atoms to predict the type of bond. For example, two atoms with identical electronegativities, such as chlorine ($\Delta EN = 3.16 - 3.16 = 0$) share their electrons equally. They are bonded covalently. When two atoms have very different electronegativities, the atom with the higher electronegativity value attracts electrons more strongly than the atom with the lower electronegativity value. In sodium chloride, for instance, chlorine ($EN = 3.16$) attracts an electron much more strongly than sodium ($EN = 0.93$). Therefore, sodium's valence electron has a very high probability of being found near chlorine, and a very low probability of being found near sodium. A high electronegativity difference is characteristic of ionic compounds.

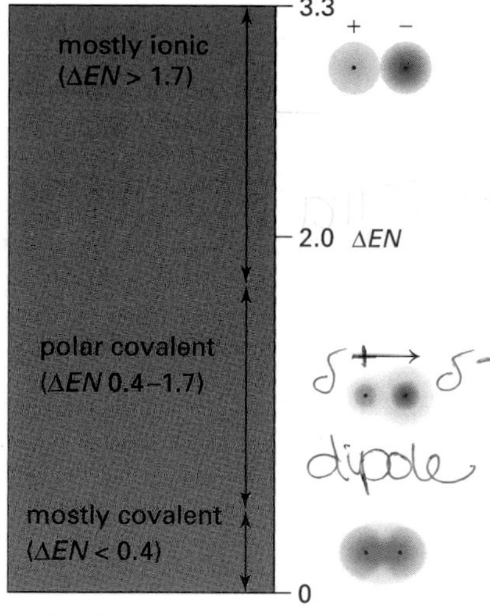

Figure 4.6 The relationship between bonding character and electronegativity difference. The arrow with the "t" shape near the tail is a vector that indicates the magnitude and direction of polarity. You will see this symbol again in section 4.2.

It is common for chemists to describe the formation of ionic compounds in terms of a transferring of electrons from one atom to another. For example, in the formation of sodium chloride, you often read that sodium loses its valence electron and that chlorine gains an electron. Terms such as these—electron transfer, losing electrons, gaining electrons—make it easier to discuss and model the behaviour of atoms during ionic bonding and the formation of ionic compounds. Figure 4.6 provides a guide to classifying chemical bonds based on their electronegativity difference. Notice that this "bonding continuum," as it is sometimes called, avoids definitive classifications. Instead, the character of chemical bonds changes gradually with ΔEN.

Chemists consider bonded atoms with ΔEN between 0 and 0.4 as being *mostly* covalent; electrons are shared equally or nearly equally. For bonded atoms that have ΔEN between 0.4 and 1.7, the bond is termed polar covalent. A **polar covalent bond** is a covalent bond with an unequally shared pair of electrons between two atoms. This unequal sharing results in a bond that has partially positive and partially negative poles. (You will learn about the significance of molecular polarity in section 4.2.) Bonded atoms with ΔEN between 1.7 and 3.3 are *mostly* ionic; electrons are shared so unequally that one bonded atom has a strong negative charge and the other has a strong positive charge.

The gradual change in the magnitude of ΔEN between two atoms is paralleled by a gradual change in the physical properties of the compounds they form. For example, chlorine forms compounds with each of the main-group elements of period 3, except for argon. Figure 4.7 shows the relationship between ΔEN and the bonding, structure, and properties of the period 3 chlorides.

Web LINK

www.mcgrawhill.ca/links/chemistry12

Another way to classify bonding involves calculating the percent ionic character of the bond. Search for information about this approach on the Internet, and write a summary of your findings. In particular, answer this question: Can a bond ever be 100% covalent or 100% ionic? To start your search, access the web site above. Click on **Web Links** to find out where to go next.

Figure 4.7 How does ΔEN affect bond type and properties of the Period 3 chlorides?

Practice Problems

5. List the following compounds in order of decreasing bond energy: H—Br, H—I, H—Cl. Use Appendix E to verify your answer.

6. Rank the following compounds in sequence from lowest melting point to highest melting point, and give reasons for your decisions: $AsBr_3$, KBr, $CaBr_2$.

Continued ...

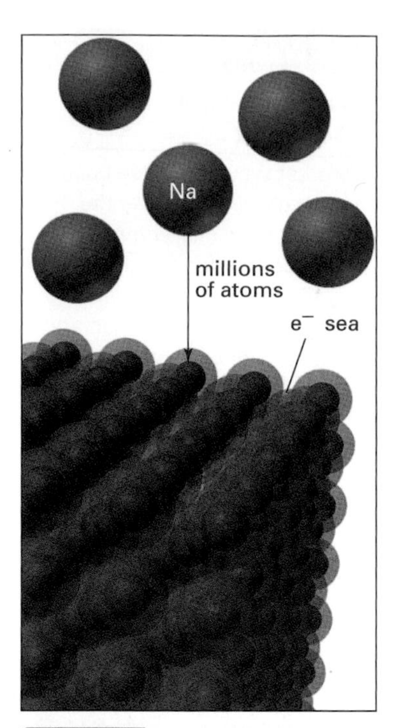

Figure 4.8 A representation of the free-electron model of metallic bonding. This model applies to metal alloys as well as to metallic elements.

Continued ...

7. From their position in the periodic table, predict which bond in the following groups is the most polar. Verify your predictions by calculating the ΔEN.
   (a) C—H, Si—H, Ge—H
   (b) Sn—Br, Sn—I, Sn—F
   (c) C—O, C—H, C—N
8. Classify the bonding in each of the following as covalent (non-polar), polar covalent, or ionic. Afterwards, rank the polar covalent compounds in order of increasing polarity.
   (a) S_8
   (b) RbCl
   (c) PF_3
   (d) SCl_2
   (e) F_2
   (f) SF_2

Metallic Bonding

About two-thirds of all the naturally occurring elements are metals. Despite this great diversity, metals have many properties in common. Metals conduct electricity and heat in both their solid and liquid states. Most metals are malleable and ductile; that is, they can be easily stretched, bent, and deformed without shattering of the whole solid, or can be easily made to be with the addition of heat or pressure. While metals show a broad range of melting and boiling points, in general, metals change state at moderate to high temperatures. Most metals have either one or two valence electrons. What kind of bonding model can account for these properties?

Based on electronegativity differences, you would not expect metals to form ionic bonds with themselves or with other metals—and they do not. Similarly, metals do not have a sufficient number of valence electrons to form covalent bonds with one another. Metals do, however, share electrons. Unlike the electron sharing in covalent compounds, however, electron sharing in metals occurs throughout the entire structure of the metal. The **free-electron model** shown in Figure 4.8 pictures metals as being composed of a densely packed core of metallic cations, within a delocalized region of shared, mobile valence electrons. The force of attraction between the positively charged cations and the pool of valence electrons that moves among them constitutes a **metallic bond**.

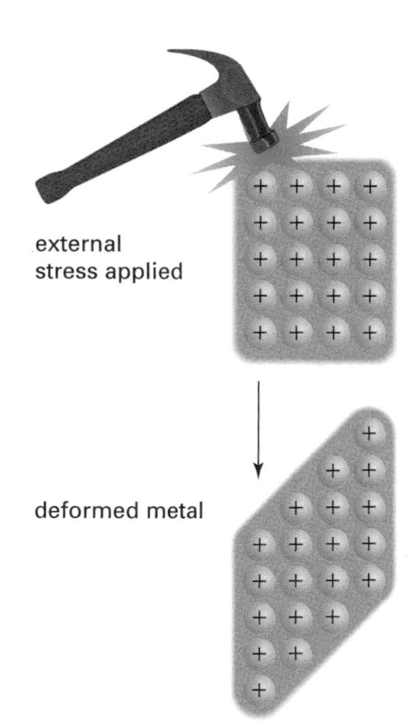

Figure 4.9 Metals are easily deformed because one layer of positive ions can slide over another. At the same time, the free electrons (shown as a yellow cloud) continue to bind the metal ions together.

Properties Explained by the Free-Electron Model

The free-electron model explains many properties of metals. For example:

- Conductivity: Metals are good conductors of electricity and heat because electrons can move freely throughout the metallic structure. This freedom of movement is not possible in solid ionic compounds, because the valence electrons are held within the individual ionic bonds in the lattice.
- Malleability and Ductility: The malleability of metals can be explained by viewing metallic bonds as being non-directional. The positive ions are often layered as fixed arrays (like soldiers lined up for inspection). When stress is applied to a metal, one layer of positive ions can slide over another layer. The layers move without breaking the array, which is the reason why metals do not shatter immediately along a clearly

defined point of stress. The delocalized electrons continue to exert a uniform attraction on the positive ions, as you can see in Figure 4.9.

- Melting and Boiling Points: The melting and boiling points of Group 1 (IA) metals (with an s^1 electron in their condensed electron configuration) are generally lower than the melting and boiling points of Group 2 (IIA) metals (with an s^2 in their condensed electron configuration). You would expect this trend, because the greater number of valence electrons and the larger positive charge of Group 2 (IIA) atoms result in stronger metallic bonding forces. This trend is also consistent with the fact that Group 2 metals are smaller in size than Group 1 metals. Trends involving the transition metals are more complex, in part because they have a greater number of unpaired electrons in their *d*-orbitals. Because these *d* electrons and those of the valence energy level have similar magnitudes, the *d* electrons can participate in metallic bonding. In general, however, transition elements have very high melting and boiling points.

Section Summary

In this section, you have used Lewis structures to represent bonding in ionic and covalent compounds, and have applied the quantum mechanical theory of the atom to enhance your understanding of bonding. All chemical bonds—whether their predominant character is ionic, covalent, or between the two—result from the atomic structure and properties of the bonding atoms. In the next section, you will learn how the positions of atoms in a compound, and the arrangement of the bonding and lone pairs of electrons, produce molecules with characteristic shapes. These shapes, and the forces that arise from them, are intimately linked to the physical properties of substances, as you will see in the final section of the chapter.

Section Review

1. (a) C Distinguish between an ionic bond, a covalent bond, and a polar covalent bond.
   (b) C Draw Lewis structures to illustrate how the element bromine can take part in an ionic bond, a covalent bond, and a polar covalent bond.
   (c) C How would the properties differ for the compounds you have shown?
2. K/U Arrange the following bonded pairs in order of decreasing electronegativity difference: H—Br, Cl—I, O—H, N—Cl, S—O.
3. K/U Select a physical property of metals. Predict and explain how this property would vary from left to right across a period in the periodic table.
4. Identify the property shown in Figure 4.10, and name the type of bonding that accounts for this property.

Web LINK

www.mcgrawhill.ca/links/chemistry12

The free-electron model is a simplified representation of metallic bonding. While it is helpful for visualizing metals at the atomic level, this model cannot sufficiently explain the properties of all metals. Quantum mechanics offers a more comprehensive model for metallic bonding. Go to the web site above, and click on **Web Links**. This will launch you into the world of molecular orbitals and band theory. Use a graphic organizer or write a brief report that compares the free-electron and band-theory models of metallic bonding.

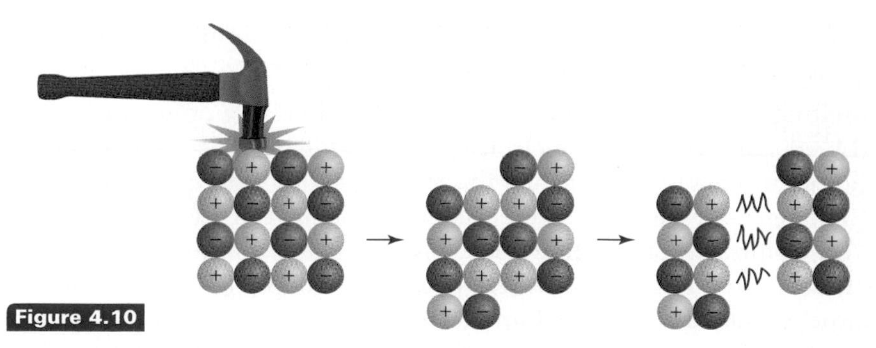

Figure 4.10

5 MC Tungsten is used to make the filament in an electric light bulb, while copper is the element of choice for electric wiring. What properties do these two metals have in common and in what way(s) do they differ to warrant their specific uses? What other metals could be used to substitute for tungsten and copper? Why are tungsten and copper used more often?

6 MC $MgCl_2$ and MgO are being considered for use in a high-temperature industrial environment. What properties of these two compounds would you check to determine which would be the better choice for this application? Refer to a reference text and determine which of the two compounds would be the better choice.

7 (a) C Iron is a hard substance, but it is easily corroded by oxygen in the presence of water to form iron(III) oxide, commonly called rust. Rust is chemically very stable but, structurally speaking, it has very little strength. Explain this difference in properties of iron and iron(III) oxide, in terms of what you have learned about the type of bonding in these substances.

(b) C Aluminum reacts with oxygen in air to form aluminum oxide. Unlike rust, which flakes off, aluminum oxide remains bound to the aluminum metal, forming a protective layer that prevents further reaction. What might account for the difference in behaviour between iron(III) oxide and aluminum oxide and their respective metals?

Molecular Shape and Polarity

4.2

A molecule is a discrete chemical entity, in which atoms are held together by the electrostatic attractions of covalent bonds. In previous chemistry courses, you used Lewis structures to represent the bonding in simple molecules. In a Lewis structure, you assume that each of the atoms in the molecule achieves a noble gas configuration through the sharing of electrons in single, double, or triple bonds. This section introduces you to a systematic approach to drawing Lewis structures for a broader range molecules and for polyatomic ions. You will see that, in some cases, there are exceptions to the octet rule. Afterwards, you will use bonding theory to help you predict the shapes of various molecules. From the shapes of these molecules (and the electronegativity difference of their atoms), you will determine the polarity of the substances that are comprised of these molecules.

Why is polarity important? It explains why, for example, water is a liquid at room temperature, rather than a gas. In other words, it helps explain why Earth is capable of supporting life! You will explore the significance of molecular polarity in greater detail in section 4.3.

Section Preview/ Specific Expectations

In this section, you will

- **draw** Lewis structures for molecules and polyatomic ions
- **represent** molecules and polyatomic ions that are exceptions to the octet rule
- **predict** molecular shape and polarity of molecules using Valence-Shell Electron-Pair Repulsion (VSEPR) theory
- **describe** advances in molecular theory as a result of Canadian research
- **communicate** your understanding of the following terms: *co-ordinate covalent bond, resonance structure, expanded valence energy level, linear, trigonal planar, tetrahedral, trigonal bipyramidal, octahedral, dipole*

Lewis Structures for Molecules and Polyatomic Ions

You can use the procedure outlined below to draw the Lewis structures for molecules and ions that have a central atom, with other atoms around it. The Sample Problems and additional text that follow show how to apply these steps for several molecules and polyatomic ions that obey the octet rule. Afterwards, use Practice Problems 9 to 13 to practice drawing Lewis structures.

How To Draw Lewis Structures for Simple Molecules and Ions with a Central Atom

Step 1 Position the least electronegative atom in the centre of the molecule or polyatomic ion. Write the other atoms around this central atom, with each atom bonded to the central atom by a single bond. Always place a hydrogen atom or a fluorine atom at an end position in the structure.

Step 2 (a) Determine the total number of valence electrons in the molecule or ion. For polyatomic ions, pay close attention to the charge. For example, if you are drawing a polyatomic anion such as CO_3^{2-}, add two electrons to the total number of valence electrons calculated for the structure CO_3. For a polyatomic cation such as NH_4^+, subtract one electron from the total number of valence electrons calculated for the structure NH_4.

(b) Once you have the total number of valence electrons, determine the total number of electrons needed for each atom to achieve a noble gas electron configuration.

(c) Subtract the first total (number of valence electrons) from the second total (number of electrons needed to satisfy the octet rule) to get the number of shared electrons. Then divide this number by 2 to give the number of bonds. Double or triple bonds may be needed to account for this number of bonds. Double bonds count as two bonds. Triple bonds count as three bonds.

Step 3 Subtract the number of shared electrons from the number of valence electrons to get the number of non-bonding electrons. Add these electrons as lone pairs to the atoms surrounding the central atom so that you achieve a noble gas electron configuration for each atom.

Sample Problem

Drawing the Lewis Structure of a Molecule

Problem

Draw the Lewis structure for formaldehyde (methanal), CH_2O.

Solution

The molecular formula, CH_2O, tells you the number of each kind of atom in the molecule. Following steps 1 to 3 from the procedure outlined above:

Step 1 Since H is always placed at the end position and C is less electronegative than O, C should be the central atom. Place the other atoms around C, attached by single bonds.

```
     O
     |
H—C—H
```

Step 2 Determine the total number of valence electrons:

$(1 \text{ C atom} \times 4 \text{ e}^-/\text{C atom}) + (1 \text{ O atom} \times 6 \text{ e}^-/\text{O atom})$
$+ (2 \text{ H atoms} \times 1 \text{ e}^-/\text{H atom})$
$= 4 \text{ e}^- + 6 \text{ e}^- + 2 \text{ e}^-$
$= 12 \text{ e}^-$

Then, determine the total number of electrons required for each atom in the structure to achieve a noble gas configuration. This would be 8 e^- for C and 8 e^- for O (to fill their respective valence energy levels) and 2 e^- for each H (to fill its 1*s* orbital). So the total number for CH_2O is:

$(2 \text{ atoms} \times 8 \text{ e}^-/\text{atom}) + (2 \text{ atoms} \times 2 \text{ e}^-/\text{atom})$
$= 20 \text{ e}^-$

To find the number of shared electrons, subtract the first total from the second,

$20 \text{ e}^- - 12 \text{ e}^- = 8 \text{ e}^-$

Now divide the number of shared electrons by two to obtain the number of bonds.

$20 \text{ e}^- - 12 \text{ e}^- = 8 \text{ e}^- \div 2 = 4$ covalent bonds

Since there are only three atoms surrounding the central atom, you can infer that one of the bonds is likely a double bond.

Step 3 Determine the number of non-bonding electrons by subtracting the number of shared electrons from the total number of valence electrons:
12 valence electrons − 8 shared electrons
= 4 non-bonding valence electrons (that is, 2 lone pairs)

Since H can only form a single bond, a possible structure is:

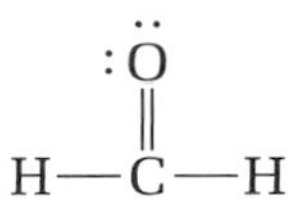

```
    ..
   :O
    ||
H—C—H
```

Check Your Solution

Each atom has achieved a noble gas electron configuration. Thus, you can be confident that this is a reasonable Lewis structure.

Co-ordinate Covalent Bonds

You know that a covalent bond involves the sharing of a pair of electrons between two atoms; each atom contributes one electron to the shared pair. In some cases, such as the hydronium ion, H_3O^+, one atom contributes both of the electrons to the shared pair. The bond in these cases is called a **co-ordinate covalent bond**. In terms of the quantum mechanical model, a co-ordinate covalent bond forms when a filled atomic orbital overlaps with an empty atomic orbital. Once a co-ordinate bond is formed, it behaves in the same way as any other single covalent bond. The next Sample Problem involves a polyatomic ion with a co-ordinate covalent bond.

Sample Problem

Lewis Structures with a Co-ordinate Covalent Bond

Problem

Draw the Lewis structure for the ammonium ion, NH_4^+.

Solution

The formula, NH_4^+, tells you the number of each kind of atom in the ion. Following steps 1 to 3 from the procedure outlined above:

Step 1 Since H is always placed at the end position, N is the central atom.

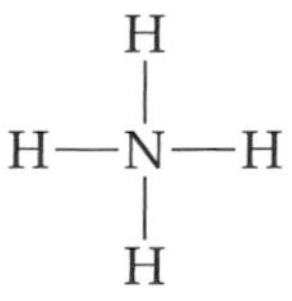

```
    H
    |
H—N—H
    |
    H
```

Step 2 Total number of valence electrons:

$(1 \text{ N atom} \times 5 \text{ e}^-/\text{N atom}) + (4 \text{ H atom} \times 1 \text{ e}^-/\text{H atom}) - 1 \text{ e}^-$
$= 8 \text{ e}^-$

Total number of electrons required for noble gas configuration:
$(1 \text{ atom} \times 8 \text{ e}^-/\text{atom}) + (4 \text{ atoms} \times 2 \text{ e}^-/\text{atom})$
$= 16 \text{ e}^-$

Continued ...

PROBLEM TIP

Because you are working with a polyatomic cation, you must account for the positive charge. Remember to subtract 1 electron from the total number of valence electrons.

Continued ...

Number of shared electrons, and the resulting number of bonds:
16 e⁻ − 8 e⁻ = 8 e⁻
8 e⁻ ÷ 2 = 4 covalent bonds

Step 3 Number of non-bonding electrons:
8 valence electrons − 8 shared electrons
= 0 non-bonding electrons (0 lone pairs)

A possible structure for NH_4^+ is:

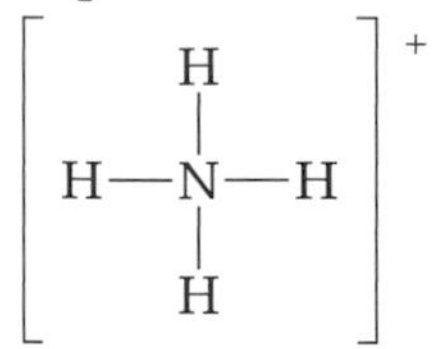

Check Your Solution

Each atom has achieved a noble gas electron configuration. The positive charge on the ion is included. This is a reasonable Lewis structure for NH_4^+.

Note that the Lewis structure for NH_4^+ does *not* indicate which atom provides each shared pair of electrons around the central nitrogen atom. However, the quantum mechanical model of the atom can explain the bonding around this nitrogen atom. The condensed electron configuration for nitrogen is [Ne] $2s^22p^3$. Each nitrogen atom has only three unpaired $2p$ electrons in three half-filled orbitals available for bonding. Since there are four covalent bonds shown around nitrogen in the Lewis structure, electrons in one of the bonds must have come from the filled orbitals of nitrogen. Therefore, one of the bonds around the central nitrogen atom must be a co-ordinate covalent bond.

Resonance Structures: More Than One Possible Lewis Structure

Imagine that you are asked to draw the Lewis structure for sulfur dioxide, SO_2. A typical answer would look like this:

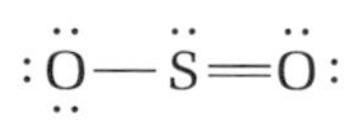

:Ö—S̈=Ö:

This Lewis structure suggests that SO_2 contains a single bond and a double bond. However, experimental measurements of bond lengths indicate that the bonds between the S and each O are identical. The two bonds have properties that are somewhere between a single bond and a double bond. In effect, the SO_2 molecule contains two "one-and-a-half" bonds.

To communicate the bonding in SO_2 more accurately, chemists draw two Lewis structures and insert a double-headed arrow between them. Each of these Lewis structures is called a resonance structure. **Resonance structures** are models that give the same relative position of atoms as in Lewis structures, but show different places for their bonding and lone pairs.

:Ö—S̈=Ö: ⟷ :Ö=S̈—Ö:

Many molecules and ions—especially organic ones—require resonance structures to represent their bonding. It is essential to bear in mind, however, that resonance structures do *not* exist in reality. For example, SO_2 does not shift back and forth from one structure to the other. An actual SO_2 molecule is a combination—a hybrid—of its two resonance structures. It is more properly referred to as a *resonance hybrid.* You could think of resonance hybrids as being like "mutts." For example, a dog that is a cross between a spaniel and a beagle (a "speagle") is part spaniel and part beagle. It does not exist as a spaniel one moment and a beagle the next. Instead, a speagle is an average, of sorts, between the two breeds. Similarly, a resonance hybrid is a weighted average between two or more resonance structures.

Practice Problems

9. Draw Lewis structures for each of the following molecules.

(a) NH_3 **(c)** CF_4 **(e)** BrO^- **(g)** H_2O_2

(b) CH_4 **(d)** AsH_3 **(f)** H_2S **(h)** ClNO

10. Draw Lewis structures for each of the following ions. (**Note:** Consider resonance structures.)

(a) CO_3^{2-} **(b)** NO^+ **(c)** ClO_3^- **(d)** SO_3^{2-}

11. Dichlorofluoroethane, CH_3CFCl_2, has been proposed as a replacement for chlorofluorocarbons (CFCs). The presence of hydrogen in CH_3CFCl_2 markedly reduces the ozone-depleting ability of the compound. Draw a Lewis structure for this molecule.

12. Draw Lewis structures for the following molecules. (**Note:** Neither of these molecules has a single central atom.)

(a) N_2H_4 **(b)** N_2F_2

13. Although Group 18 (VIIIA) elements are inactive, chemists are able to synthesize compounds of several noble gases, including Xe. Draw a Lewis structure for the XeO_4 molecule. Indicate if co-ordinate covalent bonding is likely a part of the bonding in this molecule.

Central Atoms with an Expanded Valence Level

The octet rule allows a maximum of four bonds (a total of eight electrons) to form around an atom. Based upon bond energies, however, chemists have suggested that the bonding in some molecules (and polyatomic ions) is best explained by a model that shows more than eight electrons in the valence energy level of the central atom. This central atom is said to have an **expanded valence energy level.** One example of a molecule with an expanded valence energy level is phosphorus pentachloride, PCl_5. This substance is a pungent, pale-yellow powder that is used in the agricultural, pharmaceutical, and dyeing industries. Organic chemists also use reactions with PCl_5 to identify compounds that contain hydroxyl groups.

```
         ..
        :Cl:   ..
         |   :Cl:
  ..     |  /
:Cl — P
  ..     |  \
         |   :Cl:
        :Cl:   ..
         ..
```

Until recently, chemists explained bonding in molecules such as PCl5 by assuming that empty d orbitals of the central atom were involved. However, experimental evidence does not support this idea. Current thinking suggests that larger atoms can accommodate additional valence electrons because of their size. The theory behind this idea goes beyond the scope of this textbook. Assume that the octet rule *usually* applies when drawing Lewis structures for simple molecules. Sometimes, however, you must violate the rule to allow for more than four bonds around a central atom.

Practice Problems

14. Draw Lewis structures for the following molecules.

(a) SF_6 **(b)** BrF_5

15. Draw Lewis structures for the following molecules.

(a) XeF_4 **(b)** PF_5

16. How does the arrangement of electrons around the central atom differ in PI_3 and ClI_3? Draw the Lewis structures for these compounds to answer this question.

17. Draw a Lewis structure for the ion, IF_4^-.

Shapes and Polarity of Molecules

Up to now, you have represented molecules using Lewis structures. These two-dimensional, formula-like diagrams help you count and keep track of valence electrons, and communicate essential information about the placement and bonding of atoms or ions in a molecule. Chemists often draw Lewis structures in a way that suggests the shape of a molecule. However, this is not their function. It is important to remember that Lewis structures do *not* communicate any information about a molecule's shape. To represent the shapes of real molecules, you need a model that depicts them in three-dimensions.

In 1957, an English chemist, Ronald Gillespie, joined Hamilton, Ontario's McMaster University for a long, distinguished career as a professor of chemistry. Earlier that year, Gillespie, in collaboration with a colleague, Ronald Nyholm, developed a model for predicting the shape of molecules. Like chemistry students around the world since that time, you are about to learn the central features of this model, called the Valence-Shell Electron-Pair Repulsion theory. This is usually abbreviated to VSEPR (pronounced "vesper") theory. (Note that some chemists use the term "shell" instead of energy level to describe atomic orbitals.)

Introducing Valence-Shell Electron-Pair Repulsion (VSEPR) Theory

The fundamental principle of the **Valence-Shell Electron-Pair Repulsion theory** is that the bonding pairs and lone, non-bonding pairs of electrons in the valence level of an atom repel one another. As you know, electron pairs of atoms are localized in orbitals, which are shapes that describe the space in which electrons are most likely to be found around a nucleus. The orbital for each electron pair is positioned as far apart from the other orbitals as possible.

The effect of this positioning minimizes the forces of repulsion between electron pairs. A lone pair (LP) will spread out more than a bond pair. Therefore, the repulsion is greatest between lone pairs (LP—LP). Bonding pairs (BP) are more localized between the atomic nuclei, so they spread out less than lone pairs. Therefore, the BP-BP repulsions are smaller than the LP—LP repulsions. The repulsion between a bond pair and a lone-pair (BP—LP) has a magnitude intermediate between the other two. In other words, in terms of decreasing repulsion:

$$LP—LP > LP—BP > BP—BP.$$

In Unit 1, you encountered the tetrahedral shape around a single-bonded carbon atom, the trigonal planar shape around a carbon atom with one double bond, and the linear shape around a carbon atom with a triple bond. These shapes result from repulsions between lone pairs and bonding pairs of electrons. Now you will use your understanding of LP and BP repulsions to explain the shapes you learned in Unit 1, and to predict the shapes of other molecules. Figure 4.11 shows the five basic geometrical arrangements that result from the interactions of lone pairs and bonding pairs around a central atom. These arrangements involve up to six electron groups. An electron group is usually one of the following:

- a single bond
- a double bond
- a triple bond
- a lone pair

Each of the electron-group arrangements in Figure 4.11 results in a minimum total energy for the molecule. Variations in these arrangements are possible, depending on whether a bonding pair or a lone pair occupies a specific position. Therefore, it is important to distinguish between *electron-group arrangement* and *molecular shape.* Electron-group arrangement refers to the ways in which groups of valence electrons are positioned around the central atom. Molecular shape refers to the relative positions of the atomic nuclei in a molecule. *When all the electron groups are bonding pairs, a molecule will have one of the five geometrical arrangements shown in Figure 4.11. If one or more of the electron groups include a lone pair of electrons, variations in one of the five geometric arrangements result.*

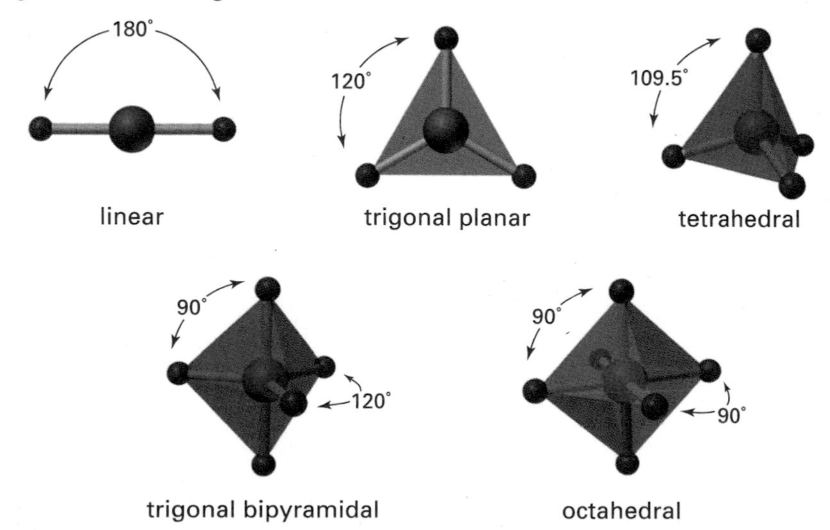

Figure 4.11 **The five basic electron-group arrangements and their bond angles**

In the following ExpressLab, you will make models of the five electron-group arrangements, and measure their bond angles. Afterwards, you will consider some of the variations in molecular shapes that can occur

ExpressLab  Using Soap Bubbles to Model Molecular Shape

You can use soap bubbles to simulate the molecular shapes that are predicted by VSEPR theory. The soap bubbles represent the electron clouds surrounding the central atom in a molecule.

Safety Precautions

- Ensure that each person uses a clean straw.
- Clean up all spills immediately.

Materials

soap solution (mixture of 80 mL distilled water, 15 mL detergent, and 5 mL glycerin)
100 mL beaker
straw
protractor
hard, flat surface
paper towels

Procedure

1. Obtain approximately 25 mL of the prepared soap solution in a 100 mL beaker.
CAUTION Clean up any spills immediately.
2. Wet a hard, flat surface, about 10 cm × 10 cm in area, with the soap solution.
3. Dip a straw into the soap solution in the beaker and blow a bubble on the wetted surface.
CAUTION Each person must use a clean straw.
Blow a second bubble of the same size onto the same surface to touch the first bubble. Record the shape of this simulated molecule and measure the bond angle between the centres of the two bubbles where the nuclei of the atoms would be located.
4. Repeat step 3 with three bubbles of the same size (to simulate three bonding pairs). Record the shape and the bond angle between the centres of the bubbles.
5. Repeat step 3, but this time make one bubble slightly larger than the other two. This will simulate a lone pair of electrons. Record what happens to the bond angles.
6. On top of the group of three equal-sized bubbles, blow a fourth bubble of the same size that touches the centre. Estimate the angle that is formed where bubbles meet.
7. On top of the group of bubbles you made in step 5, make a fourth bubble that is larger than the other three (to simulate a lone pair). Record what happens to the bond angles.
8. Have five students each blow a bubble of the same size at the same time to join and float in the air. Identify the shape that results. If successful, have six partners join six bubbles in the air and identify the shape.
9. Clean your work area using dry paper towels first. Then wipe the area with wet paper towels.

Analysis

1. What shapes and bond angles were associated with two, three, and four same-sized bubbles? Give an example of a molecule that matches each of these shapes.
2. Give an example of a molecule that matches each of the shapes in steps 5 and 7, where you made one bubble larger than the others.
3. What molecular shapes match the arrangements of the bubbles in step 8?
4. If they are available, use inflated balloons or molecule modeling kits to construct each of the arrangements in Figure 4.11. How do these models compare with your bubble models?

Molecular Geometry

Each of the molecules in Figure 4.12 has four pairs of electrons around the central atom. Observe the differences in the number of bonding and lone pairs in these molecules. Methane, CH_4, has four BPs. Ammonia, NH_3, has three BPs and one LP. Water, H_2O, has two BPs and two LPs. Notice the effect of these differences on the shapes and bond angles of the molecules. Methane, with four BPs, has a molecular shape that is the same as the electron-group arrangement in which four electron groups surround a central atom: tetrahedral. The angle between any two bonding pairs in the tetrahedral electron-group arrangement is 109.5°. This angle corresponds to the most favourable arrangement of electron groups to minimize the forces of repulsion among them.

Electronic Learning Partner

Your Chemistry 12 Electronic Learning Partner can help you visualize the geometry of molecular shapes.

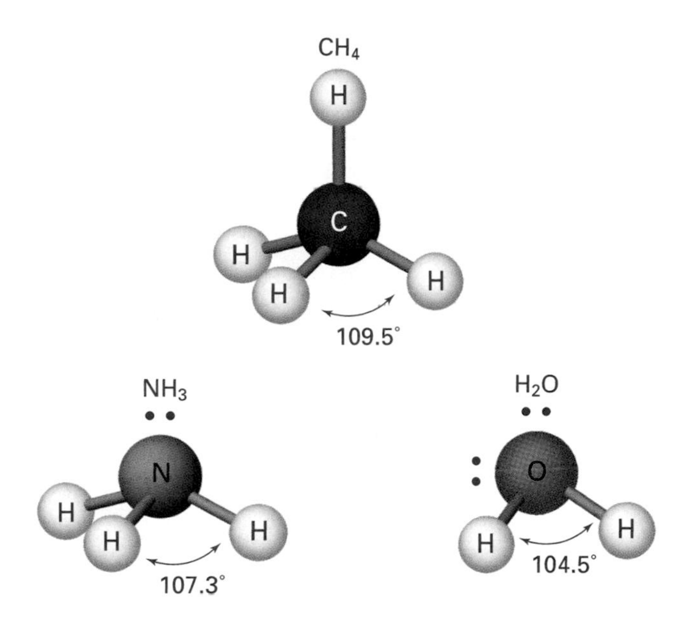

Figure 4.12 Comparing the bond angles in the molecules of CH_4, NH_3, and H_2O.

When there are 1 LP and 3 BPs around a central atom, there are two types of repulsions: LP—BP and BP—BP. Since LP—BP repulsions are greater than BP—BP repulsions, the bond angle between the bond pairs in NH_3 is reduced from 109.5° to 107.3°. When you draw the shape of a trigonal pyramidal molecule, without the lone pair, you can see that the three bonds form the shape of a pyramid with a triangular base. In a molecule of H_2O, there are two BPs and two LPs. The strong LP—LP repulsions, in addition to the LP—BP repulsions, cause the angle between the bonding pairs to be reduced further to 104.5°. You encountered this variation of the tetrahedral shape in Unit 1: the bent shape around an oxygen atom with 2 LPs and two single bonds.

Table 4.2 summarizes the molecular shapes that commonly occur. The VSEPR notation used for these shapes adopts the letter "A" to represent the central atom, the letter "X" to represent a bonding pair, and the letter "E' to represent a lone pair of electrons. For example, the VSEPR notation for NH_3 is AX_3E. This indicates that ammonia has three BPs around its central atom, and one LP.

A trigonal bipyramid has two electron regions where the repulsions are different. An electron pair in the *axial* region (green) experiences three repulsions at 90°, and one at 180°. An electron pair in the *equatorial* region (yellow) experiences two repulsions at 120°, and two at 90°. Repulsions between electron pairs are much greater when 90° apart as compared to 120°. Therefore, repulsions in the equatorial plane are less than those in the axial positions. *Because repulsions are less in the equatorial region, lone pairs and double bonds always occupy equatorial positions in trigonal bipyramidal arrangements.*

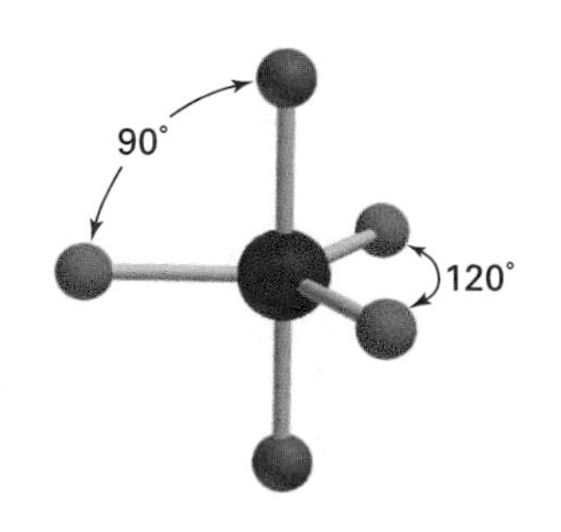

Table 4.2 Common Molecular Shapes and Their Electron Group Arrangements

Number of electron groups	Geometric arrangement of electron groups	Type of electron pairs	VSEPR notation	Name of Molecular shape	Example
2	linear	2 BP	AX_2	X—A—X linear	BeF_2
3	trigonal planar	3 BP	AX_3	trigonal planar	BF_3
3	trigonal planar	2 BP, 1 LP	AX_2E	angular	$SnCl_2$
4	tetrahedral	4 BP	AX_4	tetrahedral	CF_4
4	tetrahedral	3 BP, 1LP	AX_3E	trigonal pyramidal	PCl_3
4	tetrahedral	2 BP, 2LP	AX_2E_2	angular	H_2S
5	trigonal bipyramidal	5 BP	AX_5	trigonal bipyramidal	$SbCl_5$
5	trigonal bipyramidal	4 BP, 1LP	AX_4E	seesaw	$TeCl_4$

5	trigonal bipyramidal	3 BP, 2LP	AX_3E_2	T-shaped	BrF_3
5	trigonal bipyramidal	2 BP, 3LP	AX_2E_3	linear	XeF_2
6	octahedral	6 BP	AX_6	octahedral	SF_6
6	octahedral	5 BP, 1LP	AX_5E	square pyramidal	BrF_5
6	octahedral	4 BP, 2LP	AX_4E_2	square planar	XeF_4

Predicting Molecular Shape

You can use the steps below to help you predict the shape of a molecule (or polyatomic ion) that has one central atom. Refer to these steps as you work through the Sample Problems and the Practice Problems that follow.

1. Draw a preliminary Lewis structure of the molecule based on the formula given.
2. Determine the total number of electron groups around the central atom (bonding pairs, lone pairs and, where applicable, account for the charge on the ion). Remember that a double bond or a triple bond is counted as one electron group.
3. Determine which one of the five geometric arrangements will accommodate this total number of electron groups.
4. Determine the molecular shape from the positions occupied by the bonding pairs and lone pairs.

Sample Problem

Predicting Molecular Shape for a Simpler Compound

Problem

Determine the molecular shape of the hydronium ion, H_3O^+.

Plan Your Strategy

Follow the four-step procedure that helps to predict molecular shape. Use Table 4.2 for names of the electron-group arrangements and molecular shapes.

Act on Your Strategy

Step 1 A possible Lewis structure for H_3O^+ is:

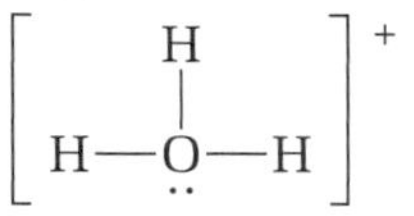

```
[      H      ] +
       |
  H —  O  — H
       ..
```

Step 2 The Lewis structure shows 3 BPs and 1 LP. That is, there are a total of four electron groups around the central O atom.

Step 3 The geometric arrangement of the electron groups is tetrahedral.

Step 4 For 3 BPs and 1 LP, the molecular shape is trigonal pyramidal.

Check Your Solution

This molecular shape corresponds to the VSEPR notation for this ion, AX_3E.

Sample Problem

Predicting Molecular Shape for a Complex Compound

Problem

Determine the shape of SiF_6^{2-} using VSEPR theory.

Plan Your Strategy

Follow the four-step procedure that helps to predict molecular shape apply. Use Table 4.2 for names of the electron group arrangements and molecular shapes.

Act on Your Strategy

Step 1 Draw a preliminary Lewis structure for SiF_6^{2-}.

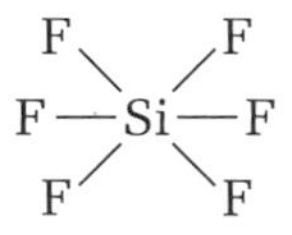

```
 F     F
   \  /
F — Si — F
   /  \
 F     F
```

This polyatomic ion has six bonds around the central Si atom, an obvious exception to the octet rule, so the central atom needs an expanded valence shell.

Total number of valence electrons

$= 1 \text{ Si atom} \times 4 \text{ e}^-/\text{Si atom} + 6 \text{ F atom} \times 7 \text{ e}^-/\text{F atom} + 2 \text{ e}^-$ (ionic charge)

$= 48 \text{ e}^-$

For all seven atoms, 48 e^- are available for bonding. Si forms 6 single bonds to each F atom, 12 electrons are used. There remains 36 e^- (18 lone pairs) to be placed around the F atoms. Each F atom uses 3 lone pairs (6 e^-) to complete an octet. Therefore, a possible structure is:

```
 [  ..     ..  ]2−
 [ :F:    :F:  ]
 [ ..  \  / .. ]
 [ :F — Si — F:]
 [ ..  /  \ .. ]
 [ :F:    :F:  ]
 [  ..     ..  ]
```

Step 2 From the structure, there are 6 BPs and 0 LP, a total of six electron groups around the central Si atom.

Step 3 The geometric arrangement of the electron group is octahedral.

Step 4 For 6 BPs, the molecular shape is also octahedral.

Check Your Solution

This molecular shape corresponds to the VSEPR notation for this ion, AX_6.

Keep in mind that the need for an expanded valence level for the central atom may not always be as obvious as in the previous Sample Problem. For example, what if you were asked to predict the molecular shape of the polyatomic ion, $BrCl_4^-$? Drawing the Lewis structure enables you to determine that the central atom has an expanded valence level.

Total number of valence electrons = 36 e-
Total number of electrons required for each atom to have an octet = 40 e^-

There are not enough electrons for bonding. Therefore, the central atom requires an expanded valence level. A possible Lewis structure for $BrCl_4^-$ is:

```
 [  ..      ..  ]−
 [ :Cl:    :Cl: ]
 [     \  /     ]
 [     :Br:     ]
 [     /  \     ]
 [ :Cl:    :Cl: ]
 [  ..      ..  ]
```

Now you can see that the central atoms has 4 BPs and 2 LPs, a total of 6 electron groups around it. The geometric arrangement of the electron group is octahedral. For 4 BPs and 2 LPs, the molecular shape is square planar.

Practice Problems

18. Use VSEPR theory to predict the molecular shape for each of the following:
(a) HCN **(b)** SO_2 **(c)** SO_3 **(d)** SO_4^{2-}

19. Use VSEPR theory to predict the molecular shape for each of the following:
(a) CH_2F_2 **(b)** $AsCl_5$ **(c)** NH_4^+ **(d)** BF_4^-

20. Use VSEPR theory to predict the molecular shapes of NO_2^+ and NO_2^-.

Continued ...

Continued ...

21. Draw Lewis structures for the following molecules and ions, and use VSEPR theory to predict the molecular shape. Indicate the examples in which the central atom has an expanded octet.
    (a) XeI_2 **(b)** PF_6^- **(c)** AsF_3 **(d)** AlF_4^-
22. Given the general formula and the shape of the molecule or ion, suggest possible elements that could be the central atom, X, in each of the following:
    (a) XF_3^+ (trigonal pyramidal) **(c)** XF_3 (T-shaped)
    (b) XF_4^+ (tetrahedral)

Tools 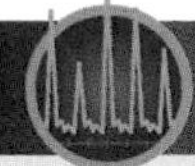 & Techniques

AIM Theory and Electron Density Maps

The microscopic world of atoms is difficult to imagine, let alone visualize in detail. Chemists and chemical engineers employ different molecular modelling tools to study the structure, properties, and reactivity of atoms, and the way they bond to one another. Richard Bader, a chemistry professor at McMaster University, has invented an interpretative theory that is gaining acceptance as an accurate method to describe molecular behaviour and predict molecular properties. According to Dr. Bader, shown below, small molecules are best represented using topological maps, where contour lines (which are commonly used to represent elevation on maps) represent the electron density of molecules.

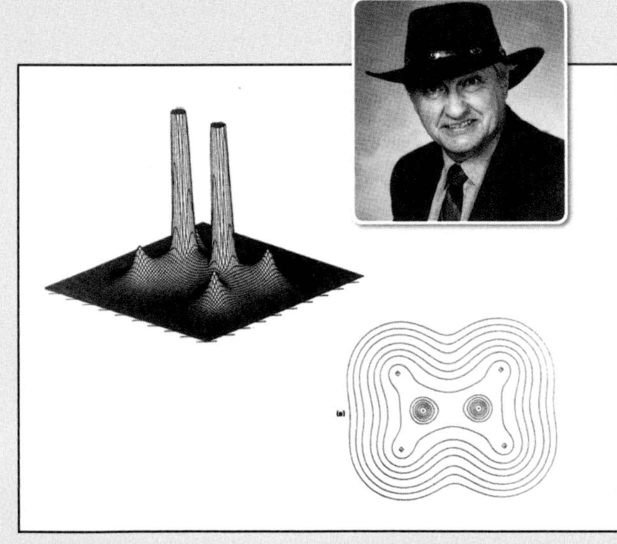

The diagrams above show two different ways to model the electron density in a plane of a molecule of ethylene .

In the early 1970s, Dr. Bader invented the theory of "Atoms in Molecules," otherwise known as AIM theory. This theory links the mathematics of quantum mechanics to the atoms and bonds in a molecule. AIM theory adopts electron density, which is related to the Schrödinger description of the atom, as a starting point to mapping molecules.

Chemists often focus on the energetic, geometric, and spectroscopic properties of molecules. However, since the electron density exists in ordinary three-dimensional space, electron density maps of molecules can be used as tools to unearth a wealth of information about the molecule. This information includes, but is not limited to, a molecule's magnetic properties, per-atom electron population, and bond types.

Using diagrams and computer graphics, chemists use AIM theory to construct three-dimensional electron density maps of molecules. Chemists can then use the maps as tools to perform experiments on the computer as if they were performing the same experiments in the laboratory, but with higher efficiency. Many chemists are excited by the prospect of using computer graphics, both inside and outside of the classroom, as teaching tools for students and the public at large.

In addition to being able to simulate very simple isolated molecules and their reactions, AIM theory provides a physical basis for the theory of Lewis electron pairs and the VSEPR model of molecular geometry. Equipped with computers and computer-generated, three-dimensional electron density maps, scientists are able to view molecules and predict molecular phenomena without even having to get off their chairs!

The Relationship Between Molecular Shape and Molecular Polarity

A molecule's shape and polarity are directly related. As you know, the electronegativity difference between two atoms is the principal factor that determines the type of chemical bonding between them. As you saw in section 4.1, when atoms of two different elements have ΔEN between 0.5 and 1.7, the bond between them is polar covalent. The electrons are shared unequally between the two types of atoms. For example, in the HCl molecule, the Cl atom is more electronegative than the H atom. Thus, the Cl atom exerts a stronger force of attraction on the shared pair of electrons. As a result, the Cl end of the bond develops a partial negative charge (δ^-) and the H end develops a partial positive charge (δ^+). This polarity in a bond is called a *bond dipole*. The vector used to represent a bond dipole points toward the more electronegative element. For example, you could represent the polarity of the bond in hydrogen chloride as:

$$\overset{\delta^+}{H}:\overset{\delta^-}{\ddot{\underset{..}{Cl}}}: \text{ or } \overset{\longmapsto}{HCl}$$

For a diatomic molecule such as HCl, or the bond polarity is also the molecule's polarity. For polyatomic molecules, such as the examples that follow, molecular polarity depends on the polarity of all of the bonds and the angles at which the dipoles come together. Thus, molecular dipole—or just **dipole**—is a term used to describe the charge separation for the entire molecule.

Consider, for example, BeF_2, which is a symmetrical, linear molecule. Each Be—F bond is polar (because fluorine has a greater electronegativity than beryllium). Due to the linear shape of this molecule, however, the polarities of the two bonds are directly opposite each other. The two bonding polarities exactly counteract each other, so that BeF_2 is a non-polar molecule. The shape of a molecule, combined with the polarity of its individual bonds, therefore, determine polarity.

$$\overset{\delta^-}{F} \overset{\longleftarrow\!\!+\!\!\longrightarrow}{—\overset{\delta^+}{Be}—} \overset{\delta^-}{F}$$

CONCEPT CHECK

Look back at Chapter 1, pages 8–10, to review how to predict the molecular polarity of organic compounds.

Compare two molecules with the same molecular shape: CCl_4 and CCl_3H. Since the polarity of the C—Cl bond is different from that of the C—H bond, the polarities of these two molecules are different. The CCl_4 molecule is symmetrical about any axis joining the two atoms of the C—Cl bond. The polarities of the C—Cl bonds counteract one another. Thus, the carbon tetrachloride molecule is non-polar. In the case of CCl_3H, the polarity of the H—Cl bond is different from that of the three C—Cl bonds. Thus, the molecule is polar.

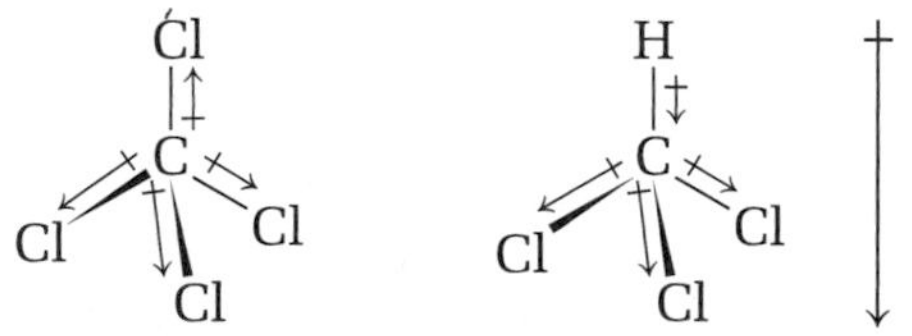

Cl, C, Cl, Cl, Cl — non-polar

H, C, Cl, Cl, Cl — polar

Table 4.3 summarizes the relationship between molecular polarity and molecular shape. The letter "A" represents the central atom and the letter "X" represent a more electronegative atom. An atom represented by the letter "Y" is more electronegative than the atom represented by the letter "X." After you examine this table, do the Practice Problems that follow.

Table 4.3 Effect of Molecular Shape on Molecular Polarity

Molecular shape	Bond polarity	Molecular polarity
linear	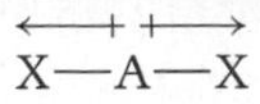 X—A—X	non-polar
linear	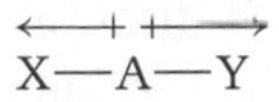 X—A—Y	polar
bent	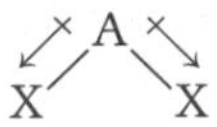 X—A—X (bent)	polar
trigonal planar	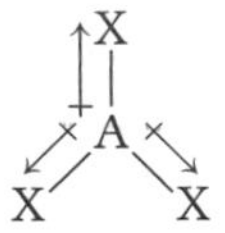 AX_3	non-polar
trigonal planar	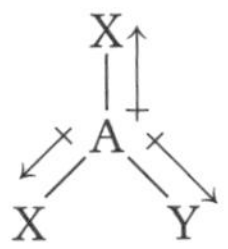 AX_2Y	polar
tetrahedral	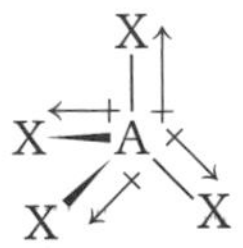 AX_4	non-polar
tetrahedral	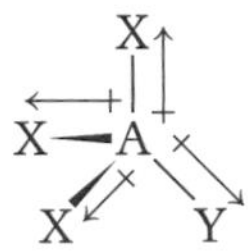 AX_3Y	polar

Practice Problems

23. Use VSEPR theory to predict the shape of each of the following molecules. From the molecular shape and the polarity of the bonds, determine whether or not the molecule is polar.

(a) CH_3F **(b)** CH_2O **(c)** AsI_3

24. Freon-12, CCl_2F_2, was used as a coolant in refrigerators until it was suspected to be a cause of ozone depletion. Determine the molecular shape of CCl_2F_2 and discuss the possibility that the molecule will be a dipole.

25. Which of the following is more polar? Justify your answer in each case.

(a) NF_3 or NCl_3 **(b)** $SeCl_4$ or $TeCl_4$

26. A hypothetical molecule with the formula XY_3 is discovered, through experiment, to exist. It is polar. Which molecular shapes are possible for this molecule? Which shapes are impossible? Explain why in each case.

Section Summary

In this section, you have pieced together the main components that determine the structure and polarity of molecules. Why is the polarity of a molecule important? Polar molecules attract one another more than non-polar molecules do. Because of this attraction, many physical properties of substances are affected by the polarity of their molecules. In the next section, you will consider some of these physical properties for liquid and solid substances, and learn about other forces that have a significant effect on the interactions within and among molecules.

Section Review

1. K/U For each of the following molecules, draw a Lewis structure, and determine if the molecule is polar or non-polar.

   (a) AsH_3 **(b)** CH_3CN **(c)** Cl_2O

2. C Discuss the validity of the statement: "All polar molecules must have polar bonds and all non-polar molecules must have non-polar bonds."

3. K/U What similarities and what differences would you expect in the molecular shape and the polarity of CH_4 and CH_3OH? Explain your answer.

4. I The molecules BF_3 and NH_3 are known to undergo a combination reaction in which the boron and nitrogen atoms of the respective molecules join together. Sketch a Lewis structure of the molecule that would be expected to be formed from this reaction. Would the product molecule be expected to be more or less reactive than the starting materials? Give a reason for your answer.

5. I Use VSEPR theory to predict the molecular shape and the polarity of SF_2. Explain your answer.

6. I Draw a Lewis structure for PCl_4^+. Use VSEPR theory to predict the molecular shape of this ion.

7. C Identify and explain the factors that determine the structure and polarity of molecules.

4.3 Intermolecular Forces in Liquids and Solids

Section Preview/ Specific Expectations

In this section, you will

- **distinguish** between intermolecular and intramolecular forces
- **explain** how the properties of solids and liquids are connected to their component particles and the forces between and among them
- **conduct** experiments to **observe** and **analyze** the physical properties of different substances and **determine** the type of bonding involved
- **communicate** your understanding of the following terms: *intramolecular forces, intermolecular forces, dipole-dipole force, ion-dipole force, ion-induced dipole force, dipole-induced dipole force, dispersion force, hydrogen bonding, crystalline solids, amorphous solids, network solids, allotrope*

Water and hydrogen sulfide have the same molecular shape: bent. Both are polar molecules. However, H_2O, with a molar mass of 18 g, is a liquid at room temperature, while H_2S, with a molar mass of 34 g, is a gas. Water's boiling point is 100°C, while hydrogen sulfide's boiling point is −61°C. These property differences are difficult to explain using the bonding models you have learned so far. These models focused on the forces of attraction between ions in ionic solids, between individual atoms in covalent and polar covalent compounds, and between positive metal ions and free electrons in metals. These are **intramolecular forces**—forces exerted *within* a molecule or polyatomic ion. During a chemical change, intramolecular forces are overcome and chemical bonds are broken. Therefore, these forces influence the chemical properties of substances.

Forces that influence the physical properties of substances are called **intermolecular forces**. These are forces of attraction and repulsion that act *between* molecules or ions.

Intermolecular forces were extensively studied by the Dutch physicist Johannes van der Waals (1837–1923). To mark his contributions to the understanding of intermolecular attractions, these forces are often called van der Waals forces. They are categorized into the following groups: dipole-dipole forces, ion-dipole forces, induced dipole forces, dispersion (London) forces, and hydrogen bonding. In the pages that follow, you will see how each of these types of intermolecular forces affects the molecular properties of liquids. Following that, you will apply your understanding of intermolecular forces to help you explain some physical properties of substances in the solid state.

Dipole-Dipole Forces

In the liquid state, polar molecules (dipoles) orient themselves so that oppositely charged ends of the molecules are near to one another. The attractions between these opposite charges are called **dipole-dipole forces**. Figure 4.13 shows the orientation of polar molecules due to these forces in a liquid.

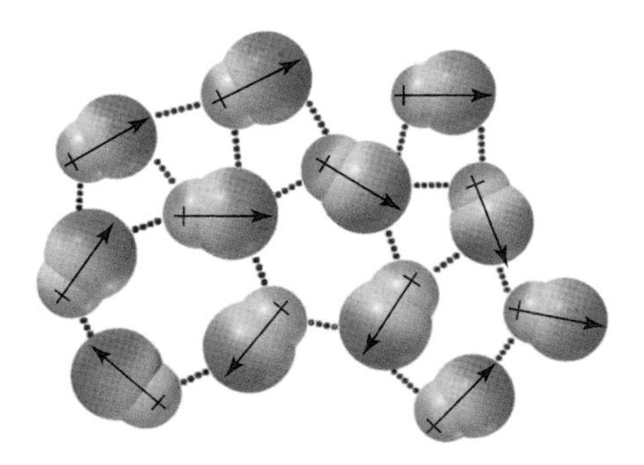

Figure 4.13 Dipole-dipole forces among polar molecules in the liquid state

As a result of these dipole-dipole forces of attraction, polar molecules will tend to attract one another more at room temperature than similarly sized non-polar molecules would. The energy required to separate polar molecules from one another is therefore greater than that needed to separate non-polar molecules of similar molar mass. This is indicated by the extreme difference in melting and boiling points of these two types of molecular substances. (Recall that melting and boiling points are physical properties of substances.)

CONCEPT CHECK

Acetaldehyde (ethanal), CH_3CHO, is a polar molecule that has a boiling point of 20°C. Propane, on the other hand, is a non-polar molecule of similar size, number of electrons, and molar mass. The boiling point of propane, $CH_3CH_2CH_3$, is −42°C. Use the concept of dipole-dipole forces to explain these property differences.

Ion-Dipole Forces

Sodium chloride and other soluble ionic solids dissolve in polar solvents such as water because of ion-dipole forces. An **ion-dipole force** is the force of attraction between an ion and a polar molecule (a dipole). For example, NaCl dissolves in water because the attractions between the Na^+ and Cl^- ions and the water molecules provide enough energy to overcome the forces that bind the ions together. Figure 4.14 shows how ion-dipole forces dissolve any type of soluble ionic compound.

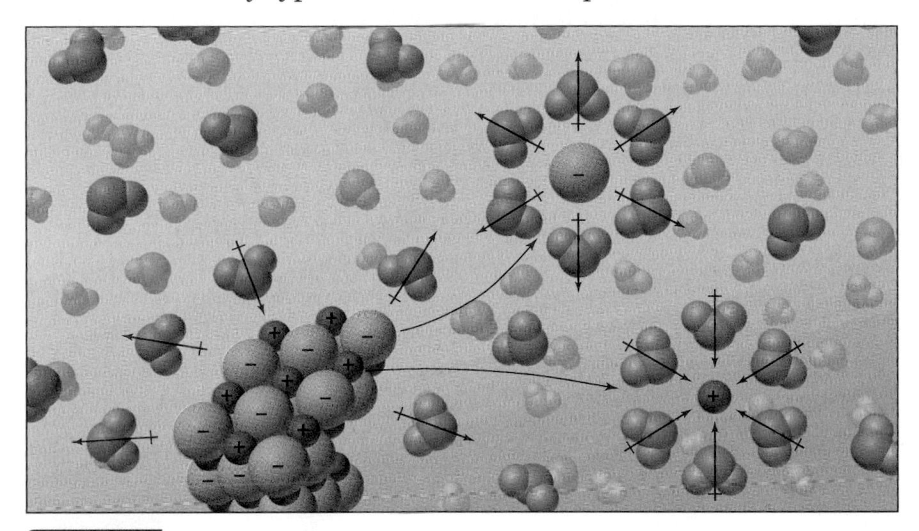

Figure 4.14 In aqueous solution, ionic solids dissolve as the negative ends and the positive ends of the water molecules become oriented with the corresponding oppositely charged ions that make up the ionic compound, pulling them away from the solid into solution.

Induced Intermolecular Forces

Induction of electric charge occurs when a charge on one object causes a change in the distribution of charge on a nearby object. Rubbing a balloon to make it "stick" to a wall is an example of charging by induction. There are two types of charge-induced dipole forces.

An **ion-induced dipole force** results when an ion in close proximity to a non-polar molecule distorts the electron density of the non-polar molecule. The molecule then becomes momentarily polarized, and the two species are attracted to each other. This force is active during every moment of your life, in the bonding between non-polar O_2 molecules and the Fe^{2+} ion in hemoglobin. Ion-induced dipole forces, therefore, are part of the process that transports vital oxygen throughout your body.

A **dipole-induced dipole force** is similar to that of an ion-induced dipole force. In this case, however, the charge on a polar molecule is responsible for inducing the charge on the non-polar molecule. Non-polar gases such as oxygen and nitrogen dissolve, sparingly, in water because of dipole-induced dipole forces.

Dispersion (London) Forces

The shared pairs of electrons in covalent bonds are constantly vibrating. While this applies to all molecules, it is of particular interest for molecules that are non-polar. The bond vibrations, which are part of the normal condition of a non-polar molecule, cause momentary, uneven distributions of charge. In other words, a non-polar molecule becomes slightly polar for an instant, and continues to do this on a random but on-going basis. At the instant that one non-polar molecule is in a slightly polar condition, it is capable of inducing a dipole in a nearby molecule. An intermolecular force of attraction results. This force of attraction is called a **dispersion force**. Chemists also commonly refer to it as a London force, in honour of the German physicist, Fritz London, who studied this force. Unlike other intermolecular forces, dispersion (London) forces act between any particles, polar or otherwise. They are the main intermolecular forces that act between non-polar molecules, as shown in Figure 4.15.

Two factors affect the magnitude of dispersion forces. One is the number of electrons in the molecule and the other is the shape of the molecule. Vibrations within larger molecules that have more electrons than smaller molecules can easily cause an uneven distribution of charge. The dispersion forces between these larger molecules are thus stronger, which has the effect of raising the boiling point for larger molecules. A molecule with a spherical shape has a smaller surface area than a straight chain molecule that has the same number of electrons. The smaller surface area allows less opportunity for the molecule to induce a charge on a nearby molecule. Therefore, for two substances with molecules that have a similar number of electrons, the substance with molecules that have a more spherical shape will have weaker dispersion forces and a lower boiling point.

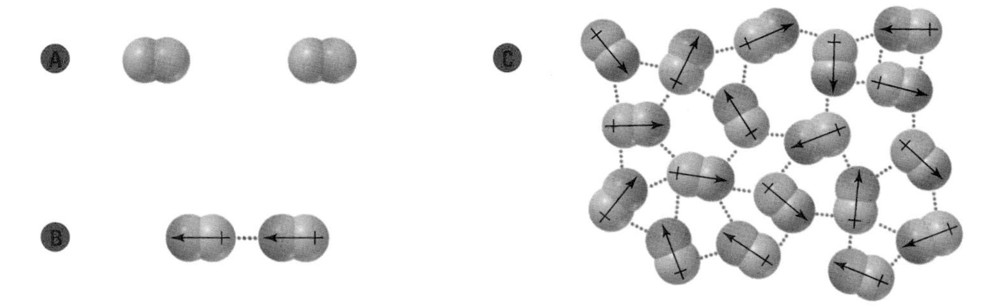

Figure 4.15 How dispersion forces develop between identical non-polar molecules. In A, neither molecule interacts with the other. In B, one molecule becomes, instantaneously, a dipole. At that moment, the dipolar molecule is able to induce a temporary charge separation in the other molecule, resulting in a force of attraction between the two. All the molecules within a sample undergo this same process, as shown in C.

Hydrogen Bonding

A dipole-dipole interaction that is very significant in many biological molecules and some inorganic molecules, such as $H_2O_{(\ell)}$, is hydrogen bonding. **Hydrogen bonding** is a particularly strong form of dipole-dipole attraction that exists between a hydrogen atom in a polar-bonded molecule that contains bonds such as H—O, H—N, or H—F, and an unshared pair of electrons on another small, electronegative atom such as O, N, or F. The small, electronegative atom can be on its own, but is usually bonded in a molecule.

When hydrogen is bonded to oxygen, fluorine, or nitrogen, the strong force of attraction exerted by any of these three very electronegative atoms draws the electron density from the hydrogen atom in the polar-bonded molecule, leaving the hydrogen atom with a partial positive charge. This dipole is easily attracted to the partially negative lone pair on a nearby electronegative atom: N, O, or F.

You can see the effect of hydrogen bonding clearly in the boiling point data of the binary hydrides of Groups 14 to 17 (IVA to VIIA), shown in Figure 4.16. In Group 14, the trend in boiling point is as expected. The increase in boiling point from methane, CH_4, to tin(IV) hydride, SnH_4, is due to an increase in the number of electrons. Therefore, there are larger dispersion forces, since more electrons are able to temporarily shift from one part of the molecule to another. In SnH_4, this results in greater intermolecular forces of attraction. There is no hydrogen bonding in these substances, because there are no lone pairs on the molecules.

However, the hydrides of Groups 15, 16, and 17 (VA to VIIA) only show the expected trend based on dispersion forces for the larger mass hydrides. The smallest mass hydrides, NH_3, H_2O, and HF, have relatively high boiling points, indicating strong intermolecular forces caused by hydrogen bonding.

Why is there no hydrogen bonding in H_2S and H_2Se, which have unshared electron pairs on their central atoms? Neither S nor Se is small enough or electronegative enough to support hydrogen bonding. Thus, H_2O (with its small and very electronegative O atom) has a much higher boiling point than H_2S and H_2Se, as you can see in Figure 4.16.

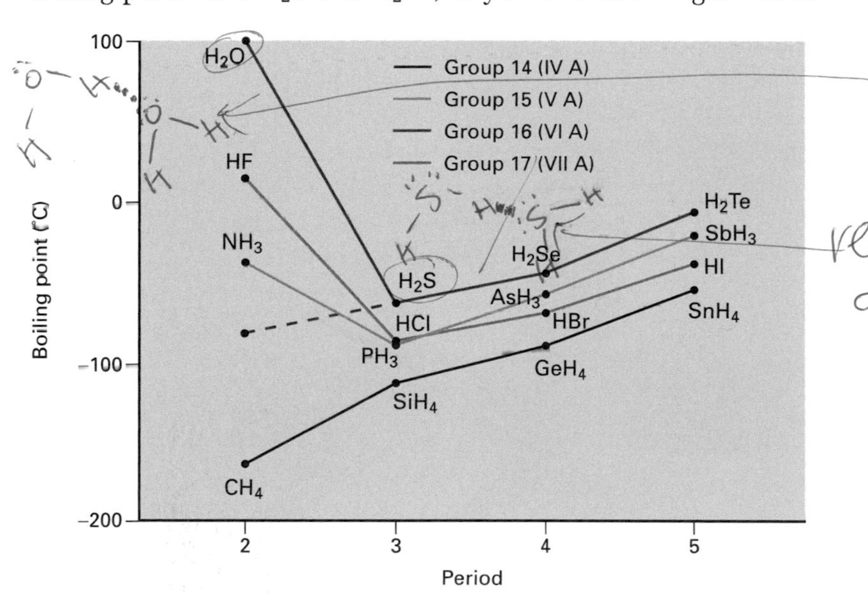

Figure 4.16 The boiling points of four groups of hydrides. The break in the trends for NH_3, H_2O, and HF is due to hydrogen bonding. The dashed line shows the likely boiling point of water, if no hydrogen bonding were present.

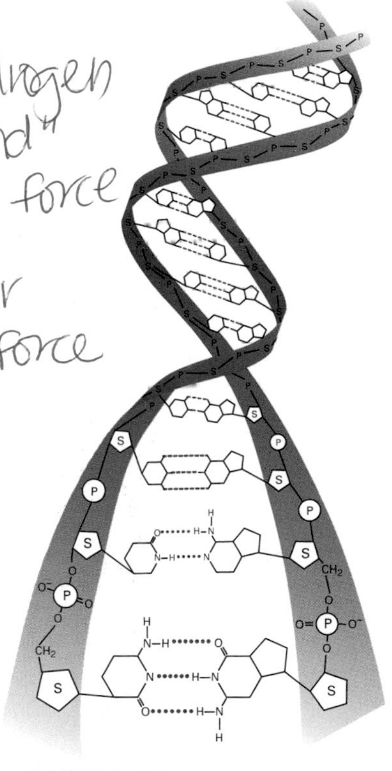

A hydrogen bond is only about 5% as strong as a single covalent bond. However, in substances that contain numerous hydrogen bonds the impact of hydrogen bonding can be significant. For example, the double-helix structure of DNA occurs because of hydrogen bonds, as shown in Figure 4.17. Another substance that contains many hydrogen bonds, water, is discussed on the following page.

Figure 4.17 A DNA (deoxyribonucleic acid) molecule is comprised of two chains, each formed from covalently bonded atoms. Hydrogen bonds hold the two chains together, producing DNA's familiar, spiral-like, double helix structure.

Hydrogen Bonding and Properties of Water

Hydrogen bonding also explains water's unusual property of being less dense in the solid state than in the liquid state. As you know from kinetic molecular theory, lower temperatures cause liquid molecules to move slower than in their gaseous states. When liquids solidify, the kinetic energy of the particles is no longer enough to prevent the intermolecular forces of attraction from causing particles to stick together. For most substances, their solid states are denser than their liquid states. When water freezes, however, the water molecules pack in such a way that the solid state is *less* dense than the liquid state. Water molecules align in a specific pattern so that hydrogen atoms of one molecule are oriented toward the oxygen atom of another molecule. You can see this clearly in Figure 4.18. If water molecules behaved as most molecules do, then lake water would freeze from the bottom up, rather than from the top down. The consequences for aquatic life, and on organisms (like us) that depend on aquatic ecosystems, would be profound. A lake that freezes from the bottom up would solidify completely, killing most of the life it contains. Because ice floats on water, it insulates the water beneath it, preventing it from freezing.

Hydrogen bonding also explains the solubility in water of polar covalent compounds. As you learned in Unit 1, the O—H functional group in alcohols is polar. Because of this polar group, alcohols can form hydrogen bonds with water molecules. Consequently, methanol and ethanol are soluble in water in all proportions, and every alcohol is more soluble than its corresponding alkane with the same number of carbon atoms. (Recall that alkanes do not have the O—H functional group.) Ammonia, NH_3, is a very soluble gas because it hydrogen-bonds with water. The smaller mass amines, such as methylamine, CH_3NH_2, are also soluble in water, for the same reason.

CONCEPT CHECK

How is hydrogen bonding different from a covalent bond such as O—H?

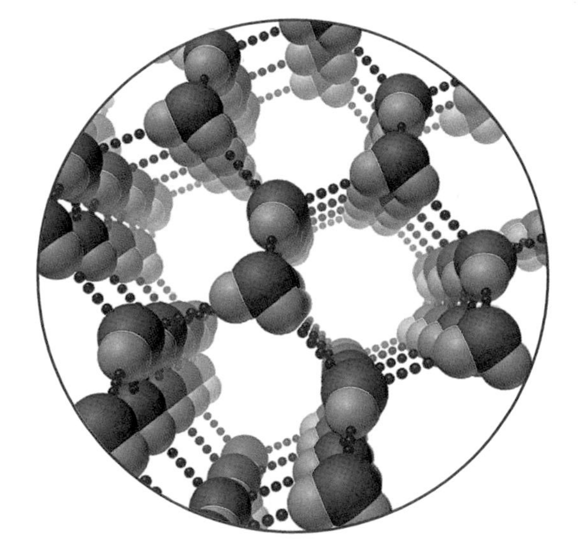

Figure 4.18 In its solid state, water has an open, hexagonal, crystalline structure. As a result of these spaces, ice is less dense than liquid water.

Summarizing Intramolecular and Intermolecular Forces

Table 4.4 compares key characteristics of intramolecular and intermolecular forces. Once you have reviewed this information, you will have an opportunity to expand the table by doing the ThoughtLab that follows.

Table 4.4 Comparing Intramolecular and Internolecular Forces

	Force	Model	Nature of Attraction	Energy (kJ/mol)	Example
Intramolecular	ionic		cation-anion	400–4000	NaCl
	covalent		nuclei-shared electron pair	150–1100	H—H
	metallic		cations-delocalized electrons	75–1000	Fe
Intermolecular	ion-dipole		ion charge-dipole charge	40-600	Na^+----OH_2
	hydrogen bond	$\overset{\delta^-}{—A}—\overset{\delta^+}{H}\cdots\cdots:\overset{\delta^-}{B}—$	polar bond to hydrogen-dipole charge (lone pair, high *EN* of N, O, F)	10–40	H_2O—H----:OH_2
	dipole-dipole		dipole charges	5–25	I—Cl----I—Cl
	ion-induced dipole		ion charge-polarizable electrons	3–15	Fe^{2+}----O_2
	dipole-induced dipole		dipole charge-polarizable electrons	2–10	H—Cl----Cl—Cl
	dispersion (London)		polarizable electrons	0.05–40	F—F----F—F

ThoughtLab  Properties of Liquids

As you know, intermolecular forces act between both like and unlike molecules. *Adhesive forces* are intermolecular forces between two different kinds of molecules. *Cohesive forces* are intermolecular forces between two molecules of the same kind. Adhesive and cohesive forces affect the physical properties of many common phenomena in the world around you.

Procedure

1. Consider each of the following observations. For each, infer the intermolecular (and, where appropriate, intramolecular) forces that are active, and explain how they account for the observations.
   - Why are bubbles in a soft drink spherical?
   - How can water striders walk on water?
   - Why does water form beaded droplets on surfaces such as leaves, waxed car hoods, and waxed paper?
   - How does a ballpoint pen work?
   - How does a towel work?
   - Why does water form a concave meniscus in a tube, while mercury forms a convex meniscus?
   - Why does liquid honey flow so slowly from a spoon?
   - Why do some motorists change to different motor oils over the course of a year?
2. Based on your inferences and explanations, develop definitions for the following terms: surface tension, capillarity, and viscosity.
3. Verify your inferences and definitions by consulting print and electronic resources. Then identify the property or properties that best explain each of the observation statements.

Analysis

1. Compare your inferences and explanations with those that you found in reference sources. Identify those cases in which your ideas matched or closely matched what you found. Explain where your reasoning may have led you astray, for those cases in which your ideas did not match closely with reference-source information.
2. Re-design Table 4.4 on page 195 so that it includes practical applications and explanations such as those you have investigated in this activity.

Bonding in Solids

In Investigation 4-A, you classified solid substances on the basis of the types of bonds that hold their molecules or ions together. This is not, however, the only way to classify solids. Chemists also divide solids into two categories based on the macroscopic properties that arise from the *arrangements* of their component particles. **Crystalline solids** have organized particle arrangements, so they have distinct shapes. Gemstones such as amethyst and garnet are examples of crystalline solids. **Amorphous solids** have indistinct shapes, because their particle arrangements lack order. Glass and rubber are common amorphous solids. In the next few pages, you will learn about the structure and properties of crystalline and amorphous solids.

Bonding and Properties of Crystalline Solids

Chemists further classify crystalline solids into five types, based on their composition. These types are atomic, molecular, network, ionic, and metallic. Since you have already learned about metallic solids, the material that follows concentrates on the other four crystalline solids. You will find a summary of their properties in Table 4.5, following these descriptions.

Atomic Solids

Atomic solids are made up of individual atoms that are held together solely by dispersion forces. The number of naturally occurring atomic solids is quite small. In fact, the noble gases in their solid state are the only examples. Since the only forces holding atomic solids together are dispersion forces, these solids have very low melting and boiling points.

Molecular Solids

Molecular solids are, as their name implies, made up of molecules. Like atomic solids, the molecules that make up molecular solids, such as frozen methane, are held together mainly by dispersion forces. Consequently, they tend to have low melting and boiling points. In polar molecular solids, however, stronger dipole-dipole forces, and sometimes hydrogen bonding, give these compounds higher melting points and boiling points than non-polar molecular solids. For example, the melting point of non-polar propane, C_3H_8, is –190°C. A molecule of polar formic acid, HCOOH, which has a similar number of electrons and the same molar mass as propane, melts at 8.4°C. The large difference in boiling point is due to the hydrogen bonding and dipole-dipole forces in formic acid. You will often see molecular solids classified as covalent solids. However, it is clearer to classify them as molecular solids, because network solids, which you will see below, also contain covalent bonds.

Network Solids

Unlike the intramolecular covalent bonds that hold atoms together in discrete molecules, it is possible for atoms to bond covalently into continuous two- or three-dimensional arrays, called **network solids.** A wide range of properties can be found among network solids.

Carbon-Based Network Solids

Figure 4.19 shows four allotropic forms of carbon. **Allotropes** are different crystalline or molecular forms of the same element that differ in physical and chemical properties.

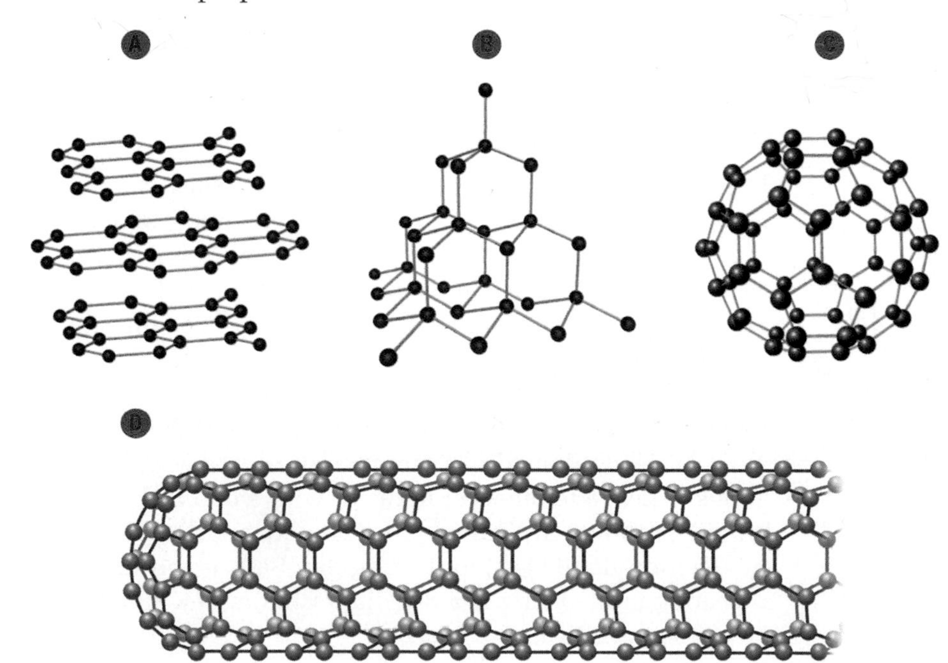

Figure 4.19 Allotropic forms of carbon: (A) graphite; (B) diamond; (C) C_{60}, buckminsterfullerene; (D) fullerene nanotube

At temperatures around 600°C and extremely low pressures, methane will decompose and deposit a thin film of carbon, in the form of diamond, on a surface. The process, known as chemical vapour deposition, or CVD, can be used to coat a wide variety of surfaces. The hardness of this diamond film can be used for applications that require non-scratch surfaces such as cookware, eyeglasses, and razor blades.

The graphite shown in Figure 4.19A is a soft, two-dimensional network solid that is a good conductor of electricity. It has a high melting point, which suggests strong bonds, but its softness indicates the presence of weak bonds as well. Each carbon forms three strong covalent bonds with three of its nearest neighbours in a trigonal planar pattern. This structure gives stability to the layers. The fourth, non-bonding pair of electrons is delocalized and free to move within the layers. This gives graphite the ability to conduct an electric current. There are no intramolecular forces between the layers. Dispersion forces attract one layer to another, enabling layers to slide by one another. Graphite feels slippery as a result of this characteristic. In fact, many industrial processes use graphite as a lubricant.

In diamond, each carbon atom forms four strong covalent bonds with four other carbon atoms. This three-dimensional array makes diamond the hardest naturally occurring substance. The valence electrons are in highly localized bonds between carbon atoms. Therefore, diamond does not conduct electricity. The planes of atoms within the diamond reflect light, which gives diamond its brilliance and sparkle. The regular arrangement of the carbon atoms in a crystalline structure, coupled with the strength of the covalent C—C bonds, gives diamond its extreme hardness, and makes it inert to corrosive chemicals.

Fullerenes make up a group of spherical allotropes of carbon. Figure 4.19C shows C_{60}, which is one type of fullerene discovered in 1985. It was given the name buckminsterfullerene because it resembles the geodesic-domed structure designed by architect R. Buckminster Fuller. Also known as "buckyballs," C_{60} is just one of several fullerenes that have been discovered. Others have been shown to have the formula C_{70}, C_{74}, and C_{82}. Because of their spherical shape, researchers have speculated that fullerenes might make good lubricants.

Recently, microscopic-level research has developed very small carbon networks called nanotubes. As you can see in Figure 4.19D, nanotubes are like a fullerene network that has been stretched into a cylinder shape. Nanotubes of C_{400} and higher may have applications in the manufacture of high-strength fibres. In the year 2000, researchers built a nanotube with a diameter of 4×10^{-8} m. Up to that time, this nanotube was the smallest structure assembled.

Other Network Solids

Many network solids are compounds, not elements. Examples include silica, SiO_2, and various metal silicates. If you were asked to draw a Lewis structure representing SiO_2, you might draw a molecule that looks like CO_2. However, unlike the bonding that occurs in CO_2, the silicon atom in silica does not form a double bond with an oxygen atom. The likely reason is that silicon's $3p$ orbital is too large to overlap effectively with a $2p$ orbital of the oxygen atom. Instead, each silicon atom bonds to four oxygen atoms to give a tetrahedral network with shared oxygen atoms, as shown in Figure 4.20. The network is represented by the formula $(SiO_2)_n$, indicating that SiO_2 is the simplest repeating unit. Quartz is a familiar example of a material composed of silica.

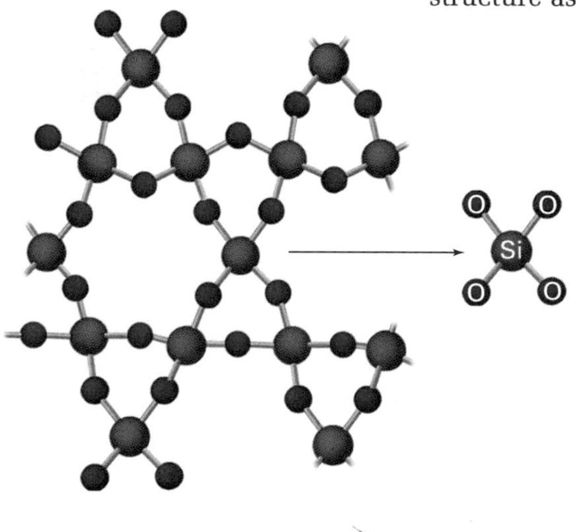

Figure 4.20 A representation of quartz, $(SiO_2)_n$, a tetrahedral network with shared oxygen atoms

Ionic Crystals

You know that an **ionic crystal** is an array of ions, arranged at regular positions in a crystal lattice. However, a closer look at the structure of ionic crystals shows that they have different and distinct arrangements, depending on the ions involved—specifically, their charge and the size.

You can compare the effect of size on crystal packing to fruits stacked in a grocery store. For example, if you were trying to stack two kinds of fruit to attain the most efficient use of space, kiwi may fit in among grapefruit more efficiently than oranges could fill the same space. Similarly, in a crystal, pairs of ions of the same or similar size are arranged differently compared with ions of different sizes.

A crystal lattice is made up of identical repeating *unit cells* that give the crystal its characteristic shape. Figure 4.21 shows a three-dimensional representation of a *simple cubic* arrangement of unit cells.

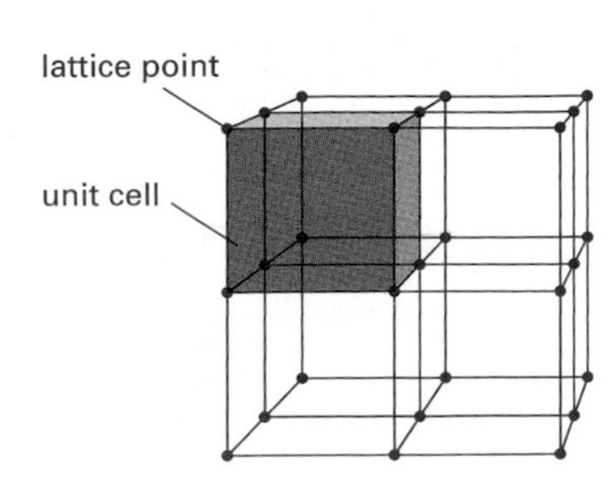

Figure 4.21 Simple cubic crystal array indicating lattice points and unit cell.

In most ionic crystals, the anion is larger than the cation and, therefore, the packing of the anions determines the arrangement of ions in the crystal lattice. There are several possible arrangements for ionic crystals in which the anions are larger than cations, and cations and anions are present in equal molar amounts. For example, Figure 4.22 shows two different arrangements found in the structures of sodium chloride, NaCl, and cesium chloride, CsCl.

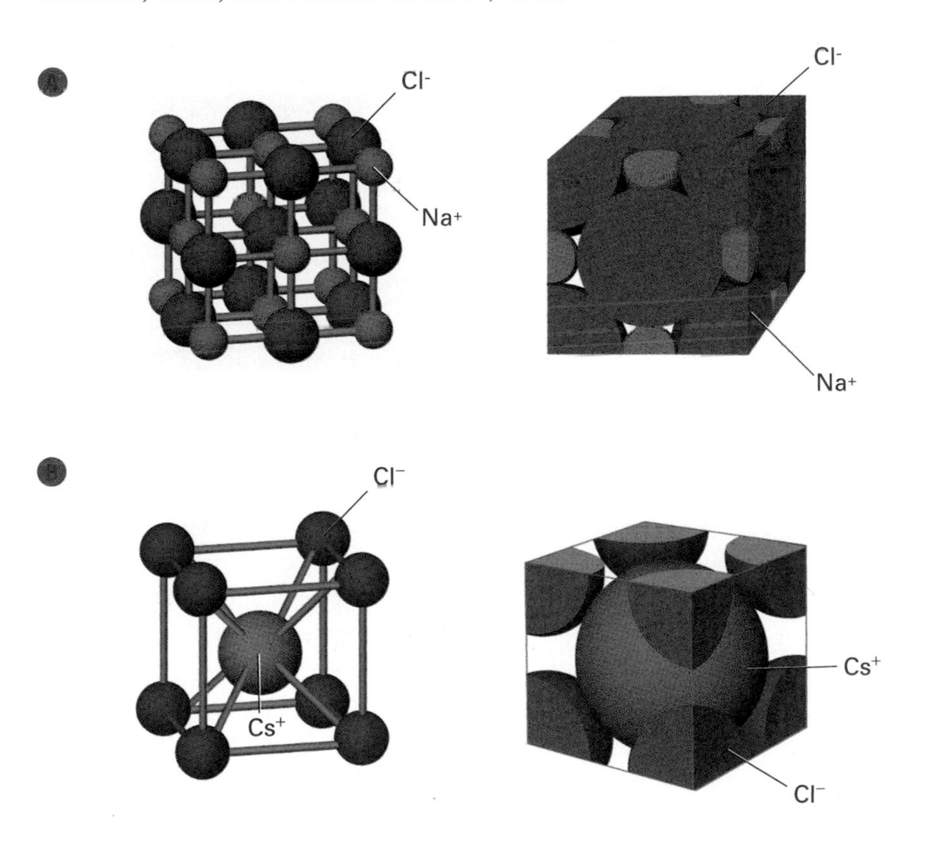

Figure 4.22 (A) Face-centred cubic structure of NaCl; (B) Body-centred cubic structure of CsCl

In the sodium chloride crystal, each chloride ion (Cl^-) is surrounded by six sodium ions (Na^+). Similarly, each Na^+ ion is surrounded by six Cl^- ions. In CsCl, each cesium ion (Cs^+) ion is surrounded by eight Cl^- ions.

Within a crystal lattice, definite lines of cleavage are present along which a crystal may be split. This is an identifying property of ionic solids.

Technology LINK

Chemists commonly use a process called X-ray diffraction to determine the structure of a compound. Two pioneers of X-ray diffraction were Rosalind Franklin and Dorothy Crowfoot Hodgkin. Franklin's work on the DNA molecule was instrumental in the discovery of its helical shape. Nobel Prize winner Hodgkin discovered the structure of complex molecules, such as cholesterol, penicillin, insulin, and vitamin B-12. Find out about X-ray diffraction: how it works and the types of scientists who employ this technology.

Canadians 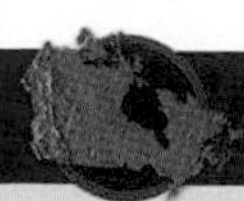 in Chemistry

Dr. R.J. Le Roy

You usually study science as discrete fields of inquiry: chemistry, physics, and biology, for example. Each of these fields has numerous sub-fields, such as geophysics and biochemistry. One of the scientists at the forefront of chemical physics is Canada's Dr. R.J. Le Roy.

During a fourth-year undergraduate research project at the University of Toronto, Dr. R.J. Le Roy pursued an answer to the following question: If you use photodissociation to separate iodine molecules into atoms, what fraction of the light is absorbed, and how does that fraction depend on the colour (frequency) of the light and on the temperature of the molecules? (Photodissociation is a process that involves the use of light to decompose molecules.)

The light causes a bonded molecule in its ground state to undergo a transition into an excited state. In this new state, the atoms repel one another, causing the molecule to break apart. The fraction of light absorbed, as measured spectroscopically, depends on the population of molecules in the initial state.

Chemists can use the fraction of light absorbed by a sample to monitor the concentration of specific molecules in a closed, controlled environment. To do this, chemists also need to know the value for a key proportionality constant that is unique to each substance, known as the *molar absorption coefficient*. These coefficients can be measured at room temperature, but not at high temperatures.

"In principle, we have known the quantum mechanics to describe these processes," Dr. Le Roy explains, "But to use that information, we have to understand the forces atoms experience as the atoms come together or fall apart."

For his master's thesis at the University of Toronto, Dr. Le Roy developed computer programs to perform the exact quantum mechanical calculation of the absorption coefficients governing photodissociation processes. Later, when doing his doctoral thesis at the University of Wisconsin, he and his thesis supervisor, Dr. Richard Bernstein, developed the Le Roy-Bernstein theory. This theory describes the patterns of energies and other properties for vibrational levels of molecules lying very close to dissociation.

To determine when to trust the equations of the Le Roy-Bernstein theory, Dr. Le Roy developed what he calls "a little rule of thumb," more generally known as the Le Roy radius. A calculation for the Le Roy radius, R_{LR}, is:

$$R > R_{LR} = 2[< r_A^2 >^{1/2} + < r_B^2 >^{1/2}]$$

where R is the internuclear distance between two atoms A and B.

The Le Roy radius is the minimum distance at which the equation describing the vibrations of molecules close to dissociation is thought to be valid. The $< r_A^2 >$ and $< r_B^2 >$ values are the squares of the radii of the atoms of the two elements (or of the two atoms of the same element) in their ground states.

Is there a practical outcome to mathematical chemistry? Dr. Le Roy strongly encourages "curiosity-oriented science," as opposed to a focus on short-term, practical results, partly because practical results do come when processes are fully understood.

For example, the study of global warming is hindered, in his view, because the intermolecular interactions are not well known. "To properly model the processes, we have to understand the exchange of energy during molecular collisions. If we know them, we can predict an immense amount."

Making Connections

1. In what ways are chemistry and physics related to each other? Design a graphic organizer to outline their similarities and differences.
2. Research and development funding often encourages chemists and chemical engineers to focus on practical applications. How does this differ from Dr. Le Roy's perspective? With which perspective do you agree more? Write an editorial outlining your opinions.

Determining the Type of Bonding in Substances

The type of bonding in a substance depends upon the kinds of atoms it contains and the forces of attraction between those atoms. If you know the physical properties of a substance, you can often predict the type of bonding in the substance. Table 4.5 summarizes the types of forces of attraction and the physical properties of solids with different types of bonding discussed in this section. In Investigation 4-B, you will design an experiment in which you predict the type of bonding in various substances, then test your prediction.

Table 4.5 Summary of Properties of Different Types of Solids

Type of Crystalline Solid	Particles Involved	Primary Forces of Attraction Between Particles	Boiling Point	Electrical Conductivity in Liquid State	Other Physical Properties of Crystals	Conditions Necessary for Formation	Examples
Atom	atoms	dispersion	low	very low	very soft	formed between atoms with no electronegativity difference	all Group 18 (VIIIA) atoms
Molecular	non-polar molecules or polar molecules	dispersion, dipole-dipole, hydrogen bonds	generally low (non-polar); intermediate polar	very low	non-polar: very soft; soluble in non-polar solvents	non-polar: formed from symmetrical molecules containing covalent bonds between atoms with small electronegativity differences	non-polar: Br_2, CH_4, CO_2, N_2
					polar: somewhat hard, but brittle; many are soluble in water	polar: formed from asymmetrical molecules containing polar covalent bonds. Electronegativity difference between atoms < 1.7	polar: H_2O, NH_3, $CHCl_3$., CH_3COOH, CO, SO_2, many organic compounds
Covalent Network	atoms	covalent bonds	very high	low	hard crystals that are insoluble in most liquids	formed usually from elements belonging to Group 14 (IV A)	graphite, diamond, SiO_2
Ionic	cations and anions	electrostatic attraction between oppositely charged ions	high	high	hard and brittle; many dissolve in water	formed between atoms with electronegativity difference > 1.7	$NaCl$, CaF_2, Cs_2S, MgO, MgF_2
Metallic	atoms	metallic bonds	most high	very high	all have a lustre, are malleable and ductile, and are good electrical and thermal conductors; they dissolve in other metals to form alloys	formed by metals with low electronegativity	Hg, Cu, Fe, Ca, Zn, Pb

Investigation 4-B

SKILL FOCUS

Initiating and planning

Performing and recording

Analyzing and interpreting

Determining the Type of Bonding in Substances

This investigation challenges you to design an experiment to observe and analyze the physical properties of different substances, and to determine the types of bonding present.

Question

How can you determine the type of bonding in an unknown substance?

Safety Precautions

- Tie back long hair and any loose clothing.
- Before you light the candle or Bunsen burner, check that there are no flammable solvents nearby.
- Ensure you have a bucket of sand nearby in case of fire.

Materials

a substance of each of the following types: non-polar molecular, polar molecular, ionic, network, and metallic

equipment or apparatus, depending on how you design your experiment

Procedure

1. Your teacher will give you five samples stored in bottles. Record the visible properties that you can observe. Based on these properties, predict the main type of bonding in each.
2. Design an experiment to test your predictions. Ensure that you have adequately addressed all possible safety considerations. Obtain approval from your teacher before you begin your investigation.
3. Carry out your investigation and record your observations.

Analysis

1. Based on your observations, classify the solids. Identify the type of bonding in each substance. Explain your decisions about classification in detail.

Conclusion

2. How did your results compare with your predictions? If some results were inconclusive, explain the source of this uncertainty and suggest one way to rectify it.

Application

3. Suppose you want to challenge a classmate to identify a set of substances.
   (a) Suggest five solids (different from the ones you used in this investigation) that you could use.
   (b) What tests could your classmate carry out to classify the substances successfully? Explain why these tests would work.
   (c) Suggest several tests that might provide inconclusive results. Explain why.

Chemistry Bulletin

Science | Technology | Society | Environment

Ionic Liquids: A Solution to the Problem of Solutions

Although organic solvents are commonplace in industry, they are highly volatile and dangerous in the large quantities that manufacturing processes require. Disposing of used solvents is a major environmental concern.

Until recently, the idea of a recyclable, non-volatile solvent seemed like the stuff of science fiction. But in the 1980s, researchers in the United States were trying to create a new electrolyte for batteries. Instead, they created a colourless, odorless liquid that was composed of nothing but ions—an *ionic liquid*.

An ionic liquid is a salt in liquid form. Ionic liquids do not usually find practical applications, since ionic compounds have such high boiling points. For example, sodium chloride does not begin to melt until it reaches a temperature of about 800°C.

In an ionic liquid, the cations are much larger than in typical ionic substances. As a result, the anions and cations cannot be packed together in an orderly way that balances both the sizes and distances of the ions with the charges between them. As a result, they remain in a loosely-packed, liquid form. Because the charges on the cations have the capacity to bind more anions, an ionic liquid has a net positive charge. Notice the large cation of the ionic liquid molecule shown below. This is what makes ionic liquids suitable solvents. The net charge can attract molecules of other compounds, dissolving them.

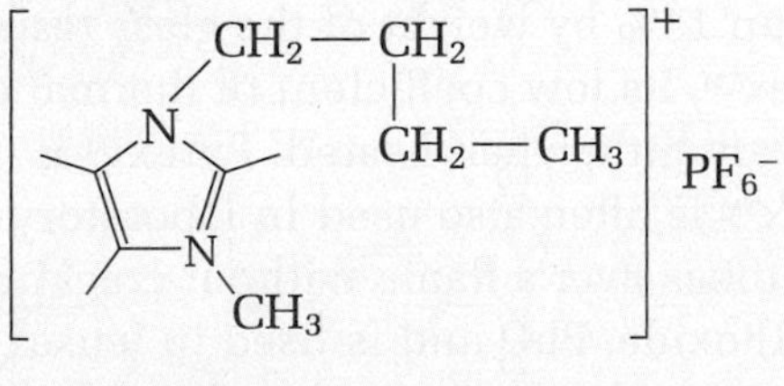

CH_2—CH_2 | N | CH_2—CH_3 | N | CH_3 | $]^+$ PF_6^-

Many properties of ionic liquids make them more desirable solvents than organic solvents. For example, ionic liquids require a lot of energy to change their state—they remain as liquids even at temperatures of 200°C. In addition, ionic liquids evaporate sparingly. Organic solvents, on the other hand, easily release fumes that are harmful, often toxic, and that can eventually oxidize and create carbon dioxide, a notorious greenhouse gas.

As solvents, ionic liquids can be uniquely "tuned" to a particular purpose by adjusting the anion/cation ratio. To decaffeinate coffee, for example, you could create an ionic liquid that would just dissolve caffeine and nothing else. Current research suggests that ionic liquids can be recovered from solution and reused.

Yet, there are still many hurdles that must be overcome before these substances can be put to widespread use. Because ionic liquids boil at high temperatures, conventional methods of separating reaction products from solution cannot be used. Pharmaceutical manufacturing processes, for instance, often involve the process of distillation. The higher temperatures required to distill using ionic liquids (as opposed to organic liquids) would decompose many of the desired pharmaceutical compounds.

The discovery of ionic liquids is proving a fertile field for researchers. Ionic solvents are revealing new mechanisms of reaction and enhancing our understanding of the molecular world. As well, their potential applications in both industry and the home have encouraged several companies, world-wide, to further explore and develop these unusual substances.

Making Connections

1. Investigate some of the industrial processes where solvents are currently used. Could ionic liquids be used in any of these industries instead? Why or why not? Look into some of the ideas that ionic liquid researchers have proposed for dealing with the unique challenges posed by different industries.

16. Draw Lewis structures and indicate the molecular shape of the following.
    (a) NOCl **(b)** AlF_6^{3-} **(c)** XeO_3

17. Draw Lewis diagrams to illustrate the following.
    (a) a compound of chlorine that is diatomic and has only dispersion forces between molecules when in the liquid state
    (b) a compound of chlorine that is diatomic and has dipole-dipole attractions when in the liquid state

18. In what cases is the name of the molecular shape the same as the name of the electron group arrangement?

19. How can a molecule with polar covalent bonds *not* be a polar molecule?

Inquiry

20. Draw a Lewis structure for PF_2Cl_3 to show that it may be polar or non-polar.

21. Water reaches its maximum density at 4°C. At temperatures above or below 4°C, the density is lower. Discuss what is happening to the water molecules at temperatures around 4°C.

22. The melting point of NaCl is 801°C, of $CaCl_2$ is 782°C, and of $AlCl_3$ is 190°C. The electrostatic forces of attraction between ions increase with an increase in the charge. In these ionic solids, the charge on the cations Na^+, Ca^{2+}, and Al^{3+} is increasing, but the melting point of the corresponding compound is decreasing. Suggest an explanation for this trend in the melting point.

23. Transition metal ions can be multivalent. For example, gold(I), Au^+, and gold(III), Au^{3+}, are the two ionic forms for gold. Decomposition occurs at 240°C for Au_2S and at 197°C for Au_2S_3.
    (a) Based upon the given information, which of the two compounds has bonds with greater ionic character?
    (b) Predict which compound would have the greater lattice energy. Outline the reasoning for your answers.

24. Determine if the molecule XeF_6 will have an octahedral shape.

25. In which liquid, $HF_{(\ell)}$ or $H_2O_{(\ell)}$, would the hydrogen bonding be stronger? Based upon your prediction, which of these two liquids would have a higher boiling point? Refer to a reference book to find the boiling points of these two liquids to check your predictions. Account for any difference between your predictions and the actual boiling points.

Communication

26. Compare the molecules SF_4 and SiF_4 with respect to molecular shape and molecular polarity.

27. Compare the forces of attraction that must be overcome to melt samples of NaI and HI.

28. **(a)** Contrast the physical properties of diamond, a covalent network solid with the molecular compound 2,2-dimethylpropane.
    (b) In diamond, covalent bonds join the carbons atoms. In a molecule of 2,2-dimethylpropane, a central carbon atom is covalently bonded to four $—CH_3$ groups. How are the covalent bonds in the two substances different from one another?
    (c) How are the physical properties of these two substances related to the covalent bonds in them?

29. At room temperature, carbon dioxide, CO_2, is a gas, while silica, SiO_2, is a hard solid. Compare the bonding in these two compounds to account for this difference in physical states.

30. Compare the bonding and molecular polarity of SeO_3 and SeO_2.

31. Discuss the intermolecular and intramolecular forces in N_2H_4 and C_2H_4. Based upon the bonding between molecules, which of these two compounds would have a lower boiling point?

Making Connections

32. Many advanced materials, such as KEVLAR®, have applications that depend on their chemical inertness. What hazards do such materials pose for the environment? In your opinion, do the benefits of using these materials outweigh the long-term risks associated with their use? Give reasons to justify your answer.

33. Considering the changes of states that occur with water in the environment, suggest how intermolecular forces of attraction influence the weather.

Answers to Practice Problems and Short Answers to Section Review Questions

1.(a) $1s^2$ **(b)** $1s^2 2s^2 2p^6 3s^2 3p^6$ **(c)** $1s^2 2s^2 2p^6 3s^2 3p^6 4s^2 3d^{10} 4p^6$ **(d)** $1s^2 2s^2 2p^6$ **2.(a)** $[Li]^+$ **(b)** $[Ca]^{2+}$ **(c)** $[:Br:]^-$ **(d)** $[:O:]^{2-}$

3.(a)
Li
+ :S: → $[Li]^+ [:S:]^{2-} [Li]^+$
Li

(b)
:Cl:
Ca + → $[:Cl:]^- [Ca]^{2+} [:Cl:]^-$
:Cl:

(c) K + :Cl: → $[K]^- [Cl]^+$

(d)
Na
Na + ·N· → $[Na]^+ [:N:]^{3-} [Na]^+$
Na $[Na]^+$

4.(a) 6 (VIA) **(b)** 1 (IA) **(c)** 3 (IIIA) **(d)** 2 (IIA) **5.** H—Cl, H—Br, H—I **6.** $AsBr_3$ $\Delta EN = 0.78$, $CaBr_2$ $\Delta EN = 1.96$, KBr $\Delta EN = 2.14$: increasing ionic character **7.(a)** C—H **(b)** Sn—F **(c)** C—O **8.** covalent: S_8, F_2; polar covalent $SCl_2 < SF_2 < PF_3$; ionic: RbCl **9.(a)** H—N—H
H

(b)
H
H—C—H
H

(c)
:F:
:F—C—F:
:F:

(d) H—As—H
H

(e) $:Br—O:^-$ **(f)** H—S—H

(g) H—O—O—H **(h)** :Cl—N=O:

10.(a)
$[$:O: / C / :O :O: $]^{2-}$ ⟷ $[$:O: / C / :O: :O: $]^{2-}$ ⟷ $[$ O: / C / :O: :O: $]^{2-}$

(b) $:N{\equiv}O:^+$ **(c)**
:O:
:O—Cl—O: $^-$

(d)
:O :O:
:S
:O:

11.
H :F:
H—C—C—Cl:
H :Cl:

12.(a)
H H
:N—N:
H H

(b) :F—N=N—F: **13.**
:O:
:O—Xe—O: (all co-ordinate covalent)
:O:

14.(a)
:F :F: :F:
S
:F :F: F:

(b)
:F: :F:
:F—Br:
:F: F:

15.(a)
:F :F:
Xe
:F: :F:

(b)
:F: :F:
:F—P
:F: F:

16. In PI_3, one lone pair around P; in ClI_3, two lone pairs around the Cl.

17.
$[$:F F:
I
:F: F: $]^-$

18.(a) linear **(b)** bent **(c)** trigonal planar **(d)** tetrahedral **19.(a)** tetrahedral **(b)** trigonal bipyramidal **(c)** tetrahedral **(d)** tretrahedral **20.** NO_2^+ linear; NO_2^- bent

21.(a) linear:
I
:Xe:
I

(b) octahedral:
$[$:F :F: F:
P
:F :F: F: $]^-$

(c) trigonal pyrimidal:
As
:F F:
:F:

(d) tetrahedral:
$[$:F:
Al—F:
:F :F: $]^-$

22.(a) O **(b)** P **(c)** Cl **23.(a)** tetrahedral, polar **(b)** trigonal planar, polar **(c)** trigonal planar, non-polar **24.** tetrahedral, polar **25.(a)** NF_3 **(b)** $TeCl_4$ **26.** trigonal pyramidal, T-shaped

Section 4.1: 1. (b) NaBr ionic, Br—F polar covalent, Br—Cl covalent **2.** O—H > S—O > H—Br > Cl—I > N—Cl **3.** property will tend to increase **4.** cleavage; ionic **4.2: 1.(a)** non-polar **(b)** polar **(c)** polar **3.** both tetrahedral; CH_4 is non-polar and CH_3OH is polar **4.** less reactive since B and N now have octet in valence energy level **5.** bent, polar **6.** tetrahedral **4.3: 2.** more opportunity for H-bonding and greater dispersion forces in H_2O_2 **3.(a)** molecular non-polar, ionic **(c)** metallic **(d)** network **6.** trigonal pyramidal, polar **7.** tetrahedral, non-polar **8.** CsF > CsCl > CsBr > CsI

Materials Convention

Background

The qualities and characteristics of a material are determined by the bonding between the particles that make up the material. For example, pure carbon exists in several different forms, or allotropes. Each allotrope contains different bonding, and therefore has different properties. The "lead" of a pencil is really made of graphite, a silver-grey allotrope of carbon. Graphite is a good writing tool because the layered bonding arrangement creates a soft material that is easily transferred to paper by rubbing. The softness and slipperiness of graphite also make it a valuable industrial lubricant. Diamond, another allotrope of carbon, is one of the hardest materials on Earth. Diamonds are useful in cutting and drilling applications. Because they are beautiful, sparkling stones, diamonds are also used in jewellery.

Humans have used the materials in their environment for thousands of years. In addition to natural materials, new materials are continually being invented. Some inventions take place by accident, and others by design. These new materials are often more effective, efficient, or cheaper than traditional materials.

Optical fibres, each made of two different types of glass, pass through the eye of a steel needle.

Whale oil was once a popular choice for lamp fuel. Later, a new, "cutting-edge" fuel was developed from petroleum to replace whale oil. What was this new fuel? How did it compare to whale oil?

Copper has long been prized for its beauty, malleability, and conductivity. However, in recent fibre-optic technology, intricate atom-layering production processes are being used to produce a material that has thousands of times the conductive ability of copper. What is this new material?

Challenge

In this project, you will choose and investigate a material. You may be interested in a common material or substance that you encounter every day. Alternatively, you may want to investigate one of the many new, cutting-edge materials that are currently being developed. You will research the structure, properties, and applications of your material, and then prepare a presentation to be set up in the classroom.

Materials

Try to obtain a sample of your material, to examine its physical properties. If this is not possible, find illustrations or photos of the material from books, magazines, or the Internet. You will need materials for your presentation, which may include charts, a poster, models, pictures, or pamphlets.

Design Criteria

A. With your class, prepare a rubric that you will use for self- and peer-assessments. Your teacher will use the same rubric to assess your work.

B. Your presentation should be visually attractive, interesting, and educational. It should examine the relationship between the bonding and physical properties of your material.

Action Plan

1. Examine or research the physical properties of the material you have chosen to study. This step may include a hands-on investigation of a sample of the material, if available. Record a detailed description of the properties of your material. Include tests to determine any of the following applicable properties.

thermal conductivity
electrical conductivity
hardness
boiling point
melting point
solubility in various solvents
surface tension
density

2. Research the chemical composition of the material, and, if applicable, the processes used to make it. Is your material natural or synthetic? If it is synthetic, what substances are used to produce it? Do these substances come from a natural source?

3. If possible for your material, draw a Lewis structure of the molecule or molecules on which your material is based. Predict the molecular shape using VSEPR theory.

4. Build a three-dimensional model showing the composition of your material. Prepare a detailed drawing to explain the bonding that occurs between particles in your material.

5. Use your model and drawing to discuss the following topics.
   (a) the polarity of your material
   (b) intermolecular forces affecting the properties of your material
   (c) the types of intramolecular bonding present in your material

6. Describe and explain any relationship you observe between the polarity and bonding of your material (step 5) and its physical properties (step 1).

7. (a) Decide on some possible functions of the material you have chosen, based on its properties.
   (b) Research applications of your material in industry, science, and medicine. How have inventors used the specific characteristics of your material to benefit society?
   (c) Compare historical, current, and possible future applications of this material. Has this material been replaced by a new material? Are there economic or other reasons why this material may no longer be used?

Assessment

After you complete this project, answer the following questions to assess your work.

- Assess the success of your project based on the response from your peers.
- Assess the completeness of your research. Could you have found out more?
- Assess the visual presentation of your project. Was your information laid out in a clear, interesting, and easy-to-understand manner?

Evaluate

1. Prepare your presentation, and set it up in the classroom. During one or two class periods, proceed around the classroom, taking notes on the materials presented by your classmates. While studying all of the materials together, note patterns, similarities, and differences in structure and function.

2. Use the rubric that your class designed to evaluate your presentation.

3. As directed by your teacher, use the rubric your class designed to evaluate the presentation of another student in the class.

Extension

What methods of analysis would provide scientists with the most information on your material? What scientific instruments would be helpful in studying your material? Explain your answer in terms of the structure of your material and the function of the instrument.

UNIT 2 Review

Knowledge/Understanding

Multiple Choice

In your notebook, write the letter for the best answer to each question.

1. Which part of the atom did Rutherford discover by bombarding gold foil with alpha particles?
   (a) protons
   (b) neutrons
   (c) electrons
   (d) nucleus
   (e) a uniform, positively charged sphere
2. Which of the following did the Bohr model of the atom help explain?
   (a) nuclear structure
   (b) atomic spectra
   (c) structure of the electron
   (d) structure of alpha particles
   (e) wave property of the atom
3. Which of the following can help explain atomic emission spectra?
   (a) A photon of a specific wavelength of light is absorbed when an atom's electron jumps from a lower energy level to a higher energy level.
   (b) A photon of a specific wavelength of light is released when an atom's electron falls from a higher energy level to a lower energy level.
   (c) Energy of electromagnetic radiation is continuous.
   (d) Electrons continuously release energy as they orbit the nucleus of the atom.
   (e) Energy is released when an electron jumps from a lower energy level to a higher energy level.
4. Which of the following is an example of a network solid?
   (a) sodium, $Na_{(s)}$
   (b) sucrose, $C_{12}H_{22}O_{11(s)}$
   (c) graphite, $C_{(s)}$
   (d) silica, $SiO_{2(s)}$
   (e) magnesium fluoride, $MgF_{2(s)}$
5. Which of the following electron configurations is impossible for an atom, regardless of whether it is in the ground state or an excited state?
   (a) $1s^22s^22p^3$
   (b) $1s^2s^12p^1$
   (c) $1s^32s^12p^4$
   (d) $1s^22s^12p^3$
   (e) $1s^22s^22p^6$
6. A crystalline substance that is hard, unmalleable, has a high melting point, and is a non-conductor of electricity could be
   I an ionic crystal
   II a polar covalent solid
   III a metal
   IV a network solid
   (a) I or II
   (b) I or III
   (c) III or IV
   (d) II or IV
   (e) I or IV
7. Which of the following molecules does not have a linear shape?
   (a) Cl_2O
   (b) CO_2
   (c) XeF_2
   (d) OCS
   (e) BeF_2
8. Which of the following electron configurations shows a transition metal atom?
   (a) $1s^22s^22p^5$
   (b) $1s^22s^22p^63s^23p^1$
   (c) $1s^22s^22p^63s^23p^63d^{10}4s^1$
   (d) $1s^22s^22p^63s^23p^63d^{10}4s^24p^1$
   (e) $1s^22s^22p^63s^23p^63d^{10}4s^24p^3$
9. A substance that has a melting point of 1850°C and a boiling point of 2700°C is insoluble in water and is a good insulator of electricity. The substance is most likely
   (a) an ionic solid
   (b) a polar covalent solid
   (c) a metal
   (d) a network solid
   (e) a molecular solid
10. The main factor that leads to the formation of a chemical bond between atoms is
    (a) the lowering of melting point and boiling point when a compound is formed
    (b) a tendency in nature to achieve a minimum energy for a system
    (c) the formation of a shape consistent with VSEPR theory
    (d) a tendency to make the attractions equal to the repulsions between atoms
    (e) a tendency to reduce the number of existing particles

Short Answers

In your notebook, write a sentence or a short paragraph to answer each question.

11. The emission spectrum of an element has spectral lines at wavelengths of 620 nm and 640 nm. Sketch the absorption spectrum for this element.

12. Briefly account for the two spectra referred to in question 11.

13. Use the aufbau principle to write the condensed ground state electron configurations for nitrogen, phosphorus, and arsenic.

14. List these elements in order of increasing atomic size: Na, Mg, K, and Ca.

15. What VSEPR notations correspond to a molecule that have a bent shape?

16. How do the lone pairs of electrons around the central atom of a molecule affect the bond angle between two bond pairs of electrons.

17. List these elements in order of increasing first ionization energies:
(a) Li, Na, K, Rb
(b) Li, Be, B, C

18. Name the two elements in period 4 whose electron configurations are not accurately predicted by the aufbau principle.

19. Which physical properties of an ionic solid are related to the size of the lattice energy of the crystal?

20. Explain why, at room temperature, CO_2 is a gas but CS_2 is a liquid.

21. Sketch an outline of the periodic table and use it to compare the trends in atomic size, first ionization energy, and electron affinity.

22. Why do the properties of elements in the periodic table recur periodically?

23. Use hydrogen and lithium to compare the concepts of nuclear charge and effective nuclear charge.

24. Explain how to determine the maximum number of electrons in any principal energy level.

25. What is the fourth quantum number, and why does it have only two possible values?

26. What is the uncertainty principle, and what are its implications for the atomic model?

27. Distinguish between dipole-dipole attraction forces and dispersion (London) forces. Is one type of these intermolecular forces stronger than the other? Explain.

Inquiry

28. A metal, *X*, reacts with chlorine to form the compound XCl_2. The metal's third ionization energy is significantly larger than its first and second ionization energies.
(a) To what group does the metal likely belong?
(b) What is the valence electron configuration of an atom of the metal in its ground state?

29. The last electron to enter an atom, following the aufbau principle, has quantum numbers $n = 3$, $l = 2$, $m_l = -1$, and $m_s = +\frac{1}{2}$. To which block of elements and which period on the periodic table does this element belong? Identify the element, assuming that its orbitals are filled in order of increasing m_l.

30. If the first five ionization energies of an element are 1.09, 2.35, 4.62, 6.22, and 37.83 MJ/mol respectively, to which group in the periodic table does this element belong? Explain your reasoning.

31. Which ionic solid would you expect to have a higher melting point: LiI or LiBr? Justify your answer.

32. Which element in each of the following pairs would you expect to be less metallic? Explain your choices.
(a) Sb and As **(b)** Si and P **(c)** Be and Na

33. The first ionization energy for boron is lower than what you would predict, based on the general trend for ionization energy across a period. Explain this exception to the trend.

34. For any molecules that have the general formula AX_n, where n is greater than 2, how can you decide whether a given molecule is polar or not?

35. All three metals (Zn, Cd, Hg) in Group 12 (IIB) have a ground state electron configuration ending with $d^{10}s^2$.
(a) What electrons will be available to participate in metallic bonding in these elements?
(b) What is the most likely charge to be carried by stable ions of these metals?

(c) Since Hg is a liquid at room temperature, what does this indicate about the strength of the metallic bonding in Hg compared to the other two metals in this group?

36. Copy the following bond pairs into your notebook. Use the electronegativity values in the periodic table in Appendix E to indicate the polarity of the bond in each case.
   (a) N—H
   (b) F—N
   (c) I—Cl

37. Arrange the following sets of bond pairs in order of increasing polarity. Identify the direction of bond polarity for the bond pairs in each set.
   (a) Cl—F, Br—Cl, Cl—Cl
   (b) Si—Cl, P—Cl, S—Cl, Si—Si

38. If a molecule with the formula AY_3 is polar, identify the molecular shapes that this molecule *could* and *could not* have.

Communication

39. Draw a Lewis structure for the ionic compound $Ba(OH)_2$.

40. Draw a Lewis structure for the molecule POF_3 and predict if it is a dipole.

41. Use VSEPR theory to predict the molecular shape of CH_2Cl_2. Draw a sketch to indicate the polarity of the bonds around the central atom to verify that this is a polar molecule.

42. Draw and compare the Lewis structures of ClO_4^- and $OSCl_2$. In which of these cases, if any, does the central atom have an expanded valence energy level?

43. Write the electron configuration and draw an orbital diagram to show the first excited state of a sodium atom. Assume that the outermost electron is excited.

44. Write a full set of quantum numbers for the following:
   (a) the outermost electron in a Li atom
   (b) the electron that causes a chlorine atom to become a chloride ion with a charge of 1–.

45. Summarize the trend in metallic character, and compare it to the trends for atomic size and ionization energy.

46. The group trend for boiling point is the same as the trend for atomic radius. For the compounds formed between hydrogen and the first three elements of group 16 (VIA), H_2S has a lower boiling point than both H_2O and H_2Se. Explain this diversion from the trend you would expect.

47. Define the following terms, and identify the term that best describes the H—O bond in a water molecule: non-polar covalent, polar covalent, ionic.

48. For a solid metallic element:
   (a) Identify four physical properties.
   (b) Identify two chemical properties that would enable you to classify the element as metallic.

49. (a) Explain why the successive ionization energies for an element always increase.
   (b) Identify an element for which you would predict the difference between two successive ionization energies to be very large. Explain what you can infer about the electron configuration of the atoms of this element.

50. In your own words, state Hund's rule. Explain how it applies in drawing the orbital diagram for a carbon atom.

51. You can use electronegativity differences to think of chemical bonds as having a percent ionic or a percent covalent character. The graph below plots percent ionic character versus ΔEN for a number of gaseous binary molecules. Use this graph to answer the questions on the next page.

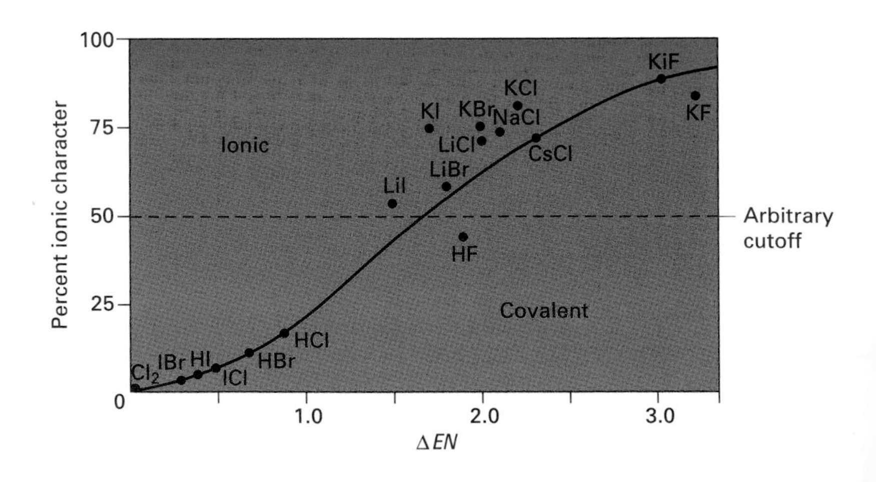

(a) Describe the ionic character of the molecules as a function of ΔEN.
(b) Which molecule has a 0 percent ionic character? What can you infer about the interactions among the electrons of this molecule?
(c) Do any molecules have a 100 percent ionic character? What can you infer about the interactions among the electrons of ionic compounds? (Hint: Can atoms give up or gain electrons in the way the ionic bonding model suggests they do?)
(d) Chemists often assign the value of 50 percent as an arbitrary cutoff for separating ionic compounds from covalent compounds. Based on your answer in part (c), what does the use of the term "arbitrary" suggest about the nature of chemical bonds?

52. The periodic trend for electronegativity is the inverse of the trend for atomic radius.
(a) Make a sketch of the periodic table to show the group and period trends for electronegativity.
(b) Explain, using an example, why you would expect this inverse relationship between electronegativity and atomic radius.

53. Name the two scientists whose ideas showed that the location of an electron in an atom is uncertain. Explain how their ideas did, in fact, suggest this fact about electrons.

54. In 1906, the Nobel Prize in chemistry was awarded to a French chemist, Ferdinand Moissan, for isolating fluorine in its pure elemental form. Why would this achievement be deserving of such a prestigious honour? Use your understanding of atomic properties as well as chemical bonds to explain your answer.

55. Use a graphic organizer such as a Venn diagram to compare ionic bonding with metallic bonding.

56. Lead and carbon belong to the same group of the periodic table. One is a metal, while the other is a non-metal. How can two elements whose atoms have the same number of valence electrons be so different in their properties?

Making Connections

57. Researchers estimate that known reserves of industrially important transition metals, such as manganese and iron, will be depleted in less than a human lifetime. Scattered about the ocean floor are billions of tonnes of small chunks of rock ranging in diameter from several millimetres to several metres. These nodules, as they are called, consist mainly of manganese and iron oxides, but contain smaller amounts of other transition-metal compounds as well. The ocean floor is designated as international property. Therefore, no single country may claim ownership of this vast potential resource. Write an editorial that outlines your recommendations for how this resource may be shared.

58. Many useful products, such as superglue and fire-retardant materials, are domestic "spin-offs" from research that is related to military applications. Military expense budgets are large, in part, because of the money required to support this research. In your opinion, should military budgets be exempt from cost-conservation measures that governments typically must consider? Debate this question with your classmates, and issue a statement that summarizes your decisions.

COURSE CHALLENGE

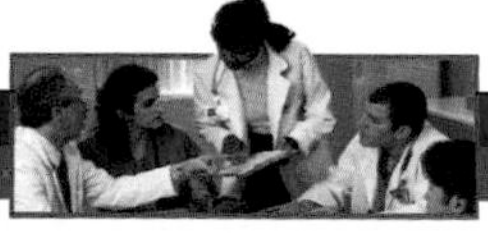

The Chemistry of Human Health

Think about these questions as you plan for your Chemistry Course Challenge.

- What are the properties characteristic of metals such as aluminum and calcium based on their electronic configurations?
- How would you test for these properties in the laboratory?
- How would you determine the type of bonding or intermolecular forces within a substance?

UNIT 3

Energy Changes and Rates of Reaction

UNIT 3 CONTENTS

CHAPTER 5
Energy and Change

CHAPTER 6
Rates of Chemical Reactions

UNIT 3 PROJECT
Developing a Bulletin About Catalysts and Enzymes

UNIT 3 OVERALL EXPECTATIONS

- What energy transformations and mechanisms are involved in chemical change?
- What skills are involved in determining energy changes for physical and chemical processes and rates of reaction?
- How do chemical technologies and processes depend on the energetics of chemical reactions?

Unit Project Prep

Look ahead to the project at the end of Unit 3. Start preparing for the project now by listing what you already know about catalysts and enzymes. Think about how catalysts and enzymes affect chemical reactions. As you work through the unit, plan how you will investigate and present a bulletin about the uses of catalysts and enzymes in Canadian industries.

In the nineteenth century, railway tunnels were blasted through the Rocky Mountains to connect British Columbia with the rest of Canada. Workers used nitroglycerin to blast through the rock. This compound is so unstable, however, that accidents were frequent and many workers died. Alfred B. Nobel found a way to stabilize nitroglycerin, and make it safer to use, when he invented dynamite.

What makes nitroglycerin such a dangerous substance? First, nitroglycerin, $C_3H_5(NO_3)_{3(\ell)}$, gives off a large amount of energy when it decomposes. In fact, about 1500 kJ of energy is released for every mole of nitroglycerin that reacts. Second, the decomposition of nitroglycerin occurs very quickly—in a fraction of a second. This fast, exothermic reaction is accompanied by a tremendous shock wave, which is caused by the expansion of the gaseous products. Finally, nitroglycerin is highly shock-sensitive. Simply shaking or jarring it can cause it to react.

Thus, nitroglycerin's explosive properties are caused by three factors: the energy that is given off by its decomposition, the rate at which the reaction occurs, and the small amount of energy that is needed to initiate the reaction. In this unit, you will learn about the energy and rates of various chemical reactions.

CHAPTER

5 Energy and Change

Chapter Preview

5.1 The Energy of Physical, Chemical, and Nuclear Changes

5.2 Determining Enthalpy of Reaction by Experiment

5.3 Hess's Law of Heat Summation

5.4 Energy Sources

Prerequisite Concepts and Skills

Before you begin this chapter, review the following concepts and skills:

- writing balanced chemical equations (Concepts and Skills Review)
- performing stoichiometric calculations (Concepts and Skills Review)

Think about a prehistoric family group building a fire. It may seem as though this fire does not have much in common with a nuclear power plant. Both the fire and the nuclear power plant, however, are technologies that harness energy-producing processes.

As you learned in Unit 2, humans continually devise new technologies that use chemical reactions to produce materials with useful properties. Since the invention of fire, humans have also worked to devise technologies that harness energy. These technologies depend on the fact that every chemical, physical, and nuclear process is accompanied by a characteristic energy change. Consider the melting of an ice cube to cool a drink, the combustion of natural gas to cook a meal, and the large-scale production of electricity via a nuclear power plant. All societies depend on the energy changes that are associated with these physical, chemical, and nuclear processes.

In this chapter, you will study the causes and magnitude of the energy changes that accompany physical changes, chemical reactions, and nuclear reactions. You will see that different processes involve vastly different amounts of energy. You will learn how to calculate the amount of energy that is absorbed or released by many simple physical changes and chemical reactions. This will allow you to predict energy changes without having to carry out the reaction—an important skill to have when dealing with dangerous reactions. Finally, you will examine the efficiency and environmental impact of traditional and alternative energy sources.

Enough radiant energy reaches Earth every day to meet the world's energy needs many times over. Since this is the case, why do fossil fuels provide most of Canada's energy, while solar power supplies only a tiny fraction?

The Energy of Physical, Chemical, and Nuclear Processes

5.1

Most physical changes, chemical reactions, and nuclear reactions are accompanied by changes in energy. These energy changes are crucial to life on Earth. For example, chemical reactions in your body generate the heat that helps to regulate your body temperature. Physical changes, such as evaporation, help to keep your body cool. On a much larger scale, there would be no life on Earth without the energy from the nuclear reactions that take place in the Sun.

The study of energy and energy transfer is known as **thermodynamics**. Chemists are interested in the branch of thermodynamics known as **thermochemistry:** the study of energy involved in chemical reactions. In order to discuss energy and its interconversions, thermochemists have agreed on a number of terms and definitions. You will learn about these terms and definitions over the next few pages. Then you will examine the energy changes that accompany chemical reactions, physical changes, and nuclear reactions.

Section Preview/ Specific Expectations

In this section, you will

- **write** thermochemical equations, expressing the energy change as a heat term in the equation or as ΔH
- **represent** energy changes using diagrams
- **compare** energy changes that result from physical changes, chemical reactions, and nuclear reactions
- **communicate** your understanding of the following terms: *thermodynamics, thermochemistry, law of conservation of energy, system, surroundings, heat (Q), temperature (T), enthalpy (H), enthalpy change (ΔH), endothermic reaction, exothermic reaction, enthalpy of reaction (ΔH_{rxn}), standard enthalpy of reaction (ΔH°_{rxn}), thermochemical equation, mass defect, nuclear binding energy, nuclear fission, nuclear fusion*

Studying Energy Changes

The **law of conservation of energy** states that the total energy of the universe is constant. In other words, energy can be neither destroyed nor created. This idea can be expressed by the following equation:

$$\Delta E_{universe} = 0$$

Energy can, however, be transferred from one substance to another. It can also be converted into various forms. In order to interpret energy changes, scientists must clearly define what part of the universe they are dealing with. The **system** is defined as the part of the universe that is being studied and observed. In a chemical reaction, the system is usually made up of the reactants and products. By contrast, the **surroundings** are everything else in the universe. The two equations below show the relationship between the universe, a system, and the system's surroundings.

$$\text{Universe} = \text{System} + \text{Surroundings}$$

$$\Delta E_{universe} = \Delta E_{system} + \Delta E_{surroundings} = 0$$

From the relationship, we know that any change in the system is accompanied by an equal and opposite change in the surroundings.

$$\Delta E_{system} = -\Delta E_{surroundings}$$

Look at the chemical reaction that is taking place in the flask in Figure 5.1. A chemist would probably define the system as the contents of the flask—the reactants and products. Technically, the rest of the universe is the surroundings. In reality, however, the entire universe changes very little when the system changes. Therefore, the surroundings are considered to be only the part of the universe that is likely to be affected by the energy changes of the system. In Figure 5.1, the flask, the lab bench, the air in the room, and the student who is carrying out the reaction all make up the surroundings. The system is more likely to significantly influence its immediate surroundings than, say, a mountaintop in Japan (also, technically, part of the surroundings).

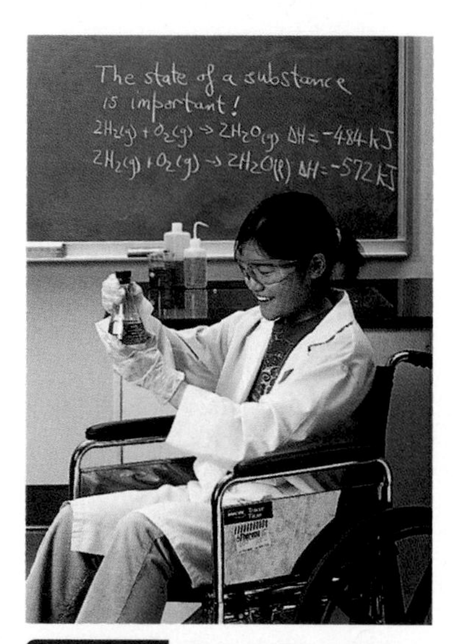

Figure 5.1 The solution in the flask is the system. The flask, the laboratory, and the student are the surroundings.

Heat and Temperature

Heat, *Q*, refers to the transfer of kinetic energy. Heat is expressed in the same units as energy—joules (J). Heat is transferred spontaneously from a warmer object to a cooler object. When you close the door of your home on a cold day to "prevent the cold from getting in," you are actually preventing the heat from escaping. You are preventing the kinetic energy in your warm home from transferring to colder objects, including the cold air, outside.

Temperature, *T*, is a measure of the average kinetic energy of the particles that make up a substance or system. You can think of temperature as a way of quantifying how hot or cold a substance is, relative to another substance.

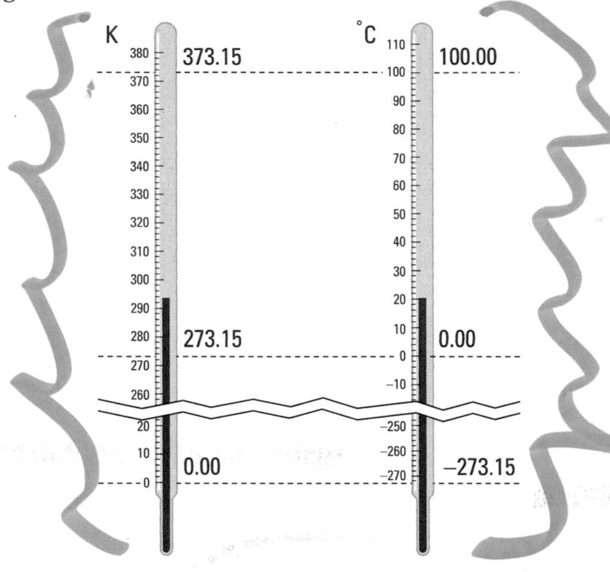

Figure 5.2 Celsius degrees and Kelvin degrees are the same size. The Kelvin scale begins at absolute zero. This is the temperature at which the particles in a substance have no kinetic energy. Therefore, Kelvin temperatures are never negative. By contrast, 0°C is set at the melting point of water. Celsius temperatures can be positive or negative.

Temperature is measured in either Celsius degrees (°C) or kelvins (K). The Celsius scale is a relative scale. It was designed so that water's boiling point is at 100°C and water's melting point is at 0°C. The Kelvin scale, on the other hand, is an absolute scale. It was designed so that 0 K is the temperature at which a substance possesses no kinetic energy. The relationship between the Kelvin and Celsius scales is shown in Figure 5.2, and by the following equation.

$$\text{Temperature in Kelvin degrees} = \text{Temperature in Celsius degrees} + 273.15$$

Enthalpy and Enthalpy Change

Chemists define the total internal energy of a substance at a constant pressure as its **enthalpy**, *H*. Chemists do not work with the *absolute* enthalpy of the reactants and products in a physical or chemical process. Instead, they study the **enthalpy change**, ΔH, that accompanies a process. That is, they study the *relative* enthalpy of the reactants and products in a system. This is like saying that the distance between your home and your school is 2 km. You do not usually talk about the *absolute* position of your home and school in terms of their latitude, longitude, and elevation. You talk about their *relative* position, in relation to each other.

The enthalpy change of a process is equivalent to its heat change at constant pressure.

Enthalpy Changes in Chemical Reactions

In chemical reactions, enthalpy changes result from chemical bonds being broken and formed. Chemical bonds are sources of stored energy. *Breaking a bond is a process that requires energy. Creating a bond is a process that releases energy.* For example, consider the combustion reaction that takes place when nitrogen reacts with oxygen.

$$N_{2(g)} + O_{2(g)} \rightarrow 2NO_{(g)}$$

In this reaction, one mole of nitrogen-nitrogen triple bonds and one mole of oxygen-oxygen double bonds are broken. Two moles of nitrogen-oxygen bonds are formed. This reaction absorbs energy. In other words, more energy is released to form two nitrogen-oxygen bonds than is used to break one nitrogen-nitrogen bond and one oxygen-oxygen bond. When a reaction results in a net *absorption* of energy, it is called an **endothermic reaction**.

On the other hand, when a reaction results in a net *release* of energy, it is called an **exothermic reaction**. In an exothermic reaction, more energy is released to form bonds than is used to break bonds. Therefore, energy is released. Figure 5.3 shows the relationship between bond breaking, bond formation, and endothermic and exothermic reactions.

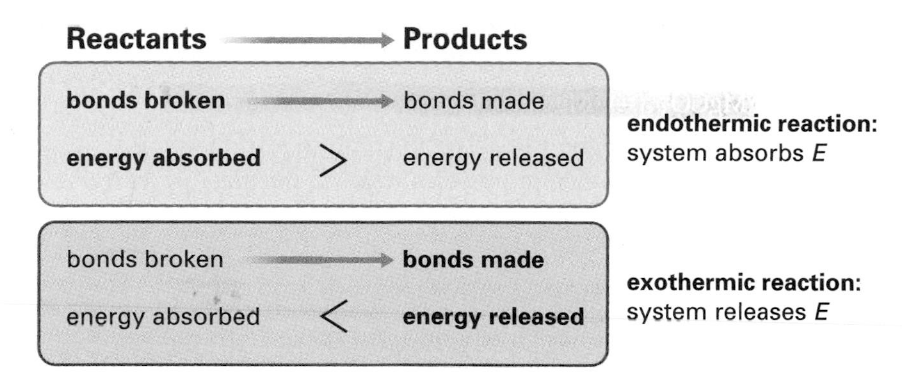

Figure 5.3 The energy changes that result from bonds breaking and forming determine whether a reaction is exothermic or endothermic.

Representing Enthalpy Changes

The enthalpy change of a chemical reaction is known as the **enthalpy of reaction, ΔH_{rxn}**. The enthalpy of reaction is dependent on conditions such as temperature and pressure. Therefore, chemists often talk about the **standard enthalpy of reaction, ΔH°_{rxn}**: the enthalpy change of a chemical reaction that occurs at SATP (25°C and 100 kPa). Often, ΔH°_{rxn} is written simply as ΔH°. The ° symbol is called "nought." It refers to a property of a substance at a standard state or under standard conditions. You may see the enthalpy of reaction referred to as the *heat of reaction* in other chemistry books.

Chemists use different subscripts to represent enthalpy changes for specific kinds of reactions. For example, ΔH_{comb} represents the enthalpy change of a combustion reaction.

Representing Exothermic Reactions

There are three different ways to represent the enthalpy change of an exothermic reaction. The simplest way is to use a **thermochemical equation**: a balanced chemical equation that indicates the amount of heat that is absorbed or released by the reaction it represents. For example, consider the exothermic reaction of one mole of hydrogen gas with half a mole of oxygen gas to produce liquid water. For each mole of hydrogen gas that reacts, 285.8 kJ of heat is produced. Notice that the heat term is included with the products because heat is produced.

$$H_{2(g)} + \frac{1}{2}O_{2(g)} \rightarrow H_2O_{(\ell)} + 285.8\ \text{kJ}$$

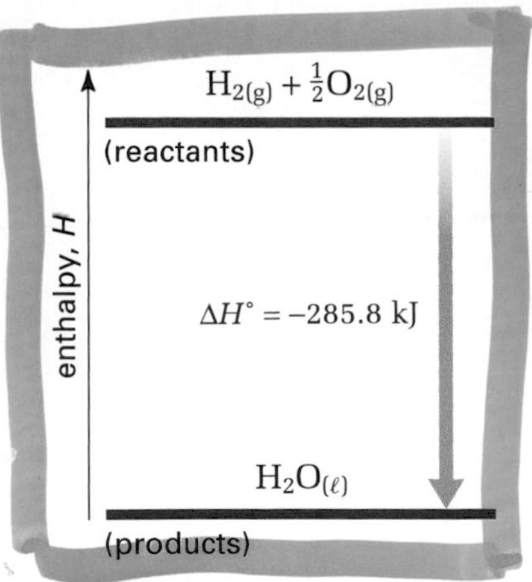

Figure 5.4 In an exothermic reaction, the enthalpy of the system decreases as energy is released to the surroundings.

You can also indicate the enthalpy of reaction as a separate expression beside the chemical equation. For exothermic reactions, ΔH° is always negative.

$$H_{2(g)} + \frac{1}{2}O_{2(g)} \rightarrow H_2O_{(\ell)} \quad \Delta H^\circ_{rxn} = -285.8 \text{ kJ}$$

A third way to represent the enthalpy of reaction is to use an enthalpy diagram. Examine Figure 5.4 to see how this is done.

Representing Endothermic Reactions

The endothermic decomposition of solid magnesium carbonate produces solid magnesium oxide and carbon dioxide gas. For each mole of magnesium carbonate that decomposes, 117.3 kJ of energy is absorbed. As for an exothermic reaction, there are three different ways to represent the enthalpy change of an endothermic reaction.

You can include the enthalpy of reaction as a heat term in the chemical equation. Because heat is absorbed in an endothermic reaction, the heat term is included on the reactant side of the equation.

$$117.3 \text{ kJ} + MgCO_{3(s)} \rightarrow MgO_{(s)} + CO_{2(g)}$$

You can also indicate the enthalpy of reaction as a separate expression beside the chemical reaction. For endothermic reactions, the enthalpy of reaction is always positive.

$$MgCO_{3(s)} \rightarrow MgO_{(s)} + CO_{2(g)} \quad \Delta H^\circ_{rxn} = 117.3 \text{ kJ}$$

Finally, you can use a diagram to show the enthalpy of reaction. Figure 5.5 shows how the decomposition of solid magnesium carbonate can be represented graphically.

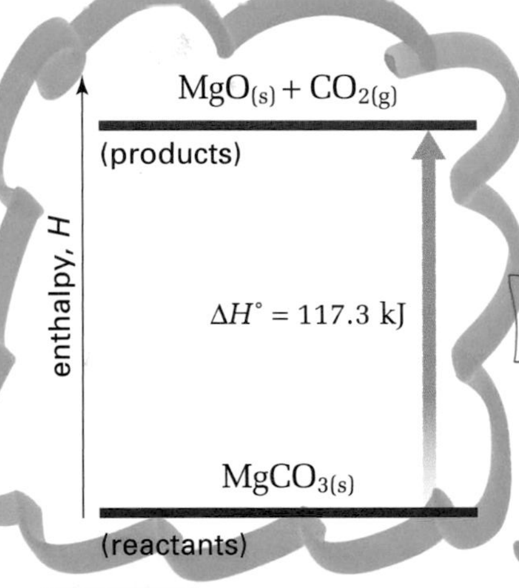

Figure 5.5 In an endothermic reaction, the enthalpy of the system increases as heat energy is absorbed from the surroundings.

Stoichiometry and Thermochemical Equations 

The thermochemical equation for the decomposition of magnesium carbonate, shown above, indicates that 117.3 kJ of energy is absorbed when one mole, or 84.32 g, of magnesium carbonate decomposes. The decomposition of two moles of magnesium carbonate absorbs twice as much energy, or 234.6 kJ.

$$MgCO_{3(s)} \rightarrow MgO_{(s)} + CO_{2(g)} \quad \Delta H^\circ_{rxn} = 117.3 \text{ kJ}$$
$$2MgCO_{3(s)} \rightarrow 2MgO_{(s)} + 2CO_{2(g)} \quad \Delta H^\circ_{rxn} = 234.6 \text{ kJ}$$

Enthalpy of reaction is *linearly dependent* on the quantity of products. That is, if the amount of products formed doubles, the enthalpy change also doubles. Figure 5.6 shows the relationship between the stoichiometry of a reaction and its enthalpy change. Because of this relationship, an exothermic reaction that is relatively safe on a small scale may be extremely dangerous on a large scale. One of the jobs of a chemical engineer is to design systems that allow exothermic reactions to be carried out safely on a large scale. For example, the blast furnaces used in steel making must withstand temperatures of up to 2000°C, produced by the exothermic combustion reaction of coal with oxygen.

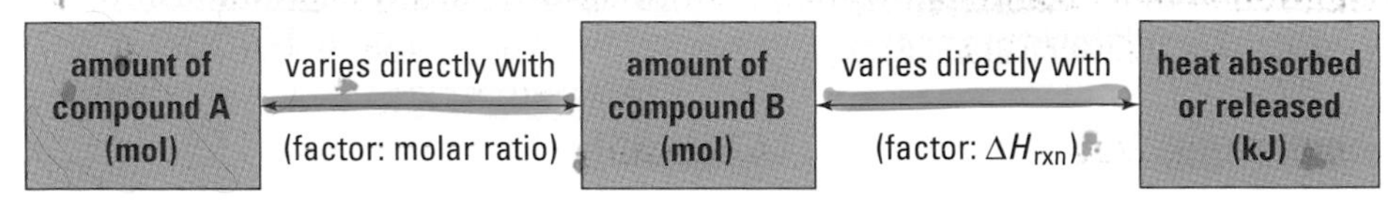

Figure 5.6 This diagram summarizes the relationship between the stoichiometry of a reaction and ΔH.

Sample Problem

Stoichiometry and Thermochemical Reactions

Problem

Aluminum reacts readily with chlorine gas to produce aluminum chloride. The reaction is highly exothermic.

$2Al_{(s)} + 3Cl_{2(g)} \rightarrow 2AlCl_{3(s)} \quad \Delta H^{\circ}_{rxn} = -1408$ kJ

What is the enthalpy change when 1.0 kg of Al reacts completely with excess Cl_2?

What Is Required?

You need to calculate the enthalpy change, ΔH, when the given amount of Al reacts.

What Is Given?

You know the enthalpy change for the reaction of two moles of Al with one mole of Cl_2. From the periodic table, you know the molar mass of Al.

$2Al_{(s)} + 3Cl_{2(g)} \rightarrow 2AlCl_{3(s)} \quad \Delta H^{\circ}_{rxn} = -1408$ kJ

$M_{Al} = 26.98$ g/mol

Plan Your Strategy

Convert the given mass of Al to moles. The enthalpy change is linearly dependent on the quantity of reactants. Therefore, you can use a ratio to determine the enthalpy change for 1.0 kg of Al reacting with Cl_2.

Act on Your Strategy

Determine the number of moles of Al in 1 kg. Remember to convert to grams.

$$n \text{ mol Al} = \frac{m_{Al}}{M_{Al}}$$

$$= \frac{1.0 \times 10^3 \text{ g}}{26.98 \text{ g/mol}}$$

$$= 37 \text{ mol}$$

Use ratios to compare the reference reaction with the known enthalpy change (ΔH_1) to the reaction with the unknown enthalpy change (ΔH_2).

$$\frac{\Delta H_2}{\Delta H_1} = \frac{n_2 \text{ mol Al}}{n_1 \text{ mol Al}}$$

$$\frac{\Delta H_2}{-1408 \text{ kJ}} = \frac{37 \text{ mol Al}}{2 \text{ mol Al}}$$

$$\Delta H_2 = -2.6 \times 10^4 \text{ kJ}$$

Check Your Solution

The sign of the answer is negative, which corresponds to an exothermic reaction. The 1 kg sample contained about 20 times more moles of Al. Therefore, the enthalpy change for the reaction should be about 20 times greater, and it is.

Practice Problems

1. Consider the following reaction.
   $N_{2(g)} + O_{2(g)} \rightarrow 2NO_{(g)} \quad \Delta H^{\circ}_{rxn} = +180.6 \text{ kJ}$
   (a) Rewrite the thermochemical equation, including the standard enthalpy of reaction as either a reactant or a product.
   (b) Draw an enthalpy diagram for the reaction.
   (c) What is the enthalpy change for the formation of one mole of nitrogen monoxide?
   (d) What is the enthalpy change for the reaction of 1.000×10^2 g of nitrogen with sufficient oxygen?

2. The reaction of iron with oxygen is very familiar. You can see the resulting rust on buildings, vehicles, and bridges. You may be surprised, however, at the large amount of heat that is produced by this reaction.
   $4Fe_{(s)} + 3O_{2(g)} \rightarrow 2Fe_2O_{3(s)} + 1.65 \times 10^3 \text{ kJ}$
   (a) What is the enthalpy change for this reaction?
   (b) Draw an enthalpy diagram that corresponds to the thermochemical equation.
   (c) What is the enthalpy change for the formation of 23.6 g of iron(III) oxide?

3. Consider the following thermochemical equation.
   $25.9 \text{ kJ} + \frac{1}{2}H_{2(g)} + \frac{1}{2}I_{2(g)} \rightarrow HI_{(g)}$
   (a) What is the enthalpy change for this reaction?
   (b) How much energy is needed for the reaction of 4.57×10^{24} molecules of iodine, I_2, with excess hydrogen, H_2?
   (c) Draw and label an enthalpy diagram that corresponds to the given thermochemical equation.

4. Tetraphosphorus decoxide, P_4O_{10}, is an acidic oxide. It reacts with water to produce phosphoric acid, H_3PO_4, in an exothermic reaction.
   $P_4O_{10(s)} + 6H_2O_{(\ell)} \rightarrow 4H_3PO_{4(aq)} \quad \Delta H^{\circ}_{rxn} = -257.2 \text{ kJ}$
   (a) Rewrite the thermochemical equation, including the enthalpy change as a heat term in the equation.
   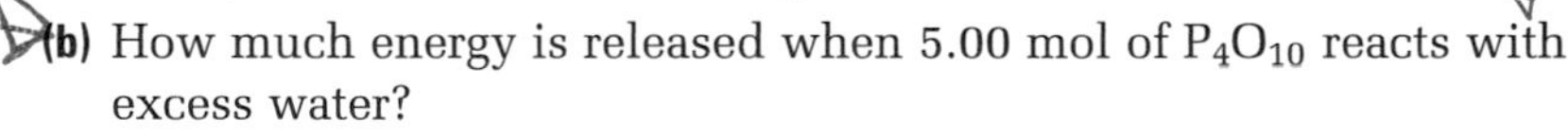
   (b) How much energy is released when 5.00 mol of P_4O_{10} reacts with excess water?
   (c) How much energy is released when 235 g of $H_3PO_{4(aq)}$ is formed?

Heat Changes and Physical Changes

Enthalpy changes are associated with physical changes as well as with chemical reactions. You have observed examples of these enthalpy changes in your daily life. Suppose that you want to prepare some pasta. You put an uncovered pot of water on a stove element. The heat from the element causes the water to become steadily hotter, until it reaches 100°C (the boiling point of water at 100 kPa). At this temperature, heat is still being added to the water. The average kinetic energy of the liquid water molecules does not increase, however. Instead, the energy is used to break the intermolecular bonds between the water molecules as they change from liquid to vapour. The temperature of the liquid water remains at

100°C until all the water has been vaporized. If you add heat to the vapour, the temperature of the vapour will increase steadily.

When you heat ice that is colder than 0°C, a similar process occurs. The temperature of the ice increases until it is 0°C (the melting point of water). If you continue to add heat, the ice remains at 0°C but begins to melt, as the bonds between the water molecules in the solid state begin to break.

Figure 5.7 shows the relationship between temperature and heat for a solid substance that melts and then vaporizes as heat is added to it.

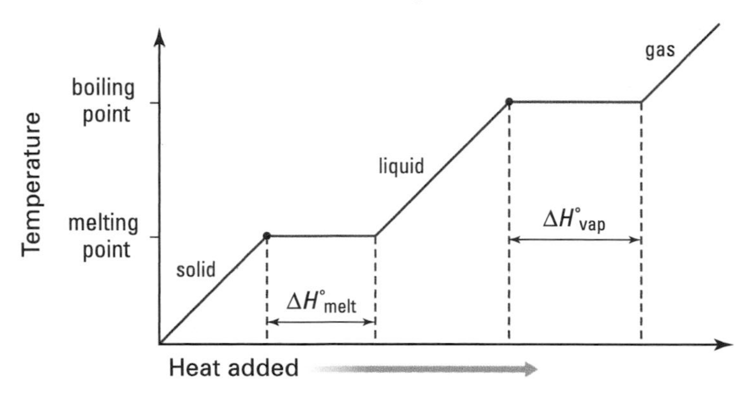

Figure 5.7 As heat is added to a substance, the temperature of the substance steadily increases until it reaches its melting point or boiling point. The temperature then remains steady as the substance undergoes a phase change.

You can represent the enthalpy change that accompanies a phase change—from liquid to solid, for example—just like you represented the enthalpy change of a chemical reaction. You can include a heat term in the equation, or you can use a separate expression of enthalpy change. For example, when one mole of water melts, it absorbs 6.02 kJ of energy.

$$H_2O_{(s)} + 6.02\ kJ \rightarrow H_2O_{(\ell)}$$

$$H_2O_{(s)} \rightarrow H_2O_{(\ell)} \quad \Delta H = 6.02\ kJ$$

Normally, however, chemists represent enthalpy changes associated with phase changes using modified ΔH symbols. These symbols are described below.

- *enthalpy of vaporization*, ΔH_{vap} : the enthalpy change for the phase change from liquid to gas
- *enthalpy of condensation*, ΔH_{cond}: the enthalpy change for the phase change of a substance from gas to liquid
- *enthalpy of melting*, ΔH_{melt} : the enthalpy change for the phase change of a substance from solid to liquid
- *enthalpy of freezing*, ΔH_{fre} : the enthalpy change for the phase change of a substance from liquid to solid

The process of melting is also known as *fusion*. Therefore, you will sometimes see the enthalpy of melting referred to as the *enthalpy of fusion*.

Vaporization and condensation are opposite processes. Thus, the enthalpy changes for these processes have the same value but opposite signs. For example, 6.02 kJ is needed to vaporize one mole of water. Therefore, 6.02 kJ of energy is released when one mole of water freezes.

$$\Delta H_{vap} = -\Delta H_{cond}$$

Similarly, melting and freezing are opposite processes.

$$\Delta H_{melt} = -\Delta H_{fre}$$

Several enthalpies of melting and vaporization are shown in Table 5.1. Notice that the same units (kJ/mol) are used for the enthalpies of melting, vaporization, condensation, and freezing. Also notice that energy changes associated with phase changes can vary widely.

Table 5.1 Enthalpies of Melting and Vaporization for Several Substances

Substance	Enthalpy of melting, ΔH_{melt} (kJ/mol)	Enthalpy of vaporization, ΔH_{vap} (kJ/mol)
argon	1.3	6.3
diethyl ether	7.3	29
ethanol	5.0	40.5
mercury	23.4	59
methane	8.9	0.94
sodium chloride	27.2	207
water	6.02	40.7

Hot Packs and Cold Packs: Using the Energy of Physical Changes

You just learned about the enthalpy changes that are associated with phase changes. Another type of physical change that involves a heat transfer is dissolution. When a solute dissolves in a solvent, the enthalpy change that occurs is called the *enthalpy of solution*, ΔH_{soln}. Dissolution can be either endothermic or exothermic.

Manufacturers take advantage of endothermic dissolution to produce cold packs that athletes can use to treat injuries. One type of cold pack contains water and a salt, such as ammonium nitrate, in separate compartments. When you crush the pack, the membrane that divides the compartments breaks, and the salt dissolves. This dissolution process is endothermic. It absorbs heat for a short period of time, so the cold pack feels cold. Figure 5.8 shows how a cold pack works.

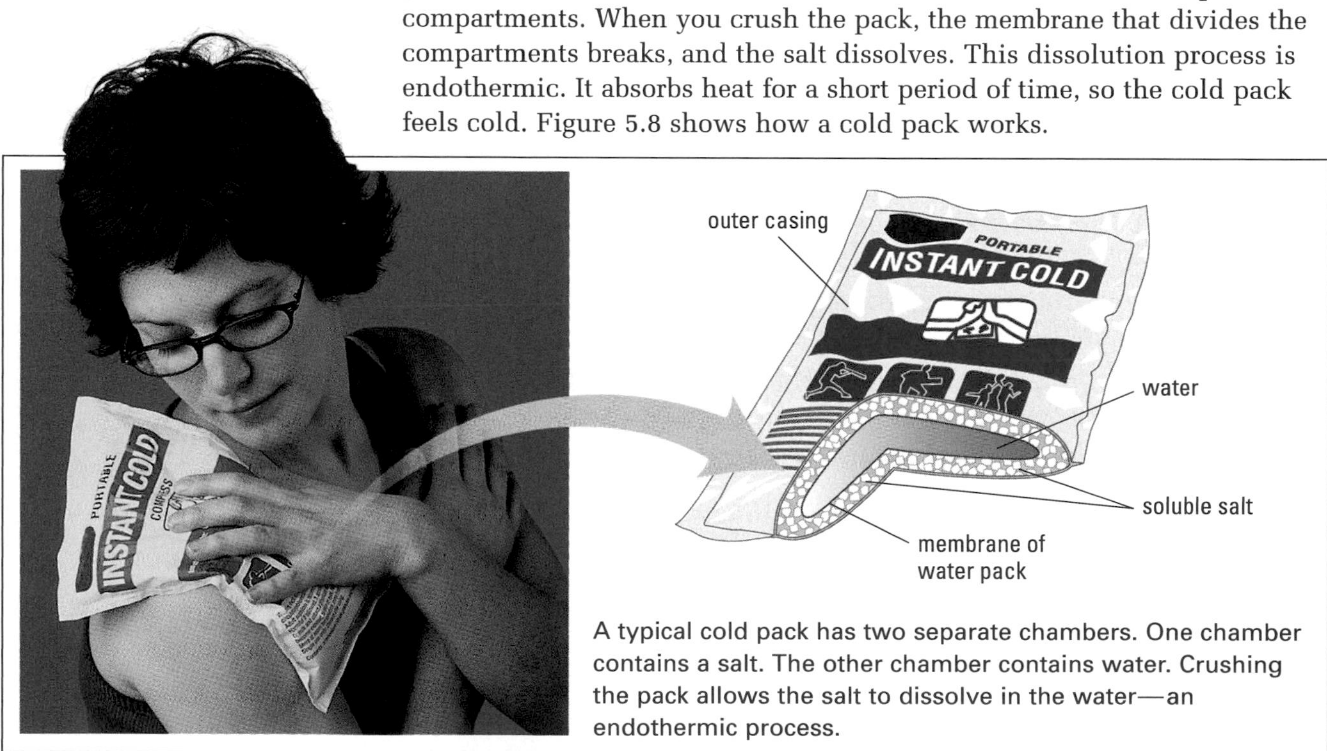

A typical cold pack has two separate chambers. One chamber contains a salt. The other chamber contains water. Crushing the pack allows the salt to dissolve in the water—an endothermic process.

Figure 5.8 **This person's shoulder was injured. Using a cold pack helps to reduce the inflammation of the joint.**

Some types of hot packs are constructed in much the same way as the cold packs described above. They have two compartments. One compartment contains a salt, such as calcium chloride. The other compartment contains water. In hot packs, however, the dissolution process is exothermic. It releases heat to the surroundings.

Energy and Nuclear Reactions

The energy that is released by a physical change, such as the dissolution of calcium chloride, can warm your hands. The energy that is released by a chemical reaction, such as the formation of water, can power a rocket. The energy that is released by a nuclear reaction, such as the nuclear reactions in the Sun, however, can provide enough heat to fry an egg on a sidewalk that is 150 000 000 km away from the surface of the Sun.

From previous science courses, you will recall that nuclear reactions involve changes in the nuclei of atoms. Often nuclear reactions result in the transformation of one or more elements into one or more different elements.

Like physical changes and chemical reactions, nuclear reactions are accompanied by energy changes. Nuclear reactions, however, produce significantly more energy than physical and chemical processes. *In nuclear reactions, a significant amount of the mass of the reactants is actually converted into energy.*

Ever since Albert Einstein devised his famous equation, $E = mc^2$, we have known that mass and energy are interconvertible. In Einstein's equation, E is energy in $kg \cdot m^2/s^2$ (J), m is the mass in kg, and c^2 is the square of the speed of light.

$$\begin{aligned} c^2 &= (3.0 \times 10^8 \text{ m/s})^2 \\ &= 9.0 \times 10^{16} \text{ m}^2/\text{s}^2 \end{aligned}$$

As you can see, c^2 is an enormous number. Therefore, even a very tiny amount of matter is equivalent to a significant amount of energy.

For example, compare the mass of 1 mol of carbon-12 atoms with the mass of the individual nucleons in 1 mol of carbon-12 atoms. The mass of 6 mol of hydrogen-1 atoms (one proton and one electron each) and 6 mol of neutrons is 12.098 940 g. The mass of 1 mol of carbon-12 atoms is exactly 12 g. Note that the mass of the electrons does not change in a nuclear reaction.

$$\begin{array}{r} 12.098\ 940 \text{ g/mol} \\ -\ 12.000\ 000 \text{ g/mol} \\ \hline 0.098\ 940 \text{ g/mol} \end{array}$$

The difference in mass is significant. It would show up on any reasonably precise balance. Thus, the mass of the nucleus of carbon-12 is significantly less than the mass of its component nucleons. The difference in mass between a nucleus and its nucleons is known as the **mass defect**. What causes this mass defect? It is caused by the **nuclear binding energy**: the energy associated with the strong force that holds a nucleus together.

$$\text{Nucleus} + \text{Nuclear binding energy} \rightarrow \text{Nucleons}$$

You can use Einstein's equation to calculate the nuclear binding energy for carbon-12.

$$\begin{aligned} \Delta E &= \Delta mc^2 \\ &= (9.89 \times 10^{-5} \text{ kg/mol})(9.0 \times 10^{16} \text{ m}^2/\text{s}^2) \\ &= 8.9 \times 10^{12} \text{ J/mol} \\ &= 8.9 \times 10^9 \text{ kJ/mol} \end{aligned}$$

Clearly, the energy associated with the bonds that hold a nucleus together is much greater than the energy associated with chemical bonds, which are usually only a few hundred kJ/mol.

The higher the binding energy of a nucleus, the more stable the nucleus is. Nuclei with mass numbers (A) that are close to 60 are the most stable. Nuclear reactions, in which nuclei break apart or fuse, tend to form nuclei that are more stable than the reactant nuclei. Figure 5.9 illustrates the relative stability of various nuclei.

Web LINK

www.mcgrawhill.ca/links/chemistry12

Not all hot packs use dissolution processes. For example, one kind of hot pack exploits the crystallization of sodium thiosulfate or sodium acetate. Another kind uses the oxidation of iron (rusting). On the Internet, investigate different kinds of hot packs. Are they all used for the same purpose? What are the pros and cons of their designs? To start your search, go to the web site above and click on **Web Links**.

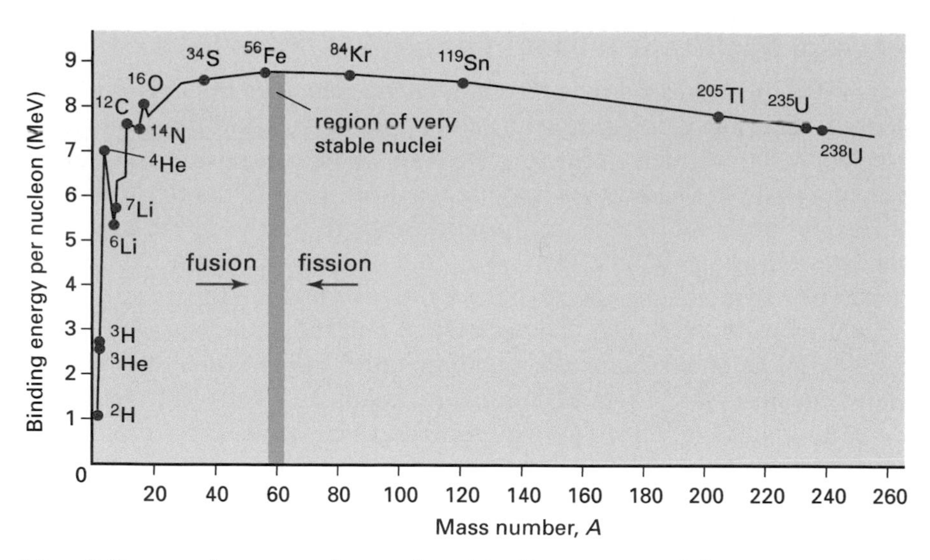

Figure 5.9 This graph shows the relative stability of nuclei. It indicates whether nuclei are more likely to split or fuse in a nuclear reaction. Notice that the helium-4 nucleus is unusually stable.

The difference between the nuclear binding energy of the reactant nuclei and the product nuclei represents the energy change of the nuclear reaction.

Nuclear Fission

A heavy nucleus can split into lighter nuclei by undergoing **nuclear fission**. Nuclear power plants use controlled nuclear fission to provide energy. Uncontrolled nuclear fission is responsible for the massive destructiveness of an atomic bomb.

The most familiar fission reactions involve the splitting of uranium atoms. In these reactions, a uranium-235 atom is bombarded with neutrons. The uranium nucleus then splits apart into various product nuclei. Two examples of fission reactions that involve uranium-235 are shown in Figure 5.10.

Electronic Learning Partner

To learn more about nuclear fission, go to the Chemistry 12 Electronic Learning Partner.

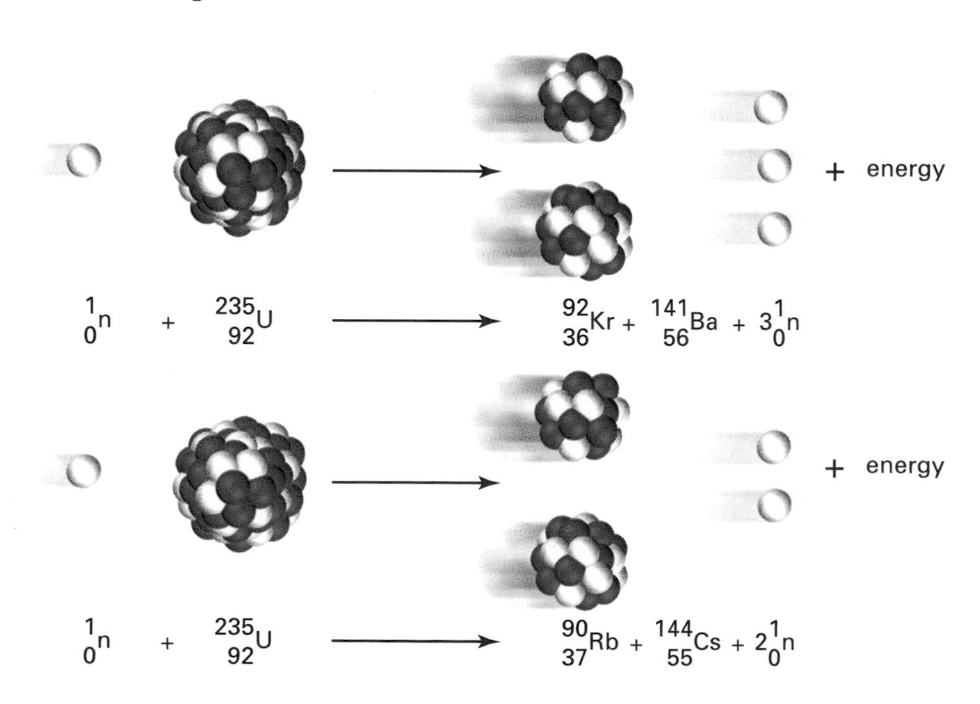

Figure 5.10 Uranium can undergo fission in numerous different ways, producing various product nuclei. Two examples are shown here.

Fission reactions produce vast quantities of energy. For example, when one mole of uranium-235 splits, it releases 2.1×10^{13} J. By contrast, when one mole of coal burns, it releases about 3.9×10^{6} J. Thus, the combustion of coal releases about five million times fewer joules of energy per mole than the fission of uranium-235.

> **Math LINK**
>
> The fission of one mole of uranium-235 produces more energy than the fusion of one mole of deuterium with one mole of tritium. What if you compare the energy that is produced in terms of *mass* of reactants? Calculate a ratio to compare the energy that is produced from fusion and fission, per gram of fuel. What practical consequences arise from your result?

Nuclear Fusion

Two smaller nuclei can fuse to form a larger nucleus, in what is called a **nuclear fusion** reaction. You and all other life on Earth would not exist without nuclear fusion reactions. These reactions are the source of the energy produced in the Sun.

One example of a fusion reaction is the fusion of deuterium and tritium.

$$ {}^{2}_{1}\text{H} + {}^{3}_{1}\text{H} \rightarrow {}^{4}_{2}\text{He} + {}^{1}_{0}\text{n} $$

The seemingly simple reaction between deuterium and tritium produces 1.7×10^{12} J of energy for each mole of deuterium. This is about 10 times fewer joules of energy than are produced by the fission of one mole of uranium. It is still, however, 500 000 times more energy than is produced by burning one mole of coal.

Scientists are searching for a way to harness the energy from fusion reactions. Fusion is a more desirable way to produce energy than fission. The main product of fusion, helium, is relatively harmless compared with the radioactive products of fission. Unfortunately, fusion is proving more difficult than fission to harness. Fusion will not proceed at a reasonable rate without an enormous initial input of energy. This is not a problem in the core of the Sun, where the temperature ranges from 7 500 000°C to 15 000 000°C. It is a problem in industry. Scientists are working on safe and economical ways to provide the high-temperature conditions that are needed to make fusion a workable energy source.

Comparing the Energy of Physical, Chemical, and Nuclear Processes

In this section, you learned that physical changes, chemical reactions, and nuclear reactions all involve energy changes. You also learned that the energy changes have some striking differences in magnitude. Figure 5.11 shows energy changes for some physical, chemical, and nuclear processes. Some other interesting energy statistics are included for reference.

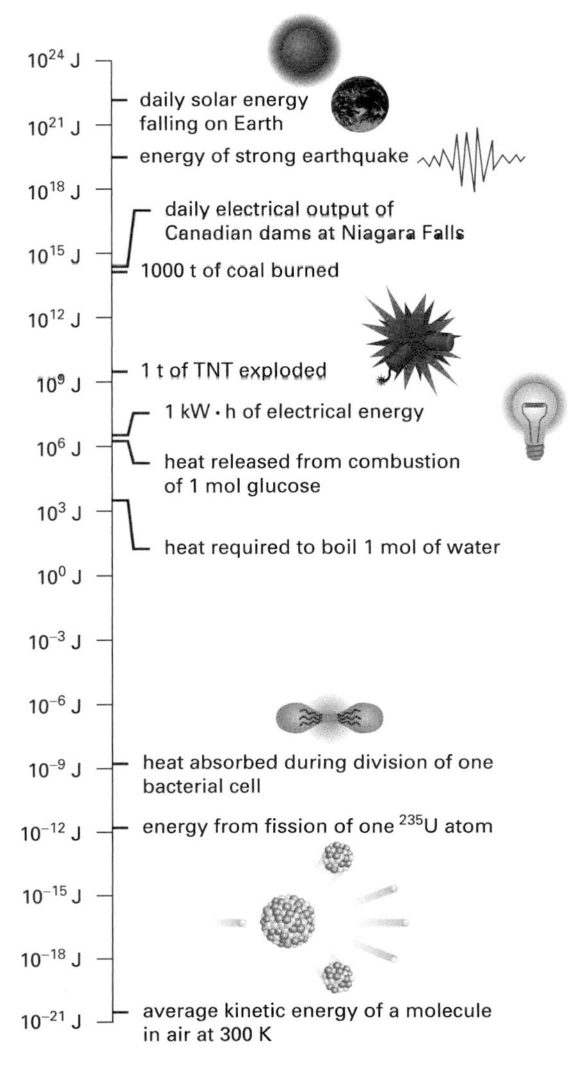

Figure 5.11 The energy changes of physical, nuclear, and chemical processes vary widely. In general, however, chemical reactions are associated with greater energy changes than physical changes. Nuclear reactions are associated with far greater energy changes than chemical reactions.

5.2 Determining Enthalpy of Reaction by Experiment

In this section, you will

- **determine** the heat that is produced by a reaction using a calorimeter, and use the data obtained to calculate the enthalpy change for the reaction
- **communicate** your understanding of the following terms: *specific heat capacity (c), heat capacity (C), calorimeter, coffee-cup calorimeter, constant-pressure calorimeter*

Chemical and physical processes, such as the ones you studied in section 5.1, are associated with characteristic enthalpy changes. How do chemists measure these enthalpy changes?

To measure the enthalpy of a chemical or physical process, chemists insulate the system from the surroundings. They can then determine the heat change by measuring the temperature change of the system. What is the relationship between the heat change and the temperature change? As you learned in previous science courses, each substance has a characteristic property that dictates how its temperature will change when heat is lost or gained.

Specific Heat Capacity

The amount of energy that is needed to raise the temperature of one gram of a substance 1°C (or 1 K) is the **specific heat capacity**, *c*, of the substance. Specific heat capacity is usually expressed in units of J/g • °C. The specific heat capacities of several substances are given in Table 5.2. Figure 5.12 shows that you can often predict the relative specific heat capacities of familiar substances.

Figure 5.12 On a sunny day, you would probably prefer to sit on a bench made of wood rather than a bench made of aluminum. Wood has a higher heat capacity than aluminum. Therefore, more heat is needed to increase its temperature.

Table 5.2 Specific Heat Capacities of Selected Substances

Substance	Specific heat capacity (J/g • °C at 25°C)
Element	
aluminum	0.900
carbon (graphite)	0.711
hydrogen	14.267
iron	0.444
Compound	
ammonia (liquid)	4.70
ethanol	2.46
ethylene glycol	2.42
water (liquid)	4.184
Other material	
air	1.02
concrete	0.88
glass	0.84
wood	1.76

You can use the specific heat capacity of a substance to calculate the amount of energy that is needed to heat a given mass a certain number of degrees. You can also use the specific heat capacity to determine the amount of heat that is released when the temperature of a given mass decreases. The specific heat capacity of liquid water, as shown in Table 5.2, is 4.184 J/g • °C. This relatively large value indicates that a considerable amount of energy is needed to raise or lower the temperature of water.

CONCEPT CHECK

Explain why water is sometimes used as a coolant for automobile engines.

All samples of the same substance have the same specific heat capacity. In contrast, **heat capacity, *C***, relates the heat of a sample, object, or system to its change in temperature. Heat capacity is usually expressed in units of kJ/°C.

Consider a bathtub full of water and a teacup full of water at room temperature. All the water has the same specific heat capacity, but the two samples have different heat capacities. It would take a great deal more heat to raise the temperature of the water in the bathtub by 10°C than it would take to raise the temperature of the water in the teacup by 10°C. Therefore, the water in the bathtub has a higher heat capacity.

Specific Heat Capacity and Heat Transfer

You can use the following equation to calculate the heat change of a substance, based on the mass of the substance. You can also use this equation to calculate the specific heat capacity of the substance and the change in its temperature.

$Q = m \bullet c \bullet \Delta T$

where Q = heat (J)

m = mass (g)

c = specific heat capacity (J/g • °C)

$\Delta T = T_f$(final temperature) – T_i(initial temperature)(°C or K)

$Q = 1.00 \times 10^2 g (4.18)(25°C)$
$= 100(4.18)(25)$
$= 10500$ J

Water is often used in controlled surroundings to measure the heat of a reaction. For example, you can use the equation above to determine the amount of energy that is needed to heat 1.00×10^2 g of water from 20.0°C to 45.0°C.

$$Q = m \bullet c \bullet \Delta T$$

The mass of the water is 1.00×10^2 g. The specific heat capacity of water is 4.184 J/g • °C. The temperature of the water increases by 25.0°C.

$$\therefore Q = (1.00 \times 10^2 \text{ g})(4.184 \text{ J/g} \bullet {}^\circ\text{C})(25.0{}^\circ\text{C})$$
$$= 1.05 \times 10^4 \text{ J}$$

To raise the temperature of 1.00×10^2 g of water by 25°C, 1.05×10^4 J of heat is needed.

Try the following problems to practise working with specific heat capacity and temperature change.

Practice Problems

5. A sample of ethylene glycol, used in car radiators, has a mass of 34.8 g. The sample liberates 783 J of heat. The initial temperature of the sample is 22.1°C. What is the final temperature?

6. A sample of ethanol, C_2H_5OH, absorbs 23.4 kJ of energy. The temperature of the sample increases from 5.6°C to 19.8°C. What is the mass of the ethanol sample? The specific heat capacity of ethanol is 2.46 J/g • °C.

7. A child's swimming pool contains 1000 L of water. When the water is warmed by solar energy, its temperature increases from 15.3°C to 21.8°C. How much heat does the water absorb?

8. What temperature change results from the loss of 255 kJ from a 10.0 kg sample of water?

Measuring Heat Transfer in a Laboratory

From previous science courses, you will probably remember that a **calorimeter** is used to measure enthalpy changes for chemical and physical reactions. A calorimeter works by insulating a system from its surroundings. By measuring the temperature change of the system, you can determine the amount of heat that is released or absorbed by the reaction. For example, the heat that is released by an exothermic reaction raises the temperature of the system.

$$Q_{\text{reaction}} = -Q_{\text{insulated system}}$$

In your previous chemistry course, you learned about various types of calorimeters. For instance, you learned about a *bomb calorimeter*, which allows chemists to determine energy changes under conditions of constant volume.

In section 5.1, however, you learned that an enthalpy change represents the heat change between products and reactants *at a constant pressure*. Therefore, the calorimeter you use to determine an enthalpy change should allow the reaction to be carried out at a constant pressure. In other words, it should be open to the atmosphere.

To determine enthalpy changes in high school laboratories, a **coffee-cup calorimeter** provides fairly accurate results. A coffee-cup calorimeter is composed of two nested polystyrene cups ("coffee cups"). They can be placed in a 250 mL beaker for added stability. Since a coffee-cup calorimeter is open to the atmosphere, it is also called a **constant-pressure calorimeter**.

As with any calorimeter, each part of the coffee-cup calori-meter has an associated heat capacity. Because these heat capacities are very small, however, and because a coffee-cup calorimeter is not as accurate as other calorimeters, the heat capacity of a coffee-cup calorimeter is usually assumed to be negligible. It is assumed to have a value of 0 J/°C.

Web LINK

www.mcgrawhill.ca/links/chemistry12

The famous French scientist Antoine Laviosier (1743-1794) is considered by many to be the first modern chemist. Lavoisier created a calorimeter to study the energy that is released by the metabolism of a guinea pig. To learn about Lavoisier's experiment, go to the web site above and click on **Web Links**. What do you think about using animals in experiments? Write an essay to explain why you agree or disagree with this practice.

Using a Calorimeter to Determine the Enthalpy of a Reaction

A coffee-cup calorimeter is well-suited to determining the enthalpy changes of reactions in dilute aqueous solutions. The water in the calorimeter absorbs (or provides) the energy that is released (or absorbed) by a chemical reaction. When carrying out an experiment in a dilute solution, the solution itself absorbs or releases the energy. You can calculate the amount of energy that is absorbed or released by the solution using the equation mentioned earlier.

$$Q = m \cdot c \cdot \Delta T$$

The mass, m, is the mass of the *solution*, because the solution absorbs the heat. When a dilute aqueous solution is used in a calorimeter, you can assume that the solution has the same density and specific heat capacity as pure water. As you saw above, you can also assume that the heat capacity of the calorimeter is negligible. In other words, you can assume that all the heat that is released or absorbed by the reaction is absorbed or released by the solution.

The following Sample Problem illustrates how calorimetry can be used to determine ΔH of a chemical reaction.

Sample Problem

Determining the Enthalpy of a Chemical Reaction

Problem

Copper(II) sulfate, $CuSO_4$, reacts with sodium hydroxide, NaOH, in a double displacement reaction. A precipitate of copper(II) hydroxide, $Cu(OH)_2$, and aqueous sodium sulfate, Na_2SO_4, is produced.

$$CuSO_{4(aq)} + 2NaOH_{(aq)} \rightarrow Cu(OH)_{2(s)} + Na_2SO_{4(aq)}$$

50.0 mL of 0.300 mol/L $CuSO_4$ solution is mixed with an equal volume of 0.600 mol/L NaOH. The initial temperature of both solutions is 21.4°C. After mixing the solutions in the coffee-cup calorimeter, the highest temperature that is reached is 24.6°C. Determine the enthalpy change of the reaction. Then write the thermochemical equation.

What Is Required?

You need to calculate ΔH of the given reaction.

What Is Given?

You know the volume of each solution. You also know the initial temperature of each solution and the final temperature of the reaction mixture.

Volume of $CuSO_4$ solution, $V_{CuSO_4} = 50.0$ mL

Volume of NaOH solution, $V_{NaOH} = 50.0$ mL

Initial temperature, $T_i = 21.4$°C

Final temperature, $T_f = 24.6$°C

Plan Your Strategy

Step 1 Determine the total volume by adding the volumes of the two solutions. Determine the total mass of the reaction mixture, assuming a density of 1.00 g/mL (the density of water).

Step 2 Use the equation $Q = m \cdot c \cdot \Delta T$ to calculate the amount of heat that is absorbed by the solution (in J). Assume that the reaction mixture has the same specific heat capacity as water ($c = 4.184$ J/g • °C).

Step 3 Use the equation $Q_{reaction} = -Q_{solution}$ to determine the amount of heat that is released by the reaction.

Step 4 Determine the number of moles of $CuSO_4$ and NaOH that reacted. If necessary, determine the limiting reactant. Use the amount of limiting reactant to get ΔH of the reaction (in kJ/mol).

Step 5 Use your ΔH to write the thermochemical equation for the reaction.

Act on Your Strategy

Step 1 The total volume of the reaction mixture is

50.0 mL + 50.0 mL = 100.0 mL

The mass of the reaction mixture, assuming a density of 1.00 g/mL, is

$$\begin{aligned} m &= DV \\ &= (1.00 \text{ g/}\cancel{\text{mL}})(100.0 \ \cancel{\text{mL}}) \\ &= 1.00 \times 10^2 \text{ g} \end{aligned}$$

Continued ...

Continued ...

Step 2 The amount of heat, Q, that is absorbed by the solution is

$$Q_{solution} = m_{solution} \cdot c_{solution} \cdot \Delta T_{solution}$$
$$= (100\ g)(4.184\ J/g \cdot {}^\circ C)(24.6^\circ C - 21.4^\circ C)$$
$$= 1.3 \times 10^3\ J$$

Step 3 Based on the value of Q in step 2, the heat change for the reaction is -1.3×10^3 J.

Step 4 Calculate the number of moles of $CuSO_4$ as follows.

$$n = c \cdot V$$
$$= (0.300\ mol/L)(50.0 \times 10^{-3}\ L)$$
$$= 0.0150\ mol$$

Calculate the number of moles of NaOH.

$$n\ mol\ NaOH = (0.600\ mol/L)(50.0 \times 10^{-3}\ L)$$
$$= 0.300\ mol$$

The reactants are present in stoichiometric amounts. (There is no limiting reactant.)

ΔH of the reaction, in kJ/mol $CuSO_4$, is

$$\Delta H = \frac{-1.3 \times 10^3\ J}{0.0150\ mol\ CuSO_4}$$
$$= -8.9 \times 10^3\ J/mol\ CuSO_4$$
$$= -89\ kJ/mol\ CuSO_4$$

The enthalpy change of the reaction is −89 kJ/mol $CuSO_4$.

Step 5 The thermochemical equation is

$$CuSO_{4(aq)} + 2NaOH_{(aq)} \rightarrow Cu(OH)_{2(s)} + Na_2SO_{4(aq)} \quad \Delta H = -89\ kJ$$

Check Your Solution

The solution has the correct number of significant digits. The units are correct. You know that the reaction was exothermic, because the temperature of the solution increased. The calculated ΔH is negative, which is correct for an exothermic reaction.

Practice Problems

9. A chemist wants to determine the enthalpy of neutralization for the following reaction.

$$HCl_{(aq)} + NaOH_{(aq)} \rightarrow NaCl_{(aq)} + H_2O_{(\ell)}$$

The chemist uses a coffee-cup calorimeter to neutralize completely 61.1 mL of 0.543 mol/L $HCl_{(aq)}$ with 42.6 mL of $NaOH_{(aq)}$. The initial temperature of both solutions is 17.8°C. After neutralization, the highest recorded temperature is 21.6°C. Calculate the enthalpy of neutralization, in units of kJ/mol of HCl. Assume that the density of both solutions is 1.00 g/mL. Also assume that the specific heat capacity of both solutions is the same as the specific heat capacity of water.

10. A chemist wants to determine empirically the enthalpy change for the following reaction.

$Mg_{(s)} + 2HCl_{(aq)} \rightarrow MgCl_{2(aq)} + H_{2(g)}$

The chemist uses a coffee-cup calorimeter to react 0.50 g of Mg ribbon with 100 mL of 1.00 mol/L $HCl_{(aq)}$. The initial temperature of the $HCl_{(aq)}$ is 20.4°C. After neutralization, the highest recorded temperature is 40.7°C.

(a) Calculate the enthalpy change, in kJ/mol of Mg, for the reaction.

(b) State any assumptions that you made in order to determine the enthalpy change.

11. Nitric acid is neutralized with potassium hydroxide in the following reaction.

$HNO_{3(aq)} + KOH_{(aq)} \rightarrow KNO_{3(aq)} + H_2O_{(\ell)} \quad \Delta H = -53.4 \text{ kJ/mol}$

55.0 mL of 1.30 mol/L solutions of both reactants, at 21.4°C, are mixed in a calorimeter. What is the final temperature of the mixture? Assume that the density of both solutions is 1.00 g/mL. Also assume that the specific heat capacity of both solutions is the same as the specific heat capacity of water. No heat is lost to the calorimeter itself.

12. A student uses a coffee-cup calorimeter to determine the enthalpy of reaction for hydrobromic acid and potassium hydroxide. The student mixes 100.0 mL of 0.50 mol/L $HBr_{(aq)}$ at 21.0°C with 100.0 mL of 0.50 mol/L $KOH_{(aq)}$, also at 21.0°C. The highest temperature that is reached is 24.4°C. Write a thermochemical equation for the reaction.

In Practice Problems 9, 11, and 12, you used experimental data to determine the enthalpy of reaction for neutralization reactions. Neutralization reactions are particularly well suited to analysis involving the use of a coffee-cup calorimeter for a number of reasons:

- When using dilute solutions of acids and bases, you can assume their density is close to the density of water. Therefore, you can easily measure the volume of the solutions and calculate their mass.
- Neutralization reactions between dilute strong acids and dilute strong bases tend to cause temperature changes in the reaction mixture that are large enough to be measurable using a standard thermometer, but small enough for safety.
- Neutralization reactions take place very quickly. Therefore, the peak temperature change also occurs very quickly. There is little time for heat transfer between the insulated system and the surroundings to take place.

In the following investigation, you will construct a coffee-cup calorimeter and use it to determine the enthalpy of a neutralization reaction.

Investigation 5-A

SKILL FOCUS

Predicting

Performing and recording

Analyzing and interpreting

Determining the Enthalpy of a Neutralization Reaction

The neutralization of hydrochloric acid with sodium hydroxide solution is represented by the following equation.

$HCl_{(aq)} + NaOH_{(aq)} \rightarrow NaCl_{(aq)} + H_2O_{(\ell)}$

Using a coffee-cup calorimeter, you will determine the enthalpy change for this reaction.

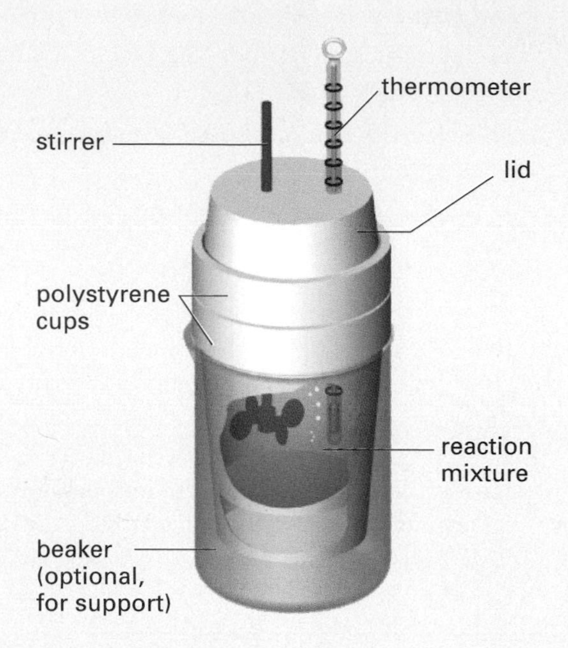

Question

What is the heat of neutralization for hydrochloric acid and sodium hydroxide solution?

Prediction

Will the neutralization reaction be endothermic or exothermic? Record your prediction, and give reasons.

Safety Precautions

If you get any hydrochloric acid or sodium hydroxide solution on your skin, flush your skin with plenty of cold water.

Materials

100 mL graduated cylinder
400 mL beaker
2 polystyrene cups that are the same size
polystyrene lid
thermometer
stirring rod
1.00 mol/L $HCl_{(aq)}$
1.00 mol/L $NaOH_{(aq)}$

Procedure

1. Your teacher will allow the hydrochloric acid and sodium hydroxide solution to come to room temperature overnight.
2. Read the rest of this Procedure carefully before you continue. Set up a graph to record your temperature observations.
3. Build a coffee-cup calorimeter, using the diagram above as a guide. You will need to make two holes in the polystyrene lid—one for the thermometer and one for the stirring rod. The holes should be as small as possible to minimize heat loss to the surroundings.
4. Rinse the graduated cylinder with a small quantity of 1.00 mol/L $NaOH_{(aq)}$. Use the cylinder to add 50.0 mL of 1.00 mol/L $NaOH_{(aq)}$ to the calorimeter. Record the initial temperature of the $NaOH_{(aq)}$. (This will also represent the initial temperature of the $HCl_{(aq)}$.) **CAUTION** The $NaOH_{(aq)}$ can burn your skin.

5. Rinse the graduated cylinder with tap water. Then rinse it with a small quantity of 1.00 mol/L $HCl_{(aq)}$. Quickly and carefully, add 50.0 mL of 1.00 mol/L $HCl_{(aq)}$ to the $NaOH_{(aq)}$ in the calorimeter. **CAUTION** The $HCl_{(aq)}$ can burn your skin.
6. Cover the calorimeter. Record the temperature every 30 s, stirring gently and continuously.
7. When the temperature levels off, record the final temperature, T_f.
8. If time permits, repeat steps 4 to 7.

Analysis

1. Determine the amount of heat that is absorbed by the solution in the calorimeter.
2. Use the following equation to determine the amount of heat that is released by the reaction:

   $-Q_{reaction} = Q_{solution}$
3. Determine the number of moles of $HCl_{(aq)}$ and $NaOH_{(aq)}$ that were involved in the reaction.
4. Use your knowledge of solutions to explain what happens during a neutralization reaction. Use equations in your answer. Was heat released or absorbed during the neutralization reaction? Explain your answer.

Conclusion

5. Use your results to determine the enthalpy change of the neutralization reaction, in kJ/mol of NaOH. Write the thermochemical equation for the neutralization reaction.

Applications

6. When an acid gets on your skin, why must you flush the area with plenty of water rather than neutralizing the acid with a base?
7. Suppose that you had added solid sodium hydroxide pellets to hydrochloric acid, instead of adding hydrochloric acid to sodium hydroxide solution?
   (a) Do you think you would have obtained a different enthalpy change?
   (b) Would the enthalpy change have been higher or lower?
   (c) How can you test your answer? Design an investigation, and carry it out with the permission of your teacher.
   (d) What change do you need to make to the thermochemical equation if you perform the investigation using solid sodium hydroxide?
8. In Investigation 5-A, you assumed that the heat capacity of your calorimeter was 0 J/°C.
   (a) Design an investigation to determine the actual heat capacity of your coffee-cup calorimeter, $C_{calorimeter}$. Include equations for any calculations you will need to do. If time permits, have your teacher approve your procedure and carry out the investigation. **Hint:** If you mix hot and cold water together and no heat is absorbed by the calorimeter itself, then the amount of heat absorbed by the cold water should equal the amount of heat released by the hot water. If more heat is released by the hot water than is absorbed by the cold water, the difference must be absorbed by the calorimeter.
   (b) Include the heat capacity of your calorimeter in your calculations for $\Delta H_{neutralization}$. Use the following equation:

$$-Q_{reaction} = (m_{solution} \cdot c_{solution} \cdot \Delta T) + (C_{calorimeter} \cdot \Delta T)$$

PROBEWARE

If you have access to probeware, do Probeware Investigation 5-A, or a similar investigation from a probeware company.

Section Summary

In this section, you measured the enthalpy change of a reaction by calorimetry. You may have noticed that the reactions you studied in this section involved relatively small energy changes. How do chemists work quantitatively with some of the large energy changes you examined in section 5.1? In the next section, you will learn how to calculate the heat of reaction for virtually any chemical reaction or physical change. This powerful skill will allow you to find heats of reaction without carrying out experiments.

Section Review

1. K/U Distinguish between heat capacity and specific heat capacity.

2. K/U What properties of polystyrene make it a suitable material for a constant-pressure calorimeter? Why are polystyrene coffee cups not suitable for a constant-volume calorimeter?

3. K/U Suppose that you use concentrated reactant solutions in an experiment with a coffee-cup calorimeter. Will you make the same assumptions that you did when you used dilute solutions? Explain.

4. I Concentrated sulfuric acid can be diluted by adding it to water. The reaction is extremely exothermic. In this question, you will design an experiment to measure the enthalpy change (in kJ/mol) for the dilution of concentrated sulfuric acid. Assume that you have access to any equipment in your school's chemistry laboratory. Do not carry out this experiment.
   (a) State the equipment and chemicals that you need.
   (b) Write a step-by-step procedure.
   (c) Set up an appropriate data table.
   (d) State any information that you need.
   (e) State any simplifying assumptions that you will make.

5. I A chemist mixes 100.0 mL of 0.050 mol/L potassium hydroxide with 100.0 mL of 0.050 mol/L nitric acid in a constant-pressure calorimeter. The temperature of the reactants is 21.01°C. The temperature of the products is 21.34°C.
   (a) Write a thermochemical equation for the reaction.
   (b) If you performed this investigation, would you change the procedure? If so, how?

6. C Explain why a bomb calorimeter may not provide accurate results for determining the enthalpy of a reaction.

7. MC From experience, you know that you produce significantly more heat when you are exercising than when you are resting. Scientists can study the heat that is produced by human metabolism reactions using a "human calorimeter." Based on what you know about calorimetry, how would you design a human calorimeter? What variables would you control and study in an investigation using your calorimeter? Write a brief proposal outlining the design of your human calorimeter and the experimental approach you would take.

Hess's Law of Heat Summation

5.3

In section 5.2, you used a coffee-cup calorimeter to determine the quantity of heat that was released or absorbed in a chemical reaction. Coffee-cup calorimeters are generally used only for dilute aqueous solutions. There are many non-aqueous chemical reactions, however. There are also many reactions that release so much energy they are not safe to perform using a coffee-cup calorimeter. Imagine trying to determine the enthalpy of reaction for the detonation of nitroglycerin, an unstable and powerfully explosive compound. Furthermore, there are reactions that occur too slowly for the calorimetric method to be practical. (You will learn more about rates of reactions in the next chapter.)

Chemists can determine the enthalpy change of any reaction using an important law, known as **Hess's law of heat summation**. This law states that *the enthalpy change of a physical or chemical process depends only on the beginning conditions (reactants) and the end conditions (products). The enthalpy change is independent of the pathway of the process and the number of intermediate steps in the process.* It is the sum of the enthalpy changes of all the individual steps that make up the process.

For example, carbon and oxygen can form carbon dioxide via two pathways.

1. Carbon can react with oxygen to form carbon monoxide. The carbon monoxide then reacts with oxygen to produce carbon dioxide. The two equations below represent this pathway.

$$C_{(s)} + \frac{1}{2}O_{2(g)} \rightarrow CO_{(g)} \quad \Delta H^\circ = -110.5 \text{ kJ}$$
$$CO_{(g)} + \frac{1}{2}O_{2(g)} \rightarrow CO_{2(g)} \quad \Delta H^\circ = -283.0 \text{ kJ}$$

2. Carbon can also react with oxygen to produce carbon dioxide directly.

$$C_{(s)} + O_{2(g)} \rightarrow CO_{2(g)} \quad \Delta H^\circ = -393.5 \text{ kJ}$$

In both cases, the net result is that one mole of carbon reacts with one mole of oxygen to produce one mole of carbon dioxide. (In the first pathway, all the carbon monoxide that is produced reacts with oxygen to form carbon dioxide.) Notice that the sum of the enthalpy changes for the first pathway is the same as the enthalpy change for the second pathway.

Examine Figure 5.13 to see how to represent the two pathways using one enthalpy diagram.

Section Preview/ Specific Expectations

In this section, you will

- **explain** Hess's law of heat summation, using examples
- **apply** Hess's law to solve problems, including problems that involve data obtained through experimentation
- **calculate** heat of reaction using given enthalpies of formation
- **communicate** your understanding of the following terms: *Hess's law of heat summation, formation reactions, standard molar enthalpy of formation (ΔH°_f)*

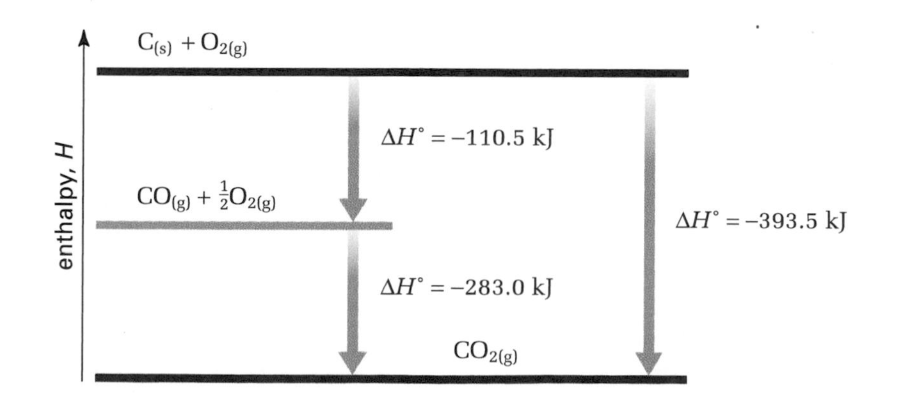

Figure 5.13 Carbon dioxide can be formed by the reaction of oxygen with carbon to form carbon monoxide, followed by the reaction of carbon monoxide with oxygen. Carbon dioxide can also be formed directly from carbon and oxygen. No matter which pathway is used, the enthalpy change of the reaction is the same.

One way to think about Hess's law is to compare the energy changes that occur in a chemical reaction with the changes in the potential energy of a cyclist on hilly terrain. This comparison is shown in Figure 5.14.

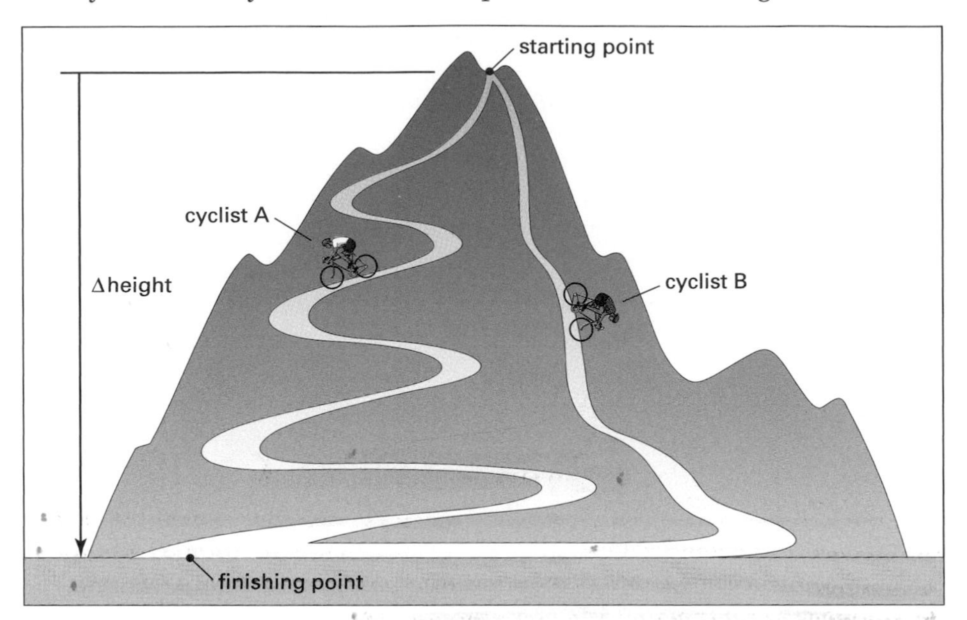

Figure 5.14 The routes that cyclists take to get from the starting point to the finishing point has no effect on the *net* change in the cyclists' gravitational potential energy.

Hess's law allows you to determine the energy of a chemical reaction without directly measuring it. In this section, you will examine two ways in which you can use Hess's law to calculate the enthalpy change of a chemical reaction:

1. by combining chemical equations algebraically
2. by using the enthalpy of a special class of reactions called formation reactions

Combining Chemical Equations Algebraically

According to Hess's law, the pathway that is taken in a chemical reaction has no effect on the enthalpy change of the reaction. How can you use Hess's law to calculate the enthalpy change of a reaction? One way is to add equations for reactions with known enthalpy changes, so that their net result is the reaction you are interested in.

For example, you can combine thermochemical equations (1) and (2) below to find the enthalpy change for the decomposition of hydrogen peroxide, equation (3).

$$\textbf{(1)} \quad H_2O_{2(\ell)} \rightarrow H_{2(g)} + O_{2(g)} \qquad \Delta H^\circ = +188 \text{ kJ}$$

$$\textbf{(2)} \quad H_{2(g)} + \frac{1}{2}O_{2(g)} \rightarrow H_2O_{(\ell)} \qquad \Delta H^\circ = -286 \text{ kJ}$$

$$\textbf{(3)} \quad H_2O_{2(\ell)} \rightarrow H_2O_{(\ell)} + \frac{1}{2}O_{2(g)} \qquad \Delta H^\circ = ?$$

Carefully examine equation (3), the *target* equation. Notice that H_2O_2 is on the left (reactant) side, while H_2O and $\frac{1}{2}O_2$ are on the right (product) side. Now examine equations (1) and (2). Notice which sides of the equations H_2O_2 and H_2O are on. They are on the correct sides, based on equation (3). Also notice that hydrogen does not appear in equation (3). Therefore, it must cancel out when equations (1) and (2) are added. Since there is one mole of $H_{2(g)}$ on the product side of equation (1) and one mole of $H_{2(g)}$ on the reactant side of equation (2), these two terms cancel. Set up equations (1) and (2) as shown on the next page. Add the products and the reactants. Then cancel any substances that appear on opposite sides.

(1) $H_2O_{2(\ell)} \rightarrow H_{2(g)} + O_{2(g)}$ $\Delta H^\circ = +188$ kJ

(2) $H_{2(g)} + \frac{1}{2}O_{2(g)} \rightarrow H_2O_{(\ell)}$ $\Delta H^\circ = -286$ kJ

$$H_2O_{2(\ell)} + \cancel{H_{2(g)}} + \frac{1}{2}O_{2(g)} \rightarrow H_2O_{(\ell)} + O_{2(g)} + \cancel{H_{2(g)}} \quad \Delta H^\circ = ?$$

or

(3) $H_2O_{2(\ell)} \rightarrow H_2O_{(\ell)} + \frac{1}{2}O_{2(g)}$ $\Delta H^\circ = ?$

Equations (1) and (2) add to give equation (3). Therefore, you know that the enthalpy change for the decomposition of hydrogen peroxide is the sum of the enthalpy changes of equations (1) and (2).

$$H_2O_{2(\ell)} \rightarrow H_2O_{(\ell)} + \frac{1}{2}O_{2(g)} \quad \Delta H^\circ = 188 \text{ kJ} - 286 \text{ kJ} = -98 \text{ kJ}$$

Figure 5.15 illustrates this combination of chemical equations in an enthalpy diagram.

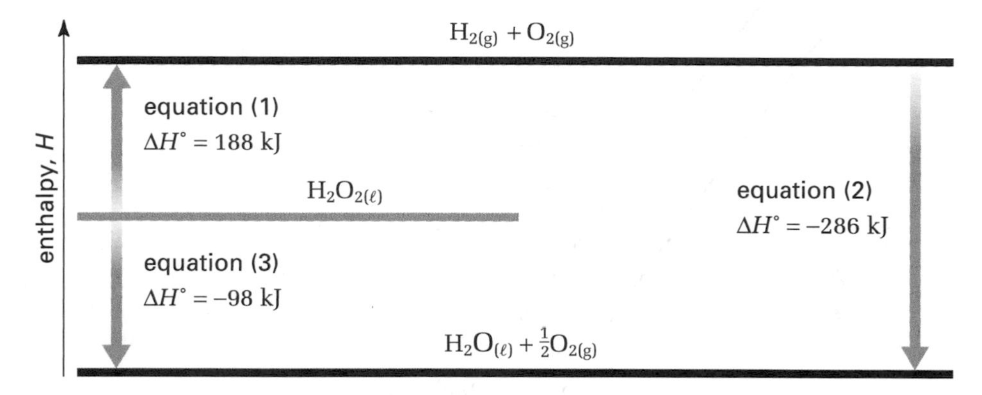

Figure 5.15 The algebraic combination of chemical reactions can be represented in an enthalpy diagram.

In the previous example, you did not need to manipulate the two equations with known enthalpy changes. They added to the target equation as they were written. In many cases, however, you will need to manipulate the equations before adding them. There are two key ways in which you can manipulate an equation:

1. *Reverse an equation* so that the products become reactants and the reactants become products. When you reverse an equation, you need to change the sign of ΔH°(multiply by -1).
2. *Multiply each coefficient in an equation* by the same integer or fraction. When you multiply an equation, you need to multiply ΔH° by the same number.

Examine the following Sample Problem to see how to manipulate equations so that they add to the target equation. Try the problems that follow to practise finding the enthalpy change by adding equations.

Sample Problem

Using Hess's Law to Determine Enthalpy Change

Problem

One of the methods that the steel industry uses to obtain metallic iron is to react iron(III) oxide, Fe_2O_3, with carbon monoxide, CO.

$Fe_2O_{3(s)} + 3CO_{(g)} \rightarrow 3CO_{2(g)} + 2Fe_{(s)}$

Continued ...

Continued ...

PROBLEM TIP

Before adding chemical equations, be sure to line up the equation arrows.

Electronic Learning Partner

Go to the Chemistry 12 Electronic Learning Partner for more information about aspects of material covered in this section of the chapter.

Determine the enthalpy change of this reaction, given the following equations and their enthalpy changes.

(1) $CO_{(g)} + \frac{1}{2}O_{2(g)} \rightarrow CO_{2(g)} \quad \Delta H^\circ = -283.0 \text{ kJ}$

(2) $2Fe_{(s)} + \frac{3}{2}O_{2(g)} \rightarrow Fe_2O_3 \quad \Delta H^\circ = -822.3 \text{ kJ}$

What Is Required?

You need to find ΔH° of the target reaction.
$Fe_2O_{3(s)} + 3CO_{(g)} \rightarrow 3CO_{2(g)} + 2Fe_{(s)}$

What Is Given?

You know the chemical equations for reactions (1) and (2), and their corresponding enthalpy changes.

Plan Your Strategy

Step 1 Examine equations (1) and (2) to see how they compare with the target equation. Decide how you need to manipulate equations (1) and (2) so that they add to the target equation. (Reverse the equation, multiply the equation, do both, or do neither). Remember to adjust ΔH° accordingly for each equation.

Step 2 Write the manipulated equations so that their equation arrows line up. Add the reactants and products on each side, and cancel substances that appear on both sides.

Step 3 Ensure that you have obtained the target equation. Add ΔH° for the combined equations.

Act on Your Strategy

Step 1 Equation (1) has CO as a reactant and CO_2 as a product, as does the target reaction. The stoichiometric coefficients do not match the coefficients in the target equation, however. To achieve the same coefficients, you must multiply equation (1) by 3.
Equation (2) has the required stoichiometric coefficients, but Fe and Fe_2O_3 are on the wrong sides of the equation. You need to reverse equation (2) and change the sign of ΔH°.

Step 2 Multiply each equation as required, and add them.

3 × (1) $3CO_{(g)} + \frac{3}{2}O_{2(g)} \rightarrow 3CO_{2(g)} \quad \Delta H^\circ = 3(-283.0 \text{ kJ})$

−1 × (2) $Fe_2O_{3(s)} \rightarrow 2Fe_{(s)} + \frac{3}{2}O_{2(g)} \quad \Delta H^\circ = -1(-824.2 \text{ kJ})$

$$Fe_2O_{3(s)} + \cancel{\frac{3}{2}O_{2(g)}} + 3CO_{(g)} \rightarrow 3CO_{2(g)} + 2Fe_{(s)} + \cancel{\frac{3}{2}O_{2(g)}}$$

or

$$Fe_2O_{3(s)} + 3CO_{(g)} \rightarrow 3CO_{2(g)} + 2Fe_{(s)}$$

Step 3 The desired equation is achieved. Therefore, you can calculate the enthalpy change of the target reaction by adding the heats of reaction for the manipulated equations.

$\Delta H^\circ = 3(-283.0 \text{ kJ}) + 824.2 \text{ kJ} = -24.8 \text{ kJ}$
$\therefore Fe_2O_{3(s)} + 3CO_{(g)} \rightarrow 3CO_{2(g)} + 2Fe_{(s)} \quad \Delta H^\circ = -24.8 \text{ kJ}$

Check Your Solution

Because the equations added correctly to the target equation, you know you manipulated the equations with known enthalpy changes correctly. Check to ensure that you adjusted ΔH° accordingly for each equation.

Practice Problems

13. Ethene, C_2H_4, reacts with water to form ethanol, $CH_3CH_2OH_{(\ell)}$.

$C_2H_{4(g)} + H_2O_{(\ell)} \rightarrow CH_3CH_2OH_{(\ell)}$

Determine the enthalpy change of this reaction, given the following thermochemical equations.

(1) $CH_3CH_2OH_{(\ell)} + 3O_{2(g)} \rightarrow 3H_2O_{(\ell)} + 2CO_{2(g)} \quad \Delta H^\circ = -1367 \text{ kJ}$

(2) $C_2H_{4(g)} + 3O_{2(g)} \rightarrow 2H_2O_{(\ell)} + 2CO_{2(g)} \quad \Delta H^\circ = -1411 \text{ kJ}$

14. A typical automobile engine uses a lead-acid battery. During discharge, the following chemical reaction takes place.

$Pb_{(s)} + PbO_{2(s)} + 2H_2SO_{4(\ell)} \rightarrow 2PbSO_{4(aq)} + 2H_2O_{(\ell)}$

Determine the enthalpy change of this reaction, given the following equations.

(1) $Pb_{(s)} + PbO_{2(s)} + 2SO_{3(g)} \rightarrow 2PbSO_{4(s)} \quad \Delta H^\circ = -775 \text{ kJ}$

(2) $SO_{3(g)} + H_2O_{(\ell)} \rightarrow H_2SO_{4(\ell)} \quad \Delta H^\circ = -133 \text{ kJ}$

15. Mixing household cleansers can result in the production of hydrogen chloride gas, $HCl_{(g)}$. Not only is this gas dangerous in its own right, but it also reacts with oxygen to form chlorine gas and water vapour.

$4HCl_{(g)} + O_{2(g)} \rightarrow 2Cl_{2(g)} + 2H_2O_{(g)}$

Determine the enthalpy change of this reaction, given the following equations.

(1) $H_{2(g)} + Cl_{2(g)} \rightarrow 2HCl_{(g)} \quad \Delta H^\circ = -185 \text{ kJ}$

(2) $H_{2(g)} + \frac{1}{2}O_{2(g)} \rightarrow H_2O_{(\ell)} \quad \Delta H^\circ = -285.8 \text{ kJ}$

(3) $H_2O_{(g)} \rightarrow H_2O_{(\ell)} \quad \Delta H^\circ = -40.7 \text{ kJ}$

16. Calculate the enthalpy change of the following reaction between nitrogen gas and oxygen gas, given thermochemical equations (1), (2), and (3).

$2N_{2(g)} + 5O_{2(g)} \rightarrow 2N_2O_{5(g)}$

(1) $2H_{2(g)} + O_{2(g)} \rightarrow 2H_2O_{(\ell)} \quad \Delta H^\circ = -572 \text{ kJ}$

(2) $N_2O_{5(g)} + H_2O_{(\ell)} \rightarrow 2HNO_{3(\ell)} \quad \Delta H^\circ = -77 \text{ kJ}$

(3) $\frac{1}{2}N_{2(g)} + \frac{3}{2}O_{2(g)} + \frac{1}{2}H_{2(g)} \rightarrow HNO_{3(\ell)} \quad \Delta H^\circ = -174 \text{ kJ}$

Sometimes it is impractical to use a coffee-cup calorimeter to find the enthalpy change of a reaction. You can, however, use the calorimeter to find the enthalpy changes of other reactions, which you can combine to arrive at the desired reaction. In the following investigation, you will determine the enthalpy changes of two reactions. Then you will apply Hess's law to determine the enthalpy change of a third reaction.

Investigation 5-B

SKILL FOCUS
- Performing and recording
- Analyzing and interpreting
- Communicating results

Hess's Law and the Enthalpy of Combustion of Magnesium

Magnesium ribbon burns in air in a highly exothermic combustion reaction. (See equation (1).) A very bright flame accompanies the production of magnesium oxide, as shown in the photograph below. It is impractical and dangerous to use a coffee-cup calorimeter to determine the enthalpy change for this reaction.

(1) $Mg_{(s)} + \frac{1}{2}O_{2(g)} \rightarrow MgO_{(s)}$

Instead, you will determine the enthalpy changes for two other reactions (equations (2) and (3) below). You will use these enthalpy changes, along with the known enthalpy change for another reaction (equation (4) below), to determine the enthalpy change for the combustion of magnesium.

(2) $MgO_{(s)} + 2HCl_{(aq)} \rightarrow MgCl_{2(aq)} + H_2O_{(\ell)}$

(3) $Mg_{(s)} + 2HCl_{(aq)} \rightarrow MgCl_{2(aq)} + H_{2(g)}$

(4) $H_{2(g)} + \frac{1}{2}O_{2(g)} \rightarrow H_2O_{(\ell)}\ \Delta H^\circ = -285.8$ kJ/mol

Notice that equations (2) and (3) occur in aqueous solution. You can use a coffee-cup calorimeter to determine the enthalpy changes for these reactions. Equation (4) represents the formation of water directly from its elements in their standard state.

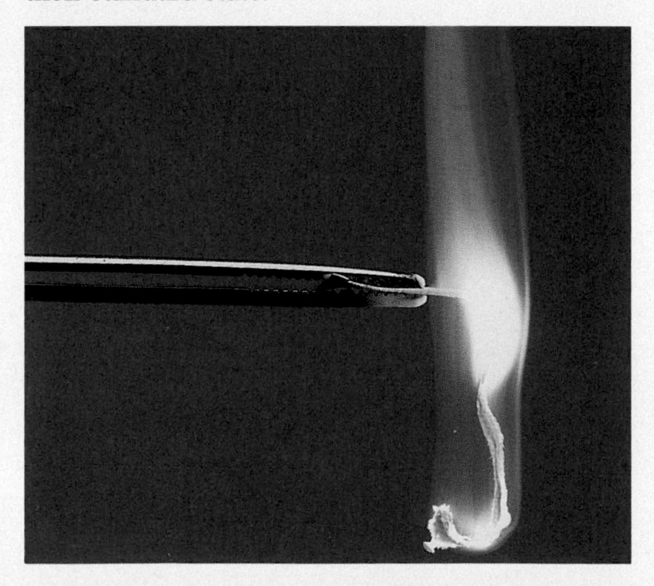

Question

How can you use equations (2), (3), and (4) to determine the enthalpy change of equation (1)?

Prediction

Predict whether reactions (2) and (3) will be exothermic or endothermic.

Materials

coffee cup calorimeter (2 nested coffee cups sitting in a 250 mL beaker)
thermometer
100 mL graduated cylinder
scoopula
electronic balance
MgO powder
Mg ribbon (or Mg turnings)
sandpaper or emery paper
1.00 mol/L $HCl_{(aq)}$

Safety Precautions

- Hydrochloric acid is corrosive. Use care when handling it.
- Be careful not to inhale the magnesium oxide powder.

Procedure

Part 1 Determining ΔH of Equation (2)

1. Read the Procedure for Part 1. Prepare a fully-labelled set of axes to graph your temperature observations.
2. Set up the coffee-cup calorimeter. (Refer to Investigation 5-A) Using a graduated cylinder, add 100 mL of 1.00 mol/L $HCl_{(aq)}$ to the calorimeter. CAUTION $HCl_{(aq)}$ can burn your skin.

3. Record the initial temperature, T_i, of the $HCl_{(aq)}$, to the nearest tenth of a degree.
4. Find the mass of no more than 0.80 g of MgO. Record the exact mass.
5. Add the MgO powder to the calorimeter containing the $HCl_{(aq)}$. Swirl the solution gently, recording the temperature every 30 s until the highest temperature, T_f, is reached.
6. Dispose of the reaction solution as directed by your teacher.

Part 2 Determining ΔH of Equation (3)

1. Read the Procedure for Part 2. Prepare a fully-labelled set of axes to graph your temperature observations.
2. Using a graduated cylinder, add 100 mL of 1.00 mol/L $HCl_{(aq)}$ to the calorimeter.
3. Record the initial temperature, T_i, of the $HCl_{(aq)}$, to the nearest tenth of a degree.
4. If you are using magnesium ribbon (as opposed to turnings), sand the ribbon. Accurately determine the mass of no more than 0.50 g of magnesium. Record the exact mass.
5. Add the Mg to the calorimeter containing the $HCl_{(aq)}$. Swirl the solution gently, recording the temperature every 30 s until the highest temperature, T_f, is reached.
6. Dispose of the solution as directed by your teacher.

Analysis

1. Use the equation $Q = m \cdot c \cdot \Delta T$ to determine the amount of heat that is released or absorbed by reactions (2) and (3). List any assumptions you make.
2. Convert the mass of MgO and Mg to moles. Calculate ΔH of each reaction in units of kJ/mol of MgO or Mg. Remember to put the proper sign (+ or –) in front of each ΔH value.
3. Algebraically combine equations (2), (3), and (4), and their corresponding ΔH values, to get equation (1) and ΔH of the combustion of magnesium.
4. **(a)** Your teacher will tell you the accepted value of ΔH of the combustion of magnesium. Based on the accepted value, calculate your percent error.
   (b) Suggest some sources of error in the investigation. In what ways could you improve the procedure?
5. What assumption did you make about the amount of heat that was lost to the calorimeter? Do you think that this is a fair assumption? Explain.
6. Why was it fair to assume that the hydrochloric acid solution has the same density and specific heat capacity as water?

Conclusion

7. Explain how you used Hess's law of heat summation to determine ΔH of the combustion of magnesium. State the result you obtained for the thermochemical equation that corresponds to chemical equation (1).

Extension

8. Design an investigation to verify Hess's law, using the following equations.

   (1) $NaOH_{(s)} \rightarrow Na^{+}_{(aq)} + OH^{-}_{(aq)}$

   (2) $NaOH_{(s)} + H^{+}_{(aq)} + Cl^{-}_{(aq)} \rightarrow Na^{+}_{(aq)} + Cl^{-}_{(aq)} + H_2O_{(\ell)}$

   (3) $Na^{+}_{(aq)} + OH^{-}_{(aq)} + H^{+}_{(aq)} + Cl^{-}_{(aq)} \rightarrow Na^{+}_{(aq)} + Cl^{-}_{(aq)} + H_2O_{(\ell)}$

   Assume that you have a coffee-cup calorimeter, solid NaOH, 1.00 mol/L $HCl_{(aq)}$, 1.00 mol/L $NaOH_{(aq)}$, and standard laboratory equipment. Write a step-by-step procedure for the investigation. Then outline a plan for analyzing your data. Be sure to include appropriate safety precautions. If time permits, obtain your teacher's approval and carry out the investigation.

Using Standard Molar Enthalpies of Formation

You have learned how to add equations with known enthalpy changes to obtain the enthalpy change for another equation. This method can be time-consuming and difficult, however, because you need to find reactions with known enthalpy changes that will add to give your target equation. There is another way to use Hess's law to find the enthalpy of an equation.

Formation Reactions

In Investigation 5-B, you used the reaction of oxygen with hydrogen to form water. Reactions like this one are known as **formation reactions**. In a formation reaction, a substance is formed from elements in their standard states. The enthalpy change of a formation reaction is called the **standard molar enthalpy of formation, ΔH°_f**. *The standard molar enthalpy of formation is the quantity of energy that is absorbed or released when one mole of a compound is formed directly from its elements in their standard states.*

Some standard molar enthalpies of formation are listed in Table 5.3. Notice that the standard enthalpies of formation of most compounds are negative. Thus, most compounds are more stable than the elements they are made from.

Table 5.3 Selected Standard Molar Enthalpies of Formation

Compound	ΔH°_f	Formation equations
$CO_{(g)}$	−110.5	$C_{(s)} + \frac{1}{2}O_{2(g)} \rightarrow CO_{(g)}$
$CO_{2(g)}$	−393.5	$C_{(s)} + O_{2(g)} \rightarrow CO_{2(g)}$
$CH_{4(g)}$	−74.6	$C_{(s)} + 2H_{2(g)} \rightarrow CH_{4(g)}$
$CH_3OH_{(\ell)}$	−238.6	$C_{(s)} + 2H_{2(g)} + \frac{1}{2}O_{2(g)} \rightarrow CH_3OH_{(\ell)}$
$C_2H_5OH_{(\ell)}$	−277.6	$2C_{(s)} + 3H_{2(g)} + \frac{1}{2}O_{2(g)} \rightarrow C_2H_5OH_{(\ell)}$
$C_6H_{6(\ell)}$	+49.0	$6C_{(s)} + 3H_{2(g)} \rightarrow C_6H_{6(\ell)}$
$C_6H_{12}O_{6(s)}$	−1274.5	$6C_{(s)} + 6H_{2(g)} + 3O_{2(g)} \rightarrow C_6H_{12}O_{6(s)}$
$H_2O_{(\ell)}$	−285.8	$H_{2(s)} + \frac{1}{2}O_{2(g)} \rightarrow H_2O_{(\ell)}$
$H_2O_{(g)}$	−241.8	$H_{2(s)} + \frac{1}{2}O_{2(g)} \rightarrow H_2O_{(g)}$
$CaCl_{2(s)}$	−795.4	$Ca_{(s)} + Cl_{2(g)} \rightarrow CaCl_{2(s)}$
$CaCO_{3(s)}$	−1206.9	$Ca_{(s)} + C_{(s)} + \frac{3}{2}O_{2(g)} \rightarrow CaCO_{3(s)}$
$NaCl_{(s)}$	−411.1	$Na_{(s)} + \frac{1}{2}Cl_{2(g)} \rightarrow NaCl_{(g)}$
$HCl_{(g)}$	−92.3	$\frac{1}{2}H_{2(s)} + \frac{1}{2}Cl_{2(g)} \rightarrow HCl_{(g)}$
$HCl_{(aq)}$	−167.5	$\frac{1}{2}H_{2(s)} + \frac{1}{2}Cl_{2(g)} \rightarrow HCl_{(ag)}$

By definition, the enthalpy of formation of an element in its standard state is zero. The standard state of an element is usually its most stable form under standard conditions. Recall, from section 5.1, that standard conditions are 25°C and 100 kPa (close to room temperature and pressure). Therefore, the standard state of nitrogen is $N_{2(g)}$. The standard state of magnesium is $Mg_{(s)}$.

Some elements exist in more than one form under standard conditions. For example, carbon can exist as either graphite or diamond, as shown in Figure 5.16. Graphite is defined as the standard state of carbon. Therefore, the standard enthalpy of formation of graphite carbon is 0 kJ/mol. The standard enthalpy of formation of diamond is 1.9 kJ/mol. Another example is oxygen, $O_{2(g)}$. Oxygen also exists in the form of ozone,

$O_{3(g)}$, under standard conditions. The diatomic molecule is defined as the standard state of oxygen, however, because it is far more stable than ozone. Therefore, the standard enthalpy of formation of oxygen gas, $O_{2(g)}$, is 0 kJ/mol. The standard enthalpy of formation of ozone is 143 kJ/mol.

Figure 5.16 Carbon can exist as graphite or diamond under standard conditions. It can, however, have only one standard state. Carbon's standard state is graphite.

When writing a formation equation, always write the elements in their standard states. For example, examine the equation for the formation of water directly from its elements under standard conditions.

$$H_{2(g)} + \frac{1}{2}O_{2(g)} \rightarrow H_2O_{(\ell)} \quad \Delta H^\circ_f = -285.8 \text{ kJ}$$

A formation equation should show the formation of exactly one mole of the compound of interest. The following equation shows the formation of benzene, C_6H_6 under standard conditions.

$$6C_{(graphite)} + 3H_{2(g)} \rightarrow C_6H_{6(\ell)} \quad \Delta H^\circ_f = 49.1 \text{ kJ}$$

Practice Problems

17. Write a thermochemical equation for the formation of each substance. Be sure to include the physical state of all the elements and compounds in the equation. You can find the standard enthalpy of formation of each substance in Appendix E.

(a) CH_4 **(b)** NaCl **(c)** MgO **(d)** $CaCO_3$

18. Liquid sulfuric acid has a very large negative standard enthalpy of formation (−814.0 kJ/mol). Write an equation to show the formation of liquid sulfuric acid. The standard state of sulfur is rhombic sulfur ($S_{(s)}$).

19. Write a thermochemical equation for the formation of gaseous cesium. The standard enthalpy of formation of $Cs_{(g)}$ is 76.7 kJ/mol.

20. Solid phosphorus is found in two forms: white phosphorus (P_4) and red phosphorus (P). White phosphorus is the standard state.

(a) The enthalpy of formation of red phosphorus is −17.6 kJ/mol. Write a thermochemical equation for the formation of red phosphorus.

(b) 32.6 g of white phosphorus reacts to form red phosphorus. What is the enthalpy change?

Calculating Enthalpy Changes

You can calculate the enthalpy change of a chemical reaction by adding the heats of formation of the products and subtracting the heats of formation of the reactants. The following equation can be used to determine the enthalpy change of a chemical reaction.

$$\Delta H^\circ = \Sigma(n\Delta H^\circ_f \text{ products}) - \Sigma(n\Delta H^\circ_f \text{ reactants})$$

In this equation, n represents the molar coefficient of each compound in the balanced chemical equation and Σ means "the sum of."

As usual, you need to begin with a balanced chemical equation. If a given reactant or product has a molar coefficient that is not 1, you need to multiply its ΔH°_f by the same molar coefficient. This makes sense because the units of ΔH°_f are kJ/mol. Consider, for example, the complete combustion of methane, $CH_{4(g)}$.

$$CH_{4(g)} + 2O_{2(g)} \rightarrow CO_{2(g)} + 2H_2O_{(g)}$$

Using the equation for the enthalpy change, and the standard enthalpies of formation in Appendix E, you can calculate the enthalpy change of this reaction.

$$\Delta H^\circ = [(\Delta H^\circ_f \text{ of } CO_{2(g)}) + 2(\Delta H^\circ_f \text{ of } H_2O_{(g)})] - [1(\Delta H^\circ_f \text{ of } CH_{4(g)}) + 2(\Delta H^\circ_f \text{ of } O_{2(g)})]$$

Substitute the standard enthalpies of formation from Appendix E to get the following calculation.

$$\begin{aligned}\Delta H^\circ &= [(-393.5 \text{ kJ/mol}) + 2(-241.8 \text{ kJ/mol})] - [(-74.8 \text{ kJ/mol}) + 2(0 \text{ kJ/mol})]\\ &= -802.3 \text{ kJ/mol of } CH_4\end{aligned}$$

How does this method of adding heats of formation relate to Hess's law? Consider the equations for the formation of each compound that is involved in the reaction of methane with oxygen.

$$\begin{aligned}&\textbf{(1)}\quad H_{2(g)} + \tfrac{1}{2}O_{2(g)} \rightarrow H_2O_{(g)} \quad \Delta H^\circ_f = -241.8 \text{ kJ}\\ &\textbf{(2)}\quad C_{(s)} + O_{2(g)} \rightarrow CO_{2(g)} \quad \Delta H^\circ_f = -393.5 \text{ kJ}\\ &\textbf{(3)}\quad C_{(s)} + 2H_{2(g)} \rightarrow CH_{4(g)} \quad \Delta H^\circ_f = -74.6 \text{ kJ}\end{aligned}$$

CONCEPT CHECK

"Using enthalpies of formation is like a shortcut for adding equations to obtain ΔH°." Do you agree with this statement? Explain your answer.

There is no equation for the formation of oxygen, because oxygen is an element in its standard state.

By adding the formation equations, you can obtain the target equation. Notice that you need to reverse equation (3) and multiply equation (1) by 2.

$$\begin{aligned}&\textbf{2 × (1)}\quad 2H_{2(g)} + O_{2(g)} \rightarrow 2H_2O_{(g)} \quad \Delta H^\circ_f = 2(-241.8) \text{ kJ}\\ &\textbf{(2)}\quad C_{(s)} + O_{2(g)} \rightarrow CO_{2(g)} \quad \Delta H^\circ_f = -393.5 \text{ kJ}\\ &\textbf{−1 × (3)}\quad CH_{4(g)} \rightarrow C_{(s)} + 2H_{2(g)} \quad \Delta H^\circ_f = -1(-74.6) \text{ kJ}\end{aligned}$$

$$CH_{4(g)} + 2O_{2(g)} + \cancel{C}_{(s)} + 2\cancel{H}_{2(g)} \rightarrow 2H_2O_{(g)} + CO_{2(g)} + \cancel{C}_{(s)} + 2\cancel{H}_{2(g)}$$

or

$$CH_{4(g)} + 2O_{2(g)} \rightarrow 2H_2O_{(g)} + CO_{2(g)}$$

Add the manipulated ΔH°_f values: $\Delta H^\circ = 2(-241.8) \text{ kJ} - 393.5 \text{ kJ} + 74.6 \text{ kJ}$.
$= -802.3 \text{ kJ}$

This value of ΔH° is the same as the value you obtained using ΔH°_f data. When you used the addition method, you performed the same operations on the enthalpies of formation before adding them. Therefore, using enthalpies of formation to determine the enthalpy of a reaction is consistent with Hess's law. Figure 5.17 shows the general process for determining the enthalpy of a reaction from enthalpies of formation.

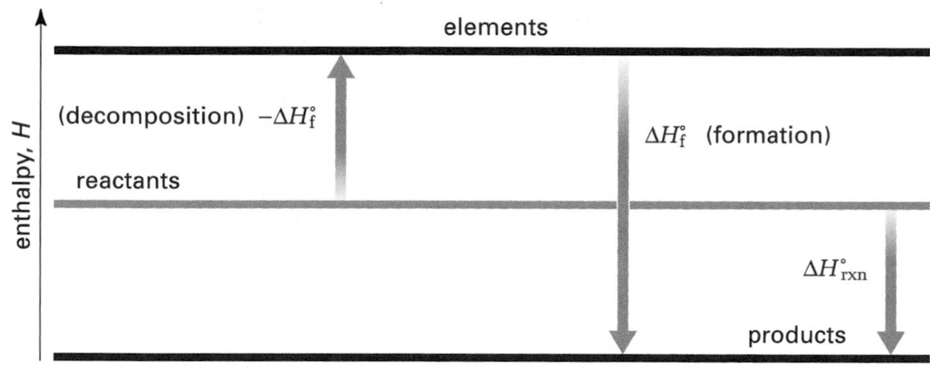

Figure 5.17 **The overall enthalpy change of any reaction is the sum of the enthalpy change of the decomposition of the reactants to their elements and the enthalpy change of the formation of the products from their elements.**

It is important to realize that, in most reactions, *the reactants do not actually break down into their elements and then react to form products.* Since there is extensive data about enthalpies of formation, however, it is useful to calculate the overall enthalpy change this way. Moreover, according to Hess's law, the enthalpy change is the same, regardless of the pathway. (In Chapter 6, you will learn more about the mechanisms by which compounds and elements react to form different elements and compounds.) Examine the following Sample Problem to see how to use enthalpies of formation to determine the enthalpy change of a reaction. Then try the Practice Problems that follow.

Sample Problem

Using Enthalpies of Formation

Problem

Iron(III) oxide reacts with carbon monoxide to produce elemental iron and carbon dioxide. Determine the enthalpy change of this reaction, using known enthalpies of formation.

$Fe_2O_{3(s)} + 3CO_{(g)} \rightarrow 3CO_{2(g)} + 2Fe_{(s)}$

What Is Required?

You need to find ΔH° of the given chemical equation, using ΔH°_f data.

What Is Given?

From Appendix E, you can obtain the enthalpies of formation.

ΔH°_f of $Fe_2O_{3(s)} = -824.2$ kJ/mol

ΔH°_f of $CO_{(g)} = -110.5$ kJ/mol

ΔH°_f of $CO_{2(g)} = -393.5$ kJ/mol

ΔH°_f of $Fe_{(s)} = 0$ kJ/mol (by definition)

Plan Your Strategy

Multiply each ΔH°_f value by its molar coefficient from the balanced chemical equation. Substitute into the following equation, and then solve.

$\Delta H^\circ = \Sigma(n\Delta H^\circ_f \text{ products}) - \Sigma(n\Delta H^\circ_f \text{ reactants})$

Continued ...

CONCEPT CHECK

You saw the reaction between iron(III) oxide and carbon monoxide in the Sample Problem on page 245. Which method for determining the enthalpy of reaction do you prefer? Explain your answer.

Continued ...

Act on Your Strategy

$$\Delta H^\circ = \Sigma(n\Delta H^\circ_f \text{ products}) - \Sigma(n\Delta H^\circ_f \text{ reactants})$$
$$= [3(\Delta H^\circ_f CO_{2(g)}) + 2(\Delta H^\circ_f Fe_{(s)})] - [(\Delta H^\circ_f Fe_2O_{3(s)}) + 3(\Delta H^\circ_f CO_{(g)})]$$
$$= [(-393.5 \text{ kJ/mol}) + 2(0 \text{ kJ/mol})] - [(-824.2 \text{ kJ/mol}) + 3(-110.5 \text{ kJ/mol})]$$
$$= -24.8 \text{ kJ/mol}$$

$$\therefore Fe_2O_{3(s)} + 3CO_{(g)} \rightarrow 3CO_{2(g)} + 2Fe_{(s)} \quad \Delta H^\circ = -24.8 \text{ kJ/mol}$$

Check Your Solution

A balanced chemical equation was used in the calculation. The number of significant digits is correct. The units are also correct.

Practice Problems

21. Hydrogen can be added to ethene, C_2H_4, to obtain ethane, C_2H_6.

$C_2H_{4(g)} + H_{2(g)} \rightarrow C_2H_{6(g)}$

Show that the equations for the formation of ethene and ethane from their elements can be algebraically combined to obtain the equation for the addition of hydrogen to ethene.

22. Zinc sulfide reacts with oxygen gas to produce zinc oxide and sulfur dioxide.

$2ZnS_{(s)} + 3O_{2(g)} \rightarrow 2ZnO_{(s)} + 2SO_{2(g)}$

Write the chemical equation for the formation of the indicated number of moles of each compound from its elements. Algebraically combine these equations to obtain the given equation.

23. Small amounts of oxygen gas can be produced in a laboratory by heating potassium chlorate, $KClO_3$.

$2KClO_{3(s)} \rightarrow 2KCl_{(s)} + 3O_{2(g)}$

Calculate the enthalpy change of this reaction, using enthalpies of formation from Appendix E.

24. Use the following equation to answer the questions below.

$CH_3OH_{(\ell)} + 1.5O_{2(g)} \rightarrow CO_{2(g)} + 2H_2O_{(g)}$

(a) Calculate the enthalpy change of the complete combustion of one mole of methanol, using enthalpies of formation.

(b) How much energy is released when 125 g of methanol undergoes complete combustion?

Section Summary

In this section, you learned how to calculate the enthalpy change of a chemical reaction using Hess's law of heat summation. Enthalpies of reaction can be calculated by combining chemical equations algebraically or by using enthalpies of formation. Hess's law allows chemists to determine enthalpies of reaction without having to take calorimetric measurements. In the next section, you will see how the use of energy affects your lifestyle and your environment.

Section Review

1. K/U Explain why you need to reverse the sign of ΔH° when you reverse an equation. Use an example in your answer.

2. C In section 5.3, you learned two methods for calculating enthalpy changes using Hess's law. If you had only this textbook as a reference, which method would allow you to calculate enthalpy changes for the largest number of reactions? Explain your answer.

3. I In the early 1960s, Neil Bartlett, at the University of British Columbia, was the first person to synthesize compounds of the noble gas xenon. A number of noble gas compounds (such as XeF_2, XeF_4, XeF_6, and XeO_3) have since been synthesized. Consider the reaction of xenon difluoride with fluorine gas to produce xenon tetrafluoride.

   $XeF_{2(g)} + F_{2(g)} \rightarrow XeF_{4(s)}$

   Use the following standard molar enthalpies of formation to calculate the enthalpy change for this reaction.

Compound	ΔH°_f (kJ/mol)
$XeF_{2(g)}$	−108
$XeF_{4(s)}$	−251

4. I Calculate the enthalpy change of the following reaction, given equations (1), (2), and (3).

   $2H_3BO_{3(aq)} \rightarrow B_2O_{3(s)} + 3H_2O_{(\ell)}$

   (1) $H_3BO_{3(aq)} \rightarrow HBO_{2(aq)} + H_2O_{(\ell)}$ $\Delta H^{\circ} = -0.02$ kJ

   (2) $H_2B_4O_{7(s)} + H_2O_{(\ell)} \rightarrow 4HBO_{2(aq)}$ $\Delta H^{\circ} = -11.3$ kJ

   (3) $H_2B_4O_{7(s)} \rightarrow 2B_2O_{3(s)} + H_2O_{(\ell)}$ $\Delta H^{\circ} = 17.5$ kJ

5. I The standard molar enthalpy of formation of calcium carbonate is −1207.6 kJ/mol. Calculate the enthalpy of formation of calcium oxide, given the following equation.

   $CaO_{(g)} + CO_{2(g)} \rightarrow CaCO_{3(s)}$ $\Delta H^{\circ} = -178.1$ kJ

6. C A classmate is having difficulty understanding Hess's law. Write a few paragraphs to explain the law. Include examples, diagrams, and an original analogy.

7. C In your own words, explain why using enthalpies of formation to determine enthalpy of reaction depends on Hess's law. Include an example.

5.4 Energy Sources

Section Preview/ Specific Expectations

In this section, you will

- **compare** the efficiency and environmental impact of conventional and alternative sources of energy
- **communicate** your understanding of the following terms: *non-renewable*, *renewable*

Canadians depend on energy sources, such as those listed in Figure 5.18, to power vehicles, light and heat buildings, and manufacture products that support our lives and lifestyles. As society's needs for energy and energy-using products grow, scientists and technologists search for more economical and environmentally responsible ways to meet these needs. In this section, you will compare energy sources based on their efficiency and environmental impact.

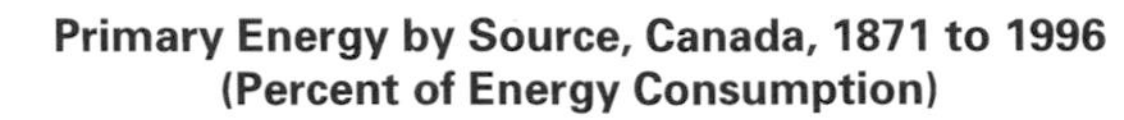

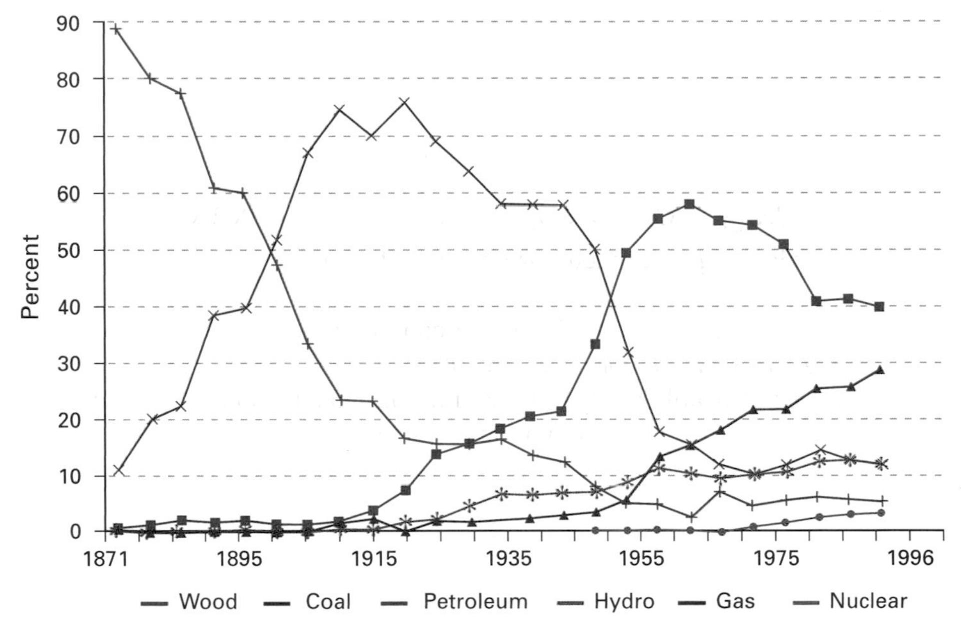

Figure 5.18 The energy that Canadians use comes from a variety of sources. What factors account for the changes you can see in this graph? How do you think energy use has changed since 1996?

Energy and Efficiency

When you think about energy efficiency, what comes to mind? You may think about taking the stairs instead of the elevator, choosing to drive a small car instead of a sport utility vehicle, or turning off lights when you are not using them. What, however, does efficiency really mean? How do you quantify it?

There are several ways to define efficiency. One general definition says that energy efficiency is the ability to produce a desired effect with minimum energy expenditure. For example, suppose that you want to bake a potato. You can use a microwave oven or a conventional oven. Both options achieve the same effect (baking the potato), but the first option uses less energy. According to the general definition above, using the microwave oven is more energy-efficient than using the conventional oven. The general definition is useful, but it is not quantitative.

Another definition of efficiency suggests that it is *the ratio of useful energy produced to energy used in its production, expressed as a percent.* This definition quantitatively compares input and output of energy. When you use it, however, you need to be clear about what you mean by "energy used." Figure 5.19 shows factors to consider when calculating efficiency or analyzing efficiency data.

"Useful energy" is
- energy delivered to consumer in usable form
- actual work done

"Energy used" could include
- ideal energy content of fuel
- energy used to extract and transport fuel
- solar energy used to create fuel (e.g. biomass)
- energy used to build and maintain power plant

$$\text{Efficiency} = \frac{\text{Useful energy produced}}{\text{Energy used}} \times 100\%$$

Figure 5.19 Efficiency is expressed as a percent. Always specify what is included in the "energy used" part of the ratio.

It is often difficult to determine how much energy is used to produce useful energy. Often an efficiency percent only takes into account the "ideal" energy output of a system, based on the energy content of the fuel.

Efficiency and Natural Gas

When discussing the efficiency of a fuel such as natural gas, you need to specify how that fuel is being used. Consider, for example, natural gas. Natural gas is primarily methane. Therefore, you can estimate an ideal value for energy production using the enthalpy of combustion of methane.

$$CH_{4(g)} + 2O_{2(g)} \rightarrow CO_{2(g)} + 2H_2O_{(g)} \quad \Delta H^\circ = -802 \text{ kJ}$$

In other words, 16 g of methane produces 802 kJ of heat (under constant pressure conditions).

When natural gas is used directly in cooking devices, its efficiency can be as high as 90%. Thus, for every 16 g of gas burned, you get about 720 kJ (0.90×802 kJ) of usable energy as heat for cooking. This is a much higher fuel efficiency than you can get with appliances that use electrical energy produced in a power plant that runs on a fuel such as coal.

If natural gas is used to produce electricity in a power plant, however, the efficiency is much lower—around 37%. Why? The heat from the burning natural gas is used to boil water. The kinetic energy of the resulting steam is transformed to mechanical energy for turning a turbine. The turbine generates the electrical energy. Each of these steps has an associated efficiency that is less than 100%. Thus, at each step, the overall efficiency of the fuel decreases.

Thinking About the Environment

Efficiency is not the only criterion for selecting an energy source. Since the 1970s, society has become increasingly conscious of the impact of energy technologies on the environment.

Suppose that you want to analyze the environmental impact of an energy source. You can ask the following questions:

- Are any waste products or by-products of the energy production process harmful to the environment? For example, any process in which a hydrocarbon is burned produces carbon dioxide. Carbon dioxide is a known greenhouse gas, which contributes to global warming. Any combustion process provides the heat required to form oxides of nitrogen from nitrogen gas. Nitrogen oxides contribute to acid precipitation.
- Is obtaining or harnessing the fuel harmful to the environment? For example, oil wells and strip coal mines destroy habitat. Natural gas pipelines, shown in Figure 5.20, are visually unappealing. They also split up habitat, which harms the ecosystem.

Figure 5.20 This gas pipeline harms the ecosystem by splitting up habitat.

- Will using the energy source permanently remove the fuel from the environment? A **non-renewable** energy source (such as coal, oil, or natural gas) is effectively gone once we have used it up. Non-renewable energy sources take millions of years to form. We use them up at a much faster rate than they can be replenished. An energy source that is clearly **renewable** is solar energy. The Sun will continue to radiate energy toward Earth over its lifetime—many millions of years. A somewhat renewable energy source is wood. Trees can be grown to replace those cut down. It takes trees a long time to grow, however, and habitat is often destroyed in the meantime.

Comparing Energy Sources

Both efficiency and environmental impact are important factors to consider when comparing energy sources. In the following ThoughtLab, you will research and compare alternative and conventional energy sources.

ThoughtLab

In this ThoughtLab, you will work as a class to compare two different energy sources.

Procedure

1. On your own, or with a group, choose an energy source from the following list. Other energy sources may be discussed and added in class.

solar (radiant) energy	wood
petroleum	biomass
hydrogen fuel cell	nuclear fission
natural gas fuel cell	natural gas
wind energy	coal
hydroelectric power	tar sands
geothermal energy	

2. Before beginning your research, record your current ideas about the efficiency and environmental impact of your chosen energy source.
3. Research the efficiency and environmental impact of your energy source. If possible, determine what the efficiency data means. For example, suppose that a source tells you that natural gas is 90% efficient. Is the source referring to natural gas burned directly for heat or for cooking? Is the energy being converted from heat to electricity in a power plant? Be as specific as possible.
4. Ensure that you use a variety of sources to find your data. Be aware of any bias that might be present in your sources.
5. Trace the energy source as far back as you can. For example, you can trace the energy in fossil fuels back to solar energy that powered the photosynthesis in the plants that eventually became the fossil fuel. Write a brief outline of your findings.
6. Your teacher will pair you (or your group) with another student (or group) that has researched a different energy source. Work together to analyze the comparative merits and drawbacks of the two energy sources, based on your research.
7. Write a conclusion that summarizes the benefits and risks of both energy sources, in terms of their efficiency and environmental impact.
8. Present your findings to the class.

Analysis

1. Discuss the presentations as a class.
   (a) Decide which energy sources are most efficient. Also decide which energy sources are least damaging to the environment.
   (b) Decide which energy source is best overall in terms of both efficiency and environmental impact.
2. Could the "best overall" energy source be used to provide a significant portion of Canada's energy needs? What obstacles would need to be overcome for this to happen?
3. Besides efficiency and environmental impact, what other factors are involved in developing and delivering an energy source to consumers?

Careers in Chemistry

Nuclear Safety Supervisor

Jennifer Noronha

In some ways, nuclear power is an appealing power source. Nuclear reactions create large amounts of energy from minimal material, and they generate none of the carbon dioxide and other emissions that cause acid rain and global warming. The products and reactants of nuclear reactions, however, are dangerously radioactive. Therefore, special measures are needed to protect nuclear power station employees from daily exposure to radiation. That is where Jennifer Noronha comes in. Noronha is the supervisor of Radiological Services at Darlington. Employee safety—especially from high radiation doses—is her first priority.

The Darlington Nuclear Generating Station is located 70 km east of Toronto. It uses a fuel of natural uranium to produce enough electricity to provide power for a city the size of Toronto. Noronha and her radiation protection team plan and implement safety programs that minimize dose rates, or the amount of radiation that station employees are exposed to.

Station employees must undergo four weeks of radiation protection training. This training was designed by Noronha's department, based on an extensive investigation of radiation fields within the station, as well as a thorough evaluation of past safety programs and approaches. Through this training, employees learn how to measure existing dose rates with survey equipment, assess what kinds of tools and protective clothing are needed, and take appropriate action to lower radiation doses. For example:

- Airborne hazards, such as tritium (present in radiated water vapour), can be reduced by running the station's dryer system. The dryer system catches the radiated vapour and dries it out of the air.
- Non-airborne radiation can be countered by shielding the affected area with lead blankets or sheeting material.

Noronha's strong mathematics skills were evident from an early age. When she moved to Canada from Kenya at age 11, she was immediately put ahead a grade. Her mathematics skills and her father's engineering profession were what propelled her toward engineering. Noronha earned her engineering physics degree from McMaster University. Her courses included general chemistry, biomedical theory, and nuclear theory. She worked as a commissioning engineer at Darlington during its start-up. She tested the station's safety shut-down systems and helped to bring the station's first reactor on-line. "It was pretty amazing," Noronha says. "At the time, it was still relatively new technology, and it was Canadian technology."

Noronha got her MBA from the University of Toronto in 1998. Soon after, she moved to her current position, which allows her to combine her people skills and technical expertise.

Making Career Connections

1. Are you interested in the different safety concerns related to Canadian nuclear reactors, and the steps that are being taken to counter these concerns? Contact the Canadian Nuclear Safety Commission (CNSC) or explore their web site. (The CNSC is the Ottawa-based government watchdog for the use of nuclear energy in Canada.)
2. To learn more about the wide variety of careers in nuclear power generation, Ontario Power Generation is a good place to start. Their web site has a helpful career page that lists opportunities for students and recent graduates, as well as experienced professionals.

Chemistry Bulletin

Science | Technology | Society | Environment

Hot Ice

When engineers first began extending natural gas pipelines through regions of bitter cold, they noticed that their lines plugged with a dangerous slush of ice and gas. The intense pressure of the lines, combined with the cold, led to the formation of *methane hydrates*, a kind of gas-permeated ice. More than a mere nuisance, methane hydrate plugs were a potential threat to pipelines. The build-up of gas pressure behind a methane hydrate plug could lead to an explosion. Now, however, this same substance may hold the key to a vast fuel supply.

Methane hydrates form when methane molecules become trapped within an ice lattice as water freezes. They can form in very cold conditions or under high-pressure conditions. Both of these conditions are met in deep oceans and in permafrost. In Canada, hydrates have already been found in large quantities in the Canadian Arctic. Methane hydrate has a number of remarkable properties. For example, when brought into an oxygen atmosphere, the methane fumes can be ignited, making it appear that the ice is burning!

Methane releases 25% less carbon dioxide per gram than coal, and it emits none of the oxides of nitrogen and sulfur that contribute to acid precipitation. Therefore, using methane in place of other fossil fuels is very desirable. Methane hydrates seem to be an ideal and plentiful "pre-packaged" source of natural gas. Estimates of the exact amount of methane stored in hydrates suggest there could be enough to serve our energy needs anywhere from 350 years to 3500 years, based on current levels of energy consumption. This would constitute a significant source of fossil fuels, if we can find a way to extract the gas safely and economically.

Unfortunately, hydrates become unstable when the pressure or temperature changes. Even small changes in these conditions can cause hydrates to degrade rapidly. Methane hydrates are stable at ocean depths greater than 300 m, but offshore drilling at these depths has been known to disturb the hydrate formations, causing large, uncontrolled releases of flammable methane gas. Also, methane hydrates often hold sediment layers together. Therefore, in addition to the danger of a gas explosion, there is the danger of the sea floor collapsing where drilling occurs.

Methane is a significant greenhouse gas. A massive release of methane could cause catastrophic global climate change. Some researchers believe that the drastic climate change that occurred during the Pleistocene era was due to methane hydrate destabilization and widespread methane release.

Nonetheless, Canada, Japan, the United States, and Russia all have active research and exploration programs in this area. As global oil supplies dwindle, using methane hydrates might increasingly be seen as worth the risk and cost.

Making Connections

1. Compare using methane from natural gas with using methane from methane hydrates in terms of environmental impact and efficiency. You will need to do some research to find out extraction methods for each source of methane.
2. On the Internet, research one possible structure of methane hydrate. Create a physical model or a three-dimensional computer model to represent it. Use your model to explain why methane hydrates are unstable at temperatures that are warmer than 0°C.

Emerging Energy Sources

In the ThoughtLab on page 258, you probably noticed that all energy sources have drawbacks as well as benefits. Scientists and engineers are striving to find and develop new and better energy sources. One energy source that engineers are trying to harness is nuclear fusion. As you learned in section 5.1, nuclear fusion provides a great deal of energy from readily available fuel (isotopes of hydrogen). In addition, nuclear fusion produces a more benign waste product than nuclear fission (helium). Unfortunately, fusion is not yet practical and controllable on a large scale because of the enormous temperatures involved.

Chemists are also striving to find new sources for existing fuels that work well. The Chemistry Bulletin on the facing page discusses a new potential source of methane.

Section Summary

In this section, you learned about efficiency. You learned how it can be defined in different ways for different purposes. You used your understanding of processes that produce energy to investigate the efficiency and environmental impact of different energy sources.

In Chapter 5, you learned about the energy that is associated with chemical reactions. You used a calorimeter to measure heat changes, and you used these heat changes to write thermochemical equations. You probably already realize that adding heat to reactants often speeds up a reaction. In other words, raising the temperature of a system consisting of a chemical reaction often increases the speed of the reaction. A familiar example is cooking. You increase the temperature of a heating element to speed up the reactions that are taking place in the food as it cooks. How does increasing temperature speed up a reaction? Is the enthalpy of a reaction related to its speed? Chapter 6 addresses these questions.

Section Review

1. **C** Your friend tells you about an energy source that is supposed to be 46% efficient. What questions do you need to ask your friend in order to clarify this claim?

2. **C** Efficiency and environmental concerns are not always separate. In fact, they are often closely linked. Give three examples of energy sources in which changes in efficiency affect environmental impact, or vice versa.

3. **I** Design an experiment to determine the efficiency of a laboratory burner. You will first need to decide how to define the efficiency, and you will also need to find out what fuel your burner uses. Include a complete procedure and safety precautions.

4. **MC** Some high-efficiency gas furnaces can heat with an efficiency of up to 97%. These gas furnaces work by allowing the water vapour produced during combustion to condense. Condensation is an exothermic reaction that releases further energy for heating. Use the information in this section to demonstrate the increased heat output, using Hess's law. The enthalpy of condensation of water is 44 kJ/mol.

5 **MC** "Nuclear energy is an energy source that is less hazardous to the environment than energy derived from burning coal." Write a brief essay, explaining why you agree or disagree with this statement.

6 **MC** The label on an electric kettle claims that the kettle is 95% efficient.

(a) What definition of efficiency is the manufacturer using?

(b) Write an expression that shows how the manufacturer might have arrived at an efficiency of 95% for the kettle.

(c) Design a detailed experiment to test the manufacturers' claim. Include safety precautions.

7 **K/U** Read the Chemistry Bulletin on page 260. How does the efficiency of using methane as a fuel source compare to using methane hydrates? Justify your answer.

8 **I** Hydrogen is a very appealing fuel, in part because burning it produces only non-polluting water. One of the challenges that researchers face in making hydrogen fuel a reality is how to produce hydrogen economically. Researchers are investigating methods of producing hydrogen indirectly. The following series of equations represent one such method.

$$3FeCl_{2(s)} + 4H_2O_{(g)} \rightarrow Fe_3O_{4(s)} + 6HCl_{(g)} + H_{2(g)} \qquad \Delta H^\circ = 318 \text{ kJ}$$

$$Fe_3O_{4(s)} + \frac{3}{2}Cl_{2(g)} + 6HCl_{(g)} \rightarrow 3FeCl_{3(s)} + 3H_2O_{(g)} + \frac{1}{2}O_{2(g)} \qquad \Delta H^\circ = -249 \text{ kJ}$$

$$3FeCl_{3(s)} \rightarrow 3FeCl_{2(s)} + \frac{3}{2}Cl_{2(g)} \qquad \Delta H^\circ = 173 \text{ kJ}$$

(a) Show that the net result of the three reactions is the decomposition of water to produce hydrogen and oxygen.

(b) Use Hess's law and the enthalpy changes for the reactions to determine the enthalpy change for the decomposition of one mole of water. Check your answer, using the enthalpy of formation of water.

CHAPTER 5 Review

Reflecting on Chapter 5

Summarize this chapter in the format of your choice. Here are a few ideas to use as guidelines:

- Compare the magnitude of energy changes resulting from physical changes, chemical reactions, and nuclear reactions.
- Compare the processes that are responsible for the energy changes resulting from physical changes, chemical reactions, and nuclear reactions.
- Explain the different ways to represent the energy changes of physical and chemical processes.
- Give examples of important exothermic and endothermic processes.
- Explain how a calorimeter is used to determine enthalpy of reaction.
- Use examples and analogies to explain Hess's law.
- Show how to use Hess's law and experimentally determined enthalpies of reaction to calculate unknown enthalpies of reaction.
- Show how to calculate enthalpy of reaction using known enthalpies of formation, and explain how this calculation relates to Hess's law.
- Explain the concept of efficiency, and discuss the efficiency and environmental impact of conventional and alternative energy sources.

Reviewing Key Terms

For each of the following terms, write a sentence that shows your understanding of its meaning.

thermodynamics	thermochemistry
law of conservation of energy	system
surroundings	heat (Q)
temperature (T)	enthalpy (H)
enthalpy change (ΔH)	endothermic reaction
exothermic reaction	enthalpy of reaction
standard enthalpy of reaction (ΔH°)	thermochemical equation
mass defect	nuclear binding energy
nuclear fission	nuclear fusion
specific heat capacity (c)	heat capacity (C)
calorimeter	coffee-cup calorimeter
constant-pressure calorimeter	Hess's law of heat summation
formation reactions	standard molar enthalpy of formation (ΔH°_f)
non-renewable	renewable

Knowledge/Understanding

1. In your own words, describe the relationship between a system and its surroundings. Use an example to illustrate your description.
2. The vaporization of liquid carbon disulfide, CS_2, requires an energy input of 29 kJ/mol.
   (a) Is this reaction exothermic or endothermic? What is the enthalpy change of this reaction?
   (b) Write a thermochemical equation of this reaction. Include 29 kJ as either a reactant or a product.
   (c) Draw and label an enthalpy diagram for the vaporization of liquid carbon disulfide.
3. A given chemical equation is tripled and then reversed. What effect, if any, will there be on the enthalpy change of the reaction?
4. Explain why two nested polystyrene coffee cups, with a lid, make a good constant-pressure calorimeter.
5. Write the balanced equation for the formation of each substance.
   (a) $LiCl_{(s)}$ (b) $C_2H_5OH_{(\ell)}$ (c) $NH_4NO_{3(s)}$
6. If the enthalpy of formation of an element in its standard state is equal to zero, explain why the heat of formation of iodine gas, $I_{2(g)}$, is 21 kJ/mol.

Inquiry

7. In an oxygen-rich atmosphere, carbon burns to produce carbon dioxide, CO_2. Both carbon monoxide, CO, and carbon dioxide are produced when carbon is burned in an oxygen-deficient atmosphere. This makes the direct measurement of the enthalpy of formation of CO difficult. CO, however, also burns in oxygen, O_2, to produce pure carbon dioxide. Explain how you would experimentally determine the enthalpy of formation of carbon monoxide.
8. Two 30.0 g pieces of aluminium, Al, are placed in an insulated container.
   (a) One piece of Al has an initial temperature of 100.0°C. The other piece has an initial temperature of 20.0°C. What is the temperature inside the container after the system has equilibrated? Assume that no heat is lost to the container or the surroundings.

(b) Repeat the calculation in part (a) with the following change: The piece of Al at 20.0°C has a mass of 50.0 g.

9. The complete combustion of 1.00 mol of sucrose, $C_{12}H_{22}O_{11}$, releases –5641 kJ of energy (at 25°C and 100 kPa).
$C_{12}H_{22}O_{11(s)} + 12O_{2(g)} \rightarrow 12CO_{2(g)} + 11H_2O_{(\ell)}$
(a) Use the enthalpy change of this reaction, and enthalpies of formation from Appendix E, to determine the enthalpy of formation of sucrose.
(b) Draw and label an enthalpy diagram for this reaction.

10. A 10.0 g sample of pure acetic acid, CH_3CO_2H, is completely burned. The heat released warms 2.00 L of water from 22.3°C to 39.6°C. Assuming that no heat was lost to the calorimeter, what is the enthalpy change of the complete combustion of acetic acid? Express your answer in units of kJ/g and kJ/mol.

11. Use equations (1), (2), and (3) to find the enthalpy change of the formation of methane, CH_4, from chloroform, $CHCl_3$.
$CHCl_{3(\ell)} + 3HCl_{(g)} \rightarrow CH_{4(g)} + 3Cl_{2(g)}$
(1) $\frac{1}{2}H_{2(g)} + \frac{1}{2}Cl_{2(g)} \rightarrow HCl_{(g)} \quad \Delta H^\circ = -92.3 \text{ kJ}$
(2) $C_{(s)} + 2H_{2(g)} \rightarrow CH_{4(g)} \quad \Delta H^\circ = -74.8 \text{ kJ}$
(3) $C_{(s)} + \frac{1}{2}H_{2(g)} + \frac{3}{2}Cl_{2(g)} \rightarrow CHCl_{3(\ell)}$
$\Delta H^\circ = -134.5 \text{ kJ}$

12. The following equation represents the combustion of ethylene glycol, $(CH_2OH)_2$.
$(CH_2OH)_{2(\ell)} + \frac{5}{2}O_{2(g)} \rightarrow 2CO_{2(g)} + 3H_2O_{(\ell)}$
$\Delta H^\circ = -1178 \text{ kJ}$
Use known enthalpies of formation and the given enthalpy change to determine the enthalpy of formation of ethylene glycol.

13. Most of us associate the foul smell of hydrogen sulfide gas, $H_2S_{(g)}$, with the smell of rotten eggs.
$H_2S_{(g)} + \frac{3}{2}O_{2(g)} \rightarrow SO_{2(g)} + H_2O_{(g)} \quad \Delta H = -519 \text{ kJ}$
How much energy is released when 15.0 g of $H_2S_{(g)}$ burns?

14. Hydrogen peroxide, H_2O_2, is a strong oxidizing agent. It is used as an antiseptic in a 3.0% aqueous solution. Some chlorine-free bleaches contain 6.0% hydrogen peroxide.
(a) Write the balanced chemical equation for the formation of one mole of $H_2O_{2(\ell)}$.
(b) Using the following equations, determine the enthalpy of formation of H_2O_2.
(1) $2H_2O_{2(\ell)} \rightarrow 2H_2O_{(\ell)} + O_{2(g)} \quad \Delta H^\circ = -196 \text{ kJ}$
(2) $H_{2(g)} + \frac{1}{2}O_{2(g)} \rightarrow H_2O_{(\ell)} \quad \Delta H^\circ = -286 \text{ kJ}$

15. Hydrogen cyanide is a highly poisonous gas. It is produced from methane and ammonia.
$CH_{4(g)} + NH_{3(g)} \rightarrow HCN_{(g)} + 3H_{2(g)}$
Find the enthalpy change of this reaction, using the following thermochemical equations.
(1) $H_{2(g)} + 2C_{(graphite)} + N_{2(g)} \rightarrow 2HCN_{(g)}$
$\Delta H^\circ = 270 \text{ kJ}$
(2) $N_{2(g)} + 3H_{2(g)} \rightarrow 2NH_{3(g)} \quad \Delta H^\circ = -92 \text{ kJ}$
(3) $C_{(graphite)} + 2H_{2(g)} \rightarrow CH_{4(g)} \quad \Delta H^\circ = -75 \text{ kJ}$

16. The following equation represents the complete combustion of butane, C_4H_{10}.
$C_4H_{10(g)} + 6.5O_{2(g)} \rightarrow 4CO_{2(g)} + 5H_2O_{(g)}$
(a) Using known enthalpies of formation, calculate the enthalpy change of the complete combustion of C_4H_{10}. (The enthalpy of formation of C_4H_{10} is –126 kJ/mol.)
(b) Using known enthalpies of formation, calculate the enthalpy change of the complete combustion of ethane, C_2H_6, to produce carbon dioxide and water vapour. Express your answer in units of kJ/mol and kJ/g.
(c) A 10.0 g sample that is 30% C_2H_6 and 70% C_4H_{10}, by mass, is burned in excess oxygen. How much energy is released?

17. The caloric content (energy content) of foods is measured using a bomb calorimeter. A sample of food is burned in oxygen, O_2, inside the calorimeter. You can make a simple food calorimeter using an empty food can or pop can. Design an investigation in which you make your own calorimeter and then use your calorimeter to measure the energy content of a piece of cheese or a cracker. Include a diagram of your proposed calorimeter. List the data, and other observations, that you plan to record. Get your teacher's approval before carrying out your investigation. **Note:** Some students have a severe nut allergy. Do not use peanuts, or any other nuts, in your investigation.

18. Design an investigation to determine the enthalpy change of the combustion of ethanol using a wick-type burner, similar to that in a kerosene lamp.
(a) Draw and label a diagram of the apparatus.

(b) Write a step-by-step procedure.
(c) Prepare a table to record your data and other observations.
(d) State any assumptions that you will make when carrying out the calculations.

Communication

19. Suppose that you need to find the enthalpy change of a chemical reaction. Unfortunately, you are unable to carry out the reaction in your school laboratory. Does this mean that you cannot find the enthalpy change of the reaction? Explain.
20. Acetylene, C_2H_2, and ethylene, C_2H_4, are both used as fuels. They combine with oxygen gas to produce carbon dioxide and water in an exothermic reaction. Acetylene also reacts with hydrogen to produce ethylene, as shown.
$C_2H_{2(g)} + H_{2(g)} \rightarrow C_2H_{4(g)} \quad \Delta H^\circ = -175.1 \text{ kJ}$
(a) Without referring to any tables or doing any calculations, explain why C_2H_2 has a more positive enthalpy of formation than C_2H_4.
(b) Do you think C_2H_2 or C_2H_4 is a more energetic fuel? Explain.

Making Connections

21. When a vehicle is parked in the sunlight on a hot summer day, the temperature inside can approach 55°C. One company has patented a non-CFC propelled aerosol that can be sprayed inside a vehicle to reduce the temperature to 25°C within seconds. The spray contains a mixture of two liquids: 10% ethanol, C_2H_5OH, and 90% water by mass.
(a) Use thermochemical equations, and the corresponding enthalpy changes, to explain how the spray works.
(b) 1.0 g of the aerosol is sprayed into a hot vehicle. How much heat (in kJ) can be absorbed due to vaporization of the aerosol? **Note:** ΔH_{vap} of water = 44.0 kJ/mol and ΔH_{vap} of ethanol = 38.56 kJ/mol
(c) Do you think there are any risks associated with using a spray like the one described above? Explain your answer.
22. Consider methane, CH_4, and hydrogen, H_2, as possible fuel sources.
(a) Write the chemical equation for the complete combustion of each fuel. Then find the enthalpy of combustion, ΔH_{comb}, of each fuel. Express your answers in kJ/mol and kJ/g. Assume that water *vapour*, rather than liquid water, is formed in both reactions.
(b) Which is the more energetic fuel, per unit mass?
(c) Consider a fixed mass of each fuel. Which fuel would allow you to drive a greater distance? Explain briefly.
(d) Describe how methane and hydrogen could be obtained. Which of these methods do you think is less expensive? Explain.
(e) Which fuel do you think is more environmentally friendly? Explain.

Answers to Practice Problems and Short Answers to Section Review Questions

Practice Problems: 1.(a) $N_{2(g)} + O_{2(g)} + 180.6 \text{ kJ} \rightarrow 2NO_{(g)}$ (b) Reactants have lower energy than products. (c) +90.3 kJ (d) +644.5 kJ 2.(a) -1.65×10^3 kJ (b) Reactants have higher energy than products. (c) -1.22×10^2 kJ 3.(a) +25.9 kJ (b) +393 kJ (c) Reactants have lower energy than products. 4.(a) $P_4O_{10(s)} + 6H_2O_{(\ell)} \rightarrow 4H_3PO_{4(aq)} + 257.2 \text{ kJ}$ (b) -1.29×10^3 kJ (c) -1.54×10^2 kJ 5. 12.8°C 6. 6.70×10^2 g 7. 2.72×10^4 kJ 8. 6.09°C 9. 53 kJ/mol 10.(a) -4.1×10^2 kJ/mol Mg (b) The density and heat capacity of the solutions are the same as the density and heat capacity of water. No heat is lost to the calorimeter. 11. 29.7°C 12. $HBr_{(aq)} + KOH_{(aq)} \rightarrow H_2O_{(\ell)} + KBr_{(aq)} + 57 \text{ kJ}$ 13. −44 kJ 14. −509 kJ 15. −120 kJ 16. +30 kJ 17.(a) $C_{(s)} + 2H_{2(g)} \rightarrow CH_{4(g)} + 74.6 \text{ kJ}$ (b) $Na_{(s)} + \frac{1}{2}Cl_{2(g)} \rightarrow NaCl_{(s)} + 411.2 \text{ kJ}$ (c) $Mg_{(s)} + \frac{1}{2}O_{2(g)} \rightarrow MgO_{(s)} + 601.2 \text{ kJ}$ (d) $Ca_{(s)} + C_{(s)} + \frac{3}{2}O_{2(g)} \rightarrow CaCO_{3(s)} + 1207.6 \text{ kJ}$ 18. $H_{2(g)} + S_{(s)} + 2O_{2(g)} \rightarrow H_2SO_{4(\ell)} + 814.0 \text{ kJ}$ 19. $Cs_{(s)} + 76.7 \text{ kJ} \rightarrow Cs_{(g)}$ 20.(a) $\frac{1}{4}P_{4(s)} \rightarrow P_{(s)} + 17.6 \text{ kJ}$ (b) −18.5 kJ 21. Get the target equation from the two formation equations. 22. Get the target equation from the formation equations. 23. −77.6 kJ/mol 24.(a) −637.9 kJ (b) 2.49×10^3 kJ

Section Review: 5.1: 2.(a) $\Delta H^\circ_{rxn} = -127.0$ kJ (b) $\Delta H^\circ_{vap} = +44.0$ kJ (c) $\Delta H^\circ_{rxn} = \Delta H^\circ_{comb} < 0$ 4.(a) 34.21 g (b) 599 kJ (c) 1.06×10^3 kJ 5.(a) $C_2H_{2(g)} + \frac{5}{2}O_{2(g)} \rightarrow H_2O_{(\ell)} + 2CO_{2(g)} + 1.3 \times 10^3 \text{ kJ}$ (c) 108 kJ 7.(a) $H_{2(g)} + \frac{1}{2}O_{2(g)} \rightarrow H_2O_{(g)} + 241.8 \text{ kJ}$ **5.2:** 5.(a) $KOH_{(aq)} + HNO_{3(aq)} \rightarrow H_2O_{(\ell)} + KNO_{3(aq)} + 55 \text{ kJ}$ **5.3:** 3. −143 kJ 4. 14.3 kJ 5. −636.0 kJ

CHAPTER

6 Rates of Chemical Reactions

Chapter Preview

6.1 Expressing and Measuring Reaction Rates

6.2 The Rate Law: Reactant Concentration and Rate

6.3 Theories of Reaction Rates

6.4 Reaction Mechanisms and Catalysts

Prerequisite Concepts and Skills

Before you begin this chapter, review the following concepts and skills:

- balancing chemical equations (Concepts and Skills Review)
- expressing concentration in units of mol/L (Concepts and Skills Review)

Racing cars, such as the one shown below, can reach speeds that are well above 200 km/h. In contrast, the maximum speed of many farm tractors is only about 25 km/h. Just as some vehicles travel more quickly than others, some chemical reactions occur more quickly than others. For example, compare the two reactions that occur in vehicles: the decomposition of sodium azide in an air bag and the rusting of iron in steel.

When an automobile collision activates an air bag, sodium azide, $NaN_{3(g)}$, decomposes to form sodium, $Na_{(s)}$, and nitrogen gas, $N_{2(g)}$. (The gas inflates the bag.) This chemical reaction occurs almost instantaneously. It inflates the air bag quickly enough to cushion a driver's impact in a collision.

On the other hand, the reaction of iron with oxygen to form rust proceeds quite slowly. Most Canadians know that the combination of road salt and wet snow somewhat increases the rate of the reaction. Even so, it takes several years for a significant amount of rust to form on the body of a car. This is a good thing for car owners—if rusting occurred as fast as the reaction in an inflating air bag, cars would flake to pieces in seconds.

Why do some reactions occur slowly while others seem to take place instantaneously? How do chemists measure, compare, and express the rates at which chemical reactions occur? Can chemists predict and control the rate of a chemical reaction? These questions will be answered in Chapter 6.

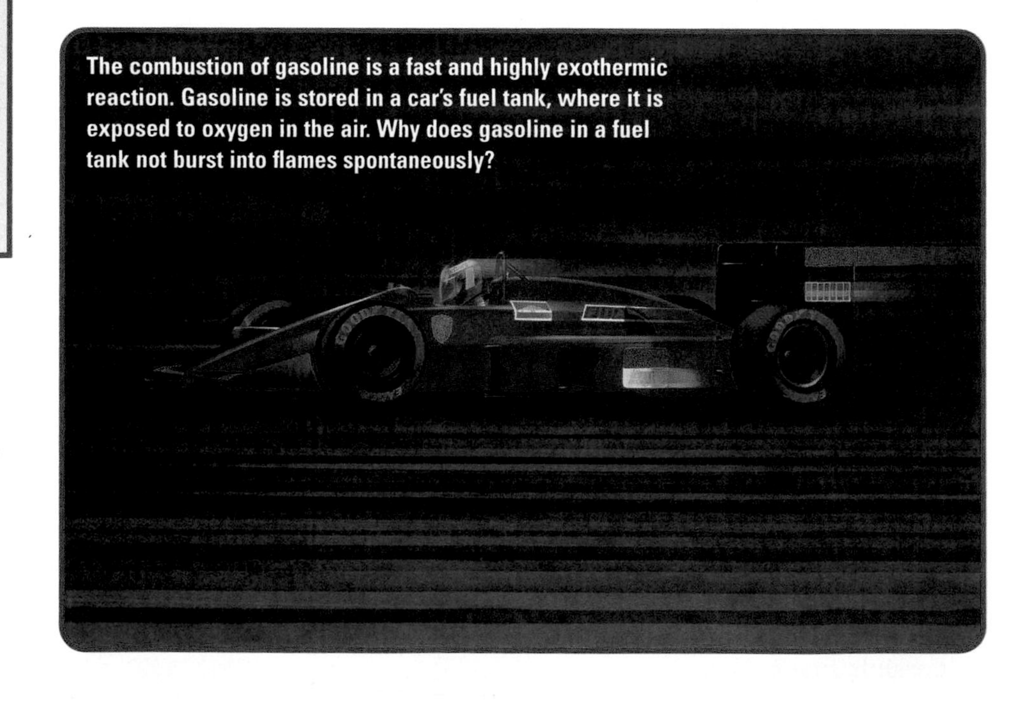

The combustion of gasoline is a fast and highly exothermic reaction. Gasoline is stored in a car's fuel tank, where it is exposed to oxygen in the air. Why does gasoline in a fuel tank not burst into flames spontaneously?

Expressing and Measuring Reaction Rates

6.1

Section Preview/ Specific Expectations

In this section, you will

- **describe**, with the help of a graph, reaction rate as a function of the change of concentration of a reactant or product with respect to time
- **examine** various methods that are used to monitor the rate of a chemical reaction
- **determine** and **distinguish** between the average rate and the instantaneous rate of a chemical reaction
- **review** the factors that affect reaction rate
- **communicate** your understanding of the following terms: *reaction rate, average rate, instantaneous rate, catalyst*

As you learned in the Unit 3 opener, nitroglycerin is an explosive that was used to clear the way for railroads across North America. It decomposes instantly. The reactions that cause fruit to ripen, then rot, take place over a period of days. The reactions that lead to human ageing take place over a lifetime.

How quickly a chemical reaction occurs is a crucial factor in how the reaction affects its surroundings. Therefore, knowing the rate of a chemical reaction is integral to understanding the reaction.

Expressing Reaction Rates

The change in the amount of reactants or products over time is called the **reaction rate**. How do chemists express reaction rates? Consider how the rates of other processes are expressed. For example, the Olympic sprinter in Figure 6.1 can run 100 m in about 10 s, resulting in an average running rate of 100 m/10 s or about 10 m/s.

The running rate of a sprinter is calculated by dividing the distance travelled by the interval of time the sprinter takes to travel this distance. In other words, running rate (speed) is expressed as a change in distance divided by a change in time. In general, a change in a quantity with respect to time can be expressed as follows.

$$\begin{aligned}\text{Rate} &= \frac{\text{Change in quantity}}{\text{Change in time}}\\ &= \frac{\text{Quantity}_{\text{final}} - \text{Quantity}_{\text{initial}}}{t_{\text{final}} - t_{\text{initial}}}\\ &= \frac{\Delta\ \text{Quantity}}{\Delta t}\end{aligned}$$

Chemists express reaction rates in several ways. For example, a reaction rate can be expressed as a change in the amount of reactant consumed or product made per unit of time, as shown below. (The letter A represents a compound.)

$$\begin{aligned}\text{Rate of reaction} &= \frac{\text{Amount of A}_{\text{final}} - \text{Amount of A}_{\text{initial}}\ \text{(in mol)}}{t_{\text{final}} - t_{\text{initial}}\ \text{(in s)}}\\ &= \frac{\Delta\ \text{Amount of A}}{\Delta t}\ \text{(in mol/s)}\end{aligned}$$

When a reaction occurs between gaseous species or in solution, chemists usually express the reaction rate as a change in the concentration of the reactant or product per unit time. Recall, from your previous chemistry course, that the concentration of a compound (in mol/L) is symbolized by placing square brackets, [ ], around the chemical formula. The equation below is the equation you will work with most often in this section.

$$\begin{aligned}\text{Rate of reaction} &= \frac{\text{Concentration of A}_{\text{final}} - \text{Concentration of A}_{\text{initial}}\ \text{(in mol/L)}}{t_{\text{final}} - t_{\text{initial}}\ \text{(in s)}}\\ &= \frac{\Delta[\text{A}]}{\Delta t}\ \text{(in mol/(L} \cdot \text{s))}\end{aligned}$$

Figure 6.1 The running rate (speed) of a sprinter is expressed as a change in distance over time.

Reaction rates are always positive, by convention. A rate that is expressed as the change in concentration of a product is the rate at which the concentration of the product is increasing. The rate that is expressed in terms of the change in concentration of a reactant is the rate at which the concentration of the reactant is decreasing.

Average and Instantaneous Rates of Reactions

If reactions always proceeded at a constant rate, it would be straightforward to find reaction rates. You would just need the initial and final concentrations and the time interval. Reaction rates, however, are not usually constant. They change with time. How does this affect the way that chemists determine reaction rates?

Consider the following reaction.

$$A_{(g)} \rightarrow C_{(g)} + D_{(g)}$$

Now examine the graph in Figure 6.2. The blue line on the graph shows the concentration of product C as the reaction progresses, based on the data in Table 6.1.

Table 6.1 Concentration of C During a Reaction at Constant Temperature

Time (s)	[C] (mol/L)
0.0	0.00
5.0	3.12×10^{-3}
10.0	4.41×10^{-3}
15.0	5.40×10^{-3}
20.0	6.24×10^{-3}

The **average rate** of a reaction is the average change in the concentration of a reactant or product per unit time over a given time interval. For example, using the data in Table 6.1, you can determine the average rate of the reaction from $t = 0.0$ s to $t = 5.0$ s.

$$\begin{aligned}\text{Average rate} &= \frac{\Delta[\text{C}]}{\Delta t} \\ &= \frac{(3.12 \times 10^{-3}\ \text{mol/L}) - 0.00\ \text{mol/L}}{5.0\ \text{s} - 0.0\ \text{s}} \\ &= 6.2 \times 10^{-4}\ \text{mol/(L} \bullet \text{s)}\end{aligned}$$

You can see this calculation in Figure 6.2. On a concentration-time graph, the average rate of a reaction is represented by the slope of a line that is drawn between two points on the curve. This line is called a *secant*.

The average rate of a reaction gives an overall idea of how quickly the reaction is progressing. It does not, however, tell you how fast the reaction is progressing at a specific time. For example, suppose that someone asked you how fast the reaction in Figure 6.2 was progressing over 20.0 s. You would probably calculate the average rate from $t = 0.0$ s to $t = 20.0$ s. You would come up with the answer 3.12×10^{-3} mol/(L • s). (Try this calculation yourself.) What would you do, however, if you were asked how fast the reaction was progressing at exactly $t = 10.0$ s?

The **instantaneous rate** of a reaction is the rate of the reaction at a particular time. To find the instantaneous rate of a reaction using a concentration-time graph, draw a tangent line to the curve and find the slope of the tangent. A *tangent* line is like a secant line, but it touches the curve at only one point. It does not intersect the curve.

The slope of the tangent is the instantaneous rate of the reaction. Figure 6.2 shows the tangent line at $t = 10.0$ s. As shown on the graph, the slope of the tangent (therefore the instantaneous rate) at $t = 10.0$ s is 2.3×10^{-4} mol/(L • s).

Notice that near the beginning of the reaction, when the concentration of the reactants is relatively high, the slope of the tangent is greater (steeper). This indicates a faster reaction rate. As the reaction proceeds, the reactants are used up and the slope of the tangent decreases.

Physics LINK

In your previous courses in science or physics, you probably learned the difference between *instantaneous* velocity and *average* velocity. How did you use a displacement-time graph to determine instantaneous velocity and average velocity? Write a memo that explains instantaneous rate and average rate to a physicist, by comparing reaction rate with velocity.

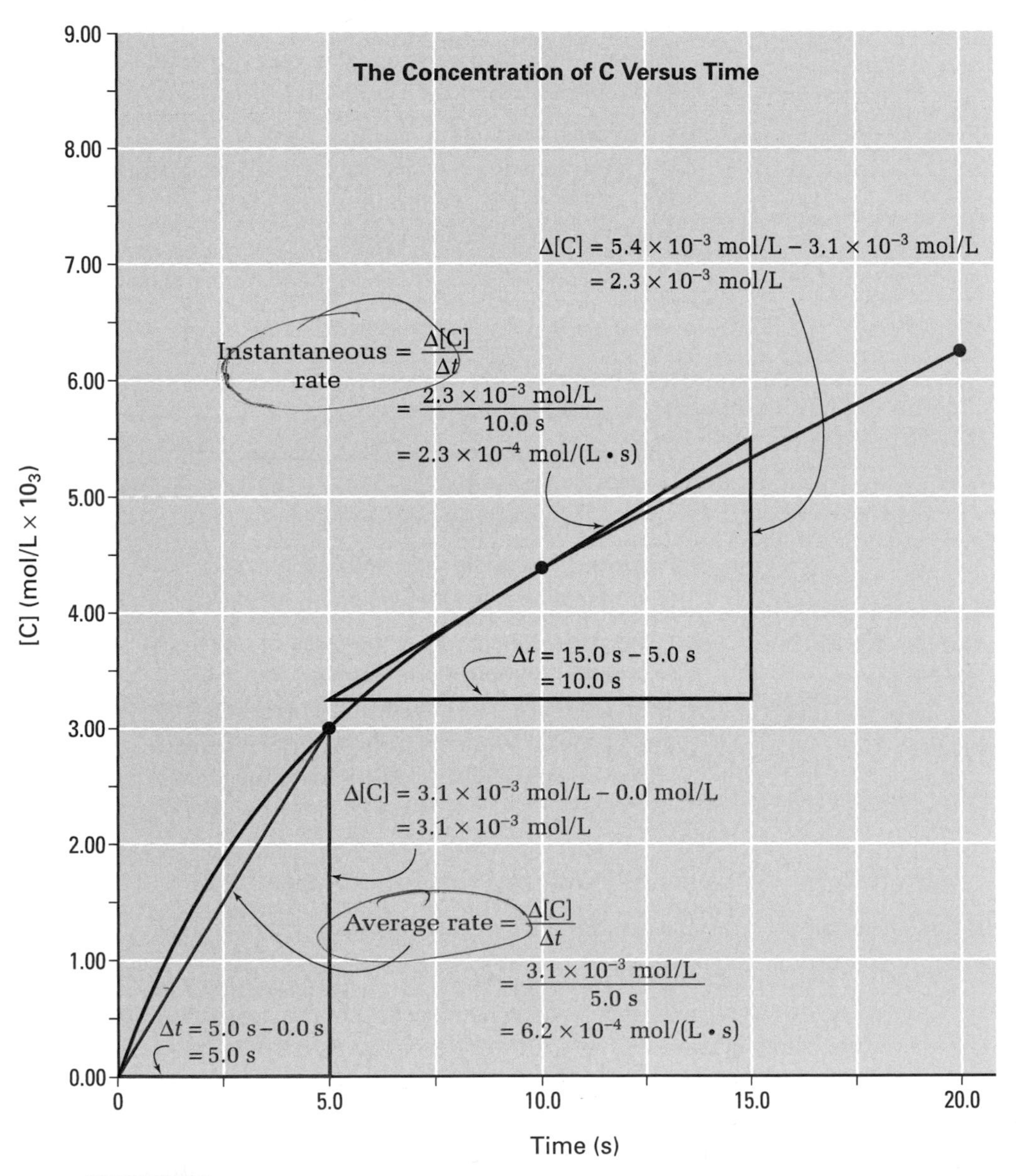

Figure 6.2 The slope of a tangent drawn to a concentration-time curve represents the instantaneous rate of the reaction. The slope of a secant is used to determine the average rate of a reaction.

In the following ThoughtLab, you will use experimental data to draw a graph that shows the change in concentration of the product of a reaction. Then you will use the graph to help you determine the instantaneous rate and average rate of the reaction.

ThoughtLab

A chemist carried out a reaction to trace the rate of decomposition of dinitrogen pentoxide.

$$2N_2O_{5(g)} \rightarrow 4NO_{2(g)} + O_{2(g)}$$

The chemist collected the following data at a constant temperature.

Time (s)	$[O_2]$ (mol/L)
0.00	0.0
6.00×10^2	2.1×10^{-3}
1.20×10^3	3.6×10^{-3}
1.80×10^3	4.8×10^{-3}
2.40×10^3	5.6×10^{-3}
3.00×10^3	6.4×10^{-3}
3.60×10^3	6.7×10^{-3}
4.20×10^3	7.1×10^{-3}
4.80×10^3	7.5×10^{-3}
5.40×10^3	7.7×10^{-3}
6.00×10^3	7.8×10^{-3}

Procedure

1. Using graph paper or spreadsheet software, plot and label a graph that shows the rate of formation of oxygen gas. The concentration of O_2 (in mol/L) is the dependent variable and time (in s) is the independent variable.
2. Draw a secant to the curve in the interval from $t = 0$ s to $t = 4800$ s.
3. Draw a tangent to the curve at $t = 1200$ s and at $t = 4800$ s.
4. Determine the slope of the secant. What is the average rate of the reaction over the given time interval? Include proper units, and pay attention to significant digits.
5. Determine the slope of each tangent. What is the instantaneous reaction rate at $t = 1200$ s and at $t = 4800$ s? Include proper units, and pay attention to significant digits.

Analysis

1. Why are the units for the average rate and the instantaneous rate the same?
2. For a given set of data, two students determined different average reaction rates. If neither student made an error in the calculations, account for the difference in their reaction rates.
3. Propose a reason for the difference in the instantaneous rates at 1200 s and 4800 s.
4. When chemists compare the rates of reactions carried out under different conditions, they often compare the rates near the beginning of the reactions. What advantage(s) do you see in this practice? **Hint:** Think of slow reactions.

Reaction Rates in Terms of Products and Reactants

In the ThoughtLab, you analyzed the rate of the following reaction in terms of the production of oxygen.

$$2N_2O_{5(g)} \rightarrow 4NO_{2(g)} + O_{2(g)}$$

There are two other ways to represent the rate of this reaction:

- in terms of the rate of the disappearance of dinitrogen pentoxide
- in terms of the production of nitrogen dioxide

For every 1 mol of O_2 that is produced, 4 mol of NO_2 are also produced. This means that the rate of production of NO_2 is four times greater than the rate of production of O_2. Therefore, the rate of production of O_2 is one quarter the rate of production of NO_2. You can express the relationship between O_2 production and NO_2 production as follows.

$$\frac{\Delta[O_2]}{\Delta t} = \frac{1}{4}\frac{\Delta[NO_2]}{\Delta t}$$

When 1 mol of O_2 is produced, 2 mol of N_2O_5 are consumed. Therefore, the rate of production of O_2 is half the rate of disappearance of N_2O_5. You can represent this relationship as follows:

$$\frac{\Delta[O_2]}{\Delta t} = -\frac{1}{2}\frac{\Delta[N_2O_5]}{\Delta t}$$

Notice that the expression involving N_2O_5 (the reactant) has a negative sign. A change in concentration is calculated using the expression below.

$$\text{Change in concentration} = \text{Concentration}_{\text{final}} - \text{Concentration}_{\text{initial}}$$

Since the concentration of a reactant always decreases as a reaction progresses, the change in concentration is always negative. By convention, however, *a rate is always expressed as a positive number.* Therefore, expressions that involve reactants must be multiplied by –1 to become positive.

Examine the Sample Problem below to see how to express reaction rates in terms of products and reactants. Then try the Practice Problems that follow.

Electronic Learning Partner

To learn more about reaction rates expressed as changes in concentration over time, go to the Chemistry 12 Electronic Learning Partner.

Sample Problem

Expressing Reaction Rates

Problem

Dinitrogen pentoxide, N_2O_5, decomposes to form nitrogen dioxide and oxygen.

$$2N_2O_{5(g)} \rightarrow 4NO_{2(g)} + O_{2(g)}$$

NO_2 is produced at a rate of 5.0×10^{-6} mol/(L • s). What is the corresponding rate of disappearance of N_2O_5 and rate of formation of O_2?

What Is Required?

Since N_2O_5 is a reactant, you need to calculate its rate of disappearance. O_2 is a product, so you need to find its rate of formation.

What Is Given?

You know the rate of formation of NO_2 and the balanced chemical equation.

Plan Your Strategy

First check that the chemical equation is balanced. Then use the molar coefficients in the balanced equation to determine the relative rates of disappearance and formation.

Since 4 mol of NO_2 are produced for every 2 mol of N_2O_5 that decompose, the rate of disappearance of N_2O_5 is $\frac{2}{4}$, or $\frac{1}{2}$, the rate of formation of NO_2. Similarly, 1 mol of O_2 is formed for every 4 mol of NO_2. Therefore, the rate of production of O_2 is $\frac{1}{4}$ the rate of NO_2 production.

Act on Your Strategy

$$\text{Rate of disappearance of } N_2O_5 = \frac{1}{2} \times 5.0 \times 10^{-6} \text{ mol/(L • s)}$$
$$= 2.5 \times 10^{-6} \text{ mol/(L • s)}$$

$$\text{Rate of production of } O_2 = \frac{1}{4} \times 5.0 \times 10^{-6} \text{ mol/(L • s)}$$
$$= 1.2 \times 10^{-6} \text{ mol/(L • s)}$$

Check Your Solution

From the coefficients in the balanced chemical equation, you can see that the rate of decomposition of N_2O_5 is $\frac{2}{4}$, or $\frac{1}{2}$, the rate of formation of NO_2. The rate of production of O_2 is $\frac{1}{2}$ the rate of decomposition of N_2O_5.

Practice Problems

1. Cyclopropane, C_3H_6, is used in the synthesis of organic compounds and as a fast-acting anesthetic. It undergoes rearrangement to form propene. If cyclopropane disappears at a rate of 0.25 mol/s, at what rate is propene being produced?

2. Ammonia, NH_3, reacts with oxygen to produce nitric oxide, NO, and water vapour.

   $4NH_{3(g)} + 5O_{2(g)} \rightarrow 4NO_{(g)} + 6H_2O_{(g)}$

   At a specific time in the reaction, ammonia is disappearing at rate of 0.068 mol/(L • s).

   What is the corresponding rate of production of water?

3. Hydrogen bromide reacts with oxygen to produce bromine and water vapour.

   $4HBr_{(g)} + O_{2(g)} \rightarrow 2Br_{2(g)} + 2H_2O_{(g)}$

   How does the rate of decomposition of HBr (in mol/(L • s)) compare with the rate of formation of Br_2 (also in mol/(L • s))? Express your answer as an equation.

4. Magnesium metal reacts with hydrochloric acid to produce magnesium chloride and hydrogen gas.

   $Mg_{(s)} + 2HCl_{(aq)} \rightarrow MgCl_{2(aq)} + H_{2(g)}$

   Over an interval of 1.00 s, the mass of $Mg_{(s)}$ changes by −0.011 g.

   (a) What is the corresponding rate of consumption of $HCl_{(aq)}$ (in mol/s)?

   (b) Calculate the corresponding rate of production of $H_{2(g)}$ (in L/s) at 20°C and 101 kPa.

PROBLEM TIP

Recall, from your previous chemistry course, that 1.00 mol of any gas occupies a volume of 24.0 L at 20°C and 101 kPa.

Methods for Measuring Reaction Rates

How do chemists collect the data they need to determine a reaction rate? To determine empirically the rate of a chemical reaction, chemists must monitor the concentration or amount of at least one reactant or product. There are a variety of techniques available. The choice of technique depends on the reaction under study and the equipment available.

Monitoring Mass, pH, and Conductivity

Consider the reaction of magnesium with hydrochloric acid.

$$Mg_{(s)} + 2HCl_{(aq)} \rightarrow MgCl_{2(aq)} + H_{2(g)}$$

Hydrogen gas is released in the reaction. You can track the decrease in mass, due to the escaping hydrogen, by carrying out the reaction in an open vessel on an electric balance. The decrease in mass can be plotted against time. Some electronic balances can be connected to a computer, with the appropriate software, to record mass and time data automatically as the reaction proceeds.

Another technique for monitoring the reaction above involves pH. Since HCl is consumed in the reaction, you can record changes in pH with respect to time. Figure 6.3 shows a probe being used to monitor the changing pH of a solution.

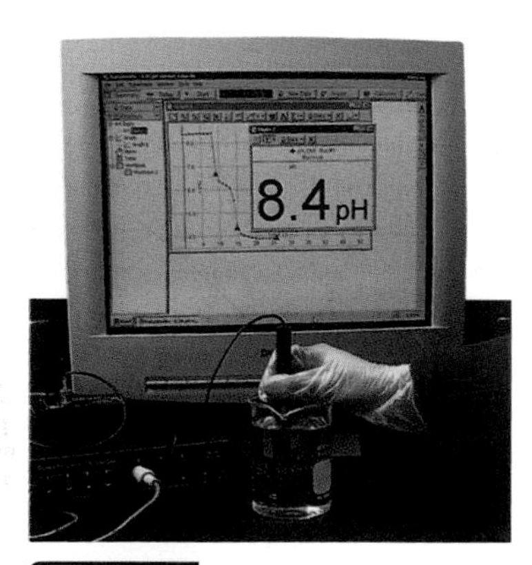

Figure 6.3 If the concentration of H_3O^+ or OH^- ions changes over the course of a reaction, a chemist can use a pH meter to monitor the reaction.

A third technique involves electrical conductivity. Dissolved ions in aqueous solution conduct electricity. The electrical conductivity of the solution is proportional to the concentration of ions. Therefore, reactions that occur in aqueous solution, and involve a change in the quantity of dissolved ions, undergo a change in electrical conductivity. In the reaction above, hydrochloric acid is a mix of equal molar amounts of two ions: hydronium, H_3O^+, and chloride, Cl^-. The $MgCl_2$ that is produced exists as three separate ions in solution: one Mg^{2+} ion and two Cl^- ions. Since there is an increase in the concentration of ions as the reaction proceeds, the conductivity of the solution also increases with time.

PROBEWARE

If you have access to probeware, do Probeware Investigation 6-A, or a similar investigation from a probeware company.

Monitoring Pressure

When a reaction involves gases, the pressure of the system often changes as the reaction progresses. Chemists can monitor this pressure change. For example, consider the decomposition of dinitrogen pentoxide, shown in the following chemical reaction.

$$2N_2O_{5(g)} \rightarrow 4NO_{2(g)} + O_{2(g)}$$

When 2 mol of N_2O_5 gas decompose, they form 5 mol of gaseous products. Therefore, the pressure of the system increases as the reaction proceeds, provided that the reaction is carried out in a closed container. Chemists use a pressure sensor to monitor pressure changes.

Monitoring Colour

Colour change can also be used to monitor the progress of a reaction. The absorption of light by a chemical compound is directly related to the concentration of the compound. For example, suppose you add several drops of blue food colouring to a litre of water. If you add a few millilitres of bleach to the solution, the intensity of the colour of the food dye diminishes as it reacts. You can then monitor the colour change. (Do not try this experiment without your teacher's supervision.)

For accurate measurements of the colour intensity of a solution, chemists use a device called a *spectrophotometer*. (See Figure 6.4.)

Figure 6.4 This photograph shows a simple spectrophotometer, which measures the amount of visible light that is absorbed by a coloured solution. More sophisticated devices can measure the absorption of ultraviolet and infrared radiation.

Monitoring Volume

When a reaction generates gas, chemists can monitor the volume of gas produced. In Investigation 6-A, you will determine average reaction rates by recording the time taken to produce a fixed volume of gas. You will perform several trials of the same reaction to investigate the effects that temperature, concentration of reactants, and surface area of reactants have on the reaction rate. You will also perform one trial using a different reactant.

SKILL FOCUS

Predicting

Performing and recording

Analyzing and interpreting

Investigation 6-A

Studying Reaction Rates

You have probably already encountered the reaction of vinegar with baking soda. The carbon dioxide that is produced can be used to simulate a volcano, for example, or to propel a toy car or rocket.

$$CH_3COOH_{(aq)} + NaHCO_{3(s)} \rightarrow NaCH_3COO_{(aq)} + CO_{2(g)} + H_2O_{(\ell)}$$

Other carbonate-containing compounds, such as calcium carbonate, can also react with vinegar to produce CO_2.

In this investigation, you will determine reaction rates by recording the time taken to produce a fixed volume of CO_2. You will collect the CO_2 by downward displacement of water.

Question

How do factors such as concentration, temperature, a different reactant, and surface area affect the rate of this reaction?

Prediction

Read the Procedure. *Quantitatively* predict the effects of changes to the concentration and temperature. *Qualitatively* predict the effects of changes to the reactant and surface area.

Materials

electronic balance
pneumatic trough
stopwatch
250 mL Erlenmeyer flask
retort stand and clamp
one-holed rubber stopper, fitted with a piece of glass tubing (must be airtight)
1 m rubber hose to fit glass tubing (must be airtight)
25 or 100 mL graduated cylinder
large test tube
weighing paper, weighing boat, or small beaker
100 mL vinegar (at room temperature)
10 g baking soda, $NaHCO_3$
2.0 g powdered $CaCO_3$
2.0 g solid $CaCO_3$ (marble chips)
scoopula
thermometer
wash bottle with distilled water (at room temperature)
warm-water bath (prepared using a large beaker and a hot plate or electric kettle)
ice bath (ice cubes and water)
paper towel

Safety Precautions

- Beware of shock hazard if an electric kettle or hot plate is used.
- Wear safety glasses at all times.

Procedure

Part 1 The Effect of Concentration

1. The distilled water and vinegar that you are going to use should be at room temperature. Measure and record the temperature of either the vinegar or the distilled water.
2. Assemble the apparatus for the collection of CO_2, by downward displacement of water, as shown below.

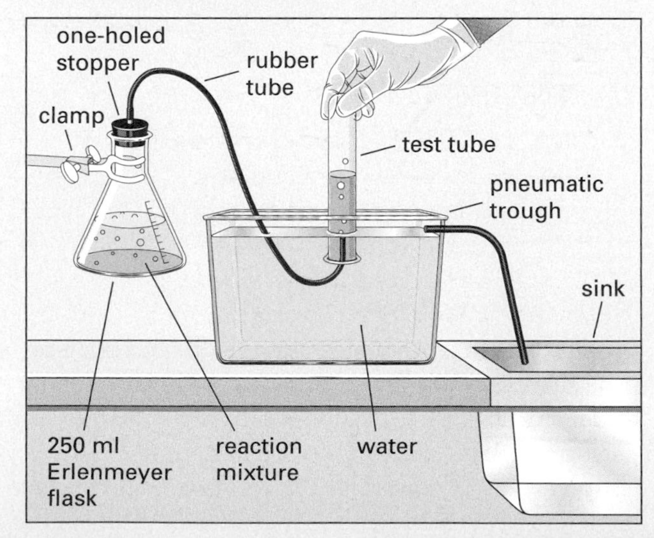

Note: To invert a test tube filled with water, place a piece of paper over the mouth of the filled test tube before inverting it.

3. Copy the table below into your notebook, to record your data.

Trial	Mass of $NaHCO_3$ (g)	Volume of vinegar (mL)	Volume of distilled water (mL)	Time to fill test tube with CO_2 (s)	Average reaction rate (mL/s)
1	1.00	20.0	0.0		
2	1.00	15.0	5.0		
3	1.00	10.0	10.0		
4	1.00	5.0	15.0		

4. For trial 1, add 20.0 mL of vinegar to the flask. Have the stopwatch ready. The end of the rubber tubing should be in place under the water-filled test tube in the pneumatic trough. Quickly add 1.00 g of $NaHCO_3$ to the flask, and put in the stopper. Record the time taken to fill the tube completely with CO_2.

5. Complete trials 2 to 4 with the indicated quantities.

6. Determine the volume of CO_2 you collected by filling the gas collection test tube to the top with water and then pouring the water into a graduated cylinder.

Part 2 The Effect of Temperature

1. Repeat trial 3 in Part 1 using a mixture of 10 mL of water and 10 mL of vinegar that has been cooled to about 10°C below room temperature in an ice bath. Measure and record the temperature of the mixture immediately before the reaction. Record the time taken to fill the test tube with CO_2. Determine the average rate of production of CO_2 (in mL/s).

2. Use a hot-water bath to warm a mixture of 10 mL of distilled water and 10 mL of vinegar to about 10°C above room temperature. Repeat trial 3 in Part 1 using this heated mixture. Measure and record the temperature of the vinegar-water mixture immediately before the reaction. Record the time taken to fill the test tube with CO_2. Determine the average rate of production of CO_2 (in mL/s).

Part 3 The Effect of Reactants and Surface Area

1. Repeat trial 3 in Part 1, using 1.00 g of powdered calcium carbonate, $CaCO_3$, instead of $NaHCO_3$. All the reactants should be at room temperature. Record the time taken to fill the tube with CO_2. Determine the average rate of production of CO_2 (in mL/s).

2. Repeat step 1, using 1.00 g of solid $CaCO_3$ (marble chips) instead of powdered $CaCO_3$.

Analysis

1. Draw a graph to show your results for Part 1. Plot average reaction rate (in mL CO_2/s) on the *y*-axis. Plot [CH_3COOH] (in mol/L) on the *x*-axis. Vinegar is 5.0% (m/v) CH_3COOH.

2. As quantitatively as possible, state a relationship between [CH_3COOH] and the average rate of the reaction.

3. Compare the average reaction rate for corresponding concentrations of vinegar at the three different temperatures tested.

4. What effect did a 10°C temperature change have on the reaction rate? Be as quantitative as possible.

5. What effect did using $CaCO_3$ instead of $NaHCO_3$ have on the reaction rate?

6. What effect did the surface area of the $CaCO_3$ have on the reaction rate?

Conclusion

7. State the effect of each factor on the reaction rate. Compare your results with your prediction.

Application

8. The rate of this reaction can be expressed by the following equation:
Rate = $k[CH_3COOH]^m$, where k is a constant and m is usually equal to 0, 1, or 2.
What value of m do your results suggest?

Factors That Affect Reaction Rate

Chemists have made the following observations about factors that affect reaction rate. You may already be familiar with some of these observations, from your previous studies of chemical reactions and from Investigation 6-A.

Summary of Some Factors That Affect Reaction Rate

1. The rate of a reaction can be increased by increasing the temperature.
2. Increasing the concentrations of the reactants usually increases the rate of the reaction.
3. A **catalyst** is a substance that increases the rate of a reaction. The catalyst is regenerated at the end of the reaction and can be re-used. You will learn more about catalysts in section 6.4.
4. Increasing the available surface area of a reactant increases the rate of a reaction.
5. The rate of a chemical reaction depends on what the reactants are.

Chemists and engineers use these and other factors to manipulate the rate of a particular reaction to suit their needs. For example, consider the synthesis of ammonia, NH_3, from nitrogen and hydrogen.

$$N_{2(g)} + 3H_{2(g)} \rightarrow 2NH_{3(g)}$$

This reaction must be carried out with high concentrations of reactants, at a temperature of 400°C to 500°C, in the presence of a catalyst. Otherwise, the rate of production of ammonia is too slow to be economically feasible.

Section Summary

In this section, you learned how to express reaction rates and how to analyze reaction rate graphs. You also learned how to determine the average rate and instantaneous rate of a reaction, given appropriate data. Then you examined different techniques for monitoring the rate of a reaction. Finally, you carried out an investigation to review some of the factors that affect reaction rate. In the next section, you will learn how to use a rate law equation to show the quantitative relationships between reaction rate and concentration.

Section Review

1. (C) In your own words, explain why the rate of a chemical reaction is fastest at the beginning of the reaction.
2. (K/U) Show why the expression for the rate of disappearance of a reactant is always negative, even though rates are always positive, by convention.
3. (K/U) Under what circumstances is the rate at which the concentration of a reactant decreases numerically equal to the rate at which the concentration of a product increases?

4 **I** In the following reaction, the rate of production of sulfate ions is 1.25×10^{-3} mol/(L • s).

$$2HCrO_4^- + 3HSO_3^- + 5H^+ \rightarrow 2Cr^{3+} + 3SO_4^{2-} + 5H_2O$$

(a) What is the corresponding rate at which $[HSO_3^-]$ decreases over the same time interval?

(b) What is the corresponding rate at which $[HCrO_4^-]$ decreases over the same time interval?

5 **K/U** In this section, you examined the following techniques for monitoring the progress of a reaction. State the conditions that must change as a reaction proceeds, to allow each technique to be used.

(a) using a pH meter

(b) using a spectrophotometer

(c) using a conductivity meter

(d) monitoring pressure

6 **I** Refer to the table in the ThoughtLab on page 270. Redraw the table, adding one column for $[N_2O_5]$ and one column for $[NO_2]$.

(a) Calculate $[NO_2]$ at each time interval, based on $[O_2]$. Assume initial $[NO_2] = 0.0$ mol/L.

(b) On the same set of axes, draw and label a concentration-time graph with two curves: one for $[NO_2]$ and one for $[O_2]$.

(c) What do you notice about the shapes of the two curves?

(d) Determine the instantaneous rate at $t = 1200$ and $t = 4800$ for each compound. How do the instantaneous rates compare?

(e) Is it possible to use the information in the table to calculate $[N_2O_5]$ at each time interval? Explain your answer.

(f) On a concentration-time graph, sketch the shape of a curve that represents $[N_2O_5]$ versus time.

7 **C** For each reaction, suggest one or more techniques for monitoring the progress of the reaction. Explain your answers.

(a) $CH_3CH_2CH_2CH_2CH{=}CH_{2(\ell)} + Br_{2(aq)} \rightarrow CH_3CH_2CH_2CH_2CHBrCH_2Br_{(\ell)}$
Hint: Br_2 is brownish-orange. The other compounds are colourless.

(b) $H_2O_{2(aq)} \rightarrow H_2O_{(\ell)} + \frac{1}{2}O_{2(g)}$

(c) $CaCO_{3(s)} + H_2SO_{4(aq)} \rightarrow CaSO_{4(aq)} + CO_{2(g)} + H_2O_{(\ell)}$

(d) $5Fe^{2+}{}_{(aq)} + MnO_4^-{}_{(aq)} + 8H^+{}_{(aq)} \rightarrow Mn^{2+}{}_{(aq)} + 5Fe^{3+}{}_{(aq)} + 4H_2O_{(\ell)}$

Hint: The permanganate ion, MnO_4^-, is purple. All the other species in the reaction are colourless.

(e) $N_{2(g)} + 3H_{2(g)} \rightarrow 2NH_{3(g)}$

6.2 The Rate Law: Reactant Concentration and Rate

Section Preview/ Specific Expectations

In this section, you will

- **express** the rate of a reaction as a rate law equation
- **explain** the concept of half-life for a reaction
- **determine** the rate law equation and rate constant for a reaction from experimental data
- **communicate** your understanding of the following terms: *rate law equation, rate constant, overall reaction order, first-order reaction, second-order reaction, initial rates method, half-life* ($t_{1/2}$)

The rate of a chemical reaction depends on several factors, as you learned in section 6.1. One of the factors that affect reaction rate is the concentrations of the reactants. You know that the rates of most chemical reactions increase when the concentrations of the reactants increase. Is there a more specific relationship? In this section, you will explore the quantitative relationships between the rate of a reaction and the concentrations of the reactants.

Relating Reactant Concentrations and Rate

Consider the general reaction below.

$$aA + bB \rightarrow \text{products}$$

This reaction occurs at a constant temperature. The reactant formulas are represented by A and B. The stoichiometric coefficients are represented by a and b. In this section, you will study reaction rates that are not affected by the concentrations of the products. Therefore, you do not need to use symbols for the products.

In general, the rate of a reaction increases when the concentrations of the reactants increase. The dependence of the rate of a reaction on the concentrations of the reactants is given by the following relationship.

$$\text{Rate} \propto [A]^m[B]^n$$

This relationship can be expressed in a general equation called the **rate law equation**. The rate law equation is shown in Figure 6.5. For any reaction, the rate law equation expresses the relationship between the concentrations of the reactants and the rate of the reaction. The letter k represents a proportionality constant called the **rate constant**. There is a different rate constant for each reaction at any given temperature.

The exponents m and n must be determined by experiment. They do not necessarily correspond to the stoichiometric coefficients of their reactants. They are usually 1 or 2, but values of 0, 3, and even fractions can occur.

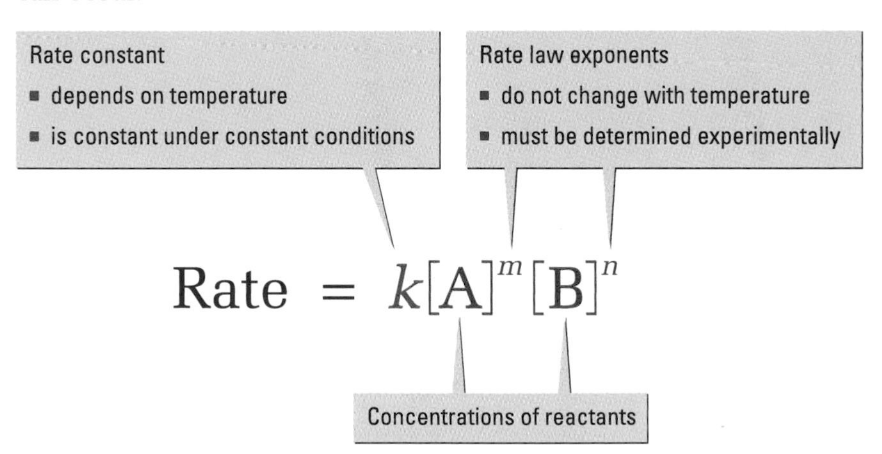

Figure 6.5 This diagram explains the components of the rate law.

The values of the exponents in the rate law equation establish the order of the reaction. If a given reactant is found to have an exponent of 1, the reaction is said to be *first order* in this reactant. Similarly, if the exponent of a reactant is 2, the reaction is *second order* in this reactant. The sum of the exponents ($m + n$) is known as the **overall reaction order**. For example, the rate law equation below represents a reaction that is first order in A, second order in B, and third order (1 + 2) overall.

$$\text{Rate} = k[\text{A}]^1[\text{B}]^2$$

Chemists carry out experiments to determine the rate law equation for a given reaction at a given temperature. Later in the section, you will work with data and do this yourself. First, however, it is necessary to examine further the rate constant and reaction order.

The Rate Constant

The magnitude of the rate constant, k, indicates the speed of a reaction. A small rate constant indicates a slow reaction. A large rate constant indicates a fast reaction. For example, a first-order reaction with a rate constant of 10^2 s^{-1} will be essentially complete in less than 0.10 s. By contrast, a first-order reaction with a rate constant of 10^{-3} s^{-1} will take about 2 h.

As a reaction proceeds, the reaction rate decreases because the concentrations of the reactants decrease. The value of the rate constant, however, remains the same throughout the reaction. In other words, for a given reaction under constant conditions, the value of k remains constant.

What happens if the temperature changes? Since the rate of a chemical reaction depends on temperature, so does k. Because a chemical reaction depends on temperature, chemists who study reaction rates must work at constant temperature. Furthermore, the value of k must be accompanied by the temperature at which the reaction occurred. For example, for the second order decomposition of hydrogen iodide, $k = 2.7 \times 10^{-3}$ L/(mol • s) at 440°C. When the temperature is increased to 500°C, $k = 3.9 \times 10^{-3}$ L/(mol • s).

Defining First-Order Reactions

A **first-order reaction** has an overall order of 1. The decomposition of dinitrogen pentoxide, N_2O_5, is an example of a first-order reaction.

$$2N_2O_{5(g)} \rightarrow 4NO_{2(g)} + 5O_{2(g)}$$

Experiments have shown that this reaction is first order in N_2O_5. In other words, the rate law equation is written as follows:

$$\text{Rate} = k[N_2O_5]^1$$

Because the overall order of the reaction is one, it is a first-order reaction.

Reactions with more than one reactant can also be first-order reactions. For example, consider the following reaction.

$$(CH_3)_3CBr_{(aq)} + H_2O_{(\ell)} \rightarrow (CH_3)_3OH_{(aq)} + H^+_{(aq)} + Br^-_{(aq)}$$

Experiments have shown that the reaction is first order in $(CH_3)_3CBr$, and zero order in water. In other words, the rate of the reaction does not depend, at all, on the concentration of water. The rate law equation is written as follows:

$$\text{Rate} = k[(CH_3)_3CBr_{(\ell)}]^1[H_2O_{(\ell)}]^0$$

The overall reaction order is 1 (1 + 0). The reaction is first order.

CONCEPT CHECK

Chemists can trace the progress of the chemical reaction shown on the left by monitoring its conductivity. Explain why.

You frequently encounter first-order relationships in everyday life. For example, if one soft drink costs \$1.50, two soft drinks cost \$3.00, three cost \$4.50, and so on. This first-order relationship can be represented by the following equation.

$$\text{Cost} = \$1.50\ (\text{Number of soft drinks})^1$$

Defining Second-Order Reactions

For the decomposition reaction of hydrogen iodide, the value of the exponent of [HI] in the rate law equation is the same as its molar coefficient in the balanced chemical equation. This is not always the case. The values of the exponents in a rate law equation *must* be determined *by experiment*.

A **second-order reaction** has an overall reaction order of 2. An example of a second-order reaction is the decomposition of hydrogen iodide.

$$2HI_{(g)} \rightarrow H_{2(g)} + I_{2(g)}$$

Chemists have determined, by experiment, that this reaction is second order in hydrogen iodide. Therefore, the rate law equation is written as follows:

$$\text{Rate} = k[HI]^2$$

Another example of a second-order reaction is the reaction between nitric oxide and ozone.

$$NO_{(g)} + O_{3(g)} \rightarrow NO_{2(g)} + O_{2(g)}$$

By experiment, chemists have determined that the reaction is first order in nitric oxide and first order in ozone, as shown below:

$$\text{Rate} = k[NO]^1[O_3]^1$$

The overall order of the reaction is 2 (1 + 1). Therefore, the reaction is a second order reaction.

For a non-chemistry example of a second-order relationship, consider the equation for the simple parabola, $y = x^2$. When $x = 3$, $y = 9$. When $x = 12$, the value of y increases to $12^2 = 144$. In other words, when the value of x increases by 4, the value of y increases by a factor of 4^2 $(16 \times 9 = 144)$.

Electronic Learning Partner

Go to the Chemistry 12 Electronic Learning Partner for more information about aspects of material covered in this section of the chapter.

Other Reactions

There are also other types of reactions, besides first-order and second-order reactions. For example, consider the decomposition of ammonia on a tungsten, W, catalyst.

$$2NH_{3(g)} \xrightarrow{W_{(s)}} N_{2(g)} + 3H_{2(g)}$$

The rate law equation for this reaction has been found experimentally.

$$\text{Rate} = k[NH_3]^0 = k$$

The rate of decomposition of ammonia on a tungsten catalyst is *independent* of the concentration of ammonia, since $[NH_3]^0 = 1$. In other words, the rate of the reaction is a constant.

Many reactions have overall orders that are higher than 2. Some reactions have overall orders that are fractions. In this text, however, you will concentrate on first-order and second-order reactions.

The Initial Rates Method

As you have learned, the values of the exponents in a rate law equation must be determined experimentally. Chemists determine the values of m and n by carrying out a series of experiments. Each experiment has a different, known set of initial concentrations. All other factors, such as temperature, remain constant. Chemists measure and compare the initial rate of each reaction. Thus, this method is called the **initial rates method**.

There are several reasons for using initial rates. One reason is that chemists do not need to follow each reaction to its end. This saves time, especially when studying slow reactions.

Rate Experiments

To see how the initial rates method works, consider the following reaction:

$$2N_2O_{5(g)} \rightarrow 2NO_{2(g)} + O_{2(g)}$$

The general rate law equation for this reaction is given below.

$$\text{Rate} = k[N_2O_5]^m$$

To determine the value of m, a chemist performs three experiments. A different initial concentration of dinitrogen pentoxide, $[N_2O_5]_0$, is used for each experiment. The subscript 0 represents $t = 0$. Table 6.2 shows the results of these experiments.

Table 6.2 Data for Rate Experiments

Experiment	Initial $[N_2O_5]_0$ (mol/L)	Initial rate (mol/(L • s))
1	0.010	4.8×10^{-6}
2	0.020	9.6×10^{-6}
3	0.030	1.5×10^{-5}

The value of m can be determined using at least two different methods. By inspection, you can see that when the concentration of N_2O_5 is doubled (experiments 1 and 2), the rate also doubles. When the concentration of N_2O_5 is tripled (experiments 1 and 3), the rate also triples. This indicates a first-order relationship, as follows:

$$\text{Rate} = k[N_2O_5]^1$$

Alternatively, you can compare the rate law equation for each experiment using ratios. This method is useful when the relationship between concentration and rate is not immediately obvious from the data.

Write the rate expressions for experiments 1 and 2 as follows:

$$\begin{aligned}\text{Rate}_1 &= k(0.010)^m \\ &= 4.8 \times 10^{-6} \text{ mol/(L} \cdot \text{s)}\end{aligned}$$

$$\begin{aligned}\text{Rate}_2 &= k(0.020)^m \\ &= 9.6 \times 10^{-6} \text{ mol/(L} \cdot \text{s)}\end{aligned}$$

Create a ratio to compare the two rates.

$$\frac{\text{Rate}_1}{\text{Rate}_2} = \frac{k(0.010 \text{ mol/L})^m}{k(0.020 \text{ mol/L})^m} = \frac{4.8 \times 10^{-6} \text{ mol/(L} \cdot \text{s)}}{9.6 \times 10^{-6} \text{ mol/(L} \cdot \text{s)}}$$

Since k is a constant for reactions that occur at a constant temperature, you can cancel out k.

$$\frac{\cancel{k}(0.010 \cancel{\text{mol/L}})^m}{\cancel{k}(0.020 \cancel{\text{mol/L}})^m} = \frac{4.8 \times 10^{-6} \cancel{\text{mol/(L}} \cdot \text{s)}}{9.6 \times 10^{-6} \cancel{\text{mol/(L}} \cdot \text{s)}}$$

$$(0.5)^m = 0.5$$

$$m = 1 \text{ (by inspection)}$$

CONCEPT CHECK

Check your value of m by setting up a ratio for experiments 1 and 3 (or experiments 2 and 3). Solve for m, and compare.

CONCEPT CHECK

Notice that the units of the rate constant are determined by your calculation. For a first-order reaction, the units are always s^{-1}. Show why this is the case. What are the units for a second-order reaction?

Determining the Rate Constant

Once you know the rate law equation for a reaction, you can calculate the rate constant using the results from any of the experiments. The rate law equation for the previous rate experiments is written as follows:

$$\text{Rate} = k[N_2O_5]^1$$

You can use data from any of the three experiments to calculate k. Substituting data from experiment 1, you can solve for k.

$$4.8 \times 10^{-6}\ \text{mol/(L} \cdot \text{s)} = k(0.010\ \text{mol/L})$$

$$k = \frac{4.8 \times 10^{-6}\ \cancel{\text{mol}}/(\cancel{\text{L}} \cdot \text{s})}{0.010\ \cancel{\text{mol/L}}}$$

$$= 4.8 \times 10^{-6}\text{s}^{-1}$$

Therefore, for the temperature at which the experiments were carried out, $4.8 \times 10^{-4}\ \text{s}^{-1}$ is the rate constant. Using data from experiment 2 would give the same value of k.

Sample Problem

Finding a Rate Law Equation

Problem

Chlorine dioxide, ClO_2, reacts with hydroxide ions to produce a mixture of chlorate and chlorite ions.

$$2ClO_{2(aq)} + 2OH^-_{(aq)} \rightarrow ClO_3{}^-_{(aq)} + ClO_2{}^-_{(aq)} + H_2O_{(\ell)}$$

The rate data in the table below were determined at a constant temperature. Find the rate law equation and the value of k.

Experiment	Initial $[ClO_2]$ (mol/L)	Initial $[OH^-]$ (mol/L)	Initial rate of formation of products (mol/(L $\cdot$s))
1	0.0150	0.0250	1.30×10^{-3}
2	0.0150	0.0500	2.60×10^{-3}
3	0.0450	0.0250	1.16×10^{-2}

What Is Required?

You need to find the value of m, n, and k in the general rate law equation for the reaction.

$$\text{Rate} = k[ClO_2]^m[OH^-]^n$$

What Is Given?

You know the initial concentrations of the reactants and the initial rates for three experiments.

Plan Your Strategy

Step 1 Find two experiments in which $[ClO_2]$ remains constant and $[OH^-]$ changes. Compare the rates and concentrations to solve for n. Then find two experiments in which $[OH^-]$ remains constant and $[ClO_2]$ changes. Compare rates and concentrations to find m.

Step 2 Use the data and your calculated values for m and n to solve for k, using the following equation.

$\text{Rate} = k[ClO_2]^m[OH^-]^n$

Act on Your Strategy

Step 1 Begin with experiments 1 and 2, where $[ClO_2]$ is constant, to determine the order of the reaction with respect to OH^-.

Set up a ratio for the two rates as shown. Notice that you can cancel out $[ClO_2]$ because you chose experiments in which $[ClO_2]$ did not change.

$$\frac{Rate_2}{Rate_1} = \frac{k(\cancel{0.0150\ mol/L})^m(0.0500\ mol/L)^n}{k(\cancel{0.0150\ mol/L})^m(0.0250\ mol/L)^n} = \frac{2.60 \times 10^{-3}\ mol/(L \cdot s)}{1.30 \times 10^{-3}\ mol/(L \cdot s)}$$

Because k is a constant, you can cancel it out.

$$\frac{\cancel{k}(0.0500\ \cancel{mol/L})^n}{\cancel{k}(0.0250\ \cancel{mol/L})^n} = \frac{2.60 \times 10^{-3}\ \cancel{mol/(L} \cdot s)}{1.30 \times 10^{-3}\ \cancel{mol/(L} \cdot s)}$$

$$(2)^n = 2$$

$$n = 1 \text{ (by inspection)}$$

Therefore, the reaction is first order with respect to OH^-.

Set up a ratio for experiments 1 and 3, in which $[OH^-]$ is constant and $[ClO_2]$ changes. You can cancel out the $[OH^-]$ terms right away.

$$\frac{Rate_3}{Rate_1} = \frac{k(0.0450\ mol/L)^m(\cancel{0.0250\ mol/L})^n}{k(0.0150\ mol/L)^m(\cancel{0.0250\ mol/L})^n} = \frac{1.16 \times 10^{-2}\ mol/(L \cdot s)}{1.30 \times 10^{-3}\ mol/(L \cdot s)}$$

Since k is a constant, you can cancel it out.

$$\frac{\cancel{k}[0.0450\ \cancel{mol/L}]^m}{\cancel{k}[0.0150\ \cancel{mol/L}]^m} = \frac{1.16 \times 10^{-2}\ \cancel{mol/(L} \cdot s)}{1.30 \times 10^{-3}\ \cancel{mol/(L} \cdot s)}$$

$$(3)^m = 8.92 \approx 9$$

$$m = 2 \text{ (by inspection)}$$

The reaction is second order with respect to ClO_2.
Therefore, the rate law equation is written as follows:

$$Rate = k[ClO_2]^2[OH^-]$$

The reaction is second order in ClO_2, first order in OH^-, and third order overall.

Step 2 To find the value of the rate constant, substitute data from any of the three experiments into the rate law equation. Using the data from experiment 1 gives the following equation.

$$1.30 \times 10^{-3} mol/(L \cdot s) = k(0.0150\ mol/L)^2(0.0250\ mol/L)$$

Solve for k.

$$k = \frac{1.30 \times 10^{-3}\ mol/(L \cdot s)}{(0.0150\ mol/L)^2(0.0250\ mol/L)}$$
$$= 231\ L^2/(mol^2 \cdot s)$$

Therefore, the rate constant at the temperature of the experiment is $231\ L^2/(mol^2 \cdot s)$. The rate law equation is $Rate = 231\ L^2/(mol^2 \cdot s)[ClO_2]^2[OH^-]$.

Check Your Solution

To check your values of m and n, solve by inspection. When $[ClO_2]$ is constant and $[OH^-]$ doubles, the rate also doubles and the reaction is first order in OH^-. When $[OH^-]$ is constant and $[ClO_2]$ triples, the rate increases by a factor of 9. Recall that $3^2 = 9$, indicating a second-order relationship. To check your value of k, substitute data from experiment 2 or 3. You should get the same answer. Also check the units and the number of significant digits for k.

PROBLEM TIP

When you set up your ratio to compare rates, it does not matter which rate is the numerator, and which is the denominator. It is often easier to see the solution, however, if the larger rate is the numerator.

Practice Problems

5. When heated, ethylene oxide decomposes to produce methane and carbon monoxide.
$C_2H_4O_{(g)} \rightarrow CH_{4(g)} + CO_{(g)}$
At 415°C, the following initial rate data were recorded.

Experiment	$[C_2H_4O]_0$ (mol/L)	Initial rate (mol/(L • s))
1	0.002 85	5.84×10^{-7}
2	0.004 28	8.76×10^{-7}
3	0.005 70	1.17×10^{-6}

Determine the rate law equation and the rate constant at 415°C.

6. Iodine chloride reacts with hydrogen to produce iodine and hydrogen chloride.
$2ICl + H_2 \rightarrow I_2 + 2HCl$

At temperature *T*, the following initial rate data were recorded.

Experiment	$[ICl]_0$ (mol/L)	$[H_2]_0$ (mol/L)	Initial rate (mol/(L • s))
1	0.20	0.050	0.0015
2	0.40	0.050	0.0030
3	0.20	0.200	0.0060

Determine the rate law equation and the rate constant at temperature *T*.

7. Sulfuryl chloride (also known as chlorosulfuric acid and thionyl chloride), SO_2Cl_2, is used in a variety of applications, including the synthesis of pharmaceuticals, rubber-based plastics, dyestuff, and rayon. At a certain temperature, the rate of decomposition of sulfuryl chloride was studied.
$SO_2Cl_{2(g)} \rightarrow SO_{2(g)} + Cl_{2(g)}$

$[SO_2Cl_2]$ (mol/L)	Initial rate (mol/(L • s))
0.150	3.3×10^{-6}
0.300	6.6×10^{-6}
0.450	9.9×10^{-6}

(a) Write the rate law equation for the decomposition of sulfuryl chloride.

(b) Determine the rate constant, *k*, for the reaction, with the appropriate units.

8. Consider the following reaction.
$2A + 3B + C \rightarrow$ products

This reaction was found to obey the following rate law equation.
Rate = $k[A]^2[B][C]$

Copy the following table into your notebook. Then use the given information to predict the blank values. Do not write in this textbook.

Experiment	Initial [A] (mol/L)	Initial [B] (mol/L)	Initial [C] (mol/L)	Initial rate (mol/(L • s))
1	0.10	0.20	0.050	0.40
2	0.10	(a)	0.10	0.40
3	0.20	0.050	(b)	0.20
4	(c)	0.025	0.040	0.45
5	0.10	0.010	0.15	(d)

The Half-Life of a Reaction

You have seen that a large rate constant indicates a fast reaction, while a small rate constant indicates a slow reaction. It is not always easy, however, to determine how long a reaction will take to be completed just by looking at the value of k. How can you relate the speed of a reaction to the units of s^{-1} for a first-order reaction?

The **half-life**, $t_{1/2}$, of a reaction is the time that is needed for the reactant mass or concentration to decrease by one half of its initial value. The SI units for half-life are seconds. Usually, however, half-life is expressed in whatever units of time are appropriate to the reaction. For example, the half-life of a very fast reaction may be measured in microseconds. The half-life of a slow reaction may be measured in days. Knowing the half-life of a reaction is an easy way to tell how fast or how slow a reaction is.

Examine Figure 6.6 to see how half-life and reactant concentration are related for the first-order decomposition of dinitrogen pentoxide.

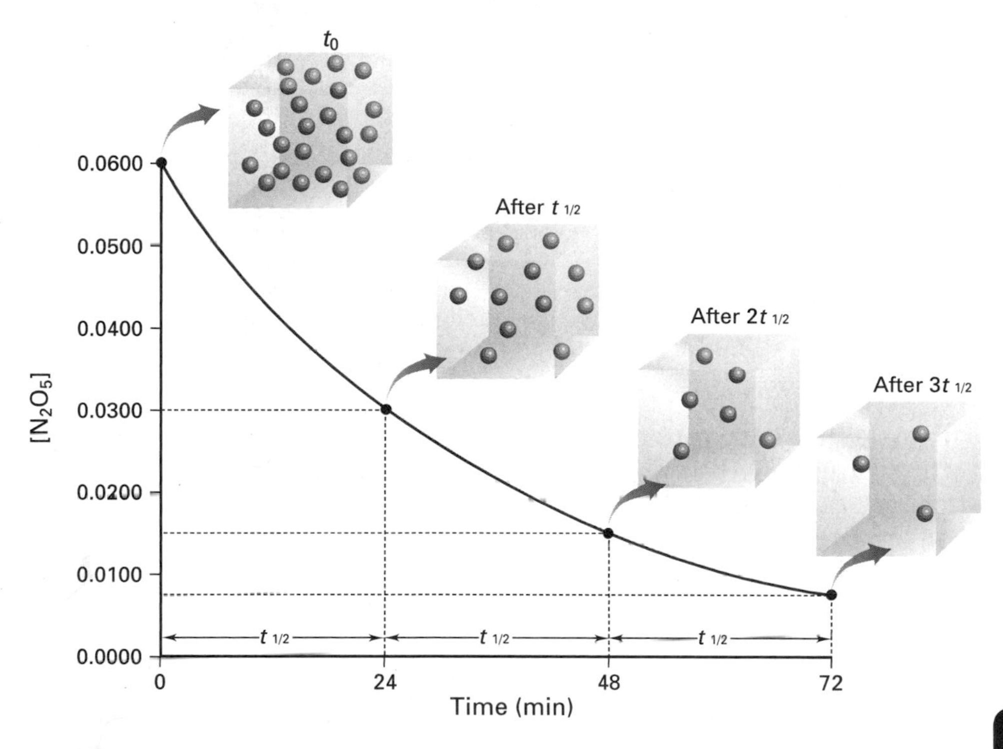

Figure 6.6 This graph shows $[N_2O_5]$ versus time for the first three half-lives of the decomposition reaction of N_2O_5 at 44°C . The spheres in the cubes represent $[N_2O_5]$.

The half-life of any first-order reaction is always a constant, and it depends on k. In other words, any first-order reaction (that is, any reaction with a rate law equation in the form Rate = k[A]) has a half-life that is independent of the initial concentration of the reactant, A. The half-life of *any first-order reaction* can be calculated using the following equation.

$$t_{1/2} = \frac{0.693}{k}$$

In this equation, k is the first-order rate constant. It has units of s^{-1}.

Examine the Sample Problem on the next page to learn how to calculate the half-life of a first-order reaction. Then try the Practice Problems that follow.

Math LINK

Where does the equation for half-life come from? Each rate law has an associated *integrated rate law.* (A calculus teacher may be able to show you how to arrive at the integrated rate law.) For first-order reactions, the integrated rate law is $\ln\left(\frac{[A]_0}{[A]_t}\right) = kt$, where A is a reactant. Show that $[A]_{t_{1/2}} = \frac{1}{2}[A]_0$ for all first-order reactions. Then use the integrated rate law to arrive at the equation for half-life.

Sample Problem

Calculating the Half-Life of a First-Order Reaction

Problem

The decomposition of sulfuryl chloride, SO_2Cl_2, is a first-order reaction.

$SO_2Cl_{2(g)} \rightarrow SO_{2(g)} + Cl_{2(g)}$

Rate $= k[SO_2Cl_2]$

At 320°C, the rate constant is 2.2×10^{-5} s^{-1}.

(a) Calculate the half-life of the reaction, in hours.

(b) How long does the sulfuryl chloride take to decrease to $\frac{1}{8}$ of its original concentration?

What Is Required?

(a) You need to find the half-life, $t_{1/2}$, of the reaction.

(b) You need to find the time required for $[SO_2Cl_2]$ to decrease to $\frac{1}{8}$ of its initial value, $[SO_2Cl_2]_0$.

What Is Given?

You know the rate equation and the rate constant.

Rate $= k[SO_2Cl_2]$

$k = 2.2 \times 10^{-5}$ s^{-1}

Plan Your Strategy

(a) To find $t_{1/2}$, use the following equation.

$$t_{1/2} = \frac{0.693}{k}$$

(b) Each half-life reduces the initial concentration of the reactant, $[A]_0$, by one half.

After 1 $t_{1/2}$, $[A] = \frac{1}{2}[A]_0$

After 2 $t_{1/2}$, $[A] = \frac{1}{2} \times (\frac{1}{2}[A]_0)$
$= \frac{1}{4}[A]_0$

After 3 $t_{1/2}$, $[A] = \frac{1}{2} \times (\frac{1}{4}[A]_0)$
$= \frac{1}{8}[A]_0$

The concentration of reactant remaining after 3 half-lives is $\frac{1}{8}$, or $(\frac{1}{2})^3$, of the original amount.

PROBLEM TIP

In general, the concentration of reactant that is left after n half-lives have passed, $[A]_t$, can be expressed as follows.

$[A]_t = x[A]_0 = (\frac{1}{2})^n \times [A]_0$

Act on Your Strategy

(a) Solve for $t_{1/2}$.

$$t_{1/2} = \frac{0.693}{2.2 \times 10^{-5}\ s^{-1}}$$
$$= 3.2 \times 10^4\ s$$

Convert the units of your answer to hours.

$$t_{1/2} = \frac{3.2 \times 10^4\ \cancel{s}}{(60\ \cancel{s}/\cancel{min})(60\ \cancel{min}/h)}$$
$$= 8.8\ h$$

The half-life of the reaction, in hours, is 8.8 h.

(b) Determine how many half-lives correspond to $\frac{1}{8} \times [SO_2Cl_2]_0$.

$$\frac{1}{8} \times [SO_2Cl_2]_0 = (\frac{1}{2})^n \times [SO_2Cl_2]_0$$

$$\frac{1}{8} = (\frac{1}{2})^n$$

$$n = 3 \text{ (by inspection)}$$

Determine the time span that corresponds to 3 half-lives.

$3 \times 8.8 \text{ h} = 26 \text{ h}$

It would take 26 h for the concentration of SO_2Cl_2 to decrease to $\frac{1}{8}[SO_2Cl_2]_0$.

Check Your Solution

The units and significant digits for half-life are correct. The small rate constant of 2.2×10^{-5} corresponds to a slow reaction with a long half-life. Three half-lives corresponds to $\frac{1}{2} \times \frac{1}{2} \times \frac{1}{2} = \frac{1}{8}$ of the initial concentration.

Practice Problems

9. Cyclopropane, C_3H_6, has a three-membered hydrocarbon ring structure. It undergoes rearrangement to propene. At 1000°C, the first-order rate constant for the decomposition of cyclopropane is 9.2 s^{-1}.

(a) Determine the half-life of the reaction.

(b) What percent of the original concentration of cyclopropane will remain after 4 half-lives?

10. Peroxyacetyl nitrate (PAN), $H_3CCO_2ONO_2$, is a constituent of photochemical smog. It undergoes a first-order decomposition reaction with $t_{1/2} = 32$ min.

(a) Calculate the rate constant in s^{-1} for the first-order decomposition of PAN.

(b) 128 min after a sample of PAN began to decompose, the concentration of PAN in the air is 3.1×10^{13} molecules/L. What was the concentration of PAN when the decomposition began?

11. In general, a reaction is essentially over after 10 half-lives. Prove that this generalization is reasonable.

12. The half-life of a certain first-order reaction is 120 s. How long do you estimate that it will take for 90% of the original sample to react?

Section Summary

In this section, you learned how to relate the rate of a chemical reaction to the concentrations of the reactants using the rate law. You classified reactions based on their reaction order. You determined the rate law equation from empirical data. Then you learned about the half-life of a first-order reaction. As you worked through sections 6.1 and 6.2, you may have wondered *why* factors such as concentration and temperature affect the rates of chemical reactions. In the following section, you will learn about some theories that have been developed to explain the effects of these factors.

Section Review

1 K/U The following reaction is second order in A and first order in B.

$2A + 1B \rightarrow 3C$

What is the rate law equation for the reaction below?
(Assume that A, B, and C are the same compounds for each reaction.)

$4A + 2B \rightarrow 6C$

2 K/U Consider the general reaction below.

$aA + bB \rightarrow cC + dD$

Based on this equation, is it correct to write the following rate law equation by inspection? Explain your answer.

$\text{Rate} = k[A]^a[B]^b$

3 K/U Consider the following rate law equation.

$\text{Rate} = k[A]^2[B]$

(a) How does the reaction rate change if [A] decreases by a factor of 2 and [B] increases by a factor of 4?

(b) How does the reaction rate change if [A] and [B] are doubled?

4 K/U Consider the following rate law equation.

$\text{Rate} = k[HCrO_4^-][HSO_3^-]^2[H^+]$

(a) What is the order with respect to each reactant?

(b) What is the overall reaction order?

(c) What are the units for the rate constant?

5 T/I The data in the table below were collected for the following reaction.

$A + B \rightarrow C + D$

Experiment	$[A]_0$ (mol/L)	$[B]_0$ (mol/L)	Initial rate (mol/(L • s))
1	0.020	0.020	5.0×10^{-3}
2	0.040	0.020	1.0×10^{-2}
3	0.040	0.060	9.0×10^{-2}

(a) What is the rate law equation for the reaction?

(b) Calculate the value of the rate constant, with proper units.

6 T/I The data in the table below were collected, at 25°C, for the following reaction.

$2A + B + 2C \rightarrow 3D$

Determine the rate law equation and the value of k at 25°C, with proper units.

Experiment	$[A]_0$ (mol/L)	$[B]_0$ (mol/L)	$[C]_0$ (mol/L)	Initial rate (mol/(L • s))
1	0.20	0.10	0.10	4.0×10^{-4}
2	0.20	0.30	0.20	1.2×10^{-3}
3	0.20	0.10	0.30	4.0×10^{-4}
4	0.60	0.30	0.40	3.6×10^{-3}

7 T/I A first-order decomposition reaction has a rate constant of 2.34×10^{-2} year^{-1}. What is the half-life of the reaction? Express your answer in years and in seconds.

Theories of Reaction Rates

6.3

In section 6.2, you explored the rate law, which defines the relationship between the concentrations of reactants and reaction rate. Why, however, does the rate of a reaction increase with increased concentrations of reactants? Why do increased temperature and surface area increase reaction rates? To try to explain these and other macroscopic observations, chemists develop theories that describe what happens as reactions proceed on the molecular scale. In this section, you will explore these theories.

Section Preview/
Specific Expectations

In this section, you will

- **explain**, using the collision theory and potential energy diagrams, how different factors, such as temperature and concentration, control the rate of a chemical reaction
- **analyze** simple potential energy diagrams of chemical reactions
- **communicate** your understanding of the following terms: *collision theory, activation energy (E_a), transition state theory, potential energy diagram, transition state, activated complex*

Collision Theory

Why do factors such as temperature and concentration increase or decrease the rate of a reaction? To answer this question, chemists must first answer another question: What causes a reaction to occur? One obvious answer is that a reaction occurs when two reactant particles collide with one another. This answer is the basis for **collision theory**: In order for a reaction to occur, reacting particles (atoms, molecules, or ions) must collide with one another.

The Effect of Concentration on Reactant Rates

You can use simple collision theory to begin to understand why factors such as concentration affect reaction rate. If a collision is necessary for a reaction to occur, then it makes sense that the rate of the reaction will increase if there are more collisions per unit time. More reactant particles in a given volume (that is, greater concentration) will increase the number of collisions between the particles per second. Figure 6.7 illustrates this idea.

Electronic Learning Partner

Go to the Chemistry 12 Electronic Learning Partner for more information about aspects of material covered in this section of the chapter.

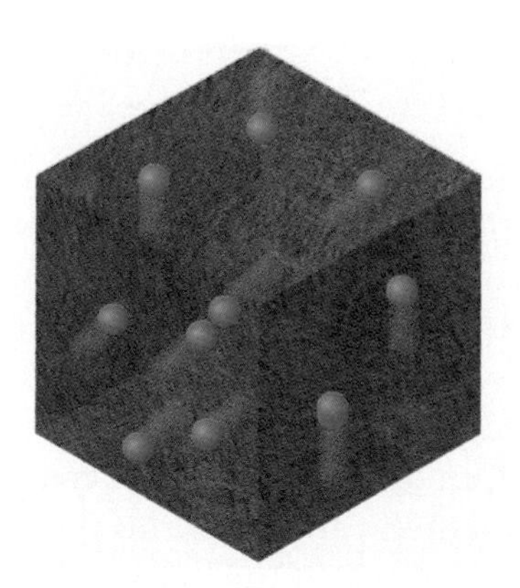

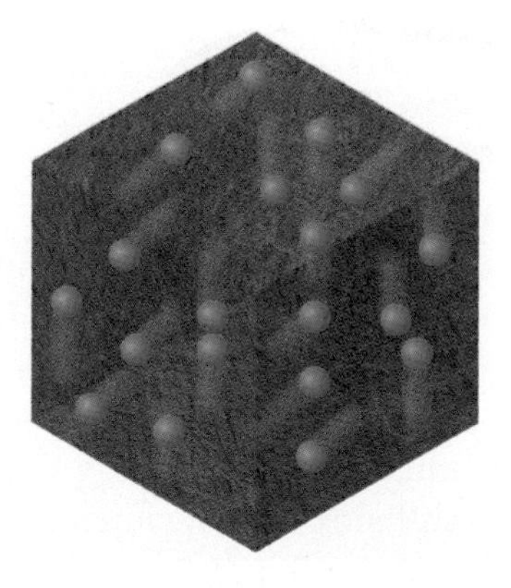

Figure 6.7 At increased reactant concentrations, there is an increased number of collisions per second.

Collision Theory and Surface Area

You can also use simple collision theory to explain why increasing the surface area of a solid-phase reactant speeds up a reaction. With greater surface area, more collisions can occur. This explains why campfires are started with paper and small twigs, rather than logs. Figure 6.8 shows an example of the effect of surface area on collision rate.

Figure 6.8 Mixtures of grain or coal dust and air are potentially explosive, as shown in this grain elevator explosion.

Beyond Collision Theory

Simple collision theory recognizes that a collision between reactants is necessary for a reaction to proceed. Does every collision result in a reaction? Consider a 1 mL sample of gas at room temperature and atmospheric pressure. In the sample, about 10^{28} collisions per second take place between gas molecules. If each collision resulted in a reaction, *all* gas phase reactions would be complete in about a nanosecond (10^{-9} s)—a truly explosive rate! As you know from section 6.2, however, gas phase reactions can occur quite slowly. This suggests that *not every collision between reactants results in a reaction.*

In order for a collision between reactants to result in a reaction, the collision must be *effective*. An effective collision—one that results in the formation of products—must satisfy the following two criteria. You will investigate these criteria over the next few pages.

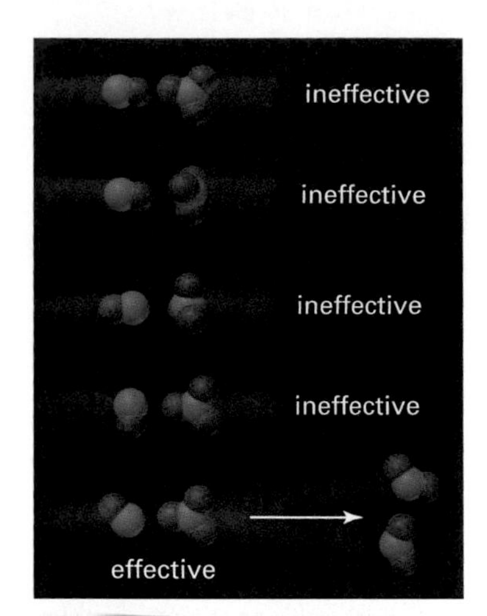

Figure 6.9 Only one of these five possible orientations of NO and NO_3 will lead to the formation of a product.

For a collision to be effective, it must satisfy both of these criteria:
1. correct orientation of reactants
2. sufficient collision energy

Orientation of Reactants

Reacting particles must collide with the proper orientation relative to one another. This is also known as having the correct *collision geometry*. The importance of proper collision geometry can be illustrated by the following reaction.

$$NO_{(g)} + NO_{3(g)} \rightarrow NO_{2(g)} + NO_{2(g)}$$

Figure 6.9 shows five of the many possible ways in which NO and NO_3 can collide. Only one of these five possibilities has the correct collision geometry for a reaction to occur. As shown in the figure, only a certain orientation of reactants prior to collision leads to the formation of two molecules of nitrogen dioxide.

Activation Energy

In addition to collision geometry, there is a second factor that determines whether a collision will result in a reaction: the energy of the collision. The reactants must collide with energy that is sufficient to begin to break the bonds in the reactants and to begin to form the bonds in the products. In most reactions, only a small fraction of collisions have sufficient energy for a reaction to occur. The **activation energy, E_a**, of a reaction is the minimum collision energy that is required for a successful reaction.

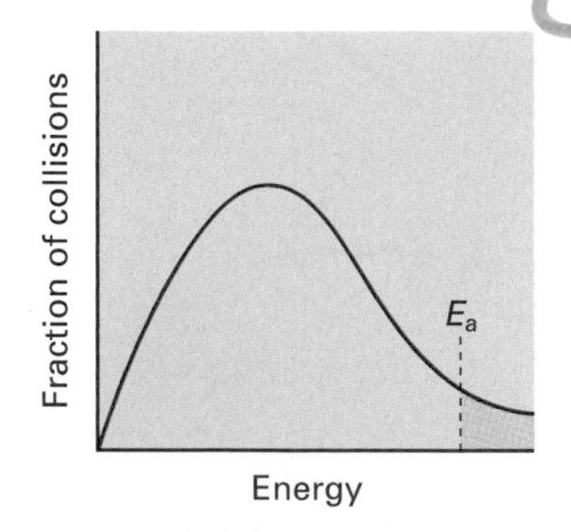

Figure 6.10 The area under a Maxwell-Boltzmann distribution graph represents the distribution of the kinetic energy of collisions at a constant temperature. At a given temperature, only a certain fraction of the molecules in a sample have enough kinetic energy to react.

The collision energy depends on the kinetic energy of the colliding particles. As you know, temperature is a measure of the average kinetic energy of the particles in a substance. If you plot the number of collisions in a substance at a given temperature against the kinetic energy of each collision, you get a curve like the one in Figure 6.10. The type of distribution that is shown by this curve is known as a *Maxwell-Boltzmann distribution*. The dotted line indicates the activation energy. The shaded part of the graph indicates the collisions with energy that is equal to or greater than the activation energy.

How does the distribution of kinetic energy change as the temperature of a substance increases? Figure 6.11 shows the distribution of kinetic energy in a sample of reacting gases at two different temperatures, T_1 and T_2, where $T_2 > T_1$. The activation energy is indicated by the dashed vertical line. Two observations are apparent from the graph:

1. At both temperatures, a relatively small fraction of collisions have sufficient kinetic energy—the activation energy—to result in a reaction.
2. As the temperature of a sample increases, the fraction of collisions with sufficient energy to cause a reaction increases significantly.

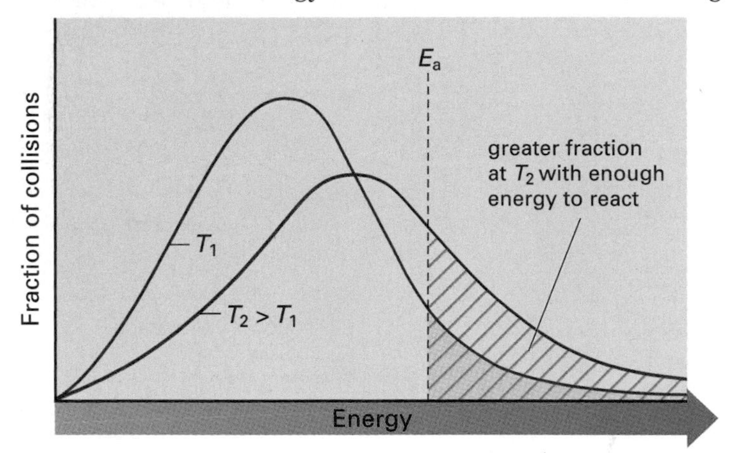

Figure 6.11 At increased temperatures, more particles collide with enough energy to react.

Go to the Chemistry 12 Electronic Learning Partner to learn more about collision geometry.

For many reactions, the rate roughly doubles for every 10°C rise in temperature.

Transition State Theory

Transition state theory is used to explain what happens when molecules collide in a reaction. It examines the transition, or change, from reactants to products. The kinetic energy of the reactants is transferred to potential energy as the reactants collide, due to the law of conservation of energy. This is analogous to a bouncing basketball. The kinetic energy of the ball is converted to potential energy, which is stored in the deformed ball as it hits the floor. The potential energy is converted to kinetic energy as the ball bounces away.

You can represent the increase in potential energy during a chemical reaction using a **potential energy diagram**: a diagram that charts the potential energy of a reaction against the progress of the reaction. Examples of potential energy diagrams are shown in Figures 6.12 and 6.13 on the next page. The *y*-axis represents potential energy. The *x*-axis, labelled "Reaction progress," represents the progress of the reaction through time.

The "hill" in each diagram illustrates the activation energy barrier of the reaction. A slow reaction has a high activation energy barrier. This indicates that relatively few reactants have sufficient kinetic energy for a successful reaction. A fast reaction, by contrast, has a low activation energy barrier.

A potential energy diagram for an exothermic reaction is shown in Figure 6.12. The reactants at the beginning of the reaction are at a higher energy level than the products. The overall difference in potential energy is the enthalpy change, ΔH. *There is no way to predict the activation energy of a reaction from its enthalpy change.* A highly exothermic reaction may be very slow because it has a high activation energy. Conversely, a reaction that is very fast may release very little heat. The enthalpy change of a reaction is determined by finding the overall energy that is transferred. The activation energy of a reaction is determined by analyzing the reaction rate.

A potential energy diagram for an endothermic reaction is shown in Figure 6.13. The reactants at the beginning of the reaction are at a lower energy level than the products. The overall difference in potential energy between reactants and products is the enthalpy change.

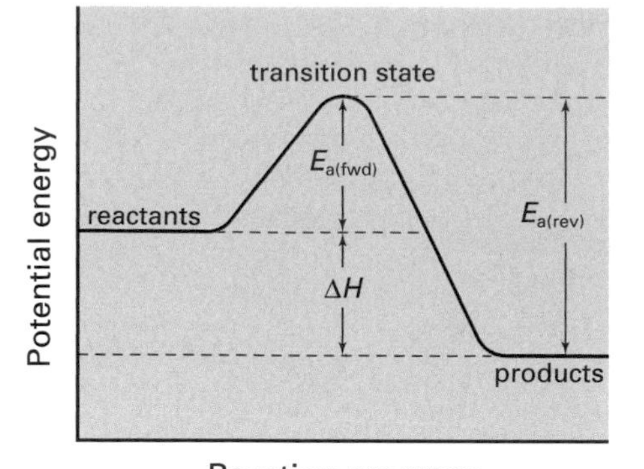

Figure 6.12 **A potential energy diagram for an exothermic reaction**

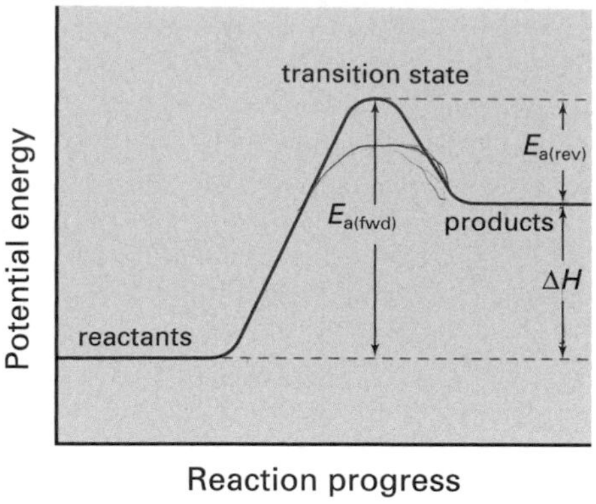

Figure 6.13 **A potential energy diagram for an endothermic reaction**

You may already know that many reactions can proceed in two directions. For example, hydrogen and oxygen react to form water. Water, however, can also undergo electrolysis, forming hydrogen and oxygen. This is the reverse of the first reaction.

$$H_{2(g)} + \frac{1}{2}O_{2(g)} \rightarrow H_2O_{(\ell)} \quad \Delta H^\circ = -285.8 \text{ kJ}$$

$$H_2O_{(\ell)} \rightarrow H_{2(g)} + \frac{1}{2}O_{2(\ell)} \quad \Delta H^\circ = 285.8 \text{ kJ}$$

The enthalpy change of the first reaction is the same as the enthalpy change of the second reaction, with the opposite sign. (You can show this using Hess's law.) How are the activation energies of forward and reverse reactions related? For an exothermic reaction, the activation energy of the reverse reaction, $E_{a(rev)}$ equals $E_{a(fwd)} + \Delta H$. For an endothermic reaction, $E_{a(rev)}$ equals $E_{a(fwd)} - \Delta H$. Figures 6.12 and 6.13 show the activation energies of forward and reverse reactions.

The top of the activation energy barrier on a potential energy diagram represents the **transition state**, or change-over point, of the reaction. The chemical species that exists at the transition state is referred to as the **activated complex**. The activated complex is a transitional species that is neither product nor reactant. It has partial bonds and is highly unstable. There is a subtle difference between the transition state and the activated complex. The transition state refers to the top of the "hill" on a potential energy diagram. The *chemical species* that exists at this transition point is called the activated complex.

Tracing a Reaction With a Potential Energy Diagram

Consider the substitution reaction between a hydroxide, OH^-, ion and methyl bromide, $BrCH_3$. Methanol, CH_3OH, and a bromide, Br^-, ion are formed.

$$BrCH_3 + OH^- \rightarrow CH_3OH + Br^-$$

Figure 6.14 is a potential energy diagram for this reaction. It includes several "snapshots" as the reaction proceeds.

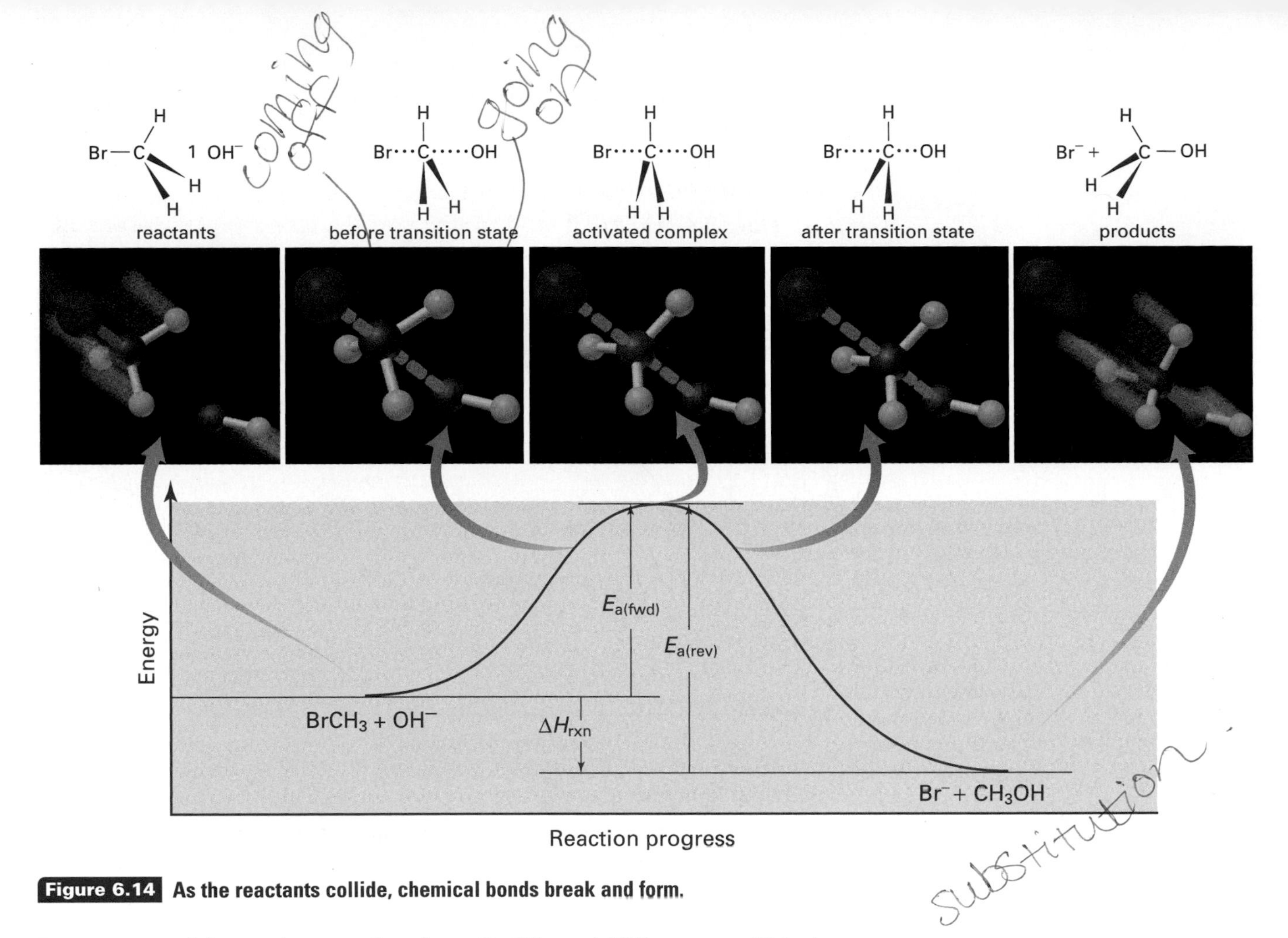

Figure 6.14 As the reactants collide, chemical bonds break and form.

For a successful reaction to take place, $BrCH_3$ and OH^- must collide in a favourable orientation. The OH^- ion must approach $BrCH_3$ from the side that is *opposite* to the Br atom. When this occurs, a partial bond is formed between the O of the OH^- ion and the C atom. Simultaneously, the C—Br bond is weakened.

Because the activated complex contains partial bonds, it is highly unstable. It can either break down to form products or it can decompose to re-form the reactants. The activated complex is like a rock teetering on top of a mountain. It could fall either way.

Sample Problem

Drawing a Potential Energy Diagram

Problem

Carbon monoxide, CO, reacts with nitrogen dioxide, NO_2. Carbon dioxide, CO_2, and nitric oxide, NO, are formed. Draw a potential energy diagram to illustrate the progress of the reaction. (You do not need to draw your diagram to scale). Label the axes, the transition state, and the activated complex. Indicate the activation energy of the forward reaction, $E_{a(fwd)} = 134$ kJ, as well as $\Delta H = -226$ kJ. Calculate the activation energy of the reverse reaction, $E_{a(rev)}$, and show it on the graph.

Continued...

Continued ...

Solution

Since $E_{a(rev)} = \Delta H + E_{a(fwd)}$

$$E_{a(rev)} = 226 \text{ kJ} + 134 \text{ kJ}$$
$$= 360 \text{ kJ}$$

The activation energy of the reverse reaction is 360 kJ.

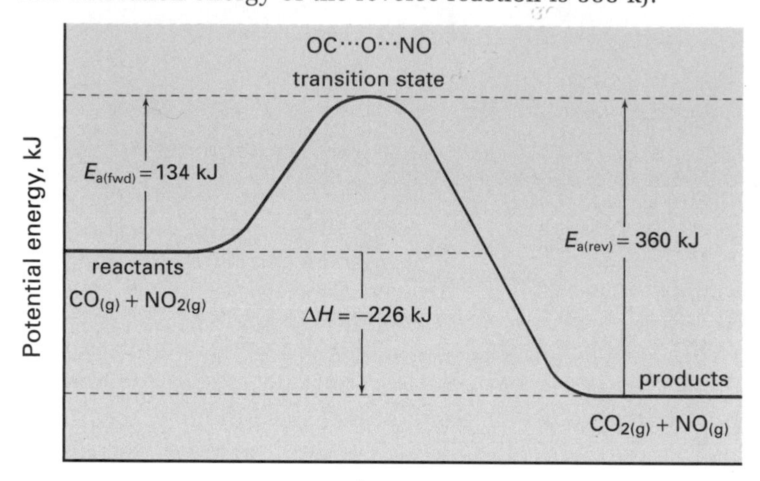

Check Your Solution

Look carefully at the potential energy diagram. Check that you have labelled it completely. Since the forward reaction is exothermic, the reactants should be at a higher energy level than the products, and they are. The value of $E_{a(rev)}$ is reasonable.

Practice Problems

13. The following reaction is exothermic.

$2ClO_{(g)} \rightarrow Cl_{2(g)} + O_{2(g)}$

Draw and label a potential energy diagram for the reaction. Propose a reasonable activated complex.

14. Consider the following reaction.

$AB + C \rightarrow AC + B \quad \Delta H = +65 \text{ kJ}, E_{a(rev)} = 34 \text{ kJ}$

Draw and label a potential energy diagram for this reaction. Calculate and label $E_{a(fwd)}$. Include a possible structure for the activated complex.

15. Consider the reaction below.

$C + D \rightarrow CD \quad \Delta H = -132 \text{ kJ}, E_{a(fwd)} = 61 \text{ kJ}$

Draw and label a potential energy diagram for this reaction. Calculate and label $E_{a(rev)}$. Include a possible structure for the activated complex.

16. In the upper atmosphere, oxygen exists in forms other than $O_{2(g)}$. For example, it exists as ozone, $O_{3(g)}$, and as single oxygen atoms, $O_{(g)}$. Ozone and atomic oxygen react to form two molecules of oxygen. For this reaction, the enthalpy change is −392 kJ and the activation energy is 19 kJ. Draw and label a potential energy diagram. Include a value for $E_{a(rev)}$. Propose a structure for the activated complex.

Temperature Dependence of Reaction Rates: Applications

As you saw in Figure 6.11, when reactions occur at higher temperatures, the percent of particles that have sufficient kinetic energy to react (E_a) increases. Since a greater percent of reactant particles have enough energy to react, the rate of the reaction increases.

Chemists exploit the temperature dependence of reaction rates by carrying out chemical reactions at elevated temperatures to speed them up. In organic chemistry, especially, reactions are commonly performed under *reflux*: that is, while boiling the reactants. To prevent reactants and products from escaping as gases, a water-cooled condenser tube is fitted to the reaction vessel. The tube condenses the vapours to liquids and returns them to the reaction vessel. Figure 6.15 shows an experiment performed under reflux.

At home, you take advantage of the temperature dependence of chemical reactions all the time. For example, to keep your food fresh, you store it in a refrigerator. If you have ever left milk or vegetables in the refrigerator for several weeks, however, you have probably observed that refrigeration does not *stop* food from spoiling. Instead, it decreases the *rate* of the reactions that cause spoilage. When you want to cook your food quickly, you increase the temperature of the stove. This increases the rate of the reactions that take place as the food cooks.

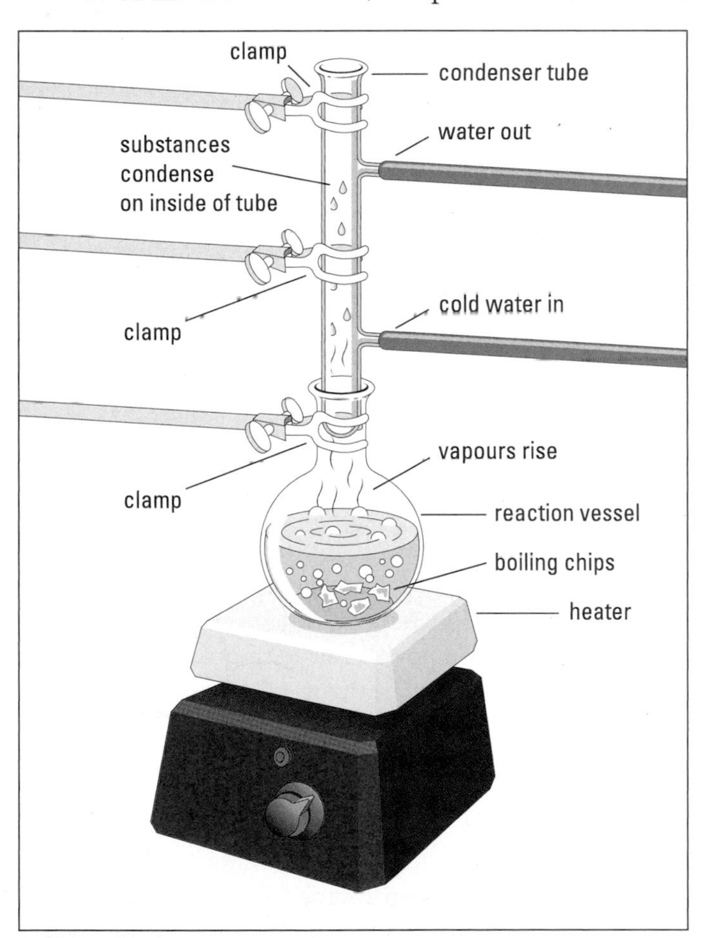

Figure 6.15 As the reactants are boiled under reflux, the vapours condense and are returned to the reaction vessel. In this way, a chemical reaction can be carried out at the boiling point of the reactants at atmospheric pressure, with no loss of reactants, products, or solvent.

Web LINK

www.mcgrawhill.ca/links/chemistry12

The chirping rate of a cricket is dependent on temperature, much like chemical reaction rates are. Why does this suggest that a cricket is a cold-blooded creature? Search the Internet to determine the mathematical relationship between ambient temperature and the chirping rate of a cricket. Start your search by going to the web site above and clicking on **Web Links**.

Section Summary

In this section, you used collision theory and transition state theory to explain how reaction rates are affected by various factors. You considered simple reactions, consisting of a single-step collision between reactants. Not all reactions are simple, however. In fact, most chemical reactions take place via several steps, occurring in sequence. In the next section, you will learn about the steps that make up reactions and discover how these steps relate to reaction rates.

Section Review

1. K/U In your own words, describe how collision theory explains that increased concentrations result in increased reaction rates.

2. I Suppose that you have a solid sample of an organic substance. You are going to burn the substance in an analytical furnace and determine whether or not any unburned matter remains. The substance is somewhat hard, and you receive it in a solid chunk.
   - **(a)** Suggest two ways to increase the rate of burning so that your analysis will take less time.
   - **(b)** Explain why your suggestions would work, using the theories you learned in this section.

3. K/U Consider the following reaction.
   $A_2 + B_2 \rightarrow 2AB \quad E_{a(fwd)} = 143$ kJ, $E_{a(rev)} = 75$ kJ
   - **(a)** Is the reaction endothermic or exothermic in the forward direction?
   - **(b)** Draw and label a potential energy diagram. Include a value for ΔH.
   - **(c)** Suggest a possible activated complex.

4. C Consider two exothermic reactions. Reaction (1) has a much smaller activation energy than reaction (2).
   - **(a)** Sketch a potential energy diagram for each reaction, showing how the difference in activation energy affects the shape of the graph.
   - **(b)** How do you think the rates of reactions (1) and (2) compare? Explain your answer.

5. C Your friend is confused about the difference between the enthalpy change and the activation energy of a chemical reaction. Write a few paragraphs, in which you define each term and distinguish between them. Use potential energy diagrams to illustrate your answer.

6. MC In a coal-fired electric generating plant, the coal is pulverized before being mixed with air in the incinerator. Use collision theory to explain why the coal is pulverized.

7. C People who have been submerged in very cold water, and presumed drowned, have sometimes been revived. By contrast, people who have been submerged for a similar period of time in warmer water have not survived. Suggest reasons for this difference.

8. C Create a graphic organizer to summarize how the theories you have studied in section 6.3 explain the effects of concentration, temperature, and surface area on reaction rates.

Reaction Mechanisms and Catalysts

6.4

Chemical reactions are like factories, where goods (products) are created from raw materials (reactants). An assembly line, like the one shown in Figure 6.16, involves many steps. An automobile is not formed from its components in just one step. Similarly, most chemical reactions do not proceed immediately from products to reactants. They take place via a number of steps. While you can go inside a factory to see all the steps that are involved in making an automobile or a piece of clothing, you cannot observe a chemical equation on a molecular scale as it proceeds. Chemists can experimentally determine the reactants and products of a reaction, but they must use indirect evidence to suggest the steps in-between.

Section Preview/
Specific Expectations

In this section, you will

- **demonstrate** an understanding that most reactions occur as a series of elementary reactions in a reaction mechanism
- **evaluate** a proposed mechanism for a chemical reaction
- **determine**, through experiment, the reaction rate and rate law equation for a chemical reaction
- **communicate** your understanding of the following terms: *reaction mechanism, elementary reaction, reaction intermediates, molecularity, bimolecular, unimolecular, termolecular, rate-determining step, homogeneous catalyst, heterogeneous catalyst, enzymes, active site, substrate*

Elementary Reactions

A **reaction mechanism** is a series of steps that make up an overall reaction. Each step, called an **elementary reaction**, involves a single molecular event, such as a simple collision between atoms, molecules, or ions. An elementary step can involve the formation of different molecules or ions, or it may involve a change in the energy or geometry of the starting molecules. It cannot be broken down into further, simpler steps.

For example, consider the following reaction.

$$2NO_{(g)} + O_{2(g)} \rightarrow 2NO_{2(g)}$$

Chemists have proposed the following two-step mechanism for this reaction. Each step is an elementary reaction.

Step 1 $NO_{(g)} + O_{2(g)} \rightarrow NO_{3(g)}$

Step 2 $NO_{3(g)} + NO_{(g)} \rightarrow 2NO_{2(g)}$

When you add the two elementary reactions, you get the overall reaction. Notice that NO_3 is present in both elementary reactions, but it is not present in the overall reaction. It is produced in the first step and consumed in the second step.

Molecules (or atoms or ions) that are formed in an elementary reaction and consumed in a subsequent elementary reaction are called **reaction intermediates**. Even though they are not products or reactants in the overall reaction, reaction intermediates are essential for the reaction to take place. They are also useful for chemists who are trying to find evidence to support a proposed reaction mechanism. You will explore this idea in Investigation 6-B.

The Molecularity of Elementary Reactions

The term **molecularity** refers to the number of reactant particles (molecules, atoms, or ions) that are involved in an elementary reaction.

In the reaction mechanism you just saw, each elementary reaction consisted of two molecules colliding and reacting. When two particles collide and react, the elementary reaction is said to be **bimolecular**.

Figure 6.16 Like most chemical reactions, building a car on an assembly line takes more than one step.

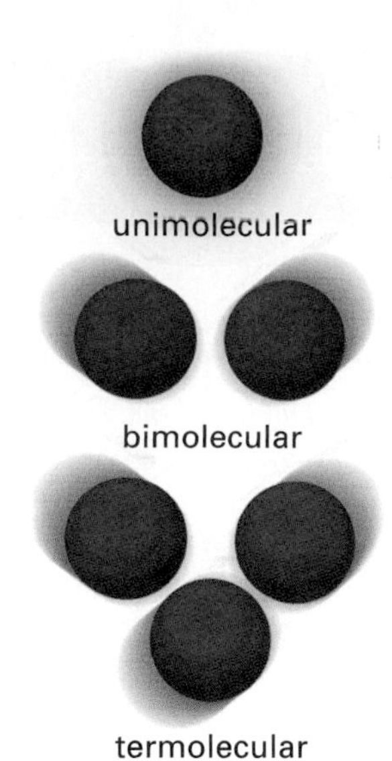

Figure 6.17 This figure shows the molecularity of elementary reactions. Termolecular reactions are rare.

A **unimolecular** elementary reaction occurs when one molecule or ion reacts. For example, when one molecule of chlorine absorbs ultraviolet light, the Cl—Cl bond breaks. The product is two chlorine atoms.

$$Cl_{2(g)} \xrightarrow{\text{UV light}} 2Cl_{(g)}$$

An elementary reaction may also involve three particles colliding in a **termolecular** reaction. Termolecular elementary steps are rare, because it is unlikely that three particles will collide all at once. Think of it this way. You have probably bumped into someone accidentally, many times, on the street or in a crowded hallway. How many times, however, have you and two other people collided at exactly the same time? Figure 6.17 models unimolecular, bimolecular, and termolecular reactions.

Rate Law Equations for Elementary Reactions

In section 6.2, you learned that the rate law equation for a chemical reaction cannot be determined just by looking at the chemical equation. It must be found by experiment. Elementary reactions are the exception to this rule. *For an elementary reaction, the exponents in the rate law equation are the same as the stoichiometric coefficients for each reactant in the chemical equation.* Table 6.3 shows how rate laws correspond to elementary reactions.

Table 6.3 Elementary Reactions and Their Rate Laws

Elementary reaction	Rate law
$A \rightarrow$ products	Rate = $k[A]$
$A + B \rightarrow$ products	Rate = $k[A][B]$
$2A \rightarrow$ products	Rate = $k[A]^2$
$2A + B \rightarrow$ products	Rate = $k[A]^2[B]$

Proposing and Evaluating Mechanisms

You can take apart a grandfather clock to take a look at its mechanism. You can physically examine and take apart its gears and springs to determine how it works.

When chemists investigate the mechanism of a reaction, they are not so lucky. Determining the mechanism of a chemical reaction is a bit like figuring out how a clock works just by looking at its face and hands. For this reason, reaction mechanisms are *proposed* rather than definitively stated. Much of the experimental evidence that is obtained to support a mechanism is indirect. Researchers need a lot of creativity as they propose and test mechanisms.

One of the ways that researchers provide evidence for proposed mechanisms is by proving the existence of a reaction intermediate. Although a reaction intermediate usually appears for a very short time and cannot be isolated, there are other ways to show its presence. For example, if a reaction intermediate is coloured while other reactants are colourless, a spectrophotometer will show when the reaction intermediate is formed. You may even be able to see a fleeting colour change without a spectrophotometer.

When chemists propose a mechanism, they must satisfy the following criteria:

- The equations for the elementary steps must combine to give the equation for the overall reaction.
- The proposed elementary steps must be reasonable.
- The mechanism must support the experimentally determined rate law.

The Rate-Determining Step

How do chemists determine whether a proposed mechanism supports the experimentally determined rate law? They must consider how the rates of the elementary reactions relate to the rate of the overall reaction. Elementary reactions in mechanisms all have different rates. Usually one elementary reaction, called the **rate-determining step**, is much slower. Hence, it determines the overall rate.

To understand the rate-determining step, consider a two-step process. Suppose that you and a friend are making buttered toast for a large group of people. The first step is toasting the bread. The second step is buttering the toast. Suppose that toasting two slices of bread takes about two minutes, but buttering the toast takes only a few seconds.

The rate at which the toast is buttered does not have any effect on the overall rate of making buttered toast, because it is much faster than the rate at which the bread is toasted. The overall rate of making buttered toast, therefore, depends only on the rate of toasting the bread. In other words, two pieces of buttered toast will be ready every two minutes. This is the same as the rate for the first step in the "buttered toast mechanism." Thus, toasting the bread is the rate-determining step in the mechanism.

Step 1 Bread → Toast (slow, rate-determining)

Step 2 Toast + Butter → Buttered toast (fast)

Now suppose that the butter is frozen nearly solid. It takes you about five minutes to scrape off enough butter for one piece of toast. Pieces of toast pile up, waiting to be buttered. In this case, the rate of making buttered toast depends on the rate of spreading the butter. The rate of toasting is fast relative to the rate of buttering.

Step 1 Bread → Toast (fast)

Step 2 Toast + Frozen butter → Buttered toast (slow, rate-determining)

The Rate-Determining Step and the Rate Law

How does the rate-determining step relate to the rate law for the overall equation? Consider the reaction of nitrogen dioxide with chlorine.

$$2NO_{2(g)} + Cl_{2(g)} \rightarrow 2NO_2Cl_{(g)}$$

Experiments show that this reaction has the following rate equation.

$$\text{Rate} = k[NO_2][Cl_2]$$

The proposed mechanism for the reaction is shown below.

Step 1 $NO_{2(g)} + Cl_{2(g)} \rightarrow NO_2Cl_{(g)} + Cl_{(g)}$ (slow)

Step 2 $NO_{2(g)} + Cl_{(g)} \rightarrow NO_2Cl_{(g)}$ (fast)

The mechanism seems reasonable, because the steps add up to give the overall reaction. Both steps are plausible, because they are bimolecular elementary reactions. Does the reaction mechanism support, however, the experimentally determined rate law?

Rate-determining steps can occur anywhere in a reaction mechanism. If a slow step is not the first step of the mechanism, however, the overall rate law is more complicated to determine. One reason is that fast reactions occurring early in a mechanism are often reversible—both the forward and reverse reactions take place. (You will learn more about these reactions in Unit 4.)

The first step in the mechanism is slow when compared with the second step. Therefore, the rate-determining step is step 1. Since the rate of the overall reaction depends on the rate of the slow step, the rate law equation for the rate-determining step should match the rate law for the overall reaction.

By definition, the rate law equation for step 1 is written as follows:

$$\text{Rate}_1 = k_1[NO_2][Cl_2]$$

The rate law of the rate-determining step in the proposed mechanism matches the experimentally determined rate law. Since the proposed mechanism is consistent with the overall rate law equation, it is a reasonable mechanism.

Sample Problem

Evaluating a Proposed Mechanism

Problem

Consider the reaction below.

$2NO_{(g)} + 2H_{2(g)} \rightarrow N_{2(g)} + 2H_2O_{(g)}$

The experimentally determined rate law is written as follows:

Rate $= k[NO]^2[H_2]$

A chemist proposes the mechanism below for the reaction.

Step 1 $2NO_{(g)} + H_{2(g)} \rightarrow N_2O_{(g)} + H_2O_{(g)}$ (slow)

Step 2 $N_2O_{(g)} + H_{2(g)} \rightarrow N_{2(g)} + H_2O_{(g)}$ (fast)

Determine whether the proposed mechanism is reasonable.

What Is Required?

You need to determine whether the proposed mechanism is reasonable. To do this, you need to answer the following questions:

- Do the steps add up to the overall reaction?
- Are the steps reasonable in terms of their molecularity?
- Is the proposed mechanism consistent with the experimentally determined rate law?

What Is Given?

You know the proposed mechanism, the overall reaction, and the rate law for the overall reaction.

Plan Your Strategy

Add the two reactions, and cancel out reaction intermediates. Check the molecularity of the steps. Determine the rate law equation for the rate-determining step, and compare it to the overall rate law equation.

Act on Your Strategy

Add the two steps.

Step 1 $2NO_{(g)} + H_{2(g)} \rightarrow N_2O_{(g)} + H_2O_{(g)}$

Step 2 $N_2O_{(g)} + H_{2(g)} \rightarrow N_2 + H_2O_{(g)}$

$$\cancel{N_2O_{(g)}} + 2H_{2(g)} + 2NO_{(g)} \rightarrow \cancel{N_2O_{(g)}} + N_{2(g)} + 2H_2O_{(g)}$$

or

$$2NO_{(g)} + 2H_{2(g)} \rightarrow N_{2(g)} + 2H_2O_{(g)}$$

The two steps add up to give the overall reaction. The second step is bimolecular, so it is chemically reasonable. The first step is termolecular, which is possible but rare. In the proposed mechanism, step 1 is the rate-determining reaction. The rate law equation for the first step is written as follows:

$$\text{Rate}_1 = k_1[NO_2]^2[H_2]$$

The rate law equation for the slow step matches the rate law equation for the overall reaction.

Based on the equations for the elementary reactions, the molecularity of these reactions, and the rate law for the rate-determining step, the reaction mechanism seems reasonable.

Check Your Solution

The reaction intermediate in the proposed mechanism is N_2O. When adding the two steps, you were able to cancel out N_2O.

Practice Problems

17. $NO_{2(g)}$ and $F_{2(g)}$ react to form $NO_2F_{(g)}$. The experimentally determined rate law for the reaction is written as follows:
Rate = $k[NO_2][F_2]$

A chemist proposes the following mechanism. Determine whether the mechanism is reasonable.

Step 1 $NO_{2(g)} + F_{2(g)} \rightarrow NO_2F_{(g)} + F_{(g)}$ (slow)

Step 2 $NO_{2(g)} + F_{(g)} \rightarrow NO_2F_{(g)}$ (fast)

18. A researcher is investigating the following overall reaction.
$2C + D \rightarrow E$

The researcher claims that the rate law equation for the reaction is written as follows:
Rate = $k[C][D]$

(a) Is the rate law equation possible for the given reaction?

(b) If so, suggest a mechanism that would match the rate law. If not, explain why not.

19. A chemist proposes the following reaction mechanism for a certain reaction.

Step 1 $A + B \rightarrow C$ (slow)

Step 2 $C + A \rightarrow E + F$ (fast)

(a) Write the equation for the chemical reaction that is described by this mechanism.

(b) Write a rate law equation that is consistent with the proposed mechanism.

20. Consider the reaction between 2-bromo-2-methylpropane and water.
$(CH_3)_3CBr_{(aq)} + H_2O_{(\ell)} \rightarrow (CH_3)_3COH_{(aq)} + H^+_{(aq)} + Br^-_{(aq)}$

Rate experiments show that the reaction is first order in $(CH_3)_3CBr$, but zero order in water. Demonstrate that the accepted mechanism, shown below, is reasonable.

Step 1 $(CH_3)_3CBr_{(aq)} \rightarrow (CH_3)_3C^+_{(aq)} + Br^-_{(aq)}$ (slow)

Step 2 $(CH_3)_3C^+_{(aq)} + H_2O_{(\ell)} \rightarrow (CH_3)_3COH_2^+{}_{(aq)}$ (fast)

Step 3 $(CH_3)_3COH_2^+{}_{(aq)} \rightarrow H^+_{(aq)} + (CH_3)_3COH_{(aq)}$ (fast)

Catalysts

A catalyst is a substance that increases the rate of a chemical reaction without being consumed by the reaction. Catalysts are of tremendous importance in all facets of chemistry, from the laboratory to industry. Many industrial reactions, for example, would not be economically viable without catalysts. Well over three million tonnes of catalysts are produced annually in North America.

How a Catalyst Works

A catalyst works by lowering the activation energy of a reaction so that a larger fraction of the reactants have sufficient energy to react. It lowers the activation energy by providing an alternative mechanism for the reaction. The potential energy diagram in Figure 6.18 shows the activation energy for an uncatalyzed reaction and the activation energy for the same reaction with the addition of a catalyst.

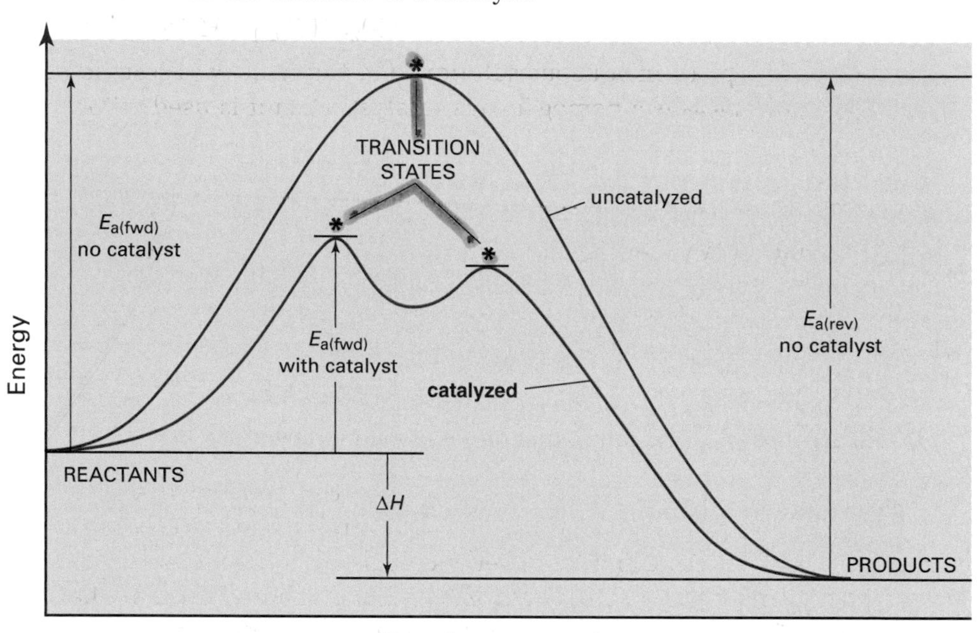

Figure 6.18 **A catalyst lowers the activation energy of a reaction by providing an alternative mechanism. A catalyst also increases the rate of the reverse reaction. What effect does a catalyst have on ΔH of a reaction?**

In Figure 6.18, the catalyzed reaction consists of a two-step mechanism. The uncatalyzed reaction consists of a one-step mechanism. To see how a catalyst works, in general, consider a simple, one-step, bimolecular reaction:

$$A + B \rightarrow AB$$

A catalyst can increase the rate of this reaction by providing an alternative mechanism with lower activation energy. A possible mechanism for the catalyzed reaction is shown below.

Step 1	A + catalyst → A—catalyst
Step 2	A—catalyst + B → AB + catalyst
Overall reaction	A + B → AB

Both steps are faster than the original, uncatalyzed reaction. Therefore, although the overall reaction of the catalyzed mechanism has the same reactants and products as the uncatalyzed reaction, the catalyzed mechanism is faster. The chemical species A—catalyst is a reaction intermediate. It is produced in step 1 but consumed in step 2. By contrast, the catalyst is *regenerated* in the reaction. It appears as a reactant in step 1 and as a product in step 2. Although the catalyst changes *during* the overall reaction, it is regenerated unchanged at the *end* of the overall reaction.

Figure 6.19 shows the reaction of sodium potassium tartrate with H_2O_2, catalyzed by cobaltous chloride, $CoCl_2$. The $CoCl_2$ is pink in solution. Notice that the contents of the beaker briefly change colour to dark green, suggesting the formation of a reaction intermediate. Also notice that you can see the regeneration of the catalyst when the reaction is over.

Catalysts are divided into two categories, depending on whether or not they are in the same phase as the reactants.

Figure 6.19 The reaction of Rochelle's salt, sodium potassium tartrate, with hydrogen peroxide is catalyzed with $CoCl_2$ at 70°C.

Photograph A shows the reactants before mixing.

Photograph B was taken immediately after $CoCl_2$ was added. Notice the pink colour.

Photograph C was taken after about 20 s.

Photograph D was taken after about 2 min.

Homogeneous Catalysts

A **homogeneous catalyst** exists in the same phase as the reactants. Homogeneous catalysts most often catalyze gaseous and aqueous reactions. For example, aqueous zinc chloride, $ZnCl_2$, is used to catalyze the following reaction.

$$(CH_3)_2CHOH_{(aq)} + HCl_{(aq)} \xrightarrow{ZnCl_{2(aq)}} (CH_3)_2CHCl_{(aq)} + H_2O_{(\ell)}$$

The reaction takes place in aqueous solution, and the catalyst is soluble in water. Therefore, $ZnCl_2$ is a homogeneous catalyst when it is used with this reaction.

Heterogeneous Catalysts

A **heterogeneous catalyst** exists in a phase that is different from the phase of the reaction it catalyzes. An important industrial use of heterogeneous catalysts is the addition of hydrogen to an organic compound that contains C=C double bonds. This process is called *hydrogenation*.

Consider the hydrogenation of ethylene, shown below.

$$H_2C{=}CH_{2(g)} + H_{2(g)} \xrightarrow{Pd_{(s)}} H_3C{-}CH_{3(g)}$$

Without a catalyst, the reaction is very slow. When the reaction is catalyzed by a metal such as palladium or platinum, however, the rate increases dramatically. The ethylene and hydrogen molecules form bonds with the metal surface. This weakens the bonds of the hydrogen and ethylene. The H—H bonds of the hydrogen molecules break, and the hydrogen atoms are somewhat stabilized because of their attraction to the metal. The hydrogen atoms react with the ethylene, forming ethane. Figure 6.20 shows the hydrogenation of ethylene to ethane.

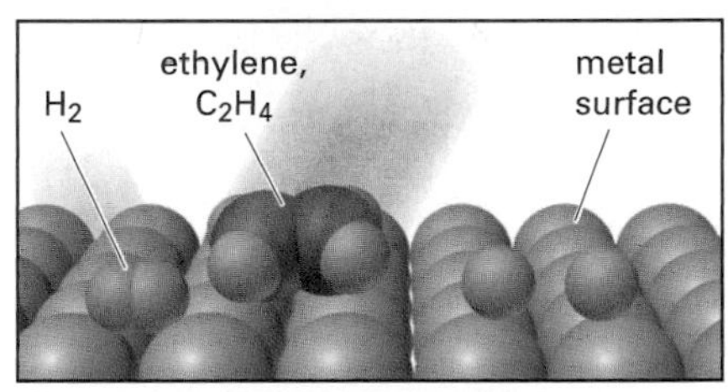

1 H_2 and C_2H_4 approach and bond to metal surface.

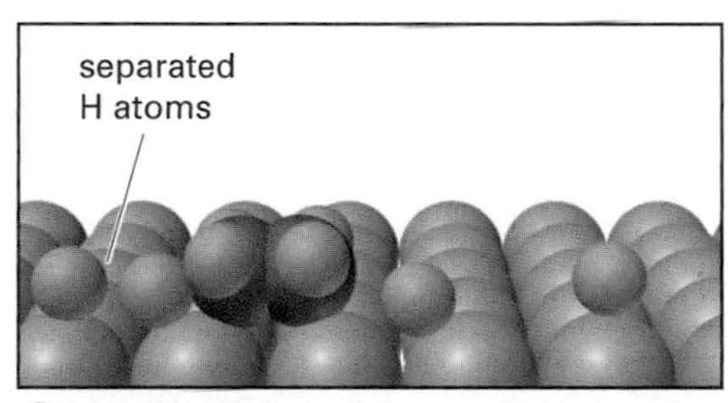

2 Rate-limiting step is H—H bond breakage.

3 One H atom bonds to C_2H_4.

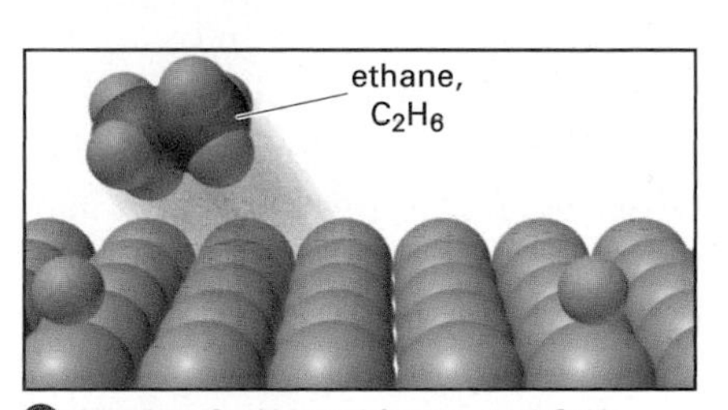

4 Another C—H bond forms, and C_2H_6 leaves the surface.

Figure 6.20 Platinum metal is used as a heterogeneous catalyst for the hydrogenation of gaseous ethylene to ethane.

Web LINK

www.mcgrawhill.ca/links/chemistry12
You probably know that compounds called chlorofluorocarbons (CFCs) are responsible for depleting the ozone layer in Earth's stratosphere. Did you know, however, that CFCs do their destructive work by acting as homogeneous catalysts? Use the Internet to find out how CFCs catalyze the decomposition of ozone in the stratosphere. To start your research, go to the web site above and click on **Web Links**. Communicate your findings as a two-page press release.

You may find the term "hydrogenation" familiar. Some food products, such as margarine and peanut butter, contain hydrogenated vegetable oils. Hydrogenation is used in the food industry to convert liquid vegetable oils, which contain carbon-carbon double bonds, to solid fats, such as shortening, which are fully saturated.

Biological Catalysts

Your body depends on reactions that are catalyzed by amazingly efficient and specific biological catalysts. Biological catalysts are enormous protein molecules called **enzymes**. Their molecular masses range from 15 000 to 1 000 000 g/mol.

Only a small portion of the enzyme, called the **active site**, is actually involved in the catalysis reaction. In terms of the enzyme's overall shape, the active site is like a nook or a fold in its surface. The reactant molecule, called the **substrate** in an enzyme reaction, binds to the active site. The enzyme works by stabilizing the reaction's transition state.

Two models currently exist to explain how an enzyme and its substrate interact. One model, called the *lock and key model*, suggests that an enzyme is like a lock, and its substrate is like a key. The shape of the active site on the enzyme exactly fits the shape of the substrate. A second model, called the *induced fit model*, suggests that the active site of an enzyme changes its shape to fit its substrate. Figure 6.21 shows both models.

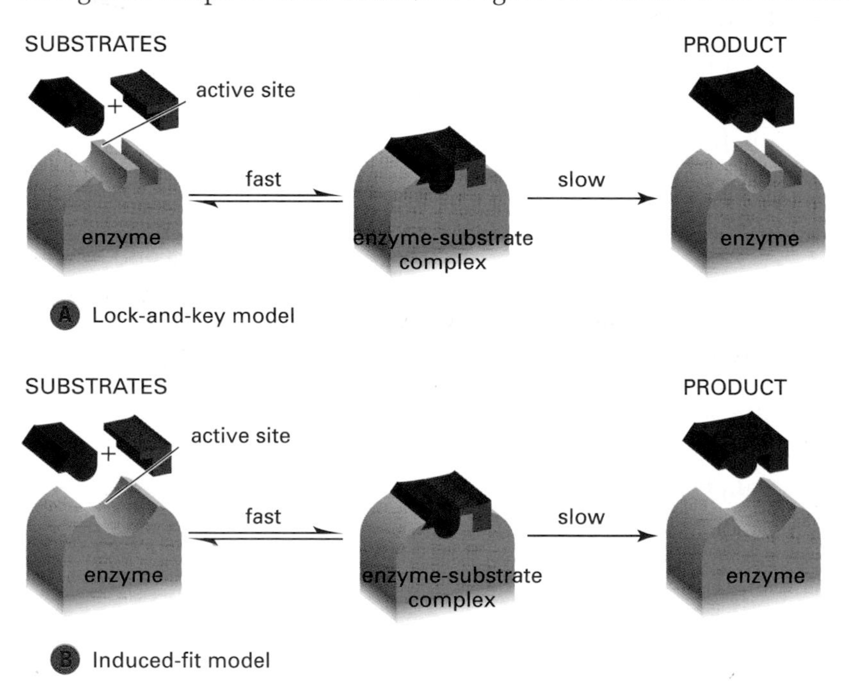

Figure 6.21 **Diagram A shows the lock and key model of enzyme function. Diagram B shows the induced-fit model of enzyme function.**

Enzymes are involved in many functions of the human body, including digestion and metabolism. For example, the enzyme lactase is responsible for catalyzing the breakdown of lactose, a sugar found in milk. People who are lactose-intolerant are usually missing lactase, or they have insufficient amounts. If you are lactose-intolerant, you can take commercially produced supplements that contain lactase.

In the following ThoughtLab, you will investigate other ways in which industries use and produce catalysts.

ThoughtLab  Researching Catalysts

Many reactions that produce useful compounds proceed too slowly to be used in industries. Some reactions need to be carried out at high temperatures or pressures to proceed quickly. These conditions, however, are often expensive to maintain. Therefore, chemists and engineers use catalysts to speed up the reactions in order to obtain products at a reasonable rate and under reasonable conditions. Similarly, many necessary biological reactions would proceed too slowly to sustain life without the presence of enzymes.

Procedure

1. As a class, hold a brainstorming session to create a list of catalysts that are used in industries and a list of enzymes. Come to class prepared with several suggestions. (You may need to do some initial research.) Two examples in each category are given below.
   - *industrial catalysts*: catalytic converter, V_2O_5 catalyst in synthesis of $H_2SO_{4(aq)}$ (contact process)
   - *enzymes*: papain, pepsin
2. Choose one catalyst and one enzyme from each brainstormed list.
3. Using electronic and print resources, research your substances.
4. Prepare a brief report about each substance. Your report should include answers to the Analysis questions below. If you prefer, present your research as a web page.
5. Include a list of the sources you used. You should include at least three electronic sources and one print source for each substance.

Analysis

1. Your report about industrial catalysts should answer the following questions.
   (a) What is the chemical formula of the catalyst? Is the catalyst heterogeneous or homogeneous?
   (b) What process does the catalyst speed up? Include products, reactants, the reaction mechanism (if possible), and information about the uncatalyzed reaction.
   (c) What technology is required to support the catalyzed reaction (for example, the design of a catalytic converter)?
   (d) What is the importance and relevance, to Canadians, of the industry associated with the catalyst?
2. Your report about enzymes should answer the questions below.
   (a) What chemical process is the enzyme involved in?
   (b) How does the enzyme fit into an overall biological process?
   (c) Are there any ways in which the enzyme can be prevented from doing its job? Explain.
   (d) Is the enzyme mass-produced by industry? If so, how? For what purpose is it mass-produced?

In the Thought Lab above, you may have included the enzyme catalase in your brainstormed list. Catalase in your blood is responsible for the "fizzing" you see when you use a dilute hydrogen peroxide solution to disinfect a cut. When dilute hydrogen peroxide is poured on a cut, it decomposes to oxygen gas and water. The decomposition reaction is catalyzed by catalase.

Have you ever opened an old bottle of hydrogen peroxide and noticed that it has lost its potency? The hydrogen peroxide has decomposed, leaving you with a bottle of water. At room temperature, in the absence of a catalyst, the decomposition of H_2O_2 occurs very slowly, over a period of months or years. With the help of a catalyst, however, the rate of the reaction can be increased to the point where it can be easily studied in a high school laboratory.

In Investigation 6-B, you will write a detailed procedure to determine the rate law for the catalyzed decomposition of hydrogen peroxide. Instead of using catalase to catalyze the reaction, you will use an inorganic catalyst.

Investigation 6-B

SKILL FOCUS

- Predicting
- Performing and recording
- Analyzing and interpreting

Determining the Rate Law for a Catalyzed Reaction

Hydrogen peroxide, H_2O_2, can be purchased as a dilute solution in a pharmacy or supermarket. It is used, among other things, as a topical antiseptic for minor cuts. In this investigation, you will use the I^- ion, in the form of aqueous NaI, to catalyze the decomposition of H_2O_2.

Questions

What are the exponents m and n, and the rate constant, in the following general rate law equation?

Rate = $k[H_2O_2]^m[I^-]^n$

Prediction

Will the catalyzed decomposition be first order or second order with respect to H_2O_2 and with respect to I^-?

Materials

100 mL beaker
250 mL Erlenmeyer flask
60 mL 6% (m/v) $H_2O_{2(aq)}$
60 mL 1.0 mol/L $NaI_{(aq)}$
3 graduated cylinders (10 mL)
2 medicine droppers or plastic pipettes
masking tape or grease pencil
electronic top-loading balance, accurate to two decimal places

Option 1
pneumatic trough
one-holed rubber stopper, fitted with a piece of glass tubing (must be airtight)
1 m rubber hose to fit glass tubing (must be airtight)
graduated cylinder to collect gas

Option 2
balance, preferably accurate to three decimal places

Safety Precautions

- The reaction mixture will get hot. Handle the Erlenmeyer flask near the top.
- 6% hydrogen peroxide solution is an irritant. Wear safety glasses, latex gloves, and a laboratory apron.

Procedure

1. Copy the following table into your notebook, to record your observations. Give your table a title.

	H_2O_2		NaI		
Experiment	Volume of 6% H_2O_2 (mL)	Volume of distilled water (mL)	Volume of 1.0 mol/L NaI (mL)	Volume of distilled water (mL)	O_2 produced in 60 s (mL or g)
1	10.0	0	10.0	0	
2	8.0	2.0	10.0	0	
3	6.0	4.0	10.0	0	
4	4.0	6.0	10.0	0	
5	10.0	0	8.0	2.0	
6	10.0	0	6.0	4.0	
7	10.0	0	4.0	6.0	

2. The rate of production of oxygen gas can be monitored in different ways. Depending on the equipment you have available, choose option 1 or option 2.
   - *Option 1*: You can collect the oxygen gas by downward displacement of water. Collect the oxygen produced in the first 60 s of the reaction in an upside-down graduated cylinder. Refer to Investigation 6-A to see how to set up an apparatus for collecting the gas.
   - *Option 2*: If you have an electronic balance that is accurate to 0.001 g, you can monitor the change in mass as oxygen is released in the first 60 s of the reaction.

Note: Gently swirl the reaction mixture for the first 5 s. To help the bubbles of oxygen leave the solution, gently swirl the reaction flask every 15 s. Stop swirling at 45 s.

3. For your chosen option, write out a complete procedure based on the table on the previous page. Include a diagram showing the experimental set-up, and detailed safety precautions. Have your teacher approve your procedure, and then carry it out.

Analysis

1. For each experiment, determine the initial rate of reaction. That is, convert the volume or mass of O_2 within the time interval to get an initial rate of mol O_2/(L • s). Using the stoichiometry of the reaction, convert this initial rate to a rate of decomposition of H_2O_2 in mol/(L • s).

2. **(a)** For the experiments in which $[H_2O_2]$ was kept constant, plot a graph of reaction rate (*y*-axis) versus [NaI]. From the shape of the graph, determine if the reaction is first order or second order with respect to NaI.

   (b) For the experiments in which [NaI] was kept constant, plot a graph of reaction rate (*y*-axis) versus $[H_2O_2]$. From the shape of the graph, determine if the reaction is first order or second order with respect to H_2O_2.

3. Determine the value of *k* for the reaction, with proper units. Pay close attention to significant digits.

Conclusion

4. Write the rate law equation for the reaction.

Applications

5. The iodide ion-catalyzed decomposition of hydrogen peroxide to oxygen gas and water is believed to occur via the following mechanism.

   Step 1 $H_2O_2 + I^- \rightarrow H_2O + OI^-$ (slow)

   Step 2 $H_2O_2 + OI^- \rightarrow H_2O + O_2 + I^-$ (fast)

   (a) Show that these two steps are consistent with the overall stoichiometry of the reaction.

   (b) Is this mechanism consistent with your experimentally obtained rate law? Explain.

   (c) How does this mechanism account for the fact that the I^- ion is a catalyst?

   (d) What is the role of the hypoiodite, IO^-, ion in the mechanism? How do you know?

   (e) Draw and label a potential energy-reaction coordinate diagram that is consistent with this mechanism. Be sure to illustrate any reaction intermediates.

   (f) Postulate a possible activated complex for step 1 of the proposed mechanism.

6. The rate of decomposition of H_2O_2 can be increased using other catalysts, such as Mn^{2+} and Fe^{2+}.

   (a) How do you expect ΔH for the decomposition reaction to change if different catalysts are used?

   (b) Do you expect the rate law to change? Explain.

Canadians in Chemistry

Dr. Maud L. Menten

Dr. Maud L. Menten, a leading researcher in the field of biochemistry, was born in Port Lambton, Ontario, in 1879. She grew up in rural British Columbia, where she had to take a canoe across the Fraser River every day to get to school. Later, Menten returned to Ontario to attend the University of Toronto. Upon receiving her medical degree in 1911, she became one of the first female doctors in Canada.

Menten soon received international recognition for her study of enzymes. From 1912 to 1913, she worked at Leonor Michaelis' lab at the University of Berlin. While conducting experiments on the breakdown of sucrose by the enzyme called invertase, Menten and Michaelis were able to refine the work of Victor Henri to explain how enzymes function. A few years earlier, Henri had proposed that enzymes bind directly to their substrates. Michaelis and Menten obtained the precise measurements that were needed to support Henri's hypothesis. Using the recently developed concept of pH, they were able to buffer their chemical reactions and thereby control the conditions of their experiments more successfully than Henri. Their findings, published in 1913, provided the first useful model of enzyme function.

In the ground-breaking scientific paper that presented their work, Menten and Michaelis also derived an important mathematical formula. This formula describes the rate at which enzymes break down their substrates. It correlates the speed of the enzyme reaction with the concentrations of the enzyme and the substrate. Called the Michaelis-Menten equation, it remains fundamental to our understanding of how enzymes catalyze reactions.

Menten remained an avid researcher all her life. She worked long days, dividing her time between teaching at the University of Pittsburgh's School of Medicine and working as a pathologist at the Children's Hospital of Pittsburgh. She also worked with her peers and students on more than 70 papers that dealt with a wide range of medical topics. In addition to her work with Michaelis, Menten's contributions to science include pioneering techniques for studying the chemical composition of tissues and for investigating the nature of hemoglobin in human red blood cells.

It was not until 1949, only a year before retiring from the University of Pittsburgh, that Menten obtained a full professorship. Shortly after retiring, Menten joined the Medical Research Institute of British Columbia to do cancer research. Sadly, ill health forced her to cease her scientific work only a few years later. She spent her remaining days in Leamington, Ontario.

Although she was a brilliant medical researcher, Menten's enthusiasm overflowed beyond the boundaries of science. She was a world traveller, a mountain climber, a musician, and a student of languages. She was even a painter, whose work made its way into art exhibitions.

In honour of her lifetime of achievements, Maud L. Menten was inducted into the Canadian Medical Hall of Fame in 1998.

Section Summary

In this section, you learned that chemical reactions usually proceed as a series of steps called elementary reactions. You related the equations for elementary reactions to rate laws. You learned how the relative speed of the steps in a reaction mechanism help to predict the rate law of an overall reaction. Finally, you learned how a catalyst controls the rate of a chemical reaction by providing a lower-energy reaction mechanism. In this chapter, you compared activation energies of forward and reverse reactions. In the next unit, you will study, in detail, reactions that proceed in both directions.

Section Review

1. K/U In your own words, describe an elementary reaction.

2. K/U Distinguish between an overall reaction and an elementary reaction. Include a discussion of rate law.

3. K/U Why do chemists say that a reaction mechanism is *proposed*?

4. K/U What three criteria must a proposed reaction mechanism satisfy, for it to be accepted?

5. C Explain the difference between a reaction intermediate and an activated complex in a reaction mechanism.

6. K/U Consider the reaction below.

   $2A + B_2 \xrightarrow{C} D + E$

   A chemist proposes the following reaction mechanism.

   Step 1 $A + B_2 \rightarrow AB_2$

   Step 2 $AB_2 + C \rightarrow AB_2C$

   Step 3 $AB_2C + A \rightarrow A_2B_2 + C$

   Step 4 $A_2B_2 \rightarrow D + E$

   (a) Show that the proposed mechanism can account for the overall reaction.

   (b) What is the role of AB_2C and AB_2?

   (c) What is the role of C?

   (d) Given a proposed reaction mechanism, how can you differentiate, in general, between a reaction intermediate and a catalyst?

7. I Chlorine gas reacts with aqueous hydrogen sulfide (also known as hydrosulfuric acid) to form elemental sulfur and hydrochloric acid.

   $Cl_{2(g)} + H_2S_{(aq)} \rightarrow S_{(s)} + 2HCl_{(aq)}$

   The experimentally obtained rate law equation is written as follows:

   Rate = $k[Cl_2][H_2S]$

   Which of the following mechanisms is consistent with this information? Explain your answer.

   Mechanism A

   Step 1 $Cl_2 + H_2S \rightarrow Cl^+ + HCl + HS^-$ (slow)

   Step 2 $Cl^+ + HS^- \rightarrow HCl + S$ (fast)

   Mechanism B

   Step 1 $Cl_2 \rightarrow Cl^+ + Cl^-$ (slow)

   Step 2 $Cl^- + H_2S \rightarrow HCl + HS^-$ (fast)

   Step 3 $Cl^+ + HS^- \rightarrow HCl + S$ (fast)

8. C Consider the following general endothermic reaction.

   $A + B \rightarrow C + D$

   (a) Explain why a catalyst has no effect on the enthalpy change. Illustrate your answer with a potential energy diagram.

   (b) Explain the effect of a catalyst on a reaction that has similar products and reactants, but is an exothermic reaction. Illustrate your answer with a potential energy diagram.

9 K/U Consider the general reaction shown below.

$2A + B \rightarrow 2E + F$

The experimentally determined rate law equation is written as follows:

Rate = $k[A][B]$

A proposed mechanism for the reaction has the first step below.

Step 1 $A + B \rightarrow C + F$ (rate-determining step)

Assuming a two-step mechanism, write a possible second elementary reaction for this mechanism.

10 MC Historians suspect that many ancient Romans suffered from lead poisoning. The ancient Romans used lead for their plumbing system (hence, the Latin name for the element). Lead is a poison because it can form strong bonds with proteins, including enzymes. Lead and other heavy metals react with enzymes by binding to functional groups on their active sites. When lead binds to a protein, such as an enzyme, the protein often precipitates out of solution.

(a) Based on your understanding of enzymes and the information above, explain why lead's ability to form strong bonds with the active sites of enzymes makes lead toxic to humans.

(b) Egg whites and milk are used as antidotes for heavy metal poisoning. The victim must ingest the egg white or milk soon after the lead has been ingested. Then the victim's stomach must be pumped. Explain why this antidote works. Why must the stomach be pumped?

(c) Would the antidote described in part (b) work for someone who suffers from long-term exposure to lead by ingestion? Explain why or why not.

CHAPTER 6 Review

Reflecting on Chapter 6

Summarize this chapter in the format of your choice. Here are a few ideas to use as guidelines:

- Use a graph to describe the rate of reaction as a function of the change of concentration of a reactant or product with respect to time.
- Distinguish between an average rate and an instantaneous rate of a reaction.
- Express the rate of a reaction as a rate law equation.
- Explain what the term "half-life" means. Calculate the half-life of a first-order reaction.
- Use collision theory and transition state theory to explain how concentration, temperature, surface area, and the nature of reactants control the rate of a chemical reaction.
- Use a potential energy diagram to demonstrate your understanding of the relationships between activation energy, reactants, products, enthalpy change, and the activated complex.
- Explain why most reactions occur as a series of elementary reactions. Show how the rate law for a rate-determining step relates to the rate law for an overall reaction.
- Explain how a catalyst increases the rate of a chemical reaction.

Reviewing Key Terms

For each of the following terms, write a sentence that shows your understanding of its meaning.

reaction rate	average rate
instantaneous rate	catalyst
rate law equation	rate constant
overall reaction order	first-order reaction
second-order reaction	initial rates method
half-life ($t_{1/2}$)	collision theory
activation energy (E_a)	transition state theory
potential energy diagram	transition state
activated complex	reaction mechanism
elementary reaction	reaction intermediates
molecularity	bimolecular
unimolecular	termolecular
rate-determining step	homogeneous catalyst
heterogeneous catalyst	enzymes
active site	substrate

Knowledge/Understanding

1. The way that chemists monitor the rate of a chemical reaction depends on the reaction. What conditions of a reacting system must change for the each instrument or method below to be useful?
   (a) balance
   (b) downward displacement of water
   (c) spectrophotometer
2. In your own words, describe how collision theory explains why increased surface area increases the rate of a reaction.
3. Agitation (stirring) often increases the rate of a reaction. Use collision theory to explain why.
4. At elevated temperatures, ammonia reacts with oxygen as follows:
   $4NH_{3(g)} + 5O_{2(g)} \rightarrow 4NO_{(g)} + 6H_2O_{(g)}$
   (a) Write an equation that shows the relationship between the rate of reaction expressed in terms of each reactant and product.
   (b) The average rate of production of nitrogen monoxide is 6.2×10^{-2} mol/(L • s). What is the average rate of change in the concentration of ammonia?
5. State two requirements for an effective collision between reactants.
6. State the difference between a homogeneous catalyst and a heterogeneous catalyst.

Inquiry

7. The rate of decomposition of hydrogen peroxide was studied at a particular temperature.
   $H_2O_{2(aq)} \rightarrow H_2O_{(\ell)} + \frac{1}{2}O_{2(g)}$
   (a) The initial concentration of hydrogen peroxide was 0.200 mol/L. 10.0 s later, it was measured to be 0.196 mol/L. What was the initial rate of the reaction, expressed in mol/(L • s)?
   (b) 0.500 L of hydrogen peroxide solution was used for the experiment. What mass was lost as O_2 bubbled out of solution in this initial 10.0 s interval?
8. The following first-order reaction has a rate of 3.8×10^{-3} mol/(L • s) when [A] = 0.38 mol/L.
   A → products
   (a) Calculate the value of k and the half-life of the reaction.

(b) What did you assume when answering part (a)?

9. The units of k, the rate constant, depend on the overall order of a reaction. For what overall reaction order does the rate constant have the same units as the reaction rate?

10. The rate constant of a first-order reaction is $4.87 \times 10^{-2}\ s^{-1}$ at a particular temperature. Calculate the half-life of this reaction.

11. Phosgene, $COCl_2$, is a highly toxic gas that is heavier than air. It can be produced by reacting carbon monoxide with chlorine in a very slow reaction.
$CO_{(g)} + Cl_{2(g)} \rightarrow COCl_{2(g)}$
The following initial rate data were collected at a particular temperature.

Experiment	Initial [CO] (mol/L)	Initial [Cl_2] (mol/L)	Initial rate (mol/(L • s))
1	0.500	0.0500	6.45×10^{-30}
2	0.0500	0.0500	6.65×10^{-31}
3	0.0500	0.500	6.50×10^{-30}
4	0.05 00	0.00500	6.60×10^{-32}

(a) Write the rate law equation for this reaction.
(b) Calculate the value of the rate constant. Make sure that you use the proper units.

12. The following reaction was studied using the method of initial rates.
$3A_{(aq)} + 4B_{(aq)} \rightarrow$ products
The following data were collected.

Experiment	Initial [A] (mol/L)	Initial [B] (mol/L)	Initial rate (mol/(L • s))
1	0.200	0.200	5.00
2	0.600	0.200	45.0
3	0.200	0.400	10.0
4	0.600	0.400	90.0

(a) Write the rate law equation.
(b) What is the overall reaction order?
(c) Calculate the value of the rate constant, with the proper units.

13. Mercury(II) chloride, $HgCl_2$, reacts with oxalate ions, $C_2O_4^{2-}$, as follows:
$2HgCl_{2(aq)} + C_2O_4^{2-}{}_{(aq)} \rightarrow$
$2Cl^-{}_{(aq)} + 2CO_{2(g)} + Hg_2Cl_{2(s)}$
The reaction rate was monitored by measuring the mass of Hg_2Cl_2 formed as a function of time.

Experiment	Initial [Hg_2Cl_2] (mol/L)	Initial [$C_2O_4^{2-}$] (mol/L)	Initial rate (mol/(L • s))
1	0.0788	0.113	1.35×10^{-6}
2	0.0788	0.225	5.33×10^{-6}
3	0.039	0.225	2.63×10^{-6}

(a) Write the rate law equation for the reaction.
(b) Calculate the value of the rate constant, with the appropriate units.
(c) Calculate the initial rate of the reaction when $[HgCl_2] = 0.0400$ mol/L and $[C_2O_4^{2-}] = 0.150$ mol/L.

14. A chemical reaction between compounds A and B is first order in A and first order in B. Find the unknown information in the table below.

Experiment	Rate (mol/(L • s))	[A] (mol/L)	[B] (mol/L)
1	0.10	0.20	0.050
2	0.40	(a)	0.050
3	0.80	0.40	(b)

15. A chemical reaction between compounds C and D is first order in C and second order in D. Find the unknown information in the table below.

Experiment	Rate (mol/(L • s))	[A] (mol/L)	[B] (mol/L)
1	0.10	1.0	0.20
2	(a)	2.0	0.20
3	(b)	2.0	0.40

16. Consider the reaction below.
$2A + B \rightarrow C + D$
At 20°C, the activation energy of the forward reaction is 59.9 kJ/mol and the activation energy of the reverse reaction is 72.0 kJ/mol.
(a) What is the enthalpy change for the reaction?
(b) Sketch a potential energy diagram for the reaction.

17. Consider the reaction below.

$$2A + B_2 \xrightarrow{C} D + E$$

A chemist proposes the following reaction mechanism.

Step 1 $A + B_2 \rightarrow AB_2$

Step 2 $AB_2 + C \rightarrow AB_2C$

Step 3 $AB_2C + A \rightarrow A_2B_2 + C$

Step 4 $A_2B_2 \rightarrow D + E$

Suggest a rate law equation corresponding to each of the following situations. Remember that a rate law equation may include only the concentration of reactants and catalyst, if any. If you think that it is impossible to predict a rate law for any of the situations, explain why. State any assumptions you make for each situation.

(a) Step 1 is the rate determining step.

(b) Step 2 is the rate determining step.

(c) Step 3 is the rate determining step.

Communication

18. Explain why chemical reactions are fastest at the beginning.

19. Explain why a rate constant must always be accompanied by the temperature at which it was determined.

20. Your friend is having trouble understanding how a rate-determining step in a reaction mechanism determines the rate of the overall reaction. Invent a new analogy to explain the idea to your friend, using a process from everyday life that involves slow and fast steps.

21. Write a few paragraphs to distinguish between an activated complex and a reaction intermediate. Include equations and diagrams in your explanation.

22. "For an endothermic reaction, the activation energy will always be greater than the enthalpy change for the reaction."

(a) Do you agree or disagree with this statement? Explain your answer.

(b) Replace the word "endothermic" with "exothermic." Do you agree or disagree with this statement now? Explain your answer.

23. **(a)** Explain, in your own words, what is meant by the term "activation energy."

(b) How can the idea of activation energy be used to explain the temperature dependence of rate?

(c) How can activation energy be used to explain why a catalyst increases the rate of a chemical reaction?

Making Connections

24. Suppose that you are using a Polaroid™ camera outdoors during the winter. You want your photographs to develop faster. Suggest a way to accomplish this.

25. Why are catalytic converters not very effective immediately after starting a vehicle in the winter? Suggest a way to correct this problem.

Answers to Practice Problems and Short Answers to Section Review Questions

Practice Problems: **1.** 0.25 mol/s **2.** 0.10 mol/(L • s) **3.** $\Delta[Br_2]/\Delta t = -0.5\Delta[HBr]/\Delta t$ **4.(a)** 9.0×10^{-4} mol/s **(b)** 0.011 L/s **5.** Rate = $(2.05 \times 10^{-4}\ s^{-1})[C_2H_4O]$ **6.** Rate = $(0.15\ L/(mol \cdot s))[ICl][H_2]$ **7.(a)** Rate = $k[SO_2Cl_2]$ **(b)** $k = 2.2 \times 10^{-5}\ s^{-1}$ **8.(a)** 0.10 **(b)** 0.025 **(c)** 0.34 **(d)** 0.060 **9.(a)** 0.075 s **(b)** 6.2% **10.(a)** $3.6 \times 10^{-4}\ s^{-1}$ **(b)** 5.0×10^{14} molecules/L **11.** Only 0.1% of the reactants are left. **12.** about 400 s **13.** Products are at lower energy than reactants. **14.** $E_{a(fwd)} = 99$ kJ **15.** $E_{a(rev)} = 193$ kJ **16.** $E_{a(rev)} = 411$ kJ **17.** Yes: Steps add to give overall reaction. Both steps are bimolecular. Rate law for slow step matches experimental rate law. **18.(a)** yes **(b)** $C + D \rightarrow B$ (slow); $B + C \rightarrow E$ (fast) **19.(a)** $A + 2B \rightarrow E + F$ **(b)** Rate = $k[A][B]$ **20.** Steps add to give overall reaction. Steps are bimolecular or unimolecular. Rate law for slow step matches experimental rate law.

Section Review: **6.1:** **4.(a)** 1.25×10^{-3} mol/(L • s) **(b)** 8.33×10^{-4} mol/(L • s) **6.2** **1.** Rate = $k[A]^2[B]$ **2.** no **3.(a)** no change **(b)** increase by a factor of 8 **4.(a)** 1, 2, 1 **(b)** 4 **(c)** $L^3/(mol^3 \cdot s)$ **5.(a)** Rate = $k[A][B]^2$ **(b)** $6.2 \times 10^2\ L^2/(mol^2 \cdot s)$ **6.** Rate = $(0.020\ L^2/(mol^2 \cdot s))[A][B]$ **7.** 29.6 years; 9.33×10^8 s **6.3:** **3.(a)** endothermic **(b)** $\Delta H = 68$ kJ **6.4:** **6.(b)** reaction intermediates **(c)** catalyst **7.** mechanism A **9.** $A + C \rightarrow 2E$ (fast)

Developing a Bulletin About Catalysts and Enzymes

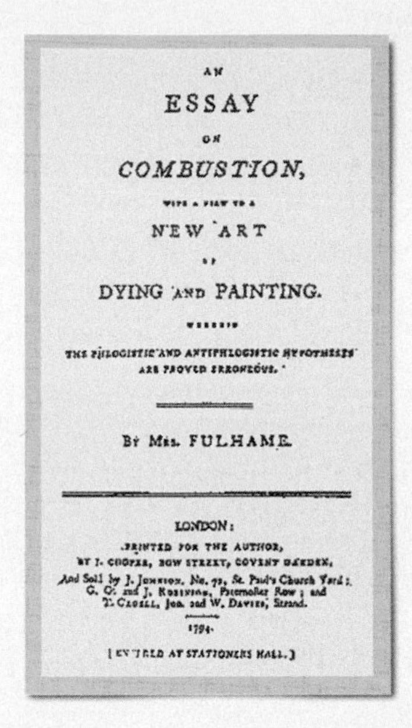

AN
ESSAY
ON
COMBUSTION,
WITH A VIEW TO A
NEW ART
OF
DYING AND PAINTING.
WHEREIN
THE PHLOGISTIC AND ANTIPHLOGISTIC HYPOTHESES
ARE PROVED ERRONEOUS.

BY MRS. FULHAME.

LONDON:
PRINTED FOR THE AUTHOR,
BY J. COOPER, BOW STREET, COVENT GARDEN,
And Sold by J. Johnson, No. 72, St. Paul's Church Yard;
G. G. and J. Robinson, Paternoster Row; and
T. Cadell, Jun. and W. Davies, Strand.
1794.
[ENTERED AT STATIONERS HALL.]

Background

Scientists have studied the amazing properties of enzymes and catalysts for more than 200 years. In 1794, the English chemist Elizabeth Fulhame published a book called *An Essay on Combustion*. Her book included many of the first recorded ideas about the role of catalysis in chemical processes. In particular, Fulhame was interested in the catalytic properties of the hydrogen and oxygen atoms in water.

Fulhame's ideas about inorganic catalysts were swept under by the fast and furious debate of her scientific era. Many of the most prominent scientists of her day were arguing over whether organic reactions needed living organisms to occur. Many scientists believed that a *vital force* was present in living organisms and was essential for the formation of all organic compounds. For example, the fermentation of sugar into alcohol takes place through the action of yeast, a living organism. Scientists argued that it was impossible to produce alcohol, an organic compound, without the action of yeast, or some other living organism.

Although no one realized it at the time, Fulhame's ideas on catalysts were directly relevant to the debate on vital force. There is no mysterious vital force in living organisms. Instead, organic reactions in living organisms depend on organic catalysts called enzymes.

In 1828, the German chemist Friedrich Wohler discovered that urea, an organic compound, could be produced in a laboratory without any help from a living organism. Wohler's discovery inspired other scientists to hypothesize that the fermentation of sugar into alcohol could also occur without the presence of living organisms, such as yeast. Some scientists went further, and insisted that yeast was not even a living organism.

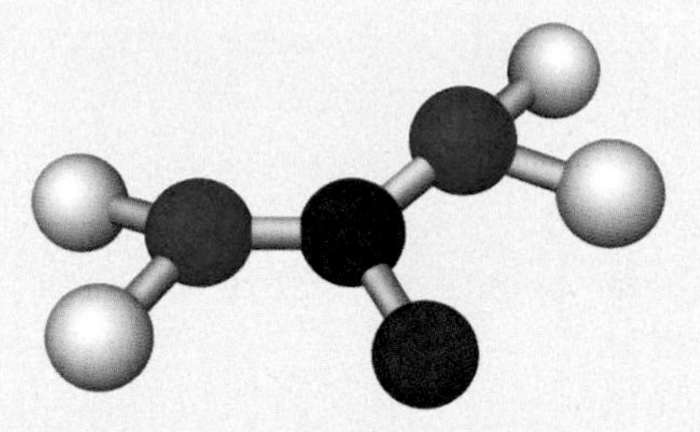

Urea

Eventually, scientists proved that fermentation did not require whole cells of yeast. Fermentation could still occur if non-living extracts of yeast were present. Through further study and experimentation, the important component in the yeast extract was identified as an enzyme. If earlier scientists had considered Fulhame's ideas on catalysts, they might have arrived at the truth about enzymes sooner!

Over time, scientists have built up a large body of knowledge about many different catalysts and enzymes. This knowledge has been put to good use in industry. Difficult and expensive industrial processes have been made faster, cheaper, and easier through the use of catalysts and enzymes. For example, enzymes are used in the pharmaceutical industry, in paper-making and recycling processes, and in the petroleum industry. Many more industrial uses of catalysts and enzymes are possible, and research into catalysts continues.

In this project, you will research some of the many catalysts and enzymes that are used in Canadian industries. Then you will create an information bulletin to present your findings.

Challenge

Design, produce, and distribute an information bulletin to inform your community about the use of catalysts or enzymes in Canadian industries. Choose one or more of the following options:

- If there is a local industry you can study, focus on different aspects of this industry's use of catalysts.
- Study a catalyst that is used by several different industries.
- Compile a variety of uses of catalysts and enzymes across Canada.
- Analyze a cutting-edge technology that involves catalysts or enzymes.
- Analyze the historical development and use of a particular enzyme or catalyst.

Materials

Set up your information bulletin in the format of a brochure or leaflet, suitable to be distributed to classmates and other students in your school. Alternatively, present your information bulletin as a web page, so that you can share your research with a wider audience.

Design Criteria

A. As a class, develop a product rubric to assess the bulletins. For example, you may wish to evaluate how interesting the bulletins are, or whether they are easy for non-chemistry students to understand.

B. Your bulletin should be interesting, visually attractive, factual, and aimed at a wide audience.

Action Plan

1. Decide whether you will work on your own, with a partner, or as part of a small group.
2. Decide what you want your bulletin to look like. The following questions will help you decide on a format.
   - How many topics will you cover in your bulletin?
   - How long will each article in your bulletin be?
   - Will you study a local industry, or cover a variety of Canadian industries?
   - Will you examine one enzyme or catalyst in detail, or will you give an overview of many different catalysts and/or enzymes?
   - How can you ensure that your bulletin is visually attractive and interesting?
   - How can you motivate readers? For example, you may want to include activities (such as puzzles and quizzes) or a brief list of interesting facts.
3. If you are working as part of a group, assign a task to each member of the group.

Evaluate

Print out your bulletin. Display it on a poster board in a hallway of your school, or in a local public building. If your bulletin is in the form of a web page, present it to the class and then post it on the Internet or on your school's Intranet.

Assessment

After you complete this project, answer the following questions to assess your work.

- Assess the interest level of the information you gathered. Did you obtain information of interest to Grade 12 chemistry students? Did you obtain information of interest to non-science students?
- Assess the scientific accuracy and complexity of the information you gathered. Did you use at least three major sources? If you used web sites, were they run by reputable organizations? Did you include information above and beyond the content of this course?
- Assess your communication skills. Did you explain detailed chemical information in a clear and understandable way?

Web LINK

www.mcgrawhill.ca/links/chemistry12

Go to the web site above, and click on **Web Links** to find appropriate sites to start your research.

Knowledge/Understanding

Multiple Choice

In your notebook, write the letter for the best answer to each question.

1. The ΔH°_f of an element in its standard state is defined to be
   (a) 0 kJ/mol
   (b) 10 kJ/mol
   (c) –10 kJ/mol
   (d) greater than 0 kJ/mol
   (e) a unique value for each element
2. 10.9 kJ of energy is needed to vaporize 60.0 g of liquid Br_2 vapour at 60°C. What is the molar heat of vaporization of Br_2 at 60°C?
   (a) 3.64 kJ/mol
   (b) 7.27 kJ/mol
   (c) 14.6 kJ/mol
   (d) 29.1 kJ/mol
   (e) 10.9 kJ/mol
3. What is the molar heat of vaporization of water, given the following thermochemical equations?
   $H_{2(g)} + \frac{1}{2}O_{2(g)} \rightarrow H_2O_{(g)} + 241.8 \text{ kJ}$
   $H_{2(g)} + \frac{1}{2}O_{2(g)} \rightarrow H_2O_{(\ell)} + 285.8 \text{ kJ}$
   (a) 44.0 kJ/mol
   (b) –527.6 kJ/mol
   (c) –44.0 kJ/mol
   (d) –527.6 kJ/mol
   (e) 241.8 kJ/mol
4. Which substance has a standard enthalpy of formation, ΔH°_f, equal to zero?
   (a) gold, $Au_{(s)}$
   (b) water, $H_2O_{(\ell)}$
   (c) carbon monoxide, $CO_{(s)}$
   (d) zinc, $Zn_{(g)}$
   (e) water, $H_2O_{(g)}$
5. Which of the following statements are true?
   I. The reaction vessel cools when an endothermic reaction occurs.
   II. An endothermic reaction has a negative value of ΔH.
   III. Heat is liberated when an exothermic reaction occurs.
   (a) I and II
   (b) I, II, and III
   (c) I and III only
   (d) II and III only
   (e) none of them
6. Which of the following processes are exothermic?
   I. boiling water
   II. freezing water
   III. condensing steam
   IV. melting ice
   (a) I and II only
   (b) II and III only
   (c) I and IV only
   (d) II, III, and IV only
   (e) II and IV only
7. Which factor does not affect the rate of a chemical reaction in aqueous solution?
   (a) the enthalpy change of the reaction
   (b) the activation energy of the reaction
   (c) the collision frequency of the reacting particles
   (d) the relative orientation of the colliding particles
   (e) the temperature of the solution
8. Which statement about an activated complex is true?
   (a) It is a stable substance.
   (b) It has lower chemical potential energy, or enthalpy, than reactants or products.
   (c) It occurs only in endothermic reactions.
   (d) It occurs at the transition state of the reaction.
   (e) It always breaks down to form product molecules.
9. A catalyst changes the
   I. mechanism of a reaction
   II. enthalpy change of a reaction
   III. activation energy of a reaction
   (a) I only
   (b) III only
   (c) I and III only
   (d) II and III only
   (e) I, II, and III
10. The overall rate of any chemical reaction is most closely related to
    (a) the number of steps in the reaction mechanism
    (b) the overall reaction
    (c) the fastest step in the reaction mechanism
    (d) the slowest step in the reaction mechanism
    (e) the average rate of all the steps in the reaction mechanism

Short Answer

In your notebook, write a sentence or a short paragraph to answer each question.

11. Distinguish between an open system and an insulated system.
12. In your own words, define the terms "system" and "surroundings." Use an example.

13. In a chemical reaction, bonds are formed and broken.
    (a) How would you characterize the enthalpy change of bond breaking?
    (b) How would you characterize the enthalpy change of bond formation?
    (c) State the relationship between the enthalpy change of the overall reaction (exothermic or endothermic) and bond breakage and formation.
14. "The reactants have more potential energy than the products." What kind of reaction does this statement describe? Justify your answer.
15. What is the relationship between the initial quantity of reactants for a reaction and the enthalpy change for a reaction? Use a thermochemical equation in your answer.
16. Compare and contrast enthalpy of vaporization and enthalpy of condensation.
17. What are the characteristics of a good constant-pressure calorimeter?
18. A wooden or plastic spoon, rather than a metal spoon, should be used to stir hot soup. Explain why, in terms of specific heat capacity.
19. Butane, C_4H_{10}, is the fuel that is used in disposable lighters. Consider the following equation for the complete combustion of butane.
    $C_4H_{10(g)} + 6.5O_{2(g)} \rightarrow 4CO_{2(g)} + 5H_2O_{(\ell)}$
    (a) Write a separate balanced chemical equation for the formation of C_4H_{10}, the formation of CO_2, and the formation of H_2O, directly from the elements in their standard states.
    (b) Algebraically combine these equations to get the balanced chemical equation for the complete combustion of C_4H_{10}.
20. The enthalpy of formation of an element in its standard state is zero. Explain why the enthalpy of formation of $I_{2(g)}$ is 21 kJ/mol, not 0 kJ/mol.
21. Briefly state why it is useful to know the half-life of a reaction.
22. The rate of a chemical reaction is dependent on its ΔH. Do you agree or disagree with this statement? Briefly justify your answer.
23. What is the relationship between the units of the rate constant, k, and the overall reaction order?
24. List the factors that affect the rate of a chemical reaction.
25. Draw and label a potential energy diagram for the following exothermic reaction. Include E_a, the activated complex, and ΔH.
    $A + B \rightarrow AB$
26. Distinguish between the instantaneous rate and the initial rate of a chemical reaction. Under what circumstances would these two rates be the same?

Inquiry

27. A student wants to determine the enthalpy change associated with dissolving solid sodium hydroxide, NaOH, in water. The student dissolves 1.96 g of NaOH in 100.0 mL of water in a coffee-cup calorimeter. The initial temperature of the water is 23.4°C. After the NaOH dissolves, the temperature of the water rises to 28.7°C.
    (a) Use these data to determine the enthalpy of dissolution of sodium hydroxide, in kJ/mol NaOH. Assume that the heat capacity of the calorimeter is negligible.
    (b) Suppose that the heat capacity of the calorimeter was not negligible. Explain how the value of ΔH that you calculated in part (a) would compare with the *actual* ΔH.
    (c) Draw and label an enthalpy diagram for this reaction.
28. Some solid ammonium nitrate, NH_4NO_3, is added to a coffee-cup calorimeter that contains water at room temperature. After the NH_4NO_3 has dissolved, the temperature of the solution drops to near 0°C. Explain this observation.
29. Consider the following chemical equations and their enthalpy changes.
    $CH_{4(g)} + 2O_{2(g)} \rightarrow CO_{2(g)} + 2H_2O_{(g)}$ $\Delta H = -8.0 \times 10^2$ kJ
    $CaO_{(s)} + H_2O_{(\ell)} \rightarrow Ca(OH)_{2(aq)}$ $\Delta H = -65$ kJ
    What volume of methane, at 20°C and 100 kPa, would have to be combusted in order to release the same amount of energy as the reaction of 1.0×10^2 g of CaO with sufficient water? (The volume of 1.00 mol of any gas at 20°C and 100 kPa is 24 L.)

30. The complete combustion of 1.00 mol of sucrose (table sugar), $C_{12}H_{22}O_{11}$, yields 5.65×10^3 kJ.
   (a) Write a balanced thermochemical equation for the combustion of sucrose.
   (b) Calculate the amount of energy that is released when 5.00 g of sucrose (about one teaspoon) is combusted.

31. Carbon monoxide reacts with hydrogen gas to produce a mixture of methane, carbon dioxide, and water. (This mixture is known as substitute natural gas.)
$4CO_{(g)} + 8H_{2(g)} \rightarrow 3CH_{4(g)} + CO_{2(g)} + 2H_2O_{(\ell)}$
Use the following thermochemical equations to determine the enthalpy change of the reaction.
$C_{(graphite)} + 2H_{2(g)} \rightarrow CH_{4(g)} + 74.8 \text{ kJ}$
$CO_{(g)} + \frac{1}{2}O_{2(g)} \rightarrow CO_{2(g)} + 283.1 \text{ kJ}$
$H_{2(g)} + \frac{1}{2}O_{2(g)} \rightarrow H_2O_{(g)} + 241.8 \text{ kJ}$
$C_{(graphite)} + \frac{1}{2}O_{2(g)} \rightarrow CO_{(g)} + 110.5 \text{ kJ}$
$H_2O_{(\ell)} + 44.0 \text{ kJ} \rightarrow H_2O_{(g)}$

32. The decomposition of aqueous hydrogen peroxide, H_2O_2, can be catalyzed by different catalysts, such as aqueous sodium iodide, NaI, or aqueous iron(II) nitrate, $Fe(NO_3)_2$.
   (a) The enthalpy change, in kJ/mol of H_2O_2, would be the same for this reaction, regardless of the catalyst. Explain why, with the help of a potential energy diagram.
   (b) Design an investigation to verify your explanation in part (a). Do not attempt to carry out the investigation without the supervision of your teacher.

33. Acetone, CH_3COCH_3, reacts with iodine in acidic solution.
$$CH_3COCH_{3(aq)} + I_{2(aq)} \xrightarrow{H^+} CH_3COCH_2I_{(aq)} + HI_{(aq)}$$
The experimentally observed rate law equation is written as follows:
Rate = $k[CH_3COCH_3][H^+]$
   (a) What is the effect on the reaction rate if the concentration of CH_3COCH_3 is doubled?
   (b) What is the effect on the reaction rate if the concentration of I_2 is doubled?

34. Consider the heterogeneous reaction of solid magnesium with hydrochloric acid.
$Mg_{(s)} + 2HCl_{(aq)} \rightarrow MgCl_{2(aq)} + H_{2(g)}$
The general rate law of this reaction is given below.
Rate = $k[Mg]^m[HCl]^n$
Several trials of the reaction are carried out, each using a piece of magnesium ribbon of a fixed length. This is analogous to constant concentration. The general rate law equation can now be written as follows:
Rate = $k'[HCl]^n$, where $k' = k[Mg]^m$
Design an investigation to determine the value of n in the rate law equation. Assume that you have a stock solution of 6.0 mol/L HCl, magnesium ribbon, a stopwatch, and any standard laboratory glassware available. If you want to carry out your investigation, have your teacher approve the procedure first.

35. Thioacetamide, CH_3CSNH_2, reacts with water as shown in the balanced chemical equation below.
$$CH_3C(S)NH_{2(aq)} + H_2O_{(\ell)} \rightarrow H_2S_{(aq)} + CH_3CONH_{2(aq)}$$
The experimentally observed rate law equation is written as follows:
Rate = $k[H_3O^+][CH_3C(S)NH_2]$
   (a) Do you expect the reaction rate to change if some solid NaOH is added to the reaction vessel? Explain.
   (b) Do you expect the reaction rate to change if 250 mL of water is added to 250 mL of the reacting solution? Explain.
   (c) How will the reaction rate change if the reaction is carried out at 40°C instead of at 20°C?

Communication

36. If a solution of acid accidentally comes in contact with your skin, you are told to run the affected area under cold water for several minutes. Explain why it is not advisable to simply neutralize the acid with a basic solution.

37. A classmate is having difficulty understanding how the concepts of system, insulated system, and surroundings are related to exothermic and endothemic reactions. Write a note to explain to your classmate how the concepts are related. Use diagrams to help clarify your explanation.

38. Consider the following data for the complete combustion of the C_1 to C_8 alkanes.

Name	Formula	ΔH_{comb} (kJ/mol of alkane)
methane	CH_4	-8.90×10^2
ethane	C_2H_6	-1.56×10^3
propane	C_3H_8	-2.22×10^3
butane	C_4H_{10}	-2.88×10^3
pentane	C_5H_{12}	-3.54×10^3
hexane	C_6H_{14}	-4.16×10^3
heptane	C_7H_{16}	-4.81×10^3
octane	C_8H_{18}	-5.45×10^3

(a) Using either graph paper or spreadsheet software, plot a graph of ΔH_{comb} (y-axis) versus the number of C atoms in the fuel (x-axis).

(b) Extrapolate your graph to predict ΔH_{comb} of decane, $C_{10}H_{22}$.

(c) From your graph, develop an equation to determine ΔH_{comb} of a straight-chain alkane, given the number of carbons. Your equation should be of the form $\Delta H = \ldots$.

(d) Use the equation you developed in part (c) to determine ΔH_{comb} of $C_{10}H_{22}$. How does this value compare with the value of ΔH you determined by extrapolation from the graph? Explain why.

(e) Methane, ethane, propane, and butane are all gases at room temperature. You know that equal volumes of different gases contain the same number of moles under identical conditions of temperature and pressure. Which of these gases do you think would make the best fuel? Explain your answer.

39. A chemistry student wrote the following sentences to remind herself of important concepts. Write a paragraph to expand on each concept. Use examples and diagrams where appropriate.

(a) Hess's law can be used to determine the enthalpy change of a reaction, instead of measuring the enthalpy change in a laboratory.

(b) A pH meter can be used to monitor the progress of a reaction.

(c) The rate law equation for a reaction is dictated by the rate-determining step in the reaction's mechanism.

(d) A catalyst speeds up a reaction, but it does not affect the enthalpy change of the reaction.

Making Connections

40. Suppose that you are having a new home built in a rural area, where natural gas is not available. You have two choices for fuelling your furnace:
- propane, C_3H_8, delivered as a liquid under pressure and stored in a tank
- home heating oil, delivered as a liquid (not under pressure) and stored in a tank

What factors do you need to consider in order to decide on the best fuel? What assumptions do you need to make?

41. Suppose that you read the following statement in a magazine: 0.95 thousand cubic feet of natural gas is equal to a gigajoule, GJ, of energy. Being a media-literate student, you are sceptical of this claim and wish to verify it. The following assumptions/information may be useful.
- Natural gas is pure methane.
- Methane undergoes complete combustion.
- $H_2O_{(\ell)}$ is formed, rather than $H_2O_{(g)}$.
- 1.00 mol of any gas occupies 24 L at 20°C and 100 kPa.
- 1 foot = 12 inches; 1 inch = 2.54 cm; 1 L = 1 dm^3

42. Many taxis and delivery vehicles, especially in large cities, have been converted to burn propane, C_3H_8, rather than gasoline. This is done to save money. For most vehicles, the conversion to propane must be done at the owner's expense. Suppose that a taxi owner wants to calculate how far a taxi needs to be driven in order to recoup the conversion cost. List the information that the owner needs in order to do the calculation. Also list any simplifying assumptions.

COURSE CHALLENGE

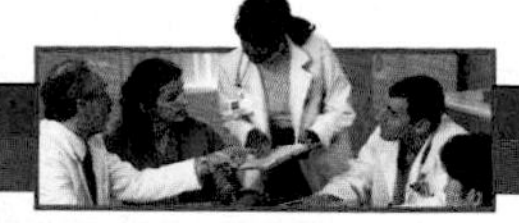

The Chemistry of Human Health

Consider the following points as you plan your Chemistry Course Challenge:
- How can you determine and compare rates of chemical reactions?
- What are some key differences between inorganic catalysts and enzymes?
- How are enzymes important for maintaining human health?

UNIT 4

Chemical Systems and Equilibrium

UNIT 4 CONTENTS

CHAPTER 7
Reversible Reactions and Chemical Equilibrium

CHAPTER 8
Acids, Bases, and pH

CHAPTER 9
Aqueous Solutions and Solubility Equilibria

UNIT 4 ISSUE
Earth in Equilibrium

UNIT 4 OVERALL EXPECTATIONS

- What are the properties of a system at equilibrium? What factors affect the extent of a chemical reaction?
- How can you measure the position of equilibrium experimentally and predict the concentrations of chemicals in a system at equilibrium?
- How can understanding chemical equilibrium help you explain systems in biology, ecology, and chemical industries?

Unit Issue Prep

Why is an understanding of Earth's geochemical cycles important? Look at the issue for Unit 4, on page 456. Start preparing now by listing questions that you would like answered. Record additional questions as you progress through this unit.

Photochromic lenses darken in sunlight and then gradually clear in shade and dim light. The change is permanent in constant light conditions, yet reversible if the light intensity changes. How?

The molten glass contains dissolved silver chloride. As the solution cools, silver chloride precipitates, forming tiny silver chloride crystals. Light striking these crystals produces silver and chlorine atoms. The silver atoms tend to clump together to form particles that are big enough to block light and darken the glass.

Adding copper(I) chloride to the molten glass makes the process reversible. When light intensity diminishes, copper ions remove electrons from silver atoms, converting the silver atoms into silver ions. The silver ions then migrate back to the silver chloride crystals. The glass becomes transparent again.

Most of the reactions that you have studied have essentially gone to completion. You can use stoichiometry to calculate the amounts of products formed, assuming that the chemical reaction proceeds only from left to right. In many chemical reactions, however, a mixture of reactants and products form because the reaction can also proceed from right to left. In this unit, you will study the concepts that describe reversible reactions.

CHAPTER 7

Reversible Reactions and Chemical Equilibrium

Chapter Preview

7.1 Recognizing Equilibrium

7.2 Thermodynamics and Equilibrium

7.3 The Equilibrium Constant

7.4 Predicting the Direction of a Reaction

Prerequisite Concepts and Skills

Before you begin this chapter, review the following concepts and skills:

- using the relationship between rate of reaction and concentration (Chapter 6, section 6.2)
- using stoichiometry with chemical equations (previous studies)
- calculating molar concentrations (previous studies)

Your body contains over 1 kg of nitrogen, mainly in the proteins and nucleic acids that make up your cells. Your body, however, cannot absorb nitrogen directly from the atmosphere. You get the nitrogen-containing compounds you need by eating plants and other foods.

Plants cannot absorb nitrogen directly from the atmosphere, either. Their roots absorb nitrogen compounds—mainly nitrates—from the soil. Some of these nitrates result from atmospheric chemical reactions that involve nitrogen and oxygen. First, lightning supplies the energy that is needed to form nitric oxide from nitrogen and oxygen. The resulting nitric oxide combines with oxygen to form nitrogen dioxide, which reacts with rainwater to form nitric acid. Then, in the seas and the soil at Earth's surface, nitric acid ionizes to produce nitrate, NO_3^-. Plants and plant-like organisms can readily absorb nitrate.

Scientists call this chemical "circle of life" the *nitrogen cycle*. There are, as you know, similar, equally important cyclic processes in nature. Each process involves *reversible changes*: changes that may proceed in either direction, from reactants to products or from products to reactants.

In this chapter, you will study factors that affect reversible changes, notably those in chemical reactions. You will learn how to determine the amounts of reactants and products that are present when their proportions no longer change. You will also learn how to make qualitative predictions about the ways that chemists can change these proportions. Finally, you will see how industrial chemists apply their knowledge of reversible changes to increase the yield of chemicals that are important to society.

How is this dynamic situation similar to the activity of reactants and products in a chemical reaction?

Recognizing Equilibrium

7.1

In this section, you will

- **identify** and **illustrate** equilibrium in various systems and the conditions that lead to equilibrium
- **describe**, in terms of equilibrium, the behaviour of ionic solutes in unsaturated, saturated, and supersaturated solutions
- **communicate** your understanding of the following terms: *equilibrium, homogeneous equilibrium, heterogeneous equilibrium*

In Unit 3, you learned that the rate of any reaction depends on the concentration of the reacting chemicals. As a reaction proceeds, the concentrations of the "product" chemicals increases, and the reverse reaction may re-form "reactants." Under certain conditions, the rate of the reverse reaction increases as the rate of the forward reaction decreases. Eventually, the rate of the forward reaction equals the rate of the reverse reaction. **Equilibrium** occurs when opposing changes, such as those just described, are occurring simultaneously at the same rate.

How can you recognize equilibrium? In the following paragraphs, you will read about physical changes that may or may not involve a system at equilibrium. As you consider these changes, note which systems are at equilibrium and what changes are taking place. Also note the conditions that would be needed for the changes to occur in the opposite direction at the same rate. The ExpressLab that follows these descriptions will add to your understanding of equilibrium. It will help you recognize systems that are at equilibrium and systems that are not.

- A puddle of water remains after a summer shower. The puddle evaporates because some of the water molecules near the surface have enough kinetic energy to escape from the liquid. At the same time, some water molecules in the air may condense back into the liquid. The chance of this happening is small, however. The puddle soon evaporates completely, as shown in Figure 7.1(A).
- Even if you put a lid on a jar of water, some of the water in the jar evaporates. Careful measurements show that the level of liquid water in the jar initially decreases because more water evaporates than condenses. As the number of water vapour molecules increases, however, some condense back into the liquid. Eventually, the number of evaporating molecules and the number of condensing molecules are equal. The level of water inside the jar remains constant. At this point, shown in Figure 7.1(B), equilibrium has been reached.

Figure 7.1 Why is (B) an example of equilibrium, while (A) is not?

- Crystals of copper(II) sulfate pentahydrate, $CuSO_4 \cdot 5H_2O$, are blue. If you place a few small crystals in a beaker of water, water molecules break apart the ions and they enter into the solution. A few ions in the solution may re-attach to the crystals. Because more ions enter the solution than re-attach to the crystals, however, all the solid eventually dissolves. If you keep adding crystals of copper(II) sulfate pentahydrate, the solution eventually becomes saturated. Crystals remain at the bottom of the beaker, as shown in Figure 7.2.

Figure 7.2 Which of these systems, (A) or (B), is at equilibrium? Will changing the temperature affect the concentration of the solution in the equilibrium system?

- Figure 7.3(A) shows a supersaturated solution of sodium acetate. It was prepared by adding sodium acetate to a saturated aqueous solution. The mixture was heated to dissolve the added crystals. Finally, the solution was left to return slowly to room temperature. In Figure 7.3(B), a single crystal of sodium acetate has been added to the supersaturated solution. As you can see, solute ions rapidly leave the solution, and solid forms.

Figure 7.3 These photographs show a supersaturated solution of sodium acetate and the effect of adding a crystal to it. Is either of these systems at equilibrium?

ExpressLab  Modelling Equilibrium

In this ExpressLab, you will model what happens when forward and reverse reactions occur. You will take measurements to gain quantitative insight into an equilibrium system. Then you will observe the effect of introducing a change to the equilibrium.

Materials

2 graduated cylinders (25 mL)
2 glass tubes with different diameters (for example, 10 mm and 6 mm)
2 labels or a grease pencil
supply of water, coloured with food dye

Procedure

1. Copy the following table into your notebook, to record your observations.

Transfer number	Volume of water in reactant cylinder (mL)	Volume of water in product cylinder (mL)
0	25.0	0.0
1		
2		

2. Label one graduated cylinder "reactant." Label the other "product."
3. Fill the reactant cylinder with coloured water, up to the 25.0 mL mark. Leave the product cylinder empty.
4. With your partner, transfer water simultaneously from one cylinder to the other as follows: Lower the larger-diameter glass tube into the reactant cylinder. Keep the top of the tube open. When the tube touches the bottom of the cylinder, cover the open end with a finger. Then transfer the liquid to the product cylinder. *At the same time* as you are transferring liquid into the product cylinder, your partner must use the smaller-diameter tube to transfer liquid from the product cylinder into the reactant cylinder.
5. Remove the glass tubes. Record the volume of water in each graduated cylinder, to the nearest 0.1 mL.
6. Repeat steps 4 and 5 until there is no further change in the volumes of water in the graduated cylinders.
7. Add approximately 5 mL of water to the reactant cylinder. Record the volume in each cylinder. Then repeat steps 4 and 5 until there is no further change in volume.

Analysis

1. Plot a graph of the data you collected. Put transfer number on the *x*-axis and volume of water on the *y*-axis. Use different symbols or colours to distinguish between the reactant volume and the product volume. Draw the best smooth curve through the data.
2. In this activity, the volume of water is a model for concentration. How can you use your graph to compare the rate of the forward reaction with the rate of the reverse reaction? What happens to these rates as the reaction proceeds?
3. At the point where the two curves cross, is the rate of the forward reaction equal to the rate of the reverse reaction? Explain.
4. How can you recognize when the system is at equilibrium?
5. Were the volumes of water (that is, concentrations of reactants and products) in the two tubes equal at equilibrium? How do you know?
6. In a chemical reaction, what corresponds to the addition of more water to the reactant cylinder? How did the final volume of water in the product cylinder change as a result of adding more water to the reactant cylinder?
7. Determine the ratio $\frac{\text{Volume of product}}{\text{Volume of reactant}}$ at the end of the first equilibrium and at the end of the second equilibrium. Within experimental error, were these two ratios the same or different?
8. In this activity, what do you think determined the relative volumes of water in the graduated cylinders? In a real chemical reaction, what factors might affect the relative concentrations of reactants and products at equilibrium?
9. Explain why the system in this activity is a closed system.

Conditions That Apply to All Equilibrium Systems

The fundamental requirement for equilibrium is that opposing changes must occur at the same rate. There are many processes that reach equilibrium. Three physical processes that reach equilibrium are

- a solid in contact with a solution that contains this solid: for example, sugar crystals in a saturated aqueous sugar solution
- the vapour above a pure liquid: for example, a closed jar that contains liquid water
- the vapour above a pure solid: for example, mothballs in a closed drawer

Two chemical processes that reach equilibrium are

- a reaction with reactants and products in the same phase: for example, a reaction between two gases to produce a gaseous product. In this chapter, you will focus on reactions in which the reactants and products are in the same phase. The equilibrium they reach is called **homogeneous equilibrium**.
- a reaction in which reactants and products are in different phases: for example, an aqueous solution of ions, in which the ions combine to produce a slightly soluble solid that forms a precipitate. In Chapters 8 and 9, you will work extensively with reactions in which the reactants and products are in different phases. The equilibrium they reach is called **heterogeneous equilibrium**.

CONCEPT CHECK

A decomposition reaction is taking place in a closed container, $R \rightarrow P$. At equilibrium, does the concentration of reactant have to equal the concentration of product? Explain your answer.

What is it that equilibrium systems like these have in common? What conditions are necessary for equilibrium to become established? As outlined in the box below, there are four conditions that apply to all equilibrium systems.

The Four Conditions That Apply to All Equilibrium Systems

1. Equilibrium is achieved in a reversible process when the rates of opposing changes are equal. A double arrow, $\rightleftharpoons$, indicates reversible changes. For example:

$$H_2O_{(\ell)} \rightleftharpoons H_2O_{(g)}$$

$$H_{2(g)} + Cl_{2(g)} \rightleftharpoons 2HCl_{(g)}$$

2. The observable (macroscopic) properties of a system at equilibrium are constant. At equilibrium, there is no overall change in the properties that depend on the total quantity of matter in the system. Examples of these properties include colour, pressure, concentration, and pH.

You can summarize the first two equilibrium conditions by stating that *equilibrium involves dynamic change at the molecular level but no change at the macroscopic level.*

3. Equilibrium can only be reached in a closed system. A closed system is a system that does not allow the input or escape of any component of the equilibrium system, including energy. For this reason, a system can be at equilibrium only if it is at constant temperature. A common example of a closed system is carbon dioxide gas that is in equilibrium with dissolved carbon dioxide in a soda drink.

The system remains at equilibrium as long as the container is not opened. Note that small changes to the components of a system are sometimes negligible. Thus, equilibrium principles can be applied if a system is not physically closed. For example, consider the equilibrium of a solid in a saturated aqueous solution, such as $CaO_{(s)} + H_2O_{(\ell)} \rightleftharpoons Ca^{2+}_{(aq)} + 2OH^{-}_{(aq)}$. You can neglect the small amount of water that vaporizes from the open beaker during an experiment.

4. Equilibrium can be approached from either direction. For example, the proportions of $H_{2(g)}$, $Cl_{2(g)}$, and $HCl_{(g)}$ (in a closed container at constant temperature) are the same at equilibrium, regardless of whether you started with $H_{2(g)}$ and $Cl_{(g)}$ or whether you started with $HCl_{(g)}$.

Section Summary

In this section, you learned how to recognize equilibrium. As well, you learned about the conditions that are needed for equilibrium to be reached. Later in this chapter, you will examine what happens when some equilibrium conditions in a system are changed. You will learn how chemists can control conditions to increase the yield of reactions.

Energy is an important component of most equilibrium systems. The input or output of energy in a system causes the temperature to change. Thus, the requirement that an equilibrium system be closed means that the temperature of the system must remain constant. In the next section, you will examine more closely the effects of thermodynamics on equilibrium systems. In particular, you will examine the factors that affect the amount of reactant and product in a reaction and the factors that determine whether or not a reaction is spontaneous.

Section Review

1 K/U Give two physical and two chemical processes that are examples of reversible changes that are not at equilibrium.

2 K/U Explain, in terms of reaction rates, how the changes that take place in a reversible reaction approach equilibrium.

3 K/U A sealed carbonated-drink bottle contains a liquid drink with a space above it. The space contains carbon dioxide at a pressure of about 405 kPa.

(a) What changes are taking place at the molecular level?

(b) Which macroscopic properties are constant?

4 K/U Agree or disagree with the following statement, and give reasons to support your answer: "In a sealed jar of water at equilibrium, the quantity of water molecules in the liquid state equals the quantity of water vapour molecules."

5 I Ice and slush are a feature of Canadian winters. Under what conditions do ice and water form an equilibrium mixture?

Electronic Learning Partner

Observe a molecular view of dynamic equilibrium in your Chemistry 12 Electronic Learning Partner.

7.2 Thermodynamics and Equilibrium

Section Preview/ Specific Expectations

In this section, you will

- **identify** qualitatively entropy changes that are associated with physical and chemical processes
- **describe** how reactions have a tendency to achieve minimum energy and maximum entropy
- **explain** the effect of changes in enthalpy, entropy, and temperature on a chemical reaction
- **communicate** your understanding of the following terms: *favourable change, entropy (S), second law of thermodynamics, free energy*

You can easily predict what will happen in a number of physical and chemical processes. What will happen if you let go of a pencil that you are holding tip-down on a table? What will you observe if you add a few drops of food colouring to some water in a glass? If a piece of paper starts to burn in a plentiful supply of air, what are the products of the reaction? How will an iron nail change if it is left outside? These are all examples of favourable (or *spontaneous*) changes. A **favourable change** is a change that has a natural tendency to happen under certain conditions.

What Conditions Favour a Change?

What conditions determine whether or not a change is favourable? How are different conditions related to equilibrium, where forward and reverse changes occur at the same rate? The answers to these questions are linked to two important concepts in thermodynamics: enthalpy and entropy.

Enthalpy and Favourable Changes

Energy is a key condition in favourable changes. For example, you know that a ball always rolls downhill. Its gravitational potential energy is lower at the bottom of the hill than at the top. For the ball to roll uphill—an unfavourable (or non-spontaneous) change—you have to apply energy to the ball. Chemical reactions are similar. When products have less enthalpy than reactants, the reaction releases energy. Therefore, the reaction is exothermic. Are favourable changes those that are exothermic under certain conditions?

Certainly many favourable physical and chemical changes are exothermic. Some favourable changes, however, involve no release of energy. Others are actually endothermic. For example, a favourable endothermic chemical reaction occurs when barium hydroxide and ammonium thiocyanate are placed in an Erlenmeyer flask, as shown in Figure 7.4.

$$Ba(OH)_2 \cdot 8H_2O_{(s)} + 2NH_4SCN_{(s)} \rightarrow Ba(SCN)_{2(aq)} + 10H_2O_{(\ell)} + 2NH_{3(g)} \quad \Delta H = +170 \text{ kJ}$$

When the flask is stoppered and the contents are shaken, the solids intermingle. The reaction produces a slush-like mixture. This favourable, but endothermic, reaction absorbs enough energy to freeze a thin layer of water under the flask.

Figure 7.4 A favourable endothermic reaction occurs when barium hydroxide is mixed with ammonium thiocyanate. If you removed the stopper in photograph (B), you would detect the characteristic odour of ammonia.

Temperature and Favourable Changes

Another condition that determines whether or not a particular reaction is favoured is temperature. Consider the reversible synthesis reaction between mercury and oxygen, and the decomposition of mercury(II) oxide.

$$Hg_{(\ell)} + \frac{1}{2}O_{2(g)} \rightleftharpoons HgO_{(s)} \quad \Delta H = \pm 90.8 \text{ kJ}$$

From left to right, the reaction is exothermic. Therefore, the enthalpy change, ΔH, is negative. If the enthalpy change was the only condition that determined whether a reaction is favourable, then the synthesis reaction would take place. The synthesis reaction does take place—but only at relatively moderate temperatures. Above 400°C, the reverse reaction is favourable. The decomposition of $HgO_{(s)}$ occurs. Thus, the direction in which this reaction proceeds depends on temperature. This is fundamentally different from the dependence of reaction rate on temperature that you learned about in Unit 3. The reaction between $Hg_{(\ell)}$, $O_{2(g)}$, and $HgO_{(s)}$ does not just change its rate with a change in temperature, it changes its *direction*.

Temperature and Enthalpy Alone Cannot Explain Favourable Reactions

Temperature and enthalpy are not the only conditions that determine whether a change is favourable. Consider the process shown in Figure 7.5. A closed valve links two flasks together. The left flask contains an ideal gas. The right flask is evacuated. When the valve is opened, you expect the gas to diffuse into the evacuated flask until the pressure in both flasks is equal. You do not expect to see the reverse process—with all the gas molecules ending up in one of the flasks—unless work is done on the system. The process shown in Figure 7.5 is certainly a favourable change. Yet no exchange of energy is involved. The condition that influences this change is called entropy. It is an important condition in all physical and chemical changes.

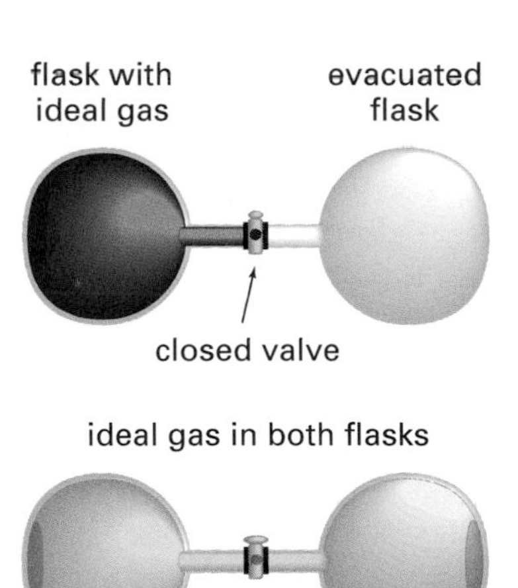

Figure 7.5 If the gas is ideal, this favourable change involves no change in energy. Why must the gas be ideal?

Entropy and Favourable Changes

Energy is conserved, but it tends to spread out and become less concentrated. Hot objects always dissipate heat to cooler surroundings. **Entropy, S**, is a way to measure the energy that is spread out or dispersed during a process.

The entropy of a system depends on the number of energy states the molecules in the system possess. The entropy of a system increases with temperature because the molecules possess more energy states at the higher temperature. When a solid melts the entropy again increases because the molecules are able to move about in more ways than before. See Figure 7.6 on the next page.

An ordered arrangement of particles (atoms, ions, or molecules) has lower entropy (smaller disorder) than the same number of particles in random arrangements. Thus, the entropy of a pure substance depends on its state. The entropy of a system increases (becomes more disordered) with temperature, because the motion of particles becomes more chaotic at higher temperatures. See Figure 7.6 on the next page.

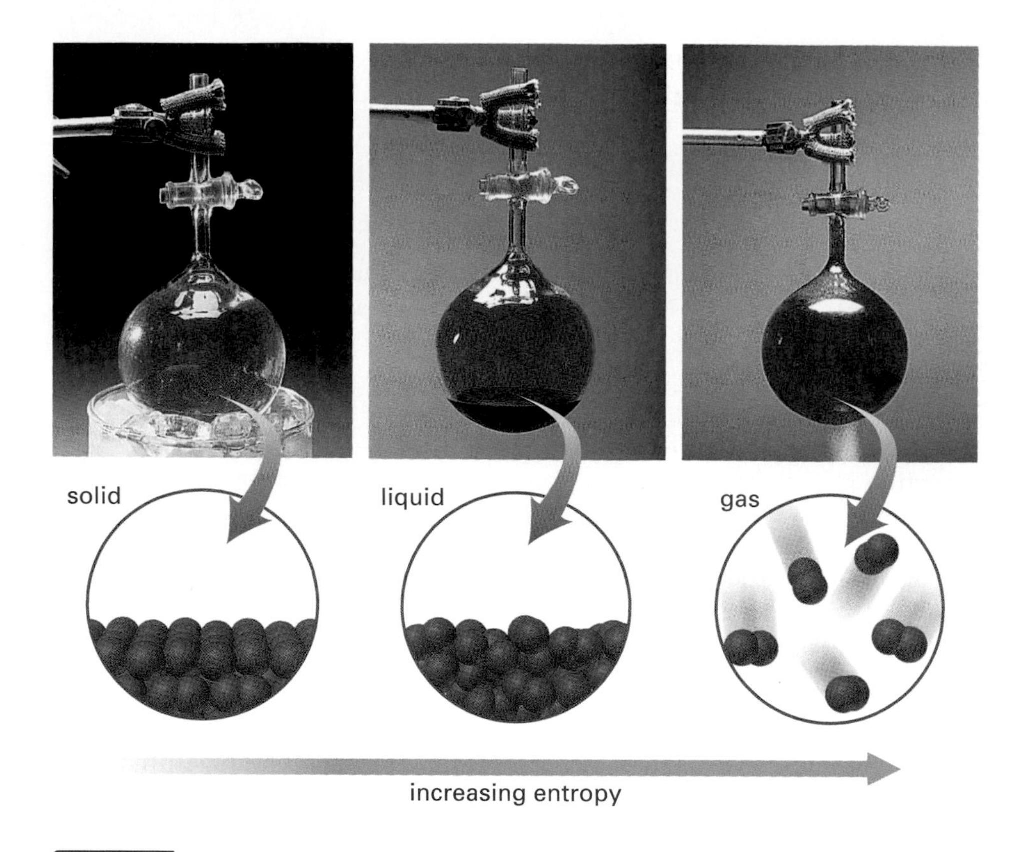

Figure 7.6 The entropy of the particles in a sample of matter increases as the matter changes state from solid to liquid to gas.

The mathematical definition of entropy describes the probability of a given arrangement of particles. A system of moving particles with low entropy *probably* has greater order than a similar system with higher entropy. Why say "probably"? Consider gas molecules in a container at room temperature. The gas molecules are moving rapidly, colliding with other molecules and the walls of the container. They have relatively large entropy. What if two different gases occupy the same container? It is possible, in principle, for the gases to separate so they are on opposite sides of the container. Given the huge number of particles, however, the probability of this happening is so close to zero that chemists say it will never happen. Nevertheless, it remains possible. The outcome is not certain.

As you know from Unit 3, chemists often use the terms "system" and "surroundings" to describe a chemical reaction. The system is the reaction itself—the reactants and products. The surroundings are everything else in the universe. According to the first law of thermodynamics (the law of conservation of energy) the algebraic sum of the energy changes in a system and its surroundings is zero. Unlike energy, entropy is not conserved. The total amount of entropy is increasing. This is part of the **second law of thermodynamics**: the total entropy of the universe is constantly increasing. The key word here is "total." If you are studying a system, you must add together changes in the entropy of the system, ΔS_{sys}, and changes in the entropy of the surroundings, ΔS_{surr}.

Consider what happens to you and your classmates (the particles in a "system") on your way to class. Initially, you and your classmates are in different places in the school, walking with different velocities toward the classroom. Going to class represents a relatively large amount of entropy. The final state of the system, when everyone is seated, represents a system that appears to have the same, or even less entropy. How, then, did the entropy of the universe increase?

The entropy of you and your classmates changed on your way to class because each of you metabolised energy. The movement and heat from your bodies and the warm moist air you exhaled on your way to class added to the entropy of the air particles around you, increasing the entropy of the universe. The total amount of entropy always increases when energy is dispersed.

What does entropy have to do with favourable chemical changes and equilibrium systems? *All favourable changes involve an increase in the total amount of entropy.* Recall the endothermic reaction in Figure 7.4.

$$Ba(OH)_2 \bullet 8H_2O_{(s)} + 2NH_4SCN_{(s)} \rightarrow Ba(SCN)_{2(aq)} + 10H_2O_{(\ell)} + 2NH_{3(g)} \quad \Delta H = +170 \text{ kJ}$$

The reaction is favourable, partly because the reactants are two solids and the products are a solution, water, and a gas. The products have a much larger entropy than the reactants. This helps to make the reaction favourable.

Now recall the reaction between mercury and oxygen. It favours the formation of HgO below about 400°C, but the decomposition of HgO above 400°C. This reaction highlights the importance of temperature to favourable change. Enthalpy, entropy, and temperature are linked in a concept called free energy.

Free Energy and Equilibrium

Imagine a perfect engine—an engine without any friction between its moving parts. This engine would still be unable to convert all the energy in a fuel into useful work. The reason is that some energy, such as the energy of the hot exhaust, is used to increase the entropy of the surroundings. The term **free energy** means available energy. In a chemical change, free energy is a measure of the useful work that can be obtained from a reaction. Free energy is often called *Gibbs free energy*, and given the symbol G, in honour of Josiah Willard Gibbs (1839–1903). Gibbs, a professor of chemistry at Harvard University, developed the concept of free energy and the equations that describe it.

The change in the free energy of a system, ΔG, at constant temperature is given by the equation $\Delta G = \Delta H - T\Delta S$. In this equation, as in most chemistry equations, the temperature, T, must be in Kelvin degrees. The symbol ΔH represents the change in enthalpy. The symbol ΔS represents the change in entropy.

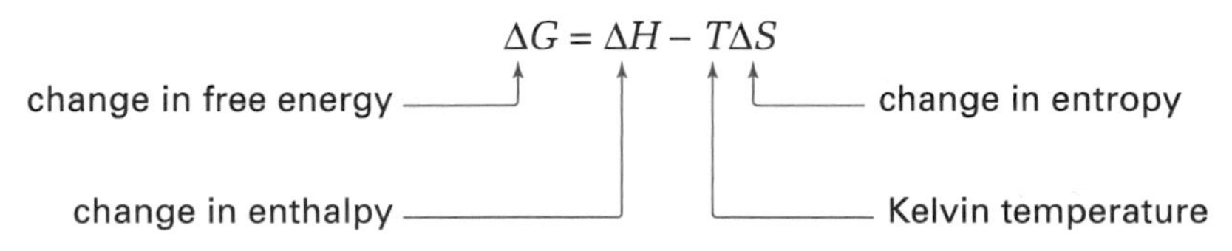

The value of ΔG for a chemical reaction tells us whether the reaction is a source of useful energy. Such energy can be converted into another form, such as mechanical energy to drive machinery or electrical energy in a battery. The rest of the energy from the reaction enters the environment as heat. It increases the entropy of the surroundings.

Biology LINK

The concept of free energy appears in other areas of science, notably biology. Find out the meanings of the terms "exergonic" and "endergonic." How are these terms related to free energy?

The equation $\Delta G = \Delta H - T\Delta S$, and the sign of ΔG, can help to explain why some chemical reactions are favourable at room temperature, why others are not, and how temperature affects the direction of a reaction. The change in the free energy of a chemical reaction is related to the direction of the reaction and to equilibrium as follows:

- When ΔG is negative, the forward reaction is favourable.
- When ΔG is zero, the reaction is at equilibrium.
- When ΔG is positive, the reaction is favourable in the reverse direction but not in the forward direction.

Exothermic reactions are often favourable because the sign of ΔH is negative. The equation shows that exothermic reactions are unfavourable only when ΔS is negative and the value of $-T\Delta S$ is large enough to make ΔG positive. This is what happens in the reaction between mercury and oxygen.

$$Hg_{(\ell)} + \frac{1}{2}O_{2(g)} \rightarrow HgO_{(s)} \quad \Delta H = -90.8 \text{ kJ}$$

From left to right, the reaction is exothermic. There is a decrease in entropy because gas molecules react to form a solid. At temperatures greater than about 400°C (673 K), the value of $-T\Delta S$ is positive and greater than 90.8 kJ. Thus, ΔG becomes positive. The forward reaction is no longer favourable, as experiments confirm.

In most reactions, the entropy of the system increases (S is positive), so the value of $-T\Delta S$ is negative. Therefore, most reactions tend to be more favourable at higher temperatures. At low temperatures, reactions are likely to be favourable only if they are highly exothermic.

Section Summary

Table 7.1 shows how you can use the signs of ΔH and ΔS to determine whether a chemical reaction is favourable. It also shows how ΔH and ΔS may vary with temperature. Keep in mind that a favourable reaction may be fast or slow. Thermodynamics makes no prediction about the rate of a reaction, only whether or not it can take place. Also, before any reaction begins, the activation energy must be supplied.

Table 7.1 How the Signs of ΔH and ΔS Affect the Favourability of a Reaction

ΔH	ΔS	$-T\Delta S$	Comments and examples
−	+	−	Both ΔH and ΔS favour the reaction. The reaction is favourable at all temperatures. Example: $2O_{3(g)} \rightarrow 3O_{2(g)}$
+	−	+	Neither ΔH nor ΔS favours the reaction. The reaction is unfavourable at all temperatures. Example: $3O_{2(g)} \rightarrow 2O_{3(g)}$
−	−	+	ΔH is favourable, but ΔS is not. The reaction is likely to be favourable at relatively low temperatures. Example: $Hg_{(\ell)} + \frac{1}{2}O_{2(g)} \rightarrow HgO_{(s)}$
+	+	−	ΔS is favourable, but ΔH is not. The reaction is likely to be favourable at higher temperatures. Example: $HgO_{(s)} \rightarrow Hg_{(\ell)} + \frac{1}{2}O_{2(g)}$

Note that the entropy of a system cannot increase forever. Eventually, a maximum state of disorder is reached. When this happens, the system appears to have constant properties, even though changes are still taking place at the molecular level. We say that a chemical system is at equilibrium when it has constant observable properties. Therefore, equilibrium occurs when a system has reached its maximum entropy. In the next section, you will look more closely at the reactants and products of chemical systems and learn how equilibrium is measured.

Section Review

1. K/U Explain the difference between a favourable chemical change and an unfavourable chemical change. Give two examples of each type of change.

2. C Write a short paragraph, or use a graphic organizer, to show the relationship among the following concepts: favourable chemical change, temperature, enthalpy, entropy, free energy.

3. K/U In each process, how does the entropy of the system change?

   (a) ice melting

   (b) water vapour condensing

   (c) sugar dissolving in water

   (d) $HCl_{(g)} + NH_{3(g)} \rightarrow NH_4Cl_{(s)}$

   (e) $CaCO_{3(s)} \rightarrow CaO_{(s)} + CO_{2(g)}$

4. K/U What is the sign of the entropy change in each chemical reaction?

   (a) $N_2O_{4(g)} \rightarrow 2NO_{2(g)}$

   (b) $PCl_{3(g)} + Cl_{2(g)} \rightarrow PCl_{5(g)}$

   (c) $2Al_{(s)} + \frac{3}{2}O_{2(g)} \rightarrow Al_2O_{3(s)}$

   (d) $N_{2(g)} + 3H_{2(g)} \rightarrow 2NH_{3(g)}$

   (e) $S_{(s)} + O_{2(g)} \rightarrow SO_{2(g)}$

5. I When water freezes, the phase change that occurs is exothermic.

   $H_2O_{(\ell)} \rightarrow H_2O_{(s)} \quad \Delta H = -6.02 \text{ kJ}$

   Based on the change in enthalpy, you would expect that water would always freeze. Use the concepts of entropy and free energy to explain why this phase change is favourable only below 0°C.

7.3 The Equilibrium Constant

Section Preview/ Specific Expectations

In this section, you will

- **express** your understanding of the law of chemical equilibrium as it applies to concentrations of reactants and products at equilibrium
- **collect** experimental data to determine an equilibrium constant for concentration
- **solve** equilibrium problems involving concentrations of reactants and products
- **communicate** your understanding of the following terms: *law of chemical equilibrium, equilibrium constant (K_c), ICE table*

Chemists use both thermodynamics and rate to study chemical reactions. Thermodynamics determines *whether* a reaction will occur at a certain temperature and *when* equilibrium will be reached. The rate of a reaction determines the *time it takes* for a certain concentration of product to form. In this section, you will learn about the *extent* of a reaction: the relative concentrations of products to reactants at equilibrium.

Opposing Rates and the Law of Chemical Equilibrium

In 1864, two Norwegian chemists, Cato Guldberg and Peter Waage, summarized their experiments on chemical equilibrium in the **law of chemical equilibrium**: *At equilibrium, there is a constant ratio between the concentrations of the products and reactants in any change.* Figure 7.7 shows how the law of chemical equilibrium applies to one chemical system. Chemists have studied this system extensively. It involves the reversible reaction between two gases: dinitrogen tetroxide, which is colourless, and nitrogen dioxide, which is dark brown.

$$N_2O_{4(g)} \rightleftharpoons 2NO_{2(g)}$$

colourless **brown**

By observing the intensity of the brown colour in the mixture, chemists can determine the concentration of nitrogen dioxide.

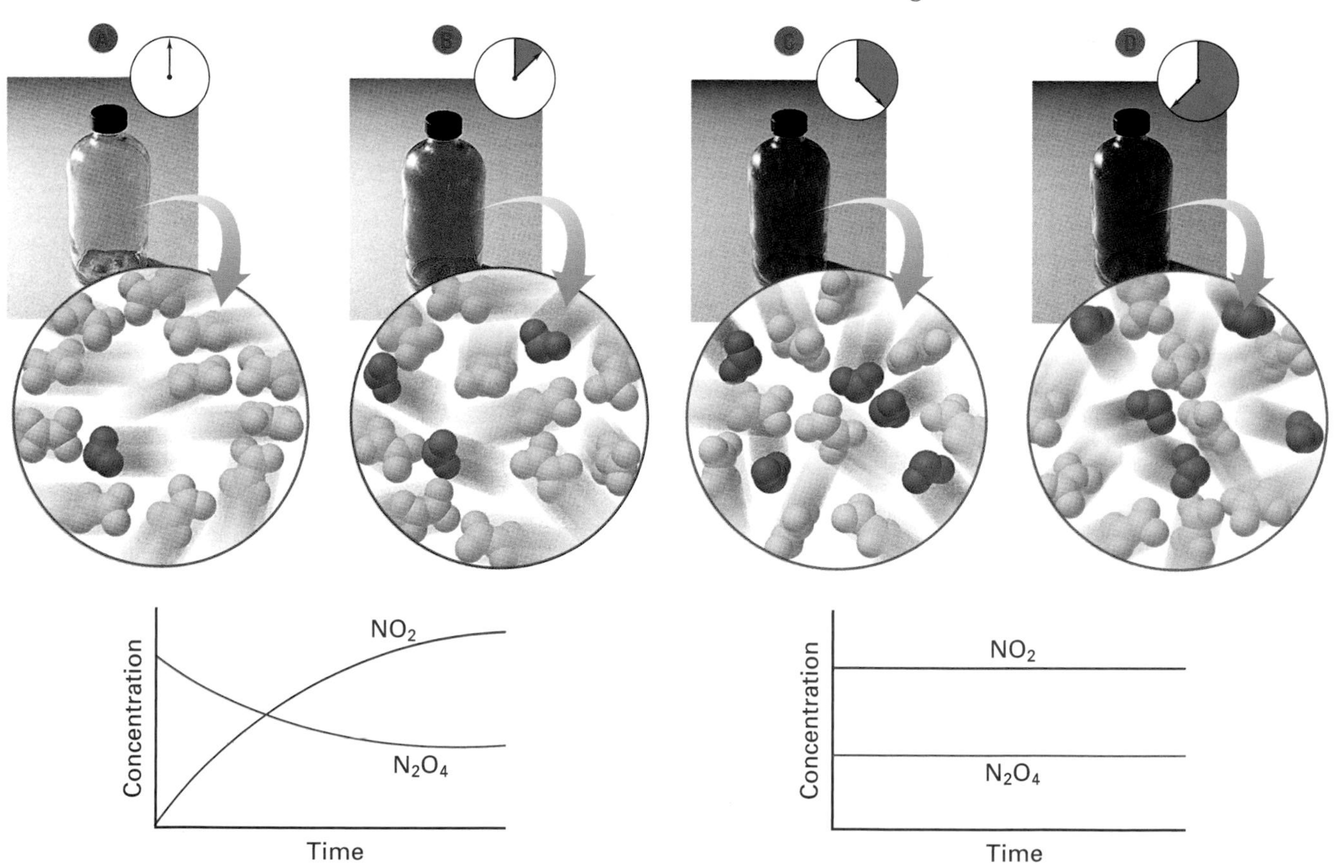

Figure 7.7 As this system nears equilibrium, the rate of the forward reaction decreases and the rate of the reverse reaction increases. At equilibrium, the macroscopic properties of this system are constant. Changes at the molecular level take place at equal rates.

Dinitrogen tetroxide gas is produced by vaporizing dinitrogen tetroxide liquid. Dinitrogen tetroxide liquid boils at 21°C. If a small quantity is placed in a sealed flask at 100°C, it vaporizes, filling the flask with dinitrogen tetroxide gas.

Suppose that the initial concentration of $N_2O_{4(g)}$ is 0.0200 mol/L. The initial concentration of $NO_{2(g)}$ is zero. The initial rate of the forward reaction, k_f, is relatively large, while the initial rate of the reverse reaction, k_r, is zero. The initial conditions correspond to the first exchange in the ExpressLab on page 325, where the reactant cylinder was full and the product cylinder was empty. As the reaction proceeds, the rate of the forward reaction decreases because the concentration of $N_2O_{4(g)}$ decreases. At the same time, the rate of the reverse reaction increases because the concentration of $NO_{2(g)}$ increases. At equilibrium, $k_f = k_r$. There are no further changes in the relative amounts of N_2O_4 and NO_2.

CONCEPT CHECK

How did the ExpressLab on page 325 model a reaction approaching equilibrium? How did it model the situation when equilibrium was reached?

The Equilibrium Constant

From other experiments involving the reaction between nitrogen tetroxide and nitrogen dioxide, chemists know that both the forward and reverse reactions involve elementary steps. Thus, you can write rate equations for the reactions.

Forward reaction: $N_2O_{4(g)} \rightarrow 2NO_{2(g)}$ Reverse reaction: $2NO_{2(g)} \rightarrow N_2O_{4(g)}$

Forward rate: $k_f[N_2O_4]$ Reverse rate: $k_f[NO_2]^2$

At equilibrium,

$$\text{Foward rate} = \text{Reverse rate}$$

$$k_f[N_2O_4] = k_r[NO_2]^2$$

The ratio of rate constants is another constant. The forward rate constant, k_f, divided by the reverse rate constant, k_r, is called the **equilibrium constant**, K_{eq}.

$$\frac{k_f}{k_r} = K_{eq} = \frac{[NO_2]^2}{[N_2O_4]}$$

You can write rate equations for these reactions because they are elementary mechanisms. Guldberg and Waage, however, showed that similar results are found for any reaction, regardless of the mechanism. For any general equilibrium equation, let P, Q, R, and S represent chemical formulas and *a*, *b*, *c*, and *d* represent their respective coefficients in the chemical equation.

$$a\text{P} + b\text{Q} \rightleftharpoons c\text{R} + d\text{S}$$

Recall that molar concentrations are indicated by square brackets. The equilibrium expression is usually expressed in terms of molar concentrations. Thus, the subscript "c" is usually used instead of "eq" in the equilibrium constant.

$$K_c = \frac{[\text{R}]^c[\text{S}]^d}{[\text{P}]^a[\text{Q}]^b}$$

This equilibrium expression depends only on the stoichiometry of the reaction. By convention, chemists always write the concentrations of the products in the numerator and the concentrations of the reactants in the denominator. Each concentration term is raised to the power of the coefficient in the chemical equation. The terms are multiplied, never added.

The following Sample Problem shows how to find the equilibrium expression for a reaction. In this chapter, you will use equilibrium expressions for homogeneous reactions (mostly reactions between gases). In Chapters 8 and 9, you will learn how to use equilibrium expressions for heterogeneous systems.

The law of chemical equilibrium is sometimes known as the law of mass action. Before the term "concentration" was used, the concept of amount per unit volume was called "active mass."

Sample Problem

Writing Equilibrium Expressions

Problem

One of the steps in the production of sulfuric acid involves the catalytic oxidation of sulfur dioxide.

$$2SO_{2(g)} + O_{2(g)} \rightarrow 2SO_{3(g)}$$

Write the equilibrium expression.

What Is Required?

You need to find an expression for K_c.

What Is Given?

You know the balanced chemical equation.

Plan Your Strategy

The expression for K_c is a fraction. The concentration of the product is in the numerator, and the concentrations of the reactants are in the denominator. Each concentration term must be raised to the power of the coefficient in the balanced equation.

Act on Your Strategy

$$K_c = \frac{[SO_3]^2}{[SO_2]^2[O_2]}$$

Check Your Solution

The square brackets indicate concentrations. The product is in the numerator, and each term is raised to the power of the coefficient in the chemical equation. The coefficient or power of 1 is not written, thus following chemistry conventions.

Practice Problems

Write the equilibrium expression for each homogeneous reaction.

1. The reaction between ethanol and ethanoic acid to form ethyl ethanoate and water:
   $CH_3CH_2OH_{(\ell)} + CH_3COOH_{(\ell)} \rightleftharpoons CH_3CHOOCH_2CH_{3(\ell)} + H_2O_{(\ell)}$
2. The reaction between nitrogen gas and oxygen gas at high temperatures:
   $N_{2(g)} + O_{2(g)} \rightleftharpoons 2NO_{(g)}$
3. The reaction between hydrogen gas and oxygen gas to form water vapour:
   $2H_{2(g)} + O_{2(g)} \rightleftharpoons 2H_2O_{(g)}$
4. The reduction-oxidation equilibrium of iron and iodine ions in aqueous solution:
   $2Fe^{3+}_{(aq)} + 2I^{-}_{(aq)} \rightleftharpoons 2Fe^{2+}_{(aq)} + I_{2(aq)}$
   Note: You will learn about reduction-oxidation reactions in the next unit.
5. The oxidation of ammonia (one of the reactions in the production of nitric acid):
   $4NH_{3(g)} + 5O_{2(g)} \rightleftharpoons 4NO_{(g)} + 6H_2O_{(g)}$

The Equilibrium Constant and Temperature

Adding a chemical that is involved in a reaction at equilibrium increases the rate at which this chemical reacts. The rate, however, decreases as the concentration of the added chemical decreases. Eventually, equilibrium is re-established with the same equilibrium constant. For a given system at equilibrium, the value of the equilibrium constant depends only on temperature. Changing the temperature of a reacting mixture changes the rate of the forward and reverse reactions by different amounts, because the forward and reverse reactions have different activation energies. A reacting mixture at one temperature has an equilibrium constant whose value changes if the mixture is allowed to reach equilibrium at a different temperature.

The numerical value of the equilibrium constant does not depend on whether the starting point involves reactants or products. These are just labels that chemists use to identify particular chemicals in the reaction mixture. Also, at a given temperature, the value of K_c does not depend on the starting concentrations. The reaction gives the same ratio of products and reactants according to the equilibrium law. Remember, however, that K_c is calculated using *concentration* values when the system is at *equilibrium*.

Sample Problem

Calculating an Equilibrium Constant

Problem

A mixture of nitrogen and chlorine gases was kept at a certain temperature in a 5.0 L reaction flask.

$$N_{2(g)} + 3Cl_{2(g)} \rightleftharpoons 2NCl_{3(g)}$$

When the equilibrium mixture was analyzed, it was found to contain 0.0070 mol of $N_{2(g)}$, 0.0022 mol of $Cl_{2(g)}$, and 0.95 mol of $NCl_{3(g)}$. Calculate the equilibrium constant for this reaction.

What Is Required?

You need to calculate the value of K_c.

What Is Given?

You have the balanced chemical equation and the amount of each substance at equilibrium.

Plan Your Strategy

Step 1 Calculate the molar concentration of each compound at equilibrium.

Step 2 Write the equilibrium expression. Then substitute the equilibrium molar concentrations into the expression.

Act on Your Strategy

Step 1 The reaction takes place in a 5.0 L flask. Calculate the molar concentrations at equilibrium.

$$[N_2] = \frac{0.0070 \text{ mol}}{5.0 \text{ L}} = 1.4 \times 10^{-3} \text{ mol/L}$$

Continued ...

Continued ...

$$[Cl_2] = \frac{0.0022\text{ mol}}{5.0\text{ L}} = 1.9 \times 10^{-1}\text{ mol/L}$$

$$[NCl_3] = \frac{0.95\text{ mol}}{5.0\text{ L}} = 1.9 \times 10^{-1}\text{ mol/L}$$

Step 2 Write the equilibrium expression. Substitute the equilibrium molar concentrations into the expression.

$$K_c = \frac{[NCl_3]^2}{[N_2][Cl_2]^3}$$

$$= \frac{(1.9 \times 10^{-1})^2}{(1.4 \times 10^{-3}) \times (4.4 \times 10^{-4})^3}$$

$$= 3.0 \times 10^{11}$$

PROBLEM TIP

Notice that units are not included when using or calculating the value of K_c. This is the usual practice. The units do not help you check your solution.

Check Your Solution

The equilibrium expression has the product terms in the numerator and the reactant terms in the denominator. The exponents in the equilibrium expression match the corresponding coefficients in the chemical equation. The molar concentrations at equilibrium were substituted into the expression.

Practice Problems

6. The following reaction took place in a sealed flask at 250°C.

$PCl_{5(g)} \rightleftharpoons PCl_{3(g)} + Cl_{2(g)}$

At equilibrium, the gases in the flask had the following concentrations: $[PCl_5] = 1.2 \times 10^{-2}$ mol/L, $[PCl_3] = 1.5 \times 10^{-2}$ mol/L, and $[Cl_2] = 1.5 \times 10^{-2}$ mol/L. Calculate the value of K_c at 250°C.

7. Iodine and bromine react to form iodine monobromide, IBr.

$I_{2(g)} + Br_{2(g)} \rightleftharpoons 2IBr_{(g)}$

At 250°C, an equilibrium mixture in a 2.0 L flask contained 0.024 mol of $I_{2(g)}$, 0.050 mol of $Br_{2(g)}$, and 0.38 mol of $IBr_{(g)}$. What is the value of K_c for the reaction at 250°C?

8. At high temperatures, carbon dioxide gas decomposes into carbon monoxide and oxygen gas. At equilibrium, the gases have the following concentrations: $[CO_{2(g)}] = 1.2$ mol/L, $[CO_{(g)}] = 0.35$ mol/L, and $[O_{2(g)}] = 0.15$ mol/L. Determine K_c at the temperature of the reaction.

9. Hydrogen sulfide is a pungent, poisonous gas. At 1400 K, an equilibrium mixture was found to contain 0.013 mol/L hydrogen, 0.046 mol/L sulfur in the form of $S_{2(g)}$, and 0.18 mol/L hydrogen sulfide. Calculate the equilibrium constant, at 1400 K, for the following reaction.

$2H_2S_{(g)} \rightleftharpoons 2H_{2(g)} + S_{2(g)}$

10. Methane, ethyne, and hydrogen form the following equilibrium mixture.

$2CH_{4(g)} \rightleftharpoons C_2H_{2(g)} + 3H_{2(g)}$

While studying this reaction mixture, a chemist analyzed a 4.0 L sealed flask at 1700°C. The chemist found 0.46 mol of $CH_{4(g)}$, 0.64 mol of $C_2H_{2(g)}$, and 0.92 mol of $H_{2(g)}$. What is the value of K_c for the reaction at 1700°C?

Measuring Equilibrium Concentrations

The equilibrium constant, K_c, is calculated by substituting equilibrium concentrations into the equilibrium expression. Experimentally, this means that a reaction mixture must come to equilibrium. Then one or more properties are measured. The properties that are measured depend on the reaction. Common examples for gaseous reactions include colour, pH, and pressure. From these measurements, the concentrations of the reacting substances can be determined. Thus, you do not need to measure all the concentrations in the mixture at equilibrium. You can determine some equilibrium concentrations if you know the initial concentrations of the reactants and the concentration of one product at equilibrium.

For example, a mixture of iron(III) nitrate and potassium thiocyanate, in aqueous solution, react to form the iron(III) thiocyanate ion, $Fe(SCN)^{2+}{}_{(aq)}$. The reactant solutions are nearly colourless. The product solution ranges in colour from orange to blood-red, depending on its concentration. The nitrate and potassium ions are spectators. Therefore, the net ionic equation is

$$Fe^{3+}{}_{(aq)} + SCN^{-}{}_{(aq)} \rightleftharpoons Fe(SCN)^{2+}{}_{(aq)}$$

colourless **red**

CONCEPT CHECK

Think about a chemical system at equilibrium. Are the concentrations of the reactants and products in the same ratio as the coefficients in the chemical equation? Explain your answer.

Because the reaction involves a colour change, you can determine the concentration of $Fe(SCN)^{2+}{}_{(aq)}$ by measuring the intensity of the colour. You will find out how to do this in Investigation 7-A. For now, assume that it can be done. From the measurements of colour intensity, you can calculate the equilibrium concentration of $Fe(SCN)^{2+}{}_{(aq)}$. Then, knowing the initial concentration of each solution, you can calculate the equilibrium concentration of each ion using the chemical equation.

Suppose, for instance, that the initial concentration of $Fe^{3+}{}_{(aq)}$ is 0.0064 mol/L and the initial concentration of $SCN^{-}{}_{(aq)}$ is 0.0010 mol/L. When the solutions are mixed, the red complex ion forms. By measuring the intensity of its colour, you can determine that the concentration of $Fe(SCN)^{2+}{}_{(aq)}$ is 4.5×10^{-4} mol/L. From the stoichiometry of the equation, each mole of $Fe(SCN)^{2+}{}_{(aq)}$ forms when equal amounts of $Fe^{3+}{}_{(aq)}$ and $SCN^{-}{}_{(aq)}$ react. So, if there is 4.5×10^{-4} mol/L of $Fe(SCN)^{2+}{}_{(aq)}$ at equilibrium, then the same amounts of both $Fe^{3+}{}_{(aq)}$ and $SCN^{-}{}_{(aq)}$ must have reacted. This represents the change in their concentrations. The equilibrium concentration of a reacting species is the sum of its initial concentration and the change that results from the reaction. Therefore, the concentration of $Fe^{3+}{}_{(aq)}$ at equilibrium is (0.0064 – 0.000 45) mol/L = 0.005 95 mol/L, or 0.0060 mol/L. You can calculate the equilibrium concentration of $SCN^{-}{}_{(aq)}$ the same way, and complete the table below.

Concentration (mol/L)	$Fe^{3+}{}_{(aq)}$	+	$SCN^{-}{}_{(aq)}$	$\rightleftharpoons$	$Fe(SCN)^{2+}{}_{(aq)}$
Initial	0.0064		0.0010		0
Change	-4.5×10^{-4}		-4.5×10^{-4}		4.5×10^{-4}
Equilibrium	0.0060		5.5×10^{-4}		4.5×10^{-4}

Finally, you can calculate K_c by substituting the equilibrium concentrations into the equilibrium expression.

In the following investigation, you will collect experimental data to determine an equilibrium constant.

Investigation 7-A

SKILL FOCUS

- Predicting
- Performing and recording
- Analyzing and interpreting
- Communicating results

Measuring an Equilibrium Constant

The colour intensity of a solution is related to the concentration of the ions and the depth of the solution. By adjusting the depth of a solution with unknown concentration until it has the same intensity as a solution with known concentration, you can determine the concentration of the unknown solution. For example, if the concentration of a solution is lower than the standard, the depth of the solution has to be greater in order to have the same colour intensity. Thus, the ratio of the concentrations of two solutions with the same colour intensity is inversely proportional to the ratio of their depths.

In this investigation, you will examine the homogeneous equilibrium between iron(III) (ferric) ions, thiocyanate ions, and ferrithiocyanate ions, $Fe(SCN)^{2+}$.

$$Fe^{3+}{}_{(aq)} + SCN^{-}{}_{(aq)} \rightleftharpoons Fe(SCN)^{2+}{}_{(aq)}$$

You will prepare four equilibrium mixtures with different initial concentrations of $Fe^{3+}{}_{(aq)}$ ions and $SCN^{-}{}_{(aq)}$ ions. You will calculate the initial concentrations of these reacting ions from the volumes and concentrations of the stock solutions used, and the *total* volumes of the equilibrium mixtures. Then you will determine the concentration of $Fe(SCN)^{2+}$ ions in each mixture by comparing the colour intensity of the mixture with the colour intensity of a solution that has known concentration. After you find the concentration of $Fe(SCN)^{2+}$ ions, you will use it to calculate the concentrations of the other two ions at equilibrium. You will substitute the three concentrations for each mixture into the equilibrium expression to determine the equilibrium constant.

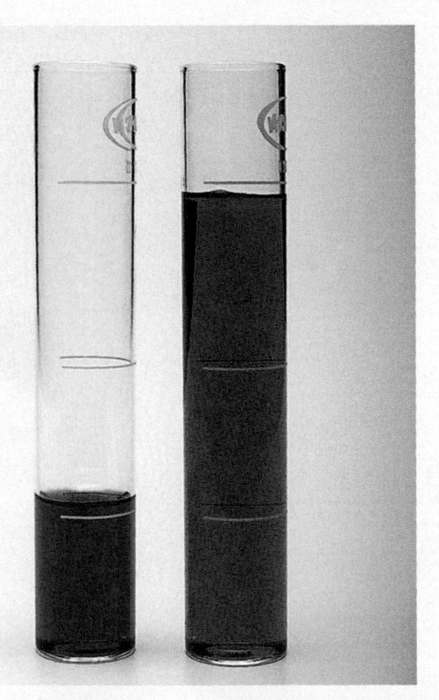

Which solution is the least concentrated? Why is the colour intensity the same when you look vertically through the solutions?

Question

What is the value of the equilibrium constant at room temperature for the following reaction?

$$Fe^{3+}_{(aq)} + SCN^{-}_{(aq)} \rightleftharpoons Fe(SCN)^{2+}_{(aq)}$$

Prediction

Write the equilibrium expression for this reaction.

Safety Precautions

- The $Fe(NO_3)_3$ solution is acidified with nitric acid. It should be handled with care. Wash any spills on your skin or clothing with plenty of water, and inform your teacher.
- All glassware must be clean and dry before using it.

Materials

3 beakers (100 mL)
5 test tubes (18 mm × 150 mm)
5 flat-bottom vials
test tube rack
labels or grease pencil
thermometer
stirring rod
strip of paper
diffuse light source, such as a light box
medicine dropper
3 Mohr pipettes (5.0 mL)
20.0 mL Mohr pipette
pipette bulb
30 mL 0.0020 mol/L $KSCN_{(aq)}$
30 mL 0.0020 mol/L $Fe(NO_3)_{3(aq)}$ (acidified)
25 mL 0.200 mol/L $Fe(NO_3)_{3(aq)}$ (acidified)
distilled water

Procedure

1. Copy the following tables into your notebook, and give them titles. You will use the tables to record your measurements and calculations.
2. Label five test tubes and five vials with the numbers 1 through 5. Label three beakers with the names and concentrations of the stock solutions: 2.00×10^{-3} mol/L KSCN, 2.00×10^{-3} mol/L $Fe(NO_3)_3$, and 0.200 mol/L $Fe(NO_3)_3$. Pour about 30 mL of each stock solution into its labelled beaker. Be sure to distinguish between the different concentrations of the iron(III) nitrate solutions. Make sure that you choose the correct solution when needed in the investigation. Measure the volume of each solution as carefully as possible to ensure the accuracy of your results.

Test tube	$Fe(NO_3)_3$ (mL)	H_2O (mL)	KSCN (mL)	Initial $[SCN^-]$ (mol/L)
2	5.0	3.0	2.0	
3	5.0	2.0	3.0	
4	5.0	1.0	4.0	
5	5.0	0	5.0	

Vial	Depth of solution in vial (mm)	Depth of standard solution (mm)	$\frac{\text{Depth of standard solution}}{\text{Depth of solution in vial}}$
2			
3			
4			
5			

3. Prepare the standard solution of $FeSCN^{2+}$ in test tube 1. Use the 20 mL Mohr pipette to transfer 18.0 mL of 0.200 mol/L $Fe(NO_3)_{3(aq)}$ into the test tube. Then use a 5 mL pipette to add 2.0 mL of 2.00×10^{-3} mol/L KSCN. The large excess of $Fe^{3+}_{(aq)}$ is to ensure that all the $SCN^-_{(aq)}$ will react to form $Fe(SCN)^{2+}_{(aq)}$.
4. Pipette 5.0 mL of 2.0×10^{-3} mol/L $Fe(NO_3)_{3(aq)}$ into each of the other four test tubes (labelled 2 to 5).
5. Pipette 3.0 mL, 2.0 mL, 1.0 mL, and 0 mL of distilled water into test tubes 2 to 5.
6. Pipette 2.0 mL, 3.0 mL, 4.0 mL, and 5.0 mL of 2.0×10^{-3} mol/L $KSCN_{(aq)}$ into test tubes 2 to 5. Each of these test tubes should now contain 10.0 mL of solution. Notice that the first table you prepared (in step 1) shows the volumes of the liquids you added to the test tubes. Use a stirring rod to mix each solution. (Remember to rinse the stirring rod with water and then dry it with a paper towel before you stir each solution. Measure and record the temperature of one of the solutions. Assume that all the solutions are at the same temperature.
7. Pour about 5 mL of the standard solution from test tube 1 into vial 1.
8. Pour some of the solution from test tube 2 into vial 2. Look down through vials 1 and 2. Add enough solution to vial 2 to make its colour intensity about the same as the colour intensity of the solution in vial 1. Use a sheet of white paper as background to make your rough colour intensity comparison.
9. Wrap a sheet of paper around vials 1 and 2 to prevent light from entering the sides of the solutions. Looking down through the vials over a diffuse light source, adjust the volume of the standard solution in vial 1 until the colour intensity in the vials is the same. Use a medicine dropper to remove or add standard solution. Be careful not to add standard solution to vial 2.
10. When the colour intensity is the same in both vials, measure and record the depth of solution in each vial as carefully as possible.
11. Repeat steps 9 and 10 using vials 3 to 5.
12. Discard the solutions into the container supplied by your teacher. Rinse the test tubes and vials with distilled water. Then return all the equipment. Wash your hands.
13. Copy the following table into your notebook, to summarize the results of your calculations.
   (a) Calculate the equilibrium concentration of $Fe(SCN)^{2+}$ in the standard solution you prepared in test tube 1. The $[Fe(SCN)^{2+}]_{standard}$ is essentially the same as the starting concentration of $SCN^-_{(aq)}$ in test tube 1. The large excess of $Fe^{3+}_{(aq)}$ ensured that the reaction of SCN^- was almost complete. Remember, however, to include the volume of $Fe(NO_3)_{3(aq)}$ in the total volume of the solution for your calculation.

	Initial concentrations		Equilibrium concentrations			
Test tube	$[Fe^{3+}]_i$	$[SCN^-]_i$	$[Fe^{3+}]_{eq}$	$[SCN^-]_{eq}$	$[Fe(SCN)^{2+}]_{eq}$	Equilibrium constant, K_c
2						
3						
4						
5						

(b) Calculate the initial concentration of $Fe^{3+}_{(aq)}$ in test tubes 2 to 5. $[Fe^{3+}]_i$ is the same in these four test tubes. They all contained the same volume of $Fe(NO_3)_{3(aq)}$, and the total final volume was the same. Remember to use the total volume of the solution in your calculation.

(c) Calculate the initial concentration of $SCN^{-}_{(aq)}$ in test tubes 2 to 5. $[SCN^-]_i$ is different in each test tube.

(d) Calculate the equilibrium concentration of $Fe(SCN)^{2+}$ in test tubes 2 to 5. Use the following equation:

$$[FeSCN^{2+}]_{eq} = \frac{\text{Depth of standard solution}}{\text{Depth of solution in vial}} \times [FeSCN^{2+}]_{standard}$$

(e) Based on the stoichiometry of the reaction, each mole of $Fe(SCN)^{2+}$ is formed by the reaction of one mole of Fe^{3+} with one mole of SCN^-. Thus, you can find the equilibrium concentrations of these ions by using the equations below:

$[Fe^{3+}]_{eq} = [Fe^{3+}]_i - [Fe(SCN)^{2+}]_{eq}$

$[SCN^-]_{eq} = [SCN^-]_i - [Fe(SCN)^{2+}]_{eq}$

(f) Calculate four values for the equilibrium constant, K_c, by substituting the equilibrium concentrations into the equilibrium expression. Find the average of your four values for K_c.

Analysis

1. How did the colour intensity of the solutions in test tubes 2 to 5 vary at equilibrium? Explain your observation.
2. How consistent are the four values you calculated for K_c? Suggest reasons that could account for any differences.
3. Use your four sets of data to evaluate the following expressions:

   $[Fe(SCN)^{2+}][Fe^{3+}][SCN^-]$

   $$\frac{[Fe^{3+}] + [SCN^-]}{[Fe(SCN)^{2+}]}$$

   Do these expressions give a value that is more constant than your equilibrium constant? Make up another expression, and evaluate it using your four sets of data. How constant is its value?
4. Suppose that the following reaction was the equilibrium reaction.

   $Fe^{3+}_{(aq)} + 2SCN^{-}_{(aq)} \rightleftharpoons Fe(SCN)_2{}^{+}_{(aq)}$

   (a) Would the equilibrium concentration of the product be different from the concentration of the product in the actual reaction? Explain.

   (b) Would the value of K_c be different from the value you calculated earlier? Explain.

Conclusion

5. Write a short conclusion, summarizing your results for this investigation.

Continued ...

What Is Required?

You need to find $[H_2]$, $[I_2]$, and $[HI]$.

What Is Given?

You have the balanced chemical equation. You know the equilibrium constant for the reaction, $K_c = 25$. You also know the concentrations of the reactants and product: $[H_2]_i = 2.00$ mol/L, $[I_2]_i = 3.00$ mol/L, and $[HI]_i = 0$.

Plan Your Strategy

Step 1 Set up an ICE table. Let the change in molar concentrations of the reactants be x. Use the stoichiometry of the chemical equation to write expressions for the equilibrium concentrations. Record these expressions in your ICE table.

Step 2 Write the equilibrium expression. Substitute the expressions for the equilibrium concentrations into the equilibrium expression. Rearrange the equilibrium expression into the form of a quadratic equation. Solve the quadratic equation for x.

Step 3 Substitute x into the equilibrium line of the ICE table to find the equilibrium concentrations.

Act on Your Strategy

Step 1 Set up an ICE table.

Concentration (mol/L)	$H_{2(g)}$	+	$I_{2(g)}$	$\rightleftharpoons$	$2HI_{(g)}$
Initial	2.00		3.00		0
Change	$-x$		$-x$		$+2x$
Equilibrium	$2.00 - x$		$3.00 - x$		$+2x$

PROBLEM TIP

In the equilibrium expression, the numerator is $[HI]^2$. The concentration is $2x$, so the substitution is $(2x)^2 = 4x^2$. A common mistake is to write $2x^2$. You can avoid this mistake by using brackets.

Ge
x r
the
ch
to
in
Fra
the

Step 2 Write and solve the equilibrium expression.

$$K_c = \frac{[HI]^2}{[H_2][I_2]}$$

$$25.0 = \frac{(2x)^2}{(2.00 - x)(3.00 - x)}$$

This equation does not involve a perfect square. It must be re-arranged into a quadratic equation.

$$0.840x^2 - 5.00x + 6.00 = 0$$

Recall that a quadratic equation of the form $ax^2 + bx + c = 0$ has the following solution:

$$x = \frac{-b \pm \sqrt{b^2 - 4ac}}{2a}$$

$$\therefore x = \frac{-(-5.00) \pm \sqrt{25.0 - 20.16}}{1.68}$$

$$= \frac{5.00 \pm 2.2}{1.68}$$

$x = 4.3$ and $x = 1.7$

Step 3 The value $x = 4.3$ is not physically possible. It would result in negative concentrations of H_2 and I_2 at equilibrium. The concentration of each substance at equilibrium is found by substituting $x = 1.7$ into the last line of the ICE table.

Concentration (mol/L)	$H_{2(g)}$	+	$I_{2(g)}$	$\rightleftharpoons$	$2HI_{(g)}$
Equilibrium	2.00 − 1.7		3.00 − 1.7		3.4

Applying the rule for subtraction involving measured values,
$[H_2] = 0.3$ mol/L

$[I_2] = 1.3$ mol/L

$[HI] = 3.4$ mol/L

Check Your Solution

The coefficients in the chemical equation match the exponents in the equilibrium expression. To check your concentrations, substitute them back into the equilibrium expression.

$$K_c = \frac{3.4^2}{0.3 \times 1.3} = 30$$

Solving the quadratic equation gives a value of x, correct to one decimal place. As a result, $[H_2]$ can have only one significant figure. The calculated value of K_c is equal to the given value, within the error introduced by rounding.

Practice Problems

11. At 25°C, the value of K_c for the following reaction is 82.

$I_{2(g)} + Cl_{2(g)} \rightleftharpoons 2ICl_{(g)}$

0.83 mol of $I_{2(g)}$ and 0.83 mol of $Cl_{2(g)}$ are placed in a 10 L container at 25°C. What are the concentrations of the three gases at equilibrium?

12. At a certain temperature, $K_c = 4.0$ for the following reaction.

$2HF_{(g)} \rightleftharpoons H_{2(g)} + F_{2(g)}$

A 1.0 L reaction vessel contained 0.045 mol of $F_{2(g)}$ at equilibrium. What was the initial amount of HF in the reaction vessel?

13. A chemist was studying the following reaction.

$SO_{2(g)} + NO_{2(g)} \rightleftharpoons NO_{(g)} + SO_{3(g)}$

In a 1.0 L container, the chemist added 1.7×10^{-1} mol of $SO_{2(g)}$ to 1.1×10^{-1} mol of $NO_{2(g)}$. The value of K_c for the reaction at a certain temperature is 4.8. What is the equilibrium concentration of $SO_{3(g)}$ at this temperature?

14. Phosgene, $COCl_{2(g)}$, is an extremely toxic gas. It was used during World War I. Today it is used to manufacture pesticides, pharmaceuticals, dyes, and polymers. It is prepared by mixing carbon monoxide and chlorine gas.

$CO_{(g)} + Cl_{2(g)} \rightleftharpoons COCl_{2(g)}$

Continued ...

Continued ...

0.055 mol of $CO_{(g)}$ and 0.072 mol of $Cl_{2(g)}$ are placed in a 5.0 L container. At 870 K, the equilibrium constant is 0.20. What are the equilibrium concentrations of the mixture at 870 K?

15. Hydrogen bromide decomposes at 700 K.

$2HBr_{(g)} \rightleftharpoons H_{2(g)} + Br_{2(g)} \quad K_c = 4.2 \times 10^{-9}$

0.090 mol of HBr is placed in a 2.0 L reaction vessel and heated to 700 K. What is the equilibrium concentration of each gas?

Qualitatively Interpreting the Equilibrium Constant

The Sample Problem on page 344 showed how to calculate the amounts of gases at equilibrium for the following reaction.

$$CO_{(g)} + H_2O_{(g)} \rightleftharpoons H_{2(g)} + CO_{2(g)}$$

Recall that the equilibrium constant for this reaction was 8.3 at 700 K. 1.0 mol of $CO_{(g)}$ and 1.0 mol of $H_2O_{(g)}$ were put in a 5.0 L container. At equilibrium, 0.48 mol of $H_{2(g)}$ and 0.48 mol of $CO_{2(g)}$ were present. The starting amounts were in stoichiometric ratios. Consequently, if the reaction had gone to completion, there would have been no limiting reactant. You would have expected 1.0 mol of both $H_{2(g)}$ and $CO_{2(g)}$. Thus, you can calculate the percent yield of the reaction as follows:

$$\begin{aligned}\text{Percent yield} &= \frac{\text{Equilibrium } [H_2] \times 100\%}{\text{Maximum } [H_2]}\\ &= \frac{0.48 \times 100\%}{1.0}\\ &= 48\%\end{aligned}$$

This value indicates the extent of the reaction. It is linked to the size of K_c. As you know, the equilibrium expression is always written with the product terms over the reactant terms. Therefore, a large K_c means that the concentration of products is larger than the concentration of reactants at equilibrium. When referring to a reaction with a large K_c, chemists often say that the position of equilibrium lies to the right, or that it favours the products. Similarly, if K_c is small, the concentration of reactants is larger than the concentration of products. Chemists say that the position of equilibrium lies to the left, or that it favours reactants. Thus, the following general statements are true:

- *When $K > 1$, products are favoured.* The equilibrium lies far to the right. Reactions where K is greater than 10^{10} are usually regarded as going to completion.
- *When $K \approx 1$, there are approximately equal concentrations of reactants and products at equilibrium.*
- *When $K < 1$, reactants are favoured.* The equilibrium lies far to the left. Reactions in which K is smaller than 10^{-10} are usually regarded as not taking place at all.

Notice that the subscript "c" has been left off K in these general statements. This reflects the fact that there are other equilibrium constants that these statements apply to, not just equilibrium constants involving concentrations. You will learn about other types of equilibrium constants in the next two chapters. For the rest of this chapter, though, you will continue to see the subscript "c" used.

In the Sample Problem and Practice Problems below, you will consider how temperature affects the extent of a reaction. Keep in mind that the size of K_c is not related to the time that a reaction takes to achieve equilibrium. Very large values of K_c may be associated with reactions that take place extremely slowly. The time that a reaction takes to reach equilibrium depends on the rate of the reaction. This is determined by the size of the activation energy.

Sample Problem

Temperature and the Extent of a Reaction

Problem

Consider the reaction of carbon monoxide and chlorine to form phosgene, $COCl_{2(g)}$.

$$CO_{(g)} + Cl_{2(g)} \rightleftharpoons COCl_{2(g)}$$

At 870 K, the value of K_c is 0.20. At 370 K, the value of K_c is 4.6×10^7. Based on only the values of K_c, is the production of $COCl_{2(g)}$ more favourable at the higher or lower temperature?

What Is Required?

You need to choose the temperature that favours the greater concentration of product.

What Is Given?

$K_c = 0.2$ at 870 K

$K_c = 4.6 \times 10^7$ at 370 K

Plan Your Strategy

Phosgene is a product in the chemical equation. A greater concentration of product corresponds to a larger value of K_c.

Act on Your Strategy

The value of K_c is larger at 370 K. The position of equilibrium lies far to the right. It favours the formation of $COCl_2$ at the lower temperature.

Check Your Solution

The problem asked you to choose the temperature for a larger concentration of the product. K_c is expressed as a fraction with product terms in the numerator. Therefore, the larger value of K_c corresponds to the larger concentration of product.

CONCEPT CHECK

Does a small value of K_c indicate a slow reaction? Use an example to justify your answer.

Practice Problems

16. For the reaction $H_{2(g)} + I_{2(g)} \rightleftharpoons 2HI_{(g)}$, the value of K_c is 25.0 at 1100 K and 8.0×10^2 at room temperature, 300 K. Which temperature favours the dissociation of $HI_{(g)}$ into its component gases?

17. Three reactions, and their equilibrium constants, are given below.

I. $N_{2(g)} + O_{2(g)} \rightleftharpoons 2NO_{(g)}$ $K_c = 4.7 \times 10^{-31}$

II. $2NO_{(g)} + O_{2(g)} \rightleftharpoons 2NO_{2(g)}$ $K_c = 1.8 \times 10^{-6}$

III. $N_2O_{4(g)} \rightleftharpoons 2NO_{2(g)}$ $K_c = 0.025$

Continued ...

Continued ...

Arrange these reactions in the order of their tendency to form products.

18. Identify each reaction as essentially going to completion or not taking place.

(a) $N_{2(g)} + 3Cl_{2(g)} \rightleftharpoons 2NCl_{3(g)}$ $K_c = 3.0 \times 10^{11}$

(b) $2CH_{4(g)} \rightleftharpoons C_2H_{6(g)} + H_{2(g)}$ $K_c = 9.5 \times 10^{-13}$

(c) $2NO_{(g)} + 2CO_{(g)} \rightleftharpoons N_{2(g)} + 2CO_{2(g)}$ $K_c = 2.2 \times 10^{59}$

19. Most metal ions combine with other ions in solution. For example, in aqueous ammonia, silver(I) ions are at equilibrium with different complex ions.

$[Ag(H_2O)_2]^+_{(aq)} + 2NH_{3(aq)} \rightleftharpoons [Ag(NH_3)_2]^+_{(aq)} + 2H_2O_{(\ell)}$

At room temperature, K_c for this reaction is 1×10^7. Which of the two silver complex ions is more stable? Explain your reasoning.

20. Consider the following reaction.

$H_{2(g)} + Cl_{2(g)} \rightleftharpoons 2HCl_{(g)}$ $K_c = 2.4 \times 10^{33}$ at 25°C

$HCl_{(g)}$ is placed in a reaction vessel. To what extent do you expect the equilibrium mixture to dissociate into $H_{2(g)}$ and $Cl_{2(g)}$?

The Meaning of a Small Equilibrium Constant

Understanding the meaning of a small equilibrium constant can sometimes help to simplify a calculation that would otherwise involve a quadratic equation. When K_c is small compared with the initial concentration, the value of the initial concentration minus x is approximately equal to the initial concentration. Thus, you can ignore x. Of course, if the initial concentration of a substance is zero, any equilibrium concentration of the substance, no matter how small, is significant. In general, values of K_c are not measured with accuracy better than 5%. Therefore, making the approximation is justified if the calculation error you introduce is less than 5%.

To help you decide whether or not the approximation is justified, divide the initial concentration by the value of K_c. If the answer is greater than 500, the approximation is justified. If the answer is between 100 and 500, it may be justified. If the answer is less than 100, it is not justified. The equilibrium expression must be solved in full.

Sample Problem

Using the Approximation Method

Problem

The atmosphere contains large amounts of oxygen and nitrogen. The two gases do not react, however, at ordinary temperatures. They do react at high temperatures, such as the temperatures produced by a lightning flash or a running car engine. In fact, nitrogen oxides from exhaust gases are a serious pollution problem.

A chemist is studying the following equilibrium reaction.

$$N_{2(g)} + O_{2(g)} \rightleftharpoons 2NO_{(g)}$$

The chemist puts 0.085 mol of $N_{2(g)}$ and 0.038 mol of $O_{2(g)}$ in a 1.0 L rigid cylinder. At the temperature of the exhaust gases from a particular engine, the value of K_c is 4.2×10^{-8}. What is the concentration of $NO_{(g)}$ in the mixture at equilibrium?

What Is Required?

You need to find the concentration of NO at equilibrium.

What Is Given?

You have the balanced chemical equation. You know the value of K_c and the following concentrations: $[N_2] = 0.085$ mol/L and $[O_2] = 0.038$ mol/L.

Plan Your Strategy

Step 1 Divide the smallest initial concentration by K_c to determine whether you can ignore the change in concentration.

Step 2 Set up an ICE table. Let x represent the change in $[N_2]$ and $[O_2]$.

Step 3 Write the equilibrium expression. Substitute the equilibrium concentrations into the equilibrium expression. Solve the equilibrium expression for x.

Step 4 Calculate [NO] at equilibrium.

Act on Your Strategy

Step 1

$$\frac{\text{Smallest initial concentration}}{K_c} = \frac{0.038}{4.2 \times 10^{-8}}$$
$$= 9.0 \times 10^5$$

Because this is well above 500, you can ignore the changes in $[N_2]$ and $[O_2]$.

Step 2

Concentration (mol/L)	$N_{2(g)}$	+	$O_{2(g)}$	$\rightleftharpoons$	$2NO_{(g)}$
Initial	0.085		0.038		0
Change	$-x$		$-x$		$+2x$
Equilibrium	$0.085 - x \approx 0.085$		$0.038 - x \approx 0.038$		$2x$

Step 3

$$K_c = \frac{[NO]^2}{[N_2][O_2]}$$
$$4.2 \times 10^{-8} = \frac{(2x)^2}{0.085 \times 0.038}$$
$$= \frac{4x^2}{0.003\,23}$$
$$x = \sqrt{3.39 \times 10^{-11}}$$
$$= 5.82 \times 10^{-6}$$

Step 4 $[NO] = 2x$

Therefore, the concentration of $NO_{(g)}$ at equilibrium is 1.2×10^{-5} mol/L.

Continued ...

Continued ...

Check Your Solution

First, check your assumption that x is negligible compared with the initial concentrations. Your assumption is valid because, using the rules for subtracting measured quantities, $0.038 - (5.8 \times 10^{-6}) = 0.038$. Next, check the equilibrium values:

$$\frac{(1.2 \times 10^{-5})^2}{0.0085 \times 0.038} = 4.5 \times 10^{-8}$$

This is equal to the equilibrium constant, within rounding errors in the calculation.

Practice Problems

21. The following equation represents the equilibrium reaction for the dissociation of phosgene gas.

$COCl_{2(g)} \rightleftharpoons CO_{(g)} + Cl_{2(g)}$

At 100°C, the value of K_c for this reaction is 2.2×10^{-8}. The initial concentration of $COCl_{2(g)}$ in a closed container at 100°C is 1.5 mol/L. What are the equilibrium concentrations of $CO_{(g)}$ and $Cl_{2(g)}$?

22. Hydrogen sulfide is a poisonous gas with a characteristic, offensive odour. It dissociates at 1400°C, with K_c equal to 2.4×10^{-4}.

$H_2S_{(g)} \rightleftharpoons 2H_{2(g)} + S_{2(g)}$

4.0 mol of H_2S is placed in a 3.0 L container. What is the equilibrium concentration of $H_{2(g)}$ at 1400°C?

23. At a certain temperature, the value of K_c for the following reaction is 3.3×10^{-12}.

$2NCl_{3(g)} \rightleftharpoons N_{2(g)} + 3Cl_{2(g)}$

A certain amount of nitrogen trichloride, $NCl_{3(g)}$, is put in a 1.0 L reaction vessel at this temperature. At equilibrium, 4.6×10^{-4} mol of $N_{2(g)}$ is present. What amount of $NCl_{3(g)}$ was put in the reaction vessel?

24. At a certain temperature, the value of K_c for the following reaction is 4.2×10^{-8}.

$N_{2(g)} + O_{2(g)} \rightleftharpoons 2NO_{(g)}$

0.45 mol of $N_{2(g)}$ and 0.26 mol of $O_{2(g)}$ are put in a 6.0 L reaction vessel. What is the equilibrium concentration of $NO_{(g)}$ at this temperature?

25. At a particular temperature, K_c for the decomposition of carbon dioxide gas is 2.0×10^{-6}.

$2CO_{2(g)} \rightleftharpoons 2CO_{(g)} + O_{2(g)}$

3.0 mol of CO_2 is put in a 5.0 L container. Calculate the equilibrium concentration of each gas.

Section Summary

In this section, you learned that the equilibrium constant, K_c, is a ratio of product concentrations to reactant concentrations. You used concentrations to find K, and you used K to find concentrations. You also used an ICE table to track and summarize the initial, change, and equilibrium quantities in a reaction. You found that the value of K_c is small for reactions that reach equilibrium with a high concentration of reactants, and the value of K_c is large for reactions that reach equilibrium with a low concentration of reactants. In the next section, you will learn how to determine whether or not a reaction is at equilibrium, and, if it is not, in which direction it will go to achieve equilibrium.

Section Review

1 K/U Write equilibrium expressions for each homogeneous reaction.

(a) $SbCl_{5(g)} \rightleftharpoons SbCl_{3(g)} + Cl_{2(g)}$

(b) $2H_{2(g)} + 2NO_{(g)} \rightleftharpoons N_{2(g)} + 2H_2O_{(g)}$

(c) $2H_2S_{(g)} + CH_{4(g)} \rightleftharpoons 4H_{2(g)} + CS_{2(g)}$

2 I When 1.0 mol of ammonia gas is injected into a 0.50 L flask, the following reaction proceeds to equilibrium.

$2NH_{3(g)} \rightleftharpoons N_{2(g)} + 3H_{2(g)}$

At equilibrium, 0.30 mol of hydrogen gas is present.

(a) Calculate the equilibrium concentrations of $N_{2(g)}$ and $NH_{3(g)}$.

(b) What is the value of K_c?

3 I At a certain temperature, K_c for the following reaction between sulfur dioxide and nitrogen dioxide is 4.8.

$SO_{2(g)} + NO_{2(g)} \rightleftharpoons NO_{(g)} + SO_{3(g)}$

$SO_{2(g)}$ and $NO_{2(g)}$ have the same initial concentration: 0.36 mol/L. What amount of $SO_{3(g)}$ is present in a 5.0 L container at equilibrium?

4 I Phosphorus trichloride reacts with chlorine to form phosphorus pentachloride.

$PCl_{3(g)} + Cl_{2(g)} \rightleftharpoons PCl_{5(g)}$

0.75 mol of PCl_3 and 0.75 mol of Cl_2 are placed in a 8.0 L reaction vessel at 500 K. What is the equilibrium concentration of the mixture? The value of K_c at 500 K is 49.

5 MC Hydrogen gas has several advantages and disadvantages as a potential fuel. Hydrogen can be obtained by the thermal decomposition of water at high temperatures.

$2H_2O_{(g)} \rightleftharpoons 2H_{2(g)} + O_{2(g)} \quad K_c = 7.3 \times 10^{-18}$ at 1000°C

(a) The initial concentration of water in a reaction vessel is 0.055 mol/L. What is the equilibrium concentration of $H_{2(g)}$ at 1000°C?

(b) Comment on the practicality of the thermal decomposition of water to obtain $H_{2(g)}$.

7.4 Predicting the Direction of a Reaction

Section Preview/ Specific Expectations

In this section, you will

- **explain** how to use Le Châtelier's principle to predict the direction in which a chemical system at equilibrium will shift when concentration changes
- **apply** Le Châtelier's principle to **make** and **test** predictions about how different factors affect a chemical system at equilibrium
- **communicate** your understanding of the following terms: *reaction quotient (Q_c), Le Châtelier's principle, common ion effect*

So far in this chapter, you have worked with reactions that have reached equilibrium. What if a reaction has not yet reached equilibrium, however? How can you predict the direction in which the reaction must proceed to reach equilibrium? To do this, you substitute the concentrations of reactants and products into an expression that is identical to the equilibrium expression. Because these concentrations may not be the concentrations that the equilibrium system would have, the expression is given a different name: the reaction quotient. The **reaction quotient, Q_c**, is an expression that is identical to the equilibrium constant expression, *but* its value is calculated using concentrations that are not necessarily those at equilibrium.

The Relationship Between the Equilibrium Constant and the Reaction Quotient

Recall the general reaction $aP + bQ \rightleftharpoons cR + dS$. The reaction quotient expression for this reaction is

$$Q_c = \frac{[R]^c[S]^d}{[P]^a[Q]^b}$$

You can calculate a value for Q_c by substituting the concentration of each substance into this expression. If the value of Q_c is equal to K_c, the system must be at equilibrium. If Q_c is greater than K_c, the numerator must be very large. The concentrations of the chemicals on the right side of the equation must be more than their concentrations at equilibrium. In this situation, the system attains equilibrium by moving to the left. Conversely, if Q_c is less than K_c, the system attains equilibrium by moving to the right. Figure 7.8 summarizes these relationships between Q_c and K_c.

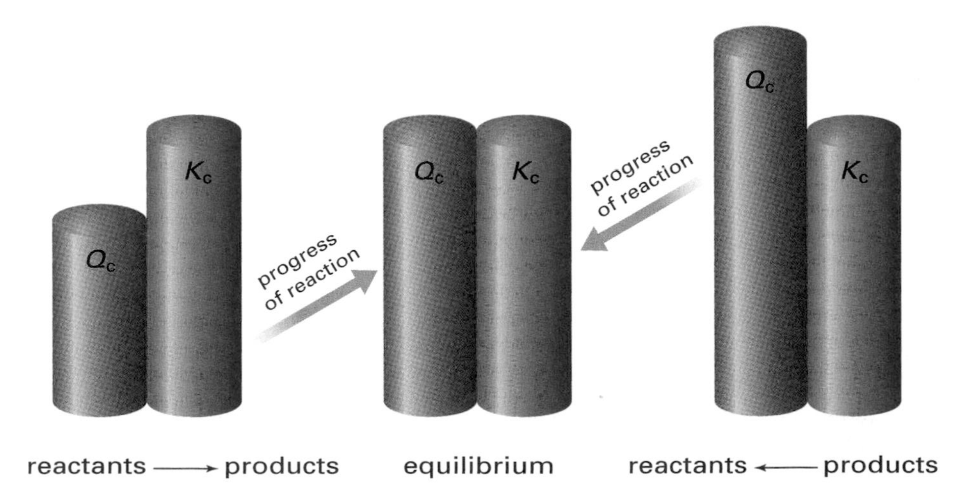

Figure 7.8 This diagram shows how Q_c and K_c determine reaction direction. When $Q_c < K_c$, the system attains equilibrium by moving to the right, favouring products. When $Q_c = K_c$, the system is at equilibrium. When $Q_c > K_c$, the system attains equilibrium by moving to the left, favouring reactants.

The next Sample Problem shows you how to calculate Q_c and interpret its value.

Sample Problem

Determining the Direction of Shift to Attain Equilibrium

Problem

Ammonia is one of the world's most important chemicals, in terms of the quantity manufactured. Some ammonia is processed into nitric acid and various polymers. Roughly 80% of ammonia is used to make fertilizers, such as ammonium nitrate. In the Haber process for manufacturing ammonia, nitrogen and hydrogen combine in the presence of a catalyst.

$$N_{2(g)} + 3H_{2(g)} \rightleftharpoons 2NH_{3(g)}$$

At 500°C, the value of K_c for this reaction is 0.40. The following concentrations of gases are present in a container at 500°C: $[N_{2(g)}] = 0.10$ mol/L, $[H_{2(g)}] = 0.30$ mol/L, and $[NH_{3(g)}] = 0.20$ mol/L. Is this mixture of gases at equilibrium? If not, in which direction will the reaction go to reach equilibrium?

What Is Required?

You need to calculate Q_c and interpret its value.

What Is Given?

You have the balanced chemical equation. You know that the value of K_c is 0.40. You also know the concentrations of the gases: $[N_2] = 0.10$ mol/L, $[H_2] = 0.30$ mol/L, and $[NH_3] = 0.20$ mol/L.

Plan Your Strategy

Write the expression for Q_c, and then calculate its value. Compare the value of Q_c with the value of K_c. Decide whether the system is at equilibrium and, if not, in which direction the reaction will go.

Act on Your Strategy

$$Q_c = \frac{[NH_3]^2}{[N_2][H_2]^3}$$

$$= \frac{(0.20)^2}{(0.10)(0.30)^3}$$

$$= 14.8$$

$\therefore Q_c > 0.40$

The system is not at equilibrium. The reaction will proceed by moving to the left.

Check Your Solution

Check your calculation of Q_c. (Errors in evaluating fractions with exponents are common.) The value of K_c is less than one, so you would expect the numerator to be less than the denominator. This is difficult to evaluate without a calculator, however, because of the powers.

Practice Problems

26. The following reaction takes place inside a cylinder with a movable piston.

$2NO_{2(g)} \rightleftharpoons N_2O_{4(g)}$

At room temperature, the equilibrium concentrations inside the cylinder are $[NO_2] = 0.0206$ mol/L and $[N_2O_4] = 0.0724$ mol/L.

(a) Calculate the value of K_c.

(b) Calculate the concentration of each gas at the moment that the piston is used to halve the volume of the reacting mixture. Assume that the temperature remains constant.

(c) Determine the value of Q_c when the volume is halved.

(d) Predict the direction in which the reaction will proceed to re-establish equilibrium.

27. Ethyl acetate is an ester that can be synthesized by reacting ethanoic acid (acetic acid) with ethanol. At room temperature, the equilibrium constant for this reaction is 2.2.

$CH_3COOH_{(\ell)} + CH_3CH_2OH_{(\ell)} \rightleftharpoons CH_3COOCH_2CH_{3(l)} + H_2O_{(\ell)}$

Various samples were analyzed. The concentrations are given in the table below. Decide whether each sample is at equilibrium. If it is not at equilibrium, predict the direction in which the reaction will proceed to establish equilibrium.

Sample	$[CH_3COOH]$ (mol/L)	$[CH_3CH_2OH]$ (mol/L)	$[CH_3COOCH_2CH_3]$ (mol/L)	$[H_2O]$ (mol/L)
(a)	0.10	0.10	0.10	0.10
(b)	0.084	0.13	0.16	0.28
(c)	0.14	0.21	0.33	0.20
(d)	0.063	0.11	0.15	0.17

28. In the past, methanol was obtained by heating wood without allowing the wood to burn. The products were collected, and methanol (sometimes called "wood alcohol") was separated by distillation. Today methanol is manufactured by reacting carbon monoxide with hydrogen gas.

$CO_{(g)} + 2H_{2(g)} \rightleftharpoons CH_3OH_{(g)}$

At 210°C, K_c for this reaction is 14.5. A gaseous mixture at 210°C contains the following concentrations of gases: $[CO] = 0.25$ mol/L, $[H_2] = 0.15$ mol/L, and $[CH_3OH] = 0.36$ mol/L. What will be the direction of the reaction if the gaseous mixture reaches equilibrium?

Le Châtelier's Principle

What happens to a system at equilibrium if the concentration of one of the reacting chemicals is changed? This question has practical importance, because many manufacturing processes are continual. Products are removed and more reactants are added without stopping the process. For example, consider the Haber process that was mentioned in the previous Sample Problem.

$$N_{2(g)} + 3H_{2(g)} \rightleftharpoons 2NH_{3(g)}$$

At 500°C, K_c is 0.40. The gases have the following concentrations: $[N_2] = 0.10$ mol/L, $[H_2] = 0.10$ mol/L, and $[NH_3] = 0.0063$ mol/L. (You can check this by calculating Q_c and making sure that it is equal to K_c.)

Ammonia can be removed from the equilibrium mixture by cooling because ammonia liquefies at a higher temperature than either nitrogen or hydrogen. What happens to an equilibrium mixture if some ammonia is removed? To find out, you need to calculate Q_c.

$$Q_c = \frac{[NH_3]^2}{[N_2][H_2]^3}$$

If ammonia is removed from the equilibrium mixture, the numerator will decrease, so Q_c will be smaller than K_c. When $Q_c < K_c$, the system attains equilibrium by moving to the right. Why? The system must respond to the removal of ammonia by forming more ammonia. Similarly, if nitrogen, hydrogen, or both were added, the system would re-establish equilibrium by shifting to the right. As long as the temperature remains constant, the re-established equilibrium will have the same K_c as the original equilibrium.

A French chemist, Henri Le Châtelier, experimented with various chemical equilibrium systems. (See Figure 7.9.) In 1888, Le Châtelier summarized his work on changes to equilibrium systems in a general statement called **Le Châtelier's principle**. It may be stated as follows: *A dynamic equilibrium tends to respond so as to relieve the effect of any change in the conditions that affect the equilibrium.*

Figure 7.9 Henri Le Châtelier (1850–1936) specialized in mining engineering. He studied the thermodynamics involved in heating ores to obtain metals. His results led to important advances in the understanding of equilibrium systems.

Le Châtelier's principle predicts the way that an equilibrium system responds to change. For example, when the concentration of a substance in a reaction mixture is changed, Le Châtelier's principle qualitatively predicts what you can show quantitatively by evaluating the reaction quotient. If products are removed from an equilibrium system, more products must be formed to relieve the change. This is just as you would predict, because Q_c will be less than K_c.

Le Châtelier's principle also predicts what will happen when other changes are made to an equilibrium system. For example, you can use Le Châtelier's principle to predict the effects of changing the volume of a cylinder that contains a mixture of gases, or the effects of changing the temperature of an equilibrium system. In the next investigation, you will use Le Châtelier's principle to predict the effects of making various changes to different equilibrium systems. The Chemistry Bulletin that follows the investigation considers applications of Le Châtelier's principle to areas outside of chemistry. Afterward, you will continue your studies of Le Châtelier's principle and equilibrium systems in chemistry.

Electronic Learning Partner

Your Chemistry 12 Electronic Learning Partner can help you reinforce your understanding of Le Châtelier's principle.

SKILL FOCUS
Predicting
Performing and recording
Analyzing and interpreting

Investigation 7-B

Perturbing Equilibrium

In this investigation, you will use Le Châtelier's principle to predict the effect of changing one factor in each system at equilibrium. Then you will test your prediction using a colour change or the appearance (or disappearance) of a precipitate.

Question

How can Le Châtelier's principle qualitatively predict the effect of a change in a chemical equilibrium?

Predictions

Your teacher will give you a table that lists four equilibrium systems and the changes you will make to each system. In the appropriate column, record your predictions for each test. If you predict that the change will cause the system to re-attain equilibrium by shifting toward the reactants, record "left." If you predict that the system will re-establish equilibrium by shifting toward the products, record "right."

Safety Precautions

- Potassium chromate and barium chloride are toxic.
- Hydrochloric acid and sodium hydroxide are corrosive. Wash any spills on your skin or clothing with plenty of cool water. Inform your teacher immediately.

Materials

Part 1

3 test tubes
0.1 mol/L $K_2CrO_{4(aq)}$
1 mol/L $HCl_{(aq)}$
1 mol/L $NaOH_{(aq)}$
1 mol/L $Fe(NO_3)_{3(aq)}$
1 mol/L $BaCl_{2(aq)}$

Part 2

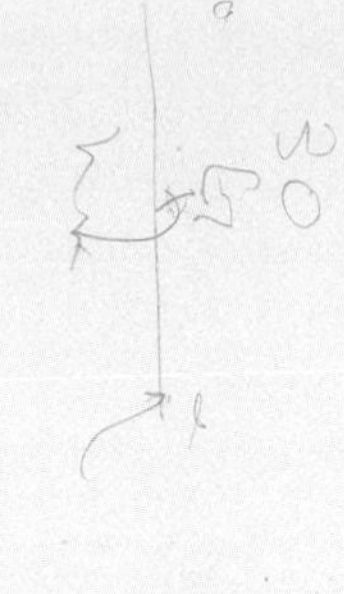

25 mL beaker
2 test tubes
test tube rack
scoopula
0.01 mol/L $NH_{3(aq)}$
6.0 mol/L $HCl_{(aq)}$
phenolphthalein solution
$NH_4Cl_{(s)}$

Part 3

4 small test tubes
$CoCl_{2(s)}$
ethanol
concentrated $HCl_{(aq)}$ in a dropper bottle
0.1 mol/L $AgNO_{3(aq)}$ in a dropper bottle
distilled water in a dropper bottle
25 mL or 50 mL beaker
test tube rack
test tube holder
hot-water bath (prepared by your teacher)
cold-water bath (prepared by your teacher)

Part 4

small piece of copper
concentrated nitric acid
test tube
test tube rack
one-hole stopper
glass delivery tube
short length of rubber tubing
syringe with a cap or rubber stopper to seal the tip
$NO_{2(g)}/N_2O_{4(g)}$ tubes
boiling water
ice water

Procedure

Part 1 The Chromate/Dichromate Equilibrium

Equilibrium System

$$H^+_{(aq)} + 2CrO_4^{2-}{}_{(aq)} \rightleftharpoons Cr_2O_7^{2-}{}_{(aq)} + OH^-_{(aq)}$$

yellowish orange

1. Pour about 5 mL of 0.1 mol/L $K_2CrO_{4(aq)}$ into each of three test tubes. You will use test tube 1 as a colour reference.
2. Add 5 drops of $HCl_{(aq)}$ to test tube 2. Record any colour change.
3. Add 5 drops of $NaOH_{(aq)}$ to test tube 2. Record what happens.
4. Finally, add 5 drops of $Fe(NO_3)_{3(aq)}$ to test tube 2. A precipitate of $Fe(OH)_{3(s)}$ should form. Note the colour of the solution above the precipitate.
5. Add 5 drops of $BaCl_{2(aq)}$ to test tube 3. A precipitate of $BaCrO_{4(s)}$ should form. Record the colour of the solution above the precipitate.
6. Dispose of the chemicals as instructed by your teacher.

Part 2 Changes to a Base Equilibrium System

Safety Precautions

Ammonia and hydrochloric acid are corrosive. Wash any spills on your skin or clothing with plenty of cool water. Inform your teacher immediately.

Equilibrium System

$$NH_{3(aq)} + H_2O_{(\ell)} \rightleftharpoons NH_4^+{}_{(aq)} + OH^-_{(aq)}$$

1. Pour about 10 mL of $NH_{3(aq)}$ into a small beaker. Place the beaker on a sheet of white paper. Add 2 drops of phenolphthalein indicator.
2. Divide the solution equally into two small test tubes. To one of the test tubes, add a few small crystals of $NH_4Cl_{(s)}$ on the end of a scoopula. Record your observations.
3. To the other test tube, add a few drops of $HCl_{(aq)}$ until you see a change. Again note the colour change. (The H^+ ions combine with the OH^- ions, removing them from the equilibrium mixture.)
4. Dispose of the chemicals as instructed by your teacher.

Part 3 Concentration and Temperature Changes

Safety Precautions

- Concentrated hydrochloric acid is hazardous to your eyes, skin, and clothing. Treat spills with baking powder and copious amounts of cool water. Inform your teacher immediately.
- Ethanol is flammable. Keep samples of ethanol and the supply bottle away from open flames.

Equilibrium System

$$Co(H_2O)_6^{2+}{}_{(aq)} + 4Cl^-_{(aq)} \rightleftharpoons CoCl_4^{2-}{}_{(aq)} + 6H_2O_{(\ell)}$$

pink **blue or purple**

$\Delta H = +50$ kJ/mol

1. Measure about 15 mL of ethanol into a small beaker.
2. Record the colour of the $CoCl_2$. Dissolve a small amount (about half the size of a pea) in the beaker of ethanol. The solution should be blue or purple. If it is pink, add drops of concentrated $HCl_{(aq)}$ until the solution is blue-purple.
3. Divide the cobalt solution equally among the four small test tubes. Put one of the test tubes aside as a control.
4. To each of the other three test tubes, add 3 drops of distilled water, one drop at a time. Stir after you add each drop. Record any change in colour that occurs with each drop.
5. To one of the test tubes from step 4, add 5 drops of concentrated HCl, one drop at a time with stirring. Record the results.

6. Silver and chloride ions combine to form a precipitate of AgCl. To the third test tube, add $AgNO_{3(aq)}$, one drop at a time, until no more precipitate appears. Record the colour of the solution as the chloride ions precipitate.
7. Record the colour of the liquid mixture in the fourth test tube. Use a test tube holder to immerse this test tube in the hot-water bath. Record any colour change.
8. Place the test tube from step 7 in the cold-water bath. Record any colour change.
9. Dispose of the chemicals as instructed by your teacher.

Part 4 Investigating Gaseous Equilibria

CAUTION These are not student tests. Your teacher may demonstrate this equilibrium if a suitable fume hood is available for the first test, and if sealed tubes containing a mixture of nitrogen dioxide, $NO_{2(g)}$, and dinitrogen tetroxide, $N_2O_{4(g)}$, are available for the second test. If either or both tests are not demonstrated, refer to the photographs that show the changes.

Safety Precautions

- Concentrated nitric acid is highly corrosive and a strong oxidizing agent.
- Nitrogen dioxide and dinitrogen tetroxide are poisonous gases.

Equilibrium System

$N_2O_{4(g)} \rightleftharpoons 2NO_{2(g)} \quad \Delta H = +59$ kJ/mol

colourless **brown**

1. Your teacher will use sealed tubes that contain a mixture of $N_2O_{4(g)}$ and $NO_{2(g)}$. One tube will be placed in boiling water, and a second tube will be placed in ice water. A third tube (if available) will remain at room temperature as a control. Compare and record the colour of the gas mixture at each temperature.
2. $NO_{2(g)}$ can be prepared by reacting copper with concentrated nitric acid. The gas is poisonous. The reaction, if your teacher performs it, *must* take place in a fume hood.
3. By using a one-hole stopper, glass delivery tube, and a short length of rubber tubing, some $NO_{2(g)}$ can be collected in a syringe. The syringe is then sealed by attaching a cap or by pushing the needle into a rubber stopper.
4. Observe what happens when the syringe plunger is pressed down sharply, changing the volume of the equilibrium mixture. You will observe an immediate change in colour. Then, if the plunger is held in a fixed position, the colour will change over a few seconds as the system re-establishes equilibrium. Carefully record these colour changes.

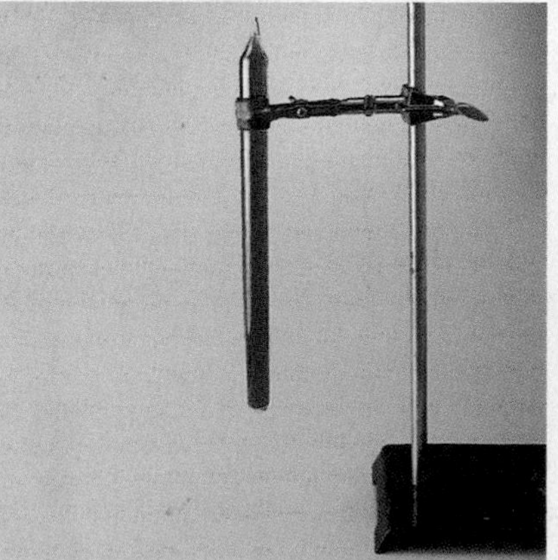

These three tubes contain a mixture of $NO_{2(g)}$ and $N_2O_{4(g)}$. The tube on the left is in an ice-water mixture. The centre tube is at room temperature. The tube on the right is in boiling water. Given that $NO_{2(g)}$ is brown, can you explain the shift in equilibrium? Think about Le Châtelier's principle and the enthalpy of the reaction between the two gases.

Analysis

1. Compare the predictions you made using Le Châtelier's principle with the observations you made in your tests. Account for any differences.

2. In which tests did you increase the concentration of a reactant or a product? Did your observations indicate a shift in equilibrium to form more or less of the reactant or product?

3. In which tests did you decrease the concentration of a reactant or product? Did your observations indicate a shift in equilibrium to form more or less of the reactant or product?

4. In two of the systems you studied, the enthalpy changes were given.

   $N_2O_{4(g)} \rightleftharpoons 2NO_{2(g)} \quad \Delta H = +59 kJ/mol$

   $Co(H_2O)_6{}^{2+}{}_{(aq)} + 4Cl^-{}_{(aq)} \rightleftharpoons CoCl_4{}^{2-}{}_{(aq)} + 6H_2O_{(\ell)}$

   $\Delta H = +50 kJ/mol$

   (a) Are these systems endothermic or exothermic when read from left to right?

   (b) When heated, did these systems shift to the left or to the right? In terms of the energy change, was the observed shift in equilibrium toward the endothermic or exothermic side of the reaction?

   (c) Do you think the value of K_c changed or remained the same when the equilibrium mixture was heated? Explain your answer.

5. Think about the $N_2O_{4(g)}/NO_{2(g)}$ equilibrium.

   (a) How was the total pressure of the mixture affected when the plunger was pushed down?

   (b) How was the pressure of the mixture affected by the total number of gas molecules in the syringe?

   (c) Explain the observed shift in equilibrium when the plunger was pushed down. In your explanation, refer to Le Châtelier's principle and the total amount of gas in the syringe.

   (d) What would be the effect, if any, on the following equilibrium system if the volume was reduced? Explain.

   $2IBr_{(g)} \rightleftharpoons I_{2(g)} + Br_{2(g)}$

Conclusion

6. How did your results compare with your predictions? Discuss and resolve any discrepancies with your class.

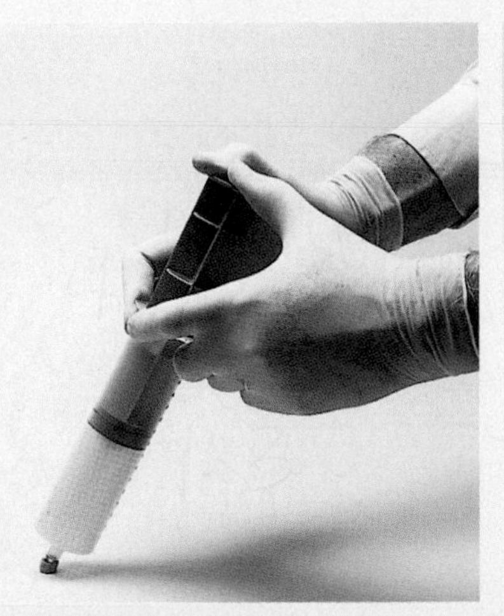
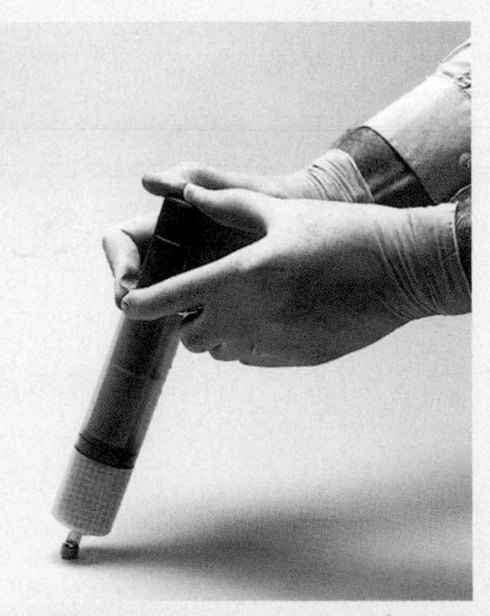
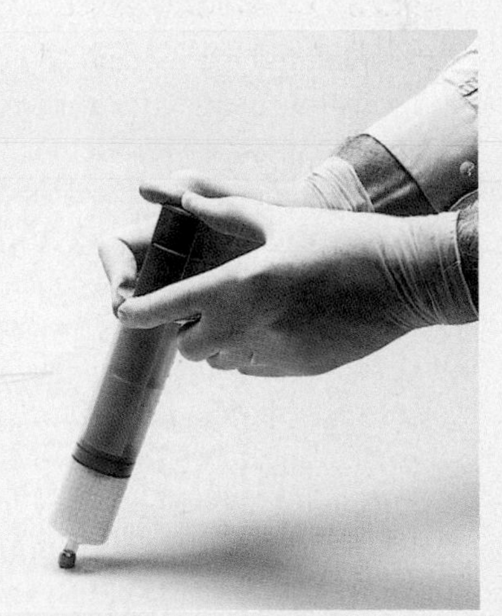

The sealed syringe contains a mixture of $NO_{2(g)}$ and $N_2O_{4(g)}$. The photograph on the left shows an equilibrium mixture at atmospheric pressure. The middle photograph shows that the plunger has been pushed down, increasing the pressure. Two changes cause the darker appearance of the gas mixture. First, the concentration of gases is greater. Second, decreasing the volume heats the gas. This causes a shift toward $NO_{2(g)}$. The photograph on the right shows the result a few seconds after the plunger was pushed down. The gas has cooled back to room temperature. The colour of the mixture is less brown, indicating a shift toward $N_2O_{4(g)}$.

Chemistry Bulletin

Science | Technology | Society | Environment

Le Châtelier's Principle: Beyond Chemistry

How is Le Châtelier's principle related to predator-prey interactions in ecosystems?

Le Châtelier's principle is a general statement about how any system in equilibrium—not just a chemical system—responds to change. Le Châtelier principle concerns the conservation of energy or matter. There are corresponding laws in several other areas of science.

Heinrich Lenz studied the direction of the current that is induced in a conductor as a result of changing the magnetic field near it. You can think of this as the change in a system in electromagnetic equilibrium. Lenz published his law in 1834. It states that when a conductor interacts with a magnetic field, there must be an induced current that opposes the interaction, because of the law of conservation of energy. Lenz's law is used to explain the direction of the induced current in generators, transformers, inductors, and many other systems.

In geology, John Pratt and George Airy introduced the idea of isostacy in 1855. Earth's crust is in gravitational equilibrium in almost all places. The crust responds to changes in the load on it, however, by slowly moving vertically. Examples of how the load on Earth's crust can change include the weathering of mountain ranges and the melting of ice caps. Geologists estimate that the ice over parts of Scandinavia and northern Europe was well over 2 km thick during the last ice age. As a result of the ice melting and the removal of the weight pushing down on Earth's crust, the central part of Scandinavia has risen over 500 m. This uplift is continuing. Here gravitational potential energy is being conserved.

In ecology, ecosystems are really examples of steady state systems. Ecosystems involve a flow of energy in only one direction, from the Sun. Nevertheless, populations of plants and animals develop stable numbers that react to external changes, such as disease and variations in the weather. For example, if the number of carnivores in an ecosystem increases, more herbivores are eaten and the herbivore population decreases. The carnivores then have difficulty finding food. *Their* populations decrease due to competition, emigration, and starvation until a new balance, however delicate, is attained.

Le Châtelier's principle also applies to biology. Homeostasis is the tendency of a body system to remain in a state of equilibrium. Examples of homeostasis include the maintenance of body temperature (homeothermy) and the pH balance of blood.

In economics, the law of supply and demand is similar to Le Châtelier's principle. When the price of a commodity, such as the price of a kilogram of apples, is constant, the market for the commodity is at equilibrium. If the supply of the commodity falls, the equilibrium is changed. The market adjusts by increasing the price, which tends to increase the supply.

Making Connections

1. Identify and describe two other examples, in science or other fields, that illustrate Le Châtelier's principle.
2. Draw diagrams to show the above examples of Le Châtelier's principle in human physiology, ecology, and economics. Show how different conditions affect the equilibrium and how the systems react to establish a new equilibrium.

The Effect of Ions on Aqueous Equilibrium Systems

Many important equilibrium systems involve ions in aqueous solution. The **common ion effect** applies Le Châtelier's principle to ions in aqueous solution. As its name suggests, the common ion effect involves adding an ion to a solution in which the ion is already present in solution. It is really a concentration effect. The equilibrium shifts away from the added ion, as predicted by Le Châtelier's principle.

An aqueous solution that contains certain ions can be added to another solution to form a precipitate, if the ions have low solubility. For example, adding silver nitrate solution to test for $Cl^-_{(aq)}$ is effective due to the very low solubility of silver chloride. You can use the precipitation of an insoluble salt to remove almost all of a particular ion from a solution and, as a result, cause a shift in the position of equilibrium of the original solution. The common ion effect is important in the solubility of salts. The precipitation of insoluble salts is used to identify the presence of unknown ions. You will learn more about the common ion effect in Chapter 9.

The Effect of Temperature Changes on the Position of Equilibrium

As you know, the value of the equilibrium constant changes with temperature, because the rates of the forward and reverse reactions are affected. Le Châtelier's principle still holds, however. It can be used to predict the effect of temperature changes on a system when the sign of the enthalpy change for the reaction is known. For example, the dissociation of sulfur trioxide is endothermic.

$$2SO_{3(g)} \rightleftharpoons 2SO_{2(g)} + O_{2(g)} \quad \Delta H = 197 \text{ kJ}$$

As the reaction proceeds from left to right, energy is absorbed by the chemical system and converted to chemical potential energy. If an equilibrium mixture of these gases is heated, energy is added to the system. Le Châtelier's principle predicts a shift that will relieve the imposed change. Therefore, the shift will tend to remove the added energy. This will happen if the equilibrium shifts to the right, increasing the amount of $SO_{2(g)}$ and $O_{2(g)}$ formed. As a result, the value of K_c will increase. The shift in equilibrium is consistent (as it must be) with the law of conservation of energy.

Suppose that the equilibrium of an endothermic reaction shifted to the left when the system was heated. From right to left, the reaction is exothermic. Thus, a shift to the left would release energy to the surroundings. This would make the mixture hotter. The equilibrium would shift still more to the left, releasing more and more energy. Clearly, this cannot happen without violating the law of conservation of energy.

The effect of temperature on the position of equilibrium can be summarized as follows:

- *Endothermic change* ($\Delta H > 0$): An increase in temperature shifts the equilibrium to the right, forming more products. The value of K_c increases. A decrease in temperature shifts the equilibrium to the left, forming more reactants. The value of K_c decreases.
- *Exothermic change* ($\Delta H < 0$): An increase in temperature shifts the equilibrium to the left, forming more reactants. The value of K_c decreases. A decrease in temperature shifts the equilibrium to the right, forming more products. The value of K_c increases.

Chickens have no sweat glands. When the temperature rises, they tend to breathe faster. This lowers the concentration of carbonate ions in their blood. Because eggshells are mostly calcium carbonate, faster-breathing chickens lay eggs with thinner shells. Rather than installing expensive air conditioning, chicken farmers can supply carbonated water for their chickens to drink when the temperature reaches "fowl" highs. How does this relate to Le Châtelier's principle?

The Effects of Volume and Pressure Changes on Equilibrium

When the volume of a mixture of gases decreases, the pressure of the gases must increase. Le Châtelier's principle predicts a shift in equilibrium to relieve this change. Therefore, the shift must tend to reduce the pressure of the gases. Molecules striking the walls of a container cause gas pressure, so a reduction in gas pressure at constant temperature must mean fewer gas molecules. Consider the following reaction again.

$$2SO_{3(g)} \rightleftharpoons 2SO_{2(g)} + O_{2(g)}$$

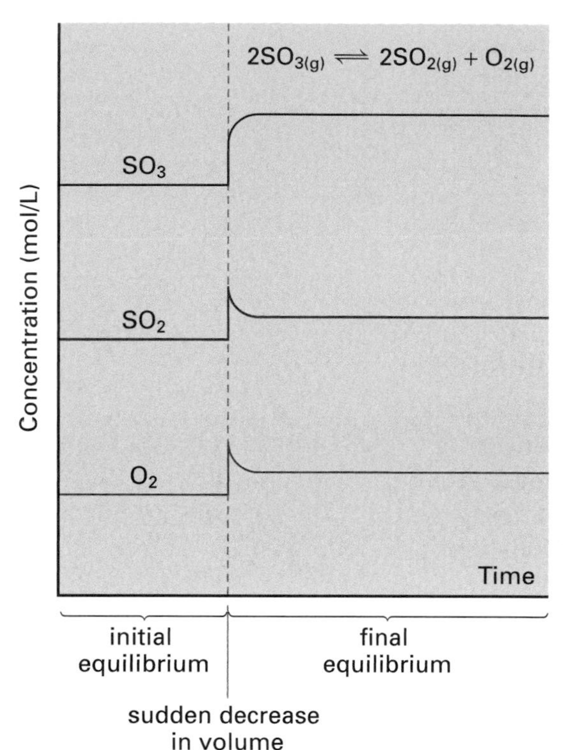

Figure 7.10 Note the changes in the concentrations of the reacting gases. When the new equilibrium is established, the concentrations are different. Compare the concentrations of the gases in the first equilibrium and second equilibrium.

There are two gas molecules on the left side of the equation. There are three gas molecules on the right side. Consequently, if the equilibrium shifts to the left, the pressure of the mixture will decrease. *Reducing the volume of an equilibrium mixture of gases, at constant temperature, causes a shift in equilibrium in the direction of fewer gas molecules.* What if there is the same number of gas molecules on both sides of the reaction equation? Changing the volume of the container, as shown in Figure 7.10, has no effect on the position of equilibrium. The value of K_c will be unchanged, as long as there is no change in temperature.

It is possible to increase the total pressure of gases in a rigid container by injecting more gas. If the injected gas reacts with the other gases in the equilibrium mixture, the effect is the same as increasing the concentration of a reactant. The equilibrium shifts to reduce the amount of added gas. If the injected gas does not react with the other gases (if an "inert" gas is added), there is no effect on the equilibrium. This is because the added gas is not part of the equilibrium system. Its addition causes no change in the volume of the container. Nitrogen (because of its low reactivity) and the noble gases are often used as inert gases. If the container expands when an inert gas is added, the effect is the same as increasing the volume and therefore decreasing the pressure of the reacting gases.

The Effect of a Catalyst on Equilibrium

A catalyst speeds up the *rate* of a reaction, either by allowing a different reaction mechanism or by providing additional mechanisms. The overall effect is to lower the activation energy, which increases the rate of reaction. The activation energy is lowered the same amount for the forward and reverse reactions, however. There is the same increase in reaction rates for both reactions. As a result, a catalyst does not affect the position of equilibrium. It only affects the time that is taken to achieve equilibrium.

Table 7.2 summarizes the effect of a catalyst, and other effects of changing conditions, on a system at equilibrium. The Sample Problem that follows provides an opportunity for you to use Le Châtelier's principle to predict the equilibrium shift in response to various conditions.

Table 7.2 The Effects of Changing Conditions on a System at Equilibrium

Type of reaction	Change to system	Effect on K_c	Direction of change
all reactions	increasing any reactant concentration, or decreasing any product concentration	no effect	toward products
	decreasing any reactant concentration, or increasing any product concentration	no effect	toward reactants
	using a catalyst	no effect	no change
exothermic	increasing temperature	decreases	toward reactants
	decreasing temperature	increases	toward products
endothermic	increasing temperature	increases	toward products
	decreasing temperature	decreases	toward reactants
equal number of reactant and product gas molecules	changing the volume of the container, or adding a non-reacting gas	no effect	no change
more gaseous product molecules than reactant gaseous molecules	decreasing the volume of the container at constant temperature	no effect	toward reactants
	increasing the volume of the container at constant temperature, or adding a non-reacting gas at contstant pressure	no effect	toward products
fewer gaseous product molecules than reactant gaseous molecules	decreasing the volume of the container at constant temperature	no effect	toward products
	increasing the volume of the container at constant temperature	no effect	toward reactants

Sample Problem

Using Le Châtelier's Principle

Problem

The following equilibrium takes place in a rigid container.

$$PCl_{5(g)} \rightleftharpoons PCl_{3(g)} + Cl_{2(g)} \quad \Delta H = 56 \text{ kJ}$$

In which direction does the equilibrium shift as a result of each change?

(a) adding phosphorus pentachloride gas

(b) removing chlorine gas

(c) decreasing the temperature

(d) increasing the pressure by adding helium gas

(e) using a catalyst

What Is Required?

You need to determine whether each change causes the equilibrium to shift to the left or the right, or whether it has no effect.

What Is Given?

You have the balanced chemical equation. You know that the enthalpy change is 56 kJ. Therefore, the reaction is endothermic.

Continued ...

Continued ...

Plan Your Strategy

Identify the change. Then use the chemical equation to determine the shift in equilibrium that will minimize the change.

Act on Your Strategy

(a) $[PCl_5]$ increases. Therefore, the equilibrium must shift to minimize $[PCl_5]$. The reaction shifts to the right.

(b) $[Cl_2]$ is reduced. Therefore, the equilibrium must shift to increase $[Cl_2]$. The reaction shifts to the right.

(c) The temperature decreases. Therefore, the equilibrium must shift in the direction in which the reaction is exothermic. From left to right, the reaction is endothermic. Thus, the reaction must be exothermic from right to left. The reaction shifts to the left if the temperature is decreased.

(d) Helium does not react with any of the gases in the mixture. The position of equilibrium does not change.

(e) A catalyst has no effect on the position of equilibrium.

Check Your Solution

Check the changes. Any change that affects the equilibrium reaction must result in a shift that minimizes it.

Practice Problems

29. Consider the following reaction.

$H_{2(g)} + I_{2(g)} + 52\text{ kJ} \rightleftharpoons 2HI_{(g)}$

In which direction does the equilibrium shift if there is an increase in temperature?

30. A decrease in the pressure of each equilibrium system below is caused by increasing the volume of the reaction container. In which direction does the equilibrium shift?

(a) $CO_{2(g)} + H_{2(g)} \rightleftharpoons CO_{(g)} + H_2O_{(g)}$

(b) $2NO_{2(g)} \rightleftharpoons N_2O_{4(g)}$

(c) $2CO_{2(g)} \rightleftharpoons 2CO_{(g)} + O_{2(g)}$

(d) $CH_{4(g)} + 2H_2S_{(g)} \rightleftharpoons CS_{2(g)} + 4H_{2(g)}$

31. The following reaction is exothermic.

$2NO_{(g)} + 2H_{2(g)} \rightleftharpoons N_{2(g)} + 2H_2O_{(g)}$

In which direction does the equilibrium shift as a result of each change?

(a) removing the hydrogen gas

(b) increasing the pressure of gases in the reaction vessel by decreasing the volume

(c) increasing the pressure of gases in the reaction vessel by pumping in argon gas while keeping the volume of the vessel constant

(d) increasing the temperature

(e) using a catalyst

32. In question 31, which changes affect the value of K_c? Which changes do not affect the value of K_c?

33. Toluene, C_7H_8, is an important organic solvent. It is made industrially from methyl cyclohexane.

$C_7H_{14(g)} \rightleftharpoons C_7H_{8(g)} + 3H_{2(g)}$

The forward reaction is endothermic. State three different changes to an equilibrium mixture of these reacting gases that would shift the equilibrium toward greater production of toluene.

Applying Le Châtelier's Principle: Manufacturing Ammonia

Industrial processes must be run at optimal conditions to be economic. This means more than simply manipulating the reaction conditions to maximize the extent of the reaction. Business people, and the professionals they hire to advise them, must consider other factors as well. These factors include the rate of the reaction, safety, the location of the plant, the cost to build and operate the plant, the cost of acquiring reactant materials, the cost of transporting products, and the cost of hiring, educating, and maintaining plant workers.

The chapter introduction mentioned the importance of nitrogen to plants and animals. Despite nitrogen's abundance in the atmosphere, its low reactivity means that there is a limited supply in a form that organisms can use. In nature, bacteria and the energy of lightning supply the nitrates and other nitrogen compounds that plants need to survive. Humans use technology to produce these compounds for our extensive agricultural industry.

In this chapter, you learned about the Haber process for manufacturing ammonia. You used this process to help you understand various concepts related to equilibrium. As you can see in Figure 7.11, ammonia is a valuable industrial chemical. Its annual global production is well over 100 million tonnes. The vast majority of ammonia, roughly 80%, is used to make fertilizers. You will now examine how the equilibrium concepts you have been studying work together to provide society with a reliable, cost-effective supply of ammonia.

Before World War I, the main source of nitrates for human use was from large deposits of bird droppings in Peru and sodium nitrate from Chile. These sources were becoming scarce and expensive. Then Fritz Haber (1868–1934), a lecturer in a technical college in Germany, began to experiment with ways to manufacture ammonia. Haber knew that ammonia could be easily converted to nitrates and other useful nitrogen

Unit Issue Prep

To begin preparing for the Unit 4 Issue, make a simplified sketch of the carbon cycle,

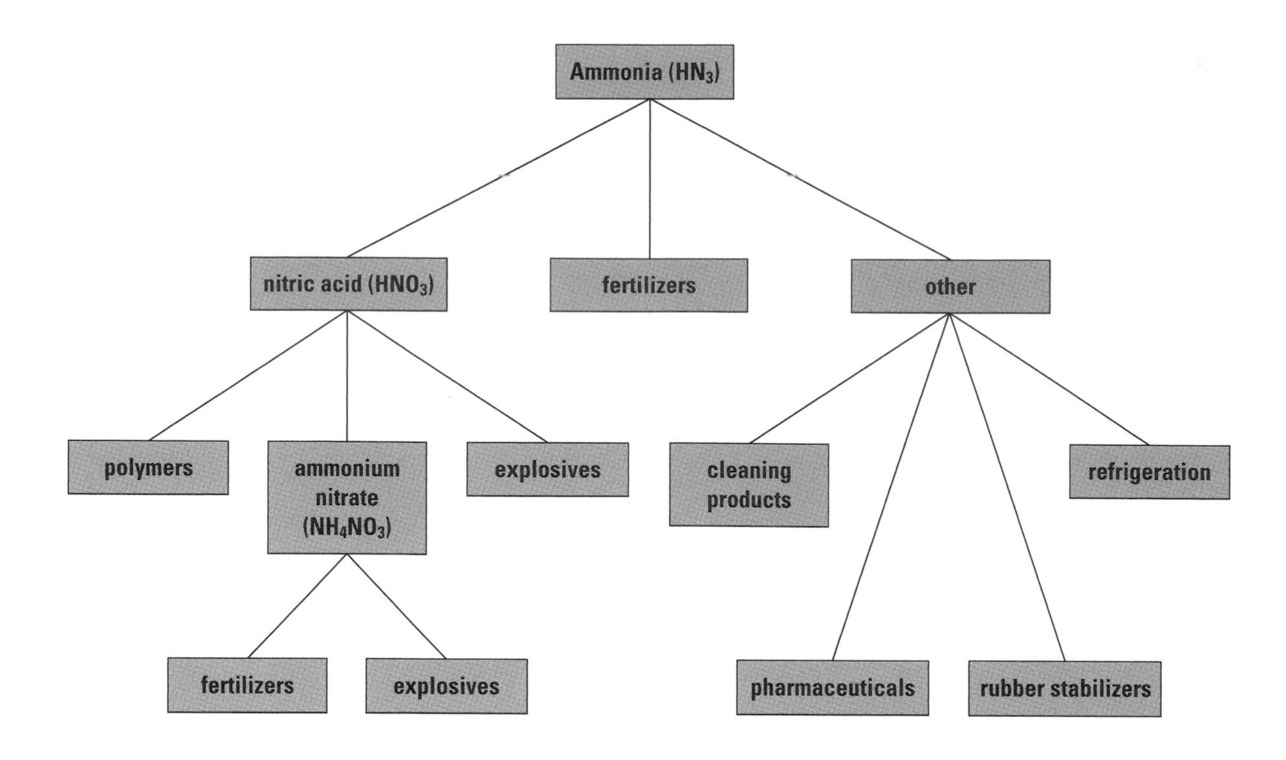

Figure 7.11 **Because ammonia is used to manufacture many essential products, it is one of the top five industrial chemicals in the world.**

compounds. What he needed was a method for producing large quantities of ammonia at minimal cost.

Haber experimented with the direct synthesis of ammonia.

$$N_{2(g)} + 3H_{2(g)} \rightleftharpoons 2NH_{3(g)} \quad \Delta H^{\circ} = -92 \text{ kJ/mol}$$

Because the reaction is exothermic, heat is released as the reaction proceeds. Le Châtelier's principle predicts that the yield of ammonia is greater at lower temperatures. Just as in the contact process for manufacturing sulfuric acid, however, high yield is not the only important factor. The rate of reaction for ammonia synthesis is too slow at low temperatures.

Le Châtelier's principle also predicts that the yield of ammonia is greater at higher pressures. High-pressure plants are expensive to build and maintain, however. In fact, the first industrial plant that manufactured ammonia had its reaction vessel blow up. A German chemical engineer, Carl Bosch, solved this problem by designing a double-walled steel vessel that could operate at several hundred times atmospheric pressure. Modern plants operate at pressures in the range of 20 200 kPa to 30 400 kPa.

The gas mixture is kept pressurized in a condenser. (See Figure 7.12.) Ammonia is removed from the reaction vessel by cooling the gas mixture. Because of the hydrogen bonding between ammonia molecules, the gas condenses into a liquid while the nitrogen and hydrogen remain as gases. Nitrogen is removed to shift the equilibrium toward the production of more ammonia. Once the ammonia is removed, the gases are recycled back to the reaction vessel in a continuous operation.

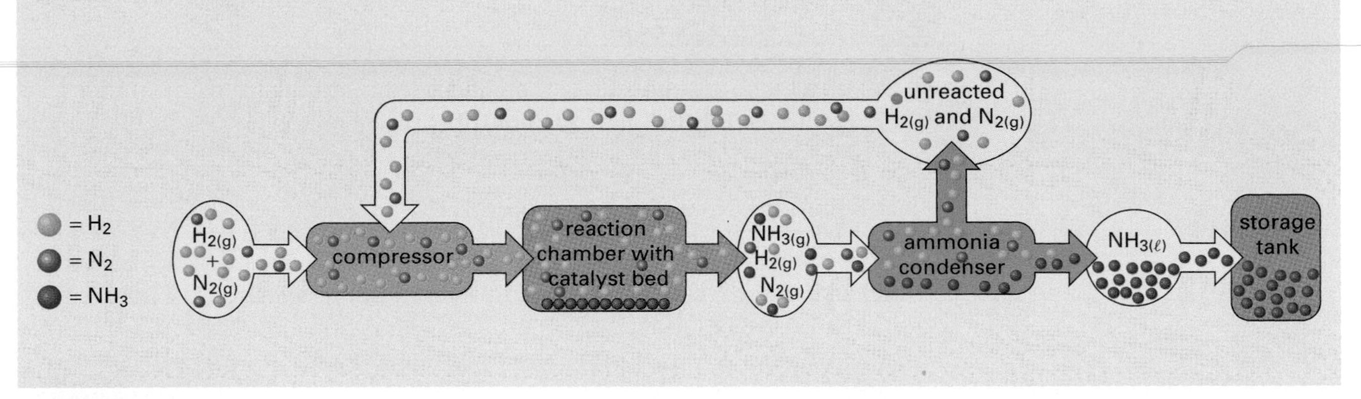

Figure 7.12 **This diagram shows the stages in the Haber-Bosch process for manufacturing ammonia. The catalyst is a mixture of MgO, Al_2O_3, and SiO_2, with embedded iron crystals.**

By manipulating the pressure and removing ammonia from the reaction vessel, Haber successfully increased the yield of ammonia. To increase the rate of the reaction, Haber needed to find a catalyst. A catalyst would allow the reaction to proceed at higher temperatures—a compromise between rate and yield. Some historians claim that Haber performed more than 6500 experiments trying to find a suitable catalyst. He finally chose an iron catalyst. This catalyst works well at the relatively moderate temperature of 400°C that is used for the reaction. It lasts about five years before losing its effectiveness.

Haber received the Nobel prize for Chemistry in 1918. Bosch received a Nobel prize in 1931 for his work on high pressure reactions. Today the Haber-Bosch process is used to manufacture virtually all the ammonia that is produced in the world. Plants are usually located near a source of natural gas, which is used to obtain the hydrogen gas.

$$CH_{4(g)} + H_2O_{(g)} \rightarrow CO_{(g)} + 3H_{2(g)}$$

The other raw material is air, which provides an inexhaustible supply of nitrogen gas.

During World War I, Haber helped to develop the technology for deploying phosgene, chlorine, and mustard gas as weapons of chemical warfare. His wife Clara, also a chemist, was disgusted by the use of science in war. When her husband refused to stop his support of the war effort, she committed suicide.

Section Summary

In this section, you learned that the expression for the reaction quotient is the same as the expression for the equilibrium constant. The concentrations that are used to solve these expressions may be different, however. When Q_c is less than K_c, the reaction proceeds to form more products. When Q_c is greater than K_c, the reaction proceeds to form more reactants. These changes continue until Q_c is equal to K_c. Le Châtelier's principle describes this tendency of a chemical system to return to equilibrium after a change moves it from equilibrium. The industrial process for manufacturing ammonia illustrates how chemical engineers apply Le Châtelier's principle to provide the most economical yield of a valuable chemical product.

K_c is only one of several equilibrium constants that chemists use to describe chemical systems. In the next chapter, you will learn how equilibrium applies to chemical systems that involve acids and bases.

Web LINK

www.mcgrawhill.ca/links/chemistry12

Sulfuric acid, methanol, and polystyrene are other industrially important chemicals that depend on equilibrium reactions for their production. Choose one of these chemicals, or another industrial chemical. Research methods that are used to produce it, as well as the products that are derived from it. To start your research, go to the web site above and click on **Web Links**. Prepare a report that outlines what you learned.

Section Review

1 **I** In which direction does the equilibrium shift as a result of the change to each homogeneous equilibrium system?

(a) Adding $Cl_{2(g)}$: $2Cl_{2(g)} + O_{2(g)} \rightleftharpoons 2Cl_2O_{(g)}$

(b) Removing $N_{2(g)}$: $2NO_{2(g)} \rightleftharpoons N_{2(g)} + 2O_{2(g)}$

(c) Using a catalyst: $CH_{4(g)} + H_2O_{(g)} \rightleftharpoons CO_{2(g)} + H_{2(g)}$

(d) Decreasing the total volume of the reaction container: $2NO_{2(g)} \rightleftharpoons N_2O_{4(g)}$

(e) Increasing the temperature: $CO_{(g)} + 3H_{2(g)} \rightleftharpoons CH_{4(g)} + H_2O_{(g)}$ $\Delta H = -230$ kJ

2 **I** For each reversible reaction, determine whether the forward reaction is favoured by high temperatures or low temperatures.

(a) $N_2O_{4(g)} \rightleftharpoons 2NO_{2(g)}$ $\Delta H = +59$ kJ

(b) $2ICl_{(g)} \rightleftharpoons I_{2(g)} + Cl_{2(g)}$ $\Delta H = -35$ kJ

(c) $2CO_{2(g)} + 566\text{ kJ} \rightleftharpoons 2CO_{(g)} + O_{2(g)}$

(d) $2HF_{(g)} \rightleftharpoons H_{2(g)} + F_{2(g)}$ $\Delta H = -536$ kJ

3 **I** In each reaction, the volume of the reaction vessel is decreased. What is the effect (if any) on the position of equilibrium?

(a) $N_{2(g)} + O_{2(g)} \rightleftharpoons 2NO_{(g)}$

(b) $4HCl_{(g)} + O_{2(g)} \rightleftharpoons 2Cl_{2(g)} + 2H_2O_{(g)}$

(c) $2H_2S_{(g)} + CH_{4(g)} \rightleftharpoons 4H_{2(g)} + CS_{2(g)}$

(d) $2CH_3COOH_{(aq)} + Ba(OH)_{2(aq)} \rightleftharpoons Ba(CH_3COO)_{2(aq)} + 2H_2O_{(\ell)}$

4 **MC** As you learned, the Haber process is used to produce ammonia.

(a) Write the chemical equation for the Haber process.

(b) Describe three changes that would increase the yield of ammonia, and explain why.

5 **MC** A key step in the production of sulfuric acid is the conversion of sulfur dioxide to sulfur trioxide. This step is exothermic.

(a) Write the chemical equation for this step, showing the sign of ΔH.

(b) Describe three changes that would increase the yield of sulfur trioxide.

(c) Explain why the reaction is carried out at a relatively high temperature.

(d) Why is a catalyst (vanadium pentoxide) used?

Careers

Anesthesiology: A Career in Pain Management

Laila Rafik Karwa

Dr. Laila Rafik Karwa is a specialist in pain. She prevents it, manages it, and alleviates it. She is an anesthesiologist.

Anesthesiologists are medical doctors who are responsible for inducing sleep before surgery, and for managing patients' pain before, during, and after surgery. Anesthesiologists are also involved in managing pain in patients who have severe illnesses, or acute or chronic pain. For example, Karwa's areas of interest include pediatric and obstetric anesthesia and acute pain management.

Anesthesiologists must have an intimate knowledge of the chemical and physical properties of gases. Many anesthetics are inhaled and are delivered to the bloodstream by diffusion. The speed at which diffusion occurs between the lungs, the blood, and other tissues of the body depends on a constant called the *partition coefficient*. This constant is a ratio that describes the equilibrium concentrations of a solute that is dissolved in two separate phases. The solute becomes separated (partitioned) between the two solvents in such a way that its concentration in one is directly proportional to its concentration in the other.

Gases diffuse from areas of high concentration to areas of low concentration. The speed at which diffusion occurs in the body depends on partition coefficients. The faster the concentration in the lung and brain tissues reaches the inhaled concentration of an anesthetic, the sooner a patient is induced.

Dr. Karwa enjoys the immediate gratification that comes with anesthesiology and the speed at which the anesthetics exact their effects. "What I like most about my job is the ability to let patients have their surgery being completely unaware of it, and to make them as pain-free as possible during and after surgery," she says. She chose medicine as a career because she was interested in biology and because she wanted to help other people with their health problems, after witnessing her father's struggles with rheumatic heart disease.

Dr. Karwa completed high school in Bombay, India, and later studied medicine at the University of Bombay. After medical school, she immigrated to Canada. She completed residency training in anesthesia at the University of Toronto and now works at St. Joseph's Health Science Centre as a staff anesthesiologist.

In Canada, students who are interested in a career as an anesthesiologist must complete two to three years of university before applying to medical school. Many medical schools in Canada require applicants to have taken courses in specific subjects in the biological and physical sciences, and to have some background in the humanities or social sciences. This education also helps students prepare for the Medical College Admissions Test (MCAT), which is required for application to most medical schools. The MCAT tests an applicant's aptitude for science, verbal reasoning, and writing. Most medical school programs in Canada are four years long.

Medical school graduates who decide to do specialty training in anesthesiology must complete a postgraduate residency program, which usually takes at least five years.

Making Career Connections

1. Although anesthesiologists tend to be associated with surgery, they also work outside the confines of the operating room. Research other locations where you might find an anesthesiologist.
2. Anesthetics are powerful medications. What risks to patients' health do anesthesiologists have to consider when they administer anesthetics?
3. Anesthetics are classified into three main groups: general, regional, and local. Research the criteria for this classification, and explain the distinctions.

CHAPTER 7 Review

Reflecting on Chapter 7

Summarize this chapter in the format of your choice. Here are a few ideas to use as guidelines:

- State the four conditions that apply to all equilibrium systems. Give examples to illustrate these conditions.
- Identify conditions that favour a reaction, and explain how they are related to equilibrium.
- Describe how enthalpy and entropy are related to chemical equilibrium.
- Compare the first and second laws of thermodynamics.
- Use experimental data to determine an equilibrium constant for concentration.
- Describe how to use an ICE table to solve problems that involve K_c.
- Compare Q_c and K_c to determine the direction of a chemical reaction.
- Explain the meaning of a small value of K_c.
- Use Le Châtelier's principle to predict the direction of a reaction.
- Outline the effects of changing conditions on a chemical system at equilibrium.
- Summarize the use of equilibrium and Le Châtelier's principle in industrial processes, such as the production of ammonia and sulfuric acid.

Reviewing Key Terms

For each of the following terms, write a sentence that shows your understanding of its meaning.

equilibrium
homogeneous equilibrium
heterogeneous equilibrium
favourable change
entropy (S)
second law of thermodynamics
free energy
law of chemical equilibrium
equilibrium constant (K_c)
ICE table
reaction quotient (Q_c)
Le Châtelier's principle
common ion effect

Knowledge/Understanding

1. Explain the difference between the rate of a reaction and the extent of a reaction.
2. At equilibrium, there is no overall change in the concentrations of reactants and products. Why, then, is this state described as dynamic?
3. For a reaction that goes to completion, is K_c very large or very small? Explain why.
4. In a chemical reaction, the change in enthalpy and the change in entropy are determined by the nature of the compounds involved.
   (a) What signs of ΔH and ΔS indicate that both factors contribute to a favourable reaction?
   (b) What signs of ΔH and ΔS indicate that both factors combine to make a reaction unfavourable?
5. Name the factors that can affect the equilibrium of a reaction.
6. Increasing temperature tends to increase the solubility of a solid in a liquid, but it tends to decrease the solubility of gases. Explain why.
7. The following reaction is at equilibrium. Which condition will produce a shift to the right: a decrease in volume or a decrease in temperature? Explain why.
   $H_{2(g)} + Cl_{2(g)} \rightleftharpoons 2HCl_{(g)} + \text{heat}$
8. The following system is at equilibrium. Will an increase in pressure result in a shift to the left or to the right? How do you know?
   $2CO_{2(g)} \rightleftharpoons 2CO + O_{2(g)}$
9. Consider the following reaction.
   $CO_{(g)} + 3H_{2(g)} \rightleftharpoons CH_{4(g)} + H_2O_{(g)}$
   (a) The volume and temperature are kept constant, but the pressure on the system is increased. Explain how this affects the concentration of the reactants and products, and the direction in which the equilibrium shifts.
   (b) When equilibrium has been re-established, which substance(s) will show an increase in concentration?

Inquiry

10. The following equation represents the dissociation of hydrogen iodide gas. At 430°C, the value of K_c is 0.20.

$2HI_{(g)} \rightleftharpoons H_{2(g)} + I_{(g)}$

Some $HI_{(g)}$ is placed in a closed container at 430°C. Analysis at equilibrium shows that the concentration of $I_{2(g)}$ is 5.6×10^{-4} mol/L. What are the equilibrium concentrations of $H_{2(g)}$ and $HI_{(g)}$?

11. The oxidation of sulfur dioxide to sulfur trioxide is an important reaction. At 1000 K, the value of K_c is 3.6×10^{-3}.

$2SO_{2(g)} + O_{2(g)} \rightleftharpoons 2SO_{3(g)}$

A closed flask originally contains 1.7 mol/L $SO_{2(g)}$ and 1.7 mol/L $O_{2(g)}$. What is $[SO_3]$ at equilibrium when the reaction vessel is maintained at 1000 K?

12. Ethanol and propanoic acid react to form the ester ethyl propanoate, which has the odour of bananas.

$CH_3CH_2OH_{(\ell)} + CH_3CH_2COOH_{(\ell)} \rightleftharpoons CH_3CH_2COOCH_2CH_3 + H_2O_{(\ell)}$

At 50°C, K_c for this reaction is 7.5. If 30.0 g of ethanol is mixed with 40.0 g of propanoic acid, what mass of ethyl propanoate will be present in the equilibrium mixture at 50°C?

Hint: Calculate the initial amounts of the reactants. Then solve the equilibrium equation using amounts instead of concentrations. The volume of the mixture does not affect the calculation.

13. 0.50 mol of $CO_{(g)}$ and 0.50 mol of $H_2O_{(g)}$ are placed in a 10 L container at 700 K. The following reaction occurs.

$CO_{(g)} + H_2O_{(g)} \rightleftharpoons H_{2(g)} + CO_{2(g)} \quad K_c = 8.3$

What is the concentration of each gas that is present at equilibrium?

14. Sulfur atoms combine to form molecules that have different numbers of atoms depending on the temperature. At about 1050°C, the following dissociation occurs.

$S_{8(g)} \rightleftharpoons 4S_{2(g)}$

The initial concentration of $S_{8(g)}$ in a flask is 9.2×10^{-3} mol/L, and the equilibrium concentration of the same gas is 2.3×10^{-3} mol/L. What is the value of K_c?

15. Perpetual motion machines seem to be a favourite project for inventors who do not understand the second law of thermodynamics. Design a simple machine that, once started, would recycle energy and, according to the first law of thermodynamics, should carry on forever. Why does the second law of thermodynamics rule out the possibility of ever making a perpetual motion machine?

16. Consider an equilibrium in which oxygen gas reacts with gaseous hydrogen chloride to form gaseous water and chlorine gas. At equilibrium, the gases have the following concentrations:
$[O_2] = 8.6 \times 10^{-2}$ mol/L,
$[HCl] = 2.7 \times 10^{-2}$ mol/L,
$[H_2O] = 7.8 \times 10^{-3}$ mol/L,
$[Cl_2] = 3.6 \times 10^{-3}$ mol/L.

(a) Write a balanced chemical equation for this reaction.

(b) Calculate the value of the equilibrium constant.

17. Sulfur trioxide gas reacts with gaseous hydrogen fluoride to produce gaseous sulfur hexafluoride and water vapour. The value of K_c is 6.3×10^{-3}.

(a) Write a balanced chemical equation for this reaction.

(b) 2.9 mol of sulfur trioxide is mixed with 9.1 mol of hydrogen fluoride in a 4.7 L flask. Set up an equation to determine the equilibrium concentration of sodium hexafluoride.

(c) Explain why you are likely unable to solve this equation.

18. The following results were collected for two experiments that involve the reaction, at 600°C, between gaseous sulfur dioxide and oxygen to form gaseous sulfur trioxide. Show that the value of K_c was the same in both experiments.

Experiment 1		Experiment 2	
Initial concentration (mol/L)	Equilibrium concentration (mol/L)	Initial concentration (mol/L)	Equilibrium concentration (mol/L)
$[SO_2] = 2.00$	$[SO_2] = 1.50$	$[SO_2] = 0.500$	$[SO_2] = 0.590$
$[O_2] = 1.50$	$[O_2] = 1.25$	$[O_2] = 0$	$[O_2] = 0.0450$
$[SO_3] = 3.00$	$[SO_3] = 3.50$	$[SO_3] = 0.350$	$[SO_3] = 0.260$

19. Write the chemical equation for the reversible reaction that has the following equilibrium expression.

$$K_c = \frac{[NO]^4[H_2O]^6}{[NH_3]^4[O_2]^5}$$

Assume that, at a certain temperature, [NO] and $[NH_3]$ are equal. Also assume that $[H_2O] = 2.0$ mol/L and $[O_2] = 3.0$ mol/L. What is the value of K_c at this temperature?

Communication

20. Discuss the following statements, which are attributed to the chemist Harry Bent:
    - The first law of thermodynamics says you cannot win, you can only break even.
    - The second law of thermodynamics says you cannot break even.
21. Equal amounts of hydrogen gas and iodine vapour are heated in a sealed flask.
    (a) Sketch a graph to show how $[H_{2(g)}]$ and $[HI_{(g)}]$ change over time.
    (b) Would you expect a graph of $[I_{2(g)}]$ and $[HI_{(g)}]$ to appear much different from your first graph? Explain why.
    (c) How does the value of Q_c change over time for this reaction?
22. A younger student in your school wants to grow a crystal of copper(II) sulfate. Write a short explanation that outlines how the student can use a small crystal and a saturated solution of copper(II) sulfate to grow a larger crystal.

Making Connections

23. The tendency of systems to reach a maximum state of entropy has been applied to the social sciences. Does the second law of thermodynamics help to explain the increase in garbage on the streets? Justify your answer.
24. At 25°C, the value of K_c for the reaction between nitrogen and oxygen is 4.7×10^{-31}:
    $N_{2(g)} + O_{2(g)} \rightleftharpoons 2NO_{(g)} \quad K_c = 4.7 \times 10^{-31}$
    Assuming that air is composed of 80% nitrogen by volume and 20% oxygen by volume, estimate the concentration of nitric oxide to be expected in the atmosphere. Why is the actual [NO] usually greater?
25. One of the steps in the Ostwald process for the production of nitric acid involves the oxidation of ammonia.
    $4NH_{3(g)} + 5O_{2(g)} \rightleftharpoons 4NO_{(g)} + 6H_2O_{(g)} \quad \Delta H = -905$ kJ
    (a) State the reaction conditions that favour the production of nitrogen monoxide.
    (b) A rhodium/platinum alloy is used as a catalyst. What effect does the catalyst have on the rate of reaction? What effect does the catalyst have on the position of equilibrium?
    (c) Explain why the reaction temperature is relatively high, typically about 900°C.
    (d) A relatively low pressure of about 710 kPa is used. Suggest why.
    (e) In the next step of the Ostwald process, nitrogen monoxide is mixed with air to form nitrogen dioxide.
    $2NO_{(g)} + O_{2(g)} \rightleftharpoons 2NO_{2(g)} \quad \Delta H = -115$ kJ
    Why are the gases cooled for this reaction? What do you think happens to the heat that is extracted?
    (f) Finally, the nitrogen dioxide reacts with water to form nitric acid.
    $3NO_{2(g)} + H_2O_{(\ell)} \rightleftharpoons 2HNO_{3(aq)} + NO_{(g)}$
    What is done with the $NO_{(g)}$ that is formed? Name three uses of this product.
26. Polystyrene is one of our most useful polymers. Polystyrene resin is manufactured from styrene, which is made as follows:
    $C_6H_5CH_2CH_{3(g)} + 123 \text{ kJ} \rightleftharpoons C_6H_5CHCH_{2(g)} + H_{2(g)}$
    (a) Predict the effects (if any) on this equilibrium if the following changes are made:
    - increasing the applied pressure
    - removing styrene
    - reducing the temperature
    - adding a catalyst
    - adding helium
27. **(a)** Based on your predictions in question 26, select the conditions that would maximize the equilibrium yield of styrene.
    (b) Why are the conditions you selected unlikely to be used by industry? What conditions must be used to maximize the yield of an industrial process?

(c) In practice, the pressure of the ethyl benzene is kept low. An inert substance—super-heated steam—is added to keep the total pressure of the mixture at atmospheric pressure. What advantage does this method have over running the reaction without the steam at a pressure that is less than atmospheric pressure?

(d) The super-heated steam supplies energy to the system. Why is this desirable? It also reacts with any carbon that is formed as a by-product at high temperatures, preventing the carbon from contaminating the catalyst. What products are formed as a result of the reaction between carbon and steam?

(e) What is the advantage of using super-heated steam at 600°C rather than ordinary steam at 100°C?

(f) In practice, an iron oxide catalyst is used. Explain why this is desirable.

(g) At 600°C, with an iron oxide catalyst, the conversion is only about 35% complete. The yield could be increased by raising the temperature. Suggest two plausible reasons why this is not done.

(h) The ethyl benzene that has not reacted is separated from the styrene by fractional distillation. The boiling point of ethyl benzene is 136°C, and the boiling point of styrene is 146°C. The styrene, of course, is used to make polystyrene. What is done with the ethyl benzene?

Answers to Practice Problems and Short Answers to Section Review Questions

Practice Problems: **1.** $K_c = \dfrac{[CH_3COOCH_2CH_3][H_2O]}{CH_3CH_2OH][CH_3COOH]}$

2. $K_c = \dfrac{[NO]^2}{[N_2][O_2]}$ **3.** $K_c = \dfrac{[H_2O]^2}{[H_2]^2[O_2]}$ **4.** $K_c = \dfrac{[Fe^{2+}]^2[I_2]}{[Fe^{3+}]^2[I^-]^2}$

5. $K_c = \dfrac{[NO]^4[H_2O]^6}{[NH_3]^4[O_2]^5}$ **6.** 1.9×10^{-2} **7.** 1.2×10^2 **8.** 0.013 **9.** 2.4×10^{-4} **10.** 0.15 **11.** $[I_2] = [Cl_2] = 0.015$ mol/L; $[ICl] = 0.14$ mol/L **12.** HF = 0.11 mol **13.** $[SO_3] = 0.089$ mol/L **14.** $[CO] = 0.011$ mol/L; $[Cl_2] = 0.014$ mol/L; $[COCl_2] = 3.1 \times 10^{-5}$ mol/L **15.** $[HBr] = 0.045$ mol/L; $[H_2] = [Br_2] = 2.9 \times 10^{-6}$ mol/L **16.** 1100 K **17.** III, II, I **18.(a)** completion **(b)** no reaction **(c)** completion **19.** $Ag(NH_3)_2]^+{}_{(aq)}$ **20.** essentially no dissociation **21.** $[CO] = [Cl_2] = 1.8 \times 10^{-4}$ mol/L **22.** $[H_2] = 8.6 \times 10^{-2}$ mol/L **23.** 6.1×10^{-1} mol **24.** $[NO] = 1.2 \times 10^{-5}$ mol/L **25.** $[CO_2] = 0.59$ mol/L; $[CO] = 1.1 \times 10^{-2}$ mol/L; $[O_2] = 5.6 \times 10^{-3}$ mol/L **26.(a)** $K_c = 171$ **(b)** $[NO_2] = 0.0412$ mol/L; $[N_2O_4] = 0.145$ mol/L **(c)** $Q_c = 85.3$ **(d)** right **27.(a)** right **(b)** left **(c)** at equilibrium **(d)** left **28.** left **29.** right **30.(a)** no change **(b)** left **(c)** right **(d)** right **31.(a)** left **(b)** right **(c)** no change **(d)** left **(e)** no change **32.** Only (d) changes the value of K_c. **33.** adding methyl cyclohexane; removing toluene; decreasing the pressure; increasing the temperature

Section Review: 7.2: **3.(a)** increases **(b)** decreases **(c)** increases **(d)** decreases **(e)** increases **4.(a)** + **(b)** – **(c)** – **(d)** – **(e)** +

7.3: **1.(a)** $\dfrac{[SbCl_3][Cl_2]}{[SbCl_5]}$ **(b)** $\dfrac{[N_2][H_2O]^2}{[H_2]^2[NO]^2}$ **(c)** $\dfrac{[H_2]^4[CS_2]}{[H_2S]^2[CH_4]}$

2.(a) $[N_2] = 0.20$ mol/L; $[NH_3] = 1.6$ mol/L **(b)** $K_c = 0.017$ **3.** 1.2 mol **4.** $[PCl_3] = [Cl] = 0.035$ mol/L; $[PCl_5] = 0.059$ mol/L **5.(a)** $[H_2] = 3.5 \times 10^{-7}$ mol/L **(b)** an impractical method **7.4:** **1.(a)** right **(b)** right **(c)** unchanged **(d)** right **(e)** left **2.(a)** high temperatures **(b)** low temperatures **(c)** high temperatures **(d)** low temperatures **3.(a)** unchanged **(b)** right **(c)** left **(d)** unchanged (because liquid volumes are largely unaffected by pressure changes) **4.(a)** $N_2 + 3H_2 \rightleftharpoons 2NH_3$ **(b)** adding N_2 or H_2; removing NH_3; increasing the pressure; lowering the temperature (because the reaction is exothermic)

CHAPTER

8 Acids, Bases, and pH

Chapter Preview

8.1 Explaining the Properties of Acids and Bases

8.2 The Equilibrium of Weak Acids and Bases

8.3 Bases and Buffers

8.4 Acid-Base Titration Curves

Prerequisite Concepts and Skills

Before you begin this chapter, review the following concepts and skills:

- writing net ionic equations (Concepts and Skills Review)
- calculating molar concentrations (Concepts and Skills Review)
- solving equilibrium problems (Chapter 7, sections 7.3 and 7.4)
- explaining the mathematical properties of logarithms (previous studies)
- performing acid-base titrations (previous studies)

For many people, the word "acid" evokes the image of a fuming, highly corrosive, dangerous liquid. This image is fairly accurate for concentrated hydrochloric acid, a strong acid. Most acids, however, are not as corrosive as hydrochloric acid, although they may still be very hazardous. For example, hydrofluoric acid can cause deep, slow-healing tissue burns if it is handled carelessly. It is used by artists and artisans who etch glass. It reacts with the silica in glass to form a compound that dissolves, leaving the glass with a brilliant surface. Hydrofluoric acid is highly corrosive. Even a 1% solution is considered to be hazardous. Yet chemists classify hydrofluoric acid as a weak acid.

You learned about acids and bases in your previous chemistry course. In this chapter, you will extend your knowledge to learn how the structure of a compound determines whether it is an acid or a base. You will use the equilibrium constant of the reaction of an acid or base with water to determine whether the acid or base is strong or weak. You will apply your understanding of dissociation and pH to investigate buffer solutions: solutions that resist changes in pH. Finally, you will examine acid-base titrations that involve combinations of strong and weak acids and bases.

The hydrohalic acids ($HX_{(aq)}$), where X represents a halogen) include HF, HCl, HBr, and HI. Only HF is a weak acid. The rest are strong acids. What factors account for this difference?

Explaining the Properties of Acids and Bases

8.1

Table 8.1 outlines properties of acids and bases that you have examined in previous courses. In this section, you will review two theories that help to explain these and other properties. As well, you will use your understanding of molecular structure to help you understand why acids and bases differ in strength.

Table 8.1 Examples and Common Properties of Acids and Bases

Example	Acids	Bases
solid	acetyl salicylic acid	sodium hydroxide
liquid	acetic acid	aniline
gas	hydrogen chloride	ammonia
Property	**Acids**	**Bases**
taste CAUTION Never taste chemicals in a lab.	Acids taste sour.	Bases taste bitter.
texture of solution CAUTION Never deliberately touch chemicals. Strong, concentrated acids and bases will burn your skin.	Acids do not have a characteristic texture.	Bases feel slippery.
reaction with phenolphthalein	Acidic phenolphthalein is colourless.	Basic phenolphthalein is pink.
reaction with litmus paper	Acids turn blue litmus red.	Bases turn red litmus blue.
reaction with metals	Acids react with metals above hydrogen in the activity series to displace $H_{2(g)}$.	Bases react with certain metals (such as Al) to form $H_{2(g)}$.
reaction with carbonates	Carbon dioxide is formed.	No reaction occurs.
reaction with ammonium chloride	No reaction occurs.	Ammonia, NH_3, a gas with a characteristic odour, is produced.
neutralization reaction	Acids neutralize basic solutions.	Bases neutralize acidic solutions.
reaction with fatty acids	No reaction occurs.	Bases react to form soap (a saponification reaction).
aqueous property of oxides	Non-metal oxides form acidic solutions: for example, $CO_{2(g)} + H_2O_{(\ell)} \rightarrow H_2CO_{3(aq)}$	Metal oxides form basic solutions: for example, $CaO_{(s)} + H_2O_{(\ell)} \rightarrow Ca(OH)_{2(aq)}$
amount of dissociation in aqueous solution (strength)	Strong acids dissociate completely. Weak acids dissociate only partially.	Strong bases dissociate completely. Weak bases dissociate only partially.

Section Preview/ Specific Expectations

In this section, you will

- **compare** strong acids and bases, and weak acids and bases, in terms of equilibrium
- **identify** conjugate acid-base pairs
- **solve** problems that involve strong acids and strong bases
- **communicate** your understanding of the following terms: *hydronium ion ($H_3O^+_{(aq)}$), conjugate acid-base pair, monoprotic acids, polyprotic acids*

The following ExpressLab highlights concepts that you will examine in this section, as well as later in the chapter.

ExpressLab Comparing Acid-Base Reactions

You will perform three acid-base reactions. Before you begin, read the Procedure and make a prediction about the relative rates of these reactions.

Safety Precautions

The solutions that are used in this lab are irritants and should be handled with care. Wash any spills on your skin or clothing with plenty of water. Inform your teacher immediately.

Materials

powdered calcium carbonate, $CaCO_{3(s)}$
3 squeeze bottles, each containing one of the following solutions: 2.0 mol/L $HCl_{(aq)}$; 2.0 mol/L $CH_3COOH_{(aq)}$; mixture of 2.0 mol/L $CH_3COOH_{(aq)}$ and 2.0 mol/L $NaCH_3COO_{(aq)}$
scoopula
3 test tubes
test tube rack
labels or grease pencil

Procedure

1. Label each test tube to identify the solution it will contain. Then fill each test tube with the corresponding solution, to a depth of about 2 cm.
2. Add a small amount of $CaCO_{3(s)}$ (enough to cover the tip of a scoopula) to each test tube. Try to add the same amount of $CaCO_{3(s)}$ to all three test tubes.
3. Record your observations. Rank the rates of the three reactions from fastest to slowest.

Analysis

1. **(a)** What rate-related change did you observe in each test tube?
   (b) If you wanted to collect quantitative data for each reaction, how could you modify the experiment?
2. In all three test tubes, the following reaction occurred.
   $CaCO_{3(s)} + 2H_3O^+_{(aq)} \rightarrow CO_{2(g)} + Ca^{2+}_{(aq)} + 3H_2O_{(\ell)}$
   The only difference between the test tubes was the concentration of $H^+_{(aq)}$ in the acidic solutions that reacted with $CaCO_{3(s)}$. Explain your ranking of the rates of reaction in terms of $[H_3O^+]$.
3. The concentrations of $HCl_{(aq)}$ and $CH_3COOH_{(aq)}$ were identical. The following dissociation reactions occurred.
   $HCl_{(aq)} + H_2O_{(\ell)} \rightleftharpoons H_3O^+_{(aq)} + Cl^+_{(aq)}$
   $CH_3COOH_{(aq)} + H_2O_{(\ell)} \rightleftharpoons H_3O^+_{(aq)} + CH_3COO^-_{(aq)}$
   Explain your ranking of the $[H_3O^+]$ in these solutions in terms of the extent of the equilibrium dissociation.
4. The third solution was a mixture of 2.0 mol/L $CH_3COOH_{(aq)}$ and 2.0 mol/L $NaCH_3COO_{(aq)}$. How did the addition of sodium acetate affect the equilibrium of the dissociation reaction of acetic acid?
5. Explain your ranking of the rate of the reaction between calcium carbonate and the solution that was a mixture of acetic acid and sodium acetate.

The Arrhenius Theory of Acids and Bases

According to the Arrhenius theory (1887), acids and bases are defined in terms of their structure and the ions produced when they dissolve in water.

- An acid is a substance that dissociates in water to form $H^+_{(aq)}$. Two examples of Arrhenius acids are hydrochloric acid, HCl, and sulfuric acid, H_2SO_4.
- A base is a substance that dissociates in water to form $OH^-_{(aq)}$. Two examples of Arrhenius bases are sodium hydroxide, NaOH, and potassium hydroxide, KOH.

The Arrhenius theory explains acid-base reactions as a combination of $H^+_{(aq)}$ and $OH^-_{(aq)}$. It provides insight into the heat of neutralization for the reaction between a strong acid and a strong base. (Strong acids and bases dissociate completely into ions in solution.) For example, consider the following reaction.

$$HCl_{(aq)} + NaOH_{(aq)} \rightarrow NaCl_{(aq)} + H_2O_{(\ell)} \quad \Delta H = -56 \text{ kJ}$$

The total ionic equation for this reaction is

$$H^+_{(aq)} + Cl^-_{(aq)} + Na^+_{(aq)} + OH^-_{(aq)} \rightarrow Na^+_{(aq)} + Cl^-_{(aq)} + H_2O_{(\ell)} \quad \Delta H = -56 \text{ kJ}$$

Subtracting spectator ions from both sides, the net ionic equation is

$$H^+_{(aq)} + OH^-_{(aq)} \rightarrow H_2O_{(\ell)} \quad \Delta H = -56 \text{ kJ}$$

Different combinations of strong Arrhenius acids and bases react with the same exothermic result. Measurements always show the release of 56 kJ of energy per mole of water formed. This makes sense, because the net ionic equation is the same regardless of the specific neutralization reaction that occurs.

The Arrhenius theory has limitations, however. For example, $H^+_{(aq)}$, a bare proton, does not exist in water. The positive charge on a proton is attracted to the region of negative charge on the lone pair of electrons on a water molecule's oxygen atom. The combination is a hydrated proton called a **hydronium ion, $H_3O^+_{(aq)}$**.

$$H^+_{(aq)} + H_2O_{(\ell)} \rightarrow H_3O^+_{(aq)}$$

The hydronium ion, itself, forms hydrogen bonds with other water molecules. (See Figure 8.1.) Thus, a better formula for the ion that is present in acidic solutions is $[H(H_2O)_n]^+$, where n is usually 4 or 5. For convenience, however, chemists usually use a single hydronium ion when writing equations.

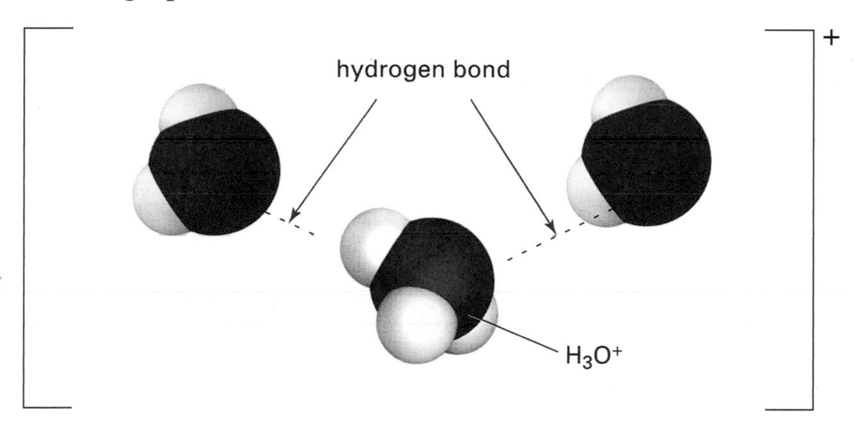

Figure 8.1 In aqueous solution, the hydronium ion, H_3O^+, forms hydrogen bonds with other water molecules.

The Arrhenius theory also has limitations for explaining certain reactions. For example, aqueous solutions of ammonia are basic. They react with acids in neutralization reactions, even though ammonia does not contain the hydroxide ion. Many aqueous solutions of salts with no hydroxide ions are basic, too. Some reactions take place without any liquid solvent. For example, ammonium chloride can be formed by the reaction between ammonia and hydrogen chloride, which are both gases:

$$NH_{3(g)} + HCl_{(g)} \rightarrow NH_4Cl_{(s)}$$

The Brønsted-Lowry Theory

The limitations of the Arrhenius theory of acids and bases are overcome by a more general theory, called the Brønsted-Lowry theory. This theory was proposed independently, in 1923, by Johannes Brønsted, a Danish chemist, and Thomas Lowry, an English chemist. It recognizes an acid-base reaction as a chemical equilibrium, having both a forward reaction and a reverse reaction that involve the transfer of a proton. The Brønsted-Lowry theory defines acids and bases as follows:

- An acid is a substance from which a proton can be removed. (Some chemists describe Brønsted-Lowry acids as "proton-donors.")
- A base is a substance that can accept a proton. (Some chemists describe Brønsted-Lowry bases as "proton-acceptors.")

Note that the word "proton" refers to the nucleus of a hydrogen atom — an H^+ ion that has been removed from the acid molecule. It does not refer to a proton removed from the nucleus of another atom, such as oxygen or sulfur, that may be present in the acid molecule. As mentioned previously, H^+ ions share electrons with any species (ion or molecule) that has a lone pair of electrons. In aqueous solution, the proton bonds with a water molecule to form the hydronium ion. Unlike the Arrhenius theory, however, the Brønsted-Lowry theory is not restricted to aqueous solutions. For example, the lone pair of electrons on an ammonia molecule can bond with H^+, and liquid ammonia can act as a base.

Definition Term	Arrhenius Theory	Brønsted-Lowry Theory
acid	a substance that contains hydrogen and dissociates in water to form $H^+_{(aq)}$	a substance from which a proton can be removed
base	a substance that contains the hydroxide group and dissociates in water to form $OH^-_{(aq)}$	a substance that can accept a proton from an acid

Conjugate Acid-Base Pairs

The dissociation of acetic acid in water is represented in Figure 8.2. This dissociation is an equilibrium reaction because it proceeds in both directions. Acetic acid is weak, so only a few ions dissociate. The position of equilibrium lies to the left, and the reverse reaction is favoured. In the reverse reaction, the hydronium ion gives up a proton to the acetate ion. Thus, these ions are an acid and a base, respectively, as shown in Figure 8.3. The acid on the left (CH_3COOH) and the base on the right (CH_3COO^-) differ by one proton. They are called a **conjugate acid-base pair**. Similarly, H_2O and H_3O^+ are a conjugate acid-base pair.

acid (acetic acid) | base (water) | acid (hydronium ion) | base (acetate ion)

Figure 8.2 The dissociation of acetic acid, a weak acid, in water

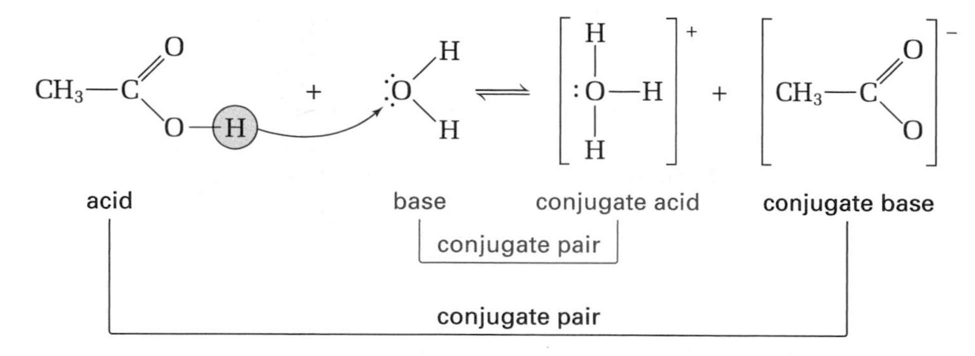

Figure 8.3 Conjugate acid-base pairs in the dissociation of acetic acid in water

Unlike the Arrhenius theory, the Brønsted-Lowry theory of acids and bases can explain the basic properties of ammonia when it dissolves in water. See Figure 8.4.

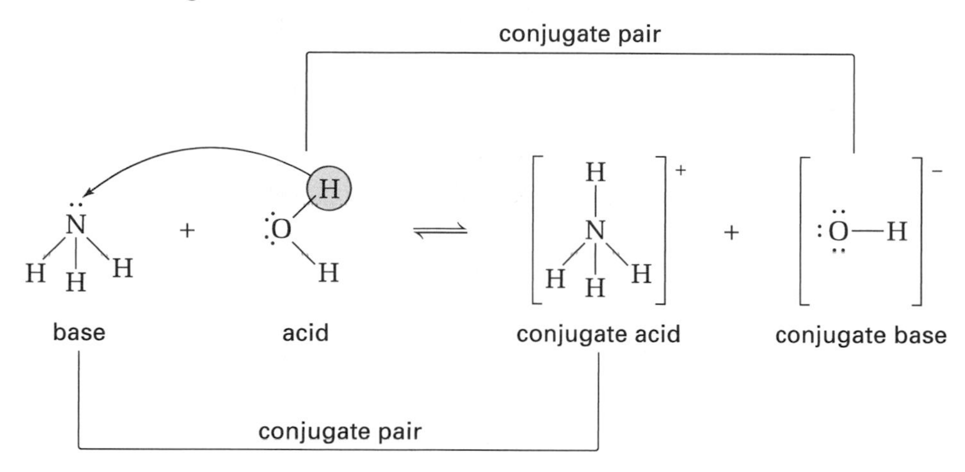

Figure 8.4 The dissociation of ammonia, a weak base, in water

Aqueous ammonia is a weak base, so relatively few hydroxide ions form. The position of equilibrium lies to the left. In the forward reaction, the water molecule gives up a proton and acts as an acid. A substance that can act as a proton donor (an acid) in one reaction and a proton acceptor (a base) in another reaction is said to be *amphoteric.* (Water acts as an acid in the presence of a stronger base, and as a base in the presence of a stronger acid.

Sample Problem

Identifying Conjugate Acid-Base Pairs

Problem

Identify the conjugate acid-base pair in each reaction.

(a) $H_3PO_{4(aq)} + H_2O_{(\ell)} \rightleftharpoons H_2PO_4^-{}_{(aq)} + H_3O^+{}_{(aq)}$

(b) $H_2PO_4^-{}_{(aq)} + OH^-{}_{(aq)} \rightleftharpoons HPO_4^{2-}{}_{(aq)} + H_2O_{(\ell)}$

Continued ...

Continued ...

Solution

On the left side of the equation, the acid is the molecule or ion that donates a proton. The base is the molecule or ion that accepts the proton. On the right side of the equation, you can identify the conjugate acid and base by the difference of a single proton from the base and acid on the left side.

(a) The conjugate acid-base pairs are $H_3PO_4/H_2PO_4^-$ and H_2O/H_3O^+.

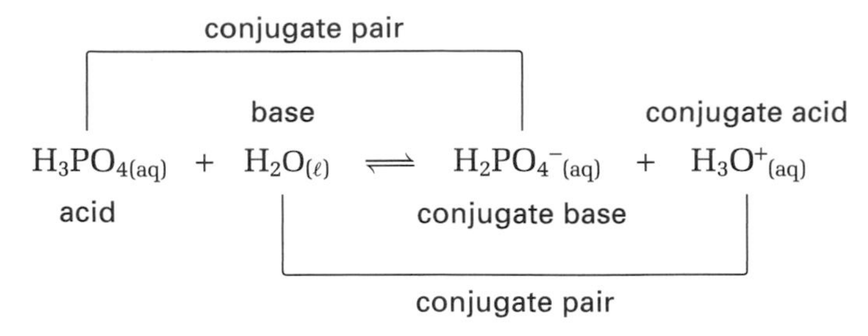

(b) The conjugate acid-base pairs are $H_2PO_4^-/HPO_4^{2-}$ and OH^-/H_2O.

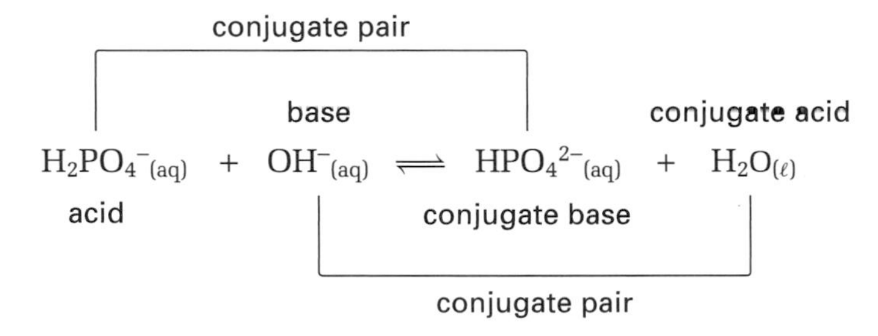

Check Your Solution

In each case, the acid has one more proton than its conjugate base.

Practice Problems

1. Name and write the formula of the conjugate base of each molecule or ion.

(a) HCl **(b)** HCO_3^- **(c)** H_2SO_4 **(d)** $N_2H_5^+$

2. Name and write the formula of the conjugate acid of each molecule or ion.

(a) NO_3^- **(b)** OH^- **(c)** H_2O **(d)** HCO_3^-

3. Identify the conjugate acid-base pairs in each reaction.

(a) $HS^-{}_{(aq)} + H_2O_{(\ell)} \rightleftharpoons H_2S_{(aq)} + OH^-{}_{(aq)}$

(b) $O^{2-}{}_{(aq)} + H_2O_{(\ell)} \rightarrow 2OH^-{}_{(aq)}$

4. Identify the conjugate acid-base pairs in each reaction.

(a) $H_2S_{(aq)} + NH_{3(aq)} \rightleftharpoons NH_4^+{}_{(aq)} + HS^-{}_{(aq)}$

(b) $H_2SO_{4(aq)} + H_2O_{(\ell)} \rightarrow H_3O^+{}_{(aq)} + HSO_4^-{}_{(aq)}$

Molecular Structure and the Strength of Acids and Bases

When a strong acid or base dissolves in water, almost every acid or base molecule dissociates. While there are many acids and bases, most are weak. Thus, the number of strong acids and strong bases is fairly small.

Strong Acids

- binary acids that have the general formula $HX_{(aq)}$, where X = Cl, Br, and I (but not F): for example, hydrochloric acid, HCl, and hydrobromic acid, HBr (HCl and HBr are *hydrohalic acids*: acids that have hydrogen bonded to atoms of the halogen elements.)
- oxoacids (acids containing oxygen atoms) in which the number of oxygen atoms exceeds, by two or more, the number of protons that can be dissociated: for example, nitric acid, HNO_3, sulfuric acid, H_2SO_4, perchloric acid, $HClO_4$, and chloric acid, $HClO_3$

The binary acids of non-metals exhibit periodic trends in their acid strength, as shown in Figure 8.5. Two factors are responsible for this trend: the electronegativity of the atom that is bonded to hydrogen, and the strength of the bond.

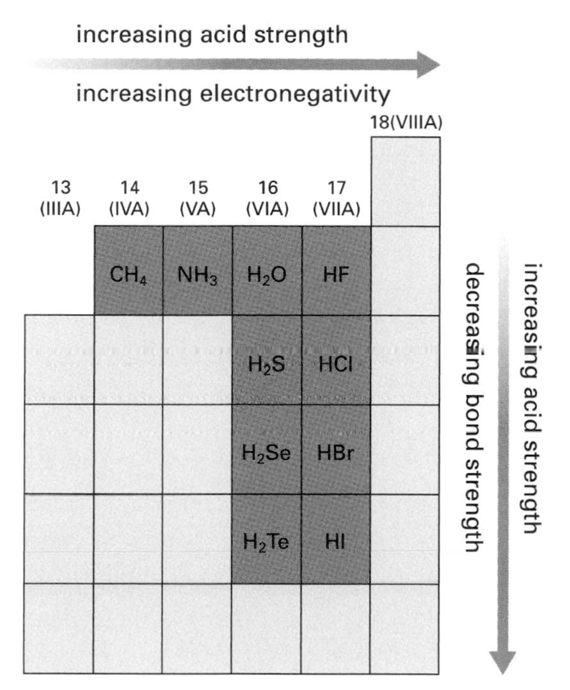

Figure 8.5 The binary acids show periodic trends, which are related to electronegativity and bond strength.

Across a period, electronegativity is the most important factor. The acid strength of hydrides increases as their electronegativity increases. This happens because an electronegative atom draws electrons away from the hydrogen atom, making it relatively positive. The negative pole of a water molecule then strongly attracts the hydrogen atom and pulls it away.

Down a group, bond strength is the most important factor. Acid strength increases as bond strength decreases. A weaker bond means that the hydrogen atom is more easily pulled away from the atom to which it is attached. For example, hydrofluoric acid is a stronger acid than water, but HF is the weakest of the hydrohalic acids because the H-F bond is relatively strong.

You might wonder how $HCl_{(aq)}$, $HBr_{(aq)}$, and $HI_{(aq)}$ can be described as increasing in strength, since each acid dissociates completely in water. This trend becomes apparent if you add equal concentrations of the acids to a solvent that is less basic than water, such as pure acetic acid. You will find that the acids dissociate to different extents.

Oxoacids increase in strength with increasing numbers of oxygen atoms, as shown in Figure 8.6. The hydrogen atoms that dissociate in water are always attached to oxygen atoms. Oxygen is more electronegative than hydrogen, so oxygen atoms draw electrons away from hydrogen atoms. The more oxygen atoms there are in a molecule, the greater is the polarity of the bond between each hydrogen atom and the oxygen atom it is attached to, and the more easily the water molecule can tear the hydrogen atom away.

increasing acid strength

H—O—Cl:	H—O—Cl(=O):	H—O—Cl(=O)=O	H—O—Cl(=O)(=O)=O
δ^+ δ^-	δ^+ δ^-	δ^+ δ^-	δ^+ δ^-
hypochlorous acid	chlorous acid	chloric acid	perchloric acid

Figure 8.6 **The relative strength of oxoacids increases with the number of oxygen atoms.**

Acids such as HCl, CH_3COOH, and HF are **monoprotic acids**. They have only a single hydrogen atom that dissociates in water. Some acids have more than one hydrogen atom that dissociates. These acids are called **polyprotic acids**. For example, sulfuric acid has two hydrogen atoms that can dissociate.

$$H_2SO_{4(aq)} + H_2O_{(\ell)} \rightleftharpoons H_3O^+_{(aq)} + HSO_4^-{}_{(aq)}$$
$$HSO_4^-{}_{(aq)} + H_2O_{(\ell)} \rightleftharpoons H_3O^+_{(aq)} + SO_4^{2-}{}_{(aq)}$$

Sulfuric acid is a far stronger acid than the hydrogen sulfate ion, because much more energy is required to remove a proton from a negatively charged ion. *The strength of a polyprotic acid decreases as the number of hydrogen atoms that have dissociated increases.*

Strong bases are confined to the oxides and hydroxides from Groups 1 (IA) and 2 (IIA).

Strong Bases

- all oxides and hydroxides of the alkali metals: for example, sodium hydroxide, NaOH, and potassium hydroxide, KOH
- alkaline earth (Group 2 (IIA)) metal oxides and hydroxides below beryllium: for example, calcium hydroxide, $Ca(OH)_2$, and barium hydroxide, $Ba(OH)_2$

The strong basic oxides have metal atoms with low electronegativity. Thus, the bond to oxygen is ionic and is relatively easily broken by the attraction of polar water molecules. The oxide ion always reacts with water molecules to produce hydroxide ions.

$$O^{2-}{}_{(aq)} + H_2O_{(\ell)} \rightarrow 2OH^-{}_{(aq)}$$

Magnesium oxide and magnesium hydroxide are not very soluble. They are strong bases, however, because the small amount that does dissolve dissociates almost completely into ions. Beryllium oxide is a weak base. (It is the exception in Group 2 (IIA).) It is a relatively small atom, so the bond to oxygen is strong and not easily broken by water molecules.

Calculations That Involve Strong Acids and Bases

When a strong acid dissociates completely into ions in water, the concentration of $H_3O^+_{(aq)}$ is equal to the concentration of the strong acid. Similarly, when a strong base dissociates completely in water, the concentration of $OH^-_{(aq)}$ is equal to the concentration of the strong base.

Sample Problem

Calculating Ion Concentrations in Acidic and Basic Solutions

Problem

During an experiment, a student pours 25.0 mL of 1.40 mol/L nitric acid into a beaker that contains 15.0 mL of 2.00 mol/L sodium hydroxide solution. Is the resulting solution acidic or basic? What is the concentration of the ion that causes the solution to be acidic or basic?

What Is Required?

You must determine the ion in excess and its concentration.

What Is Given?

You have the following data:
Volume of nitric acid = 25.0 mL
$[HNO_3] = 1.40$ mol/L
Volume of sodium hydroxide = 15.0 mL
$[NaOH] = 2.00$ mol/L

Plan Your Strategy

Step 1 Write the chemical equation for the reaction.

Step 2 Calculate the amount of each reactant using the following equation.
Amount (in mol) = Concentration (in mol/L) × Volume (in L)

Step 3 Determine the limiting reactant.

Step 4 The reactant in excess is a strong acid or base. Thus, the excess amount results in the same amount of H_3O^+ or OH^-.

Step 5 Calculate the concentration of the excess ion by using the amount in excess and the total volume of the solution.

Act on Your Strategy

Step 1 $HNO_{3(aq)} + NaOH_{(aq)} \rightarrow NaNO_{3(aq)} + H_2O_{(\ell)}$

Step 2
$$\text{Amount of } HNO_3 = 1.40 \text{ mol/L} \times 0.0250 \text{ L} = 0.0350 \text{ mol}$$
$$\text{Amount of NaOH} = 2.00 \text{ mol/L} \times 0.0150 \text{ L} = 0.0300 \text{ mol}$$

Step 3 The reactants combine in a 1:1 ratio. The amount of NaOH is less, so this reactant must be the limiting reactant.

Step 4
$$\text{Amount of excess } HNO_{3(aq)} = 0.0350 \text{ mol} - 0.0300 \text{ mol} = 0.005\,0 \text{ mol}$$

Therefore, the amount of $H_3O^+_{(aq)}$ is 5.0×10^{-3} mol.

Step 5 Total volume of solution = 25.0 mL + 15.0 mL = 40.0 mL

$$[H_3O^+] = \frac{5.0 \times 10^{-3} \text{ mol}}{0.0400 \text{ L}} = 0.12 \text{ mol/L}$$

Continued ...

Continued ...

The solution is acidic, and $[H_3O^+]$ is 0.12 mol/L.

Check Your Solution

The chemical equation has a 1:1 ratio between reactants. The amount of acid is greater than the amount of base. Therefore, the resulting solution should be acidic, which it is.

Practice Problems

5. Calculate the concentration of hydronium ions in each solution.

(a) 4.5 mol/L $HCl_{(aq)}$

(b) 30.0 mL of 4.50 mol/L $HBr_{(aq)}$ diluted to 100.0 mL

(c) 18.6 mL of 2.60 mol/L $HClO_{4(aq)}$ added to 24.8 mL of 1.92 mol/L $NaOH_{(aq)}$

(d) 17.9 mL of 0.175 mol/L $HNO_{3(aq)}$ added to 35.4 mL of 0.0160 mol/L $Ca(OH)_{2(aq)}$

6. Calculate the concentration of hydroxide ions in each solution.

(a) 3.1 mol/L $KOH_{(aq)}$

(b) 21.0 mL of 3.1 mol/L KOH diluted to 75.0 mL

(c) 23.2 mL of 1.58 mol/L $HCl_{(aq)}$ added to 18.9 mL of 3.50 mol/L $NaOH_{(aq)}$

(d) 16.5 mL of 1.50 mol/L $H_2SO_{4(aq)}$ added to 12.7 mL of 5.50 mol/L $NaOH_{(aq)}$

7. Determine whether reacting each pair of solutions results in an acidic solution or a basic solution. Then calculate the concentration of the ion that causes the solution to be acidic or basic. (Assume that the volumes in part (a) are additive. Assume that the volumes in part (b) stay the same.)

(a) 31.9 mL of 2.75 mol/L $HCl_{(aq)}$ added to 125 mL of 0.0500 mol/L $Mg(OH)_{2(aq)}$

(b) 4.87 g of $NaOH_{(s)}$ added to 80.0 mL of 3.50 mol/L $HBr_{(aq)}$

8. 2.75 g of $MgO_{(s)}$ is added to 70.0 mL of 2.40 mol/L $HNO_{3(aq)}$. Is the solution that results from the reaction acidic or basic? What is the concentration of the ion that is responsible for the character of the solution?

Section Summary

Strong acids and bases (and strong electrolytes) dissociate completely in water. Therefore, you can use the concentrations of these compounds to determine the concentrations of the ions they form in aqueous solutions. You cannot, however, use the concentrations of weak acids, bases, and electrolytes in the same way. Their solutions contain some particles that have not dissociated into ions. Nevertheless, important changes in $[H_3O^+]$ and $[OH^-]$ take place because dissolved ions affect the dissociation of water.

In the next section, you will focus on the equilibrium of water. You will discover how the pH scale is related to the concentrations of the ions that form when water dissociates. As well, you will learn how to calculate the pH values of solutions of weak acids and bases.

Section Review

1. K/U Phosphoric acid, $H_3PO_{4(aq)}$ is triprotic. It has three hydrogen ions that may be dissociated.
   (a) Write an equation to show the dissociation of each proton.
   (b) Show that $H_2PO_4^-{}_{(aq)}$ can act as either an acid or a base.
   (c) Which is the stronger acid, $H_3PO_{4(aq)}$ or $H_2PO_4^-{}_{(aq)}$? Explain your answer.

2. K/U Para-aminobenzoic acid (PABA) is a weak monoprotic acid that is used in some sunscreen lotions. Its formula is $C_6H_4NH_2COOH$. What is the formula of the conjugate base of PABA?

3. K/U Boric acid, $B(OH)_{3(aq)}$, is used as a mild antiseptic in eye-wash solutions. The following reaction takes place in aqueous solution.
   $B(OH)_{3(aq)} + 2H_2O_{(\ell)} \rightleftharpoons B(OH)_4^-{}_{(aq)} + H_3O^+{}_{(aq)}$
   (a) Identify the conjugate acid-base pairs.
   (b) Is boric acid strong or weak? How do you know?

4. K/U Classify each compound as a strong acid, weak acid, strong base, or weak base.
   (a) butyric acid, $CH_3CH_2CH_2COOH$ (responsible for the odour of rancid butter)
   (b) hydroiodic acid, $HI_{(aq)}$ (added to some cough syrups)
   (c) potassium hydroxide, KOH (used in the manufacture of soft soaps)
   (d) red iron oxide, Fe_2O_3 (used as a colouring pigment in paints)

5. C Distinguish between a concentrated solution of a weak base, and a dilute solution of a strong base. Give an example of each.

8.2 The Equilibrium of Weak Acids and Bases

In this section, you will

- **define** and **perform** calculations that involve the ion product constant for water, K_w, and the acid dissociation constant, K_a
- **compare** strong acids and bases in terms of equilibrium
- **compare** weak acids and bases in terms of equilibrium
- **communicate** your understanding of the following terms: *ion product constant for water (K_w), pH, pOH, acid dissociation constant (K_a), percent dissociation*

The dissociation of an acidic or basic compound in aqueous solution produces ions that interact with water. The pH of the aqueous solution is determined by the position of equilibrium in reactions between the ions that are present in solution and the water molecules. Pure water contains a few ions, produced by the dissociation of water molecules:

$$2H_2O_{(\ell)} \rightleftharpoons H_3O^+_{(aq)} + OH^-_{(aq)}$$

At 25°C, only about two water molecules in one billion dissociate. This is why pure water is such a poor conductor of electricity. In neutral water, at 25°C, the concentration of hydronium ions is the same as the concentration of hydroxide ions: 1.0×10^{-7} mol/L. These concentrations must be the same because the dissociation of water produces equal numbers of hydronium and hydroxide ions. Because this is an equilibrium reaction, and because the position of equilibrium of all reactions changes with temperature, $[H_3O^+]$ is not 1.0×10^{-7} mol/L at other temperatures. The same is true of $[OH^-]$.

The Ion Product Constant for Water

The equilibrium constant, K_c, for the dissociation of water is given by the following expression.

$$K_c = \frac{[H_3O^+][OH^-]}{[H_2O]^2}$$

So few ions form that the concentration of water is essentially constant. The product $K_c[H_2O]^2$ is equal to the product of the concentrations of hydronium ions and hydroxide ions. The equilibrium value of the concentration ion product $[H_3O^+][OH^-]$ at 25°C is called the **ion product constant for water**. It is given the symbol $\boldsymbol{K_w}$.

$$\begin{aligned} K_c[H_2O]^2 &= [H_3O^+][OH^-] \\ &= 1.0 \times 10^{-7} \text{ mol/L} \times 1.0 \times 10^{-7} \text{ mol/L} \\ &= 1.0 \times 10^{-14} \\ &= K_w \end{aligned}$$

The units are commonly dropped, as in other equilibrium expressions you have encountered.

The concentration of H_3O^+ in the solution of a strong acid is equal to the concentration of the dissolved acid, unless the solution is very dilute. Consider $[H_3O^+]$ in a solution of 0.1 mol/L hydrochloric acid. All the molecules of HCl dissociate in water, forming a hydronium ion concentration that equals 0.1 mol/L. The increased $[H_3O^+]$ pushes the dissociation reaction between water molecules to the left, in accordance with Le Châtelier's principle. Consequently, the concentration of hydronium ions that results from the dissociation of water is even less than 1×10^{-7} mol/L. This $[H_3O^+]$ is negligible compared with the 0.1 mol/L concentration of the hydrochloric acid. Unless the solution is very dilute (about 1×10^{-7} mol/L), the dissociation of water molecules can be ignored when determining $[H_3O^+]$ of a strong acid.

Similarly, the concentration of hydroxide ions can be determined from the concentration of the dissolved base. If the solution is a strong base, you can ignore the dissociation of water molecules when determining $[OH^-]$, unless the solution is very dilute. When either $[H_3O^+]$ or $[OH^-]$ is known, you can use the ion product constant for water, K_w, to determine the concentration of the other ion. Although the value of K_w for water is 1.0×10^{-14} at 25°C only, you can use this value unless another value is given for a different temperature.

$[H_3O^+]$ and $[OH^-]$ in Aqueous Solutions at 25°C

In an acidic solution, $[H_3O^+]$ is greater than 1.0×10^{-7} mol/L and $[OH^-]$ is less than 1.0×10^{-7} mol/L.

In a neutral solution, both $[H_3O^+]$ and $[OH^-]$ are equal to 1.0×10^{-7} mol/L.

In a basic solution, $[H_3O^+]$ is less than 1.0×10^{-7} mol/L and $[OH^-]$ is greater than 1.0×10^{-7} mol/L.

Sample Problem

Determining $[H_3O^+]$ and $[OH^-]$

Problem

Find $[H_3O^+]$ and $[OH^-]$ in each solution.

(a) 2.5 mol/L nitric acid

(b) 0.16 mol/L barium hydroxide

Solution

You know that nitric acid is a strong acid and barium hydroxide is a strong base. Since both dissociate completely in aqueous solutions, you can use their molar concentrations to determine $[H_3O^+]$ or $[OH^-]$. You can find the concentration of the other ion using K_w:

$$K_w = 1.0 \times 10^{-14} = [H_3O^+][OH^-]$$

(a) $[HNO_3] = 2.5$ mol/L, so $[H_3O^+] = 2.5$ mol/L

$$[OH^-] = \frac{1.0 \times 10^{-14}\ \text{mol/L}}{2.5} = 4.0 \times 10^{-15}\ \text{mol/L}$$

(b) $Ba(OH)_2 \xrightarrow{H_2O} Ba^{2+}{}_{(aq)} + 2OH^-{}_{(aq)}$

Each mole of $Ba(OH)_2$ in solution forms two moles of OH^- ions.

$\therefore [OH^-] = 2 \times 0.16 = 0.32$ mol/L

$$[H_3O^+] = \frac{1.0 \times 10^{-14}\ \text{mol/L}}{0.32} = 3.1 \times 10^{-14}\ \text{mol/L}$$

Check Your Solution

For a solution of a strong acid, as in part (a), $[H_3O^+]$ should be greater than 1.0×10^{-14} and $[OH^-]$ should be less than 1.0×10^{-14}. For a solution of strong base, $[OH^-]$ should be greater than, and $[H_3O^+]$ should be less than, 1.0×10^{-14}.

CONCEPT CHECK

Sulfuric acid is the only common strong diprotic acid. Explain why the concentration of hydronium ions in a solution of 1.0 mol/L $H_2SO_{4(aq)}$ is 1.0 mol/L, not 2.0 mol/L.

Practice Problems

9. Determine $[H_3O^+]$ and $[OH^-]$ in each solution.

(a) 0.45 mol/L hydrochloric acid

(b) 1.1 mol/L sodium hydroxide

10. Determine $[H_3O^+]$ and $[OH^-]$ in each solution.

(a) 0.95 mol/L hydrobromic acid

(b) 0.012 mol/L calcium hydroxide

11. $[OH^-]$ is 5.6×10^{-14} mol/L in a solution of hydrochloric acid. What is the molar concentration of the $HCl_{(aq)}$?

12. $[H_3O^+]$ is 1.7×10^{-14} in a solution of calcium hydroxide. What is the molar concentration of the $Ca(OH)_{2(aq)}$?

pH and pOH

You can describe the acidity of an aqueous solution quantitatively by stating the concentration of the hydronium ions that are present. $[H_3O^+]$ is often, however, a very small number. The pH scale was devised by a Danish biochemist named Søren Sørensen as a convenient way to represent acidity (and, by extension, basicity). The scale is logarithmic, based on 10. Think of the letter *p* as a mathematical operation representing *–log*. The **pH** of a solution is the exponential *p*ower of *h*ydrogen (or *h*ydronium) ions, in moles per litre. It can therefore be expressed as follows:

$$pH = -\log[H_3O^+]$$

The *practical* range of the pH scale, shown in Figure 8.7, is from 0 to 14. A solution of a strong acid that is more concentrated than 1.0 mol/L would give a negative pH. Since you can determine $[H_3O^+]$ of such solutions directly from the concentration of the acid, the pH scale offers no advantage. Similarly, the pH of a strong base that is more concentrated than 1.0 mol/L is greater than 14. Note that pH is a dimensionless quantity. In other words, it has no units.

You can calculate the **pOH** (the power of hydroxide ions) of a solution from the $[OH^-]$.

$$pOH = -\log[OH^-]$$
$$K_w = [H_3O^+][OH^-] = 1.0 \times 10^{-14} \text{ at } 25^\circ C$$
$$\therefore pH + pOH = 14$$

Web LINK

www.mcgrawhill.ca/links/chemistry12

Many people, for both personal and professional reasons, rely on pH meters to provide quick, reliable pH measurements. Use the Internet to find out how a pH meter works, and what jobs or tasks it is used for. To start your research, go to the web site above and click on **Web Links**. Prepare a brief report, a web page, or a brochure to present your findings.

Sample Problem

Calculating pH and pOH

Problem

A liquid shampoo has a hydroxide ion concentration of 6.8×10^{-5} mol/L at 25°C.

(a) Is the shampoo acidic, basic, or neutral?

(b) Calculate the hydronium ion concentration.

(c) What is the pH and the pOH of the shampoo?

Math LINK

Prove the relationship pH + pOH = 14 as follows. Record the ion product equation and its value at 25°C. Take the logarithm of both sides. Then reverse the sign of each term. What is your result?

Solution

(a) Compare $[OH^-]$ in the shampoo with $[OH^-]$ in neutral water at 25°C.

$[OH^-] = 6.8 \times 10^{-5}$ mol/L, which is greater than 1×10^{-7} mol/L. Therefore, the shampoo is basic.

(b) Use the equation $[H_3O^+] = \dfrac{1.0 \times 10^{-14}}{[OH^-]}$ to find the hydronium ion concentration.

$$[H_3O^+] = \frac{1.0 \times 10^{-14}}{6.8 \times 10^{-5}}$$
$$= 1.5 \times 10^{-10} \text{ mol/L}$$

(c) Substitute known values into the equations $pH = -\log[H_3O^+]$ and $pOH = -\log[OH^-]$.

$$pH = -\log(1.5 \times 10^{-10})$$
$$= 9.83$$
$$pOH = -\log(6.8 \times 10^{-5})$$
$$= 4.17$$

Check Your Solution

$pH + pOH = 14$

PROBLEM TIP

When you work with logarithms, *the number of significant digits in a number must equal the number of digits after the decimal in the number's logarithm.* Here **1.5** × 10^{-10} has two significant digits. Therefore, the calculated pH, 9.**83**, must have two significant digits after the decimal.

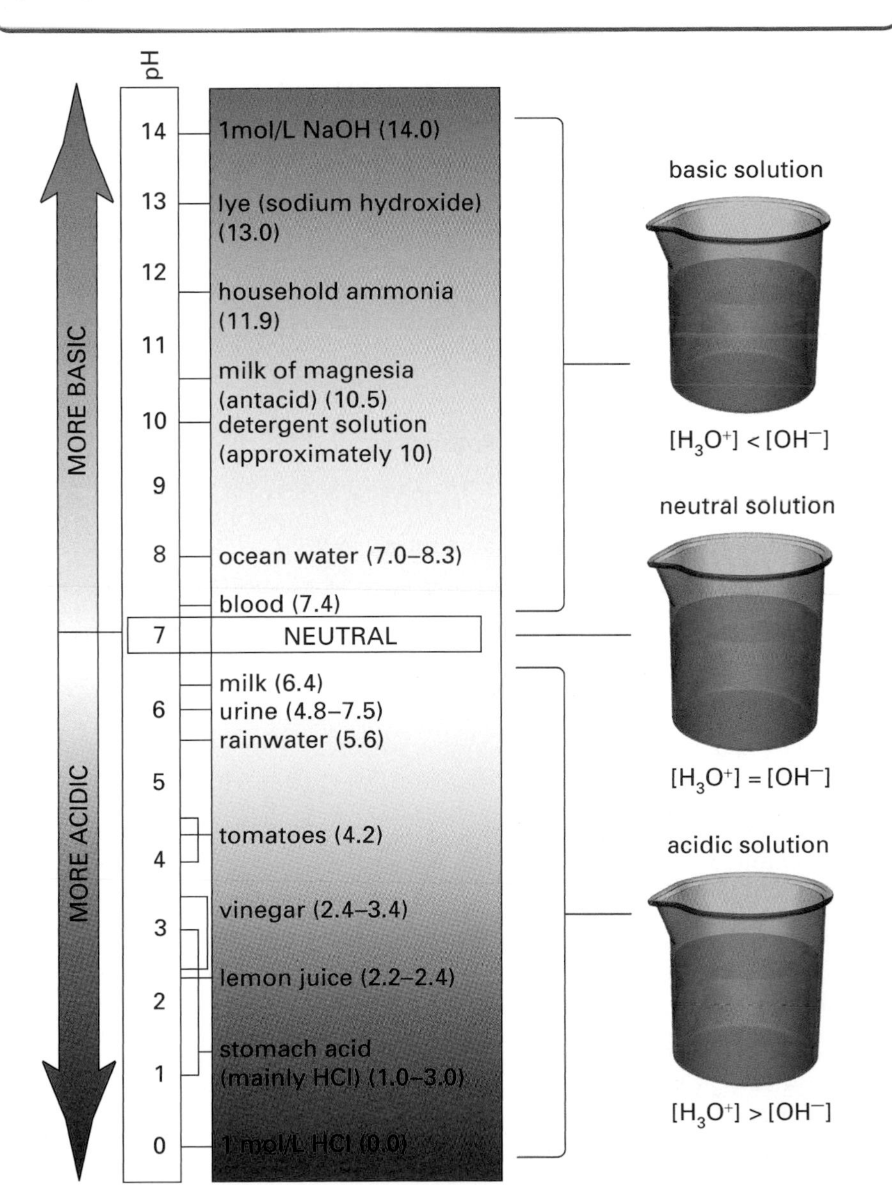

Figure 8.7 **The pH scale is logarithmic. Each change by one unit on the scale represents a change of 10 in the hydronium ion concentration of a solution.**

Another Way to Find $[H_3O^+]$ and $[OH^-]$

You can calculate $[H_3O^+]$ or $[OH^-]$ by finding the *antilog* of the pH or pOH.

$$[H_3O^+] = 10^{-pH}$$
$$[OH^-] = 10^{-pOH}$$

If you are using a calculator, you can use it to find the antilog of a number in one of two ways. If the logarithm is entered in the calculator, you can press the two keys INV and LOG in sequence. (Some calculators may have a 10^x button instead.) Alternatively, since $[H_3O^+] = 10^{-pH}$ and $[OH^-] = 10^{-pOH}$, you can enter 10, press the y^x button, enter the negative value of pH (or pOH), and then press =.

Sample Problem

Finding pOH, $[H_3O^+]$, and $[OH^-]$

Problem

If the pH of urine is outside the normal range of values, this can indicate medical problems. Suppose that the pH of a urine sample was measured to be 5.53 at 25°C. Calculate pOH, $[H_3O^+]$, and $[OH^-]$ for the sample.

Solution

You use the known value, pH = 5.53, to calculate the required values.

$$\begin{aligned} pOH &= 14.00 - 5.53 \\ &= 8.47 \end{aligned}$$

$$\begin{aligned} [H_3O^+] &= 10^{-5.53} \\ &= 3.0 \times 10^{-6} \text{ mol/L} \end{aligned}$$

$$\begin{aligned} [OH^-] &= 10^{-8.47} \\ &= 3.4 \times 10^{-9} \text{ mol/L} \end{aligned}$$

Check Your Solution

In this problem, the ion product constant is a useful check:

$$\begin{aligned} [H_3O^+][OH^-] &= (3.0 \times 10^{-6}) \times (3.4 \times 10^{-9}) \\ &= 1.0 \times 10^{-14} \end{aligned}$$

This value equals the expected value for K_w at 25°C.

Practice Problems

13. $[H_3O^+]$ of a sample of milk is found to be 3.98×10^{-7} mol/L. Is the milk acidic, neutral, or basic? Calculate the pH and $[OH^-]$ of the sample.

14. A sample of household ammonia has a pH of 11.9. What is the pOH and $[OH^-]$ of the sample?

15. Phenol, C_6H_5OH, is used as a disinfectant. An aqueous solution of phenol was found to have a pH of 4.72. Is phenol acidic, neutral, or basic? Calculate $[H_3O^+]$, $[OH^-]$, and pOH of the solution.

16. At normal body temperature, 37°C, the value of K_w for water is 2.5×10^{-14}. Calculate $[H_3O^+]$ and $[OH^-]$ at this temperature. Is pure water at 37°C acidic, neutral, or basic?

17. A sample of baking soda was dissolved in water and the pOH of the solution was found to be 5.81 at 25°C. Is the solution acidic, basic, or neutral? Calculate the pH, $[H_3O^+]$, and $[OH^-]$ of the solution.

18. A chemist dissolved some Aspirin™ in water. The chemist then measured the pH of the solution and found it to be 2.73 at 25°C. What are $[H_3O^+]$ and $[OH^-]$ of the solution?

The Acid Dissociation Constant

Many common foods (such as citrus fruits), pharmaceuticals (such as Aspirin™), and some vitamins (such as niacin, vitamin B3) are weak acids. When a weak acid dissolves in water, it does not completely dissociate. The concentration of the hydronium ions, and the concentration of the conjugate base of the acid that is formed in solution, depend on the initial concentration of the acid and the amount of acid that dissociates.

You can represent any weak monoprotic acid with the general formula HA. The equilibrium of a weak monoprotic acid in aqueous solution can be expressed as follows:

$$HA_{(aq)} + H_2O_{(aq)} \rightleftharpoons H_3O^+{}_{(aq)} + A^-{}_{(aq)}$$

The equilibrium expression for this reaction is

$$K_c = \frac{[H_3O^+][A^-]}{[HA][H_2O]}$$

In dilute solutions, the concentration of water is almost constant. Multiplying both sides of the equilibrium expression by $[H_2O]$ gives the product of two constants on the left side. This new constant is called the **acid dissociation constant, K_a**. (Some chemists refer to the acid dissociation constant as the *acid ionization constant*. With either name, the symbol is K_a.)

$$K_c[H_2O] = K_a = \frac{[H_3O^+][A^-]}{[HA]}$$

You can determine the value of K_a for a particular acid by measuring the pH of a solution. In the following investigation, you will add sodium hydroxide to acetic acid, which is a weak acid. (See Figure 8.8.) By graphing pH against the volume of sodium hydroxide that you added, you will be able to calculate the concentration of the acetic acid. Then you will be able to determine the acid dissociation constant, K_a, for this acid.

$$CH_3COOH + H_2O \rightleftharpoons CH_3COO^- + H_3O^+$$

Figure 8.8 Acetic acid is a weak monoprotic acid.

Niacin is found in many foods, including corn. The niacin in corn, however, cannot be absorbed in the intestinal tract. In regions of the world where corn is a major part of the diet, niacin deficiency can occur. If you add calcium oxide or wood ash to the water in which you boil corn, the resulting basic solution allows the niacin to be absorbed. The flour for making corn tortillas is commonly prepared using this method.

SKILL FOCUS

Predicting

Performing and recording

Analyzing and interpreting

Investigation 8-A

K_a of Acetic Acid

In your previous chemistry course, you learned how to determine the molar concentration of an acid by adding a basic solution of known concentration and measuring the volume of the basic solution required to reach the endpoint. This procedure is called a titration. The endpoint is the point at which an indicator changes colour.

In this investigation, you will be given a sample of acetic acid with an unknown concentration. Instead of measuring the volume of the basic solution required to reach the endpoint, however, you will measure the pH. Then you will graph the data you collected and use the graph to calculate the molar concentration of the acetic acid and its K_a.

Question

In a solution of acetic acid, how does the concentration of hydronium ions compare with the concentration of acetic acid?

Materials

25 mL pipette and pipette bulb
retort stand
burette and burette clamp
2 beakers (150 mL)
Erlenmeyer flask (150 mL)
labels
meniscus reader
sheet of white paper
funnel
acetic acid, CH_3COOH, solution
sodium hydroxide, NaOH, solution
dropper bottle containing phenolphthalein
pH meter or pH paper

Safety Precautions

Both CH_3COOH and NaOH are corrosive. Wash any spills on your skin or clothing with plenty of cool water. Inform your teacher immediately.

Procedure

1. Your teacher will give you the molar concentration of the NaOH solution. Record this concentration in your notebook, as well as the volume of the pipette (in mL).
2. Copy the table below into your notebook, to record your observations. Leave plenty of space. You will collect 15 to 30 sets of data, depending on the concentration of the acid.

Volume of NaOH added (mL)	pH
0.00	

3. Label a clean, dry beaker for each liquid. Obtain about 40 mL of acetic acid and about 70 mL of NaOH solution.
4. Rinse a clean burette with about 10 mL of NaOH solution. Discard the rinse with plenty of water. Then set up a retort stand, burette clamp, meniscus reader, and funnel. Fill the burette with NaOH solution. Make sure that the solution fills the tube below the tap with no air bubbles. Remove the funnel.

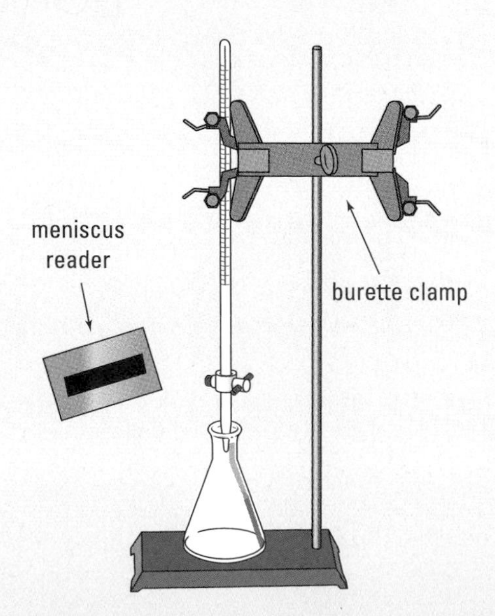

5. Obtain a clean 25 mL pipette and a suction bulb. Rinse the pipette with 5 mL to 10 mL of CH_3COOH, and discard the rinse with plenty of water. Pipette 25.00 mL of CH_3COOH into an Erlenmeyer flask. Add two or three drops of phenolphthalein indicator.

6. Record the initial pH of the solution. Make sure that the glass electrode is immersed deeply enough to get an accurate reading. If necessary, tip the flask to one side. Place a sheet of white paper under the flask.

7. Add 2 mL of NaOH from the burette. Record the volume carefully, correct to two decimal places. Swirl the contents of the Erlenmeyer flask, then measure the pH of the solution.

8. Repeat step 7 until the pH rises above 2.0. Add 1 mL amounts until the pH reaches 5.0.

9. Above pH = 5.0, add NaOH in 0.2 mL or 0.1 mL portions. Continue to swirl the contents of the flask and take pH readings. Record the volume at which the phenolphthalein changes from colourless to pink.

10. Above pH = 11, add 1 mL portions until the pH reaches at least 12.

11. Wash the liquids down the sink with plenty of water. Rinse the pipette and burette with distilled water. Leave the burette tap open.

Analysis

1. Write the chemical equation for the neutralization reaction you observed.

2. Plot a graph of your data, with pH on the vertical axis and volume of NaOH on the horizontal axis. Your graph should show a steep rise in pH as the volume of NaOH becomes enough to neutralize all the CH_3COOH. Take the midpoint on the graph (where the graph rises steeply) and read off the volume of NaOH. This is the volume of NaOH that was needed to neutralize all the CH_3COOH. Compare the volume on your graph with the volume you recorded when the phenolphthalein indicator first turned pink.

3. Calculate the molar concentration of the CH_3COOH. Use the ratio in which the acid and base react, determined from the chemical equation. You can use the following equation to find the amount of a chemical in solution.

   Amount (in mol) = Concentration (in mol/L) × Volume (in L)

   Determine the amount of NaOH added, using its concentration (given by your teacher) and the volume on your graph (from question 2).

4. Write the expression for K_a for the dissociation of CH_3COOH in water.

5. Use the initial pH of the CH_3COOH (before you added any base) to find the initial $[H_3O^+]$. What was the initial $[CH_3COO^-]$?

6. Assume that the amount of CH_3COOH that dissociates is small compared with the initial concentration of the acid. If this is true, the equilibrium value of $[CH_3COOH]$ is equal to the initial concentration of the acid. Use your values of $[H_3O^+]$, $[CH_3COO^-]$, and $[CH_3COOH]$ to calculate K_a for acetic acid. **Hint:** $[H_3O^+] = [CH_3COO^-]$

7. Refer to the volume of NaOH on your graph (from question 2). Calculate half this volume. On your graph, find the pH when the solution was half-neutralized.

8. Calculate $[H_3O^+]$ when the CH_3COOH was half-neutralized. How does this value compare with your value of K_a for CH_3COOH?

Conclusion

9. Calculate the percent difference between your value for K_a and the accepted value. State two sources of error that might account for any differences.

Application

10. Do the values you calculated for $[H_3O^+]$ and $[CH_3COOH]$ prove that CH_3COOH is a weak acid? Explain.

pH and K_a of a Weak Acid

Table 8.2 lists the acid dissociation constants for selected acids at 25°C. Notice that weak acids have K_a values that are between 1 and about 1×10^{-16}. Very weak acids have K_a values that are less than 1×10^{-16}. The smaller the value of K_a, the less the acid ionizes in aqueous solution.

Problems that involve the concentrations of ions formed in aqueous solutions are considered to be equilibrium problems. The steps for solving acid and base equilibrium problems are similar to the steps you learned in Chapter 7 for solving equilibrium problems.

Solving Equilibrium Problems That Involve Acids and Bases

The steps that you will use to solve acid and base equilibrium problems will vary depending on the problem. Below are a few general steps to guide you.

- Write the chemical equation. Use the chemical equation to set up an ICE table for the reacting substances. Enter any values that are given in the problem. (**Note:** For the problems in this textbook, you can assume that the concentrations of hydronium ions and hydroxide ions in pure water are negligible compared with the concentrations of these ions when a weak acid or weak base is dissolved in water.)
- Let x represent the change in concentration of the substance with the smallest coefficient in the chemical equation.
- For problems that give the initial concentration of the acid, [HA], compare the initial concentration of the acid with the acid dissociation constant, K_a.
- If $\frac{[HA]}{K_a} > 500$, the change in the initial concentration, x, is negligible and can be ignored.
- If $\frac{[HA]}{K_a} < 500$, the change in the initial concentration, x, may not be negligible. The equilibrium equation will be more complex, possibly requiring the solution of a quadratic equation.

Table 8.2 Some Acid Dissociation Constants for Weak Acids at 25°C

Acid	Formula	Acid dissociation constant, K_a
acetic acid	CH_3COOH	1.8×10^{-5}
chlorous acid	$HClO_2$	1.1×10^{-2}
formic acid	$HCOOH$	1.8×10^{-4}
hydrocyanic acid	HCN	6.2×10^{-10}
hydrofluoric acid	HF	6.6×10^{-4}
hydrogen oxide (water)	H_2O	1.0×10^{-14}
lactic acid	$CH_3CHOHCOOH$	1.4×10^{-4}
nitrous acid	HNO_2	7.2×10^{-4}
phenol	C_6H_5OH	1.3×10^{-10}

Percent Dissociation

The **percent dissociation** of a weak acid is the fraction of acid molecules that dissociate compared with the initial concentration of the acid, expressed as a percent. (Some chemists refer to percent dissociation as *percent of dissociation*.) The percent dissociation depends on the value of K_a for the acid, as well as the initial concentration of the weak acid. The following Sample Problems show how to solve problems that involve percent dissociation.

Sample Problem

Determining K_a and Percent Dissociation

Problem

Propanoic acid, CH_3CH_2COOH, is a weak monoprotic acid that is used to inhibit mould formation in bread. A student prepared a 0.10 mol/L solution of propanoic acid and found that the pH was 2.96. What is the acid dissociation constant for propanoic acid? What percent of its molecules were dissociated in the solution?

What Is Required?

You need to find K_a and the percent dissociation for propanoic acid.

What Is Given?

You have the following data:

Initial $[CH_3CH_2COOH] = 0.10$ mol/L

pH = 2.96

Plan Your Strategy

Step 1 Write the equation for the dissociation equilibrium of propanoic acid in water. Then set up an ICE table.

Step 2 Write the equation for the acid dissociation constant. Substitute equilibrium terms into the equation.

Step 3 Calculate $[H_3O^+]$ using $[H_3O^+] = 10^{-pH}$

Step 4 Use the stoichiometry of the equation and $[H_3O^+]$ to substitute for the unknown term, x, and calculate K_a.

Step 5 Calculate the percent dissociation by expressing the fraction of molecules that dissociate out of 100.

Act on Your Strategy

Step 1 Use the equation for the dissociation equilibrium of propanoic acid in water to set up an ICE table.

Concentration (mol/L)	$CH_3CH_2COOH_{(aq)}$	+	$H_2O_{(\ell)}$	$\rightleftharpoons$	$CH_3CH_2COO^-_{(aq)}$	+	$H_3O^+_{(aq)}$
Initial	0.10				0		~0
Change	$-x$				$+x$		$+x$
Equilibrium	$0.10 - x$				$+x$		$+x$

Continued ...

Continued ...

Step 2 $K_a = \frac{[CH_3CH_2COO^-][H_3O^+]}{CH_3CH_2COOH]}$

$= \frac{(x)(x)}{(0.10 - x)}$

Step 3 The value of x is equal to $[H_3O^+]$ and $[CH_3CH_2COOH]$.

$[H_3O^+] = 10^{-2.96}$

$= 1.1 \times 10^{-3}$ mol/L

Step 4 $K_a = \frac{(1.1 \times 10^{-3})^2}{0.10 - (1.1 \times 10^{-3})}$

$= 1.2 \times 10^{-5}$

Step 5 Percent dissociation $= \frac{1.1 \times 10^{-3} \text{ mol/L}}{0.10 \text{ mol/L}} \times 100$

$= 1.1\%$

Check Your Solution

The value of K_a and the percent dissociation are reasonable for a weak acid.

Sample Problem

Calculating pH

Problem

Formic acid, HCOOH, is present in the sting of certain ants. What is the pH of a 0.025 mol/L solution of formic acid?

What Is Required?

You need to calculate the pH of the solution.

What Is Given?

You know the concentration of formic acid:
$[HCOOH] = 0.025$ mol/L

The acid dissociation constant for formic acid is listed in Table 8.2:
$K_a = 1.8 \times 10^{-4}$

Plan Your Strategy

Step 1 Write the equation for the dissociation equilibrium of formic acid in water. Then set up an ICE table.

Step 2 Write the equation for the acid dissociation constant. Substitute equilibrium terms into the equation.

Step 3 Check the value of $\frac{[HCOOH]}{K_a}$ to see whether or not the amount that dissociates is negligible compared with the initial concentration of the acid.

Step 4 Solve the equation for x. If the amount that dissociates is not negligible compared with the initial concentration of acid, you will need to use a quadratic equation.

Step 5 $pH = -\log [H_3O^+]$

Act on Your Strategy

Step 1

Concentration (mol/L)	$HCOOH_{(aq)}$ + $H_2O_{(\ell)}$	$\rightleftharpoons$ $HCOO^-_{(aq)}$	+ $H_3O^+_{(aq)}$
Initial	0.025	0	~0
Change	$-x$	$+x$	$+x$
Equilibrium	$0.025 - x$	$+x$	$+x$

Step 2 $K_a = \frac{[HCOO^-][H_3O^+]}{[HCOOH]}$

$= \frac{(x)(x)}{(0.025 - x)}$

$= 1.8 \times 10^{-4}$

Step 3 $\frac{[HCOOH]}{K_a} = \frac{0.025}{1.8 \times 10^{-4}}$

$= 139$

Since this value is less than 500, the amount that dissociates is not negligible compared with the initial concentration of the acid.

Step 4 Rearrange the equation into a quadratic equation.

$$\frac{x^2}{(0.025 - x)} = 1.8 \times 10^{-4}$$

$x^2 + (1.8 \times 10^{-4})x - (4.5 \times 10^{-6}) = 0$

$$x = \frac{-b \pm \sqrt{b^2 - 4ac}}{2a}$$

$$= \frac{-(1.8 \times 10^{-4}) \pm \sqrt{(1.8 \times 10^{-4})^2 - 4 \times 1 \times (-4.5 \times 10^{-6})}}{2 \times 1}$$

$x = 0.0020$ or $x = -0.002$

The negative value is not reasonable, since a concentration term cannot be negative.

$\therefore x = 0.0020 \text{ mol/L} = [H_3O^+]$

Step 5 $pH = -\log 0.0020$

$= 2.70$

The pH of a solution of 0.025 mol/L formic acid is 2.70.

Check Your Solution

The pH indicates an acidic solution, as expected. Data that was given in the problem has two significant digits, and the pH has two digits following the decimal place. It is easy to make a mistake when solving a quadratic equation. You can estimate a solution to this problem, by assuming that $(0.025 - x)$ is approximately equal to 0.025.

$\frac{x^2}{0.025} = 1.8 \times 10^{-4}$

$x^2 = 4.5 \times 10^{-6}$

$x = 2.1 \times 10^{-3} \text{ mol/L}$

This answer is very close to the answer obtained by solving the quadratic equation. Therefore, the solution is probably correct.

Practice Problems

19. Calculate the pH of a sample of vinegar that contains 0.83 mol/L acetic acid. What is the percent dissociation of the vinegar?

20. In low doses, barbiturates act as sedatives. Barbiturates are made from barbituric acid, a weak monoprotic acid that was first prepared by the German chemist Adolph von Baeyer in 1864. The formula of barbituric acid is $C_4H_4N_2O_3$. A chemist prepares a 0.10 mol/L solution of barbituric acid. The chemist finds the pH of the solution to be 2.50. What is the acid dissociation constant for barbituric acid? What percent of its molecules dissociate?

21. A solution of hydrofluoric acid has a molar concentration of 0.0100 mol/L. What is the pH of this solution?

22. Hypochlorous acid, HOCl, is used as a bleach and a germ-killer. A chemist finds that 0.027% of hypochlorous acid molecules are dissociated in a 0.40 mol/L solution of the acid. What is the value of K_a for the acid?

23. The word "butter" comes from the Greek *butyros*. Butanoic acid (common name: butyric acid) gives rancid butter its distinctive odour. Calculate the pH of a 1.0×10^{-2} mol/L solution of butanoic acid ($K_a = 1.51 \times 10^{-5}$).

24. Caproic acid, $C_5H_{11}COOH$, occurs naturally in coconut and palm oil. It is a weak monoprotic acid, with $K_a = 1.3 \times 10^{-5}$. A certain aqueous solution of caproic acid has a pH of 2.94. How much acid was dissolved to make 100 mL of this solution?

Polyprotic Acids

As you know, polyprotic acids have more than one hydrogen atom that dissociates. Each dissociation has a corresponding acid dissociation constant. How can you calculate the pH of a solution of a polyprotic acid?

Problems that involve polyprotic acids can be divided into as many sub-problems as there are hydrogen atoms that dissociate. The ion concentrations that are calculated for the first dissociation are substituted as initial ion concentrations for the second dissociation, and so on. You can see this in the following Sample Problem.

Sample Problem

Calculations That Involve Polyprotic Acids

Problem

Phosphoric acid, H_3PO_4, is one of the world's most important industrial chemicals. It is mainly used to manufacture phosphate fertilizers. It is also the ingredient that gives cola drinks their tart, biting taste. Calculate the pH, $[H_2PO_4^-]$, and $[HPO_4^{2-}]$ of a 3.5 mol/L aqueous solution of phosphoric acid.

What Is Required?

You need to find pH and $[H_2PO_4^-]$.

What Is Given?

You know that $[H_3PO_4] = 3.5$ mol/L. From data tables, you can find $K_{a_1} = 7.0 \times 10^{-3}$ and $K_{a_2} = 6.3 \times 10^{-8}$.

Plan Your Strategy

Step 1 Write the equation for the dissociation equilibrium of phosphoric acid in water. Then set up an ICE table.

Step 2 Write the dissociation equation for K_{a_1}.

Step 3 Determine whether or not the dissociation of H_3PO_4 is negligible, compared with the initial concentration.

Step 4 Solve the equation for x.

Step 5 Write the equation for the dissociation equilibrium of $H_2PO_4^-$ in water. Set up an ICE table using the concentrations you calculated for the first dissociation as initial concentrations here.

Step 6 Write the dissociation equation for K_{a_2}.

Step 7 Determine whether or not the dissociation of $H_2PO_4^-$ is negligible, compared with the initial concentration.

Step 8 Solve the equation for x.

Step 9 Calculate $[H_3O^+]$ and pH.

Step 10 Calculate $[H_2PO_4^-]$.

Act on Your Strategy

Step 1

Concentration (mol/L)	$H_3PO_{4(aq)}$	+	$H_2O_{(\ell)}$	$\rightleftharpoons$	$H_2PO_4^-{}_{(aq)}$	+	$H_3O^+{}_{(aq)}$
Initial	3.5				0		~0
Change	$-x$				$+x$		$+x$
Equilibrium	$3.5 - x$				x		x

Step 2

$$K_{a_1} = \frac{[H_2PO_4^-][H_3O^+]}{[H_3PO_4]}$$
$$= 7.0 \times 10^{-3}$$
$$= \frac{(x)(x)}{(3.5 - x)}$$

Step 3

$$\frac{[H_3PO_4]}{K_{a_1}} = \frac{3.5}{7.0 \times 10^{-3}}$$
$$= 500$$

Therefore, x is probably negligible, compared with 3.5.

Step 4

$$7.0 \times 10^{-3} = \frac{x^2}{3.5}$$
$$x = 0.16 \text{ mol/L}$$

Step 5

Concentration (mol/L)	$H_2PO_4^-{}_{(aq)}$	+	$H_2O_{(\ell)}$	$\rightleftharpoons$	$HPO_4^{2-}{}_{(aq)}$	+	$H_3O^+{}_{(aq)}$
Initial	0.16				0		0.16
Change	$-x$				$+x$		$+x$
Equilibrium	$0.16 - x$				x		$(0.16 + x)$

Continued ...

Step 6 $K_{a_2} = \frac{[HPO_4^{2-}][H_3O^+]}{[H_2PO_4^-]}$

$= \frac{(x)(0.16 + x)}{(0.16 - x)}$

Step 7 $\frac{[H_2PO_4^-]}{K_{a_2}} = \frac{0.16}{6.3 \times 10^{-8}} = 2.5 \times 10^6$

This is much greater than 500, so x is negligible compared with 0.16.

Step 8 $K_{a_2} = 6.3 \times 10^{-8} = \frac{(x)\cancel{(0.16)}}{\cancel{(0.16)}} = x$

$x = [HPO_4^{2-}]$

$= 6.3 \times 10^{-8}$

Step 9 $pH = -\log 0.16$

$= 0.80$

Step 10 $[H_2PO_4^-] = 0.16 - x = 0.16$ (because x is negligible)

Check Your Solution

$[H_3O^+]$ results from the first dissociation of phosphoric acid, so the second dissociation has a negligible effect on the concentration of hydronium ions or the pH of the solution. You would expect this, because the second dissociation is much weaker than the first.

Polyprotic Acids and $[H_3O^+]$

All polyprotic acids, except sulfuric acid, are weak. Their second dissociation is much weaker than their first dissociation. For this reason, when calculating $[H_3O^+]$ and pH of a polyprotic acid, only the first dissociation needs to be considered. The calculation is then the same as the calculation for any weak monoprotic acid. In the Sample Problem, $[HPO_4^{2-}]$ was found to be the same as the second dissociation constant, K_{a_2}. *The concentration of the anions formed in the second dissociation of a polyprotic acid is equal to* K_{a_2}.

The only common strong polyprotic acid is sulfuric acid. Nevertheless, it is strong only for the first dissociation. Like the second dissociation of other polyprotic acids, the second dissociation of sulfuric acid is weak.

$$H_2SO_{4(aq)} + H_2O_{(\ell)} \rightarrow HSO_4^-{}_{(aq)} + H_3O^+{}_{(aq)} \quad 100\% \text{ dissociation}$$

$$HSO_4^-{}_{(aq)} + H_2O_{(\ell)} \rightleftharpoons SO_4^{2-}{}_{(aq)} + H_3O^+{}_{(aq)} \quad K_a = 1.0 \times 10^{-2}$$

$[H_3O^+]$ in a solution of sulfuric acid is equal to the concentration of the acid. Only in dilute solutions (less than 1.0 mol/L) does the second dissociation of sulfuric acid contribute to the hydronium ion concentration.

Practice Problems

25. Carbonated beverages contain a solution of carbonic acid. Carbonic acid is also important for forming the ions that are present in blood.

$$CO_{2(aq)} + H_2O_{(\ell)} \rightleftharpoons H_2CO_{3(aq)}$$

$$H_2CO_{3(aq)} + H_2O_{(\ell)} \rightleftharpoons HCO_3^-{}_{(aq)} + H_3O^+{}_{(aq)} \quad K_{a_1} = 4.5 \times 10^{-7}$$

$$HCO_3^-{}_{(aq)} + H_2O_{(\ell)} \rightleftharpoons CO_3^{2-}{}_{(aq)} + H_3O^+{}_{(aq)} \quad K_{a_2} = 4.7 \times 10^{-11}$$

Calculate the pH of a solution of 5.0×10^{-4} mol/L carbonic acid. What is $[CO_3^{2-}]$ in the solution?

26. Adipic acid is a diprotic acid that is used to manufacture nylon. Its formula can be abbreviated to H_2Ad. The acid dissociation constants for adipic acid are $K_{a_1} = 3.71 \times 10^{-5}$ and $K_{a_2} = 3.87 \times 10^{-6}$. What is the pH of a 0.085 mol/L solution of adipic acid?

27. Hydrosulfuric acid, $H_2S_{(aq)}$, is a weak diprotic acid that is sometimes used in analytical work. It is used to precipitate metal sulfides, which tend to be very insoluble. Calculate the pH and $[HS^-_{(aq)}]$ of a 7.5×10^{-3} mol/L solution.

28. What is the value of K_a when water acts as a Brønsted-Lowry acid? Write the expression for K_{a_2} if water acts as a diprotic acid.

Section Summary

In this section, you learned about the relationship between the pH scale and the concentrations of the ions that form when water and weak acids dissociate. In the next section, you will learn that the equilibrium of weak bases is similar to the equilibrium of weak acids. As you will see, solutions that contain a mixture of a weak acid and a salt of its conjugate base have properties with important biochemical and industrial applications.

Section Review

1 K/U Complete the following table by calculating the missing values and indicating whether each solution is acidic or basic.

$[H_3O^+]$ (mol/L)	pH	$[OH^-]$ (mol/L)	pOH	Acidic or basic?
3.7×10^{-5}	(a)	(b)	(c)	(d)
(e)	10.41	(f)	(g)	(h)
(i)	(j)	7.0×10^{-2}	(k)	(l)
(m)	(n)	(o)	8.9	(p)

2 I Lactic acid, $CH_3CHOHCOOH$, is a monoprotic acid that is produced by muscle activity. It is also produced from milk by the action of bacteria. What is the pH of a 0.12 mol/L solution of lactic acid?

3 I A 0.10 mol/L solution of a weak acid was found to be 5.0% dissociated. Calculate K_a.

4 I Oxalic acid, HOOCCOOH, is a weak diprotic acid that occurs naturally in some foods, including rhubarb. Calculate the pH of a solution of oxalic acid that is prepared by dissolving 2.5 g in 1.0 L of water. What is the concentration of hydrogen oxalate, $HOOCCOO^-$, in the solution?

5 I A sample of blood was taken from a patient and sent to a laboratory for testing. Chemists found that the blood pH was 7.40. They also found that the hydrogen carbonate ion concentration was 2.6×10^{-2} mol/L. What was the concentration of carbonic acid in the blood?

8.3 Bases and Buffers

Section Preview/ Specific Expectations

In this section, you will

- **solve** problems that involve the base dissociation constant, K_b
- **describe** the properties and components of a buffer solution
- **identify** systems in which buffer solutions are found, and **explain** how they function
- **communicate** your understanding of the following terms: *base dissociation constant (K_b), buffer solution, buffer capacity*

Many compounds that are present in plants are weak bases. Caffeine in coffee and piperidine in black pepper are two examples. A weak base, represented by B, reacts with water to form an equilibrium solution of ions.

$$B_{(aq)} + H_2O_{(\ell)} \rightleftharpoons HB^+_{(aq)} + OH^-_{(aq)}$$

The equilibrium expression for this general reaction is given as follows:

$$K_c = \frac{[HB^+][OH^-]}{[B][H_2O]}$$

The concentration of water is almost constant in dilute solutions. Multiplying both sides of the equilibrium expression by $[H_2O]$ gives the product of two constants on the left side. The new constant is called the **base dissociation constant, K_b.**

$$K_c[H_2O] = \frac{[HB^+][OH^-]}{[B]} = K_b$$

Table 8.3 lists the base dissociation constants for several weak bases at 25°C. Nitrogen-containing compounds are Brønsted-Lowry bases, because the lone pair of electrons on a nitrogen atom can bond with H^+ from water. The steps for solving problems that involve weak bases are similar to the steps you learned for solving problems that involve weak acids.

Table 8.3 Some Base Dissociation Constants at 25°C

Base	Formula	Base dissociation constant, K_b
ethylenediamine	$NH_2CH_2CH_2NH_2$	5.2×10^{-4}
dimethylamine	$(CH_3)_2NH$	5.1×10^{-4}
methylamine	CH_3NH_2	4.4×10^{-4}
trimethylamine	$(CH_3)_3N$	6.5×10^{-5}
ammonia	NH_3	1.8×10^{-5}
hydrazine	N_2H_4	1.7×10^{-6}
pyridine	C_5H_5N	1.4×10^{-9}
aniline	$C_6H_5NH_2$	4.2×10^{-10}
urea	NH_2CONH_2	1.5×10^{-14}

Sample Problem

Solving Problems Involving K_b

Problem

The characteristic taste of tonic water is due to the addition of quinine. Quinine is a naturally occurring compound that is also used to treat malaria. The base dissociation constant, K_b, for quinine is 3.3×10^{-6}. Calculate $[OH^-]$ and the pH of a 1.7×10^{-3} mol/L solution of quinine.

What Is Required?

You need to find $[OH^-]$ and pH.

What Is Given?

$K_b = 3.3 \times 10^{-6}$
Concentration of quinine = 1.7×10^{-3} mol/L

Plan Your Strategy

Step 1 Let Q represent the formula of quinine. Write the equation for the equilibrium reaction of quinine in water. Then set up an ICE table.

Step 2 Write the equation for the base dissociation constant. Substitute equilibrium terms into the equation.

Step 3 Calculate the value of $\frac{[Q]}{K_b}$ to determine whether or not the amount of quinine that dissociates is negligible compared with the initial concentration.

Step 4 Solve the equation for x. If the amount that dissociates is not negligible compared with the initial concentration of the base, you will need to use a quadratic equation.

Step 5 $pOH = -\log [OH^-]$
$pH = 14.00 - pOH$

Act on Your Strategy

Step 1

Concentration (mol/L)	$Q_{(aq)}$	+	$H_2O_{(\ell)}$	$\rightleftharpoons$	$HQ^+_{(aq)}$	+	$OH^-_{(aq)}$
Initial	1.7×10^{-3}				0		~0
Change	$-x$				$+x$		$+x$
Equilibrium	$(1.7 \times 10^{-3}) - x$				x		x

Step 2

$$K_b = \frac{[HQ^+][OH^-]}{[Q]}$$

$$3.3 \times 10^{-6} = \frac{(x)(x)}{(1.7 \times 10^{-3}) - x}$$

Step 3

$$\frac{[Q]}{K_b} = \frac{1.7 \times 10^{-3}}{3.3 \times 10^{-6}}$$
$$= 515$$

Since this value is greater than 500, the amount that dissociates is probably negligible, compared with the initial concentration of the base.

Step 4

$$3.3 \times 10^{-6} = \frac{x^2}{1.7 \times 10^{-3}}$$
$$x = \pm 7.5 \times 10^{-5}$$

The negative root is not reasonable.
$\therefore x = 7.5 \times 10^{-5}$ mol/L = $[OH^-]$

Step 5

$$pOH = -\log 7.5 \times 10^{-5}$$
$$= 4.13$$
$$pH = 14.00 - pOH$$
$$\therefore pH = 9.87$$

Check Your Solution

The pH of the solution is greater than 7, as expected for a basic solution.

Sample Problem

Calculating K_b

Problem

Pyridine, C_5H_5N, is used to manufacture medications and vitamins. Calculate the base dissociation constant for pyridine if a 0.125 mol/L aqueous solution has a pH of 9.10.

What Is Required?

You need to find K_b.

What Is Given?

$[C_5H_5N] = 0.125$ mol/L

pH = 9.10

Plan Your Strategy

Step 1 Write the equation for the equilibrium reaction of pyridine in water. Then set up an ICE table.

Step 2 Write the equation for the base dissociation constant. Substitute equilibrium terms into the equation.

Step 3 pOH = 14.0 − pH

Step 4 $[OH^-] = 10^{-pOH}$

Step 5 Substitute for x into the equilibrium equation. Calculate the value of K_b.

Act on Your Strategy

Step 1

Concentration (mol/L)	$C_5H_5N_{(aq)}$	+	$H_2O_{(\ell)}$	$\rightleftharpoons$	$C_5H_5NH^+_{(aq)}$	+	$OH^-_{(aq)}$
Initial	0.125				0		~0
Change	$-x$				$+x$		$+x$
Equilibrium	$0.125 - x$				x		x

Step 2

$$K_b = \frac{[C_5H_5NH^+][OH^-]}{[C_5H_5N]}$$

$$= \frac{(x)(x)}{(0.125 - x)}$$

Step 3 pOH = 14.00 − 9.10 = 4.90

Step 4

$$[OH^-] = 10^{-4.90}$$

$$= 1.3 \times 10^{-5} \text{ mol/L}$$

Step 5 $0.125 - (1.3 \times 10^{-5}) = 0.125$

$$K_b = \frac{(1.3 \times 10^{-5})^2}{0.125}$$

$$= 1.4 \times 10^{-9}$$

Check Your Solution

The value of K_b is reasonable for a weak organic base. The final answer has two significant digits, consistent with the two decimal places in the given pH.

Practice Problems

29. An aqueous solution of household ammonia has a molar concentration of 0.105 mol/L. Calculate the pH of the solution.

30. Hydrazine, N_2H_4, has been used as a rocket fuel. The concentration of an aqueous solution of hydrazine is 5.9×10^{-2} mol/L. Calculate the pH of the solution.

31. Morphine, $C_{17}H_{19}NO_3$, is a naturally occurring base that is used to control pain. A 4.5×10^{-3} mol/L solution has a pH of 9.93. Calculate K_b for morphine.

32. Methylamine, CH_3NH_2, is a fishy-smelling gas at room temperature. It is used to manufacture several prescription drugs, including methamphetamine. Calculate $[OH^-]$ and pOH of a 0.25 mol/L aqueous solution of methylamine.

33. At room temperature, trimethylamine, $(CH_3)_3N$, is a gas with a strong ammonia-like odour. Calculate $[OH^-]$ and the percent of trimethylamine molecules that react with water in a 0.22 mol/L aqueous solution.

34. An aqueous solution of ammonia has a pH of 10.85. What is the concentration of the solution?

Acids and Their Conjugate Bases

There is an important relationship between the dissociation constant for an acid, K_a, and the dissociation constant for its conjugate base, K_b. Consider acetic acid and its dissociation in water.

$$CH_3COOH_{(aq)} + H_2O_{(aq)} \rightleftharpoons H_3O^+_{(aq)} + CH_3COO^-_{(aq)}$$

K_a is given by the following expression.

$$K_a = \frac{[CH_3COO^-][H_3O^+]}{[CH_3COOH]}$$

The acetate ion is the conjugate base of acetic acid. A soluble salt of the conjugate base, such as sodium acetate, forms acetate ions in solution. The solution acts as a base with water.

$$CH_3COO^-_{(aq)} + H_2O_{(aq)} \rightleftharpoons CH_3COOH_{(aq)} + OH^-_{(aq)}$$

K_b is given by the expression below.

$$K_b = \frac{[CH_3COOH][OH^-]}{[CH_3COO^-]}$$

The product K_aK_b gives an interesting result.

$$K_aK_b = \frac{[H_3O^+]\cancel{[CH_3COO^-]}}{\cancel{[CH_3COOH]}} \times \frac{\cancel{[CH_3COOH]}[OH^-]}{\cancel{[CH_3COO^-]}}$$

$$= [H_3O^+][OH^-]$$

$$= K_w$$

Thus, for an acid and its conjugate base (or a base and its conjugate acid), $K_aK_b = K_w$. One interpretation of the result is the stronger an acid, the weaker its conjugate base must be. This makes sense chemically, because a strong acid gives up a proton from each molecule. Therefore, its conjugate base does not bond with the proton. In summary, then, the strength of an acid and its conjugate base are inversely related. *The conjugate of a strong acid is always a weak base, and, conversely, the conjugate of a strong base is always a weak acid.*

Sample Problem

Solving Problems Involving K_a and K_b

Problem

Sodium acetate, CH_3COONa, is used for developing photographs. Find the value of K_b for the acetate ion. Then calculate the pH of a solution that contains 12.5 g of sodium acetate dissolved in 1.00 L of water. (Only the acetate ion affects the pH of the solution.)

What Is Required?

You need to find K_b and pH.

What Is Given?

K_a for acetic acid = 1.81×10^{-5}
12.5 g of CH_3COONa is dissolved in 1.00 L of water.

Plan Your Strategy

Step 1 Find K_b using the relationship $K_aK_b = K_w$

Step 2 Calculate $[CH_3COO^-]$.

Step 3 Write the chemical equation for the acetate ion acting as a base. Then set up an ICE table.

Step 4 Write the dissociation equation for K_b, and substitute equilibrium values.

Step 5 Determine whether or not the dissociation of the acetate ions is negligible compared with its initial concentration.

Step 6 Solve the equation for x.

Step 7 Calculate $[H_3O^+]$ using $K_w = [H_3O^+][OH^-]$.

Step 8 Calculate pH from $pH = -\log[H_3O^+]$

Act on Your Strategy

Step 1 K_b for $CH_3COO^- = \dfrac{K_w}{K_a}$

$$= \frac{1.0 \times 10^{-14}}{1.8 \times 10^{-5}}$$

$$= 5.6 \times 10^{-10}$$

Step 2 $M(CH_3COONa) = 82.0$ g/mol

$$\text{Amount} = \frac{12.5\ g}{82.0\ g/mol}$$

$$= 0.152\ mol/L$$

In aqueous solution,

$CH_3COONa \rightleftharpoons Na^+_{(aq)} + CH_3COO^-_{(aq)}$

$\therefore [CH_3COO^-] = 0.152$ mol/L

Step 3

Concentration (mol/L)	$CH_3COO^-_{(aq)}$	+	$H_2O_{(\ell)}$	$\rightleftharpoons$	$CH_3COOH_{(aq)}$	+	$OH^-_{(aq)}$
Initial	0.152				0		~0
Change	$-x$				$+x$		$+x$
Equilibrium	$0.152 - x$				x		x

Step 4 $K_b = \frac{[CH_3COOH][OH^-]}{[CH_3COO^-]}$

$= \frac{(x)(x)}{0.152 - x}$

Step 5 $\frac{[CH_3COO^-]}{K_b} = \frac{0.152}{5.6 \times 10^{-10}} > 500$

Therefore, x is negligible compared with the initial concentration.

Step 6 $5.6 \times 10^{-10} = \frac{x^2}{0.152}$

Solving the equation (since only the positive root is reasonable),
$x = 9.2 \times 10^{-6} = [OH^-]$

Step 7 $[H_3O^+] = \frac{K_w}{[OH^-]}$

$= \frac{1.0 \times 10^{-14}}{9.2 \times 10^{-6}}$

$= 1.1 \times 10^{-9}$ mol/L

Step 8 $pH = -\log[H_3O^+]$

$= -\log 1.1 \times 10^{-9}$

$= 8.96$

Check Your Solution

The solution is weakly basic. The acetate ion is a very weak base, so the answer is reasonable.

Practice Problems

35. Use the table of K_a values in Appendix E to list the conjugate bases of the following acids in order of increasing base strength: formic acid, $HCOOH$; hydrofluoric acid, $HF_{(aq)}$; benzoic acid, C_6H_5COOH; phenol, C_6H_5OH.

36. K_b for ammonia, NH_3, is 1.8×10^{-5}. K_b for trimethylamine, $(CH_3)_3N$, is 6.5×10^{-5}. Which is the stronger acid, NH_4^+ or $(CH_3)_3NH^+$?

37. Sodium benzoate is used as a food preservative. Calculate the pH of a 1.0 mol/L aqueous solution of sodium benzoate. (Only the benzoate ion affects the pH of the solution.)

38. The hydrogen sulfite ion, HSO_3^-, is amphoteric. Write chemical equations to show how it acts first as an acid and then as a base.

Buffer Solutions

A solution that contains a weak acid/conjugate base mixture or a weak base/conjugate acid mixture is called a **buffer solution**. A buffer solution resists changes in pH when a moderate amount of an acid or a base is added to it. (See Figure 8.9.) For example, adding 10 mL of 1.0 mol/L hydrochloric acid to 1 L of water changes the pH from 7 to about 3, a difference of 4 units. Adding the same amount of acid to 1 L of buffered solution might change the pH by only 0.1 unit.

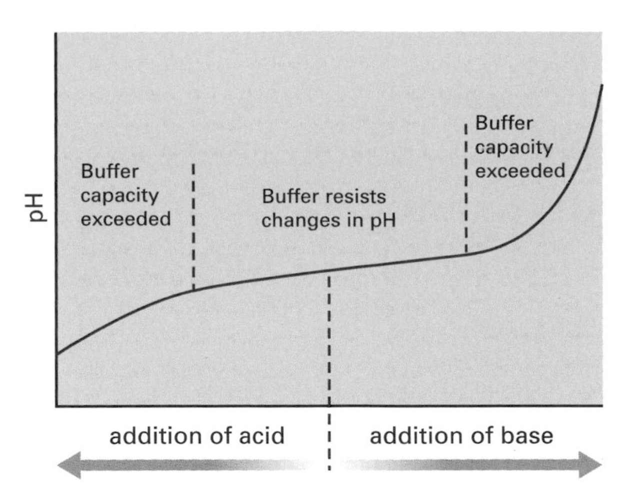

Figure 8.9 **Adding a moderate amount of an acid or a base to a buffer solution causes little change in pH.**

Buffer solutions can be made in two different ways:

1. by using a weak acid and one of its salts: for example, by mixing acetic acid and sodium acetate
2. by using a weak base and one of its salts: for example, by mixing ammonia and ammonium chloride

How does a buffer solution resist changes in pH when an acid or a base is added? Consider a buffer solution that is made using acetic acid and sodium acetate. Acetic acid is weak, so most of its molecules are not dissociated and $[CH_3COOH]$ is high. Sodium acetate is soluble and a good electrolyte, so $[CH_3COO^-]$ is also high. Adding an acid or a base has little effect because the added H_3O^+ or OH^- ions are removed by one of the components in the buffer solution. The equilibrium of the reactions between the ions in solution shifts, as predicted by Le Châtelier's principle and described below.

- Adding an acid to a buffer: Acetate ions react with the hydronium ions added to the solution.

$$CH_3COO^-_{(aq)} + H_3O^+_{(aq)} \rightleftharpoons CH_3COOH_{(aq)} + H_2O_{(\ell)}$$

  The position of equilibrium shifts to the right. Here hydronium ions are removed, by acetate ions, from the sodium acetate component.

- Adding a base to a buffer: Hydroxide ions react with the hydronium ions that are formed by the dissociation of acetic acid.

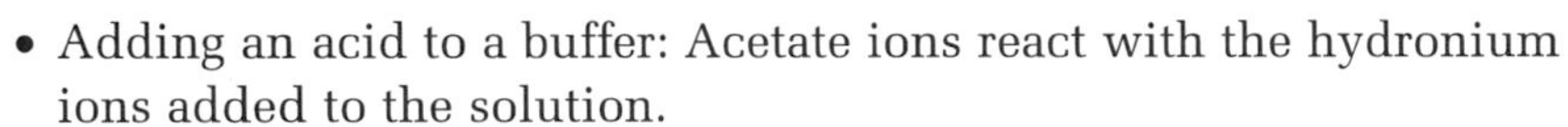

$$H_3O^+_{(aq)} + OH^-_{(aq)} \rightleftharpoons 2H_2O_{(\ell)}$$

  The position of this water equilibrium shifts to the right, replacing hydronium ions.

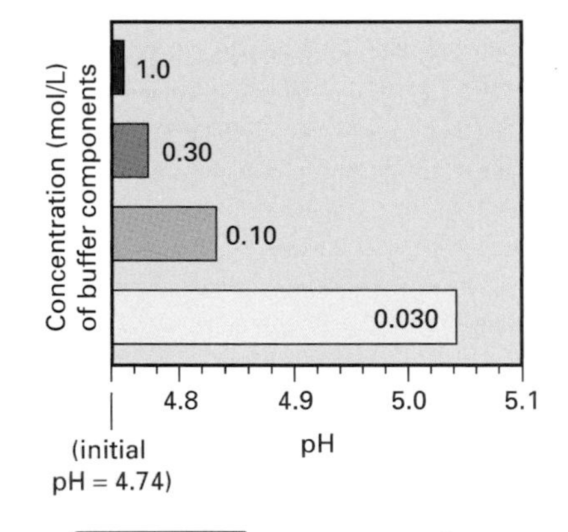

Figure 8.10 **These four buffer solutions have the same initial pH but different concentrations (shown by the numbers beside or on the bars). The pH increases with the addition of a certain amount of strong base. The more concentrated the buffer solution is (that is, the higher its buffer capacity), the smaller the change in pH is.**

Buffer solutions have two important characteristics. One of these characteristics is the pH of the solution. The other is its **buffer capacity**: the amount of acid or base that can be added before considerable change occurs to the pH. The buffer capacity depends on the concentration of the acid/conjugate base (or the base/conjugate acid) in the buffer solution. When the ratio of the concentration of the buffer components is close to 1, the buffer capacity has reached its maximum. As well, a buffer that is more concentrated resists changes to pH more than than a buffer that is more dilute. This idea is illustrated in Figure 8.10, with buffer solutions of acetic acid and acetate of different concentrations.

Buffers in the Blood

Buffers are extremely important in biological systems. The pH of arterial blood is about 7.4. The pH of the blood in your veins is just slightly less. If the pH of blood drops to 7.0, or rises above 7.5, life-threatening problems develop. To maintain its pH within a narrow range, blood contains a number of buffer systems. The most important buffer system in the blood depends on an equilibrium between hydrogen carbonate ions and carbonate ions. Dissolved carbon dioxide reacts with water to form hydrogen carbonate ions.

$$CO_{2(aq)} + 2H_2O_{(\ell)} \rightleftharpoons HCO^-_{3(aq)} + H_3O^+_{(aq)}$$

The HCO_3^- dissociates in water to form CO_3^{2-}.

$$HCO_3^-{}_{(aq)} + H_2O_{(\ell)} \rightleftharpoons CO_3^{2-}{}_{(aq)} + H_3O^+_{(aq)}$$

If metabolic changes add H_3O^+ ions to the blood, the excess H_3O^+ ions are removed by combining with HCO_3^- ions.

$$H_3O^+_{(aq)} + HCO_3^-{}_{(aq)} \rightleftharpoons CO_{2(aq)} + 2H_2O_{(\ell)}$$

If excess OH^- ions enter the blood, they are removed by reacting with the hydrogen carbonate, HCO_3^-, ions.

$$OH^-_{(aq)} + HCO_3^-{}_{(aq)} \rightleftharpoons CO_3^{2-}{}_{(aq)} + H_2O_{(\ell)}$$

Web LINK

www.mcgrawhill.ca/links/chemistry12

Some aspirin products are sold in buffered form. Infer the reasoning behind this practice. Is there clinical evidence to support it? To find out, go to the web site above and click on **Web Links**. Conduct further research to investigate whether there is more recent evidence either in support of or disproving the effectiveness of buffering aspirin.

Section Summary

In this section, you compared strong and weak acids and bases using your understanding of chemical equilibrium, and you solved problems involving their concentrations and pH. Then you considered the effect on pH of buffer solutions: solutions that contain a mixture of acid ions and base ions. In the next section, you will compare pH changes that occur when solutions of acids and bases with different strengths react together.

Section Review

1. **I** Phenol, C_6H_5OH, is an aromatic alcohol with weak basic properties. It is used as a disinfectant and cleanser. Calculate the molar concentration of OH^- ions in a 0.75 mol/L solution of phenolate, $C_6H_5O^-$, ions ($K_b = 7.7 \times 10^{-5}$). What is the pH of the solution?

2. **I** Potassium sorbate is a common additive in foods. It is used to inhibit the formation of mould. A solution contains 1.82 g of sorbate, $C_6H_7O_2^-$, ions ($K_a = 1.7 \times 10^{-5}$) dissolved in 250 mL of water. What is the pH of the solution?

3. **I** Write the chemical formula for the conjugate base of hypobromous acid, HOBr. Calculate K_b for this ion.

4. **K/U** Describe how a buffer solution differs from an aqueous acidic or basic solution.

5. **MC** Explain the function and importance of buffers in blood.

6. **C** Explain why an aqueous mixture of NaCl and HCl does not act as a buffer, but an aqueous mixture of NH_3 and NH_4Cl does.

8.4 Acid-Base Titration Curves

In this section, you will

- **interpret** acid-base titration curves and the pH at the equivalence point
- **communicate** your understanding of the following terms: *acid-base titration curve, equivalence point*

In Investigation 8-A, you performed a titration and graphed the changes in the pH of acetic acid solution as sodium hydroxide solution was added. A graph of the pH of an acid (or base) against the volume of an added base (or acid) is called an **acid-base titration curve**.

Titrations are common analytical procedures that chemists perform, with the usual goal of determining the concentration of one of the reactants. The **equivalence point** is the point in a titration when the acid and base that are present completely react with each other. If chemists know the volumes of both solutions at the equivalence point, and the concentration of one of them, they can calculate the unknown concentration.

As you can see in Figure 8.11, the equivalence point is the middle of the steep rise that occurs in a titration curve. The endpoint of a titration occurs when the indicator changes colour, which happens over a range of about 2 pH units. The pH changes rapidly near the equivalence point. Therefore, the change in colour usually takes place in a fraction of a millilitre, with the addition of a single drop of solution. Chemists have access to a variety of indicators that change colour at different pH values. The colour changes and ranges for three common indicators are given in Table 8.4.

Table 8.4 Data on the Endpoints of Three Common Acid-Base Indicators

Indicator	Colour change at endpoint	Approximate range
bromocresol green	yellow to blue	3.8–5.2
methyl red	red to yellow	4.3–6.2
phenolphthalein	colourless to pink	8.2–10.0

When an indicator is used in a titration, the range of pH values at which its endpoint occurs must include, or be close to, the equivalence point. Some representative acid-base titration curves, shown in Figures 8.11, 8.12, and 8.13, will illustrate this point.

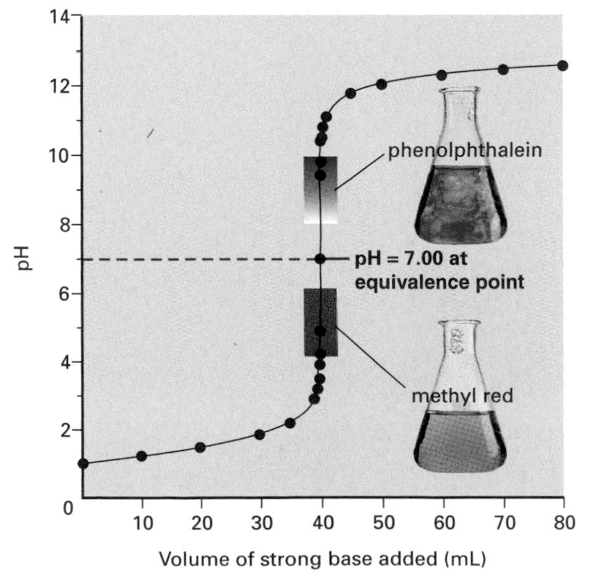

Titration Curve for a Strong Acid With a Strong Base

These titrations have a pH of 7 at equivlence. Indicators such as phenolphthalein, methyl red, and bromocresol green can be used, because their endpoints are close to the equivalence point. Many chemists prefer phenolphthalein because the change from colourless to pink is easy to see.

Figure 8.11 The curve for a strong acid-strong base titration

Titration Curve for a Weak Acid With a Strong Base

These titrations have pH values that are greater than 7 at equivalence. In the titration shown in Figure 8.12, the equivalence point occurs at a pH of 8.80. Therefore, phenolphthalein is a good indicator for this titration. Methyl red is not, because its endpoint is too far from the equivalence point.

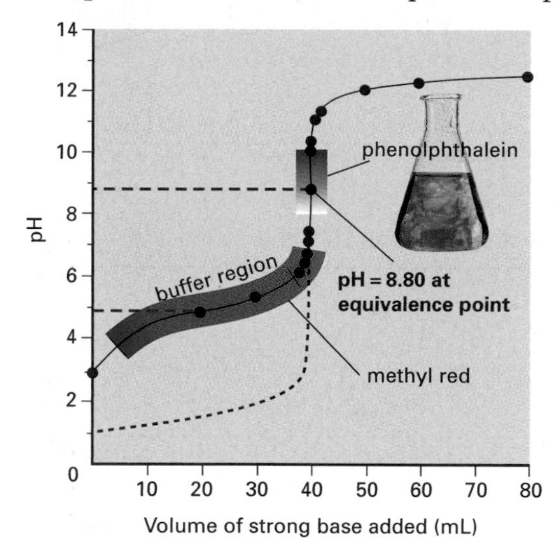

Figure 8.12 The curve for a weak acid-strong base titration: The weak acid here is propanoic acid, CH_3CH_2COOH.

Titration Curve for a Weak Base With a Strong Acid

These titrations have pH values that are less than 7 at the equivalence point. The equivalence point in the titration shown in Figure 8.13, involving ammonia and hydrochloric acid, occurs at a pH of 5.27. Either methyl red or bromocresol green could be used as an indicator, but not phenolphthalein.

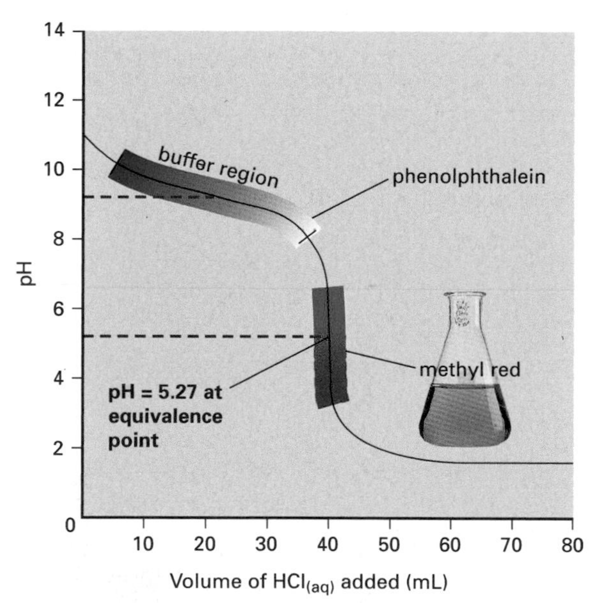

Figure 8.13 The curve for a weak base-strong acid titration

Section Summary

In this section, you examined acid-base titration curves for combinations of strong and weak acids and bases. You may have noticed the absence of a curve for the reaction of a weak acid with a weak base. A weak acid-weak base titration curve is difficult to describe quantitatively, because it has competing equilibria. You may learn about this curve in future chemistry courses.

In the next chapter, you will extend your knowledge of equilibria involving aqueous ions. You will learn how to calculate the pH at an equivalence point, so you can select an appropriate indicator for any acid-base titration. You will also learn why equilibrium is important to the solubility of compounds that are slightly soluble, and how to predict whether a precipitate will form as the result of a reaction between ions in solution.

Section Review

1. K/U In this section, you examined acid-base titration curves.
   (a) Distinguish between the equivalence point and the endpoint for a titration.
   (b) When choosing an indicator, do the pH values of the two points need to coincide exactly? Explain.

2. K/U In a titration, a basic solution is added to an acidic solution, and measurements of pH are taken. Compare a strong acid-strong base titration and a strong acid-weak base titration in terms of
   (a) the initial pH
   (b) the quantity of base that is needed to reach the equivalence point
   (c) the pH at the equivalence point

   Assume that the concentrations of the two solutions are identical.

3. C Sketch the pH curve for the titration of a weak acid with a strong base. Show the equivalence point on your sketch. Suggest an indicator that might be used, and explain your selection.

4. C Suggest an indicator that could be used for the titration of potassium hydroxide with nitrous acid. Explain your suggestion.

5. I Estimate the pH of a solution in which bromocresol green is blue, and methyl red is orange.

CHAPTER 8 Review

Reflecting on Chapter 8

Summarize this chapter in the format of your choice. Here are a few ideas to use as guidelines:

- Relate the microscopic properties of acids and bases to their macroscopic properties.
- Identify conjugate acid-base pairs for selected acid-base reactions, and compare their strengths.
- State the relationship among K_a, K_b, and K_w.
- Outline the relationship among $[H_3O^+]$, pH, $[OH^-]$, and pOH.
- Describe two examples of buffer solutions in your daily life, and explain how they function.

Reviewing Key Terms

For each of the following terms, write a sentence that shows your understanding of its meaning.

hydronium ion ($H_3O^+_{(aq)}$)	conjugate acid-base pair
monoprotic acids	polyprotic acids
ion product constant for water (K_w)	pH
acid dissociation constant (K_a)	pOH
percent dissociation	base dissociation constant (K_b)
buffer solution	buffer capacity
acid-base titration curve	equivalence point

Knowledge/Understanding

1. Give two examples of each of the following acids and bases.
   (a) Arrhenius acids
   (b) Brønsted-Lowry bases
   (c) Brønsted-Lowry bases that are not Arrhenius bases
2. Classify each compound as a strong acid, strong base, weak acid, or weak base.
   (a) phosphoric acid, H_3PO_4 (used in cola beverages and rust-proofing products)
   (b) chromic acid, H_2CrO_4 (used in the production of wood preservatives)
   (c) barium hydroxide, $Ba(OH)_2$, a white, toxic base (can be used to de-acidify paper)
   (d) CH_3NH_2, commonly called methylamine (is responsible for the characteristic smell of fish that are no longer fresh)
3. Write a chemical formula for each acid or base.
   (a) the conjugate base of OH^-
   (b) the conjugate acid of ammonia, NH_3
   (c) the conjugate acid of HCO_3^-
   (d) the conjugate base of HCO_3^-
4. Decide whether each statement is true or false, and explain your reasoning.
   (a) HBr is a stronger acid than HI.
   (b) $HBrO_2$ is a stronger acid than HBrO.
   (c) H_2SO_3 is a stronger acid than HSO_3^-.
5. Arrange the following aqueous solutions in order of pH, from lowest to highest: 2.0 mol/L $HClO_4$, 2.0 mol/L NaCl, 0.20 mol/L CH_3COOH, 0.02 mol/L HCl.
6. In each pair of bases, which is the stronger base?
   (a) $HSO_4^-{}_{(aq)}$ or $SO_4^{2-}{}_{(aq)}$
   (b) $S^{2-}{}_{(aq)}$ or $HS^-{}_{(aq)}$
   (c) $HPO_4^{2-}{}_{(aq)}$ or $H_2PO_4^-{}_{(aq)}$
   (d) $HCO_3^-{}_{(aq)}$ or $CO_3^{2-}{}_{(aq)}$
7. (a) Use Appendix E to find the values of K_a for hydrosulfuric acid, $HS^-{}_{(aq)}$, and sulfurous acid, $HSO_3^-{}_{(aq)}$.
   (b) Write equations for the base dissociation constants of $HS^-{}_{(aq)}$ and $HSO_3^-{}_{(aq)}$.
   (c) Calculate the value of K_b for each ion.
   (d) Which is the stronger base, $HS^-{}_{(aq)}$ or $HSO_3^-{}_{(aq)}$? Explain.
8. While the pH of blood must be maintained within strict limits, the pH of urine can vary. The sulfur in foods, such as eggs, is oxidized in the body and excreted in the urine. Does the presence of sulfide ions in urine tend to increase or decrease the pH? Explain.
9. Sodium methanoate, NaHCOO, and methanoic acid, HCOOH, can be used to make a buffer solution. Explain how this combination resists changes in pH when small amounts of acid or base are added.
10. Oxoacids contain an atom that is bonded to one or more oxygen atoms. One or more of these oxygen atoms may also be bonded to hydrogen. Consider the following oxoacids: $HBrO_{3(aq)}$, $HClO_{3(aq)}$, $HClO_{4(aq)}$, and $H_2SO_{3(aq)}$.
    (a) What factors are used to predict the strengths of oxoacids?
    (b) Arrange the oxoacids above in the order of increasing acid strength.

Inquiry

11. What is the pH of a mixture of equal volumes of 0.040 mol/L hydrochloric acid and 0.020 mol/L sodium hydroxide?

12. Suppose that 15.0 mL of sulfuric acid just neutralized 18.0 mL of 0.500 mol/L sodium hydroxide solution. What is the concentration of the sulfuric acid?

13. A student dissolved 5.0 g of vitamin C in 250 mL of water. The molar mass of ascorbic acid is 176 g/mol, and its K_a is 8.0×10^{-5}. Calculate the pH of the solution. **Note:** Abbreviate the formula of ascorbic acid to H_{Asc}.

14. Benzoic acid is a weak, monoprotic acid ($K_a = 6.3 \times 10^{-5}$). Its structure is shown below. Calculate the pH and the percent dissociation of each of the following solutions of benzoic acid. Then use Le Châtelier's principle to explain the trend in percent dissociation of the acid as the solution becomes more dilute.

(a) 1.0 mol/L **(b)** 0.10 mol/L **(c)** 0.01 mol/L

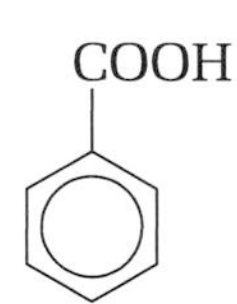

15. Hypochlorous acid, HOCl, is a weak acid that is found in household bleach. It is made by dissolving chlorine gas in water.

$Cl_{2(g)} + 2H_2O_{(\ell)} \rightleftharpoons H_3O^+_{(aq)} + Cl^-_{(aq)} + HOCl_{(aq)}$

(a) Calculate the pH and the percent dissociation of a 0.065 mol/L solution of hypochlorous acid.

(b) What is the conjugate base of hypochlorous acid? What is its value for K_b?

16. Calculate the pH of a 1.0 mol/L aqueous solution of sodium benzoate. **Note:** Only the benzoate ion affects the pH of the solution.

17. Calculate the pH of a 0.10 mol/L aqueous solution of sodium nitrite, $NaNO_2$. **Note:** Only the nitrite ion affects the pH of the solution.

18. A student prepared a saturated solution of salicylic acid and measured the pH of the solution. The student then carefully evaporated 100 mL of the solution and collected the solid. If the pH of the solution was 2.43, and 0.22 g was collected after evaporating 100 mL of the solution, what is the acid dissociation constant for salicylic acid?

COOH
OH

Communication

19. List the oxoacids of bromine (HOBr, $HBrO_2$, $HBrO_3$, and $HBrO_4$) in order of increasing strength. What is the order of increasing strength for the conjugate bases of these acids?

20. Consider the following acid-base reactions.

$HBrO_{2(aq)} + CH_3COO^-_{(aq)} \rightleftharpoons CH_3COOH_{(aq)} + BrO^-_{2(aq)}$

$H_2S_{(aq)} + OH^-_{(aq)} \rightleftharpoons HS^-_{(aq)} + H_2O_{(\ell)}$

$HS^-_{(aq)} + CH_3COOH_{(aq)} \rightleftharpoons H_2S_{(aq)} + CH_3COO^-_{(aq)}$

If each equilibrium lies to the right, arrange the following compounds in order of increasing acid strength: $HBrO_2$, CH_3COOH, H_2S, H_2O.

21. Discuss the factors that can be used to predict the relative strength of different oxoacids.

22. The formula of methyl red indicator can be abbreviated to HMr. Like most indicators, methyl red is a weak acid.

$HMr_{(aq)} + H_2O_{(\ell)} \rightleftharpoons H3O^+_{(aq)} + Mr^-_{(aq)}$

The change between colours (when the indicator colour is orange) occurs at a pH of 5.4. What is the equilibrium constant for the reaction?

23. Gallic acid is the common name for 3,4,5-trihydroxybenzoic acid.

(a) Draw the structure of gallic acid.

(b) K_a for gallic acid is 3.9×10^{-5}. Calculate K_b for the conjugate base of gallic acid. Then write the formula of the ion.

24. (a) Sketch the pH curves you would expect if you titrated

- a strong acid with a strong base
- a strong acid with a weak base
- a weak acid with a strong base

(b) Congo red changes colour over a pH range of 3.0 to 5.0. For which of the above titrations would Congo red be a good indicator to use?

Making Connections

25. Citric acid can be added to candy to give a sour taste. The structure of citric acid is shown below.

```
            .O:   .O.—H
              \\  /
     :O:  H     C    H  :O:
      ||  |     |    |   ||        ..
H—O—C—C—C—C—C—O—H
          |     |    |             ..
          H   :O:   H
                |
                H
```

(a) Identify the acidic hydrogen atoms that are removed by water in aqueous solution. Why do water molecules pull these hydrogen atoms away, rather than other hydrogen atoms in citric acid?

(b) Why does citric acid not form OH^- ions in aqueous solution, and act as a base?

(c) When citric acid and sodium hydrogen carbonate are used in bubble gum, the bubble gum foams when chewed. Suggest a reason why this happens.

26. (a) Imagine that you have collected a sample of rainwater in your community. The pH of your sample is 4.52. Unpolluted rainwater has a pH of about 5.6. How many more hydronium ions are present in your sample, compared with normal rainwater? Calculate the ratio of the concentration of hydronium ions in your sample to the concentration of hydronium ions in unpolluted rainwater.

(b) You have been invited to a community meeting to explain your findings to local residents. No one at the meeting has a background in chemistry. In a paragraph, write what you would say at this meeting.

(c) Suggest at least two possible factors that could be responsible for the pH you measured. What observations would you want to make, and what data would you want to collect, to help you gain confidence that one of these factors is responsible?

Answers to Practice Problems and Short Answers to Section Review Questions

Practice Problems: 1.(a) chloride ion, Cl^- **(b)** carbonate ion, CO_3^{2-} **(c)** hydrogen sulfate ion, HSO_4^- **(d)** hydrazine, N_2H_4 **2.(a)** nitric acid, HNO_3 **(b)** water, H_2O **(c)** hydronium ion, H_3O^+ **(d)** carbonic acid, H_2CO_3 **3.(a)** HS^-/H_2S and H_2O/OH^- **(b)** O^{2-}/OH^- and H_2O/OH^- **4.(a)** H_2S/HS^- and NH_3/NH_4^+ **(b)** H_2SO_4/HSO_4^- and H_2O/H_3O^+ **5.(a)** 4.5 mol/L **(b)** 1.35 mol/L **(c)** 0.0170 mol/L **(d)** 0.0375 mol/L **6.(a)** 3.1 mol/L **(b)** 0.87 mol/L **(c)** 0.701 mol/L **(d)** 0.692 mol/L **7.(a)** acidic solution; $[H_3O^+] = 0.479$ mol/L **(b)** acidic solution; $[H_3O^+] = 1.98$ mol/L **8.** acidic solution; $[H_3O^+] = 0.46$ mol/L **9.(a)** $[H_3O^+] = 0.45$ mol/L; $[OH^-] = 2.2 \times 10^{-14}$ mol/L **(b)** $[OH^-] = 1.1$ mol/L; $[H_3O^+] = 9.1 \times 10^{-15}$ mol/L **10.(a)** $[H_3O^+] = 0.95$ mol/L; $[OH^-] = 1.1 \times 10^{-14}$ mol/L **(b)** $[OH^-] = 0.024$ mol/L; $[H_3O^+] = 4.2 \times 10^{-13}$ mol/L **11.** $[HCl] = 0.18$ mol/L **12.** $[Ca(OH)_2] = 0.29$ mol/L **13.** acidic; pH = 6.400; $[OH^-] = 2.51 \times 10^{-8}$ mol/L **14.** pOH = 2.1; $[OH^-] = 8 \times 10^{-3}$ mol/L **15.** acidic; $[H_3O^+] = 1.9 \times 10^{-5}$ mol/L; $[OH^-] = 5.3 \times 10^{-10}$; pOH = 9.28 **16.** $[H_3O^+] = [OH^-] = 1.6 \times 10^{-7}$; neutral **17.** basic; pH = 8.19; $[H_3O^+] = 6.5 \times 10^{-9}$ mol/L; $[OH^-] = 1.5 \times 10^{-6}$ mol/L **18.** $[H_3O^+] = 1.9 \times 10^{-3}$ mol/L; $[OH^-] = 5.3 \times 10^{-12}$ mol/L **19.** pH = 2.41; 0.47% dissociation **20.** $K_a = 1.0 \times 10^{-4}$; percent dissociation = 3.2% **21.** pH = 2.65 **22.** $K_a = 2.9 \times 10^{-8}$ **23.** pH = 3.411 **24.** 1.2 g **25.** pH = 4.82; $[CO_3^{2-}] = 4.7 \times 10^{-11}$ mol/L **26.** pH = 2.74 **27.** pH = 4.59; $[HS^-] = 2.6 \times 10^{-5}$ mol/L **28.** $K_a = 1.0 \times 10^{-14}$; $K_{a_2} = \frac{[O^{2-}][H_3O^+]}{[OH^-]}$ **29.** pH = 11.14 **30.** pH = 10.50 **31.** $K_b = 1.6 \times 10^{-6}$ **32.** $[OH^-] = 1.0 \times 10^{-2}$ mol/L; pOH = 1.98 **33.** $[OH^-] = 3.8 \times 10^{-3}$ mol/L; percent dissociation = 1.7% **34.** $[NH_3] = 2.8 \times 10^{-2}$ mol/L **35.** $C_6H_5O^- > C_6H_5COO^- > HCOO^- > F^-$ **36.** NH_4^+ **37.** pH = 9.11 **38.** as an acid: $HSO_3^-{}_{(aq)} + H_2O_{(\ell)} \rightleftharpoons H_3O^+{}_{(aq)} + SO_3^{2-}{}_{(aq)}$; as a base: $HSO_3^-{}_{(aq)} + H_2O_{(\ell)} \rightleftharpoons OH^-{}_{(aq)} + H_2SO_{3(aq)}$

Section Review: 8.1:
1.(a) $H_3PO_{4(aq)} + H_2O_{(aq)} \rightleftharpoons H_2PO_4^-{}_{(aq)} + H_3O^+{}_{(aq)}$; $H_2PO_4^-{}_{(aq)} + H_2O_{(aq)} \rightleftharpoons HPO_4^{2-}{}_{(aq)} + H_3O^+{}_{(aq)}$; $HPO_4^{2-}{}_{(aq)} + H_2O_{(aq)} \rightleftharpoons PO_4^{3-}{}_{(aq)} + H_3O^+{}_{(aq)}$ **(b)** as an acid: $H_2PO_4^-{}_{(aq)} + H_2O_{(aq)} \rightleftharpoons HPO_4^{2-}{}_{(aq)} + H_3O^+{}_{(aq)}$; as a base: $H_2PO_4^-{}_{(aq)} + H_3O^+{}_{(aq)} \rightleftharpoons H_3PO_{4(aq)} + H_2O_{(aq)}$ **(c)** $H_3PO_{4(aq)}$ is much stronger (although it is still a weak acid) than $H_2PO_4^-{}_{(aq)}$. **2.** $C_6H_4NH_2COO^-$ **3.(a)** $B(OH)_3/B(OH)_4^-$ and H_2O/H_3O^+ **(b)** weak **4.(a)** weak acid **(b)** strong acid **(c)** strong base **(d)** weak base
8.2: 1.(a) 4.43 **(b)** 2.70×10^{-10} **(c)** 9.57 **(d)** acidic **(e)** 3.9×10^{-11} **(f)** 2.6×10^{-4} **(g)** 3.59 **(h)** basic **(i)** 1.4×10^{-13} **(j)** 12.85 **(k)** 1.15 **(l)** basic **(m)** 8×10^{-6} **(n)** 5.1 **(o)** 1×10^{-9} **(p)** acidic **2.** pH = 2.39 **3.** $K_a = 2.5 \times 10^{-4}$ **4.** pH = 1.70; $[HOOCCOO^-] = 5.4 \times 10^{-5}$ mol/L **5.** $[H_2CO_3] = 2.3 \times 10^{-3}$ mol/L
8.3: 1. $[OH^-] = 7.6 \times 10^{-3}$ mol/L; pH = 11.88 **2.** pH = 8.80 **3.** OBr^-; $K_b = 3.6 \times 10^{-6}$ **8.4: 4.** phenolphthalein **5.** pH ~ 5.0

CHAPTER

9 Aqueous Solutions and Solubility Equilibria

Chapter Preview

9.1 The Acid-Base Properties of Salt Solutions

9.2 Solubility Equilibria

9.3 Predicting the Formation of a Precipitate

Prerequisite Concepts and Skills

Before you begin this chapter, review the following concepts and skills:

- solving equilibrium problems (Chapter 7, section 7.3)
- identifying conjugate acid-base pairs and comparing their strengths (Chapter 8, section 8.1)
- interpreting acid-base titration curves (Chapter 8, section 8.4)

The kidneys are sometimes called the master chemists of the body. They work to maintain the constant composition of the blood by helping to balance water and the various ions that are present in the blood. A very important equilibrium in the blood, which the kidneys help to control, involves calcium ions and phosphate ions.

$$3Ca^{2+}{}_{(aq)} + 2PO_4{}^{3-}{}_{(aq)} \rightleftharpoons Ca_3(PO_4)_{2(s)}$$

This equilibrium sometimes causes problems, however. For example, calcium phosphate helps to give bones their rigidity. If the kidneys remove too many phosphate ions from the blood due to disease, the position of equilibrium shifts to the left, as predicted by Le Châtelier's principle. More calcium phosphate dissolves in the blood, reducing bone density. The loss of too much calcium phosphate can lead to osteoporosis.

If the kidneys remove too many calcium ions from the blood, the equilibrium position *in the kidneys* shifts to the right. Solid calcium phosphate can form in the kidneys, producing kidney stones. Kidney stones, which are painful, can also form as the result of calcium oxalate precipitating in the kidneys. Precipitates of other compounds can affect different areas of the body: gallstones in the gall bladder and gout in the joints are two examples.

In this chapter, you will continue your study of acid-base reactions. You will find out how ions in aqueous solution can act as acids or bases. Then, by applying equilibrium concepts to ions in solution, you will be able to predict the solubility of ionic compounds in water and the formation of a precipitate.

This kidney stone formed when potassium oxalate precipitated out of solution in a kidney. How can you predict whether a precipitate, such as potassium oxalate, will form in an aqueous solution?

The Acid-Base Properties of Salt Solutions

9.1

At the end of Chapter 8, you learned that when just enough acid and base have been mixed for a complete reaction, the solution can be acidic, neutral, or basic. The reaction between an acid and a base forms an aqueous solution of a salt. The properties of the dissolved salt determine the pH of a titration solution at equivalence. Figure 9.1 shows the acidic, basic, or neutral properties of three salt solutions.

In this section, you will learn how to predict the pH of an aqueous solution of a salt. Predicting the pH is useful when you are performing a titration experiment, because you need to choose an indicator that changes colour at a pH value that is close to the pH at equivalence.

Section Preview/ Specific Expectations

In this section, you will

- **predict** qualitatively whether a solution of a specific salt will be acidic, basic, or neutral
- **solve** problems that involve acid-base titration data and the pH at the equivalence point
- **communicate** your understanding of the following terms: *equivalence point, end-point*

Acidic and Basic Properties of Salts

The acidic or basic property of an aqueous solution of a salt results from reactions between water and the dissociated ions of the salt. Some ions do not react with water. They are neutral in solution. Ions that do react with water produce a solution with an excess of $H_3O^+{}_{(aq)}$ or $OH^-{}_{(aq)}$. The extent of the reaction determines the pH of the solution. As you will see, the reaction between an ion and water is really just another acid-base reaction.

Is there a way to classify salts so that you can predict whether their solutions in water will be basic, acidic, or neutral? In the following ExpressLab, you will determine the pH values of solutions of various salts. Then you will analyze your results to decide if there is a pattern.

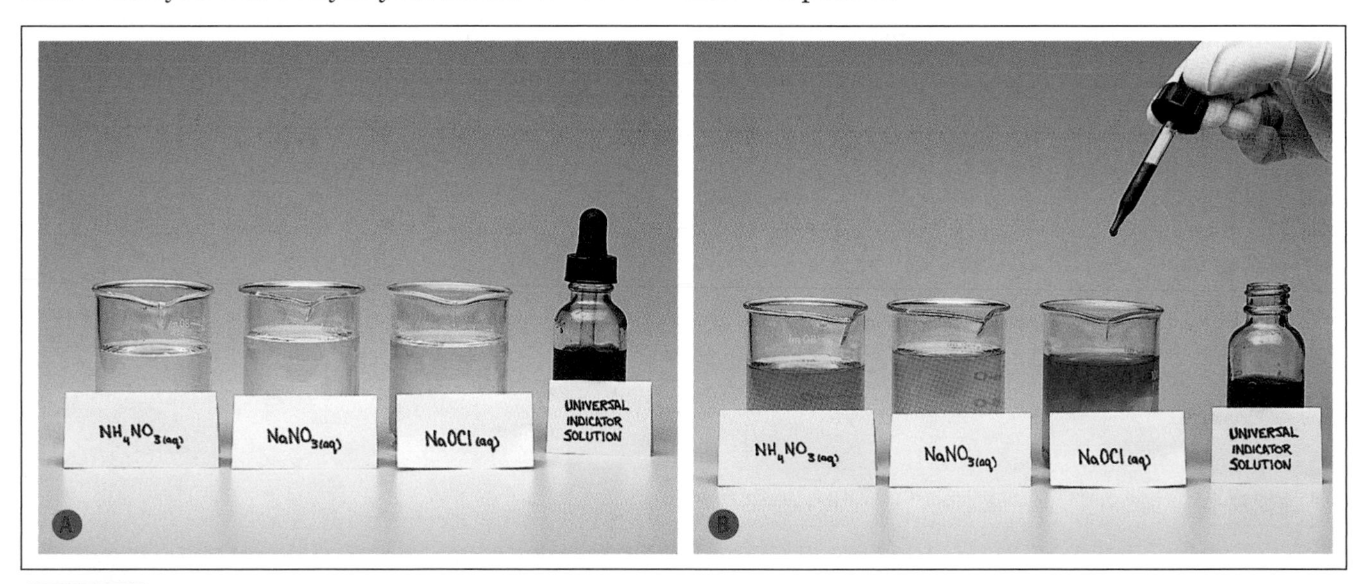

Figure 9.1 **Photograph (A) shows three aqueous salt solutions. Photograph (B) shows the solutions after a some universal indicator solution has been added to each one. The orange colour indicates an acidic solution, the yellow colour indicates a neutral solution, and the purple colour indicates a basic solution.**

ExpressLab  Testing the pH of Salt Solutions

In this ExpressLab, you will measure the pH values of solutions of different salts with the same concentration in mol/L. The pH of a solution gives the relative $[H_3O^+]$, compared with the $[OH^-]$. From this information, you can infer which ion in the salt reacted with water to the greatest extent.

Safety Precaution

- Some of the salts are irritants. Wear gloves and goggles at all times.

Materials

dropper bottles containing 0.1 mol/L solutions of each of the following salts: ammonium chloride, $NH_4Cl_{(aq)}$; sodium chloride, $NaCl_{(aq)}$; sodium acetate, $NaCH_3COO_{(aq)}$; potassium carbonate, $K_2CO_{3(aq)}$
unknown salt
universal pH paper
spot plate

Procedure

1. Copy the following table into your notebook and title it. Before you test each salt, write the formula of the salt in the appropriate box. (You will not be testing any salt that results from the reaction of a weak base and a weak acid.)

	Strong acid	Weak acid
Strong base		
Weak base		

2. Copy the following table into your notebook, to record your observations. Give your table a title.

Salt	pH of aqueous solution
NH_4Cl	
NaCl	
$NaCH_3COO$	
K_2CO_3	
unknown	

3. Pour a small volume of one of the solutions into the well of a spot plate. Test the solution using universal pH paper. By comparing the colour of the paper, estimate the pH of the solution. Record your results in you table.
4. Repeat step 3 for each of the solutions.

Analysis

1. An ion that does not react with water has no effect on the pH of water. If both the cation and the anion in a salt do not react with water, a solution of the salt will be neutral. Which salt consists of cations and anions that do not react with water? How did you classify this salt in terms of the strength of the acid and base used to form it?
2. If a salt consists of one ion that does not react with water, and another ion that does, the pH of a solution of the salt will be affected by the ion that reacts with water.
   (a) Which salt(s) formed acidic solutions? Which ion accounts for the acidic nature of the solution?
   (b) Which salt(s) formed basic solutions? Which ion accounts for the basic nature of the solution?
3. Classify your unknown salt in terms of the strength of the base and the acid from which the salt was formed.

Salts That Dissolve and Form Neutral Solutions

In the following equation, $HA_{(aq)}$ represents a strong acid, which dissociates completely in water.

$$HA_{(aq)} + H_2O_{(\ell)} \rightarrow H_3O^+{}_{(aq)} + A^-{}_{(aq)}$$

Interpreting the equation from right to left, the conjugate base of the acid, A^-, has no tendency to combine with hydronium ions in water. Therefore, it is a weak base. In fact, the anion that is formed by a strong acid is a much weaker base than water, so it does not react with water. Consequently, when a salt that contains the anion of a strong acid is dissolved in water, the anion has no effect on the pH of the solution.

For example, the conjugate base of hydrochloric acid, HCl, is Cl^-. The chloride ion is a very weak base, so it does not react significantly with water. Therefore, the chloride ion, and the conjugate bases of other strong acids, do not affect the pH of an aqueous solution.

Similarly, the cations that form strong bases (the alkali metals and the metals below beryllium in Group 2 (IIA)) do not tend to react with hydroxide ions. These cations are weaker acids than water. Therefore, when a salt contains one of these ions (for example, Na^+) the cation has no effect on the pH of an aqueous solution.

If a salt contains the cation of a strong base and the anion of a strong acid, neither ion reacts with water. Therefore, the solution has a pH of 7. Sodium chloride is an example of such a salt. It is formed by the reaction of sodium hydroxide (a strong base) and hydrochloric acid (a strong acid). *Salts of strong bases and strong acids dissolve in water and form neutral solutions.*

Salts That Dissolve and Form Acidic Solutions

A weak base, such as aqueous ammonia, dissociates very little. Therefore, the equilibrium lies to the left.

$$NH_{3(aq)} + H_2O_{(\ell)} \rightleftharpoons NH_4{}^+{}_{(aq)} + OH^-$$

Interpreting this equation from right to left, the ammonium ion, (which is the conjugate acid of aqueous ammonia) is a relatively strong acid, compared to water. In solution, ammonium ions react with water, resulting in an acidic solution:

$$NH_4{}^+{}_{(aq)} + H_2O_{(\ell)} \rightleftharpoons NH_{3(aq)} + H_3O^+{}_{(aq)}$$

If a salt consists of the cation of a weak base and the anion of a strong acid, only the cation reacts with water. The solution has a pH that is less than 7. *Salts of weak bases and strong acids dissolve in water and form acidic solutions.*

Salts That Dissolve and Form Basic Solutions

A weak acid, HA, forms a conjugate base, A^-, that is relatively strong. The reaction of such an anion with water results in a solution that is basic. For example, consider acetic acid, $CH_3COOH_{(aq)}$, a weak acid. The conjugate base of acetic acid, $CH_3COO^-{}_{(aq)}$, is relatively strong, compared to water. It reacts with water to form a basic solution.

$$CH_3COO^-{}_{(aq)} + H_2O_{(\ell)} \rightleftharpoons CH_3COOH_{(aq)} + OH^-{}_{(aq)}$$

If a salt consists of the cation of a strong base and the anion of a weak acid, such as $NaCH_3COO$, only the anion reacts significantly with water. The reaction produces hydroxide ions. Therefore, the solution will have a pH that is greater than 7. *Salts of strong bases and weak acids dissolve in water and form basic solutions.*

A reaction between an ion and water is sometimes called a *hydrolysis* reaction. An ammonium ion hydrolyzes, but a sodium ion does not. An acetate ion hydrolyzes, but a chloride ion does not.

Salts of Weak Bases and Weak Acids

If a salt consists of the cation of a weak base and the anion of a weak acid, both ions react with water. The solution is weakly acidic or weakly basic, depending on the relative strength of the ions that act as the acid or the base. You can determine which ion is stronger by comparing the values of K_a and K_b associated with the cation and the anion, respectively. *If $K_a > K_b$, the solution is acidic. If $K_b > K_a$, the solution is basic.*

Figure 9.2 shows the relationship between acid-base conjugate pairs.

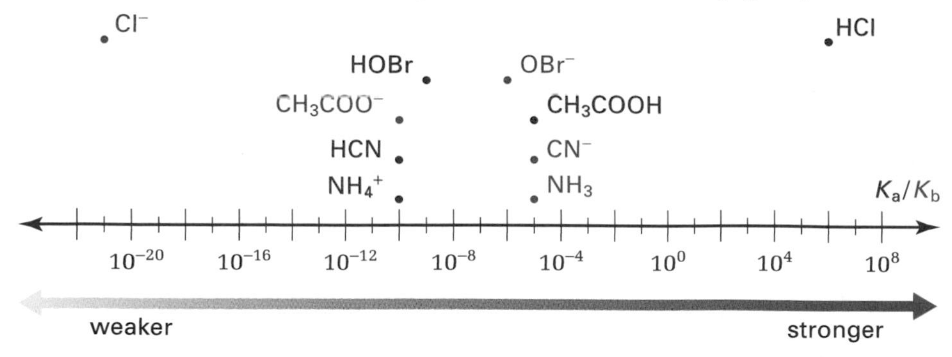

Figure 9.2 The values of K_a and K_b are shown for acid-base conjugate pairs. A strong acid, such as $HCl_{(aq)}$, forms a conjugate base, $Cl^-_{(aq)}$, that is too weak to react significantly with water. A weaker acid, such as $CH_3COOH_{(aq)}$, forms a conjugate base, $CH_3COO^-_{(aq)}$, that is strong enough to react significantly with water.

The following list summarizes the acidic and basic properties of salts. Table 9.1 shows the acid-base properties of salts in relation to the acids and bases that react to form them.

Acidic and Basic Properties of Salts

- The salt of a strong acid/weak base dissolves to form an acidic solution.
- The salt of a weak acid/strong base dissolves to form a basic solution.
- The salt of a weak acid/weak base dissolves to form an acidic solution if K_a for the cation is greater than K_b for the anion. The solution is basic if K_b for the anion is greater than K_a for the cation.

Table 9.1 The Acid-Base Properties of Various Salts

	Anion derived from a strong acid	**Anion derived from a weak acid**
Cation derived from a strong base	Reaction with water: neither ion Solution: neutral Examples: $NaCl$, K_2SO_4, $Ca(NO_3)_2$	Reaction with water: only the anion Solution: basic Examples: $NaCH_3COO$, KF, $Mg(HSO_4)_2$
Cation derived from a weak base	Reaction with water: only the cation Solution: acidic Examples: NH_4Cl, NH_4NO_3, NH_4ClO_4	Reaction with water: both ions Solution: neutral if $K_a = K_b$, acidic if $K_a > K_b$, basic if $K_b > K_a$ Examples: NH_4CN (basic), $(NH_4)_2S$ (basic), NH_4NO_2 (acidic)

Sodium Fluoride: A Basic Salt That Protects Teeth

Sugar is a common ingredient in prepared foods. When sugar remains on your teeth, bacteria in your mouth convert it into an acid. The principal constituent of tooth enamel is a mineral called hydroxyapatite, $Ca_{10}(PO_4)_6(OH)_2$. Hydroxyapatite reacts with acids to form solvated ions and water. (Solvated ions are ions surrounded by solvent particles.) Eventually, a cavity forms in the enamel.

To help prevent acid from damaging tooth enamel, many water treatment plants add small concentrations (about 1 ppm) of salts, such as NaF, to the water. You may notice that a fluoride salt has also been added to your toothpaste. Fluoride ions displace OH^- from hydroxyapatite to form fluoroapatite, $Ca_{10}(PO_4)_6F_2$, as shown in the equation below.

$$Ca_{10}(PO_4)_6(OH)_{2(s)} + 2F^-_{(aq)} \rightarrow Ca_{10}(PO_4)_6F_{2(s)} + 2OH^-_{(aq)}$$

The fluoride ion is a much weaker base than the hydroxide ion, and fluoroapatite is therefore less reactive towards acids.

The following Sample Problem shows you how to predict the acidity or basicity of aqueous salt solutions.

CHEM FACT

If you cannot brush your teeth after a sweet snack, try chewing sugar-free gum. Chewing gum stimulates the production of saliva, which is usually weakly alkaline. The saliva helps to neutralize acids on your teeth. Eating cheese for dessert also helps to neutralize acid, because cheese is basic.

Sample Problem

Predicting the Acidity or Basicity of Salts

Problem

Predict the acid/base property of an aqueous solution of each of the following salts. If you predict that the solution is not neutral, write the equation for the reaction that causes the solution to be acidic or basic.

(a) sodium phosphate, Na_3PO_4 (commonly called trisodium phosphate, TSP, used in detergents and dishwashing compounds)

(b) ammonium nitrate, NH_4NO_3 (used as a fertilizer)

(c) sodium chloride, NaCl, (used to de-ice winter roads)

(d) ammonium hydrogen carbonate, NH_4HCO_3 (used in baked foods as a leavening agent)

What Is Required?

Predict whether each aqueous solution is acidic, basic, or neutral. Write equations that represent the solutions that are acidic or basic.

What Is Given?

The formula of each salt is given.

Plan Your Strategy

Determine whether the cation is from a strong or weak base, and whether the anion is from a strong or weak acid. Ions derived from weak bases or weak acids react with water and affect the pH of the solution. If both ions react with water, compare the equilibrium constants (K_a and K_b) to determine which reaction goes farthest to completion.

Act On Your Strategy

(a) Sodium phosphate, Na_3PO_4, is the salt of a strong base (NaOH) and a weak acid (HPO_4^{2-}). Only the phosphate ions react with water.

$$PO_4^{3-}{}_{(aq)} + H_2O_{(\ell)} \rightleftharpoons HPO_4^{2-}{}_{(aq)} + OH^-_{(aq)}$$

The solution is basic.

(b) Ammonium nitrate, NH_4NO_3, is the salt of a weak base ($NH_{3(aq)}$) and a strong acid (HNO_3). Only the ammonium ions react with water.

$$NH_4^+{}_{(aq)} + H_2O_{(\ell)} \rightleftharpoons NH_{3(aq)} + H_3O^+_{(aq)}$$

The solution is acidic.

(c) Sodium chloride, NaCl, is the salt of a strong base (NaOH) and a strong acid ($HCl_{(aq)}$). Neither ion reacts with water, so the solution is neutral.

Continued ...

Continued ...

(d) Ammonium hydrogen carbonate, NH_4HCO_3, is the salt of a weak base ($NH_{3(aq)}$) and a weak acid ($H_2CO_{3(aq)}$). Both ions react with water.

$NH_4{}^+{}_{(aq)} + H_2O_{(\ell)} \rightleftharpoons NH_{3(aq)} + H_3O^+{}_{(aq)}$

$HCO_3{}^-{}_{(aq)} + H_2O_{(\ell)} \rightleftharpoons H_2CO_{3(aq)} + OH^-{}_{(aq)}$

The equilibrium that lies farther to the right has the greater influence on the pH of the solution. The equilibrium constants you need are K_a for $NH_4{}^+$ and K_b for $HCO_3{}^-$. Each ion is the conjugate of a compound for which the appropriate constant is given in tables.

From Appendix E, K_b for $NH_{3(aq)}$ is 1.8×10^{-5}. Using $K_aK_b = K_w$,

$$K_a \text{ for } NH_4{}^+ = \frac{1.0 \times 10^{-14}}{1.8 \times 10^{-5}}$$
$$= 5.6 \times 10^{-10}$$

Similarly, from Appendix E, K_a for $H_2CO = 4.5 \times 10^{-7}$.

$$K_b \text{ for } HCO_3{}^- = \frac{1.0 \times 10^{-14}}{4.5 \times 10^{-7}}$$
$$= 2.2 \times 10^{-8}$$

Because K_b for the hydrogen carbonate reaction is larger than K_a for the ammonium ion reaction, the solution is basic.

Check Your Solution

The equations that represent the reactions with water support the prediction that NH_4NO_3 dissolves to form an acidic solution and Na_3PO_4 dissolves to form a basic solution. Calcium chloride is the salt of a strong base-strong acid, so neither ion reacts with water and the solution is neutral. Both ions in ammonium hydrogen carbonate react with water. Because K_b for $HCO_3{}^-$ is greater than K_a for $NH_4{}^+$, the salt dissolves to form a weakly basic solution.

Practice Problems

1. Predict whether an aqueous solution of each salt is neutral, acidic, or basic.

(a) $NaCN$

(b) LiF

(c) $Mg(NO_3)_2$

(d) NH_4I

2. Is the solution of each salt acidic, basic, or neutral? For solutions that are not neutral, write equations that support your predictions.

(a) NH_4BrO_4

(b) $NaBrO_4$

(c) $NaOBr$

(d) NH_4Br

3. K_a for benzoic acid, C_6H_5COOH, is 6.3×10^{-5}. K_a for phenol, C_6H_5OH, is 1.3×10^{-10}. Which is the stronger base, $C_6H_5COO^-{}_{(aq)}$ or $C_6H_5O^-{}_{(aq)}$? Explain your answer.

4. Sodium hydrogen sulfite, $NaHSO_3$, is a preservative that is used to prevent the discolouration of dried fruit. In aqueous solution, the hydrogen sulfite ion can act as either an acid or a base. Predict whether $NaHSO_3$ dissolves to form an acidic solution or a basic solution. (Refer to Appendix E for ionization data.)

Calculating pH at Equivalence

In an acid-base titration, you carefully measure the volumes of acid and base that react. Then, knowing the concentration of either the acid or the base, and the stoichiometric relationship between them, you calculate the concentration of the other reactant. The **equivalence point** in the titration occurs when just enough acid and base have been mixed for a complete reaction to occur, with no excess of either reactant. As you learned in Chapter 8, you can find the equivalence point from a graph that shows pH versus volume of one solution added to the other solution. To determine the equivalence point experimentally, you need to measure the pH. Because pH meters are expensive, and the glass electrodes are fragile, titrations are often performed using an acid-base indicator.

An acid-base indicator is usually a weak, monoprotic organic acid. The indicator is in equilibrium between the undissociated acid, which is one colour, and its conjugate base, which is a different colour.

$$\underset{\textbf{colour 1}}{HIn_{(aq)}} + H_2O_{(\ell)} \rightleftharpoons H_3O^+_{(aq)} + \underset{\textbf{colour 2}}{In^-_{(aq)}}$$

For example, one indicator, called methyl red, is red when the pH is below 4.2, and yellow when the pH is above 6.2. The point in a titration at which an indicator changes colour is called the **end-point**. The colour change occurs over a range of about 2 pH units, but the pH of a solution changes rapidly near the equivalence point. Often a single drop of base causes the shift from colour 1 to colour 2. For methyl red, the end-point occurs over the pH range 4.2 to 6.2. Therefore, this indicator is used when an acid-base titration results in a moderately acidic solution at equivalence. Figure 9.3 shows the colours and end-points of various indicators.

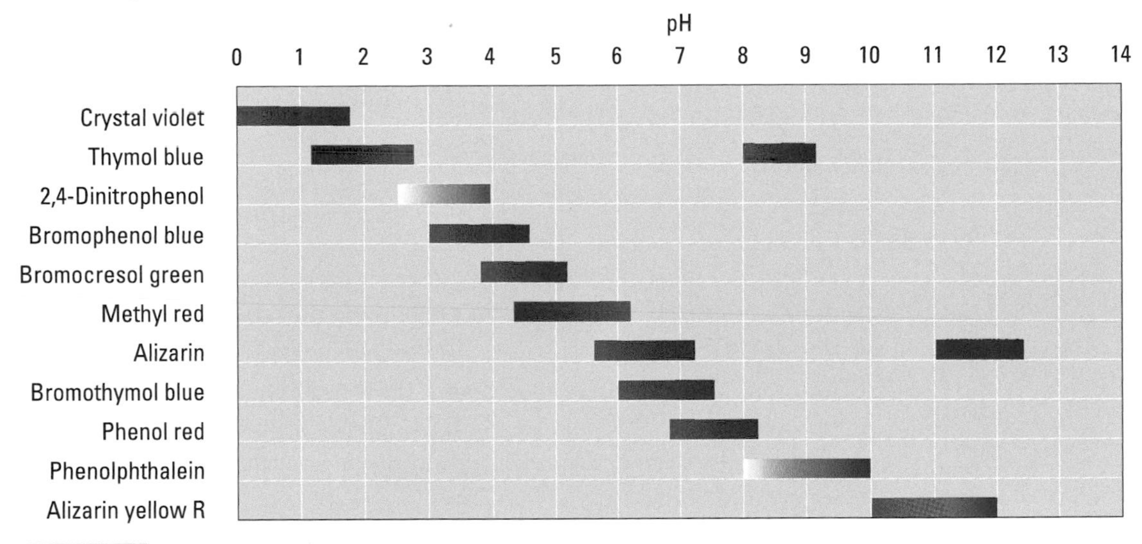

Figure 9.3 **A range of indicators are available with different end-points are available. Thymol blue and alizarin are *diprotic*. They change colour over two different ranges.**

To choose the appropriate indicator for a particular titration, you must know the approximate pH of the solution at equivalence. As you learned in Chapter 8, an acid-base titration curve provides this information. What if you do not have a titration curve? You can calculate the pH at equivalence using what you know about salts.

For example, suppose that you are titrating a strong acid solution against a weak base. At the equivalence point, the flask contains an aqueous solution of the salt that is formed by the reaction. The solution of a salt of a weak base and a strong acid is acidic, so methyl red may be an appropriate indicator. The pH at equivalence in a titration is the same as the pH of an aqueous solution of the salt formed.

Sample Problem

Calculating pH at Equivalence

Problem

2.0×10^1 mL of 0.20 mol/L $NH_{3(aq)}$ is titrated against 0.20 mol/L $HCl_{(aq)}$. Calculate the pH at equivalence. Use Figure 9.3 to select an appropriate indicator.

What Is Required?

You need to calculate the pH at equivalence. As well, you need to find an indicator that changes colour at a pH that is close to the same pH at equivalence.

What Is Given?

You know that 20 mL of 0.20 mol/L $NH_{3(aq)}$ base reacts with 0.20 mol/L $HCl_{(aq)}$. Tables of K_a and K_b values are in Appendix E. Figure 9.3 shows the pH ranges and colour changes for various indicators.

Plan Your Strategy

Step 1 Write the balanced chemical equation for the reaction

Step 2 Calculate the amount (in mol) of $NH_{3(aq)}$, using its volume and concentration. Then, determine the amount (in mol) of $HCl_{(aq)}$ that is needed to react with all the $NH_{3(aq)}$.

Step 3 Find the volume of $HCl_{(aq)}$, based on the amount and concentration of $HCl_{(aq)}$.

Step 4 Calculate the concentration of the salt formed, based on the amount (in mol) and the total volume of the solution. Decide which ion reacts with water. Write the equation that represents the reaction.

Step 5 Determine the equilibrium constant for the ion that is involved in the reaction. Then divide the $[NH_4^+]$ by K_a to determine whether the change in concentration of the ammonium ion can be ignored.

Step 6 Set up an ICE table for the ion that is involved in the reaction with water. Let x represent the change in the concentration of the ion that reacts.

Step 7 Write the equilibrium expression. Substitute the equilibrium concentrations into the expression, and solve for x.

Step 8 Use the value of x to determine the $[H_3O^+]$. Then calculate the pH of the solution. Using Figure 9.3, choose an indicator that changes colour in a range that includes the pH of the solution.

Act on Your Strategy

Step 1 The following chemical equation represents the reaction.

$NH_{3(aq)} + HCl_{(aq)} \rightarrow NH_4Cl_{(aq)}$

Step 2 Amount (in mol) of $NH_{3(aq)}$ = 0.20 mol/L × 0.020 L
= 4.0 × 10^{-3} mol

$NH_{3(aq)}$ and $HCl_{(aq)}$ react in a 1:1 ratio. Therefore, 4.0 × 10^{-3} mol of $HCl_{(aq)}$ is needed.

Step 3 Amount (in mol) $HCl_{(aq)}$ = 4.0 × 10^{-3} mol
= 0.20 mol/L × Volume $HCl_{(aq)}$ (in L)

$$\text{Volume } HCl_{(aq)} = \frac{4.0 \times 10^{-3}\text{ mol}}{0.20\text{ mol/L}} = 0.020\text{ L}$$

Step 4 From the chemical equation, the amount of NH_4Cl that is formed is equal to the amount of acid or base that reacts. Therefore, 4.0 × 10^{-3} mol of ammonium chloride is formed.

Total volume of solution = 0.020 L + 0.020 L
= 0.040 L

$$[NH_4Cl] = \frac{4.0 \times 10^{-3}\text{ mol}}{4.0 \times 10^{-2}\text{ L}} = 0.10\text{ mol/L}$$

The salt forms $NH_4{^+}_{(aq)}$ and $Cl^-_{(aq)}$ in solution. $NH_4{^+}_{(aq)}$ is the conjugate acid of a weak base, so it reacts with water. $Cl^-_{(aq)}$ is the conjugate base of a strong acid, so it does not react with water. The pH of the solution is therefore determined by the extent of the following reaction.

$NH_4{^+}_{(aq)} + H_2O_{(\ell)} \rightleftharpoons NH_{3(aq)} + H_3O^+_{(aq)}$

Step 5 From Appendix E, K_b for $NH_{3(aq)}$ is 1.8 × 10^{-5}.
K_a for the conjugate acid, $NH_4{^+}_{(aq)}$, can be calculated using the relationship $K_aK_b = K_w$.

$$K_a = \frac{K_w}{K_b} = \frac{1.0 \times 10^{-14}}{1.8 \times 10^{-5}} = 5.6 \times 10^{-10}$$

$$\frac{[NH_4^+]}{K_a} = \frac{0.10}{5.6 \times 10^{-10}} = 1.8 \times 10^{8}$$

This is well above 500, so the change in $[NH_4^+]$, x, can be ignored.

Step 6

Concentration (mol/L)	$NH_4{^+}_{(aq)}$ +	$H_2O_{(\ell)}$ ⇌	$NH_{3(aq)}$ +	$H_3O^+_{(aq)}$
Initial	0.10	—	0	~0
Change	$-x$	—	$+x$	$+x$
Equilibrium	$0.10-x$ (~0.10)	—	x	x

Continued ...

Continued ...

Step 7 $K_a = \frac{[NH_3][H_3O^+]}{[NH_4^+]}$

$5.6 \times 10^{-10} = \frac{(x)(x)}{0.10}$

$x = \sqrt{5.6 \times 10^{-11}}$

$= \pm 7.5 \times 10^{-6}$ mol/L (only the positive value makes sense)

Step 8 $x = [H_3O^+]$

$= 7.5 \times 10^{-6}$ mol/L

$pH = -\log[H_3O^+]$

$= -\log 7.5 \times 10^{-6}$

$= 5.13$

Methyl red, which changes colour over pH 4.2 to 6.2, is a good choice for an indicator. If methyl red was not available, bromocresol green, which changes colour in the pH range 3.8–5.2, could be used.

Check Your Solution

The titration forms an aqueous solution of a salt derived from a weak base and a strong acid. The solution should be acidic, which is supported by the calculation of the pH.

Practice Problems

5. After titrating sodium hydroxide with hydrofluoric acid, a chemist determined that the reaction had formed an aqueous solution of 0.020 mol/L sodium fluoride. Determine the pH of the solution.
6. Part way through a titration, 2.0×10^1 mL of 0.10 mol/L sodium hydroxide has been added to 3.0×10^1 mL of 0.10 mol/L hydrochloric acid. What is the pH of the solution?
7. 0.025 mol/L benzoic acid, C_6H_5COOH, is titrated with 0.025 mol/L sodium hydroxide solution. Calculate the pH at equivalence.
8. 50.0 mL of 0.10 mol/L hydrobromic acid is titrated with 0.10 mol/L aqueous ammonia. Determine the pH at equivalence.

PROBLEM TIP

When you are determining the concentration of ions in solution, remember to use the final volume of the solution. In Practice Problem 6, for example, the final volume of the solution is 50 mL because 20 mL of base is added to 30 mL of acid. You can assume that the total volume is the sum of the reactant volumes, because the reactant solutions are dilute.

Section Summary

In this section, you learned why solutions of different salts have different pH values. You learned how to analyze the composition of a salt to predict whether the salt forms an acidic, basic, or neutral solution. Finally, you learned how to apply your understanding of the properties of salts to calculate the pH at the equivalence point of a titration. You used the pH to determine a suitable indicator for the titration. In section 9.2, you will further investigate the equilibria of solutions and learn how to predict the solubility of ionic compounds in solution.

Section Review

1. K/U Sodium carbonate and sodium hydrogen carbonate both dissolve to form basic solutions. Comparing solutions with the same concentration, which of these salts forms the more basic solution? Explain.

2. K/U Determine whether or not each ion reacts with water. If the ion does react, write the chemical equation for the reaction. Then predict whether the ion forms an acidic solution or a basic solution.

   (a) Br^- **(b)** $CH_3NH_3^+$ **(c)** ClO_4^- **(d)** OCl^-

3. K/U Predict whether sodium dihydrogen citrate, $NaH_2C_6H_5O_7$, dissolves to form an acidic or a basic solution. (Refer to Appendix E)

4. I A chemist measures the pH of aqueous solutions of $Ca(OH)_2$, CaF_2, NH_4NO_3, KNO_3, and HNO_3. Each solution has the same concentration. Arrange the solutions from most basic to most acidic.

5. I A student uses a transfer pipette to put 25.00 mL of 0.100 mol/L acetic acid into an Erlenmeyer flask. Then the student adds sodium hydroxide from a burette to the flask, and records the following readings of volume and pH.

volume (mL)	0.00	6.00	10.00	12.00	14.00	14.40	14.60	14.80	15.20	15.40	16.00
pH	2.8	4.2	5.1	5.5	6.2	6.5	6.8	7.6	9.8	10.5	11.4

   (a) Draw a graph that shows the results. Plot pH on the vertical axis and the volume of base added on the horizontal axis. Use speadsheet software if available.

   (b) From your graph, what is the pH at the equivalence point?

   (c) Determine the initial concentration of the sodium hydroxide.

   (d) Use Figure 9.3 to suggest a suitable indicator for this titration.

6. C Congo red is an indicator that changes colour in the pH range 3.0 to 5.0. Suppose that you are titrating an aqueous sodium hydroxide solution with nitric acid added from a burette. Is congo red a suitable indicator? If so, explain why. If not, will the end-point occur before or after the equivalence point? Explain your answer.

7. K/U If you were titrating a strong base with a weak acid, which of these indicators might be suitable: bromphenol blue, bromthymol blue, or phenolphthalein? (Refer to Figure 9.3.)

8. I 0.10 mol/L hydrochloric acid is titrated with 0.10 mol/L methylamine, CH_3NH_2. Calculate the pH at the equivalence point.

9. MC Sodium phosphate, Na_3PO_4, is sold at hardware stores as TSP (trisodium phosphate). Crystals of Na_3PO_4 are dissolved in water to make an effective cleaning solution. TSP can be used, for example, to prepare a surface before painting.

   (a) Explain why TSP is an effective cleaning solution.

   (b) Suggest safety precautions you should take when using TSP.

   (c) Explain why you should never mix a solution of TSP with other household cleaning products.

9.2 Solubility Equilibria

Section Preview/ Specific Expectations

In this section, you will

- **perform** an experiment to determine the solubility product constant (K_{sp}) for calcium hydroxide
- **solve** equilibrium problems that involve concentrations of reactants and products, and K_{sp}
- **calculate** the molar solubility of a pure substance in water or in a solution of a common ion, given K_{sp}
- **communicate** your understanding of the following term: *solubility product constant* (K_{sp})

In Chapter 7, you learned that a saturated solution containing undissolved solute is an example of a system at equilibrium. Recall from your previous chemistry course that the solubility of a solute is the amount of solute that will dissolve in a given volume of solvent at a certain temperature. In other words, the solubility of a solute indicates how much of that solute is present in a solution at equilibrium. Data tables often express solubility as (g solute)/(100 mL solvent). Molar solubility, on the other hand, is always expressed in terms of (mol solute/L solvent).

The solubility of ionic solids in water covers a wide range of values. Knowing the concentration of ions in aqueous solution is important in medicine and in chemical analysis. In this section, you will continue to study equilibrium. You will examine the solubility equilibria of ionic compounds in water.

Solubility as an Equilibrium Process

In Chapter 7, you learned that three factors—change in enthalpy (ΔH), change in entropy (ΔS), and temperature (T)—determine whether or not a change is favoured. The same three factors are important for determining how much of a salt will dissolve in a certain volume of water. These factors are combined in the following equation, where ΔG is the change in free energy of the system.

$$\Delta G = \Delta H - T\Delta S$$

As you know from Chapter 7, a change is favoured when ΔG is negative. When a salt dissolves, the entropy of the system always increases, because ions in solution are more disordered than ions in a solid crystal. An increase in entropy favours the formation of a solution because the term $-T\Delta S$ is negative. Most solids dissolve to a greater extent at higher solution temperatures, because the term $-T\Delta S$ becomes more negative.

Explaining the overall enthalpy change for a dissolving salt is more complicated because it involves a number of energy changes. Cations must be separated from anions, which requires energy. Water molecules then surround each ion in solution, which releases energy. If the overall enthalpy change is negative, the formation of a solution is favoured. If the enthalpy increases, the formation of a solution is not favoured because ΔG is less negative.

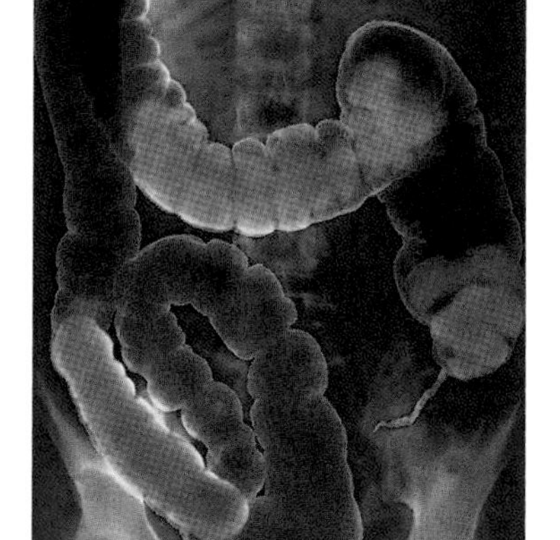

Figure 9.4 Barium sulfate, $BaSO_4$, is a sparingly soluble ionic compound that is used to enhance X-ray imaging.

Heterogeneous Equilibrium: A Solubility System

Consider barium sulfate, a sparingly soluble salt. X-ray technicians give patients a barium sulfate suspension to drink before taking an X-ray of the large intestine. A suspension of barium sulfate is opaque to X-rays, which helps to define this part of the body. (See Figure 9.4.)

When you add barium sulfate crystals to water, barium ions and sulfate ions leave the surface of the solid and enter the solution. Initially, the concentration of these ions is very low. Thus, the forward change, dissolving, occurs at a greater rate than the reverse change:

$$BaSO_{4(s)} \rightleftharpoons Ba^{2+}{}_{(aq)} + SO_4{}^{2-}{}_{(aq)}$$

As more ions enter the solution, the rate of the reverse change, recrystallisation, increases. Eventually, the rate of recrystallisation becomes equal to the rate of dissolving. As you know, when the forward rate and the backward rate of a process are equal, the system is at equilibrium. Because the reactants and the products are in different phases, the reaction is said to have reached heterogeneous equilibrium. For solubility systems of sparingly soluble ionic compounds, *equilibrium exists between the solid ionic compound and its dissociated ions in solution.*

The Solubility Product Constant

When excess solid is present in a saturated solution, you can write the equilibrium constant expression for the dissolution of the solid in the same way that you wrote the equilibrium constant expression for a homogeneous equilibrium in Chapter 7. For example, the equilibrium constant expression for barium sulfate is written as follows:

$$K = \frac{[Ba^{2+}{}_{(aq)}][SO_4{}^{2-}{}_{(aq)}]}{[BaSO_{4(s)}]}$$

The concentration of a solid, however, is itself a constant at a constant temperature. Therefore, you can combine the term for the concentration of the solid with the equilibrium constant to arrive at a new constant:

$$K[BaSO_{4(s)}] = [Ba^{2+}{}_{(aq)}][SO_4{}^{2-}{}_{(aq)}]$$

$$K_{sp} = [Ba^{2+}{}_{(aq)}][SO_4{}^{2-}{}_{(aq)}]$$

The new constant is called the **solubility product constant, K_{sp}**. The following Sample Problem shows how to write expressions for K_{sp}.

Sample Problem

Writing Solubility Product Expressions

Problem

Write the solubility product expression for each compound.

(a) barium carbonate **(b)** calcium iodate

(c) copper(II) phosphate

Solution

First write a balanced equation for the equilibrium between excess solid and dissolved ions in a saturated aqueous solution. Then use the balanced equation to write the expression for K_{sp}.

(a) $BaCO_{3(s)} \rightleftharpoons Ba^{2+}{}_{(aq)} + CO_3{}^{2-}{}_{(aq)}$

$K_{sp} = [Ba^{2+}][CO_3{}^{2-}]$

(b) $Ca(IO_3)_{2(s)} \rightleftharpoons Ca^{2+}{}_{(aq)} + 2IO_3{}^{-}{}_{(aq)}$

$K_{sp} = [Ca^{2+}][IO_3{}^{-}]^2$

(c) $Cu_3(PO_4)_{2(s)} \rightleftharpoons 3Cu^{2+}{}_{(aq)} + 2PO_4{}^{3-}{}_{(aq)}$

$K_{sp} = [Cu^{2+}]^3[PO_4{}^{3-}]^2$

Check Your Solution

The K_{sp} expressions are based on balanced equations for saturated solutions of slightly soluble ionic compounds. The exponents in the K_{sp} expressions match the corresponding coefficients in the chemical equation. The coefficient 1 is not written, following chemical convention.

PROBLEM TIP

Ion concentration terms in solubility product expressions are usually written without their phases, as you see here.

Practice Problems

9. Write the balanced chemical equation that represents the dissociation of each compound in water. Then write the corresponding solubility product expression.
   (a) copper(I) chloride
   (b) barium fluoride
   (c) silver sulfate
   (d) calcium phosphate
10. Write a balanced dissolution equation and solubility product expression for silver carbonate, Ag_2CO_3.
11. Write a balanced dissolution equation and solubility product expression for magnesium ammonium phosphate, $MgNH_4PO_4$.
12. Iron(III) nitrate has a very low solubility.
   (a) Write the solubility product expression for iron(III) nitrate.
   (b) Do you expect the value of K_{sp} of iron(III) nitrate to be larger or smaller than the K_{sp} for aluminum hydroxide, which has a slightly higher solubility?

One compound that becomes *less* soluble at higher temperatures is calcium carbonate, $CaCO_3$. If you have an electric kettle, you have probably noticed a build-up of scale on the heating element. The scale is mostly calcium carbonate, which precipitates out of the water as it heats up.

You know intuitively that solubility depends on temperature. For most ionic compounds, more solute can dissolve in a solvent at higher temperatures. Chemists determine the solubility of an ionic compound by experiment, and then use the solubility data to determine K_{sp}. Like the equilibrium constant, K_{sp} is temperature-dependent. Therefore, different experiments must be carried out to determine K_{sp} at different temperatures.

Chemical reference books list the solubility of a wide variety of compounds. In the next Sample Problem, you will learn how to use measured solubilities to determine K_{sp}.

Sample Problem

Determining K_{sp} from Measured Solubilities

Problem

A chemist finds that the solubility of silver carbonate, Ag_2CO_3, is 1.3×10^{-4} mol/L at 25°C. Calculate K_{sp} for silver carbonate.

What Is Required?

You need to find the value of K_{sp} for Ag_2CO_3 at 25°C.

What Is Given?

You know the solubility of Ag_2CO_3 at 25°C.

Plan Your Strategy

Step 1 Write an equation for the dissolution of Ag_2CO_3.

Step 2 Use the equation to write the solubility product expression.

Step 3 Find the concentration (in mol/L) of each ion.

Step 4 Substitute the concentrations into the solubility product expression, and calculate K_{sp}.

Act on Your Strategy

Step 1 $Ag_2CO_{3(s)} \rightleftharpoons 2Ag^+{}_{(aq)} + CO_3{}^{2-}{}_{(aq)}$

Step 2 $K_{sp} = [Ag^+]^2[CO_3{}^{2-}]$

Step 3 $[Ag^+] = 2 \times [Ag_2CO_3]$

$= 2 \times 1.3 \times 10^{-4}$ mol/L

$= 2.6 \times 10^{-4}$ mol/L

$[CO_3{}^{2-}] = [Ag_2CO_3] = 3.2 \times 10^{-2}$ mol/L

Step 4 $K_{sp} = [Ag^+]^2[CO_3{}^{2-}]$

$= (2.6 \times 10^{-4})^2(1.3 \times 10^{-4})$

$= 8.8 \times 10^{-12}$

Based on the given solubility data, K_{sp} for silver carbonate is 8.8×10^{-12} at 25°C.

Check Your Solution

The value of K_{sp} is less than the concentration of the salt, as expected. It has the correct number of significant digits.

Check that you wrote the balanced chemical equation and the corresponding K_{sp} equation correctly. Pay attention to molar relationships and to the exponent of each term in the K_{sp} equation.

Practice Problems

13. The maximum solubility of silver cyanide, AgCN, is 1.5×10^{-8} mol/L at 25°C. Calculate K_{sp} for silver cyanide.

14. A saturated solution of copper(II) phosphate, $Cu_3(PO_4)_2$, has a concentration of 6.1×10^{-7} g $Cu_3(PO_4)_2$ per 1.00×10^2 mL of solution at 25°C. What is K_{sp} for $Cu_3(PO_4)_2$ at 25°C?

15. A saturated solution of CaF_2 contains 1.2×10^{20} formula units of calcium fluoride per litre of solution. Calculate K_{sp} for CaF_2.

16. The concentration of mercury(I) iodide, Hg_2I_2, in a saturated solution at 25°C is 1.5×10^{-4} ppm.

(a) Calculate K_{sp} for Hg_2I_2. The solubility equilibrium is written as follows:

$Hg_2I_2 \rightleftharpoons Hg_2{}^{2+} + 2I^-$

(b) State any assumptions that you made when you converted ppm to mol/L.

Determining a Solubility Product Constant

In Investigation 9-A, you will collect solubility data and use these data to determine a K_{sp} for calcium hydroxide, $Ca(OH)_2$. When you calculate K_{sp}, you assume that the dissolved ionic compound exists as independent hydrated ions that do not affect one another. This assumption simplifies the investigation, but it is not entirely accurate. Ions *do* interfere with one another. As a result, the value of K_{sp} that you calculate will be just an approximation. K_{sp} values that are calculated from data obtained from experiments such as Investgation 9-A are generally higher than the actual values.

Investigation 9-A

MICROSCALE

SKILL FOCUS

Predicting

Performing and recording

Analyzing and interpreting

Determining K_{sp} for Calcium Hydroxide

The value of K_{sp} for a basic compound, such as $Ca(OH)_2$, can be determined by performing an acid-base titration.

Question

What is the value of K_{sp} for $Ca(OH)_2$?

Predictions

Look up the solubility of $Ca(OH)_2$ in a solubility table. Do you expect a large or a small value of K_{sp} for $Ca(OH)_2$?

Materials

50 mL beaker
white sheet of paper
2 identical microscale pipettes
wash bottle that contains distilled water
phenolphthalein in dropping bottle
glass stirring rod
Petri dish or watch glass
10 mL of recently filtered saturated solution of $Ca(OH)_2$
10 mL of 0.050 mol/L HCl

Safety Precautions

If you get acid or base on your skin, flush with plenty of cold water.

Procedure

1. Read the Procedure, and prepare an appropriate data table. Remember to give your table a title.
2. Add exactly 10 drops of 0.050 mol/L HCl to a 50 mL beaker. To ensure uniform drop size, use the same type of pipette to dispense the acid and the base. Also, hold the pipette vertically when dispensing the solution. Always record the number of drops added.
3. Place the beaker on the white sheet of paper.
4. To transfer a small amount of indicator to the acid solution, place one drop of phenolphthalein solution on a watch glass. Touch the drop with the glass rod. Transfer this small amount of indicator to the HCl in the beaker.
5. Add saturated $Ca(OH)_2$ solution, drop by drop, to the HCl solution until the solution turns a permanent pale pink. Swirl the flask, particularly as the pink begins to appear. Record the number of drops of $Ca(OH)_2$ that you added. Conduct as many trials as you can.
6. Dispose of all solutions as directed by your teacher. Do not dispose of any chemicals down the drain.

Analysis

1. Determine the average number of drops of $Ca(OH)_2$ solution that you needed to neutralize 10 drops of 0.050 mol/L HCl.
2. Write the balanced chemical equation for the dissociation of calcium hydroxide in water.
3. Write the balanced chemical equation that represents the neutralization of aqueous $Ca(OH)_2$ with $HCl_{(aq)}$.
4. Use the balanced chemical equation for the neutralization of HCl with $Ca(OH)_2$ to determine $[OH^-]$ in mol/L. **Hint:** Why do you *not* need to know the volume of one drop?
5. From $[OH^-]$, determine $[Ca^{2+}]$ in mol/L.

Conclusion

6. Calculate K_{sp} for $Ca(OH)_2$. If possible, look up the accepted value of K_{sp} for $Ca(OH)_2$. Calculate your percent error.

Using the Solubility Product Constant

You can use the value of K_{sp} for a compound to determine the concentration of its ions in a saturated solution. The following Sample Problem shows you how to do this. You will use an approach that is similar to the approach you used in section 7.2 to find equilibrium amounts using K_c for homogeneous equilibria.

Table 9.2 shows the values of K_{sp} for selected ionic compounds at 25°C. The compounds are organized from the largest K_{sp} to the smallest. You will find a more comprehensive table of K_{sp} values in Appendix E.

CONCEPT CHECK

Suppose that two different salts, AX and BY_2, have the same K_{sp}. Are the salts equally soluble at the same temperature? Explain your answer.

Table 9.2 Values of K_{sp} for Some Ionic Compounds at 25°C

Compound	K_{sp}
magnesium sulfate, $MgSO_4$	5.9×10^{-3}
lead(II) chloride, $PbCl_2$	1.7×10^{-5}
barium fluoride, BaF_2	1.5×10^{-6}
cadmium carbonate, $CdCO_3$	1.8×10^{-14}
copper(II) hydroxide, $Cu(OH)_2$	2.2×10^{-20}
silver sulfide, Ag_2S	8×10^{-48}

Sample Problem

Calculating Molar Solubility From K_{sp}

Problem

Lead(II) iodide, PbI_2, films are being investigated for their usefulness in X-ray imaging. PbI_2 is also used for decorative work, such as mosaics, because of its attractive golden yellow colour. At 25°C, K_{sp} for PbI_2 is 9.8×10^{-9}. What is the molar solubility of PbI_2 in water at 25°C?

What Is Required?

You need to determine the solubility (in mol/L) of PbI_2 at 25°C.

What Is Given?

At 25°C, K_{sp} for PbI_2 is 9.8×10^{-9}.

Plan Your Strategy

Step 1 Write the dissociation equilibrium equation.

Step 2 Use the equilibrium equation to write an expression for K_{sp}.

Step 3 Set up an ICE table. Let x represent molar solubility. Use the stoichiometry of the equilibrium equation to write expressions for the equilibrium concentrations of the ions.

Step 4 Substitute your expressions into the expression for K_{sp}, and solve for x.

Act on Your Strategy

Step 1 $PbI_{2(s)} \rightleftharpoons Pb^{2+}{}_{(aq)} + 2I^{-}{}_{(aq)}$

Step 2 $K_{sp} = [Pb^{2+}][I^{-}]^2$

Continued ...

Continued ...

Step 3

Concentration (mol/L)	$PbI_{2(s)}$	$\rightleftharpoons$	$Pb^{2+}_{(aq)}$	+	$2I^{-}_{(aq)}$
Initial	—		0		0
Change	—		$+x$		$+2x$
Equilibrium	—		x		$2x$

Step 4

$$
\begin{aligned}
K_{sp} &= [Pb^{2+}][I^-]^2 = 9.8 \times 10^{-9} \\
&= x \times (2x)^2 \\
&= x \times 4x^2 \\
&= 4x^3 \\
\therefore 4x^3 &= 9.8 \times 10^{-9} \\
x &= \sqrt[3]{\frac{9.8 \times 10^{-9}}{4}} \\
&= 1.3 \times 10^{-3} \text{ mol/L}
\end{aligned}
$$

The molar solubility of PbI_2 in water is 1.3×10^{-3} mol/L.

Check Your Solution

Recall that $x = [Pb^{2+}]_{eq}$ and $2x = [I^-]_{eq}$. Substitute these values into the K_{sp} equation. You should get the given K_{sp}.

Practice Problems

17. K_{sp} for silver chloride, AgCl, is 1.8×10^{-10} at 25°C.

(a) Calculate the molar solubility of AgCl in a saturated solution at 25°C.

(b) How many formula units of AgCl are dissolved in 1.0 L of saturated silver chloride solution?

(c) What is the percent (m/v) of AgCl in a saturated solution at 25°C?

18. Iron(III) hydroxide, $Fe(OH)_3$, is an extremely insoluble compound. K_{sp} for $Fe(OH)_3$ is 2.8×10^{-39} at 25°C. Calculate the molar solubility of $Fe(OH)_3$ at 25°C.

19. K_{sp} for zinc iodate, $Zn(IO_3)_2$, is 3.9×10^{-6} at 25°C. Calculate the solubility (in mol/L and in g/L) of $Zn(IO_3)_2$ in a saturated solution.

20. What is the maximum number of formula units of zinc sulfide, ZnS, that can dissolve in 1.0 L of solution at 25°C? K_{sp} for ZnS is 2.0×10^{-22}.

The Common Ion Effect

So far, you have considered solubility equilibria for pure substances dissolved in water. What happens to the solubility of an ionic compound when it is added to a solution that already contains one of its ions?

Consider a saturated solution of lead(II) chromate. See Figure 9.5(A) on the next page. The following equation represents this equilibrium.

$$PbCrO_{4(s)} \rightleftharpoons Pb^{2+}_{(aq)} + CrO_4{}^{2-}_{(aq)}$$

A solution of a salt that contains chromate ions, such as sodium chromate, $Na_2CrO_{4(aq)}$, is added to the mixture. More yellow lead(II) chromate precipitates out of solution, as shown in Figure 9.5(B).

As you learned in Chapter 8, this phenomenon is called the common ion effect. The observed result is predicted by Le Châtelier's principle. Adding a common ion to a solution increases the concentration of that ion in solution. As a result, equilibrium shifts away from the ion. In this example, adding chromate ions causes the equilibrium to shift to the left, and lead(II) chromate precipitates.

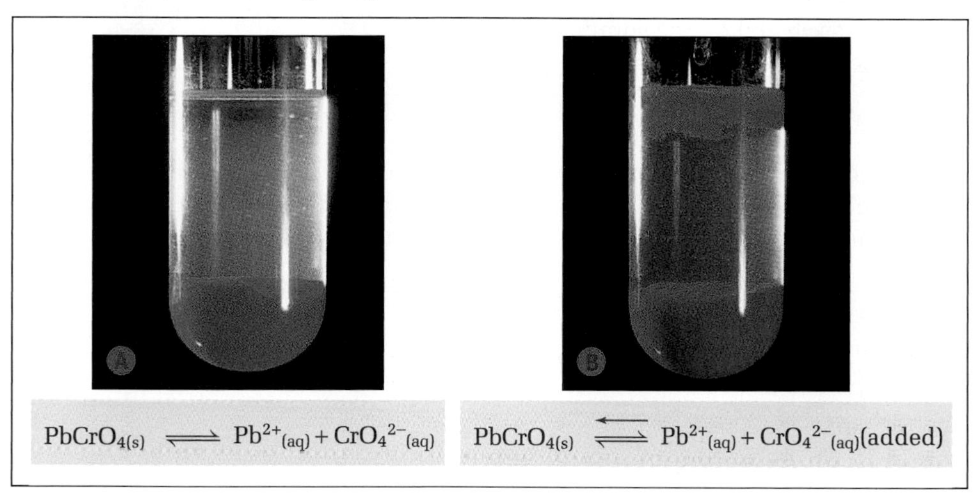

Figure 9.5 Adding chromate ions to an equilibrium system of lead(II) chromate (A) causes the equilibrium position to shift to the left. As a result, more solid lead(II) chromate precipitates (B).

Analyzing the expression for the solubility product constant gives the same result.

$$K_{sp} = [Pb^{2+}][CrO_4{}^{2-}]$$
$$= 2.3 \times 10^{-13} \text{ at } 25°C$$

Since K_{sp} is a constant at a given temperature, an increase in the concentration of one ion must be accompanied by a decrease in the concentration of the other ion, achieved by the formation of a precipitate. This explains why solid lead(II) chromate precipitates out of solution.

The next Sample Problem shows how to predict the solubility of an ionic compound when a common ion is present in solution.

Sample Problem

The Effect of a Common Ion on Solubility

Problem

The solubility of pure $PbCrO_{4(s)}$ in water is 4.8×10^{-7} mol/L.

(a) Qualitatively predict how the solubility will change if $PbCrO_{4(s)}$ is added to a 0.10 mol/L solution of sodium chromate, Na_2CrO_4.

(b) K_{sp} for $PbCrO_{4(s)}$ is 2.3×10^{-13}. Determine the solubility of $PbCrO_{4(s)}$ in a 0.10 mol/L solution of Na_2CrO_4.

What Is Required?

You need to predict, and then determine, the solubility of $PbCrO_4$ (in mol/L) in a solution of Na_2CrO_4.

What Is Given?

You know K_{sp} for $PbCrO_4$ and the concentration of the salt with the common ion.

Continued ...

Continued ...

Plan Your Strategy

(a) Use Le Châtelier's principle to make a prediction.

(b)

Step 1 Write the equilibrium equation.

Step 2 Use the equilibrium equation to write an expression for K_{sp}.

Step 3 Set up an ICE table. Let x represent the concentration of chromate (the common ion) that is contributed by $PbCrO_4$. Initial conditions are based on the solution of Na_2CrO_4.

Step 4 Solve for x, and check your prediction.

Act on Your Strategy

(a) Based on Le Châtelier's principle, the solubility of $PbCrO_4$ in a solution that contains a common ion (chromate) will be less than the solubility of $PbCrO_4$ in water.

(b)

Step 1 $PbCrO_{4(s)} \rightleftharpoons Pb^{2+}{}_{(aq)} + CrO_4{}^{2-}{}_{(aq)}$

Step 2 $K_{sp} = [Pb^{2+}][CrO_4{}^{2-}] = 2.3 \times 10^{-13}$

Step 3

Concentration (mol/L)	$PbCrO_{4(s)}$	$\rightleftharpoons$	$Pb^{2+}{}_{(aq)}$	+	$CrO_4{}^{2-}{}_{(aq)}$
Initial	—		0		0.10
Change	—		$+x$		$+x$
Equilibrium	—		x		$0.10 + x$

Step 4 Since K_{sp} is very small, you can assume that x is much smaller than 0.10. To check the validity of this assumption, determine whether or not 0.10 is more than 500 times greater than K_{sp}:

$$\frac{0.10}{K_{sp}} = \frac{0.10}{2.3 \times 10^{-13}}$$
$$= 4.3 \times 10^{11} > 500$$

Therefore, in the $(0.10 + x)$ term, x can be ignored. In other words, $(0.10 + x)$ is approximately equal to 0.10. Therefore, you can simplify the equation as follows:

$$K_{sp} = [Pb^{2+}][CrO_4{}^{2-}] = 2.3 \times 10^{-13}$$
$$= (x)(0.10 + x)$$
$$\approx (x)(0.10)$$
$$\therefore (x)(0.10) \approx 2.3 \times 10^{-13}$$
$$x \approx 2.3 \times 10^{-12} \text{ mol/L}$$

The molar solubility of $PbCrO_4$ in a solution of 0.10 mol/L Na_2CrO_4 is 2.3×10^{-12} mol/L. Your prediction was correct. The solubility of $PbCrO_4$ decreases in a solution of common ions.

Check Your Solution

The approximation in step 4 is reasonable. The solubility of x is much smaller than 0.10. To prove that the approximation works, substitute the calculated value of x into the initial expression for K_{sp}.

Practice Problems

Note: Refer to Appendix E, as necessary.

21. Determine the molar solubility of AgCl
(a) in pure water (b) in 0.15 mol/L NaCl

22. Determine the molar solubility of lead(II) iodide, PbI_2, in 0.050 mol/L NaI.

23. Calculate the molar solubility of calcium sulfate, $CaSO_4$,
(a) in pure water (b) in 0.25 mol/L Na_2SO_4

24. K_{sp} for lead(II) chloride, $PbCl_2$, is 1.6×10^{-5}. Calculate the molar solubility of $PbCl_2$
(a) in pure water (b) in 0.10 mol/L $CaCl_2$

Electronic Learning Partner

To learn more about solubility equilibria, go to the Chemistry 12 Electronic Learning Partner.

Canadians  in Chemistry

Dr. Joseph MacInnis has a medical degree, but he does not work in a clinic, a hospital, or a typical doctor's office. His office is the ocean, and he is an expert in deep-sea diving. He combines his medical expertise with his knowledge of the properties of gases to study the effects of deep-sea diving on humans.

MacInnis was born in Barrie, Ontario, in 1937. In 1963, after graduating from the University of Toronto's medical school, he received a fellowship to study diving medicine at the University of Pennsylvania. Diving medicine is a specialized field, dealing with the challenges that humans face when exposed to undersea environments.

For every 10 m below sea level that a diver descends under water, ambient pressure increases by one unit of atmospheric pressure. This increase in pressure, and its effect on heterogeneous equilibria in the body, is one of the most profound challenges that a diver encounters. As a diver descends, and is subjected to increased pressure, gases such as nitrogen dissolve in the blood and tissues. In the equilibrium between dissolved nitrogen and gaseous nitrogen, increased pressure favours dissolved nitrogen, according to Le Châtelier's principle. As the diver ascends to the ocean surface and pressure decreases, equilibrium favours the nitrogen coming out of solution.

If a diver ascends too quickly, the lungs cannot remove the nitrogen fast enough. Gas bubbles form in the blood, causing decompression sickness. Decompression sickness is also known as "the bends," because one of its symptoms is an inability to bend the joints.

MacInnis was a consultant to the team that discovered the *Titanic*, and he was the first Canadian to explore the wreck. In 1991, MacInnis and his team used small submarines with pressurized hulls to film the wreck in IMAX format. The wreck is about 4000 m below sea level, but the pressure inside the submarines was kept similar to the pressure above the surface of the sea.

MacInnis is passionate about the environment, both below and above the sea. Since 1996, MacInnis has been Chair of the Friends of the Environment Foundation, a non-profit organization that funds projects that help to protect Canada's environment. In recognition of his accomplishments, MacInnis was made a member of the Order of Canada in 1976.

Earth in Equilibrium

Background

Scientists often think about Earth as being composed of four main systems. The *biosphere* includes all living things on Earth. The *geosphere* includes non-living things, such as rocks and soil. The *hydrosphere* includes all water, such as oceans, rivers, and even puddles. The *atmosphere* includes gases, vapours, and aerosols. These four systems overlap in many ways. For example, animals are part of the biosphere, but they interact with components in the geosphere (living in a burrow or cave), the hydrosphere (drinking and excreting water), and the atmosphere (breathing).

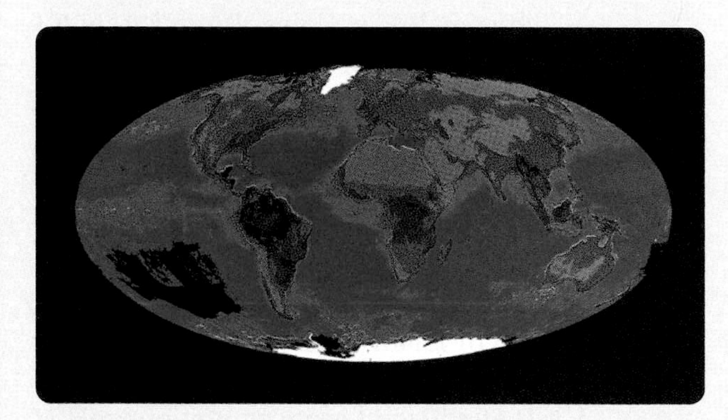

Earth, with its four spheres, is a good approximation of a closed system. As you learned in Unit 4, a closed system is subject to the principles of equilibrium. Every change to the system affects the whole equilibrium. Elements such as carbon, nitrogen, sulfur, and oxygen regularly cycle through Earth's four spheres. Thus, Earth is in a constant dynamic equilibrium.

Scientists have noticed various shifts in Earth's equilibrium in recent years. Since Earth is in a dynamic equilibrium, some shifts occur naturally. There is considerable evidence, however, that the following shifts have been caused by, or at least strongly influenced by, human activities. Some of these shifts may be familiar to you, with names such as "global warming," "ozone holes," and "loss of biodiversity."

- an increase in CO_2 and CH_4 levels in the atmosphere
- an increase in the amounts and concentrations of pollutant substances
- a decrease in ozone concentrations in certain parts of the atmosphere
- an increase in average global temperature
- a decrease in arable land and a corresponding increase in desert land
- a loss of plant and animal species, both by observable causes and by unknown causes
- an increase in extreme weather conditions, such as hurricanes and tornadoes

For this unit issue, you will research the carbon cycle, either on your own or as part of a group. Then, as a class, you will discuss the consequences of human activities on the carbon cycle.

Plan and Present

1. Research the carbon cycle. Include a general description of how carbon moves through the geosphere, biosphere, hydrosphere, and atmosphere. (Your teacher may choose to have some members of the class research different cycles, such as the phosphorus cycle or the nitrogen cycle.)
2. Identify specific chemical reactions and processes that involve the movement of carbon from one system to another. Write balanced equations for each reaction or process you identify.
3. Use the reactions and processes you identified in step 2, along with your knowledge of equilibria, to identify factors that can increase or decrease the level of carbon in each system.
4. Identify human influences on the carbon cycle.
5. Choose at least one shift in Earth's equilibrium that is listed in "Background." For each shift you choose, answer the following questions.
   (a) Is the carbon cycle involved? Explain your answer.
   (b) Where on Earth do you think this shift is most visible? Explain why you think so.
   (c) What are some possible causes of this shift?
   (d) What are some possible consequences?
6. If you come across information about a shift in Earth's equilibrium that is not listed in "Background," include a description of this shift with your research.
7. Have a class discussion on the following topics.
   (a) To what extent do human activities affect the carbon cycle? Are the consequences of any human activities serious enough to consider changing these activities?
   (b) If your answer to the second question in part (a) is yes, decide on a priority list of five human activities that your class believes should be changed.
   (c) Come up with an action plan for your school or community to address your concerns.

Assessment

After you have explored this issue, answer the following questions to assess your work.

- Assess your research skills. Was your research on this issue correct and thorough? Did you use more than one source? How have your skills improved?
- Assess your presentation of information to the class. Did you use visual aids in your explanation? Do you think everyone understood what you were trying to say?
- Assess the class discussion. Was a consensus reached? Do you think the consensus was based more on scientific research or on popular media opinion?

Evaluate the Results

1. If your class developed a priority list of five human activities that need changing, list these activities in your notebook. Explain why each activity was chosen to be on the list.
2. Describe any solutions or suggestions that your class developed. Are these suggestions likely to work? Why or why not?
3. Prepare a report that summarizes the results of your own research on the carbon cycle.

Extension

4. Do you think any other planets are in dynamic equilibrium? Do nutrient cycles occur on other planets? Explain your hypotheses. Then research the answers to these questions.

Web LINK

www.mcgrawhill.ca/links/chemistry12

Go to the web site above, and click on **Web Links** to find appropriate sites to research this unit issue.

Knowledge/Understanding

Multiple Choice

In your notebook, write the letter for the best answer to each question.

1. Which reaction represents a heterogeneous equilibrium?
   (a) $N_2O_{4(g)} \rightleftharpoons 2NO_{2(g)}$
   (b) $2SO_{2(g)} + O_{2(g)} \rightleftharpoons 2SO_{3(g)}$
   (c) $3H_{2(g)} + N_{2(g)} \rightleftharpoons 2NH_{3(g)}$
   (d) $2NaHCO_{3(s)} \rightleftharpoons Na_2CO_{3(s)} + CO_{2(g)} + H_2O_{(g)}$
   (e) $HCO_3{}^-{}_{(aq)} + H^+{}_{(aq)} \rightleftharpoons H_2CO_{3(aq)}$
2. Consider the following equilibrium.
   $N_2O_{4(g)} \rightleftharpoons 2NO_{2(g)}$ $K_c = 4.8 \times 10^{-3}$
   Which set of concentrations represents equilibrium conditions?
   (a) $[N_2O_{4(g)}] = 4.8 \times 10^{-1}$ and $[NO_{2(g)}] = 1.0 \times 10^{-4}$
   (b) $[N_2O_{4(g)}] = 1.0 \times 10^{-1}$ and $[NO_{2(g)}] = 4.8 \times 10^{-4}$
   (c) $[N_2O_{4(g)}] = 1.0 \times 10^{-1}$ and $[NO_{2(g)}] = 2.2 \times 10^{-2}$
   (d) $[N_2O_{4(g)}] = 2.2 \times 10^{-2}$ and $[NO_{2(g)}] = 1.0 \times 10^{-1}$
   (e) $[N_2O_{4(g)}] = 5.0 \times 10^{-2}$ and $[NO_{2(g)}] = 1.1 \times 10^{-2}$
3. Choose the equilibrium in which products are favoured by a decrease in pressure but reactants are favoured by a decrease in temperature.
   (a) $H_{2(g)} + I_{2(g)} + 51.8\ kJ \rightleftharpoons 2HI_{(g)}$
   (b) $2NO_{2(g)} \rightleftharpoons 2NO_{(g)} + O_{(g)}$ $\Delta H = +54\ kJ$
   (c) $N_{2(g)} + 3H_{2(g)} \rightleftharpoons 2NH_{3(g)} + 92.3\ kJ$
   (d) $PCl_{3(g)} + Cl_{2(g)} \rightleftharpoons PCl_{5(g)}$ $\Delta H = -84.2\ kJ$
   (e) $2SO_{2(g)} + O_{2(g)} \rightleftharpoons 2SO_{3(g)}$ + heat
4. A sample of lemon juice has a pH of 2. A sample of an ammonia cleaner has a pH of 10. If the two samples are combined, what ratio by volume of lemon juice to ammonia cleaner is needed to yield a neutral solution?
   (a) 1:10
   (b) 1:8
   (c) 1:100
   (d) 1:1000
   (e) 10:1
5. In which reaction is water acting as an acid?
   (a) $H_2O_{(\ell)} + NH_{3(aq)} \rightleftharpoons OH^-{}_{(aq)} + NH_4{}^+{}_{(aq)}$
   (b) $H_2O_{(\ell)} + H_3PO_{4(aq)} \rightleftharpoons H_3O^+{}_{(aq)} + H_2PO_4{}^-{}_{(aq)}$
   (c) $H_2O_{(\ell)} \rightleftharpoons H_{2(g)} + \frac{1}{2}O_{2(g)}$
   (d) $2H_2O_{(\ell)} + BaCl_{2(s)} \rightleftharpoons BaCl_2 + 2H_2O_{(g)}$
   (e) $2Na_2O_{2(s)} + 2H_2O_{(\ell)} \rightleftharpoons 4NaOH_{(aq)} + O_{2(g)}$
6. A solution of a weak acid is titrated with a solution of a strong base. What is the pH at the end-point?
   (a) 7
   (b) >7
   (c) <7
   (d) Either (b) or (c) may be correct, depending on K_a of the acid.
   (e) more information about the number of hydrogen ions is needed
7. 10 mL of 1.0 mol/L HCl is diluted by adding 990 mL of water. What is the pH of the new solution?
   (a) decrease by 2 pH units
   (b) decrease by 0.5 pH units
   (c) increase by 0.5 pH units
   (d) increase by 2 pH units
   (e) increase by a factor of 2 pH units
8. Select the correct statement about the following equilibrium.
   $HBO_3{}^{2-}{}_{(aq)} + HSiO_3{}^-{}_{(aq)} \rightleftharpoons SiO_3{}^{2-}{}_{(aq)} + H_2BO_3{}^-{}_{(aq)}$
   (a) $HBO_3{}^{2-}{}_{(aq)}$ and $HSiO_3{}^-{}_{(aq)}$ are a conjugate acid-base pair.
   (b) $HSiO_3{}^-{}_{(aq)}$ and $SiO_3{}^{2-}{}_{(aq)}$ are both acting as acids.
   (c) $HBO_3{}^{2-}{}_{(aq)}$ and $SiO_3{}^{2-}{}_{(aq)}$ are both bases.
   (d) $HSiO_3{}^-{}_{(aq)}$ and $H_2BO_3{}^-{}_{(aq)}$ are a conjugate acid-base pair.
   (e) $SiO_3{}^{2-}{}_{(aq)}$ and $H_2BO_3{}^-{}_{(aq)}$ are a conjugate acid-base pair
9. Select the equation that correctly represents the dissociation of $Al_2S_{3(s)}$ in water.
   (a) $Al_2S_{3(s)} \rightleftharpoons Al_2{}^{3+}{}_{(aq)} + S_3{}^{2-}{}_{(aq)}$
   (b) $Al_2S_{3(s)} \rightleftharpoons 2Al^{2+}{}_{(aq)} + 2S^{3-}{}_{(aq)}$
   (c) $Al_2S_{3(s)} \rightleftharpoons 2Al^{3+}{}_{(aq)} + 3S^{2-}{}_{(aq)}$
   (d) $Al_2S_{3(s)} \rightleftharpoons 3Al^{3+}{}_{(aq)} + 2S^{2-}{}_{(aq)}$
   (e) $Al_2S_{3(s)} \rightleftharpoons 2Al^{3+}{}_{(aq)} + 6S^-{}_{(aq)}$
10. A saturated solution of CH_3COOAg contains 2×10^{-3} mol Ag^+ per litre of solution. What is K_{sp} for CH_3COOAg?
    (a) 2×10^{-3}
    (b) 4×10^{-6}
    (c) 1×10^{-3}
    (d) 2×10^{-6}
    (e) 4×10^{-3}
11. In an experiment, a chemist determines that PbC_2O_4 and $Ag_2C_2O_4$ have the same K_{sp}: 1.1×10^{-11}. Based on this result, which statement can the chemist make?

(a) These compounds have the same molar solubility.
(b) $[C_2O_4^{2-}]$ in saturated solutions of these two compounds is the same.
(c) In saturated solutions of these compounds, $[Ag^+] > [Pb^{2+}]$.
(d) In equal volumes of saturated solutions, a greater mass of solute is present in PbC_2O_4.
(e) In saturated solutions of each compound, PbC_2O_4 contains 1.5 times as many ions.

Short Answer

In your notebook, write a sentence or a short paragraph to answer each question.

12. Consider the following equilibrium.
$2SO_{2(g)} + O_{2(g)} \rightleftharpoons 2SO_{3(g)}$ $\Delta H = -198.2$ kJ
Indicate if reactants or products are favoured, or if no change occurs, when
(a) the temperature is increased
(b) helium gas is added at constant volume
(c) helium gas is added at constant pressure

13. Explain the reasoning you used to answer question 12(b).

14. Describe how a decrease in volume will affect the following equilibrium.
$PCl_{3(g)} + Cl_{2(g)} \rightleftharpoons PCl_{5(g)}$

15. K_c for the equilibrium $N_2O_4 \rightleftharpoons 2NO_2$ is 4.8×10^{-3}. What is K_c for the equilibrium $\frac{1}{2}N_2O_4 \rightleftharpoons NO_2$?

16. For the following equilibrium, $K_c = 1.0 \times 10^{-15}$ at 25°C and $K_c = 0.05$ at 2200°C.
$N_{2(g)} + O_{2(g)} \rightleftharpoons 2NO_{(g)}$
Based on this information, is the reaction exothermic or endothermic? Briefly explain your answer.

17. K_c for the equilibrium system $2SO_{2(g)} + 2O_{2(g)} \rightleftharpoons SO_{3(g)}$ is 40. What is K_c for $SO_{3(g)} \rightleftharpoons SO_{2(g)} + \frac{1}{2}O_{2(g)}$?

18. 1.0 mol of $NH_{3(g)}$ is placed initially in a 500 mL flask. The following equilibrium is established.
$2NH_{3(g)} \rightleftharpoons N_{2(g)} + 3H_{2(g)}$
Write an equilibrium expression that describes the system at equilibrium.

19. A mixture is prepared for the following equilibrium.
$I_{2(aq)} + I^-_{(aq)} \rightleftharpoons I_3^-{}_{(aq)}$
The initial concentrations of $I_{2(aq)}$ and $I^-_{(aq)}$ are both 0.002 mol/L. When equilibrium is established, $[I_{2(aq)}] = 5.0 \times 10^{-4}$ mol/L. What is K_c for this equilibrium?

20. Solutions of $NaHCO_{3(aq)}$ and $CaHPO_{4(aq)}$ are mixed.
(a) Write the net ionic equation that represents the equilibrium for this system.
(b) What are the conjugate acid-base pairs?
(c) Are reactants or products favoured in this system? Give a reason for your answer.

21. What is the pH of a 100 mL sample of 0.002 mol/L H_2SO_4?

22. A 1.00 L solution contains 1.04 g of KOH. What is the pH of this solution?

23. What is $[H^+_{(aq)}]$, $[OH^-_{(aq)}]$, pH, and pOH for a 0.048 mol/L solution of benzoic acid, C_6H_5COOH?

24. How many moles of phosphate ions are present in 200 mL of 0.002 mol/L $Fe_3(PO_4)_2$?

25. Aqueous solutions of 0.2 mol/L Na_2CO_3 and 0.2 mol/L $AgNO_3$ are mixed. Write a net ionic equation that represents the overall reaction.

26. What is the highest concentration of Ca^{2+} that can be present in a solution of 0.002 00 mol/L $NH_4F_{(aq)}$?

27. 20.00 mL of 1.100×10^{-4} mol/L $Pb(NO_3)_2$ is mixed with 80.00 mL of 4.450×10^{-2} mol/L CaI_2. Show, using calculations, if a precipitate will form.

Inquiry

28. A chemist evaporates 100.0 mL of a saturated solution of lead chromate to dryness. What mass of $PbCrO_{4(s)}$ remains?

29. How can a chemist maximize the yield of $CaO_{(s)}$ in the following equilibrium system?
$CaCO_{3(s)} + \text{heat} \rightleftharpoons CaO_{(s)} + CO_{2(g)}$?

30. K_c for the following system is 1.05 at a certain temperature.
$PCl_{5(g)} + \text{heat} \rightleftharpoons PCl_{3(g)} + Cl_{2(g)}$
After 0.22 mol of $PCl_{3(g)}$ and 0.11 mol of $Cl_{2(g)}$ are added to a 500 mL flask, equilibrium is established.
(a) Explain what happens to the rates of the forward and reverse reactions as equilibrium is established.

(b) Calculate the equilibrium concentration of each component of this equilibrium.

31. Cyanide ion, CN^-, reacts with Fe^{3+} to form the blue dye that is used in blueprint paper. Hydrocyanic acid, $HCN_{(aq)}$, is a weak acid, with $K_a = 6.2 \times 10^{-10}$.
(a) Calculate K_b for the conjugate base of $HCN_{(aq)}$.
(b) Calculate the pH of 0.120 mol/L $KCN_{(aq)}$.

32. The following solutions have all been prepared at a concentration of 0.1 mol/L: $CH_3COONa_{(aq)}$, $CH_3COOH_{(aq)}$, $NH_{3(aq)}$, and $NaOH_{(aq)}$. Arrange these solutions in order of decreasing pH. For each solution, write the balanced equation that supports your answer.

33. In an experiment to estimate K_{sp}, a piece of zinc metal was left in 50.0 mL of a saturated solution of $PbCl_2$. The lead that was produced in the displacement reaction was recovered, and its mass was found to be 0.17 g.
(a) Use this information to determine the concentration of Pb^{2+} in the original solution.
(b) Write the expression for K_{sp}. Then calculate K_{sp} for $PbCl_2$.

34. For the following equilibrium, $K_c = 0.81$ at a certain temperature.
$\frac{1}{2}N_{2(g)} + \frac{3}{2}H_{2(g)} \rightleftharpoons NH_{3(g)} + \text{heat}$
(a) Explain what will happen to the magnitude of K_c if the temperature is increased.
(b) The initial concentrations of the gases are $[H_{2(g)}] = 0.76$ mol/L, $[N_{2(g)}] = 0.60$ mol/L, and $[NH_{3(g)}] = 0.48$ mol/L. Determine if the concentration of $[N_{2(g)}]$ will increase or decrease when equilibrium is established.

35. 1 L of 0.002 mol/L $Pb(NO_3)_2$ has been prepared. Given that K_{sp} for $PbCl_2$ is 1.7×10^{-5}, what is the largest number of moles of NaCl that can be added to this solution without causing a precipitate to form?

36. A 10.0 g sample of a commercial washing powder contains ammonium sulfate, $(NH_4)_2SO_4$. The sample is treated with an excess of NaOH.
$(NH_4)_2SO_{4(aq)} + 2NaOH_{(aq)} \rightarrow 2NH_{3(g)} + Na_2SO_{4(aq)} + H_2O_{(\ell)}$
The ammonia gas, $NH_{3(g)}$, that forms is dissolved in 50.0 mL of 0.250 mol/L H_2SO_4. The following reaction occurs.
$2NH_{3(g)} + H_2SO_{4(aq)} \rightarrow (NH_4)_2SO_{4(aq)}$
The $H_2SO_{4(aq)}$ that is not used up in this reaction is titrated to the end-point with 27.9 mL of 0.230 mol/L NaOH. Use this information to calculate the percent of $(NH_4)_2SO_4$ in the 10.0 g sample of washing powder.

37. Consider the following equilibrium.
$2XO_{(s)} + O_2 \rightleftharpoons 2XO_{2(g)} + \text{heat}$.
(a) Predict whether reactants or products are favoured when the volume of the container is increased. Give a reason for your answer.
(b) What do you predict will happen to the concentration of each substance when the volume of the container is increased?
(c) At a certain temperature, $K_c = 1.00 \times 10^{-4}$ and the equilibrium concentrations are $[O_2] = 1.99$ mol/L and $[XO_2] = 1.41 \times 10^{-2}$ mol/L. If the volume of the container is now doubled, calculate the concentrations of these gases when equilibrium is re-established. Assume that sufficient $XO_{(s)}$ is present to maintain equilibrium. Are the concentrations you calculate consistent with your prediction in part (b)?

38. A sample of $Ca(OH)_{2(s)}$ is allowed to stand in distilled water until a saturated solution forms. A 50.0 mL sample of this saturated solution is neutralized with 24.4 mL of 0.0500 mol/L HCl.
$2HCl_{(aq)} + Ca(OH)_{2(aq)} \rightarrow CaCl_{2(aq)} + 2H_2O_{(\ell)}$
Use this information to calculate K_{sp} for $Ca(OH)_2$.

39. Tap water that contains Cl^- at a concentration of 50 ppm is used to prepare a 0.100 mol/L $AgNO_3$ solution. Does a precipitate form?

Communication

40. Under what circumstances is the position of equilibrium not affected by a change in the pressure on the system? Give examples, and explain your answer.

41. Milk of magnesia is a suspension of $Mg(OH)_{2(s)}$ that can be used as an antacid. K_{sp} for $Mg(OH)_{2(s)}$ is only 5.6×10^{-12}. If so little of this base dissolves, how can it be effective as an antacid? Explain your answer.

42. For HSO_3^-, K_a is 6.2×10^{-8} and K_b is 7.7×10^{-13}. Your friend is confused because the

product of these two values does not equal K_w. Write a note that explains to your friend why this is correct.

43. Use equations to demonstrate how a buffer system, such as $HNO_2 : NO_2^-$, reacts with H_3O^+ and OH^-.

44. Aluminum is a self-protecting metal. It readily forms an impervious coating of Al_2O_3 that protects the aluminum from further corrosion. Does this reaction occur spontaneously at all temperatures? Explain your answer.

45. 3.50 g of CO_2 are dissolved in 375 mL of pop to carbonate the drink. Assuming that no other factors affect the acidity of the carbonated drink, explain how this information can be used to calculate its pH. (You do not need to do the calculation.)

46. Acid-base indicators are often large organic molecules. Explain why they are sold in the form of a sodium salt, for general use as an acid-base indicator.

Making Connections

47. According to guidelines published by the Ontario Ministry of the Environment in January 2001, the Maximum Acceptable Concentration (MAC) for Pb^{2+} in drinking water is 0.010 mg/L. The water in a community is contaminated with Pb^{2+} from an industrial source, at a concentration of 0.0800 mg/L.

(a) Engineers decide to clean up the water by increasing its pH. Explain why increasing the water's pH will remove the lead ions from the water.

(b) At what pH is the concentration of lead ions reduced from 0.0800 mg/L to 0.010 mg/L?

48. A patient's kidney stones are composed primarily of calcium oxalate, CaC_2O_4. The patient's physician prescribes a medicine that is basic, and advises the patient to drink plenty of water. Explain why this is a suitable treatment.

COURSE CHALLENGE

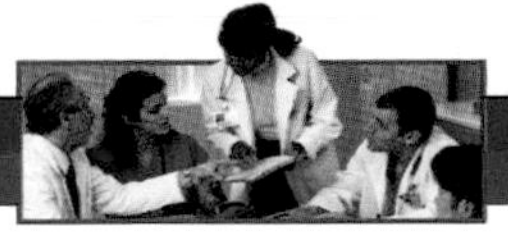

The Chemistry of Human Health

In the Course Challenge, you will explore buffer systems in the body.

- How does the body maintain a constant pH in the blood?
- How can illness affect the equilibrium in the blood?
- How can you model the buffer system in your blood?

UNIT 5

Electrochemistry

UNIT 5 CONTENTS

CHAPTER 10
Oxidation-Reduction Reactions

CHAPTER 11
Cells and Batteries

DESIGN YOUR OWN INVESTIGATION
Electroplating

UNIT 5 OVERALL EXPECTATIONS

- What are oxidation-reduction reactions? How are they involved in the interconversion of chemical and electrical energy?
- How are galvanic and electrolytic cells built, and how do they function? What equations are used to describe these types of cells? How can you solve quantitative problems related to electrolysis?
- What are the uses of batteries and fuel cells? How is electrochemical technology used to produce and protect metals? How can you assess the environmental and safety issues associated with these technologies?

Unit Issue Prep

As you progress through Unit 5, look for the skills and information you will need for the investigation at the end of the unit.

Canadian engineer Dr. John Hopps, working with a team of medical researchers in the late 1940s, developed one of the most significant medical inventions of the twentieth century: the pacemaker. The photograph on the right shows a pacemaker embedded in the body of a heart patient. A modern pacemaker is essentially a tiny computer that monitors a person's heartbeat and corrects irregularities as needed. Pacemakers are particularly useful in correcting a heartbeat that is too slow.

The pacemaker device is surgically placed in a "pocket" of tissue near the patient's collarbone. One or more wires, called "leads," are connected to the pacemaker and threaded down through a major vein to the patient's heart. By sending electrical impulses along the leads to the heart, the pacemaker can induce a heartbeat.

A pacemaker obtains electrical energy from a tiny battery that lasts for about seven years before it must be replaced. But how do batteries supply electrical energy? The answer lies in a branch of chemistry known as electrochemistry. In this unit, you will learn about the connection between chemical reactions and electricity. You will also learn about the chemical reactions that take place inside batteries.

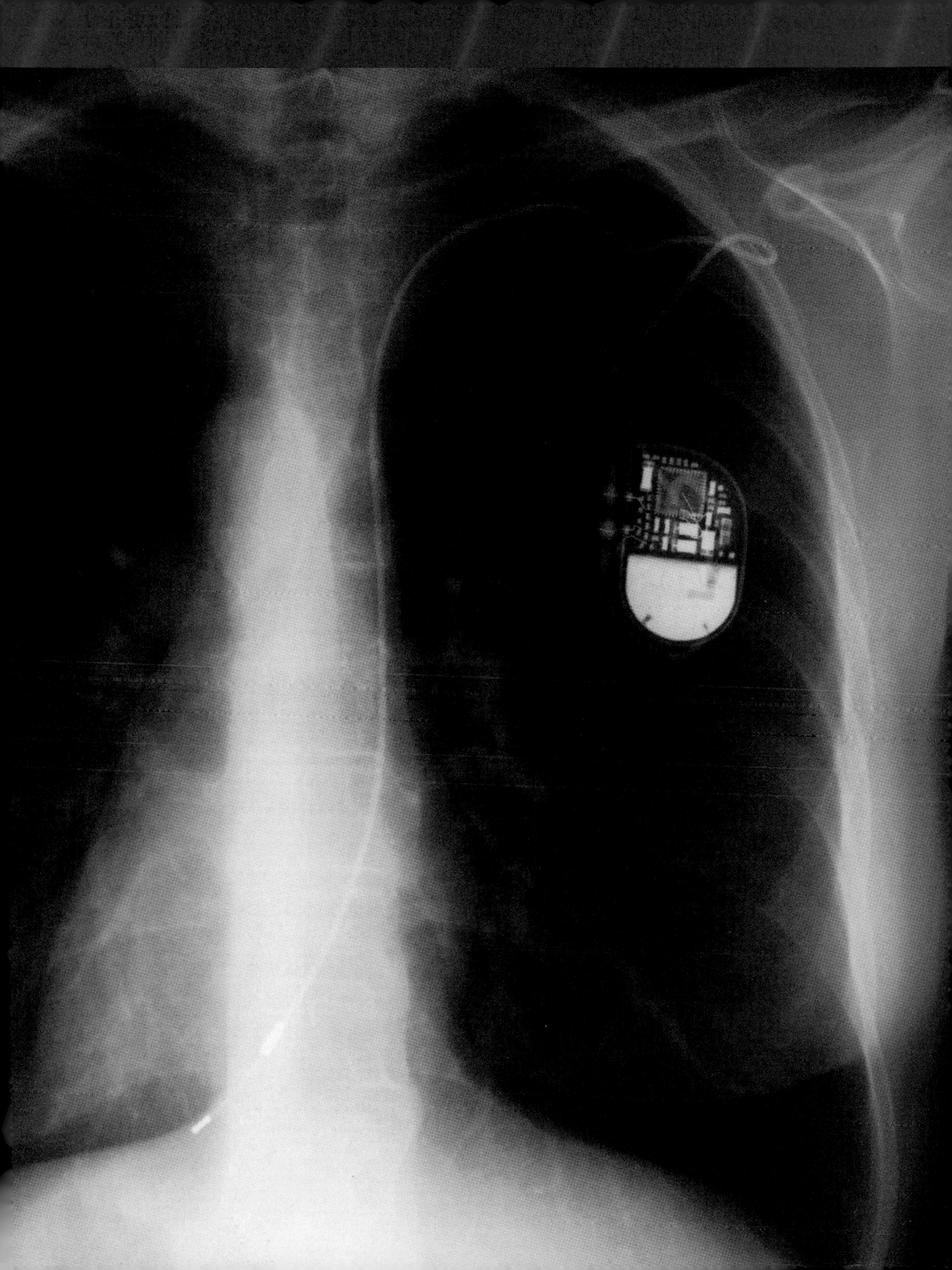

CHAPTER

10 Oxidation-Reduction Reactions

Chapter Preview

10.1 Defining Oxidation and Reduction

10.2 Oxidation Numbers

10.3 The Half-Reaction Method for Balancing Equations

10.4 The Oxidation Number Method for Balancing Equations

Prerequisite Concepts and Skills

Before you begin this chapter, review the following concepts and skills:

- balancing chemical, total ionic, and net ionic equations (Concepts and Skills Review)
- reaction types, including synthesis, decomposition, single displacement, and double displacement reactions (Concepts and Skills Review)
- the common ionic charges of metal ions and non-metal ions, and the formulas of common polyatomic ions
- drawing Lewis structures (Concepts and Skills Review)
- electronegativities and bond polarities (Chapter 4, section 4.1)

Kitchen chemistry is an important part of daily life. Cooks use chemistry all the time to prepare and preserve food. Even the simplest things you do in the kitchen can involve chemical reactions. For example, you have probably seen a sliced apple turn brown. The same thing happens to pears, bananas, avocados, and several other fruits. Slicing the fruit exposes the flesh to oxygen in the air. Compounds in the fruit react with oxygen to form brown products. An enzyme in the fruit acts as a catalyst, speeding up this reaction. How can you stop fruit from turning brown after it is sliced?

A Waldorf salad uses a simple method to prevent fruit from browning. This type of salad usually consists of diced apples, celery, and walnuts, covered with a mayonnaise dressing. The dressing keeps the air away from the food ingredients. Without air, the fruit does not turn brown.

Another way to solve this problem is to prevent the enzyme in the fruit from acting as a catalyst. Enzymes are sensitive to pH. Therefore, adding an acid such as lemon juice or vinegar to fruit can prevent the enzyme from acting. You may have noticed that avocado salad recipes often include lemon juice. In addition to hindering the enzyme, lemon juice contains vitamin C, which is very reactive toward oxygen. The vitamin C reacts with oxygen before the sliced fruit can do so.

In this chapter, you will be introduced to oxidation-reduction reactions, also called redox reactions. You will discover how to identify this type of reaction. You will also find out how to balance equations for a redox reaction.

A redox reaction causes fruit to go brown. How can you recognize other redox reactions?

Defining Oxidation and Reduction

10.1

Section Preview/ Specific Expectations

In this section, you will

- **describe** oxidation and reduction in terms of the loss and the gain of electrons
- **write** half-reactions from balanced chemical equations for oxidation-reduction systems
- **investigate** oxidation-reduction reactions by comparing the reactivities of some metals
- **communicate** your understanding of the terms *ore, oxidation, reduction, oxidation-reduction reaction, redox reaction, oxidizing agent, reducing agent, half-reaction, disproportionation*

The term *oxidation* can be used to describe the process in which certain fruits turn brown by reacting with oxygen. The original, historical definition of this term was "to combine with oxygen." Thus, oxidation occurred when iron rusted, and when magnesium was burned in oxygen gas. The term *reduction* was used historically to describe the opposite of oxidation, that is, the formation of a metal from its compounds. An **ore** is a naturally occurring solid compound or mixture of compounds from which a metal can be extracted. Thus, the process of obtaining a metal from an ore was known as a reduction. Copper ore was reduced to yield copper, and iron ore was reduced to yield iron.

As you will learn in this chapter, the modern definitions for oxidation and reduction are much broader. The current definitions are based on the idea of electron transfers, and can now be applied to numerous chemical reactions. In Unit 1, you saw the terms oxidation and reduction used to describe changes to carbon-hydrogen and carbon-oxygen bonds within organic compounds. These changes involve electron transfers, so the broader definitions that you will learn in this chapter still apply.

In your previous chemistry course, you compared the reactivities of metals. You may recall that, when a piece of zinc is placed in an aqueous solution of copper(II) sulfate, the zinc displaces the copper in a single displacement reaction. This reaction is shown in Figure 10.1. As the zinc dissolves, the zinc strip gets smaller. A dark red-brown layer of solid copper forms on the zinc strip, and some copper is deposited on the bottom of the beaker. The blue colour of the solution fades, as blue copper(II) ions are replaced by colourless zinc ions.

Figure 10.1 A solid zinc strip reacts with a solution that contains blue copper(II) ions.

CONCEPT CHECK

From your earlier work, you will recognize the sulfate ion, SO_4^{2-}, as a polyatomic ion. To review the names and formulas of common polyatomic ions, refer to Appendix E, Table E.5.

The reaction in Figure 10.1 is represented by the following equation.

$$Zn_{(s)} + CuSO_{4(aq)} \rightarrow Cu_{(s)} + ZnSO_{4(aq)}$$

This equation can be written as a total ionic equation.

$$Zn_{(s)} + Cu^{2+}{}_{(aq)} + SO_4{}^{2-}{}_{(aq)} \rightarrow Cu_{(s)} + Zn^{2+}{}_{(aq)} + SO_4{}^{2-}{}_{(aq)}$$

The sulfate ions are *spectator ions*, meaning ions that are not involved in the chemical reaction. By omitting the spectator ions, you obtain the following net ionic equation.

$$Zn_{(s)} + Cu^{2+}{}_{(aq)} \rightarrow Cu_{(s)} + Zn^{2+}{}_{(aq)}$$

Notice what happens to the reactants in this equation. The zinc atoms *lose* electrons to form zinc ions. The copper ions *gain* electrons to form copper atoms.

gains $2e^-$

$$Zn_{(s)} + Cu^{2+}{}_{(aq)} \rightarrow Cu_{(s)} + Zn^{2+}{}_{(aq)}$$

loses $2e^-$

The following chemical definitions describe these changes.

- **Oxidation** is the loss of electrons.
- **Reduction** is the gain of electrons.

Try using a mnemonic to remember the definitions for oxidation and reduction. For example, in "LEO the lion says GER," LEO stands for "Loss of Electrons is Oxidation." GER stands for "Gain of Electrons is Reduction." The mnemonic "OIL RIG" stands for "Oxidation Is Loss. Reduction Is Gain." Make up your own mnemonic to help you remember these definitions.

In the reaction of zinc atoms with copper(II) ions, the zinc atoms lose electrons and undergo oxidation. In other words, the zinc atoms are *oxidized*. The copper(II) ions gain electrons and undergo reduction. In other words, the copper(II) ions are *reduced*. Because oxidation and reduction both occur in the reaction, it is known as an **oxidation-reduction reaction** or **redox reaction**.

Notice that electrons are transferred from zinc atoms to copper(II) ions. The copper(II) ions are responsible for the oxidation of the zinc atoms. A reactant that oxidizes another reactant is called an **oxidizing agent**. The oxidizing agent accepts electrons in a redox reaction. In this reaction, copper(II) is the oxidizing agent. The zinc atoms are responsible for the reduction of the copper(II) ions. A reactant that reduces another reactant is called a **reducing agent**. The reducing agent gives or donates electrons in a redox reaction. In this reaction, zinc is the reducing agent.

A redox reaction can also be defined as a reaction between an oxidizing agent and a reducing agent, as illustrated in Figure 10.2.

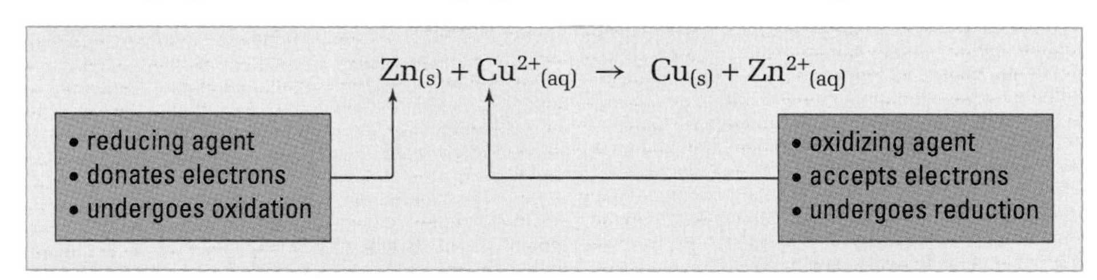

Figure 10.2 **In a redox reaction, the reducing agent is oxidized, and the oxidizing agent is reduced. Note that the oxidizing agent *does not* undergo oxidation, and that the reducing agent *does not* undergo reduction.**

Try the following practice problems to review your understanding of net ionic equations, and to work with the new concepts of oxidation and reduction.

Practice Problems

1. Write a balanced net ionic equation for the reaction of zinc with aqueous iron(II) chloride. Include the physical states of the reactants and products.
2. Write a balanced net ionic equation for each reaction, including physical states.
   (a) magnesium with aqueous aluminum sulfate
   (b) a solution of silver nitrate with metallic cadmium
3. Identify the reactant oxidized and the reactant reduced in each reaction in question 2.
4. Identify the oxidizing agent and the reducing agent in each reaction in question 2.

Half-Reactions

To monitor the transfer of electrons in a redox reaction, you can represent the oxidation and reduction separately. A **half-reaction** is a balanced equation that shows the number of electrons involved in either oxidation or reduction. Because a redox reaction involves both oxidation and reduction, two half-reactions are needed to represent a redox reaction. One half-reaction shows oxidation, and the other half-reaction shows reduction.

As you saw earlier, the reaction of zinc with aqueous copper(II) sulfate can be represented by the following net ionic equation.

$$Zn_{(s)} + Cu^{2+}{}_{(aq)} \rightarrow Cu_{(s)} + Zn^{2+}{}_{(aq)}$$

Each neutral Zn atom is oxidized to form a Zn^{2+} ion. Thus, each Zn atom must lose two electrons. You can write an oxidation half-reaction to show this change.

$$Zn_{(s)} \rightarrow Zn^{2+}{}_{(aq)} + 2e^-$$

Each Cu^{2+} ion is reduced to form a neutral Cu atom. Thus, each Cu^{2+} ion must gain two electrons. You can write a reduction half-reaction to show this change.

$$Cu^{2+}{}_{(aq)} + 2e^- \rightarrow Cu_{(s)}$$

If you look again at each half-reaction above, you will notice that the atoms and the charges are balanced. Like other types of balanced equations, half-reactions are balanced using the smallest possible whole-number coefficients. In the following equation, the atoms and charges are balanced, but the coefficients can all be divided by 2 to give the usual form of the half-reaction.

$$2Cu^{2+}{}_{(aq)} + 4e^- \rightarrow 2Cu_{(s)}$$

CONCEPT CHECK

You can write separate oxidation and reduction half-reactions to represent a redox reaction, but one half-reaction cannot occur on its own. Explain why this statement must be true.

In most redox reactions, one substance is oxidized and a different substance is reduced. In a **disproportionation** reaction, however, a single element undergoes both oxidation and reduction in the same reaction. For example, a copper(I) solution undergoes disproportionation in the following reaction.

$$2Cu^{+}{}_{(aq)} \rightarrow Cu_{(s)} + Cu^{2+}{}_{(aq)}$$

In this reaction, some copper(I) ions gain electrons, while other copper(I) ions lose electrons.

$$Cu^{+}{}_{(aq)} + Cu^{+}{}_{(aq)} \rightarrow Cu_{(s)} + Cu^{2+}{}_{(aq)}$$

gains 1e⁻

loses 1e⁻

The two half-reactions are as follows.

Oxidation: $Cu^{+}{}_{(aq)} \rightarrow Cu^{2+}{}_{(aq)} + 1e^{-}$

Reduction: $Cu^{+}{}_{(aq)} + 1e^{-} \rightarrow Cu_{(s)}$

You have learned that half-reactions can be used to represent oxidation and reduction separately. Half-reactions always come in pairs: an oxidation half-reaction is always accompanied by a reduction half-reaction, and vice versa. Try writing and balancing half-reactions using the following practice problems.

Practice Problems

5. Write balanced half-reactions from the net ionic equation for the reaction between solid aluminum and aqueous iron(III) sulfate. The sulfate ions are spectator ions, and are not included.

$Al_{(s)} + Fe^{3+}{}_{(aq)} \rightarrow Al^{3+}{}_{(aq)} + Fe_{(s)}$

6. Write balanced half-reactions from the following net ionic equations.

(a) $Fe_{(s)} + Cu^{2+}{}_{(aq)} \rightarrow Fe^{2+}{}_{(aq)} + Cu_{(s)}$

(b) $Cd_{(s)} + 2Ag^{+}{}_{(aq)} \rightarrow Cd^{2+}{}_{(aq)} + 2Ag_{(s)}$

7. Write balanced half-reactions for each of the following reactions.

(a) $Sn_{(s)} + PbCl_{2(aq)} \rightarrow SnCl_{2(aq)} + Pb_{(s)}$

(b) $Au(NO_3)_{3(aq)} + 3Ag_{(s)} \rightarrow 3AgNO_{3(aq)} + Au_{(s)}$

(c) $3Zn_{(s)} + Fe_2(SO_4)_{3(aq)} \rightarrow 3ZnSO_{4(aq)} + 2Fe_{(s)}$

8. Write the net ionic equation and the half-reactions for the disproportionation of mercury(I) ions in aqueous solution to give liquid mercury and aqueous mercury(II) ions. Assume that mercury(I) ions exist in solution as $Hg_2{}^{2+}$.

You already know that some metals are more reactive than others. You may also have carried out an investigation on the metal activity series in a previous course. In Investigation 10-A, located on page 470, you will discover how this series is related to oxidation and reduction. You will write chemical equations, ionic equations, and half-reactions for the single displacement reactions of several metals.

COURSE CHALLENGE

The Chemistry Bulletin, on the next page, introduces you to the terms *oxidant* and *antioxidant*. How may oxidants and antioxidants affect human health? Consider this question to prepare for your Chemistry Course Challenge.

Chemistry Bulletin

Science | Technology | Society | Environment

Aging: Is Oxidation a Factor?

Why do we grow old? Despite advances in molecular biology and medical research, the reasons for aging remain mysterious. One theory suggests that aging may be influenced by oxidizing agents, also known as *oxidants*.

Oxidants are present in the environment and in foods. Nitrogen oxides are oxidants present in cigarette smoke and urban smog. Other oxidants include the copper and iron salts in meat and some plants. Inhaling and ingesting oxidants such as these can increase the level of oxidants in our bodies.

Oxidants are also naturally present in the body, where they participate in important redox reactions. For example, mitochondria consume oxygen during aerobic respiration, and cells ingest and destroy bacteria. Both these processes involve oxidation and reduction.

As you have just seen, redox reactions are an essential part of your body's processes. However, these reactions can produce *free radicals*, which are highly reactive atoms or molecules with one or more unpaired electrons. Because they are so reactive, free radicals can oxidize surrounding molecules by robbing them of electrons. This process can damage DNA, proteins, and other macromolecules. Such damage may contribute to aging, and to diseases that are common among the aging, such as cancer, cardiovascular disease, and cataracts.

The study of oxidative damage has sparked a debate about the role that antioxidants might play in illness and aging. *Antioxidants* are reducing agents. They donate electrons to substances that have been oxidized, decreasing the damage caused by free radicals. Dietary antioxidants include vitamins C and E, beta-carotene, and carotenoids.

Most medical researchers agree that people with diets rich in fruits and vegetables have a lower incidence of cardiovascular disease, certain cancers, and cataracts. Although fruits and vegetables are high in antioxidants, they also contain fibre and many different vitamins and plant chemicals. It is hard to disentangle the effects of antioxidants from the beneficial effects of these other substances.

Carotenoids are pigments found in some fruits and vegetables, including spinach.

As a result, the benefits of antioxidant dietary supplements are under debate. According to one study, vitamin E supplements may lower the risk of heart disease. Another study, however, concludes that taking beta-carotene supplements does *not* reduce the risk of certain cancers.

We can be sure that a balanced diet including fruits and vegetables is beneficial to human health. Whether antioxidants confer these benefits, and whether these benefits include longevity, remain to be seen.

Making Connections

1. Research vitamins C, E, alpha- and beta-carotenes, and folic acid. How do they affect our health? What fruits and vegetables contain these vitamins?

2. Lycopene is a carotenoid that has been linked to a decreased risk of pancreatic, cervical, and prostate cancer. Find out what fruits and vegetables contain lycopene. What colour are these fruits and vegetables?

Investigation 10-A

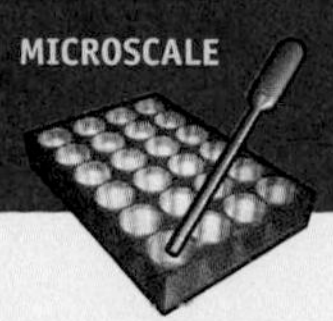

SKILL FOCUS

Predicting

Performing and recording

Analyzing and interpreting

Single Displacement Reactions

The metal activity series is shown in the table below. The more reactive metals are near the top of the series, and the less reactive metals are near the bottom. In this investigation, you will relate the activity series to the ease with which metals are oxidized and metal ions are reduced.

Activity Series of Metals

Metal	
lithium	Most Reactive
potassium	↑
barium	
calcium	
sodium	
magnesium	
aluminum	
zinc	
chromium	
iron	
cadmium	
cobalt	
nickel	
tin	
lead	
copper	
mercury	
silver	
platinum	
gold	Least Reactive

Question

How is the order of the metals in the activity series related to the ease with which metals are oxidized and metal ions are reduced?

Predictions

Predict the relative ease with which the metals aluminum, copper, iron, magnesium, and zinc can be oxidized. Predict the relative ease with which the ions of these same metals can be reduced. Explain your reasoning in both cases.

Materials

well plate
test tube rack
4 small test tubes
4 small pieces of each of these metals: aluminum foil, thin copper wire or tiny copper beads, iron filings, magnesium, and zinc
dropper bottles containing dilute solutions of aluminum sulfate, copper(II) sulfate, iron(II) sulfate, magnesium sulfate, and zinc nitrate

Safety Precautions

- Wear goggles, gloves, and an apron for all parts of this investigation.

Procedure

1. Place the well plate on a white piece of paper. Label it to match the table on the next page.
2. In each well plate, place a small piece of the appropriate metal, about the size of a grain of rice. Cover each piece with a few drops of the appropriate solution. Wait 3–5 min to observe if a reaction occurs.
3. Look for evidence of a chemical reaction in each mixture. Record the results, using "y" for a reaction, or "n" for no reaction. If you are unsure, repeat the process on a larger scale in a small test tube.

Metal \ Compound	$Al_2(SO_4)_3$	$CuSO_4$	$FeSO_4$	$MgSO_4$	$Zn(NO_3)_2$
Al					
Cu					
Fe					
Mg					
Zn					

4. Discard the mixtures in the waste beaker supplied by your teacher. Do not pour anything down the drain.

Analysis

1. For each single displacement reaction you observed, write
   (a) a balanced chemical equation
   (b) a total ionic equation
   (c) a net ionic equation
2. Write an oxidation half-reaction and a reduction half-reaction for each net ionic equation you wrote in question 1. Use the smallest possible whole-number coefficients in each half-reaction.
3. Look at each balanced net ionic equation. Compare the total number of electrons lost by the reducing agent with the total number of electrons gained by the oxidizing agent.
4. List the different oxidation half-reactions. Start with the half-reaction for the most easily oxidized metal, and end with the half-reaction for the least easily oxidized metal. Explain your reasoning. Compare your list with your first prediction from the beginning of this investigation.
5. List the different reduction half-reactions. Start with the half-reaction for the most easily reduced metal ion, and end with the half-reaction for the least easily reduced metal ion. Explain your reasoning. Compare your list with your second prediction from the beginning of this investigation.

Conclusions

6. Which list from questions 4 and 5 puts the metals in the same order as they appear in the activity series?
7. How is the order of the metals in the activity series related to the ease with which metals are oxidized and metal ions are reduced?

Applications

8. Use the activity series to choose a reducing agent that will reduce aqueous nickel(II) ions to metallic nickel. Explain your reasoning.
9. Use the activity series to choose an oxidizing agent that will oxidize metallic cobalt to form aqueous cobalt(II) ions. Explain your reasoning.

Section Summary

In this section, you learned to define and recognize redox reactions, and to write oxidation and reduction half-reactions. In Investigation 10-A, you observed the connection between the metal activity series and redox reactions. However, thus far, you have only worked with redox reactions that involve atoms and ions as reactants or products. In the next section, you will learn about redox reactions that involve covalent reactants or products.

Section Review

1 K/U Predict whether each of the following single displacement reactions will occur. If so, write a balanced chemical equation, a balanced net ionic equation, and two balanced half-reactions. Include the physical states of the reactants and products in each case.

(a) aqueous silver nitrate and metallic cadmium

(b) gold and aqueous copper(II) sulfate

(c) aluminum and aqueous mercury(II) chloride

2 (a) K/U On which side of an oxidation half-reaction are the electrons? Why?

(b) K/U On which side of a reduction half-reaction are the electrons? Why?

3 C Explain why, in a redox reaction, the oxidizing agent undergoes reduction.

4 C In a combination reaction, does metallic lithium act as an oxidizing agent or a reducing agent? Explain.

5 I Write a net ionic equation for a reaction in which

(a) Fe^{2+} acts as an oxidizing agent

(b) Al acts as a reducing agent

(c) Au^{3+} acts as an oxidizing agent

(d) Cu acts as a reducing agent

(e) Sn^{2+} acts as an oxidizing agent and as a reducing agent

6 MC The element potassium is made industrially by the single displacement reaction of molten sodium with molten potassium chloride.

(a) Write a net ionic equation for the reaction, assuming that all reactants and products are in the liquid state.

(b) Identify the oxidizing agent and the reducing agent in the reaction.

(c) Explain why the reaction is carried out in the liquid state and not in aqueous solution.

Unit Investigation Prep

In the end-of-unit investigation, you will be working with the metals zinc and copper. Which metal is more easily oxidized? Which is more easily reduced?

Oxidation Numbers

10.2

Redox reactions are very common. Some of them produce light in a process known as *chemiluminescence*. In living things, the production of light in redox reactions is known as *bioluminescence*. You can actually see the light from redox reactions occurring in some organisms, such as glowworms and fireflies, as shown in Figure 10.3.

Figure 10.3 **Fireflies use flashes of light produced by redox reactions to attract a mate.**

Section Preview/
Specific Expectations

In this section, you will

- **describe** oxidation and reduction in terms of changes in oxidation number
- **assign** oxidation numbers to elements in covalent molecules and polyatomic ions
- **identify** redox reactions using oxidation numbers
- **communicate** your understanding of the terms *oxidation numbers, oxidation, reduction*

Not all redox reactions give off light, however. How can you recognize a redox reaction, and how can you identify the oxidizing and reducing agents? In section 10.1, you saw net ionic equations with monatomic elements, such as Cu and Zn, and with ions containing a single element, such as Cu^{2+} and Zn^{2+}. In these cases, you could use ionic charges to describe the transfer of electrons. However, many redox reactions involve reactants or products with covalent bonds, including elements that exist as covalent molecules, such as oxygen, O_2; covalent compounds, such as water, H_2O; or polyatomic ions that are not spectator ions, such as permanganate, MnO_4^-. For reactions involving covalent reactants and products, you cannot use ionic charges to describe the transfer of electrons.

Oxidation numbers are actual or hypothetical charges, assigned using a set of rules. They are used to describe redox reactions with covalent reactants or products. They are also used to identify redox reactions, and to identify oxidizing and reducing agents. In this section, you will see how oxidation numbers were developed from Lewis structures, and then learn the rules to assign oxidation numbers.

CHEM FACT

Oxidation numbers are just a bookkeeping method used to keep track of electron transfers. In a covalent molecule or a polyatomic ion, the oxidation number of each element does *not* represent an ionic charge, because the elements are not present as ions. However, to assign oxidation numbers to the elements in a covalent molecule or polyatomic ion, you can *pretend* the bonds are ionic.

Oxidation Numbers from Lewis Structures

You are probably familiar with the Lewis structure of water, shown in Figure 10.4A. From the electronegativities on the periodic table in Figure 10.5, on the next page, you can see that oxygen (electronegativity 3.44) is more electronegative than hydrogen (electronegativity 2.20). The electronegativity difference is less than 1.7, so the two hydrogen-oxygen bonds are polar covalent, not ionic. In each bond, the electrons are more strongly attracted to the oxygen atom than to the hydrogen atom.

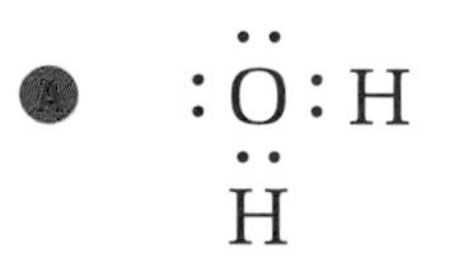

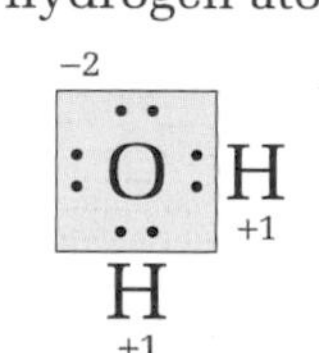

Figure 10.4 **(A) The Lewis structure of water; (B) The formal counting of electrons with the more electronegative element assigned a negative charge**

To assign oxidation numbers to the atoms in a water molecule, you can consider all the bonding electrons to be "owned" by the more electronegative oxygen atom, as shown in Figure 10.4B. Thus, each hydrogen atom in a water molecule is considered to have no electrons, as hydrogen would in a hydrogen ion, H^+. Therefore, the element hydrogen is assigned an oxidation number of +1 in water. On the other hand, the oxygen atom in a water molecule is considered to have a filled octet of electrons, as oxygen would in an oxide ion, O^{2-}. Therefore, the element oxygen is assigned an oxidation number of −2 in water. (**Note:** These are *not* ionic charges, since water is a covalent molecule. Also, note that the plus or minus sign in an oxidation number, such as −2, is written *before* the number. The plus or minus sign in an ionic charge, such as 2−, is written *after* the number.)

H 2.20																	He -
Li 0.98	Be 1.57											B 2.04	C 2.55	N 3.04	O 3.44	F 3.98	Ne -
Na 0.93	Mg 1.31											Al 1.61	Si 1.90	P 2.19	S 2.58	Cl 3.16	Ar -
K 0.82	Ca 1.00	Sc 1.36	Ti 1.54	V 1.63	Cr 1.66	Mn 1.55	Fe 1.83	Co 1.88	Ni 1.91	Cu 1.90	Zn 1.65	Ga 1.81	Ge 2.01	As 2.18	Se 2.55	Br 2.96	Kr -
Rb 0.82	Sr 0.95	Y 1.22	Zr 1.33	Nb 1.6	Mo 2.16	Tc 2.10	Ru 2.2	Rh 2.28	Pd 2.20	Ag 1.93	Cd 1.69	In 1.78	Sn 1.96	Sb 2.05	Te 2.1	I 2.66	Xe -
Cs 0.79	Ba 0.89	Lu 1.0	Hf 1.3	Ta 1.5	W 1.7	Re 1.9	Os 2.2	Ir 2.2	Pt 2.2	Au 2.4	Hg 1.9	Tl 1.8	Pb 1.8	Bi 1.9	Po 2.0	At 2.2	Rn -
Fr 0.7	Ra 0.9	Lr -	Rf -	Db -	Sg -	Bh -	Hs -	Mt -	Uun -	Uuu -	Uub -	-	Uuq -	-	Uuh -	-	Uuo -

La 1.10	Ce 1.12	Pr 1.13	Nd 1.14	Pm -	Sm 1.17	Eu -	Gd 1.20	Tb -	Dy 1.22	Ho 1.23	Er 1.24	Tm 1.25	Yb -
Ac 1.1	Th 1.3	Pa 1.5	U 1.7	Np 1.3	Pu 1.3	Am -	Cm -	Bk -	Cf -	Es -	Fm -	Md -	No -

Figure 10.5 **The periodic table, showing electronegativity values**

In a chlorine molecule, Cl_2, each atom has the same electronegativity, so the bond is non-polar covalent. Because the electrons are equally shared, you can consider each chlorine atom to "own" one of the shared electrons, as shown in Figure 10.6. Thus, each chlorine atom in the molecule is considered to have the same number of electrons as a neutral chlorine atom. Each chlorine atom is therefore assigned an oxidation number of 0.

Figure 10.6 **(A) The Lewis structure of a chlorine molecule; (B) The formal counting of electrons in a chlorine molecule for oxidation number purposes**

Figure 10.7 shows how oxidation numbers are assigned for the polyatomic cyanide ion, CN^-. The electronegativity of nitrogen (3.04) is greater than the electronegativity of carbon (2.55). Thus, the three shared pairs of electrons are all considered to belong to the nitrogen atom. As a result, the carbon atom is considered to have two valence electrons, which is two electrons less than the four valence electrons of a neutral carbon atom. Therefore, the carbon atom in CN^- is assigned an oxidation number of +2. The nitrogen atom is considered to have eight valence electrons, which is three electrons more than the five valence electrons of a neutral nitrogen atom. Therefore, the nitrogen atom in CN^- is assigned an oxidation number of –3.

Figure 10.7 **(A) The Lewis structure of a cyanide ion; (B) The formal counting of electrons in a cyanide ion for oxidation number purposes**

You have seen examples of how Lewis structures can be used to assign oxidation numbers for polar molecules such as water, non-polar molecules such as chlorine, and polar polyatomic ions such as the cyanide ion. In the following ThoughtLab, you will use Lewis structures to assign oxidation number values, and then look for patterns in your results.

ThoughtLab  Finding Rules for Oxidation Numbers

Procedure

1. Use Lewis structures to assign an oxidation number to each element in the following covalent molecules.
   (a) HI **(b)** O_2 **(c)** PCl_5 **(d)** BBr_3
2. Use Lewis structures to assign an oxidation number to each element in the following polyatomic ions.
   (a) OH^- **(b)** NH_4^+ **(c)** CO_3^{2-}
3. Assign an oxidation number to each of the following atoms or monatomic ions. Explain your reasoning.
   (a) Ne **(b)** K **(c)** I^- **(d)** Mg^{2+}

Analysis

1. For each molecule in question 1 of the procedure, find the sum of the oxidation numbers of all the atoms present. What do you notice? Explain why the observed sum must be true for a neutral molecule.
2. For each polyatomic ion in question 2 of the procedure, find the sum of the oxidation numbers of all the atoms present. Describe and explain any pattern you see.

Extension

3. Predict the sum of the oxidation numbers of the atoms in the hypochlorite ion, OCl^-.
4. Test your prediction from question 3.

Using Rules to Find Oxidation Numbers

Drawing Lewis structures to assign oxidation numbers can be a very time-consuming process for large molecules or large polyatomic ions. Instead, the results from Lewis structures have been summarized to produce a more convenient set of rules, which can be applied more quickly. Table 10.1 summarizes the rules used to assign oxidation numbers. You may have discovered some of these rules for yourself in the ThoughtLab you just completed.

Table 10.1 Oxidation Number Rules

Rules	Examples
1. A pure element has an oxidation number of 0.	Na in $Na_{(s)}$, Br in $Br_{2(\ell)}$, and P in $P_{4(s)}$ all have an oxidation number of 0.
2. The oxidation number of an element in a monatomic ion equals the charge of the ion.	The oxidation number of Al in Al^{3+} is +3. The oxidation number of Se in Se^{2-} is −2.
3. The oxidation number of hydrogen in its compounds is +1, except in metal hydrides, where the oxidation number of hydrogen is −1.	The oxidation number of H in H_2S or CH_4 is +1. The oxidation number of H in NaH or in CaH_2 is −1.
4. The oxidation number of oxygen in its compounds is usually −2, but there are exceptions. These include peroxides, such as H_2O_2, and the compound OF_2.	The oxidation number of O in Li_2O or in KNO_3 is −2.
5. In covalent compounds that do not contain hydrogen or oxygen, the more electronegative element is assigned an oxidation number that equals the negative charge it usually has in its ionic compounds.	The oxidation number of Cl in PCl_3 is −1. The oxidation number of S in CS_2 is −2.
6. The sum of the oxidation numbers of all the elements in a compound is 0.	In CF_4, the oxidation number of F is −1, and the oxidation number of C is +4. $(+4) + 4(-1) = 0$
7. The sum of the oxidation numbers of all the elements in a polyatomic ion equals the charge on the ion.	In NO_2^-, the oxidation number of O is −2, and the oxidation number of N is +3. $(+3) + 2(-2) = -1$

Some oxidation numbers found using these rules are not integers. For example, an important iron ore called magnetite has the formula Fe_3O_4. Using the oxidation number rules, you can assign oxygen an oxidation number of −2, and calculate an oxidation number of $+\frac{8}{3}$ for iron. However, magnetite contains no iron atoms with this oxidation number. It actually contains iron(III) ions and iron(II) ions in a 2:1 ratio. The formula of magnetite is sometimes written as $Fe_2O_3 \cdot FeO$ to indicate that there are two different oxidation numbers. The value $+\frac{8}{3}$ for the oxidation number of iron is an average value.

$$\frac{2(+3) + (+2)}{3} = +\frac{8}{3}$$

Even though some oxidation numbers found using these rules are averages, the rules are still useful for monitoring electron transfers in redox reactions.

In the following Sample Problem, you will find out how to apply these rules to covalent molecules and polyatomic ions.

Sample Problem

Assigning Oxidation Numbers

Problem

Assign an oxidation number to each element.

(a) $SiBr_4$ **(b)** $HClO_4$ **(c)** $Cr_2O_7^{2-}$

Solution

(a)
- Because the compound $SiBr_4$ does not contain hydrogen or oxygen, rule 5 applies. Because $SiBr_4$ is a compound, rule 6 also applies.
- Silicon has an electronegativity of 1.90. Bromine has an electronegativity of 2.96. From rule 5, therefore, you can assign bromine an oxidation number of −1.
- The oxidation number of silicon is unknown, so let it be x. You know from rule 6 that the sum of the oxidation numbers is 0. Then,

$$x + 4(-1) = 0$$
$$x - 4 = 0$$
$$x = 4$$

The oxidation number of silicon is +4. The oxidation number of bromine is −1.

(b)
- Because the compound $HClO_4$ contains hydrogen and oxygen, rules 3 and 4 apply. Because $HClO_4$ is a compound, rule 6 also applies.
- Hydrogen has its usual oxidation number of +1. Oxygen has its usual oxidation number of −2. The oxidation number of chlorine is unknown, so let it be x. You know from rule 6 that the sum of the oxidation numbers is 0. Then,

$$(+1) + x + 4(-2) = 0$$
$$x - 7 = 0$$
$$x = 7$$

The oxidation number of hydrogen is +1. The oxidation number of chlorine is +7. The oxidation number of oxygen is −2.

(c)
- Because the polyatomic ion $Cr_2O_7^{2-}$ contains oxygen, rule 4 applies. Because $Cr_2O_7^{2-}$ is a polyatomic ion, rule 7 also applies.
- Oxygen has its usual oxidation number of −2.
- The oxidation number of chromium is unknown, so let it be x. You know from rule 7 that the sum of the oxidation numbers is −2. Then,

$$2x + 7(-2) = -2$$
$$2x - 14 = -2$$
$$2x = 12$$
$$x = 6$$

The oxidation number of chromium is +6. The oxidation number of oxygen is −2.

PROBLEM TIP

When finding the oxidation numbers of elements in ionic compounds, you can work with the ions separately. For example, $Na_2Cr_2O_7$ contains two Na^+ ions, and so sodium has an oxidation number of +1. The oxidation numbers of Cr and O can then be calculated as shown in part (c) of the Sample Problem.

Practice Problems

9. Determine the oxidation number of the specified element in each of the following.

(a) N in NF_3 **(b)** S in S_8 **(c)** Cr in CrO_4^{2-}
(d) P in P_2O_5 **(e)** C in $C_{12}H_{22}O_{11}$ **(f)** C in $CHCl_3$

10. Determine the oxidation number of each element in each of the following.

(a) H_2SO_3 **(b)** OH^- **(c)** HPO_4^{2-}

11. As stated in rule 4, oxygen does not always have its usual oxidation number of −2. Determine the oxidation number of oxygen in each of the following.

(a) the compound oxygen difluoride, OF_2 **(b)** the peroxide ion, O_2^{2-}

12. Determine the oxidation number of each element in each of the following ionic compounds by considering the ions separately. **Hint:** One formula unit of the compound in part (c) contains two identical monatomic ions and one polyatomic ion.

(a) $Al(HCO_3)_3$ **(b)** $(NH_4)_3PO_4$ **(c)** $K_2H_3IO_6$

Applying Oxidation Numbers to Redox Reactions

You have seen that the single displacement reaction of zinc with copper(II) sulfate is a redox reaction, represented by the following chemical equation and net ionic equation.

$$Zn_{(s)} + CuSO_{4(aq)} \rightarrow Cu_{(s)} + ZnSO_{4(aq)}$$
$$Zn_{(s)} + Cu^{2+}{}_{(aq)} \rightarrow Cu_{(s)} + Zn^{2+}{}_{(aq)}$$

Each atom or ion shown in the net ionic equation can be assigned an oxidation number. Zn has an oxidation number of 0; Cu^{2+} has an oxidation number of +2; Cu has an oxidation number of 0; and Zn^{2+} has an oxidation number of +2. Thus, there are changes in oxidation numbers in this reaction. The oxidation number of zinc increases, while the oxidation number of copper decreases.

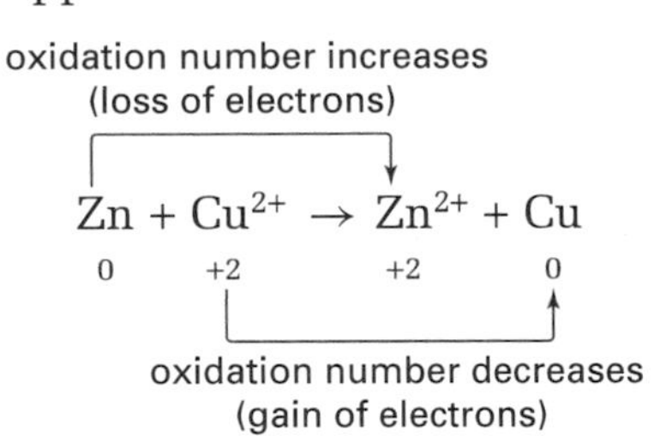

In the oxidation half-reaction, the element zinc undergoes an increase in its oxidation number from 0 to +2.

$$\underset{0}{Zn} \rightarrow \underset{+2}{Zn^{2+}} + 2e^-$$

In the reduction half-reaction, the element copper undergoes a decrease in its oxidation number from +2 to 0.

$$\underset{+2}{Cu^{2+}} + 2e^- \rightarrow \underset{0}{Cu}$$

Therefore, you can describe oxidation and reduction as follows. (Also see Figure 10.8.)

- **Oxidation** is an increase in oxidation number.
- **Reduction** is a decrease in oxidation number.

You can also monitor changes in oxidation numbers in reactions that involve covalent molecules. For example, oxidation number changes occur in the reaction of hydrogen and oxygen to form water.

$$\underset{0}{2H_{2(g)}} + \underset{0}{O_{2(g)}} \rightarrow \underset{+1\ -2}{2H_2O_{(\ell)}}$$

Because hydrogen combines with oxygen in this reaction, hydrogen undergoes oxidation, according to the historical definition given at the beginning of section 10.1. Hydrogen also undergoes oxidation according to the modern definition, because the oxidation number of hydrogen increases from 0 to +1. Hydrogen is the reducing agent in this reaction. The oxygen undergoes reduction, because its oxidation number decreases from 0 to −2. Oxygen is the oxidizing agent in this reaction.

The following Sample Problem illustrates how to use oxidation numbers to identify redox reactions, oxidizing agents, and reducing agents.

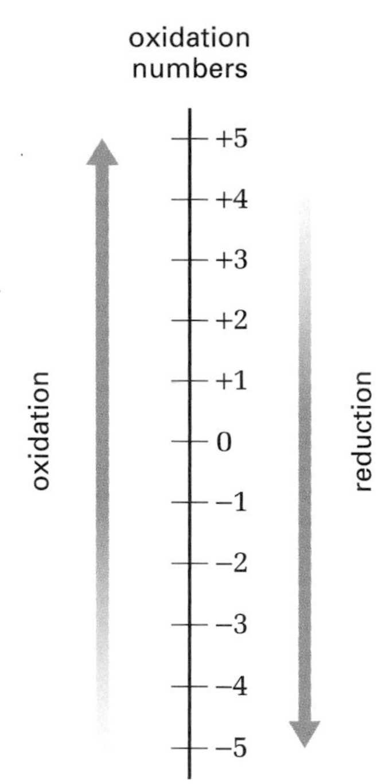

Figure 10.8 Oxidation and reduction are directly related to changes in oxidation numbers.

Sample Problem

Identifying Redox Reactions

Problem

Determine whether each of the following reactions is a redox reaction. If so, identify the oxidizing agent and the reducing agent.

(a) $CH_{4(g)} + Cl_{2(g)} \rightarrow CH_3Cl_{(g)} + HCl_{(g)}$

(b) $CaCO_{3(s)} + 2HCl_{(aq)} \rightarrow CaCl_{2(aq)} + H_2O_{(\ell)} + CO_{2(g)}$

Solution

Find the oxidation number of each element in the reactants and products. Identify any elements that undergo an increase or a decrease in oxidation number during the reaction.

(a) The oxidation number of each element in the reactants and products is as shown.

$$\underset{-4\ +1}{CH_{4(g)}} + \underset{0}{Cl_{2(g)}} \rightarrow \underset{-2\ +1\ -1}{CH_3Cl_{(g)}} + \underset{+1\ -1}{HCl_{(g)}}$$

- The oxidation number of hydrogen is +1 on both sides of the equation, so hydrogen is neither oxidized nor reduced.
- Both carbon and chlorine undergo changes in oxidation number, so the reaction is a redox reaction.
- The oxidation number of carbon increases from −4 to −2. The carbon atoms on the reactant side exist in methane molecules, $CH_{4(g)}$, so methane is oxidized. Therefore, methane is the reducing agent.
- The oxidation number of chlorine decreases from 0 to −1, so elemental chlorine, $Cl_{2(g)}$, is reduced. Therefore, elemental chlorine is the oxidizing agent.

(b) Because this reaction involves ions, write the equation in its total ionic form.

$$CaCO_{3(s)} + 2H^+_{(aq)} + 2Cl^-_{(aq)} \rightarrow Ca^{2+}_{(aq)} + 2Cl^-_{(aq)} + H_2O_{(\ell)} + CO_{2(g)}$$

Continued ...

PROBLEM TIPS

- Use the fact that the sum of the oxidation numbers in a molecule is zero to check the assignment of the oxidation numbers.
- Make sure that a reaction does not include only a reduction or only an oxidation. Oxidation and reduction must occur together in a redox reaction.

CONCEPT CHECK

In part (b) of the Sample Problem, you can assign oxidation numbers to each element in the given chemical equation *or* in the net ionic equation. What are the advantages and the disadvantages of each method?

Continued ...

The chloride ions are spectator ions, which do not undergo oxidation or reduction. The net ionic equation is as follows.

$CaCO_{3(s)} + 2H^+_{(aq)} \rightarrow Ca^{2+}_{(aq)} + H_2O_{(\ell)} + CO_{2(g)}$

For the net ionic equation, the oxidation number of each element in the reactants and products is as shown.

$\underset{+2\ +4\ -2}{CaCO_{3(s)}} + \underset{+1}{2H^+_{(aq)}} \rightarrow \underset{+2}{Ca^{2+}_{(aq)}} + \underset{+1\ -2}{H_2O_{(\ell)}} + \underset{+4\ -2}{CO_{2(g)}}$

No elements undergo changes in oxidation numbers, so the reaction is not a redox reaction.

CONCEPT CHECK

In your previous chemistry course, you classified reactions into four main types: synthesis, decomposition, single displacement, and double displacement. You also learned to recognize combustion reactions and neutralization reactions. You have now learned to classify redox reactions. In addition, you have also learned about a special type of redox reaction known as a disproportionation reaction.

1. Classify each reaction in two ways.
   - **(a)** magnesium reacting with a solution of iron(II) nitrate
     $Mg + Fe(NO_3)_2 \rightarrow Fe + Mg(NO_3)_2$
   - **(b)** hydrogen sulfide burning in oxygen
     $2H_2S + 3O_2 \rightarrow 2SO_2 + 2H_2O$
   - **(c)** calcium reacting with chlorine
     $Ca + Cl_2 \rightarrow CaCl_2$
2. Classify the formation of water and oxygen from hydrogen peroxide in three ways.
   $2H_2O_2 \rightarrow 2H_2O + O_2$

Practice Problems

13. Determine whether each reaction is a redox reaction.

(a) $H_2O_2 + 2Fe(OH)_2 \rightarrow 2Fe(OH)_3$

(b) $PCl_3 + 3H_2O \rightarrow H_3PO_3 + 3HCl$

14. Identify the oxidizing agent and the reducing agent for the redox reaction(s) in the previous question.

15. For the following balanced net ionic equation, identify the reactant that undergoes oxidation and the reactant that undergoes reduction.

$Br_2 + 2ClO_2^- \rightarrow 2Br^- + 2ClO_2$

16. Nickel and copper are two metals that are important to the Ontario economy, particularly in the Sudbury area. Nickel and copper ores usually contain the metals as sulfides, such as NiS and Cu_2S. Do the extractions of these pure elemental metals from their ores involve redox reactions? Explain your reasoning.

Section Summary

In this section, you extended your knowledge of redox reactions to include covalent reactants and products. You did this by learning how to assign oxidation numbers and how to use them to recognize redox reactions, oxidizing agents, and reducing agents. In the next section, you will extend your knowledge further by learning how to write balanced equations that represent redox reactions.

Section Review

1 **C** At the beginning of section 10.1, it was stated that oxidation originally meant "to combine with oxygen." Explain why a metal that combines with the element oxygen undergoes oxidation as we now define it. What happens to the oxygen in this reaction? Write a balanced chemical equation for a reaction that illustrates your answer.

2 **K/U** Determine whether each of the following reactions is a redox reaction.

(a) $H_2 + I_2 \rightarrow 2HI$

(b) $2NaHCO_3 \rightarrow Na_2CO_3 + H_2O + CO_2$

(c) $2HBr + Ca(OH)_2 \rightarrow CaBr_2 + 2H_2O$

(d) $PCl_5 \rightarrow PCl_3 + Cl_2$

3 **K/U** Write three different definitions for a redox reaction.

4 **C** Explain why fluorine has an oxidation number of –1 in all its compounds.

5 **C** When an element combines with another element, is the reaction a redox reaction? Explain your answer.

6 **(a)** **I** Use the oxidation number rules to find the oxidation number of sulfur in a thiosulfate ion, $S_2O_3^{2-}$.

(b) The Lewis structure of a thiosulfate ion is given here. Use the Lewis structure to find the oxidation number of each sulfur atom.

```
 [        ..          ]2–
 [       :O:          ]
 [   ..   ..   ..     ]
 [  :O :  S :  S :    ]
 [   ..   ..   ..     ]
 [       :O:          ]
 [        ..          ]
```

(c) Compare your results from parts (a) and (b) and explain any differences.

(d) What are the advantages and disadvantages of using Lewis structures to assign oxidation numbers?

(e) What are the advantages and disadvantages of using the oxidation number rules to assign oxidation numbers?

7 **(a)** **MC** The Haber Process for the production of ammonia from nitrogen and hydrogen is a very important industrial process. Write a balanced chemical equation for the reaction. Use oxidation numbers to identify the oxidizing agent and the reducing agent.

(b) When ammonia is reacted with nitric acid to make the common fertilizer ammonium nitrate, is the reaction a redox reaction? Explain. (**Hint:** Consider the two polyatomic ions in the product separately.)

8 **MC** Historically, the extraction of a metal from its ore was known as reduction. One way to reduce iron ore on an industrial scale is to use a huge reaction vessel, 30 m to 40 m high, called a blast furnace. The reactants in a blast furnace are an impure iron ore, such as Fe_2O_3, mixed with limestone, $CaCO_3$, and coke, C, which is made from coal. The solid mixture is fed into the top of the blast furnace. A blast of very hot air, at about 900°C, is blown in near the bottom of the furnace. The following reactions occur.

$2C + O_2 \rightarrow 2CO$

$Fe_2O_3 + 3CO \rightarrow 2Fe + 3CO_2$

The limestone is present to convert sand or quartz, SiO_2, which is present as an impurity in the ore, to calcium silicate, $CaSiO_3$.

$CaCO_3 \rightarrow CaO + CO_2$

$CaO + SiO_2 \rightarrow CaSiO_3$

(a) Which of the four reactions above are redox reactions?

(b) For each redox reaction that you identified in part (a), name the oxidizing agent and the reducing agent.

Redox reactions are involved in some very important industrial processes, such as iron and steel production. However, the widespread use of metals has occupied a relatively small part of human history. In the Stone Age, humans relied on stone, wood, and bone to make tools and weapons. The Stone Age ended in many parts of the world with the start of the Bronze Age, which was marked by the use of copper and then bronze (an alloy of copper and tin). In the Iron Age, bronze was replaced by the use of iron. The dates of the Bronze Age and the Iron Age vary for different parts of the world.

10.3 The Half-Reaction Method for Balancing Equations

Section Preview/ Specific Expectations

In this section, you will

- **investigate** oxidation-reduction reactions by reacting metals with acids and by combusting hydrocarbons
- **write** balanced equations for redox reactions using the half-reaction method

Did you know that redox reactions are an important part of CD manufacturing? The CDs you buy at a music store are made of Lexan®, the same plastic used for riot shields and bulletproof windows. The CDs are coated with a thin aluminum film. They are copies of a single master disc, which is made of glass coated with silver, as seen in Figure 10.9. Silver is deposited on a glass disc by the reduction of silver ions with methanal, HCHO, also known as formaldehyde. In the same reaction, formaldehyde is oxidized to methanoic acid, HCOOH, also known as formic acid. The redox reaction occurs under acidic conditions.

Figure 10.9 The production of CDs depends on a redox reaction used to coat the master disc with silver.

You have seen many balanced chemical equations and net ionic equations that represent redox reactions. There are specific techniques for balancing these equations. These techniques are especially useful for reactions that take place under acidic or basic conditions, such as the acidic conditions used in coating a master CD with silver.

In section 10.1, you learned to divide the balanced equations for some redox reactions into separate oxidation and reduction half-reactions. You will now use the reverse approach, and discover how to write a balanced equation by combining two half-reactions. To do this, you must first understand how to write a wide range of half-reactions.

Balancing Half-Reactions

In the synthesis of potassium chloride from its elements, metallic potassium is oxidized to form potassium ions, and gaseous chlorine is reduced to form chloride ions. This reaction is shown in Figure 10.10. Each half-reaction can be balanced by writing the correct formulas for the reactant and product, balancing the numbers of atoms, and then adding the correct number of electrons to balance the charges. For the oxidation half-reaction,

$$K \rightarrow K^+ + e^-$$

The atoms are balanced. The net charge on each side is 0. For the reduction half-reaction,

$$Cl_2 + 2e^- \rightarrow 2Cl^-$$

The atoms are balanced. The net charge on each side is –2.

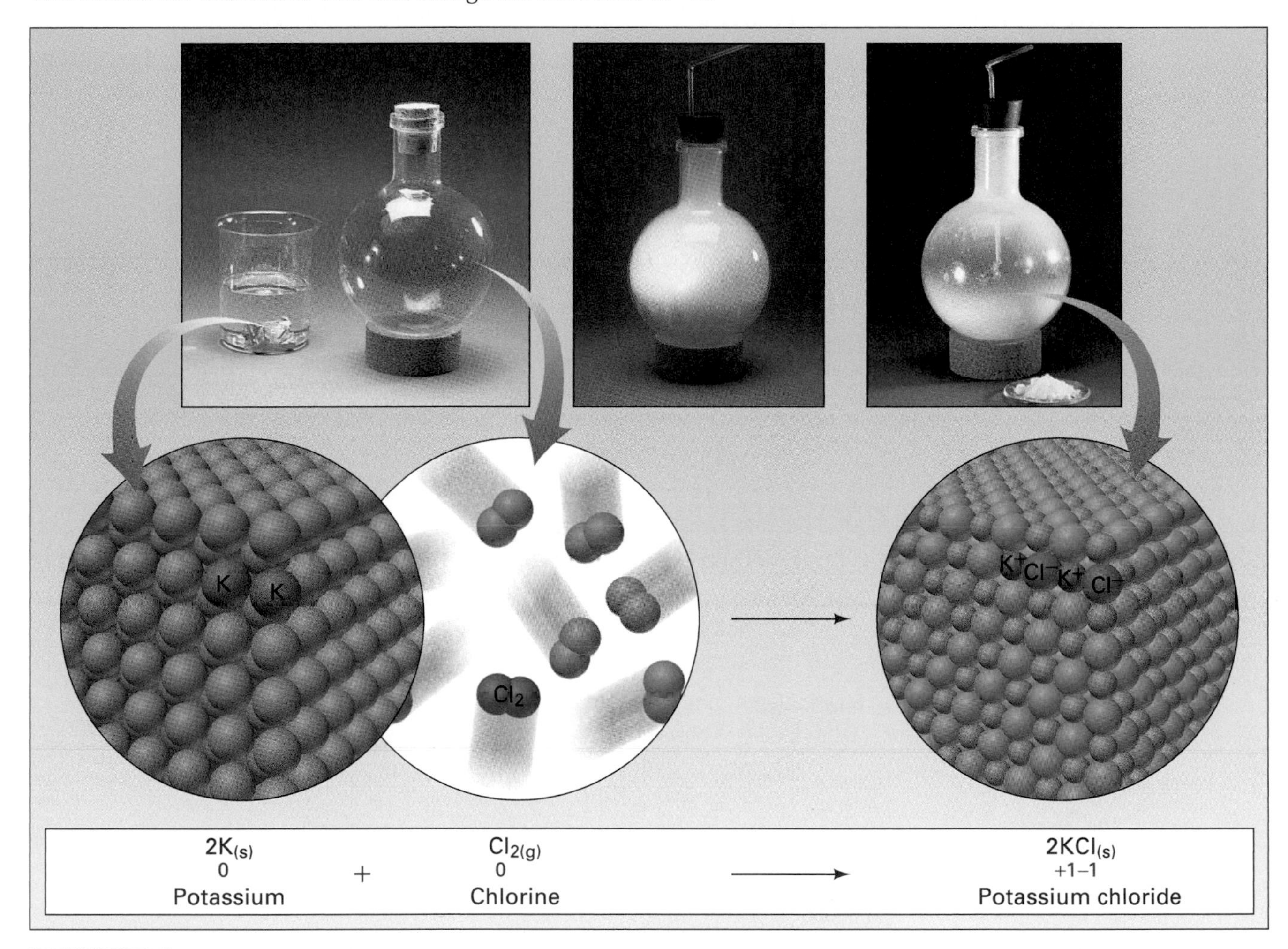

Figure 10.10 Grey potassium metal, which is stored under oil, reacts very vigorously with greenish-yellow chlorine gas to form white potassium chloride. The changes in oxidation numbers show that this synthesis reaction is also a redox reaction.

Redox reactions do not always take place under neutral conditions. Balancing half-reactions is more complicated for reactions that take place in acidic or basic solutions. When an acid or base is present, H^+ or OH^- ions must also be considered. However, the overall approach is similar. This approach involves writing the correct formulas for the reactants and products, balancing the atoms, and adding the appropriate number of electrons to one side of the half-reaction to balance the charges.

Balancing Half-Reactions for Acidic Solutions

The following steps are used to balance a half-reaction for an acidic solution. The Sample Problem that follows applies these steps.

Step 1 Write an unbalanced half-reaction that shows the formulas of the given reactant(s) and product(s).

Step 2 Balance any atoms other than oxygen and hydrogen first.

Step 3 Balance any oxygen atoms by adding water molecules.

Step 4 Balance any hydrogen atoms by adding hydrogen ions.

Step 5 Balance the charges by adding electrons.

Sample Problem

Balancing a Half-Reaction in Acidic Solution

Problem

Write a balanced half-reaction that shows the reduction of permanganate ions, MnO_4^-, to manganese(II) ions in an acidic solution.

Solution

Step 1 Represent the given reactant and product with correct formulas.
$MnO_4^- \rightarrow Mn^{2+}$

Step 2 Balance the atoms, starting with the manganese atoms. Here, the manganese atoms are already balanced.

Step 3 The reduction occurs in aqueous solution, so add water molecules to balance the oxygen atoms.
$MnO_4^- \rightarrow Mn^{2+} + 4H_2O$

Step 4 The reaction occurs in acidic solution, so add hydrogen ions to balance the hydrogen atoms.
$MnO_4^- + 8H^+ \rightarrow Mn^{2+} + 4H_2O$

Step 5 The atoms are now balanced, but the net charge on the left side is 7+, whereas the net charge on the right side is 2+. Add five electrons to the left side to balance the charges.
$MnO_4^- + 8H^+ + 5e^- \rightarrow Mn^{2+} + 4H_2O$

Practice Problems

17. Write a balanced half-reaction for the reduction of cerium(IV) ions to cerium(III) ions.

18. Write a balanced half-reaction for the oxidation of bromide ions to bromine.

19. Balance each of the following half-reactions under acidic conditions.
(a) $O_2 \rightarrow H_2O_2$ **(b)** $H_2O \rightarrow O_2$ **(c)** $NO_3^- \rightarrow N_2$

20. Balance each of the following half-reactions under acidic conditions.
(a) $ClO_3^- \rightarrow Cl^-$ **(b)** $NO \rightarrow NO_3^-$ **(c)** $Cr_2O_7^{2-} \rightarrow Cr^{3+}$

CONCEPT CHECK

The ability to balance a single half-reaction as a bookkeeping exercise does not mean that a single half-reaction can occur on its own. In a redox reaction, oxidation and reduction must both occur.

Balancing Half-Reactions for Basic Solutions

The following steps are used to balance a half-reaction for a basic solution. The Sample Problem that follows applies these steps.

Step 1 Write an unbalanced half-reaction that shows the formulas of the given reactant(s) and product(s).

Step 2 Balance any atoms other than oxygen and hydrogen first.

Step 3 Balance any oxygen and hydrogen atoms as if the conditions are acidic.

Step 4 Adjust for basic conditions by adding to both sides the same number of hydroxide ions as the number of hydrogen ions already present.

Step 5 Simplify the half-reaction by combining the hydrogen ions and hydroxide ions on the same side of the equation into water molecules.

Step 6 Remove any water molecules present on both sides of the half-reaction.

Step 7 Balance the charges by adding electrons.

Sample Problem

Balancing a Half-Reaction in Basic Solution

Problem

Write a balanced half-reaction that shows the oxidation of thiosulfate ions, $S_2O_3^{2-}$, to sulfite ions, SO_3^{2-}, in a basic solution.

Solution

Step 1 Represent the given reactant and product with correct formulas.

$S_2O_3^{2-} \rightarrow SO_3^{2-}$

Step 2 Balance the atoms, beginning with the sulfur atoms.

$S_2O_3^{2-} \rightarrow 2SO_3^{2-}$

Step 3 Balance the oxygen and hydrogen atoms as if the solution is acidic.

$S_2O_3^{2-} + 3H_2O \rightarrow 2SO_3^{2-}$

$S_2O_3^{2-} + 3H_2O \rightarrow 2SO_3^{2-} + 6H^+$

Step 4 There are six hydrogen ions present, so adjust for basic conditions by adding six hydroxide ions to each side.

$S_2O_3^{2-} + 3H_2O + 6OH^- \rightarrow 2SO_3^{2-} + 6H^+ + 6OH^-$

Step 5 Combine the hydrogen ions and hydroxide ions on the right side into water molecules.

$S_2O_3^{2-} + 3H_2O + 6OH^- \rightarrow 2SO_3^{2-} + 6H_2O$

Step 6 Remove three water molecules from each side.

$S_2O_3^{2-} + 6OH^- \rightarrow 2SO_3^{2-} + 3H_2O$

Step 7 The atoms are now balanced, but the net charge on the left side is 8–, whereas the net charge on the right side is 4–. Add four electrons to the right side to balance the charges.

$S_2O_3^{2-} + 6OH^- \rightarrow 2SO_3^{2-} + 3H_2O + 4e^-$

Practice Problems

21. Write a balanced half-reaction for the oxidation of chromium(II) ions to chromium(III) ions.

22. Write a balanced half-reaction for the reduction of oxygen to oxide ions.

23. Balance each of the following half-reactions under basic conditions.

(a) $Al \rightarrow Al(OH)_4^-$ **(b)** $CN^- \rightarrow CNO^-$ **(c)** $MnO_4^- \rightarrow MnO_2$

(d) $CrO_4^{2-} \rightarrow Cr(OH)_3$ **(e)** $CO_3^{2-} \rightarrow C_2O_4^{2-}$

24. Balance each of the following half-reactions.

(a) $FeO_4^{2-} \rightarrow Fe^{3+}$ (acidic conditions)

(b) $ClO_2^- \rightarrow Cl^-$ (basic conditions)

Half-Reaction Method for Balancing Redox Reactions

Recall that, if you consider a redox reaction as two half-reactions, electrons are lost in the oxidation half-reaction, and electrons are gained in the reduction half-reaction. For example, you know the reaction of zinc with aqueous copper(II) sulfate.

$$Zn_{(s)} + CuSO_{4(aq)} \rightarrow Cu_{(s)} + ZnSO_{4(aq)}$$

Removing the spectator ions leaves the following net ionic equation.

$$Zn_{(s)} + Cu^{2+}{}_{(aq)} \rightarrow Cu_{(s)} + Zn^{2+}{}_{(aq)}$$

- You can break the net ionic equation into two half-reactions:
  Oxidation half-reaction: $Zn \rightarrow Zn^{2+} + 2e^-$
  Reduction half-reaction: $Cu^{2+} + 2e^- \rightarrow Cu$
- You can also start with the half-reactions and use them to produce a net ionic equation. If you add the two half-reactions, the result is as follows.

$$Zn + Cu^{2+} + 2e^- \rightarrow Cu + Zn^{2+} + 2e^-$$

Removing the two electrons from each side results in the original net ionic equation.

As shown above, you can use half-reactions to write balanced net ionic equations for redox reactions. In doing so, you use the fact that *no electrons are created or destroyed in a redox reaction.* Electrons are transferred from one reactant (the reducing agent) to another (the oxidizing agent).

Balancing a Net Ionic Equation

You know from Investigation 10-A that magnesium metal, $Mg_{(s)}$, displaces aluminum from an aqueous solution of one of its compounds, such as aluminum nitrate, $Al(NO_3)_{3(aq)}$. To obtain a balanced net ionic equation for this reaction, you can start by looking at the half-reactions. Magnesium atoms undergo oxidation to form magnesium ions, which have a 2+ charge. The oxidation half-reaction is as follows.

$$Mg \rightarrow Mg^{2+} + 2e^-$$

Aluminum ions, which have a 3+ charge, undergo reduction to form aluminum atoms. The reduction half-reaction is as follows.

$$Al^{3+} + 3e^- \rightarrow Al$$

To balance the net ionic equation for this redox reaction, you can combine the two half-reactions in such a way that the number of electrons lost through oxidation equals the number of electrons gained through reduction. In other words, you can model the transfer of a certain number of electrons from the reducing agent to the oxidizing agent.

For the reaction of magnesium metal with aluminum ions, the two balanced half-reactions include different numbers of electrons, 2 and 3. The least common multiple of 2 and 3 is 6. To combine the half-reactions and give a balanced net ionic equation, multiply the balanced half-reactions by different numbers so that the results both include six electrons, as shown below.

- Multiply the oxidation half-reaction by 3. Multiply the reduction half-reaction by 2.

$$3Mg \rightarrow 3Mg^{2+} + 6e^-$$
$$2Al^{3+} + 6e^- \rightarrow 2Al$$

- Add the results.

$$3Mg + 2Al^{3+} + 6e^- \rightarrow 3Mg^{2+} + 2Al + 6e^-$$

- Remove $6e^-$ from each side to obtain the balanced net ionic equation.

$$3Mg + 2Al^{3+} \rightarrow 3Mg^{2+} + 2Al$$

To produce the balanced chemical equation, you can include the spectator ions, which are nitrate ions in this example. Include the states, if necessary. The balanced chemical equation is:

$$3Mg_{(s)} + 2Al(NO_3)_{3(aq)} \rightarrow 3Mg(NO_3)_{2(aq)} + 2Al_{(s)}$$

Math LINK

The lowest or least common multiple (LCM) of two numbers is the smallest multiple of each number. For example, the LCM of 2 and 1 is 2; the LCM of 3 and 6 is 6; and the LCM of 2 and 5 is 10. One way to find the LCM of two numbers is to list the multiples of each number and to find the smallest number that appears in both lists. For the numbers 6 and 8,

- the multiples of 6 are: 6, 12, 18, **24**, 30,...
- the multiples of 8 are: 8, 16, **24**, 32, 40,...

Thus, the LCM of 6 and 8 is 24.

What is the LCM of the numbers 4 and 12?

What is the LCM of the numbers 7 and 3?

Steps for Balancing by the Half-Reaction Method

You could balance the chemical equation for the reaction of magnesium with aluminum nitrate by inspection, instead of writing half-reactions. However, many redox equations are difficult to balance by the inspection method. In general, you can balance the net ionic equation for a redox reaction by a process known as the half-reaction method. The preceding example of the reaction of magnesium with aluminum nitrate illustrates this method. Specific steps for following the half-reaction method are given below.

Step 1 Write an unbalanced net ionic equation, if it is not already given.

Step 2 Divide the unbalanced net ionic equation into an oxidation half-reaction and a reduction half-reaction. To do this, you may need to assign oxidation numbers to all the elements in the net ionic equation to determine what is oxidized and what is reduced.

Step 3 Balance the oxidation half-reaction and the reduction half-reaction independently.

Step 4 Determine the least common multiple (LCM) of the numbers of electrons in the oxidation half-reaction and the reduction half-reaction.

Continued on the next page

Step 5 Use coefficients to write each half-reaction so that it includes the LCM of the numbers of electrons.

Step 6 Add the balanced half-reactions that include the equal numbers of electrons.

Step 7 Remove the electrons from both sides of the equation.

Step 8 Remove any identical molecules or ions that are present on both sides of the equation.

Step 9 If you require a balanced chemical equation, include any spectator ions in the chemical formulas.

Step 10 If necessary, include the states.

When using the half-reaction method, keep in mind that, in a redox reaction, *the number of electrons lost through oxidation must equal the number of electrons gained through reduction.* Figure 10.11 provides another example.

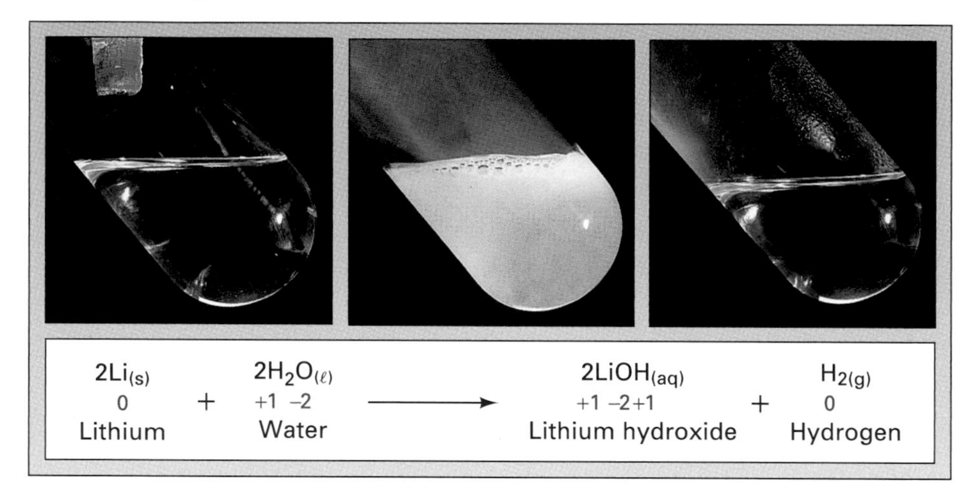

Figure 10.11 Lithium displaces hydrogen from water to form lithium hydroxide.

Oxidation half-reaction:

$Li \rightarrow Li^+ + e^-$

Reduction half-reaction:

$2H_2O + 2e^- \rightarrow H_2 + 2OH^-$

Multiply the oxidation half-reaction by 2, add the half-reactions, and simplify the result to obtain the balanced net ionic equation.

$2Li + 2H_2O \rightarrow 2Li^+ + 2OH^- + H_2$

Balancing Redox Reactions in Acidic and Basic Solutions

The half-reaction method of balancing equations can be more complicated for reactions that take place under acidic or basic conditions. The overall approach, however, is the same. You need to balance the two half-reactions, find the LCM of the numbers of electrons, and then multiply by coefficients to equate the number of electrons lost and gained. Finally, add the half-reactions and simplify to give a balanced net ionic equation for the reaction. The ten steps listed above show this process in more detail.

The Sample Problem on the next page illustrates the use of these steps for an acidic solution. To balance a net ionic equation for basic conditions by the half-reaction method, balance each half-reaction for acidic conditions, adjust for basic conditions, and then combine the half-reactions to obtain the balanced net ionic equation. The following Concept Organizer summarizes how to use the half-reaction method in both acidic and basic conditions.

CONCEPT CHECK

Explain why a balanced chemical equation or net ionic equation for a redox reaction does not include any electrons.

Concept Organizer The Half-Reaction Method for Balancing Redox Equations

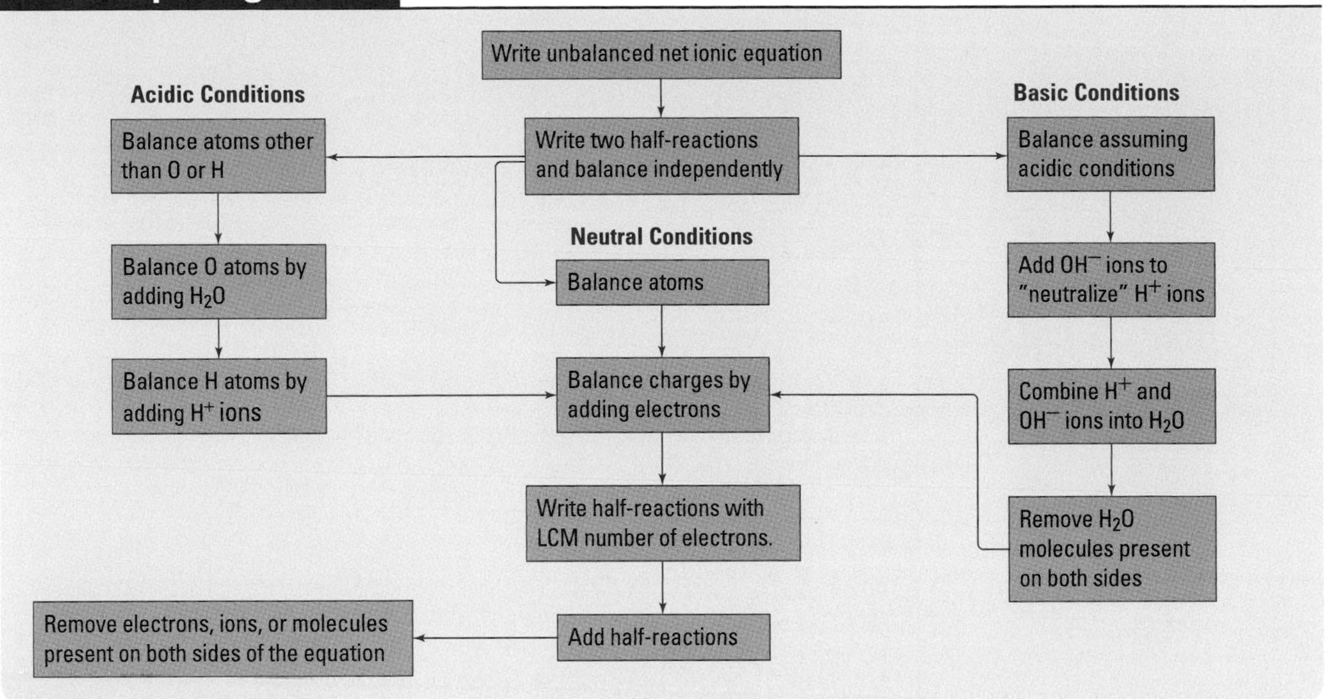

Write unbalanced net ionic equation
Acidic Conditions
Basic Conditions
Balance atoms other than O or H
Write two half-reactions and balance independently
Balance assuming acidic conditions
Neutral Conditions
Balance O atoms by adding H2O
Balance atoms
Add OH⁻ ions to "neutralize" H⁺ ions
Balance H atoms by adding H⁺ ions
Balance charges by adding electrons
Combine H⁺ and OH⁻ ions into H2O
Write half-reactions with LCM number of electrons.
Remove H2O molecules present on both sides
Remove electrons, ions, or molecules present on both sides of the equation
Add half-reactions

Sample Problem

Balancing a Redox Equation in Acidic Solution

Problem

Write a balanced net ionic equation to show the reaction of perchlorate ions, ClO_4^-, and nitrogen dioxide in acidic solution to produce chloride ions and nitrate ions.

What Is Required?

You need to write a balanced net ionic equation for the given reaction.

What Is Given?

You know the identities of two reactants and two products, and that the reaction takes place in acidic solution.

Plan Your Strategy

- Write an unbalanced ionic equation.
- Determine whether the reaction is a redox reaction.
- If it is not a redox reaction, balance by inspection.
- If it is a redox reaction, follow the steps for balancing by the half-reaction method.

Act on Your Strategy

- The unbalanced ionic equation is: $ClO_4^- + NO_2 \rightarrow Cl^- + NO_3^-$
- Assign oxidation numbers to all the elements to determine which reactant, if any, is oxidized or reduced.

$$\underset{+7\ -2}{ClO_4^-} + \underset{+4\ -2}{NO_2} \rightarrow \underset{-1}{Cl^-} + \underset{+5\ -2}{NO_3^-}$$

Continued ...

Continued ...

The oxidation number of chlorine decreases, so perchlorate ions are reduced to chloride ions.

The oxidation number of nitrogen increases, so nitrogen dioxide is oxidized to nitrate ions.

- This is a redox reaction. Use the half-reaction method to balance the equation.

Step 1 The unbalanced net ionic equation is already written.
$ClO_4^- + NO_2 \rightarrow Cl^- + NO_3^-$

Step 2 Write two unbalanced half-reactions.
Oxidation: $NO_2 \rightarrow NO_3^-$
Reduction: $ClO_4^- \rightarrow Cl^-$

Step 3 Balance the two half-reactions for acidic conditions.

Oxidation	**Reduction**
$NO_2 \rightarrow NO_3^-$	$ClO_4^- \rightarrow Cl^-$
$NO_2 + H_2O \rightarrow NO_3^-$	$ClO_4^- \rightarrow Cl^- + 4H_2O$
$NO_2 + H_2O \rightarrow NO_3^- + 2H^+$	$ClO_4^- + 8H^+ \rightarrow Cl^- + 4H_2O$
$NO_2 + H_2O \rightarrow NO_3^- + 2H^+ + e^-$	$ClO_4^- + 8H^+ + 8e^- \rightarrow Cl^- + 4H_2O$

Step 4 The LCM of 1 and 8 is 8.

Step 5 Multiply the oxidation half-reaction by 8, so that equal numbers of electrons are lost and gained.
$8NO_2 + 8H_2O \rightarrow 8NO_3^- + 16H^+ + 8e^-$

Step 6 Add the half reactions.

$$8NO_2 + 8H_2O \rightarrow 8NO_3^- + 16H^+ + 8e^-$$
$$ClO_4^- + 8H^+ + 8e^- \rightarrow Cl^- + 4H_2O$$
$$8NO_2 + 8H_2O + ClO_4^- + 8H^+ + 8e^- \rightarrow 8NO_3^- + 16H^+ + 8e^- + Cl^- + 4H_2O$$

Step 7 Simplify by removing 8 electrons from both sides.
$8NO_2 + 8H_2O + ClO_4^- + 8H^+ \rightarrow 8NO_3^- + 16H^+ + Cl^- + 4H_2O$

Step 8 Simplify by removing 4 water molecules, and 8 hydrogen ions from each side.
$8NO_2 + ClO_4^- + 4H_2O \rightarrow 8NO_3^- + 8H^+ + Cl^-$

(Steps 9 and 10 are not required for this problem.)

Check Your Solution

- The atoms are balanced.
- The charges are balanced.

Practice Problems

25. Balance each of the following redox equations by inspection. Write the balanced half-reactions in each case.

(a) $Na + F_2 \rightarrow NaF$

(b) $Mg + N_2 \rightarrow Mg_3N_2$

(c) $HgO \rightarrow Hg + O_2$

26. Balance the following equation by the half-reaction method.
$Cu^{2+} + I^- \rightarrow CuI + I_3^-$

27. Balance each of the following ionic equations for acidic conditions. Identify the oxidizing agent and the reducing agent in each case.
   (a) $MnO_4^- + Ag \rightarrow Mn^{2+} + Ag^+$
   (b) $Hg + NO_3^- + Cl^- \rightarrow HgCl_4^{2-} + NO_2$
   (c) $AsH_3 + Zn^{2+} \rightarrow H_3AsO_4 + Zn$
   (d) $I_3^- \rightarrow I^- + IO_3^-$

28. Balance each of the following ionic equations for basic conditions. Identify the oxidizing agent and the reducing agent in each case.
   (a) $CN^- + MnO_4^- \rightarrow CNO^- + MnO_2$
   (b) $H_2O_2 + ClO_2 \rightarrow ClO_2^- + O_2$
   (c) $ClO^- + CrO_2^- \rightarrow CrO_4^{2-} + Cl_2$
   (d) $Al + NO_2^- \rightarrow NH_3 + AlO_2^-$

In the next investigation, you will carry out several redox reactions, including reactions of acids with metals, and the combustion of hydrocarbons.

Tools & Techniques

The Breathalyzer Test: A Redox Reaction

The police may pull over a driver weaving erratically on the highway on suspicion of drunk driving. A police officer must confirm this suspicion by assessing whether the driver has a blood alcohol concentration over the "legal limit." The "Breathalyzer" test checks a person's breath using a redox reaction to determine blood alcohol concentration. This test was invented in 1953 by Robert Borkenstein, a former member of the Indiana State Police, and a professor of forensic studies.

What does a person's breath have to do with the alcohol in his or her blood? In fact, there is a direct correlation between the concentration of alcohol in an exhaled breath and the concentration of alcohol in the blood.

As blood moves through the lungs, it comes in close contact with inhaled gases. If the blood contains alcohol, the concentration of alcohol in the blood quickly reaches equilibrium with the concentration of alcohol in each inhaled breath. Thus, the alcohol content in an exhaled breath is a measure of the alcohol concentration in the blood itself. For example, if a person has been drinking alcohol, every 2100 mL of air exhaled contains about the same amount of alcohol as 1 mL of blood.

In the Breathalyzer test, the subject blows into a tube connected to a vial. The exhaled air collects in the vial, which already contains a mixture of sulfuric acid, potassium dichromate, water, and the catalyst silver nitrate. The alcohol reacts with the dichromate ion in the following redox reaction.

$$16H^+_{(aq)} + \underset{\text{(orange)}}{2Cr_2O_7^{2-}{}_{(aq)}} + 3C_2H_5OH_{(\ell)} \rightarrow \underset{\text{(green)}}{4Cr^{3+}{}_{(aq)}} + 3C_2H_4O_{2(aq)} + 11H_2O_{(\ell)}$$

This reaction is accompanied by a visible colour change, as orange dichromate ions become green chromium(III) ions. The concentration of alcohol in the blood is determined by measuring the intensity of the final colour.

A recent modification of the Breathalyzer test prevents drivers from starting their cars if they have been drinking. Alcohol ignition locks involve a type of Breathalyzer test that is linked to the car's ignition system. Until the driver passes the test, the car will not start. This test is useful in regulating the driving habits of people who have been previously convicted of drinking and driving.

Investigation 10-B

MICROSCALE

SKILL FOCUS

Predicting

Performing and recording

Analyzing and interpreting

Redox Reactions and Balanced Equations

A redox reaction involves the transfer of electrons between reactants. A reactant that loses electrons is oxidized and acts as a reducing agent. A reactant that gains electrons is reduced and acts as an oxidizing agent. Redox reactions can be represented by balanced equations.

Questions

How can you tell if a redox reaction occurs when reactants are mixed? Can you observe the transfer of electrons in the mixture?

Predictions

- Predict which of the metals magnesium, zinc, copper, and aluminum can be oxidized by aqueous hydrogen ions. Explain your reasoning.
- Predict whether metals that cannot be oxidized by hydrogen ions can dissolve in acids. Explain your reasoning.
- Predict whether the combustion of a hydrocarbon is a redox reaction. What assumptions have you made about the products?

Materials

well plate
4 small test tubes
test tube rack
small pieces of each of the metals magnesium, zinc, copper, and aluminum
dilute hydrochloric acid (1 mol/L)
dilute sulfuric acid (1 mol/L)
Bunsen burner
candle

Safety Precautions

- The acid solutions are corrosive. Handle them with care.
- If you accidentally spill a solution on your skin, wash the area immediately with copious amounts of cool water. If you get any acid in your eyes, wash at the eye wash station. Inform your teacher.
- Before lighting a Bunsen burner or candle, make sure that there are no flammable liquids nearby. Also, tie back long hair, and confine any loose clothing.

Procedure

Part 1 Reactions of Acids

1. Place a small piece of each metal on the well plate. Add a few drops of hydrochloric acid to each metal. Record your observations. If you are unsure of your observations, repeat the procedure on a larger scale in a small test tube.
2. Place another small piece of each metal on clean sections of the well plate. Add a few drops of sulfuric acid to each metal. Record your observations. If you are unsure of your observations, repeat the procedure on a larger scale in a small test tube.
3. Dispose of the mixtures in the beaker supplied by your teacher.

Part 2 Combustion of Hydrocarbons

4. Observe the combustion of natural gas in a Bunsen burner. Adjust the colour of the flame by varying the quantity of oxygen admitted to the burner. How does the colour depend on the quantity of oxygen?
5. Observe the combustion of a candle. Compare the colour of the flame with the colour of the Bunsen burner flame. Which adjustment of the burner makes the colours of the two flames most similar?

Analysis

Part 1 Reactions of Acids

1. Write a balanced chemical equation for each of the reactions of an acid with a metal.
2. Write each equation from question 1 in net ionic form.
3. Determine which of the reactions from question 1 are redox reactions.
4. Write each redox reaction from question 3 as two half-reactions.
5. Explain any similarities in your answers to question 4.
6. In the reactions you observed, are the hydrogen ions acting as an oxidizing agent, a reducing agent, or neither?
7. In the neutralization reaction of hydrochloric acid and sodium hydroxide, do the hydrogen ions behave in the same way as you found in question 6? Explain.
8. Your teacher may demonstrate the reaction of copper with concentrated nitric acid to produce copper(II) ions and brown, toxic nitrogen dioxide gas. Write a balanced net ionic equation for this reaction. Do the hydrogen ions behave in the same way as you found in question 6? Identify the oxidizing agent and the reducing agent in this reaction.
9. From your observations of copper with hydrochloric acid and nitric acid, can you tell whether hydrogen ions or nitrate ions are the better oxidizing agent? Explain.

Part 2 Combustion of Hydrocarbons

10. The main component of natural gas is methane, CH_4. The products of the combustion of this gas in a Bunsen burner depend on how the burner is adjusted. A blue flame indicates complete combustion. What are the products in this case? Write a balanced chemical equation for this reaction.
11. A yellow or orange flame from a Bunsen burner indicates incomplete combustion and the presence of carbon in the flame. Write a balanced chemical equation for this reaction.
12. Name another possible carbon-containing product from the incomplete combustion of methane. Write a balanced chemical equation for this reaction.
13. The fuel in a burning candle is paraffin wax, $C_{25}H_{52}$. Write a balanced chemical equation for the complete combustion of paraffin wax.
14. Write two balanced equations that represent the incomplete combustion of paraffin wax.
15. How do you know that at least one of the incomplete combustion reactions is taking place when a candle burns?
16. Are combustion reactions also redox reactions? Does your answer depend on whether the combustion is complete or incomplete? Explain.

Conclusion

17. How could you tell if a redox reaction occurred when reactants were mixed? Could you observe the transfer of electrons in the mixture?

Applications

18. Gold is very unreactive and does not dissolve in most acids. However, it does dissolve in *aqua regia* (Latin for "royal water"), which is a mixture of concentrated hydrochloric and nitric acids. The unbalanced ionic equation for the reaction is as follows.

$$Au + NO_3^- + Cl^- \rightarrow AuCl_4^- + NO_2$$

Balance the equation, and identify the oxidizing agent and reducing agent.

19. Natural gas is burned in gas furnaces. Give at least three reasons why this combustion reaction should be as complete as possible. How would you try to ensure complete combustion?

Section Summary

In this section, you learned how to use the oxidation number method to balance redox equations. You now know various techniques for recognizing and representing redox reactions. In Chapter 11, you will use these techniques to examine specific applications of redox reactions in the business world and in your daily life.

Section Review

1. K/U Is it possible to use the half-reaction method or the oxidation number method to balance the following equation? Explain your answer.

   $Al_2S_3 + H_2O \rightarrow Al(OH)_3 + H_2S$

2. I Balance each equation by the method of your choice. Explain your choice of method in each case.

   (a) $CH_3COOH + O_2 \rightarrow CO_2 + H_2O$

   (b) $O_2 + H_2SO_3 \rightarrow HSO_4^-$ (acidic conditions)

3. K/U Use the oxidation number method to balance the following equations.

   (a) $NH_3 + Cl_2 \rightarrow NH_4Cl + N_2$

   (b) $Mn_3O_4 + Al \rightarrow Al_2O_3 + Mn$

4. C Explain why, in redox reactions, the total increase in the oxidation numbers of the oxidized elements must equal the total decrease in the oxidation numbers of the reduced elements.

5. I The combustion of ammonia in oxygen to form nitrogen dioxide and water vapour involves covalent molecules in the gas phase. The oxidation number method for balancing the equation was shown in an example in this section. Devise a half-reaction method for balancing the equation. Describe the assumptions you made in order to balance the equation. Also, describe why these assumptions did not affect the final result.

CHAPTER 10 Review

Reflecting on Chapter 10

Summarize this chapter in the format of your choice. Here are a few ideas to use as guidelines:

- Give two different definitions for the term oxidation.
- Give two different definitions for the term reduction.
- Define a half-reaction. Give an example of an oxidation half-reaction and a reduction half-reaction.
- Compare the half-reaction and oxidation number methods of balancing equations.
- Practise balancing equations using both methods.
- Write an example of a balanced chemical equation for a redox reaction. Assign oxidation numbers to each element in the equation, then explain how you know it is a redox reaction.

Reviewing Key Terms

For each of the following terms, write a sentence that shows your understanding of its meaning.

ore	oxidation
reduction	oxidation-reduction reaction
redox reaction	
oxidizing agent	reducing agent
half-reaction	disproportionation
oxidation numbers	

Knowledge/Understanding

1. For each reaction below, write a balanced chemical equation by inspection.
   (a) zinc metal with aqueous silver nitrate
   (b) aqueous cobalt(II) bromide with aluminum metal
   (c) metallic cadmium with aqueous tin(II) chloride
2. For each reaction in question 1, write the total ionic and net ionic equations.
3. For each reaction in question 1, identify the oxidizing agent and reducing agent.
4. For each reaction in question 1, write the two half-reactions.
5. When a metallic element reacts with a non-metallic element, which reactant is
   (a) oxidized?
   (b) reduced?
   (c) the oxidizing agent?
   (d) the reducing agent?
6. Use a Lewis structure to assign an oxidation number to each element in the following compounds.
   (a) $BaCl_2$
   (b) CS_2
   (c) XeF_4
7. Determine the oxidation number of each element present in the following substances.
   (a) BaH_2
   (b) Al_4C_3
   (c) KCN
   (d) $LiNO_2$
   (e) $(NH_4)_2C_2O_4$
   (f) S_8
   (g) AsO_3^{3-}
   (h) VO_2^+
   (i) XeO_3F^-
   (j) $S_4O_6^{2-}$
8. Identify a polyatomic ion in which chlorine has an oxidation number of +3.
9. Determine which of the following balanced chemical equations represent redox reactions. For each redox reaction, identify the oxidizing agent and the reducing agent.
   (a) $2C_6H_6 + 15O_2 \rightarrow 12CO_2 + 6H_2O$
   (b) $CaO + SO_2 \rightarrow CaSO_3$
   (c) $H_2 + I_2 \rightarrow 2HI$
   (d) $KMnO_4 + 5CuCl + 8HCl \rightarrow KCl + MnCl_2 + 5CuCl_2 + 4H_2O$
10. Determine which of the following balanced net ionic equations represent redox reactions. For each redox reaction, identify the reactant that undergoes oxidation and the reactant that undergoes reduction.
    (a) $2Ag^+_{(aq)} + Cu_{(s)} \rightarrow 2Ag_{(s)} + Cu^{2+}_{(aq)}$
    (b) $Pb^{2+}_{(aq)} + S^{2-}_{(aq)} \rightarrow PbS_{(s)}$
    (c) $2Mn^{2+} + 5BiO_3^- + 14H^+ \rightarrow 2MnO_4^- + 5Bi^{3+} + 7H_2O$

Making Connections

33. The compound $NaAl(OH)_2CO_3$ is a component of some common stomach acid remedies.

(a) Determine the oxidation number of each element in the compound.

(b) Predict the products of the reaction of the compound with stomach acid (hydrochloric acid), and write a balanced chemical equation for the reaction.

(c) Were the oxidation numbers from part (a) useful in part (b)? Explain your answer.

(d) What type of reaction is this?

(e) Check your medicine cabinet at home for stomach acid remedies. If possible, identify the active ingredient in each remedy.

34. Two of the substances on the head of a safety match are potassium chlorate and sulfur. When the match is struck, the potassium chlorate decomposes to give potassium chloride and oxygen. The sulfur then burns in the oxygen and ignites the wood of the match.

(a) Write balanced chemical equations for the decomposition of potassium chlorate and for the burning of sulfur in oxygen.

(b) Identify the oxidizing agent and the reducing agent in each reaction in part (a).

(c) Does any element in potassium chlorate undergo disproportionation in the reaction? Explain your answer.

(d) Research the history of the safety match to determine when it was invented, why it was invented, and what it replaced.

35. Ammonium ions, from fertilizers or animal waste, are oxidized by atmospheric oxygen. The reaction results in the acidification of soil on farms and the pollution of ground water with nitrate ions.

(a) Write a balanced net ionic equation for this reaction.

(b) Why do farmers use fertilizers? What alternative farming methods have you heard of? Which farming method(s) do you support, and why?

36. One of the most important discoveries in the history of the chemical industry in Ontario was accidental. Thomas "Carbide" Willson (1860–1915) was trying to make the element calcium from lime, CaO, by heating the lime with coal tar. Instead, he made the compound calcium carbide, CaC_2. This compound reacts with water to form a precipitate of calcium hydroxide and gaseous ethyne (acetylene). Willson's discovery led to the large-scale use of ethyne in numerous applications.

(a) Was Willson trying to perform a redox reaction? How do you know? Why do you not need to know the substances in coal tar to answer this question?

(b) Write a balanced chemical equation for the reaction of calcium carbide with water. Is this reaction a redox reaction?

(c) An early use of Willson's discovery was in car headlights. Inside a headlight, the reaction of calcium carbide and water produced ethyne, which was burned to produce light and heat. Write a balanced chemical equation for the complete combustion of ethyne. Is this reaction a redox reaction?

(d) Research the impact of Willson's discovery on society, from his lifetime to the present day.

Answers to Practice Problems and Short Answers to Section Review Questions

Practice Problems: **1.** $Zn_{(s)} + Fe^{2+}_{(aq)} \rightarrow Zn^{2+}_{(aq)} + Fe_{(s)}$
2.(a) $3Mg_{(s)} + 2Al^{3+}_{(aq)} \rightarrow 3Mg^{2+}_{(aq)} + 2Al_{(s)}$
(b) $2Ag^{+}_{(aq)} + Cd_{(s)} \rightarrow 2Ag_{(s)} + Cd^{2+}_{(aq)}$
3.(a) Mg oxidized, Al^{3+} reduced
(b) Cd oxidized, Ag^{+} reduced
4.(a) Al^{3+} oxidizing agent, Mg reducing agent
(b) Ag^{+} oxidizing agent, Cd reducing agent
5. $Al_{(s)} \rightarrow Al^{3+}_{(aq)} + 3e^-$, $Fe^{3+}_{(aq)} + 3e^- \rightarrow Fe_{(s)}$
6.(a) $Fe_{(s)} \rightarrow Fe^{2+}_{(aq)} + 2e^-$, $Cu^{2+}_{(aq)} + 2e^- \rightarrow Cu_{(s)}$
(b) $Cd_{(s)} \rightarrow Cd^{2+}_{(aq)} + 2e^-$, $Ag^{+}_{(aq)} + 1e^- \rightarrow Ag_{(s)}$
7.(a) $Sn_{(s)} \rightarrow Sn^{2+}_{(aq)} + 2e^-$, $Pb^{2+}_{(aq)} + 2e^- \rightarrow Pb_{(s)}$
(b) $Ag_{(s)} \rightarrow Ag^{+} + e^-$, $Au^{3+}_{(aq)} + 3e^- \rightarrow Au_{(s)}$
(c) $Zn_{(s)} \rightarrow Zn^{2+}_{(aq)} + 2e^-$, $Fe^{3+}_{(aq)} + 3e^- \rightarrow Fe_{(s)}$
8. $Hg_2^{2+}{}_{(aq)} \rightarrow Hg_{(\ell)} + Hg^{2+}_{(aq)}$, $Hg_2^{2+}{}_{(aq)} + 2e^- \rightarrow 2Hg_{(\ell)}$, $Hg_2^{2+}{}_{(aq)} \rightarrow 2Hg^{2+}_{(aq)} + 2e^-$
9.(a) +3 **(b)** 0 **(c)** +6 **(d)** +5 **(e)** 0 **(f)** +2

10.(a) H, +1; S, +4; O, −2

(b) H, +1; O, −2 **(c)** H, +1; P, +5; O, −2

11.(a) +2 **(b)** −1 **12.(a)** Al, +3; H, +1; C, +4; O, −2

(b) N, −3; H, +1; P, +5; O, −2 **(c)** K, +1; H, +1; I, +7; O, −2

13.(a) yes **(b)** no

14.(a) H_2O_2 oxidizing agent, Fe^{2+} reducing agent

15. ClO_2^- oxidized, Br_2 reduced **16.** yes

17. $Ce^{4+} + e^- \rightarrow Ce^{3+}$ **18.** $2Br^- \rightarrow Br_2 + 2e^-$

19.(a) $O_2 + 2H^+ + 2e^- \rightarrow H_2O_2$ **(b)** $2H_2O \rightarrow O_2 + 4H^+ + 4e^-$

(c) $2NO_3^- + 12H^+ + 10e^- \rightarrow N_2 + 6H_2O$

20.(a) $ClO_3^- + 6H^+ + 6e^- \rightarrow Cl^- + 3H_2O$

(b) $NO + 2H_2O \rightarrow NO_3^- + 4H^+ + 3e^-$

(c) $Cr_2O_7^{2-} + 14H^+ + 6e^- \rightarrow 2Cr^{3+} + 7H_2O$

21. $Cr^{2+} \rightarrow Cr^{3+} + e^-$ **22.** $O_2 + 4e^- \rightarrow 2O^{2-}$

23.(a) $Al + 4OH^- \rightarrow Al(OH)_4^- + 3e^-$

(b) $CN^- + 2OH^- \rightarrow CNO^- + H_2O + 2e^-$

(c) $MnO_4^- + 2H_2O + 3e^- \rightarrow MnO_2 + 4OH^-$

(d) $CrO_4^{2-} + 4H_2O + 3e^- \rightarrow Cr(OH)_3 + 5OH^-$

(e) $2CO_3^{2-} + 2H_2O + 2e^- \rightarrow C_2O_4^{2-} + 4OH^-$

24.(a) $FeO_4^{2-} + 8H^+ + 3e^- \rightarrow Fe^{3+} + 4H_2O$

(b) $ClO_2^- + 2H_2O + 4e^- \rightarrow Cl^- + 4OH^-$

25.(a) $2Na + F_2 \rightarrow 2NaF$

ox: $Na \rightarrow Na^+ + e^-$

red: $F_2 + 2e^- \rightarrow 2F^-$

(b) $3Mg + N_2 \rightarrow Mg_3N_2$

ox: $Mg \rightarrow Mg^{2+} + 2e^-$

red: $N_2 + 6e^- \rightarrow 2N^{3-}$

(c) $2HgO \rightarrow 2Hg + O_2$

ox: $2O^{2-} \rightarrow O_2 + 4e^-$

red: $Hg^{2+} + 2e^- \rightarrow Hg$

26. $2Cu^{2+} + 5I^- \rightarrow 2CuI + I_3^-$

27.(a) $MnO_4^- + 5Ag + 8H^+ \rightarrow Mn^{2+} + 5Ag^+ + 4H_2O$

oxidizing agent, MnO_4^-; reducing agent, Ag

(b) $Hg + 2NO_3^- + 4Cl^- + 4H^+ \rightarrow HgCl_4^{2-} + 2NO_2 + 2H_2O$

oxidizing agent, NO_3^-; reducing agent, Hg

(c) $AsH_3 + 4Zn^{2+} + 4H_2O \rightarrow H_3AsO_4 + 4Zn + 8H^+$

oxidizing agent, Zn^{2+}; reducing agent, AsH_3

(d) $3I_3^- + 3H_2O \rightarrow 8I^- + IO_3^- + 6H^+$

oxidizing agent, I_3^-; reducing agent, I_3^-

28.(a) $3CN^- + 2MnO_4^- + H_2O \rightarrow 3CNO^- + 2MnO_2 + 2OH^-$

oxidizing agent, MnO_4^-; reducing agent, CN^-

(b) $H_2O_2 + 2ClO_2 + 2OH^- \rightarrow 2ClO_2^- + O_2 + 2H_2O$

oxidizing agent, ClO_2; reducing agent, H_2O_2

(c) $6ClO^- + 2CrO_2^- + 2H_2O \rightarrow 3Cl_2 + 2CrO_4^{2-} + 4OH^-$

oxidizing agent, ClO^-; reducing agent, CrO_2^-

(d) $2Al + NO_2^- + H_2O + OH^- \rightarrow NH_3 + 2AlO_2^-$

oxidizing agent, NO_2^-; reducing agent, Al

29. $CS_2 + 3O_2 \rightarrow CO_2 + 2SO_2$

30.(a) $B_2O_3 + 6Mg \rightarrow 3MgO + Mg_3B_2$

(b) $8H_2S + 8H_2O_2 \rightarrow S_8 + 16H_2O$

31.(a) $Cr_2O_7^{2-} + 6Fe^{2+} + 14H^+ \rightarrow 2Cr^{3+} + 6Fe^{3+} + 7H_2O$

(b) $I_2 + 10NO_3^- + 8H^+ \rightarrow 2IO_3^- + 10NO_2 + 4H_2O$

(c) $2PbSO_4 + 2H_2O \rightarrow Pb + PbO_2 + 2SO_4^{2-} + 4H^+$

32.(a) $3Cl^- + 2CrO_4^{2-} + H_2O \rightarrow 3ClO^- + 2CrO_2^- + 2OH^-$

(b) $3Ni + 2MnO_4^- + H_2O \rightarrow 3NiO + 2MnO_2 + 2OH^-$

(c) $I^- + 6Ce^{4+} + 6OH^- \rightarrow IO_3^- + 6Ce^{3+} + 3H_2O$

Section Review 10.1: **1.(a)** yes;

$2AgNO_{3(aq)} + Cd_{(s)} \rightarrow 2Ag_{(s)} + Cd(NO_3)_{2(aq)}$;

$2Ag^+_{(aq)} + Cd_{(s)} \rightarrow 2Ag_{(s)} + Cd^{2+}_{(aq)}$;

$Ag^+_{(aq)} + e^- \rightarrow Ag_{(s)}$; $Cd_{(s)} \rightarrow Cd^{2+}_{(aq)} + 2e^-$

(b) no **(c)** yes; $2Al_{(s)} + 3HgCl_{2(aq)} \rightarrow 2AlCl_{3(aq)} + 3Hg_{(\ell)}$;

$2Al_{(s)} + 3Hg^{2+}_{(aq)} \rightarrow 2Al^{3+}_{(aq)} + 3Hg_{(\ell)}$;

$Al_{(s)} \rightarrow Al^{3+}_{(aq)} + 3e^-$; $Hg^{2+}_{(aq)} + 2e^- \rightarrow Hg_{(\ell)}$

2.(a) right side **(b)** left side **4.** reducing agent

6.(a) $K^+ + Na_{(\ell)} \rightarrow Na^+ + K_{(\ell)}$

(b) oxidizing agent, K^+; reducing agent, $Na_{(\ell)}$

10.2: **2.(a)** yes **(b)** no **(c)** no **(d)** yes

6.(a) +2 **(b)** +5 and −1

7.(a) $N_2 + 3H_2 \rightarrow 2NH_3$ (0, 0, −3, +1); N_2 is oxidizing agent, H_2 is reducing agent **(b)** no

8.(a) $2C + O_2 \rightarrow 2CO$, $Fe_2O_3 + 3CO \rightarrow 2Fe + 3CO_2$

(b) carbon—reducing agent, oxygen—oxidizing agent; carbon monoxide—reducing agent, iron(III) oxide—oxidizing agent

10.3: **1.(a)** $2C + 2e^- \rightarrow C_2^{2-}$, reduction

(b) $2S_2O_3^{2-} \rightarrow S_4O_6^{2-} + 2e^-$, oxidation

(c) $4AsO_4^{3-} + 20H^+ + 8e^- \rightarrow As_4O_6 + 10H_2O$, reduction

(d) $Br_2 + 12OH^- \rightarrow 2BrO_3^- + 6H_2O + 10e^-$, oxidation

2.(a) $3Co^{3+} + Au \rightarrow 3Co^{2+} + Au^{3+}$

(b) $3Cu + 2NO_3^- + 8H^+ \rightarrow 3Cu^{2+} + 2NO + 4H_2O$

(c) $3NO_3^- + 8Al + 5OH^- + 2H_2O \rightarrow 3NH_3 + 8AlO_2^-$

3. $2Ag^+ + HCHO + H_2O \rightarrow 2Ag + HCOOH + 2H^+$

4.(b) $2NH_3 + 8OH^- \rightarrow N_2O + 7H_2O + 8e^-$

5.(a) $2N_2H_4 + N_2O_4 \rightarrow 3N_2 + 4H_2O$

(b) N_2O_4 is oxidizing agent, N_2H_4 is reducing agent

(c) $2NH_3 + ClO^- \rightarrow N_2H_4 + Cl^- + H_2O$

6.(a) Larissa was right

(b) $3Zn + SO_4^{2-} + 8H^+ \rightarrow 3Zn^{2+} + S + 4H_2O$

10.4: **1.** no, not a redox reaction

2.(a) $CH_3COOH + 2O_2 \rightarrow 2CO_2 + 2H_2O$

(b) $O_2 + 2H_2SO_3 \rightarrow 2HSO_4^- + 2H^+$

3.(a) $8NH_3 + 3Cl_2 \rightarrow 6NH_4Cl + N_2$

(b) $3Mn_3O_4 + 8Al \rightarrow 4Al_2O_3 + 9Mn$

CHAPTER

11 Cells and Batteries

Chapter Preview

11.1 Galvanic Cells

11.2 Standard Cell Potentials

11.3 Electrolytic Cells

11.4 Faraday's Law

11.5 Issues Involving Electrochemistry

Prerequisite Concepts and Skills

Before you begin this chapter, review the following concepts and skills:

- recognizing oxidation, reduction, and redox reactions (Chapter 10, sections 10.1, 10.2)
- writing balanced half-reactions (Chapter 10, section 10.3)
- using half-reactions to write balanced net ionic equations (Chapter 10, section 10.3)
- using stoichiometry to calculate quantities of reactants and products in a chemical reaction (Concepts and Skills Review)
- spontaneous and non-spontaneous reactions (Chapter 7, section 7.2)

Batteries come in a wide range of sizes and shapes, from the tiny button battery in a watch, to a large and heavy car battery. Although you probably use batteries every day, they may still surprise you. For example, did you know that you could make a battery from a lemon and two pieces of metal? To make a lemon battery, you could insert two different electrodes, such as copper and zinc strips, into a lemon. The battery can provide electricity for a practical use. For example, the battery can power a small light bulb, or turn a small motor.

For obvious reasons, lemon batteries are not a convenient way to power a portable device, such as a cell phone. Scientists and inventors have worked to develop a variety of batteries that are inexpensive, compact, and easy to store and to carry. Our society uses vast numbers of batteries.

What exactly is a battery, and how does it work? What is the relationship between batteries and the redox reactions studied in Chapter 10? You will find out the answers to these questions in this chapter.

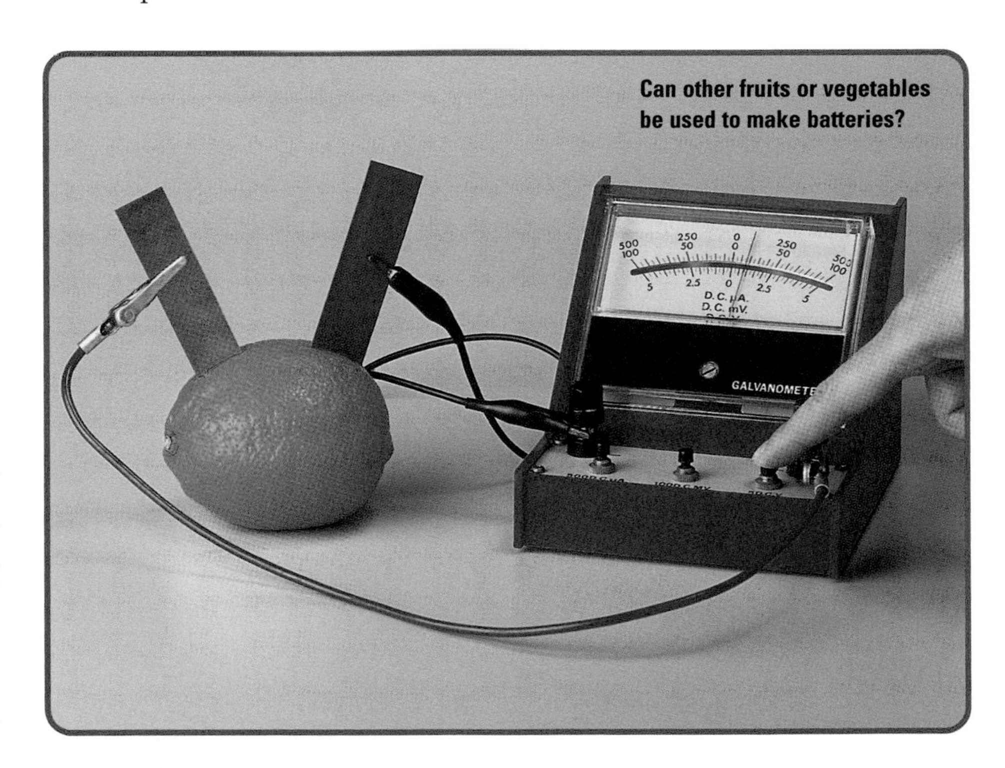

Can other fruits or vegetables be used to make batteries?

Galvanic Cells

11.1

You know that redox reactions involve the transfer of electrons from one reactant to another. You may also recall that an **electric current** is a flow of electrons in a circuit. These two concepts form the basis of **electrochemistry**, which is the study of the processes involved in converting chemical energy to electrical energy, and in converting electrical energy to chemical energy.

As you learned in Chapter 10, a zinc strip reacts with a solution containing copper(II) ions, forming zinc ions and metallic copper. The reaction is spontaneous. It releases energy in the form of heat; in other words, this reaction is exothermic.

$$Zn_{(s)} + Cu^{2+}_{(aq)} \rightarrow Zn^{2+}_{(aq)} + Cu_{(s)}$$

This reaction occurs on the surface of the zinc strip, where electrons are transferred from zinc atoms to copper(II) ions when these atoms and ions are in direct contact. A common technological invention called a galvanic cell uses redox reactions, such as the one described above, to release energy in the form of electricity.

The Galvanic Cell

A **galvanic cell**, also called a **voltaic cell**, is a device that converts chemical energy to electrical energy. The key to this invention is to prevent the reactants in a redox reaction from coming into direct contact with each other. Instead, electrons flow from one reactant to the other through an **external circuit**, which is a circuit outside the reaction vessel. This flow of electrons through the external circuit is an electric current.

An Example of a Galvanic Cell: The Daniell Cell

Figure 11.1 shows one example of a galvanic cell, called the Daniell cell. One half of the cell consists of a piece of zinc placed in a zinc sulfate solution. The other half of the cell consists of a piece of copper placed in a copper(II) sulfate solution. A *porous barrier*, sometimes called a *semi-permeable membrane*, separates these two half-cells. It stops the copper(II) ions from coming into direct contact with the zinc electrode.

Section Preview/ Specific Expectations

In this section, you will

- **identify** the components in galvanic cells and **describe** how they work
- **describe** the oxidation and reduction half-cells for some galvanic cells
- **determine** half-cell reactions, the direction of current flow, electrode polarity, cell potential, and ion movement in some galvanic cells
- **build** galvanic cells in the laboratory and **investigate** galvanic cell potentials
- **communicate** your understanding of the following terms: *electric current, electrochemistry, galvanic cell, voltaic cell, external circuit, electrodes, electrolytes, anode, cathode, salt bridge, inert electrode, electric potential, cell voltage, cell potential, dry cell, battery, primary battery, secondary battery*

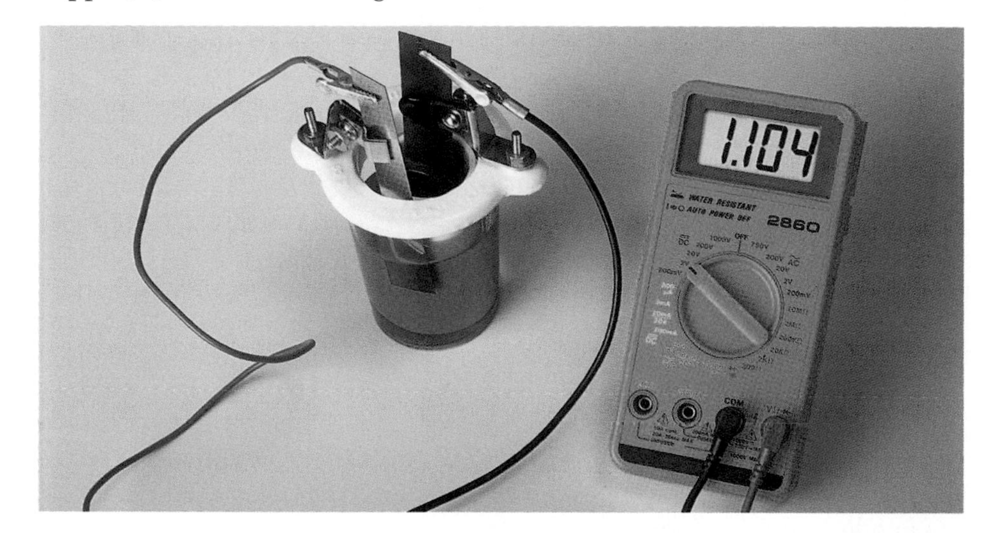

Figure 11.1 **The Daniell cell is named after its inventor, the English chemist John Frederic Daniell (1790–1845). In the photograph shown here, the zinc sulfate solution is placed inside a porous cup, which is placed in a larger container of copper sulfate solution. The cup acts as the porous barrier.**

Web LINK

www.mcgrawhill.ca/links/chemistry12

Galvanic cells are named after the Italian doctor Luigi Galvani (1737–1798), who generated electricity using two metals. These cells are also called voltaic cells, after the Italian physicist Count Alessandro Volta (1745–1827), who built the first chemical batteries. To learn more about scientists who made important discoveries in electrochemistry, such as Galvani, Volta, and Faraday, go to the web site above. Click on **Web Links** to find out where to go next.

In a Daniell cell, the pieces of metallic zinc and copper act as electrical conductors. The conductors that carry electrons into and out of a cell are named **electrodes**. The zinc sulfate and copper(II) sulfate act as electrolytes. **Electrolytes** are substances that conduct electricity when dissolved in water. (The fact that a solution of an electrolyte conducts electricity does not mean that free electrons travel through the solution. An electrolyte solution conducts electricity because of ion movements, and the loss and gain of electrons at the electrodes.) The terms *electrode* and *electrolyte* were invented by the leading pioneer of electrochemistry, Michael Faraday (1791–1867).

The redox reaction takes place in a galvanic cell when an external circuit, such as a metal wire, connects the electrodes. The oxidation half-reaction occurs in one half-cell, and the reduction half-reaction occurs in the other half-cell. For the Daniell cell:

Oxidation (loss of electrons): $Zn_{(s)} \rightarrow Zn^{2+}{}_{(aq)} + 2e^-$

Reduction (gain of electrons): $Cu^{2+}{}_{(aq)} + 2e^- \rightarrow Cu_{(s)}$

The electrode at which oxidation occurs is named the **anode**. In this example, zinc atoms undergo oxidation at the zinc electrode. Thus, the zinc electrode is the anode of the Daniell cell. The electrode at which reduction occurs is named the **cathode**. Here, copper(II) ions undergo reduction at the copper electrode. Thus, the copper electrode is the cathode of the Daniell cell.

Free electrons cannot travel through the solution. Instead, *the external circuit conducts electrons from the anode to the cathode of a galvanic cell.* Figure 11.2 gives a diagram of a typical galvanic cell.

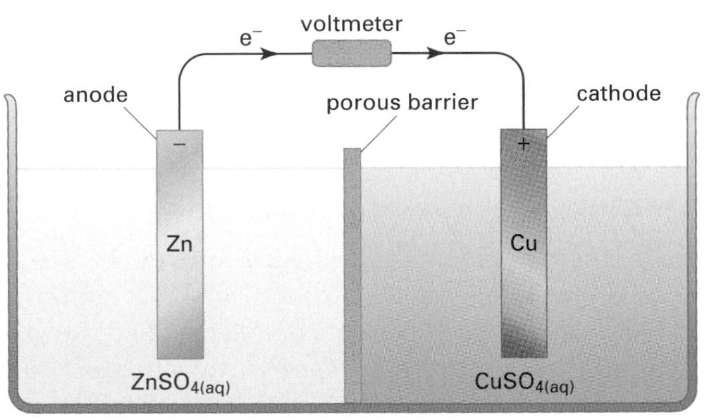

Figure 11.2 A typical galvanic cell, such as the Daniell cell shown here, includes two electrodes, electrolyte solutions, a porous barrier, and an external circuit. Electrons flow through the external circuit from the negative anode to the positive cathode.

At the anode of a galvanic cell, electrons are released by oxidation. For example, at the zinc anode of the Daniell cell, zinc atoms release electrons to become positive zinc ions. Thus, the anode of a galvanic cell is negatively charged. Relative to the anode, the cathode of a galvanic cell is positively charged. In galvanic cells, electrons flow through the external circuit from the negative electrode to the positive electrode. These electrode polarities may already be familiar to you. An example is shown in Figure 11.3.

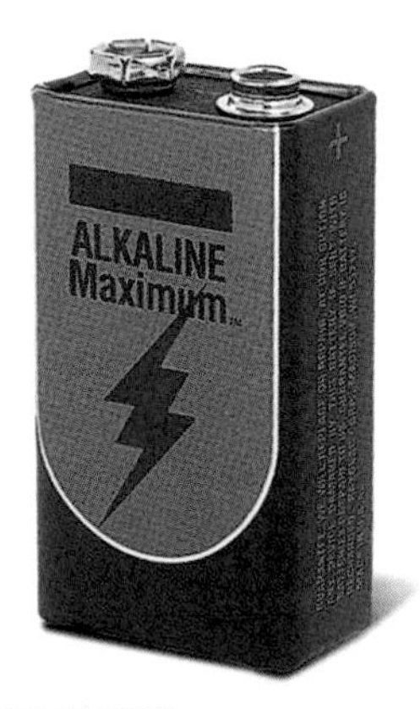

Figure 11.3 Batteries contain galvanic cells. The + mark labels the positive cathode. If there is a − mark, it labels the negative anode.

Each half-cell contains a solution of a neutral compound. In a Daniell cell, these solutions are aqueous zinc sulfate and aqueous copper(II) sulfate. How can these electrolyte solutions remain neutral when electrons are leaving the anode of one half-cell and arriving at the cathode of the other half-cell? To maintain electrical neutrality in each half-cell, some ions migrate through the porous barrier, as shown in Figure 11.4, on the next page. Negative ions (anions) migrate toward the anode, and positive ions (cations) migrate toward the cathode.

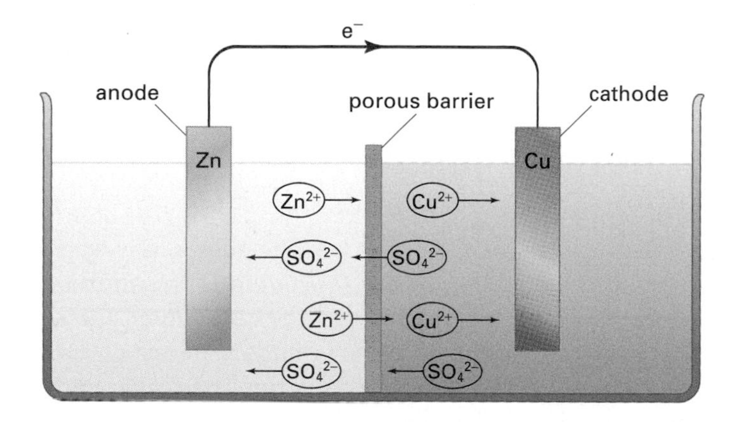

Figure 11.4 Ion migration in a Daniell cell. Some sulfate ions migrate from the copper(II) sulfate solution into the zinc sulfate solution. Some zinc ions migrate from the zinc sulfate solution into the copper(II) sulfate solution.

The separator between the half-cells does not need to be a porous barrier. Figure 11.5 shows an alternative device. This device, called a **salt bridge**, contains an electrolyte solution that does not interfere in the reaction. The open ends of the salt bridge are plugged with a porous material, such as glass wool, to stop the electrolyte from leaking out quickly. The plugs allow ion migration to maintain electrical neutrality.

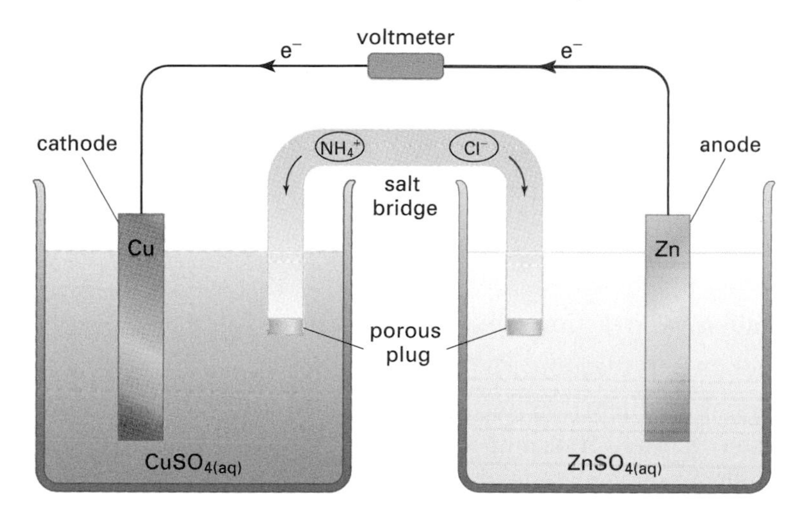

Figure 11.5 This Daniell cell includes a salt bridge instead of a porous barrier. In a diagram, the anode may appear on the left or on the right.

Suppose the salt bridge of a Daniell cell contains ammonium chloride solution, $NH_4Cl_{(aq)}$. As positive zinc ions are produced at the anode, negative chloride ions migrate from the salt bridge into the half-cell that contains the anode. As positive copper(II) ions are removed from solution at the cathode, positive ammonium ions migrate from the salt bridge into the half-cell that contains the cathode.

Other electrolytes, such as sodium sulfate or potassium nitrate, could be chosen for the salt bridge. Neither of these electrolytes interferes in the cell reaction. Silver nitrate, $AgNO_{3(aq)}$, would be a poor choice for the salt bridge, however. Positive silver ions would migrate into the half-cell that contains the cathode. Zinc displaces both copper and silver from solution, so both copper(II) ions and silver ions would be reduced at the cathode. The copper produced would be contaminated with silver.

Galvanic Cell Notation

A convenient shorthand method exists for representing galvanic cells. The shorthand representation of a Daniell cell is as follows.

$$Zn \mid Zn^{2+} \parallel Cu^{2+} \mid Cu$$

The phases or states may be included.

$$Zn_{(s)} \mid Zn^{2+}_{(aq)} \parallel Cu^{2+}_{(aq)} \mid Cu_{(s)}$$

As you saw in Figure 11.4 and Figure 11.5, the anode may appear on the left or on the right of a diagram. *In the shorthand representation, however, the anode is always shown on the left and the cathode on the right.* Each single vertical line, |, represents a phase boundary between the electrode and the solution in a half-cell. For example, the first single vertical line shows that the solid zinc and aqueous zinc ions are in different phases or states. The double vertical line, ||, represents the porous barrier or salt bridge between the half-cells. Spectator ions are usually omitted.

Inert Electrodes

The zinc anode and copper cathode of a Daniell cell are both metals, and can act as electrical conductors. However, some redox reactions involve substances that cannot act as electrodes, such as gases or dissolved electrolytes. Galvanic cells that involve such redox reactions use inert electrodes. An **inert electrode** is an electrode made from a material that is neither a reactant nor a product of the cell reaction. Figure 11.6 shows a cell that contains one inert electrode. The chemical equation, net ionic equation, and half-reactions for this cell are given below.

Chemical equation: $Pb_{(s)} + 2FeCl_{3(aq)} \rightarrow 2FeCl_{2(aq)} + PbCl_{2(aq)}$

Net ionic equation: $Pb_{(s)} + 2Fe^{3+}_{(aq)} \rightarrow 2Fe^{2+}_{(aq)} + Pb^{2+}_{(aq)}$

Oxidation half-reaction: $Pb_{(s)} \rightarrow Pb^{2+}_{(aq)} + 2e^-$

Reduction half-reaction: $Fe^{3+}_{(aq)} + e^- \rightarrow Fe^{2+}_{(aq)}$

The reduction half-reaction does not include a solid conductor of electrons, so an inert platinum electrode is used in this half-cell. The platinum electrode is chemically unchanged, so it does not appear in the chemical equation or half-reactions. However, it is included in the shorthand representation of the cell.

$$Pb \mid Pb^{2+} \parallel Fe^{3+}, Fe^{2+} \mid Pt$$

A comma separates the formulas Fe^{3+} and Fe^{2+} for the ions involved in the reduction half-reaction. The formulas are not separated by a vertical line, because there is no phase boundary between these ions. The Fe^{3+} and Fe^{2+} ions exist in the same aqueous solution.

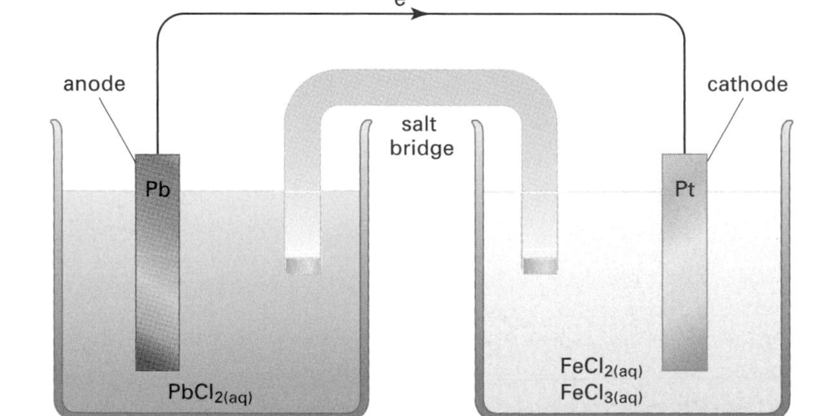

Figure 11.6 **This cell uses an inert electrode to conduct electrons. Why do you think that platinum is often chosen as an inert electrode? Another common choice is graphite.**

Practice Problems

1. (a) If the reaction of zinc with copper(II) ions is carried out in a test tube, what is the oxidizing agent and what is the reducing agent?

   (b) In a Daniell cell, what is the oxidizing agent and what is the reducing agent? Explain your answer.

2. Write the oxidation half-reaction, the reduction half-reaction, and the overall cell reaction for each of the following galvanic cells. Identify the anode and the cathode in each case. In part (b), platinum is present as an inert electrode.

   (a) $Sn_{(s)} \mid Sn^{2+}_{(aq)} \parallel Tl^{+}_{(aq)} \mid Tl_{(s)}$

   (b) $Cd_{(s)} \mid Cd^{2+}_{(aq)} \parallel H^{+}_{(aq)} \mid H_{2(g)} \mid Pt_{(s)}$

3. A galvanic cell involves the overall reaction of iodide ions with acidified permanganate ions to form manganese(II) ions and iodine. The salt bridge contains potassium nitrate.

   (a) Write the half-reactions, and the overall cell reaction.

   (b) Identify the oxidizing agent and the reducing agent.

   (c) The inert anode and cathode are both made of graphite. Solid iodine forms on one of them. Which one?

4. As you saw earlier, pushing a zinc electrode and a copper electrode into a lemon makes a "lemon cell". In the following representation of the cell, $C_6H_8O_7$ is the formula of citric acid. Explain why the representation does not include a double vertical line.

   $Zn_{(s)} \mid C_6H_8O_{7(aq)} \mid Cu_{(s)}$

Introducing Cell Potentials

You know that water spontaneously flows from a higher position to a lower position. In other words, water flows from a state of higher gravitational potential energy to a state of lower gravitational potential energy. As water flows downhill, it can do work, such as turning a water wheel or a turbine. The chemical changes that take place in galvanic cells are also accompanied by changes in potential energy. Electrons spontaneously flow from a position of higher potential energy at the anode to a position of lower potential energy at the cathode. The moving electrons can do work, such as lighting a bulb or turning a motor.

The difference between the potential energy at the anode and the potential energy at the cathode is the **electric potential**, *E*, of a cell. The unit used to measure electric potential is called the *volt*, with symbol V. Because of the name of this unit, electric potential is more commonly known as **cell voltage**. Another name for it is **cell potential**. A cell potential can be measured using an electrical device called a voltmeter.

A cell potential of 0 V means that the cell has no electric potential, and no electrons will flow. You know that you can generate electricity by connecting a zinc electrode and a copper electrode that have been inserted into a lemon. However, you cannot generate electricity by connecting two copper electrodes that have been inserted into the lemon. The two copper electrodes are the same and are in contact with the same electrolyte. There is no potential difference between the two electrodes.

Electric potentials vary from one cell to another, depending on various factors. You will examine some of these factors in the next investigation.

CONCEPT CHECK

The cell voltage is sometimes called the *electromotive force*, abbreviated *emf*. However, this term can be misleading. A cell voltage is a potential difference, not a force. The unit of cell voltage, the volt, is not a unit of force.

Investigation 11-A

SKILL FOCUS
- Predicting
- Performing and recording
- Analyzing and interpreting
- Communicating results

Measuring Cell Potentials of Galvanic Cells

In this investigation, you will build some galvanic cells and measure their cell potentials.

Question

What factors affect the cell potential of a galvanic cell?

Prediction

Predict whether the cell potentials of galvanic cells depend on the electrodes and electrolytes in the half-cells. Give reasons for your prediction.

Materials

25 cm clear aquarium rubber tubing (Tygon®), internal diameter 4–6 mm
1 Styrofoam or clear plastic egg carton with 12 wells
5 cm strip of Mg ribbon
1 cm × 5 cm strips of Cu, Al, Ni, Zn, Sn, Fe, and Ag
5 cm of thick graphite pencil lead or a graphite rod
5 mL of 0.1 mol/L solutions of each of the following: $Mg(NO_3)_2$, $Cu(NO_3)_2$, $Al(NO_3)_3$, $Ni(NO_3)_2$, $Zn(NO_3)_2$, $SnSO_4$, $Fe(NO_3)_3$, $AgNO_3$, HNO_3
15 mL of 1.0 mol/L KNO_3
5 mL of saturated NaCl solution
disposable pipette
cotton batting
sandpaper
black and red electrical leads with alligator clips
voltmeter set to a scale of 0 V to 20 V
paper towel

Safety Precautions

Handle the nitric acid solution with care. It is an irritant. Wash any spills on your skin with copious amounts of water, and inform your teacher.

Procedure

1. Use tape or a permanent marker to label the outside of nine wells of your egg carton with the nine different half-cells. Each well should correspond to one of the eight different metal/metal ion pairs: Mg/Mg^{2+}, Cu/Cu^{2+}, Al/Al^{3+}, Ni/Ni^{2+}, Zn/Zn^{2+}, Sn/Sn^{2+}, Fe/Fe^{3+}, and Ag/Ag^{+}. Label the ninth well H^{+}/H_2.
2. Prepare a 9 × 9 grid in your notebook. Label the nine columns to match the nine half-cells. Label the nine rows in the same way. You will use this chart to mark the positive cell potentials you obtain when you connect two half-cells to build a galvanic cell. You will also record the anode and the cathode for each galvanic cell you build. (You may not need to fill out the entire chart.)
3. Sand each of the metals to remove any oxides.
4. Pour 5 mL of each metal salt solution into the appropriate well of the egg carton. Pour 5 mL of the nitric acid into the well labelled H^{+}/H_2.
5. Prepare your salt bridge as follows.
   (a) Roll a small piece of cotton batting so that it forms a plug about the size of a grain of rice. Place the plug in one end of your aquarium tubing, but leave a small amount of the cotton hanging out, so you can remove the plug later.
   (b) Fill a disposable pipette as full as possible with the 1 mol/L KNO_3 electrolyte solution. Fit the tip of the pipette firmly into the open end of the tubing. Slowly inject the electrolyte solution into the tubing. Fill the tubing completely, so that the cotton on the other side becomes wet.
   (c) With the tubing completely full, insert another cotton plug into the other end. There should be no air bubbles. (You may have to repeat this step from the beginning if you have air bubbles.)

6. Insert each metal strip into the corresponding well. Place the graphite rod in the well with the nitric acid. The metal strips and the graphite rod are your electrodes.
7. Attach the alligator clip on the red lead to the red probe of the voltmeter. Attach the black lead to the black probe.
8. Choose two wells to test. Insert one end of the salt bridge into the solution in the first well. Insert the other end of the salt bridge into the solution in the second well. Attach a free alligator clip to the electrode in each well. (**Note:** The graphite electrode is very fragile. Be gentle when using it.) You have built a galvanic cell.
9. If you get a negative reading, switch the alligator clips. Once you obtain a positive value, record it in your chart. The black lead should be attached to the anode (electrons flowing into the voltmeter). Record which metal is acting as the anode and which is acting as the cathode in this galvanic cell.
10. Remove the salt bridge and wipe any excess salt solution off the outside of the tubing. Remove the alligator clips from the electrodes.
11. Repeat steps 8 to 10 for all other combinations of electrodes. Record your results.
12. Reattach the leads to the silver and magnesium electrodes, and insert your salt bridge back into the two appropriate wells. While observing the reading on the voltmeter, slowly add 5 mL of saturated NaCl solution to the Ag/Ag^+ well to precipitate AgCl. Record any changes in the voltmeter reading. Observe the Ag/Ag^+ well.
13. Rinse off the metals and the graphite rod with water. Dispose of the salt solutions into the heavy metal salts container your teacher has set aside. Rinse out your egg carton. Remove and discard the plugs of the salt bridge, and dispose of the KNO_3 solution as directed by your teacher. Return all your materials to their appropriate locations.

Analysis

1. For each cell in which you measured a cell potential, identify
   (a) the anode and the cathode
   (b) the positive and negative electrodes
2. For each cell in which you measured a cell potential, write a balanced equation for the reduction half-reaction, the oxidation half-reaction, and the overall cell reaction.
3. For any one cell in which you measured a cell potential, describe
   (a) the direction in which electrons flow through the external circuit
   (b) the movements of ions in the cell
4. Use your observations to decide which of the metals used as electrodes is the most effective reducing agent. Explain your reasoning.
5. List all the reduction half-reactions you wrote in question 2 so that the metallic elements in the half-reactions appear in order of their ability as reducing agents. Put the least effective reducing agent at the top of the list and the most effective reducing agent at the bottom.
6. In which part of your list from question 5 are the metal ions that are the best oxidizing agents? Explain.
7. (a) When saturated sodium chloride solution was added to the silver nitrate solution, what reaction took place? Explain.
   (b) Does the concentration of an electrolyte affect the cell potential of a galvanic cell? How do you know?

Conclusion

8. Identify factors that affect the cell potential of a galvanic cell.

Applications

9. Predict any other factors that you think might affect the voltage of a galvanic cell. Describe an investigation you could complete to test your prediction.

Disposable Batteries

The Daniell cell is fairly large and full of liquid. Realistically, you could not use this type of cell to power a wristwatch, a remote control, or a flashlight. Galvanic cells have been modified, however, to make them more useful.

The Dry Cell Battery

A **dry cell** is a galvanic cell with the electrolyte contained in a paste thickened with starch. This cell is much more portable than the Daniell cell. The first dry cell, invented by the French chemist Georges Leclanché in 1866, was called the Leclanché cell.

Modern dry cells are closely modelled on the Leclanché cell, and also contain electrolyte pastes. You have probably used dry cells in all kinds of applications, such as lighting a flashlight, powering a remote control, or ringing a doorbell. Dry cells are inexpensive. The cheapest AAA-, AA-, C-, and D-size 1.5-V batteries are dry cells.

A **battery** is defined as a set of galvanic cells connected in series. The negative electrode of one cell is connected to the positive electrode of the next cell in the set. *The voltage of a set of cells connected in series is the sum of the voltages of the individual cells.* Thus, a 9-V battery contains six 1.5-V dry cells connected in series. Often, the term "battery" is also used to describe a single cell. For example, a 1.5-V dry cell battery contains only a single cell.

COURSE CHALLENGE

How does electrochemistry affect your life? In the Chemistry Course Challenge, you will examine some of the roles electrochemistry plays in the human body. How is electrochemistry important in medical applications? Do research to find some specific examples.

A dry cell battery stops producing electricity when the reactants are used up. This type of battery is disposable after it has run down completely. A disposable battery is known as a **primary battery**. Some other batteries are rechargeable. A rechargeable battery is known as a **secondary battery**. The rest of this section will deal with primary batteries. You will learn about secondary batteries in section 11.3.

A dry cell contains a zinc anode and an inert graphite cathode, as shown in Figure 11.7. The electrolyte is a moist paste of manganese(IV) oxide, MnO_2, zinc chloride, $ZnCl_2$, ammonium chloride, NH_4Cl, and "carbon black," $C_{(s)}$, also known as soot.

The oxidation half-reaction at the zinc anode is already familiar to you.

$$Zn_{(s)} \rightarrow Zn^{2+}{}_{(aq)} + 2e^-$$

The reduction half-reaction at the cathode is more complicated. An approximation is given here.

$$2MnO_{2(s)} + H_2O_{(\ell)} + 2e^- \rightarrow Mn_2O_{3(s)} + 2OH^-{}_{(aq)}$$

Therefore, an approximation of the overall cell reaction is:

$$2MnO_{2(s)} + Zn_{(s)} + H_2O_{(\ell)} \rightarrow Mn_2O_{3(s)} + Zn^{2+}{}_{(aq)} + 2OH^-{}_{(aq)}$$

Figure 11.7 The D-size dry cell battery is shown whole, and cut in two. The anode is the zinc container, located just inside the outer paper, steel, or plastic case. The graphite cathode runs through the centre of the cylinder.

The Alkaline Cell Battery

The more expensive alkaline cell, shown in Figure 11.8, is an improved, longer-lasting version of the dry cell.

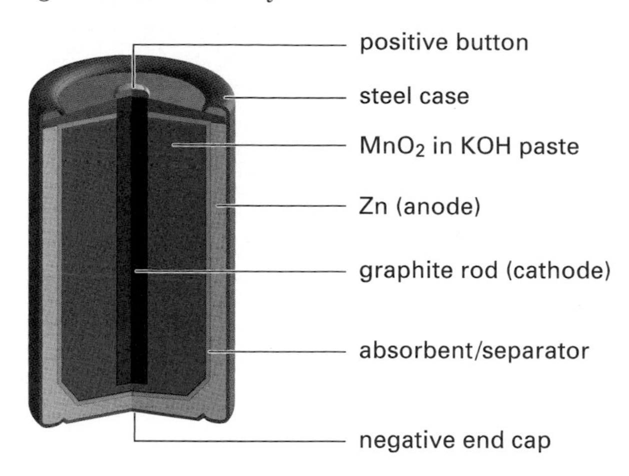

Figure 11.8 The structure of an alkaline cell is similar to the structure of a dry cell. Each type has a voltage of 1.5 V.

Billions of alkaline batteries, each containing a single alkaline cell, are made every year. The ammonium chloride and zinc chloride used in a dry cell are replaced by strongly alkaline (basic) potassium hydroxide, KOH. The half-reactions and the overall reaction in an alkaline cell are given here.

Oxidation (at the anode): $Zn_{(s)} + 2OH^-_{(aq)} \rightarrow ZnO_{(s)} + H_2O_{(\ell)} + 2e^-$

Reduction (at the cathode): $MnO_{2(s)} + 2H_2O_{(\ell)} + 2e^- \rightarrow Mn(OH)_{2(s)} + 2OH^-_{(aq)}$

Overall cell reaction: $Zn_{(s)} + MnO_{2(s)} + H_2O_{(\ell)} \rightarrow ZnO_{(s)} + Mn(OH)_{2(s)}$

Figure 11.9 Button batteries are small and long-lasting.

The Button Cell Battery

A button battery is much smaller than an alkaline battery. Button batteries are commonly used in watches, as shown in Figure 11.9. Because of its small size, the button battery is also used for hearing aids, pacemakers, and some calculators and cameras. The development of smaller batteries has had an enormous impact on portable devices, as shown in Figure 11.10.

Two common types of button batteries both use a zinc container, which acts as the anode, and an inert stainless steel cathode, as shown in Figure 11.11 on the next page. In the mercury button battery, the alkaline electrolyte paste contains mercury(II) oxide, HgO. In the silver button battery, the electrolyte paste contains silver oxide, Ag_2O. The batteries have similar voltages: about 1.3 V for the mercury cell, and about 1.6 V for the silver cell.

The reaction products in a mercury button battery are solid zinc oxide and liquid mercury. The two half-reactions and the overall equation are as follows.

Oxidation half-reaction: $Zn_{(s)} + 2OH^-_{(aq)} \rightarrow ZnO_{(s)} + H_2O_{(\ell)} + 2e^-$

Reduction half-reaction: $HgO_{(s)} + H_2O_{(\ell)} + 2e^- \rightarrow Hg_{(\ell)} + 2OH^-_{(aq)}$

Overall reaction: $Zn_{(s)} + HgO_{(s)} \rightarrow ZnO_{(s)} + Hg_{(\ell)}$

Figure 11.10 With small, long-lasting batteries, a pacemaker can now be implanted in a heart patient's chest. Early pacemakers, such as this one developed at the University of Toronto in the 1950s, were so big and heavy that patients had to wheel them around on a cart.

Figure 11.11 A common type of button battery, shown here, contains silver oxide or mercury(II) oxide. Mercury is cheaper than silver, but discarded mercury batteries release toxic mercury metal into the environment.

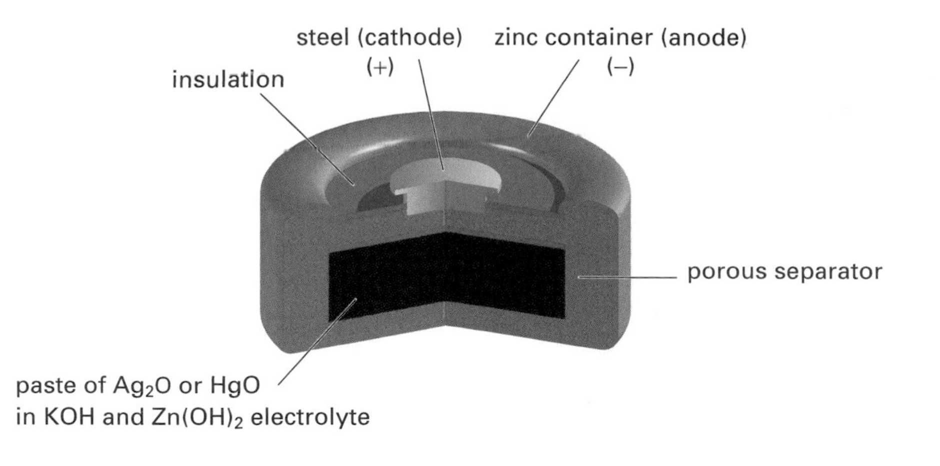

Careers  in Chemistry

Explosives Chemist

Fortunato Villamagna works as the vice-president of technology for an Australian-owned company with offices worldwide, including Canada. Villamagna's job has given him the opportunity to invent new products, build chemical plants, and conduct projects in Africa and Australia.

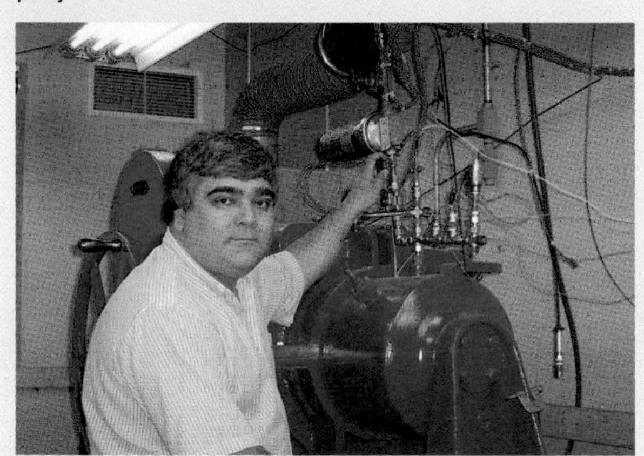

Born in Italy, Villamagna moved to Canada when he was eight. He grew up in Montréal, where his interest in chemistry was sparked by a Grade 10 teacher.

At university, Villamagna gained a greater understanding of the value of chemistry research. "Research creates new technologies and concepts, results in new products and services, creates jobs and prosperity, and in the end improves people's lives," he says. Villamagna decided to pursue graduate studies. He obtained a Masters of Science in physical chemistry at Concordia University, and a Ph.D. in physical chemistry at McGill University. Today, Villamagna leads a team of researchers responsible for developing new products and techniques involving explosives.

Redox reactions play an important role in industrial safety. Explosives are used in controlled ways in the mining, highway, and construction industries. The use of explosives allows modern workers to break up bedrock and carry out necessary demolitions from a safe distance. Chemists are involved in the development and production of explosives. They are also involved in making recommendations for the safe handling and disposal of explosives.

Many explosives are based on redox reactions. For example, the decomposition of nitroglycerin into nitrogen, carbon dioxide, water vapour, and oxygen is a redox reaction that results in a powerful explosion. The three nitrate groups of a nitroglycerin molecule act as powerful oxidizing agents, and the glycerol portion of the compound acts as a fuel. Fuels are very easily oxidized.

Nitroglycerin is highly unstable and can explode very easily. Therefore, it is difficult to manufacture and transport safely. Ammonium nitrate, an explosive that can act as both an oxidizing agent and a reducing agent, is often used to modify other explosives such as nitroglycerin. Ammonium nitrate is one of the products made by Villamagna's company.

Making Career Connections

- The chemical formula of nitroglycerin is $C_3O_9N_3H_5$. Write the balanced chemical equation for the decomposition of nitroglycerin, as described in this feature.
- Find out which Canadian companies employ chemists who specialize in safe applications of explosives. Contact those companies for more information.

Section Summary

In this section, you learned how to identify the different components of a galvanic cell. Also, you found out how galvanic cells convert chemical energy into electrical energy. You were introduced to several common primary batteries that contain galvanic cells. In the next section, you will learn more about the cell potentials of galvanic cells.

PROBEWARE

If you have access to probeware, do Probeware Investigation 11-A, or a similar investigation from a probeware company.

Section Review

1. K/U Identify the oxidizing agent and the reducing agent in a dry cell.
2. C Explain why the top of a commercial 1.5-V dry cell battery is always marked with a plus sign.
3. I The reaction products in a silver button battery are solid zinc oxide and solid silver.
   - (a) Write the two half-reactions and the equation for the overall reaction in the battery.
   - (b) Name the materials used to make the anode and the cathode.
4. I If two 1.5-V D-size batteries power a flashlight, at what voltage is the flashlight operating? Explain.
5. I How many dry cells are needed to make a 6-V dry cell battery? Explain.
6. MC Research the environmental impact of mercury pollution. Describe the main sources of mercury in the environment, the effects of mercury on human health, and at least one incident in which humans were harmed by mercury pollution.
7. K/U When a dry cell produces electricity, what happens to the container? Explain.
8. C Use the following shorthand representation to sketch a possible design of the cell. Include as much information as you can. Identify the anode and cathode, and write the half-reactions and the overall cell reaction.

$$Fe_{(s)} \mid Fe^{2+}_{(aq)} \parallel Ag^{+}_{(aq)} \mid Ag_{(s)}$$

11.2 Standard Cell Potentials

In this section, you will

- **describe** the use of the hydrogen half-cell as a reference
- **calculate** overall cell potentials under standard conditions
- **communicate** your understanding of the following terms: *reduction potential*, *oxidation potential*

In section 11.1, you learned that a cell potential is the difference between the potential energies at the anode and the cathode of a cell. In other words, a cell potential is the difference between the potentials of two half-cells. You cannot measure the potential of one half-cell, because a single half-reaction cannot occur alone. However, you can use measured cell potentials to construct tables of half-cell potentials. A table of standard half-cell potentials allows you to calculate cell potentials, rather than building the cells and measuring their potentials. Table 11.1 includes a few standard half-cell potentials. A larger table of standard half-cell potentials is given in Appendix E.

Table 11.1 Standard Half-Cell Potentials (298 K)

Half-reaction	$E°$ (V)
$F_{2(g)} + 2e^- \rightleftharpoons 2F^-_{(aq)}$	2.866
$Br_{2(\ell)} + 2e^- \rightleftharpoons 2Br^-_{(aq)}$	1.066
$I_{2(s)} + 2e^- \rightleftharpoons 2I^-_{(aq)}$	0.536
$Cu^{2+}_{(aq)} + 2e^- \rightleftharpoons Cu_{(s)}$	0.342
$2H^+_{(aq)} + 2e^- \rightleftharpoons H_{2(g)}$	0.000
$Fe^{2+}_{(aq)} + 2e^- \rightleftharpoons Fe_{(s)}$	−0.447
$Zn^{2+}_{(aq)} + 2e^- \rightleftharpoons Zn_{(s)}$	−0.762
$Al^{3+}_{(aq)} + 3e^- \rightleftharpoons Al_{(s)}$	−1.662
$Na^+_{(aq)} + e^- \rightleftharpoons Na_{(s)}$	−2.711

Table 11.1 and the larger table in Appendix E are based on the following conventions.

- Each half-reaction is written as a reduction. The half-cell potential for a reduction half-reaction is called a **reduction potential**. Look at the molecules and ions on the left side of each half-reaction. The most easily reduced molecules and ions (best oxidizing agents), such as F_2, MnO_4^-, and O_2, are near the top of the list. The least easily reduced molecules and ions (worst oxidizing agents), such as Na^+, Ca^{2+}, and H_2O, are near the bottom of the list.
- The numerical values of cell potentials and half-cell potentials depend on various conditions, so tables of *standard* reduction potentials are true when ions and molecules are in their *standard states*. These standard states are the same as for tables of standard enthalpy changes. Aqueous molecules and ions have a standard concentration of 1 mol/L. Gases have a standard pressure of 101.3 kPa or 1 atm. The standard temperature is 25°C or 298 K. Standard reduction potentials are designated by the symbol $E°$, where the superscript ° indicates standard states.
- Because you can measure potential differences, but not individual reduction potentials, all values in the table are relative. Each half-cell reduction potential is given relative to the reduction potential of the standard hydrogen electrode, which has been assigned a value of zero. The design of this electrode is shown in Figure 11.12.

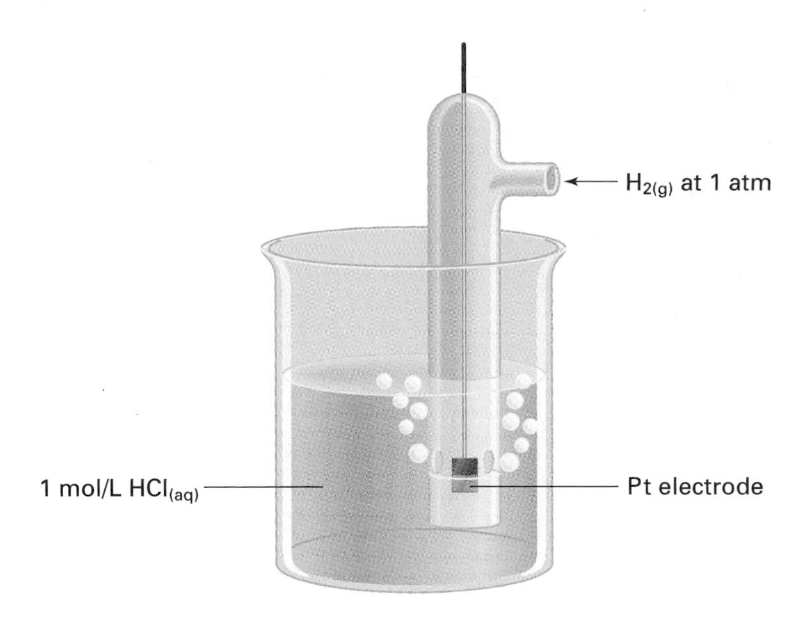

Figure 11.12 **In a standard hydrogen electrode, which is open to the atmosphere, hydrogen gas at 1 atm pressure bubbles over an inert platinum electrode. The electrode is immersed in a solution containing 1 mol/L H^+ ions.**

Calculating Standard Cell Potentials

You can use Table 11.1 to calculate the standard cell potential of the familiar Daniell cell. This cell has its standard potential when the solution concentrations are 1 mol/L, as shown in the shorthand representation below.

$$Zn \mid Zn^{2+}\ (1\ mol/L) \parallel Cu^{2+}\ (1\ mol/L) \mid Cu$$

One method to calculate the standard cell potential is to subtract the standard reduction potential of the anode from the standard reduction potential of the cathode.

$$\text{Method 1: } E^\circ_{cell} = E^\circ_{cathode} - E^\circ_{anode}$$

For a Daniell cell, you know that copper is the cathode and zinc is the anode. The relevant half-reactions and standard reduction potentials from Table 11.1 are as follows.

$$Cu^{2+}_{(aq)} + 2e^- \rightleftharpoons Cu_{(s)} \quad E^\circ = 0.342\ V$$
$$Zn^{2+}_{(aq)} + 2e^- \rightleftharpoons Zn_{(s)} \quad E^\circ = -0.762\ V$$

Use these values to calculate the cell potential.

$$\begin{aligned} E^\circ_{cell} &= E^\circ_{cathode} - E^\circ_{anode} \\ &= 0.342\ V - (-0.762\ V) \\ &= 0.342\ V + 0.762\ V \\ &= 1.104\ V \end{aligned}$$

Thus, the standard cell potential for a Daniell cell is 1.104 V.
The standard cell potentials of all galvanic cells have positive values, as explained in Figure 11.13, on the following page. Figure 11.13 shows a "potential ladder" diagram. A "potential ladder" diagram models the potential difference. The rungs on the ladder correspond to the values of the reduction potentials.

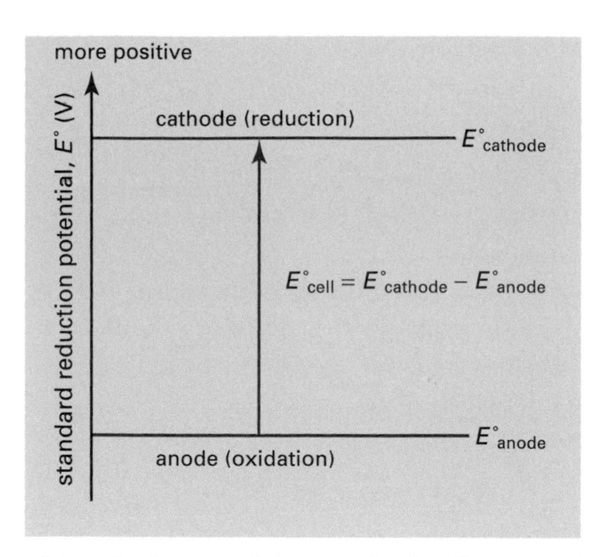

Figure 11.13 For a galvanic cell, $E^\circ_{cathode}$ is more positive (or less negative) than E°_{anode}. Thus, E°_{cell} is always positive.

This calculation of the standard cell potential for the Daniell cell used the mathematical concept that the subtraction of a negative number is equivalent to the addition of its positive value. You saw that

$$0.342 \text{ V} - (-0.762 \text{ V}) = 0.342 \text{ V} + 0.762 \text{ V}$$

In other words, the *subtraction* of the reduction potential for a half-reaction is equivalent to the *addition* of the potential for the reverse half-reaction. The reverse half-reaction of a reduction is an oxidation. The half-cell potential for an oxidation half-reaction is called an **oxidation potential**. If the reduction half-reaction is as follows,

$$Zn^{2+}_{(aq)} + 2e^- \rightleftharpoons Zn_{(s)} \quad E^\circ = -0.762 \text{ V}$$

then the oxidation half-reaction is

$$Zn_{(s)} \rightleftharpoons Zn^{2+}_{(aq)} + 2e^- \quad E^\circ_{ox} = +0.762 \text{ V}$$

To summarize, the standard cell potential can also be calculated as the sum of a standard reduction potential and a standard oxidation potential.

$$\text{Method 2: } E^\circ_{cell} = E^\circ_{red} + E^\circ_{ox}$$

As shown above, you can obtain the standard oxidation potential from a table of standard reduction potentials by reversing the reduction half-reaction, and changing the sign of the relevant potential. The reduction and oxidation half-reactions for the previous example are as follows.

$$Cu^{2+}_{(aq)} + 2e^- \rightleftharpoons Cu_{(s)} \quad E^\circ_{red} = 0.342 \text{ V}$$
$$Zn_{(s)} \rightleftharpoons Zn^{2+}_{(aq)} + 2e^- \quad E^\circ_{ox} = +0.762 \text{ V}$$

The calculation of the standard cell potential using these standard half-reaction potentials is as follows.

$$\begin{aligned} E^\circ_{cell} &= E^\circ_{red} + E^\circ_{ox} \\ &= 0.342 \text{ V} + 0.762 \text{ V} \\ &= 1.104 \text{ V} \end{aligned}$$

Finding the difference between two reduction potentials, and finding the sum of a reduction potential and an oxidation potential are exactly equivalent methods for finding a cell potential. Use whichever method you prefer. The first Sample Problem includes both methods for finding cell potentials. The second Sample Problem uses only the subtraction of two reduction potentials. Practice problems are included after the second Sample Problem.

Electronic Learning Partner

Your Chemistry 12 Electronic Learning Partner has a movie describing the operation of a galvanic cell, and the calculation of its cell potential.

Sample Problem

Calculating a Standard Cell Potential, Given a Net Ionic Equation

Problem

Calculate the standard cell potential for the galvanic cell in which the following reaction occurs.

$2I^{-}_{(aq)} + Br_{2(\ell)} \rightarrow I_{2(s)} + 2Br^{-}_{(aq)}$

What Is Required?

You need to find the standard cell potential for the given reaction.

What Is Given?

You have the balanced net ionic equation and a table of standard reduction potentials.

Plan Your Strategy

Method 1: Subtracting Two Reduction Potentials

Step 1 Write the oxidation and reduction half-reactions.

Step 2 Locate the relevant reduction potentials in a table of standard reduction potentials.

Step 3 Subtract the reduction potentials to find the cell potential, using $E^{\circ}_{cell} = E^{\circ}_{cathode} - E^{\circ}_{anode}$

Method 2: Adding an Oxidation Potential and a Reduction Potential

Step 1 Write the oxidation and reduction half-reactions.

Step 2 Locate the relevant reduction potentials in a table of standard reduction potentials.

Step 3 Change the sign of the reduction potential for the oxidation half-reaction to find the oxidation potential.

Step 4 Add the reduction potential and the oxidation potential, using $E^{\circ}_{cell} = E^{\circ}_{red} + E^{\circ}_{ox}$

Act on Your Strategy

Method 1: Subtracting Two Reduction Potentials

Step 1 The oxidation and reduction half-reactions are as follows.

Oxidation half-reaction (occurs at the anode): $2I^{-}_{(aq)} \rightarrow I_{2(s)} + 2e^{-}$

Reduction half-reaction (occurs at the cathode): $Br_{2(\ell)} + 2e^{-} \rightarrow 2Br^{-}_{(aq)}$

Step 2 The relevant reduction potentials in the table of standard reduction potentials are:

$I_{2(s)} + 2e^{-} \rightleftharpoons 2I^{-}_{(aq)} \quad E^{\circ}_{anode} = 0.536\ V$

$Br_{2(\ell)} + 2e^{-} \rightleftharpoons 2Br^{-}_{(aq)} \quad E^{\circ}_{cathode} = 1.066\ V$

Step 3 Calculate the cell potential by subtraction.

$$\begin{aligned} E^{\circ}_{cell} &= E^{\circ}_{cathode} - E^{\circ}_{anode} \\ &= 1.066\ V - 0.536\ V \\ &= 0.530\ V \end{aligned}$$

Continued ...

PROBLEM TIP

Think of a *red cat* to remember that *red*uction occurs at the *cat*hode. Think of *an ox* to remember that the *an*ode is where *ox*idation occurs.

Continued ...

Method 2: Adding an Oxidation Potential and a Reduction Potential

Step 1 The oxidation and reduction half-reactions are as follows.

Oxidation half-reaction (occurs at the anode): $2I^-_{(aq)} \rightarrow I_{2(s)} + 2e^-$

Reduction half-reaction (occurs at the cathode): $Br_{2(\ell)} + 2e^- \rightarrow 2Br^-_{(aq)}$

Step 2 The relevant reduction potentials in the table of standard reduction potentials are:

$I_{2(s)} + 2e^- \rightleftharpoons 2I^-_{(aq)} \quad E^\circ_{anode} = 0.536\ V$

$Br_{2(\ell)} + 2e^- \rightleftharpoons 2Br^-_{(aq)} \quad E^\circ_{cathode} = 1.066\ V$

Step 3 The standard electrode potential for the reduction half-reaction is $E^\circ_{red} = 1.066\ V$. Changing the sign of the potential for the oxidation half-reaction gives

$2I^-_{(aq)} \rightleftharpoons I_{2(s)} + 2e^- \quad E^\circ_{ox} = -0.536\ V$

Step 4 Calculate the cell potential by addition.

$$E^\circ_{cell} = E^\circ_{red} + E^\circ_{ox}$$
$$= 1.066\ V + (-0.536\ V)$$
$$= 0.530\ V$$

Check Your Solution

Both methods give the same answer. The cell potential is positive, as expected for a galvanic cell.

A standard cell potential depends only on the identities of the reactants and products in their standard states. As you will see in the next Sample Problem, *you do not need to consider the amounts of reactants or products present, or the reaction stoichiometry, when calculating a standard cell potential.* Since you have just completed a similar Sample Problem, only a brief solution using the subtraction method is given here. Check that you can solve this problem by adding a reduction potential and an oxidation potential.

Sample Problem

Calculating a Standard Cell Potential, Given a Chemical Equation

Problem

Calculate the standard cell potential for the galvanic cell in which the following reaction occurs.

$2Na_{(s)} + 2H_2O_{(\ell)} \rightarrow 2NaOH_{(aq)} + H_{2(g)}$

Solution

Step 1 Write the equation in ionic form to identify the half-reactions.

$2Na_{(s)} + 2H_2O_{(\ell)} \rightarrow 2Na^+_{(aq)} + 2OH^-_{(aq)} + H_{2(g)}$

Write the oxidation and reduction half-reactions.

Oxidation half-reaction (occurs at the anode): $Na_{(s)} \rightarrow Na^+_{(aq)} + e^-$

Reduction half-reaction (occurs at the cathode):

$2H_2O_{(\ell)} + 2e^- \rightarrow 2OH^-_{(aq)} + H_{2(g)}$

Step 2 Locate the relevant reduction potentials in a table of standard reduction potentials.

$Na_{(s)} + e^- \rightleftharpoons Na^+_{(aq)} \quad E^\circ_{anode} = -2.711\ V$

$2H_2O_{(\ell)} + 2e^- \rightleftharpoons H_{2(g)} + 2OH^-_{(aq)} \quad E^\circ_{cathode} = -0.828\ V$

Step 3 Subtract the standard reduction potentials to calculate the cell potential.

$$\begin{aligned} E^\circ_{cell} &= E^\circ_{cathode} - E^\circ_{anode} \\ &= -0.828\ V - (-2.711\ V) \\ &= 1.883\ V \end{aligned}$$

The standard cell potential is 1.883 V. The Problem Tip on this page illustrates this calculation.

PROBLEM TIP

A "potential ladder" diagram models the potential difference. The rungs on the ladder correspond to the values of the reduction potentials. For a galvanic cell, the half-reaction at the cathode is always on the upper rung, and the subtraction $E^\circ_{cathode} - E^\circ_{anode}$ always gives a positive cell potential.

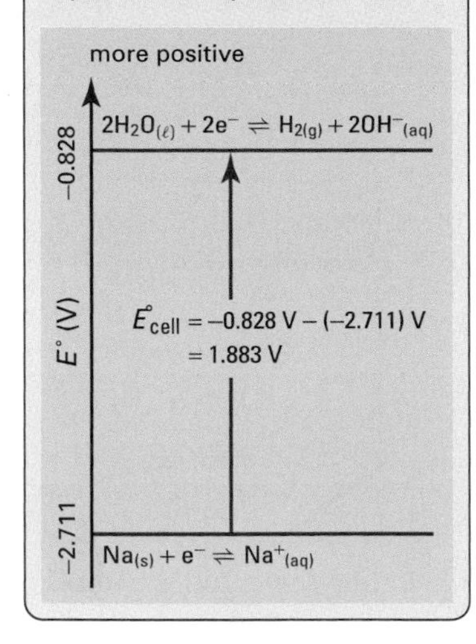

Practice Problems

(**Note:** Obtain the necessary standard reduction potential values from the table in Appendix E.)

5. Write the two half-reactions for the following redox reaction. Subtract the two reduction potentials to find the standard cell potential for a galvanic cell in which this reaction occurs.

$Cl_{2(g)} + 2Br^-_{(aq)} \rightarrow 2Cl^-_{(aq)} + Br_{2(\ell)}$

6. Write the two half-reactions for the following redox reaction. Add the reduction potential and the oxidation potential to find the standard cell potential for a galvanic cell in which this reaction occurs.

$2Cu^+_{(aq)} + 2H^+_{(aq)} + O_{2(g)} \rightarrow 2Cu^{2+}_{(aq)} + H_2O_{2(aq)}$

7. Write the two half-reactions for the following redox reaction. Subtract the two standard reduction potentials to find the standard cell potential for the reaction.

$Sn_{(s)} + 2HBr_{(aq)} \rightarrow SnBr_{2(aq)} + H_{2(g)}$

8. Write the two half-reactions for the following redox reaction. Add the standard reduction potential and the standard oxidation potential to find the standard cell potential for the reaction.

$Cr_{(s)} + 3AgCl_{(s)} \rightarrow CrCl_{3(aq)} + 3Ag_{(s)}$

Electronic Learning Partner

Go to the Chemistry 12 Electronic Learning Partner for more information about aspects of material covered in this section of the chapter.

You have learned that the standard hydrogen electrode has an assigned standard reduction potential of exactly 0 V, and is the reference for all half-cell standard reduction potentials. What would happen to cell potentials if a different reference were used? You will address this question in the following ThoughtLab.

ThoughtLab Assigning Reference Values

Many scales of measurement have zero values that are arbitrary. For example, on Earth, average sea level is often assigned as the zero of altitude. In this ThoughtLab, you will investigate what happens to calculated cell potentials when the reference half-cell is changed.

Procedure

1. Copy the following table of reduction potentials into your notebook. Change the zero on the scale by adding 1.662 V to each value to create new, adjusted reduction potentials.

Reduction half-reaction	E° (V)	$E^\circ + 1.662$ (V)
$F_{2(g)} + 2e^- \rightleftharpoons 2F^-_{(aq)}$	2.866	
$Fe^{3+}_{(aq)} + e^- \rightleftharpoons Fe^{2+}_{(aq)}$	0.771	
$2H^+_{(aq)} + 2e^- \rightleftharpoons H_{2(g)}$	0.000	
$Al^{3+}_{(aq)} + 3e^- \rightleftharpoons Al_{(s)}$	−1.662	
$Li^+_{(aq)} + e^- \rightleftharpoons Li_{(s)}$	−3.040	

2. Use the given standard reduction potentials to calculate the standard cell potentials for the following redox reactions.
   (a) $2Li_{(s)} + 2H^+_{(aq)} \rightarrow 2Li^+_{(aq)} + H_{2(g)}$
   (b) $2Al_{(s)} + 3F_{2(g)} \rightarrow 2Al^{3+}_{(aq)} + 6F^-_{(aq)}$
   (c) $2FeCl_{3(aq)} + H_{2(g)} \rightarrow 2FeCl_{2(aq)} + 2HCl_{(aq)}$
   (d) $Al(NO_3)_{3(aq)} + 3Li_{(s)} \rightarrow 3LiNO_{3(aq)} + Al_{(s)}$
3. Repeat your calculations using the new, adjusted reduction potentials.

Analysis

1. Compare your calculations from questions 2 and 3 of the procedure. What effect does changing the zero on the scale of reduction potentials have on
   (a) reduction potentials?
   (b) cell potentials?

Applications

2. Find the difference between the temperatures at which water boils and freezes on the following scales. (Assume that a difference is positive, rather than negative.)
   (a) the Celsius temperature scale
   (b) the Kelvin temperature scale
3. What do your answers for the previous question tell you about these two temperature scales?
4. The zero on a scale of masses is not arbitrary. Why not?

Section Summary

In this section, you learned that you can calculate cell potentials by using tables of half-cell potentials. The half-cell potential for a reduction half-reaction is called a reduction potential. The half-cell potential for an oxidation half-reaction is called an oxidation potential. Standard half-cell potentials are written as reduction potentials. The values of standard reduction potentials for half-reactions are relative to the reduction potential of the standard hydrogen electrode. You used standard reduction potentials to calculate standard cell potentials for galvanic cells. You learned two methods of calculating standard cell potentials. One method is to subtract the standard reduction potential of the anode from the standard reduction potential of the cathode. The other method is to add the standard reduction potential of the cathode and the standard oxidation potential of the anode. In the next section, you will learn about a different type of cell, called an electrolytic cell.

Section Review

1. **I** Determine the standard cell potential for each of the following redox reactions.

   (a) $3Mg_{(s)} + 2Al^{3+}_{(aq)} \rightarrow 3Mg^{2+}_{(aq)} + 2Al_{(s)}$

   (b) $2K_{(s)} + F_{2(g)} \rightarrow 2K^{+}_{(aq)} + 2F^{-}_{(aq)}$

   (c) $Cr_2O_7^{2-}{}_{(aq)} + 14H^{+}_{(aq)} + 6Ag_{(s)} \rightarrow 2Cr^{3+}_{(aq)} + 6Ag^{+}_{(aq)} + 7H_2O_{(\ell)}$

2. **I** Determine the standard cell potential for each of the following redox reactions.

   (a) $CuSO_{4(aq)} + Ni_{(s)} \rightarrow NiSO_{4(aq)} + Cu_{(s)}$

   (b) $4Au(OH)_{3(aq)} \rightarrow 4Au_{(s)} + 6H_2O_{(\ell)} + 3O_{2(g)}$

   (c) $Fe_{(s)} + 4HNO_{3(aq)} \rightarrow Fe(NO_3)_{3(aq)} + NO_{(g)} + 2H_2O_{(\ell)}$

3. **K/U** For which half-cell are the values of the standard reduction potential and the standard oxidation potential equal?

4. **I** Look at the half-cells in the table of standard reduction potentials in Appendix E. Could you use two of the standard half-cells to build a galvanic cell with a standard cell potential of 7 V? Explain your answer.

5. **C** Compare the positions of metals in the metal activity series with their positions in the table of standard reduction potentials. Describe the similarities and differences.

6. **I** The cell potential for the following galvanic cell is given.

   $Zn \mid Zn^{2+}$ (1 mol/L) $\parallel Pd^{2+}$ (1 mol/L) $\mid Pd \quad E^{\circ}_{cell} = 1.750$ V

   Determine the standard reduction potential for the following half-reaction.

   $Pd^{2+}_{(aq)} + 2e^{-} \rightleftharpoons Pd_{(s)}$

11.3 Electrolytic Cells

Section Preview/ Specific Expectations

In this section, you will

- **identify** the components of an electrolytic cell, and **describe** how they work
- **describe** electrolytic cells using oxidation and reduction half-cells
- **determine** oxidation and reduction half-cell reactions, direction of current flow, electrode polarity, cell potential, and ion movement in some electrolytic cells
- **build** and **investigate** an electrolytic cell in the laboratory
- **predict** whether or not redox reactions are spontaneous, using standard cell potentials
- **describe** some common rechargeable batteries, and **evaluate** their impact on the environment and on society
- **communicate** your understanding of the following terms: *electrolytic cell, electrolysis, overvoltage, electroplating*

For a galvanic cell, you have learned that the overall reaction is spontaneous, and that the cell potential has a positive value. A galvanic cell converts chemical energy to electrical energy. Electrons flow from a higher potential energy to a lower potential energy. As described earlier, the flow of electrons in the external circuit of a galvanic cell can be compared to water flowing downhill.

Although water flows downhill spontaneously, you can also pump water uphill. This process requires energy because it moves water from a position of lower potential energy to a position of higher potential energy. You will now learn about a type of cell that uses energy to move electrons from lower potential energy to higher potential energy. This type of cell, called an **electrolytic cell**, is a device that converts electrical energy to chemical energy. The process that takes place in an electrolytic cell is called **electrolysis**. The overall reaction in an electrolytic cell is non-spontaneous, and requires energy to occur. This type of reaction is the reverse of a spontaneous reaction, which generates energy when it occurs.

Like a galvanic cell, an electrolytic cell includes electrodes, at least one electrolyte, and an external circuit. Unlike galvanic cells, electrolytic cells require an external source of electricity, sometimes called the *external voltage*. This is included in the external circuit. Except for the external source of electricity, an electrolytic cell may look just like a galvanic cell. Some electrolytic cells include a porous barrier or salt bridge. In other electrolytic cells, the two half-reactions are not separated, and take place in the same container.

Electrolysis of Molten Salts

The electrolytic cell shown in Figure 11.14 decomposes sodium chloride into its elements. The cell consists of a single container with two inert electrodes dipping into liquid sodium chloride. To melt the sodium chloride, the temperature must be above its melting point of about 800°C. As in an aqueous solution of sodium chloride, the ions in molten sodium chloride have some freedom of movement. In other words, molten sodium chloride is the electrolyte of this cell.

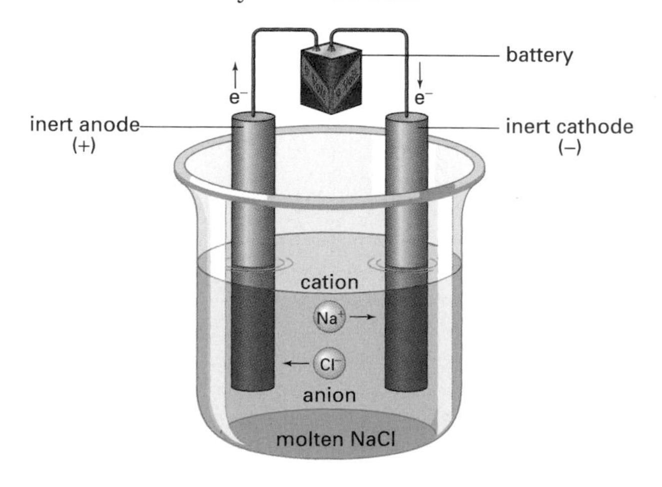

Figure 11.14 **Molten sodium chloride decomposes into sodium and chlorine in this electrolytic cell. The sodium chloride is said to undergo electrolysis, or to be *electrolyzed*.**

The external source of electricity forces electrons onto one electrode. As a result, this electrode becomes negative relative to the other electrode. The positive sodium ions move toward the negative electrode, where they gain electrons and are reduced to the element sodium. At this temperature, sodium metal is produced as a liquid. The negative chloride ions move toward the positive electrode, where they lose electrons and are oxidized to the element chlorine, a gas. *As in a galvanic cell, reduction occurs at the cathode, and oxidation occurs at the anode of an electrolytic cell.* The half-reactions for this electrolytic cell are as follows.

Reduction half-reaction (occurs at the cathode): $Na^+_{(\ell)} + e^- \rightarrow Na_{(\ell)}$

Oxidation half-reaction (occurs at the anode): $2Cl^-_{(\ell)} \rightarrow Cl_{2(g)} + 2e^-$

Because of the external voltage of the electrolytic cell, the electrodes do not have the same polarities in electrolytic and galvanic cells. In a galvanic cell, the cathode is positive and the anode is negative. In an electrolytic cell, the anode is positive and the cathode is negative.

The electrolysis of molten sodium chloride is an important industrial reaction. Figure 11.15 shows the large electrolytic cell used in the industrial production of sodium and chlorine. You will meet other industrial electrolytic processes later in this chapter.

CONCEPT CHECK

"Electrochemical cell" is a common term in electrochemistry. Some scientists include both galvanic cells and electrolytic cells as types of electrochemical cells. Other scientists consider galvanic cells, but not electrolytic cells, as electrochemical cells. If you meet the term "electrochemical cell," always check its exact meaning.

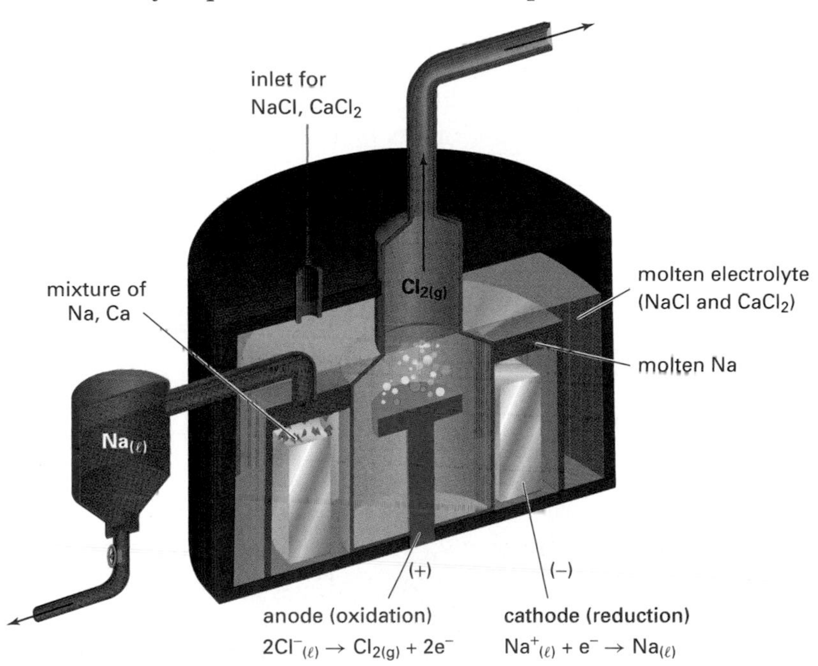

Figure 11.15 **The large cell used for the electrolysis of sodium chloride in industry is known as a *Downs cell*. To decrease heating costs, calcium chloride is added to lower the melting point of sodium chloride from about 800°C to about 600°C. The reaction produces sodium and calcium by reduction at the cathode, and chlorine by oxidation at the anode.**

Check your understanding of the introduction to electrolytic cells by completing the following practice problems.

Practice Problems

9. The electrolysis of molten calcium chloride produces calcium and chlorine. Write

(a) the half-reaction that occurs at the anode

(b) the half-reaction that occurs at the cathode

(c) the chemical equation for the overall cell reaction

Continued ...

Continued ...

10. For the electrolysis of molten lithium bromide, write
    (a) the half-reaction that occurs at the negative electrode
    (b) the half-reaction that occurs at the positive electrode
    (c) the net ionic equation for the overall cell reaction
11. A galvanic cell produces direct current, which flows in one direction. The mains supply at your home is a source of alternating current, which changes direction every fraction of a second. Explain why the external electrical supply for an electrolytic cell must be a source of direct current, rather than alternating current.
12. Suppose a battery is used as the external electrical supply for an electrolytic cell. Explain why the negative terminal of the battery must be connected to the cathode of the cell.

Electrolysis of Water

The electrolysis of aqueous solutions may not yield the desired products. Sir Humphry Davy (1778–1829) discovered the elements sodium and potassium by electrolyzing their molten salts. Before this discovery, Davy had electrolyzed aqueous solutions of sodium and potassium salts. He had not succeeded in reducing the metal ions to the pure metals at the cathode. Instead, his first experiments had produced hydrogen gas. Where did the hydrogen gas come from?

When electrolyzing an aqueous solution, there are two compounds present: water, and the dissolved electrolyte. Water may be electrolyzed as well as, or instead of, the electrolyte. The electrolysis of water produces oxygen gas and hydrogen gas, as shown in Figure 11.16.

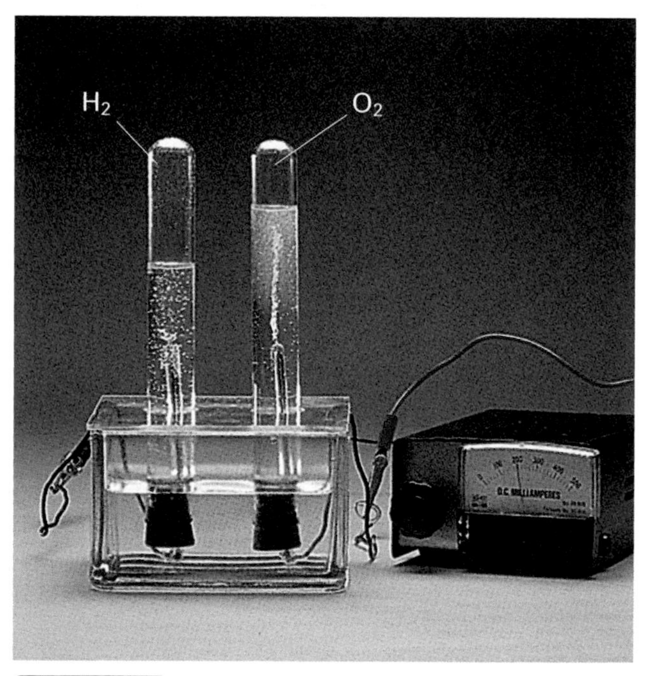

Figure 11.16 The electrolysis of water produces hydrogen gas at the cathode and oxygen gas at the anode. Explain why the volume of hydrogen gas is twice the volume of oxygen gas.

The half-reactions for the electrolysis of water are given below.

Oxidation half-reaction (occurs at the anode):
$2H_2O_{(\ell)} \rightarrow O_{2(g)} + 4H^+_{(aq)} + 4e^-$

Reduction half-reaction (occurs at the cathode):
$2H_2O_{(\ell)} + 2e^- \rightarrow H_{2(g)} + 2OH^-_{(aq)}$

Because the number of electrons lost and gained must be equal, multiply the reduction half-reaction by 2. Then add and simplify to obtain the overall cell reaction.

Overall cell reaction: $2H_2O_{(\ell)} \rightarrow 2H_{2(g)} + O_{2(g)}$

The standard reduction potentials are as follows.

$O_{2(g)} + 4H^+_{(aq)} + 4e^- \rightleftharpoons 2H_2O_{(\ell)} \quad E^\circ = 1.229 \text{ V}$

$2H_2O_{(\ell)} + 2e^- \rightleftharpoons H_{2(g)} + 2OH^-_{(aq)} \quad E^\circ = -0.828 \text{ V}$

You can use these values to calculate the E°_{cell} value for the decomposition of water.

$$\begin{aligned} E^\circ_{cell} &= E^\circ_{cathode} - E^\circ_{anode} \\ &= -0.828 \text{ V} - 1.229 \text{ V} \\ &= -2.057 \text{ V} \end{aligned}$$

Therefore,

$$2H_2O_{(\ell)} \rightarrow 2H_{2(g)} + O_{2(g)} \quad E^\circ_{cell} = -2.057 \text{ V}$$

The negative cell potential shows that the reaction is not spontaneous. Electrolytic cells are used for non-spontaneous redox reactions, so *all electrolytic cells have negative cell potentials.*

The standard reduction potentials used to calculate E°_{cell} for the decomposition of water apply only to reactants and products in their standard states. However, in pure water at 25°C, the hydrogen ions and hydroxide ions each have concentrations of 1×10^{-7} mol/L. This is not the standard state value of 1 mol/L. The reduction potential values for the non-standard conditions in pure water are given below. The superscript zero is now omitted from the E symbol, because the values are no longer standard.

$O_{2(g)} + 4H^+_{(aq)} + 4e^- \rightleftharpoons 2H_2O_{(\ell)} \quad E = 0.815 \text{ V}$

$2H_2O_{(\ell)} + 2e^- \rightleftharpoons H_{2(g)} + 2OH^-_{(aq)} \quad E = -0.414 \text{ V}$

Using these new half-cell potentials, E_{cell} for the decomposition of pure water at 25°C by electrolysis has a calculated value of −1.229 V. Therefore, the calculated value of the external voltage needed is 1.229.

In practice, the external voltage needed for an electrolytic cell is always greater than the calculated value, especially for reactions involving gases. Therefore, the actual voltage needed to electrolyze pure water is *greater than* 1.229 V. The excess voltage required above the calculated value is called the **overvoltage**. Overvoltages depend on the gases involved and on the materials in the electrodes.

When electrolyzing water, there is another practical difficulty to consider. Pure water is a very poor electrical conductor. To increase the conductivity, an electrolyte that does not interfere in the reaction is added to the water.

CONCEPT CHECK

Check that you recall how to combine the two half-reactions to obtain the overall cell reaction. You learned how to do this in section 10.3.

Electrolysis of Aqueous Solutions

As stated previously, an electrolytic cell may have the same design as a galvanic cell, except for the external source of electricity. Consider, for example, the familiar Daniell cell. (This cell was described in section 11.1 and shown in Figure 11.5.) By adding an external electrical supply, with a voltage greater than the voltage of the Daniell cell, you can push electrons in the opposite direction. By pushing electrons in the opposite direction, you reverse the chemical reaction. Figure 11.17 shows both cells, while their properties are compared in Table 11.2.

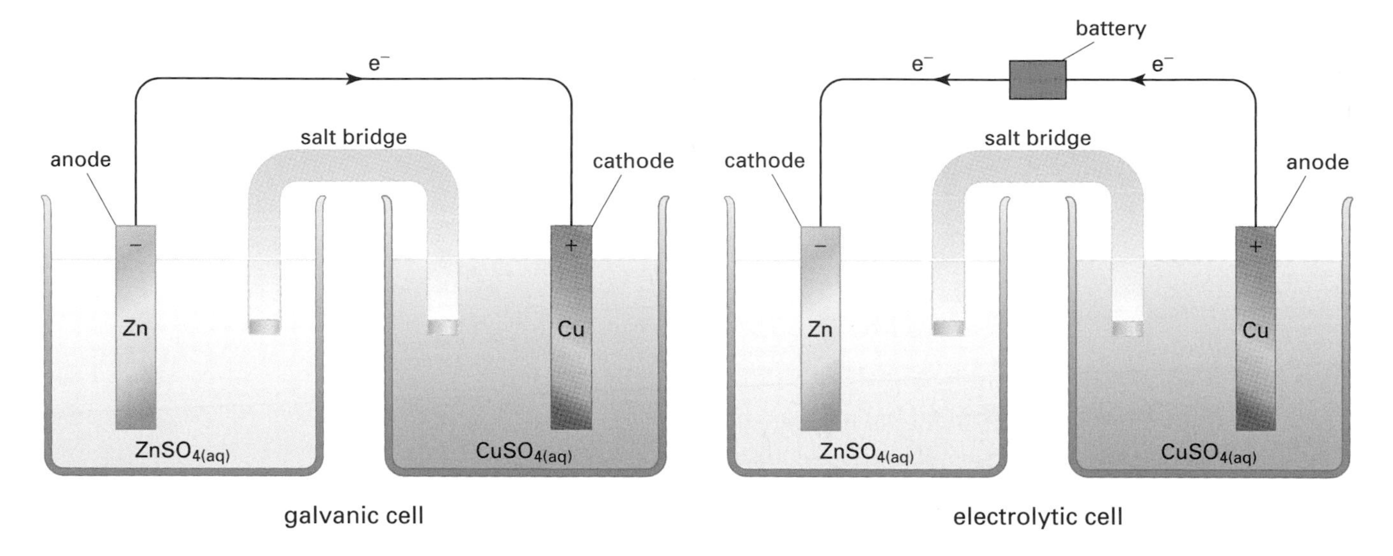

Figure 11.17 **Adding an external voltage to reverse the electron flow converts a Daniell cell from a galvanic cell into an electrolytic cell. The result is to switch the anode and cathode.**

Table 11.2 Cell Comparison

Galvanic Cell	Electrolytic Cell
Spontaneous reaction	Non-spontaneous reaction
Converts chemical energy to electrical energy	Converts electrical energy to chemical energy
Anode (negative): Zinc	Anode (positive): Copper
Cathode (positive): Copper	Cathode (negative): Zinc
Oxidation (at anode): $Zn_{(s)} \rightarrow Zn^{2+}{}_{(aq)} + 2e^-$	Oxidation (at anode): $Cu_{(s)} \rightarrow Cu^{2+}{}_{(aq)} + 2e^-$
Reduction (at cathode): $Cu^{2+}{}_{(aq)} + 2e^- \rightarrow Cu_{(s)}$	Reduction (at cathode): $Zn^{2+}{}_{(aq)} + 2e^- \rightarrow Zn_{(s)}$
Cell reaction: $Zn_{(s)} + Cu^{2+}{}_{(aq)} \rightarrow Zn^{2+}{}_{(aq)} + Cu_{(s)}$	Cell reaction: $Cu_{(s)} + Zn^{2+}{}_{(aq)} \rightarrow Cu^{2+}{}_{(aq)} + Zn_{(s)}$

In the galvanic cell, the zinc anode gradually dissolves. The copper cathode grows as more copper is deposited onto it. In the electrolytic cell, the copper anode gradually dissolves. The zinc cathode grows as more zinc is deposited onto it. The process in which a metal is deposited, or plated, onto the cathode in an electrolytic cell is known as **electroplating**. Electroplating is very important in industry, as you will learn later in this chapter.

Predicting the Products of Electrolysis for an Aqueous Solution

The comparison of the Daniell cell with the electrolytic version of the cell appears straightforward. One reaction is the reverse of the other. However, you have just learned that the electrolysis of an aqueous solution may involve the electrolysis of water. How can you predict the actual products for this type of electrolysis reaction?

To predict the products of an electrolysis involving an aqueous solution, you must examine all possible half-reactions and their reduction potentials. Then, you must find the overall reaction that requires the *lowest* external voltage. That is, you must find the overall cell reaction with a negative cell potential that is closest to zero. The next Sample Problem shows you how to predict the products of the electrolysis of an aqueous solution.

In practice, reaction products are sometimes different from the products predicted, using the method described here. Predictions are least reliable when the reduction potentials are close together, especially when gaseous products are expected. However, there are many cases in which the predictions are correct.

Sample Problem

Electrolysis of an Aqueous Solution

Problem

Predict the products of the electrolysis of 1 mol/L $LiBr_{(aq)}$.

What Is Required?

You need to predict the products of the electrolysis of 1 mol/L $LiBr_{(aq)}$.

What Is Given?

This is an aqueous solution. You are given the formula and concentration of the electrolyte. You have a table of standard reduction potentials, and you know the non-standard reduction potentials for water.

Plan Your Strategy

Step 1 List the four relevant half-reactions and their reduction potentials.

Step 2 Predict the products by finding the cell reaction that requires the lowest external voltage.

Act on Your Strategy

Step 1 The Li^+ and Br^- concentrations are 1 mol/L, so use the standard reduction potentials for the half-reactions that involve these ions. Use the non-standard values for water.

$Br_{2(\ell)} + 2e^- \rightleftharpoons 2Br^-_{(aq)} \quad E^\circ = 1.066\ V$

$O_{2(g)} + 4H^+_{(aq)} + 4e^- \rightleftharpoons 2H_2O_{(\ell)} \quad E = 0.815\ V$

$2H_2O_{(\ell)} + 2e^- \rightleftharpoons H_{2(g)} + 2OH^-_{(aq)} \quad E = -0.414\ V$

$Li^+_{(aq)} + e^- \rightleftharpoons Li_{(s)} \quad E^\circ = -3.040\ V$

Continued ...

Continued ...

There are two possible oxidation half-reactions at the anode: the oxidation of bromide ion in the electrolyte, or the oxidation of water.

$2Br^-_{(aq)} \rightarrow Br_{2(\ell)} + 2e^-$

$2H_2O_{(\ell)} \rightarrow O_{2(g)} + 4H^+_{(aq)} + 4e^-$

There are two possible reduction half-reactions at the cathode: the reduction of lithium ions in the electrolyte, or the reduction of water.

$Li^+_{(aq)} + e^- \rightarrow Li_{(s)}$

$2H_2O_{(\ell)} + 2e^- \rightarrow H_{2(g)} + 2OH^-_{(aq)}$

Step 2 Combine pairs of half-reactions to produce four possible overall reactions. (You learned how to do this in Chapter 10.)

Reaction 1: the production of lithium and bromine

$$2Li^+_{(aq)} + 2Br^-_{(aq)} \rightarrow 2Li_{(s)} + Br_{2(\ell)}$$

$$\begin{aligned} E^\circ_{cell} &= E^\circ_{cathode} - E^\circ_{anode} \\ &= -3.040 \text{ V} - 1.066 \text{ V} \\ &= -4.106 \text{ V} \end{aligned}$$

Reaction 2: the production of hydrogen and oxygen

$$2H_2O_{(\ell)} \rightarrow 2H_{2(g)} + O_{2(g)}$$

$$\begin{aligned} E_{cell} &= E_{cathode} - E_{anode} \\ &= -0.414 \text{ V} - 0.815 \text{ V} \\ &= -1.229 \text{ V} \end{aligned}$$

Reaction 3: the production of lithium and oxygen

$$4Li^+_{(aq)} + 2H_2O_{(\ell)} \rightarrow 4Li_{(s)} + O_{2(g)} + 4H^+_{(aq)}$$

$$\begin{aligned} E_{cell} &= E^\circ_{cathode} - E_{anode} \\ &= -3.040 \text{ V} - 0.815 \text{ V} \\ &= -3.855 \text{ V} \end{aligned}$$

Reaction 4: the production of hydrogen and bromine

$$2H_2O_{(\ell)} + 2Br^-_{(aq)} \rightarrow H_{2(g)} + 2OH^-_{(aq)} + Br_{2(\ell)}$$

$$\begin{aligned} E_{cell} &= E_{cathode} - E^\circ_{anode} \\ &= -0.414 \text{ V} - 1.066 \text{ V} \\ &= -1.480 \text{ V} \end{aligned}$$

The electrolysis of water requires the lowest external voltage. Therefore, the predicted products of this electrolysis are hydrogen and oxygen.

PROBLEM TIP

- Remember that spectator ions do not appear in net ionic equations. In Reaction 3, the bromide ions are spectator ions. In Reaction 4, lithium ions are spectator ions.
- As for a galvanic cell, the cell potential for an electrolytic cell is the sum of a reduction potential and an oxidation potential. Using $E_{cell} = E_{red} + E_{ox}$ gives the same predicted products as using $E_{cell} = E_{cathode} - E_{anode}$

Check Your Solution

Use a potential ladder diagram, such as the one on the next page, part A, to visualize the cell potentials. For an electrolytic cell, the half-reaction at the anode is always on the upper rung, and the subtraction $E^\circ_{cathode} - E^\circ_{anode}$ always gives a negative cell potential, as shown in part B.

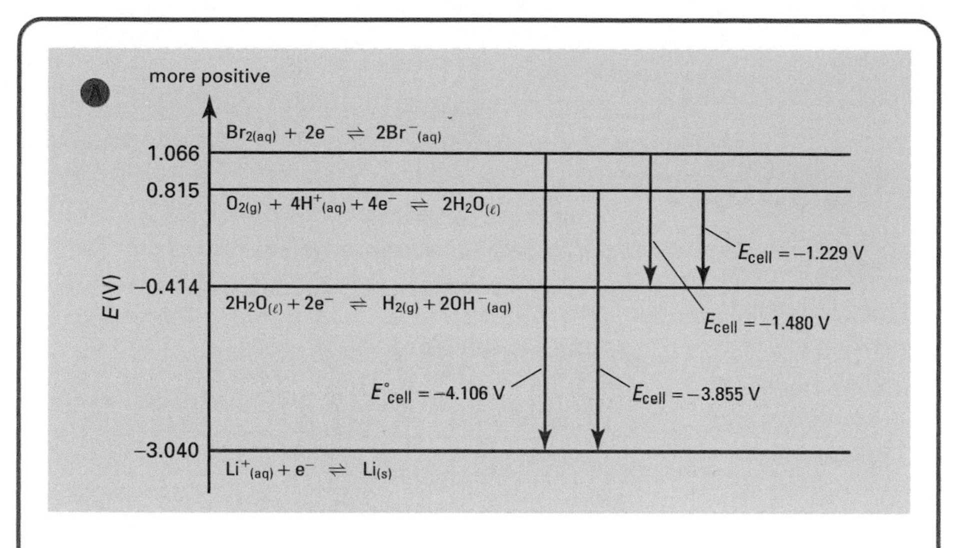

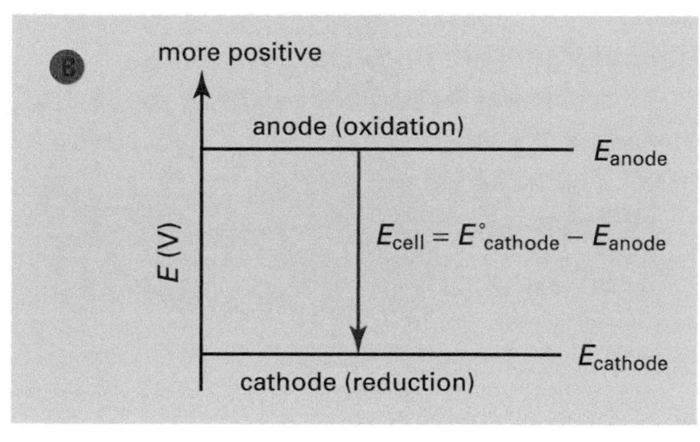

Practice Problems

13. Predict the products of the electrolysis of a 1 mol/L solution of sodium chloride.

14. Explain why calcium can be produced by the electrolysis of molten calcium chloride, but not by the electrolysis of aqueous calcium chloride.

15. One half-cell of a galvanic cell has a nickel electrode in a 1 mol/L nickel(II) chloride solution. The other half-cell has a cadmium electrode in a 1 mol/L cadmium chloride solution.

(a) Find the cell potential.

(b) Identify the anode and the cathode.

(c) Write the oxidation half-reaction, the reduction half-reaction, and the overall cell reaction.

16. An external voltage is applied to change the galvanic cell in question 15 into an electrolytic cell. Repeat parts (a) to (c) for the electrolytic cell.

In Investigation 11-B, you will build an electrolytic cell for the electrolysis of an aqueous solution of potassium iodide. You will predict the products of the electrolysis, and compare the observed products with your predictions.

Investigation 11-B

SKILL FOCUS

Predicting

Performing and recording

Analyzing and interpreting

Electrolysis of Aqueous Potassium Iodide

When an aqueous solution is electrolyzed, the electrolyte or water can undergo electrolysis. In this investigation, you will build an electrolytic cell, carry out the electrolysis of an aqueous solution, and identify the products.

Questions

What are the products from the electrolysis of a 1 mol/L aqueous solution of potassium iodide? Are the observed products the ones predicted using reduction potentials?

Predictions

Use the relevant standard reduction potentials from the table in Appendix E, and the non-standard reduction potentials you used previously for water, to predict the electrolysis products. Predict which product(s) are formed at the anode and which product(s) are formed at the cathode.

Materials

25 cm clear aquarium rubber tubing (Tygon®), internal diameter 4–6 mm
1 graphite pencil lead, 2 cm long
2 wire leads (black and red) with alligator clips
600 mL or 400 mL beaker
sheet of white paper
1 elastic band
3 toothpicks
3 disposable pipettes
2 cm piece of copper wire (20 gauge)
1 drop 1% starch solution
10 mL 1 mol/L KI
1 drop 1% phenolphthalein
9-V battery or variable power source set to 9 V

Safety Precautions

Make sure your lab bench is dry before carrying out this investigation.

Procedure

1. Fold a sheet of paper lengthwise. Curl the folded paper so that it fits inside the 600 mL beaker. Invert the beaker on your lab bench.
2. Use the elastic to strap the aquarium tubing to the side of the beaker in a U shape, as shown in the diagram.

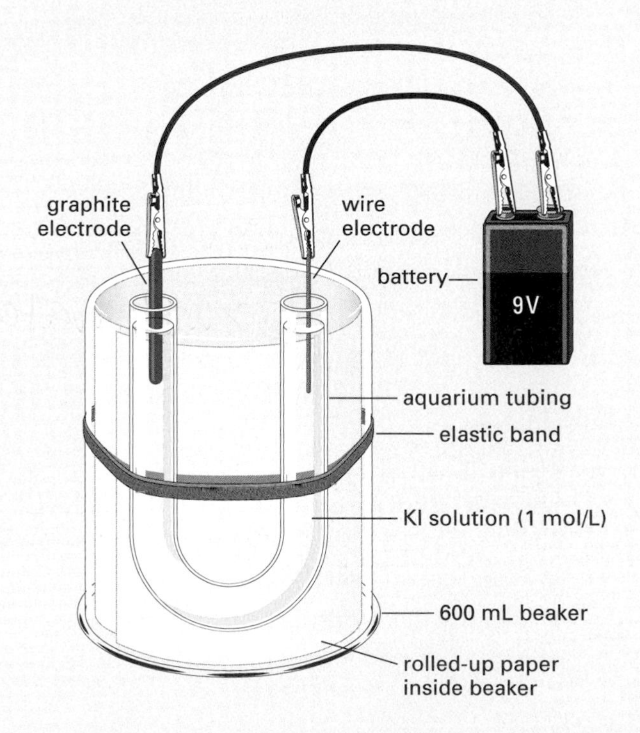

3. Fill a pipette as completely as possible with 1 mol/L KI solution. Insert the tip of the pipette firmly into one end of the aquarium tubing. Slowly inject the solution into the U-tube until the level of the solution is within 1 cm to 2 cm from the top of both ends. If air bubbles are present, try to remove them by poking them with a toothpick. You may need to repeat this step from the beginning.
4. Attach the black lead to the 2 cm piece of wire. Insert the wire into one end of the U-tube. Attach the red electrical lead to the graphite. Insert the graphite into the other end of the U-tube.
5. Attach the leads to the 9-V battery or to a variable power source set to 9 V. Attach the black lead to the negative terminal, and the red lead to the positive terminal.
6. Let the reaction proceed for three minutes, while you examine the U-tube. Record your observations. Shut off the power source and remove the electrodes. Determine the product formed around the anode by adding a drop of starch solution to the end of the U-tube that contains the anode. Push the starch solution down with a toothpick if there is an air lock. Determine one of the products around the cathode by adding a drop of phenolphthalein to the appropriate end of the U-tube.
7. Dispose of your reactants and products as instructed by your teacher. Take your apparatus apart, rinse out the tubing, and rinse off the electrodes. Return your equipment to its appropriate location.

Analysis

1. Sketch the cell you made in this investigation. On your sketch, show
   (a) the direction of the electron flow in the external circuit
   (b) the anode and the cathode
   (c) the positive electrode and the negative electrode
   (d) the movement of ions in the cell
2. Use your observations to identify the product(s) formed at the anode and the product(s) formed at the cathode.
3. Write a balanced equation for the half-reaction that occurs at the anode.
4. Write a balanced equation for the half-reaction that occurs at the cathode.
5. Write a balanced equation for the overall cell reaction.
6. Calculate the external voltage required to carry out the electrolysis. Why was the external voltage used in the investigation significantly higher than the calculated value?

Conclusion

7. What are the products from the electrolysis of a 1 mol/L aqueous solution of potassium iodide? Are the observed products the same as the products predicted using reduction potentials?

Applications

8. If you repeated the electrolysis using aqueous sodium iodide instead of aqueous potassium iodide, would your observations change? Explain your answer.
9. To make potassium by electrolyzing potassium iodide, would you need to modify the procedure? Explain your answer.

Spontaneity of Reactions

You know that galvanic cells have positive standard cell potentials, and that these cells use spontaneous chemical reactions to produce electricity. You also know that electrolytic cells have negative standard cell potentials, and that these cells use electricity to perform non-spontaneous chemical reactions. Thus, you can use the sign of the standard cell potential to predict whether a reaction is spontaneous or not under standard conditions.

Sample Problem

Predicting Spontaneity

Problem

Predict whether each reaction is spontaneous or non-spontaneous under standard conditions.

(a) $Cd_{(s)} + Cu^{2+}{}_{(aq)} \rightarrow Cd^{2+}{}_{(aq)} + Cu_{(s)}$ **(b)** $I_{2(s)} + 2Cl^{-}{}_{(aq)} \rightarrow 2I^{-}{}_{(aq)} + Cl_{2(g)}$

Solution

(a) The two half-reactions are as follows.

Oxidation (occurs at the anode): $Cd_{(s)} \rightarrow Cd^{2+}{}_{(aq)} + 2e^{-}$

Reduction (occurs at the cathode): $Cu^{2+}{}_{(aq)} + 2e^{-} \rightarrow Cu_{(s)}$

The relevant standard reduction potentials are:

$Cu^{2+}{}_{(aq)} + 2e^{-} \rightleftharpoons Cu_{(s)} \quad E^{\circ} = 0.342\ V$

$Cd^{2+}{}_{(aq)} + 2e^{-} \rightleftharpoons Cd_{(s)} \quad E^{\circ} = -0.403\ V$

$$E^{\circ}_{cell} = E^{\circ}_{cathode} - E^{\circ}_{anode}$$
$$= 0.342\ V - (-0.403\ V)$$
$$= 0.745\ V$$

The standard cell potential is positive, so the reaction is spontaneous under standard conditions.

(b) The two half-reactions are as follows.

Oxidation (occurs at the anode): $2Cl^{-}{}_{(aq)} \rightarrow Cl_{2(g)} + 2e^{-}$

Reduction (occurs at the cathode): $I_{2(s)} + 2e^{-} \rightarrow 2I^{-}{}_{(aq)}$

The relevant standard reduction potentials are:

$Cl_{2(g)} + 2e^{-} \rightleftharpoons 2Cl^{-}{}_{(aq)} \quad E^{\circ} = 1.358\ V$

$I_{2(s)} + 2e^{-} \rightleftharpoons 2I^{-}{}_{(aq)} \quad E^{\circ} = 0.536\ V$

$$E^{\circ}_{cell} = E^{\circ}_{cathode} - E^{\circ}_{anode}$$
$$= 0.536\ V - 1.358\ V$$
$$= -0.822\ V$$

The standard cell potential is negative, so the reaction is non-spontaneous under standard conditions.

Practice Problems

17. Look up the standard reduction potentials of the following half-reactions. Predict whether acidified nitrate ions will oxidize manganese(II) ions to manganese(IV) oxide under standard conditions.

$MnO_{2(s)} + 4H^{+}{}_{(aq)} + 2e^{-} \rightarrow Mn^{2+}{}_{(aq)} + 2H_2O_{(\ell)}$

$NO_3{}^{-}{}_{(aq)} + 4H^{+}{}_{(aq)} + 3e^{-} \rightarrow NO_{(g)} + 2H_2O_{(\ell)}$

18. Predict whether each reaction is spontaneous or non-spontaneous under standard conditions.

(a) $2Cr_{(s)} + 3Cl_{2(g)} \rightarrow 2Cr^{3+}{}_{(aq)} + 6Cl^{-}{}_{(aq)}$

(b) $Zn^{2+}{}_{(aq)} + Fe_{(s)} \rightarrow Zn_{(s)} + Fe^{2+}{}_{(aq)}$

(c) $5Ag_{(s)} + MnO_4{}^{-}{}_{(aq)} + 8H^{+}{}_{(aq)} \rightarrow 5Ag^{+}{}_{(aq)} + Mn^{2+}{}_{(aq)} + 4H_2O_{(\ell)}$

19. Explain why an aqueous copper(I) compound disproportionates to form copper metal and an aqueous copper(II) compound under standard conditions. (You learned about disproportionation in Chapter 10.)

20. Predict whether each reaction is spontaneous or non-spontaneous under standard conditions in an acidic solution.

(a) $H_2O_{2(aq)} \rightarrow H_{2(g)} + O_{2(g)}$

(b) $3H_{2(g)} + Cr_2O_7{}^{2-}{}_{(aq)} + 8H^{+}{}_{(aq)} \rightarrow 2Cr^{3+}{}_{(aq)} + 7H_2O_{(\ell)}$

Rechargeable Batteries

In section 11.1, you learned about several primary (disposable) batteries that contain galvanic cells. One of the most common secondary (rechargeable) batteries is found in car engines. Most cars contain a lead-acid battery, shown in Figure 11.18. When you turn the ignition, a surge of electricity from the battery starts the motor.

When in use, a lead-acid battery partially discharges. In other words, the cells in the battery operate as galvanic cells, and produce electricity. The reaction in each cell proceeds spontaneously in one direction. To recharge the battery, a generator driven by the car engine supplies electricity to the battery. The external voltage of the generator reverses the reaction in the cells. The reaction in each cell now proceeds non-spontaneously, and the cells operate as electrolytic cells. All secondary batteries, including the lead-acid battery, operate some of the time as galvanic cells, and some of the time as electrolytic cells.

As the name suggests, the materials used in a lead-acid battery include lead and an acid. Figure 11.19 shows that the electrodes in each cell are constructed using lead grids. One electrode consists of powdered lead packed into one grid. The other electrode consists of powdered lead(IV) oxide packed into the other grid. The electrolyte solution is fairly concentrated sulfuric acid, at about 4.5 mol/L.

Figure 11.18 A typical car battery consists of six 2-V cells. The cells are connected in series to give a total potential of 12 V.

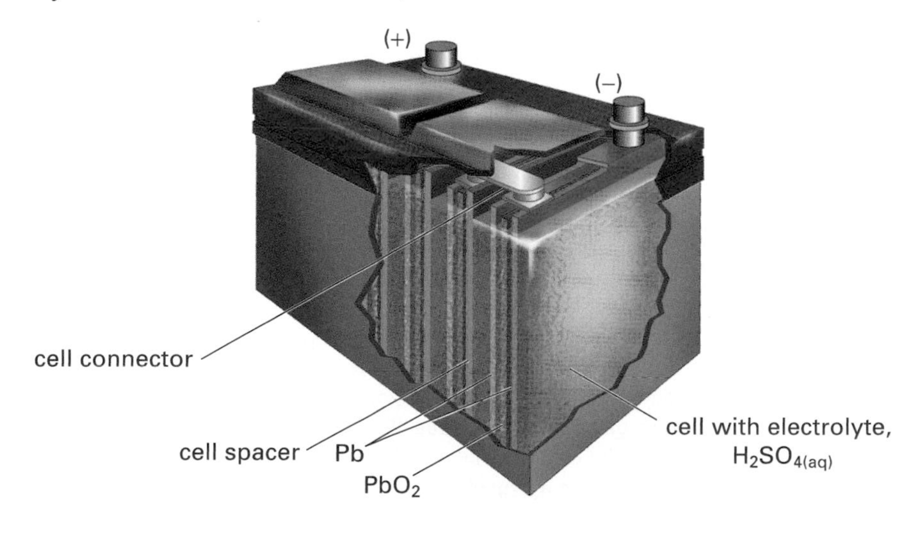

Figure 11.19 Each cell of a lead-acid battery is a single compartment, with no porous barrier or salt bridge. Fibreglass or wooden sheets are placed between the electrodes to prevent them from touching.

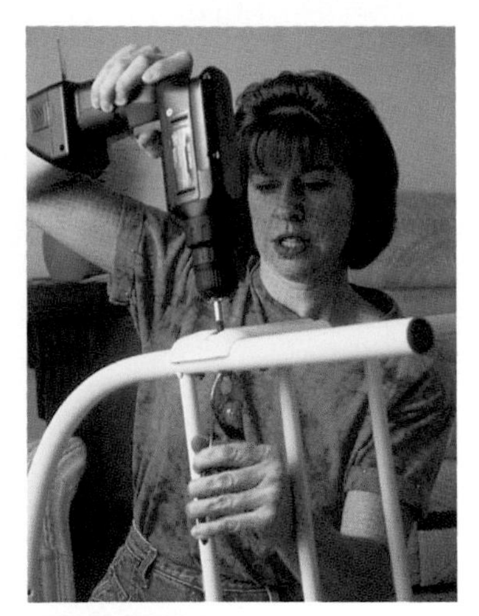

Figure 11.20 Billions of rechargeable nicad batteries are produced every year. They are used in portable devices such as cordless razors and cordless power tools.

When the battery supplies electricity, the half-reactions and overall cell reaction are as follows.

Oxidation (at the Pb anode): $Pb_{(s)} + SO_4{}^{2-}{}_{(aq)} \rightarrow PbSO_{4(s)} + 2e^-$

Reduction (at the PbO_2 cathode):

$$PbO_{2(s)} + 4H^+{}_{(aq)} + SO_4{}^{2-}{}_{(aq)} + 2e^- \rightarrow PbSO_{4(s)} + 2H_2O_{(\ell)}$$

Overall cell reaction:

$$Pb_{(s)} + PbO_{2(s)} + 4H^+{}_{(aq)} + 2SO_4{}^{2-}{}_{(aq)} \rightarrow 2PbSO_{4(s)} + 2H_2O_{(\ell)}$$

You can see that the reaction consumes some of the lead in the anode, some of the lead(IV) oxide in the cathode, and some of the sulfuric acid. A precipitate of lead(II) sulfate forms.

When the battery is recharged, the half-reactions and the overall cell reaction are reversed. In this reverse reaction, lead and lead(IV) oxide are redeposited in their original locations, and sulfuric acid is re-formed.

Reduction (at the Pb cathode): $PbSO_{4(s)} + 2e^- \rightarrow Pb_{(s)} + SO_4{}^{2-}{}_{(aq)}$

Oxidation (at the PbO_2 anode):

$$PbSO_{4(s)} + 2H_2O_{(\ell)} \rightarrow PbO_{2(s)} + 4H^+{}_{(aq)} + SO_4{}^{2-}{}_{(aq)} + 2e^-$$

Overall cell reaction:

$$2PbSO_{4(s)} + 2H_2O_{(\ell)} \rightarrow Pb_{(s)} + PbO_{2(s)} + 4H^+{}_{(aq)} + 2SO_4{}^{2-}{}_{(aq)}$$

In practice, this reversibility is not perfect. However, the battery can go through many charge/discharge cycles before it eventually wears out.

Many types of rechargeable batteries are much more portable than a car battery. For example, there is now a rechargeable version of the alkaline battery. Another example, shown in Figure 11.20, is the rechargeable nickel-cadmium (nicad) battery. Figure 11.21 shows a nickel-cadmium cell, which has a potential of about 1.4 V. A typical nicad battery contains three cells in series to produce a suitable voltage for electronic devices. When the cells in a nicad battery operate as galvanic cells, the half-reactions and the overall cell reaction are as follows.

Oxidation (at the Cd anode): $Cd_{(s)} + 2OH^-{}_{(aq)} \rightarrow Cd(OH)_{2(s)} + 2e^-$

Reduction (at the NiO(OH) cathode):

$$NiO(OH)_{(s)} + H_2O_{(\ell)} + e^- \rightarrow Ni(OH)_{2(s)} + OH^-{}_{(aq)}$$

Overall cell reaction:

$$Cd_{(s)} + 2NiO(OH)_{(s)} + 2H_2O_{(\ell)} \rightarrow Cd(OH)_{2(s)} + 2Ni(OH)_{2(s)}$$

Like many technological innovations, nickel-cadmium batteries carry risks as well as benefits. After being discharged repeatedly, they eventually wear out. In theory, worn-out nicad batteries should be recycled. In practice, however, many end up in garbage dumps. Over time, discarded nicad batteries release toxic cadmium. The toxicity of this substance makes it hazardous to the environment, as cadmium can enter the food chain. Long-term exposure to low levels of cadmium can have serious medical effects on humans, such as high blood pressure and heart disease.

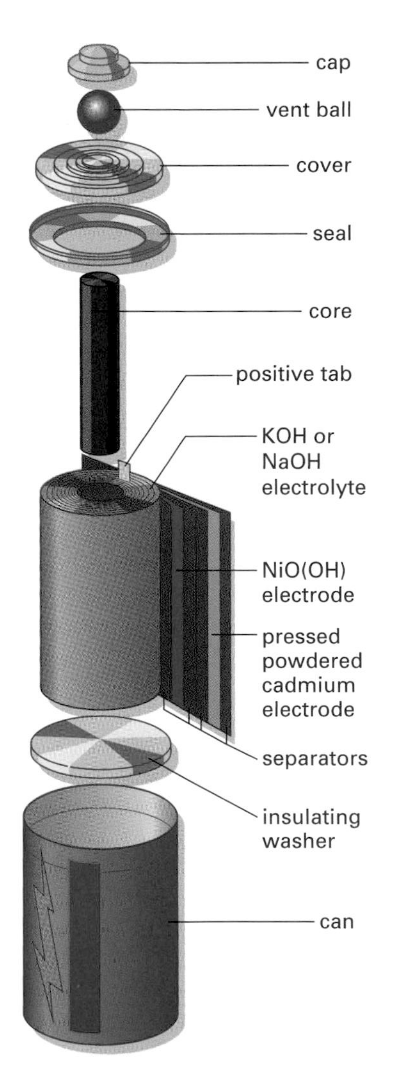

Figure 11.21 A nicad cell has a cadmium electrode and another electrode that contains nickel(III) oxyhydroxide, NiO(OH). When the cell is discharging, cadmium is the anode. When the cell is recharging, cadmium is the cathode. The electrolyte is a base, sodium hydroxide or potassium hydroxide.

Section Summary

In this section, you learned about electrolytic cells, which convert electrical energy into chemical energy. You compared the spontaneous reactions in galvanic cells, which have positive cell potentials, with the non-spontaneous reactions in electrolytic cells, which have negative cell potentials. You then considered cells that act as both galvanic cells and electrolytic cells in some common rechargeable batteries. These batteries are an important application of electrochemistry. In the next two sections, you will learn about many more electrochemical applications.

Section Review

1 I Predict the products of the electrolysis of a 1 mol/L aqueous solution of copper(I) bromide.

2 I In this section, you learned that an external electrical supply reverses the cell reaction in a Daniell cell so that the products are zinc atoms and copper(II) ions.

(a) What are the predicted products of this electrolysis reaction?

(b) Explain the observed products.

3 I Predict whether each reaction is spontaneous or non-spontaneous under standard conditions.

(a) $2FeI_{3(aq)} \rightarrow 2Fe_{(s)} + 3I_{2(s)}$

(b) $2Ag^{+}{}_{(aq)} + H_2SO_{3(aq)} + H_2O_{(\ell)} \rightarrow 2Ag_{(s)} + SO_4{}^{2-}{}_{(aq)} + 4H^{+}{}_{(aq)}$

4 K/U Write the two half-reactions and the overall cell reaction for the process that occurs when a nicad battery is being recharged.

5 K/U What external voltage is required to recharge a lead-acid car battery?

6 K/U The equation for the overall reaction in an electrolytic cell does not include any electrons. Why is an external source of electrons needed for the reaction to proceed?

7 **(a)** I Predict whether aluminum will displace hydrogen from water.

(b) I Water boiling in an aluminum saucepan does not react with the aluminum. Give possible reasons why.

8 MC Research the impact of lead pollution on the environment. Do lead-acid batteries contribute significantly to lead pollution?

9 C Lithium batteries are increasingly common. The lithium anode undergoes oxidation when the battery discharges. Various cathodes and electrolytes are used to make lithium batteries with different characteristics. Research lithium batteries. Prepare a report describing the designs, cell reactions, and uses of lithium batteries. Include a description of the advantages and disadvantages of these batteries.

11.4 Faraday's Law

Section Preview/ Specific Expectations

In this section, you will

- **describe** the relationship between time, current, and the amount of substance produced or consumed in an electrolytic process
- **solve** problems using Faraday's law
- **investigate** Faraday's law by performing an electroplating process in the laboratory
- **explain** how electrolytic processes are used to refine metals
- **research** and **assess** some environmental, health, and safety issues involving electrochemistry
- **communicate** your understanding of the following terms: *quantity of electricity, electric charge, Faraday's law, extraction, refining*

As mentioned earlier in this chapter, Michael Faraday (1791–1867) was the leading pioneer of electrochemistry. One of Faraday's major contributions was to connect the concepts of stoichiometry and electrochemistry.

Figure 11.22 A depiction of Faraday's laboratory.

You know that a balanced equation represents relationships between the quantities of reactants and products. For a reaction that takes place in a cell, stoichiometric calculations can also include the quantity of electricity produced or consumed. Stoichiometric calculations in electrochemistry make use of a familiar unit—the mole.

As a first step, you need information about measurements in electricity. You know that the flow of electrons through an external circuit is called the *electric current*. It is measured in a unit called the *ampere* (symbol A), named after the French physicist André Ampère (1775–1836). The **quantity of electricity**, also known as the **electric charge**, is the product of the current flowing through a circuit and the time for which it flows. The quantity of electricity is measured in a unit called the *coulomb* (symbol C). This unit is named after another French physicist, Charles Coulomb (1736–1806). The ampere and the coulomb are related, in that *one coulomb is the quantity of electricity that flows through a circuit in one second if the current is one ampere*. This relationship can be written mathematically.

$$\text{charge (in coulombs)} = \text{current (in amperes)} \times \text{time (in seconds)}$$

For example, suppose a current of 2.00 A flows for 5.00 min. You can use this information to find the quantity of electricity, in coulombs.

$$5.00 \text{ min} = 300 \text{ s}$$

$$2.00 \text{ A} \times 300 \text{ s} = 600 \text{ C, or } 6.00 \times 10^2 \text{ C}$$

For stoichiometric calculations, you also need to know the electric charge on a mole of electrons. This charge can be calculated by multiplying the charge on one electron and the number of electrons in one mole (Avogadro's number). The charge on a mole of electrons is known as one faraday (1 F), named after Michael Faraday.

$$\text{Charge on one mole of electrons} = \frac{1.602 \times 10^{-19}\ \text{C}}{1\cancel{e^-}} \times \frac{6.022 \times 10^{23}\ \cancel{e^-}}{1\ \text{mol}}$$
$$= 9.647 \times 10^{4}\ \text{C/mol}$$

A rounded value of 96 500 C/mol is often used in calculations. Note that this rounded value has three significant digits.

The information you have just learned permits a very precise control of electrolysis. For example, suppose you modify a Daniell cell to operate as an electrolytic cell. You want to plate 0.1 mol of zinc onto the zinc electrode. The coefficients in the half-reaction for the reduction represent stoichiometric relationships. Figure 11.23 shows that two moles of electrons are needed for each mole of zinc deposited. Therefore, to deposit 0.1 mol of zinc, you need to use 0.2 mol of electrons.

$$0.1\ \cancel{\text{mol Zn}} \times \frac{2\ \text{mol e}^-}{1\ \cancel{\text{mol Zn}}} = 0.2\ \text{mol e}^-$$

Zn^{2+} +	$2e^-$ →	Zn
1 ion	2 electrons	1 atom
$1 \times 6.02 \times 10^{23}$ ions	$2 \times 6.02 \times 10^{23}$ electrons	$1 \times 6.02 \times 10^{23}$ atoms
1 mol of ions	2 mol of electrons	1 mol of atoms

Figure 11.23 **A balanced half-reaction shows relationships between the amounts of reactants and products and the amount of electrons transferred.**

In the next Sample Problem, you will learn to apply the relationship between the amount of electrons and the amount of an electrolysis product.

Sample Problem

Calculating the Mass of an Electrolysis Product

Problem

Calculate the mass of aluminum produced by the electrolysis of molten aluminum chloride, if a current of 500 mA passes for 1.50 h.

What Is Required?

You need to calculate the mass of aluminum produced.

What Is Given?

You know the name of the electrolyte, the current, and the time.

electrolyte: $AlCl_{3(\ell)}$
current: 500 mA
time: 1.50 h

From the previous calculation, you know the charge on one mole of electrons is 96 500 C/mol.

Continued ...

Continued ...

Plan Your Strategy

Step 1 Use the current and the time to find the quantity of electricity used.

Step 2 From the quantity of electricity, find the amount of electrons that passed through the circuit.

Step 3 Use the stoichiometry of the relevant half-reaction to relate the amount of electrons to the amount of aluminum produced.

Step 4 Use the molar mass of aluminum to convert the amount of aluminum to a mass.

Act on Your Strategy

Step 1 To calculate the quantity of electricity in coulombs, work in amperes and seconds.

$$1000 \text{ mA} = 1 \text{ A}$$

$$500 \text{ mA} = 500 \cancel{\text{mA}} \times \frac{1 \text{ A}}{1000 \cancel{\text{mA}}} = 0.500 \text{ A}$$

$$1.50 \text{ h} = 1.50 \cancel{\text{h}} \times \frac{60 \cancel{\text{min}}}{1 \cancel{\text{h}}} \times \frac{60 \text{ s}}{1 \cancel{\text{min}}} = 5400 \text{ s, or } 5.40 \times 10^3 \text{ s}$$

$$\text{Quantity of electricity} = 0.500 \text{ A} \times 5400 \text{ s} = 2700 \text{ C, or } 2.70 \times 10^3 \text{ C}$$

Step 2 Find the amount of electrons. One mole of electrons has a charge of 96 500 C.

$$\text{Amount of electrons} = 2700 \cancel{\text{C}} \times \frac{1 \text{ mol e}^-}{96\,500 \cancel{\text{C}}} = 0.0280 \text{ mol e}^-$$

Step 3 The half-reaction for the reduction of aluminum ions to aluminum is $Al^{3+} + 3e^- \rightarrow Al$.

$$\text{Amount of aluminum formed} = 0.0280 \cancel{\text{mol e}^-} \times \frac{1 \text{ mol Al}}{3 \cancel{\text{mol e}^-}} = 0.00933 \text{ mol Al}$$

Step 4 Convert the amount of aluminum to a mass.

$$\text{Mass of Al formed} = 0.00933 \cancel{\text{mol}} \text{ Al} \times \frac{27.0 \text{ g Al}}{1 \cancel{\text{mol}} \text{ Al}} = 0.252 \text{ g}$$

Check Your Solution

The answer is expressed in units of mass. To check your answer, use estimation. If the current were 1 A, then 1 mol of electrons would pass in 96 500 s. In this example, the current is less than 1 A, and the time is much less than 96 500 s. Therefore, much less than 1 mol of electrons would be used, and much less than 1 mol (27 g) of aluminum would be formed.

Practice Problems

21. Calculate the mass of zinc plated onto the cathode of an electrolytic cell by a current of 750 mA in 3.25 h.

22. How many minutes does it take to plate 0.925 g of silver onto the cathode of an electrolytic cell using a current of 1.55 A?

23. The nickel anode in an electrolytic cell decreases in mass by 1.20 g in 35.5 min. The oxidation half-reaction converts nickel atoms to nickel(II) ions. What is the constant current?

24. The following two half-reactions take place in an electrolytic cell with an iron anode and a chromium cathode.

Oxidation: $Fe_{(s)} \rightarrow Fe^{2+}{}_{(aq)} + 2e^-$

Reduction: $Cr^{3+}{}_{(aq)} + 3e^- \rightarrow Cr_{(s)}$

During the process, the mass of the iron anode decreases by 1.75 g.

(a) Find the change in mass of the chromium cathode.

(b) Explain why you do not need to know the electric current or the time to complete part (a).

The preceding Sample Problem gave an example of the mathematical use of Faraday's law. **Faraday's law** states that *the amount of a substance produced or consumed in an electrolysis reaction is directly proportional to the quantity of electricity that flows through the circuit.*

To illustrate this statement, think about changing the quantity of electricity used in the Sample Problem. Suppose this quantity were doubled by using the same current, 500 mA, for twice the time, 3 h. As a result, the amount of electrons passing into the cell would also be doubled.

$$\begin{aligned} 500\text{ mA} &= 0.500\text{ A} \\ 3\text{h} &= 2 \times 1.5\text{ h} \\ &= 2 \times 5400\text{ s} \end{aligned}$$

$$\begin{aligned} \text{Quantity of electricity} &= 0.500\text{ A} \times (2 \times 5400\text{ s}) \\ &= 2 \times 2700\text{ C} \\ &= 5400\text{ C, or } 5.40 \times 10^3\text{ C} \end{aligned}$$

$$\begin{aligned} \text{Amount of electrons} &= 5400\,\cancel{\text{C}} \times \frac{1\text{ mol e}^-}{96\,500\,\cancel{\text{C}}} \\ &= 0.0560\text{ mol e}^- \end{aligned}$$

Then, as you can see from the relevant half-reaction, the mass of aluminum produced would be doubled. The mass of aluminum produced is clearly proportional to the quantity of electricity used.

In Investigation 11-C, you will apply Faraday's law to an electrolytic cell that you construct.

Investigation 11-C

SKILL FOCUS

- Predicting
- Performing and recording
- Analyzing and interpreting
- Communicating results

Electroplating

You have learned that electroplating is a process in which a metal is deposited, or plated, onto the cathode of an electrolytic cell. In this investigation, you will build an electrolytic cell and electrolyze a copper(II) sulfate solution to plate copper onto the cathode. You will use Faraday's law to relate the mass of metal deposited to the quantity of electricity used.

Question

Does the measured mass of copper plated onto the cathode of an electrolytic cell agree with the mass calculated from Faraday's law?

Prediction

Predict whether the measured mass of copper plated onto the cathode of an electrolytic cell will be greater than, equal to, or less than the mass calculated using Faraday's law.

Materials

150 mL 1.0 mol/L HNO_3 in a 250 mL beaker
120 mL acidified 0.50 mol/L $CuSO_4$ solution (with 5 mL of 6 mol/L H_2SO_4 and 3 mL of 0.1 mol/L HCl added)
drying oven, or acetone in a wash bottle
3 cm × 12 cm × 1 mm Cu strip
50 cm 16-gauge bare solid copper wire
250 mL beaker
adjustable D.C. power supply with ammeter
deionized water in a wash bottle
fine sandpaper
2 electrical leads with alligator clips
electronic balance

Safety Precautions

- Nitric acid is corrosive. Also, note that the $CuSO_4$ solution contains sulfuric acid and hydrochloric acid. Wash any spills on your skin with plenty of cold water. Inform your teacher immediately.
- Avoid touching the parts of the electrodes that have been washed with nitric acid.
- Acetone is flammable. Use acetone in the fume hood.
- Make sure your hands and your lab bench are dry before handling any electrical equipment.

Procedure

1. Clean off any tarnish on the copper strip by sanding it gently. Dip the bottom of the copper strip in the nitric acid for a few seconds, and then rinse off the strip carefully with deionized water. Avoid touching the section that has been cleaned by the acid.
2. Place the copper strip in the beaker, with the clean part of the strip at the bottom. Bend the top of the strip over the rim of the beaker so that the copper strip is secured in a vertical position. This copper strip will serve as the anode.
3. Wrap the copper wire around a pencil to make a closely spaced coil. Leave 10 cm of the wire unwrapped. Measure and record the mass of the wire. Dip the coil in the nitric acid, and rinse the coil with water. Use the 10 cm of uncoiled wire to secure the coil on the opposite side of the beaker from the anode, as shown in the diagram. This copper wire will serve as the cathode.

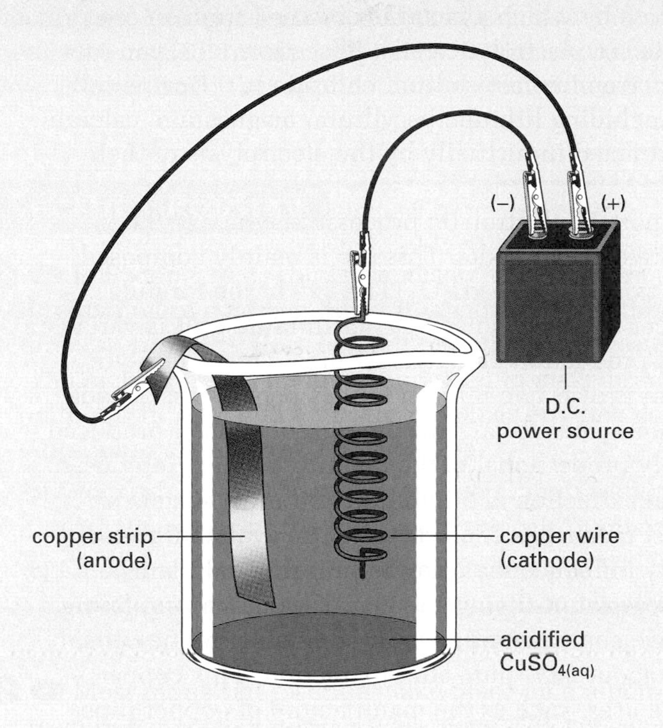

4. Pour 120 mL of the acidified $CuSO_4$ solution into the beaker. Attach the lead from the negative terminal of the power supply to the cathode. Attach the positive terminal to the anode.
5. Turn on the power supply and set the current to 1 A. Maintain this current for 20 min by adjusting the variable current knob as needed.
6. After 20 min, turn off the power. Remove the cathode and rinse it very gently with deionized water. Place the cathode in a drying oven for 20 min. Alternatively, rinse the cathode gently with acetone, and let the acetone evaporate in the fume hood for 5 min.
7. Measure and record the new mass of the cathode.
8. Dispose of all materials as instructed by your teacher.

Analysis

1. Write a balanced equation for the half-reaction that occurs at the cathode.
2. Use the measured current and the time for which the current passed to calculate the quantity of electricity used.
3. Use your answers to questions 1 and 2 to calculate the mass of copper plated onto the cathode.
4. Compare the calculated mass from question 3 with the measured increase in mass of the cathode. Give possible reasons for any difference between the two values.

Conclusion

5. How did the mass of copper electroplated onto the cathode of the electrolytic cell compare with the mass calculated using Faraday's law? Compare your answer with your prediction from the beginning of this investigation.

Applications

6. Suppose you repeated this investigation using iron electrodes, and 0.5 mol/L iron(II) sulfate solution as the electrolyte. If you used the same current for the same time, would you expect the increase in mass of the cathode to be greater than, less than, or equal to the increase in mass that you measured? Explain your answer.
7. Suppose you repeated the investigation with the copper(II) sulfate solution, but you passed the current for only half as long as before. How would the masses of copper plated onto the cathode compare in the two investigations? Explain your answer.
8. Could you build a galvanic cell without changing the electrodes or the electrolyte solution you used in this investigation? Explain your answer.

11.5 Issues Involving Electrochemistry

Section Preview/ Specific Expectations

In this section, you will

- **explain** corrosion as an electrochemical process, and **describe** some techniques used to prevent corrosion
- **evaluate** the environmental and social impact of some common cells, including the hydrogen fuel cell in electric cars
- **explain** how electrolytic processes are used in the industrial production of chlorine, and how this element is used in the purification of water
- **research** and **assess** some environmental, health, and safety issues connected to electrochemistry
- **communicate** your understanding of the following terms: *corrosion, galvanizing, sacrificial anode, cathodic protection, fuel cell, chlor-alkali process*

You have probably seen many examples of rusty objects, such as the one in Figure 11.26. However, you may not realize that rusting costs many billions of dollars per year in prevention, maintenance, and replacement costs. You will now learn more about rusting and about other issues involving electrochemistry.

Figure 11.26 Rust is a common sight on iron objects.

Corrosion

Rusting is an example of **corrosion**, which is a spontaneous redox reaction of materials with substances in their environment. Figure 11.27 shows an example of the hazards that result from corrosion.

Many metals are fairly easily oxidized. The atmosphere contains a powerful oxidizing agent: oxygen. Because metals are constantly in contact with oxygen, they are vulnerable to corrosion. In fact, the term "corrosion" is sometimes defined as the oxidation of *metals* exposed to the environment. In North America, about 20% to 25% of iron and steel production is used to replace objects that have been damaged or destroyed by corrosion. However, not all corrosion is harmful. For example, the green layer formed by the corrosion of a copper roof is considered attractive by many people.

Figure 11.27 On April 28, 1988, this Aloha Airlines aircraft was flying at an altitude of 7300 m when a large part of the upper fuselage ripped off. This accident was caused by undetected corrosion damage. The pilot showed tremendous skill in landing the plane.

Rust is a hydrated iron(III) oxide, $Fe_2O_3 \cdot xH_2O$. The electrochemical formation of rust occurs in small galvanic cells on the surface of a piece of iron, as shown in Figure 11.28. In each small cell, iron acts as the anode. The cathode is inert, and may be an impurity that exists in the iron or is deposited onto it. For example, the cathode could be a piece of soot that has been deposited onto the iron surface from the air.

Water, in the form of rain, is needed for rusting to occur. Carbon dioxide in the air dissolves in water to form carbonic acid, $H_2CO_{3(aq)}$. This weak acid partially dissociates into ions. Thus, the carbonic acid is an electrolyte for the corrosion process. Other electrolytes, such as road salt, may also be involved. The circuit is completed by the iron itself, which conducts electrons from the anode to the cathode.

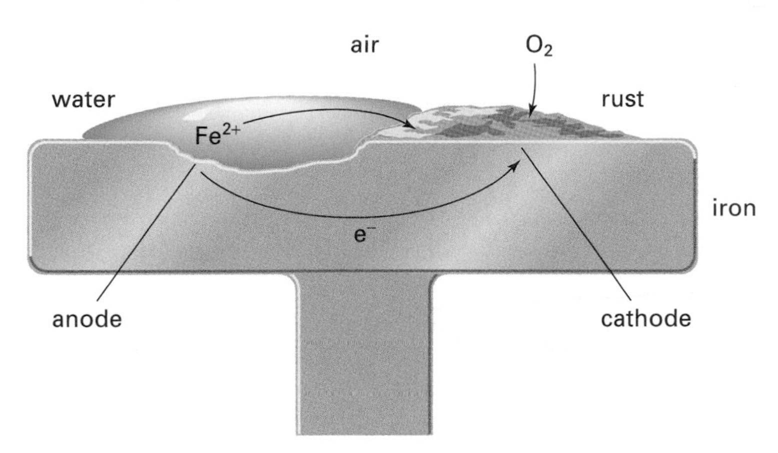

Figure 11.28 The rusting of iron involves the reaction of iron, oxygen, and water in a naturally occurring galvanic cell on the exposed surface of the metal. There may be many of these small cells on the surface of the same piece of iron.

The rusting process is complex, and the equations may be written in various ways. A simplified description of the half-reactions and the overall cell reaction is given here.

Oxidation half-reaction (occurs at the anode):

$$Fe_{(s)} \rightarrow Fe^{2+}{}_{(aq)} + 2e^-$$

Reduction half-reaction (occurs at the cathode):

$$O_{2(g)} + 2H_2O_{(\ell)} + 4e^- \rightarrow 4OH^-{}_{(aq)}$$

Multiply the oxidation half-reaction by two and add the half-reactions to obtain the overall cell reaction.

$$2Fe_{(s)} + O_{2(g)} + 2H_2O_{(\ell)} \rightarrow 2Fe^{2+}{}_{(aq)} + 4OH^-{}_{(aq)}$$

There is no barrier in the cell, so nothing stops the dissolved Fe^{2+} and OH^- ions from mixing. The iron(II) ions produced at the anode and the hydroxide ions produced at the cathode react to form a precipitate of iron(II) hydroxide, $Fe(OH)_2$. Therefore, the overall cell reaction could be written as follows.

$$2Fe_{(s)} + O_{2(g)} + 2H_2O_{(\ell)} \rightarrow 2Fe(OH)_{2(s)}$$

The iron(II) hydroxide undergoes further oxidation by reaction with the oxygen in the air to form iron(III) hydroxide.

$$4Fe(OH)_{2(s)} + O_{2(g)} + 2H_2O_{(\ell)} \rightarrow 4Fe(OH)_{3(s)}$$

Iron(III) hydroxide readily breaks down to form hydrated iron(III) oxide, $Fe_2O_3 \cdot xH_2O$, more commonly known as rust. As noted earlier, the "x" signifies a variable number of water molecules per formula unit.

$$2Fe(OH)_{3(s)} \rightarrow Fe_2O_3 \cdot 3H_2O_{(s)}$$

Both $Fe(OH)_3$ and $Fe_2O_3 \cdot xH_2O$ are reddish-brown, or "rust coloured." A rust deposit may contain a mixture of these compounds.

Fortunately, not all metals corrode to the same extent as iron. Many metals do corrode in air to form a surface coating of metal oxide. However, in many cases, the oxide layer adheres, or sticks firmly, to the metal surface. This layer protects the metal from further corrosion. For example, aluminum, chromium, and magnesium are readily oxidized in air to form their oxides, Al_2O_3, Cr_2O_3, and MgO. Unless the oxide layer is broken by a cut or a scratch, the layer prevents further corrosion. In contrast, rust easily flakes off from the surface of an iron object and provides little protection against further corrosion.

Corrosion Prevention

Corrosion, and especially the corrosion of iron, can be very destructive. For this reason, a great deal of effort goes into corrosion prevention. The simplest method of preventing corrosion is to paint an iron object. The protective coating of paint prevents air and water from reaching the metal surface. Other effective protective layers include grease, oil, plastic, or a metal that is more resistant to corrosion than iron. For example, a layer of chromium protects bumpers and metal trim on cars. An enamel coating is often used to protect metal plates, pots, and pans. *Enamel* is a shiny, hard, and very unreactive type of glass that can be melted onto a metal surface. A protective layer is effective as long as it completely covers the iron object. If a hole or scratch breaks the layer, the metal underneath can corrode.

It is also possible to protect iron against corrosion by forming an alloy with a different metal. *Stainless steel* is an alloy of iron that contains at least 10% chromium, by mass, in addition to small quantities of carbon and occasionally metals such as nickel. Stainless steel is much more resistant to corrosion than pure iron. Therefore, stainless steel is often used for cutlery, taps, and various other applications where rust-resistance is important. However, chromium is much more expensive than iron. As a result, stainless steel is too expensive for use in large-scale applications, such as building bridges.

Galvanizing is a process in which iron is covered with a protective layer of zinc. Galvanized iron is often used to make metal buckets and chain-link fences. Galvanizing protects iron in two ways. First, the zinc acts as a protective layer. If this layer is broken, the iron is exposed to air and water. When this happens, however, the iron is still protected. Zinc is more easily oxidized than iron. Therefore, zinc, not iron, becomes the anode in the galvanic cell. The zinc metal is oxidized to zinc ions. In this situation, zinc is known as a **sacrificial anode**, because it is destroyed (sacrificed) to protect the iron. Iron acts as the cathode when zinc is present. Thus, iron does not undergo oxidation until all the zinc has reacted.

Cathodic protection is another method of preventing rusting, as shown in Figure 11.29. As in galvanizing, a more reactive metal is attached to the iron object. This reactive metal acts as a sacrificial anode, and the iron becomes the cathode of a galvanic cell. Unlike galvanizing, the metal used in cathodic protection does not completely cover the iron. Because the sacrificial anode is slowly destroyed by oxidation, it must be replaced periodically.

If iron is covered with a protective layer of a metal that is *less* reactive than iron, there can be unfortunate results. A "tin" can is actually a steel can coated with a thin layer of tin. While the tin layer remains intact, it provides effective protection against rusting. If the tin layer is broken or scratched, however, the iron in the steel corrodes *faster* in contact with the tin than the iron would on its own. Since tin is less reactive than iron, tin acts as a cathode in each galvanic cell on the surface of the can. Therefore, the tin provides a large area of available cathodes for the small galvanic cells involved in the rusting process. Iron acts as the anode of each cell, which is its normal role when rusting.

Sometimes, the rusting of iron is promoted accidentally. For example, by connecting an iron pipe to a copper pipe in a plumbing system, an inexperienced plumber could accidentally speed up the corrosion of the iron pipe. Copper is less reactive than iron. Therefore, copper acts as the cathode and iron as the anode in numerous small galvanic cells at the intersection of the two pipes.

Build on your understanding of corrosion by completing the following practice problems.

Figure 11.29 Magnesium, zinc, or aluminum blocks are attached to ships' hulls, oil and gas pipelines, underground iron pipes, and gasoline storage tanks. These reactive metals provide cathodic protection by acting as a sacrificial anode.

Practice Problems

25. (a) Use the two half-reactions for the rusting process, and a table of standard reduction potentials. Determine the standard cell potential for this reaction.

(b) Do you think that your calculated value is the actual cell potential for each of the small galvanic cells on the surface of a rusting iron object? Explain.

26. Explain why aluminum provides cathodic protection to an iron object.

27. In the year 2000, Transport Canada reported that thousands of cars sold in the Atlantic Provinces between 1989 and 1999 had corroded engine cradle mounts. Failure of these mounts can cause the steering shaft to separate from the car. The manufacturer recalled the cars so that repairs could be made, where necessary. The same cars were sold across the country. Why do you think that the corrosion problems showed up in the Atlantic Provinces?

28. (a) Use a table of standard reduction potentials to determine whether elemental oxygen, $O_{2(g)}$, is a better oxidizing agent under acidic conditions or basic conditions.

(b) From your answer to part (a), do you think that acid rain promotes or helps prevent the rusting of iron?

Automobile Engines

The internal combustion engine found in most automobiles uses gasoline as a fuel. Unfortunately, this type of engine produces pollutants, such as carbon dioxide (CO_2), nitrogen oxides (NO_x), and volatile organic compounds (VOCs). These pollutants contribute to health and environmental problems, such as smog and the greenhouse effect. In addition, the internal combustion engine is very inefficient. It converts only about 25% of the chemical energy of the fuel into the kinetic energy of the car. Electric cars, such as those shown in Figure 11.30, may provide a more efficient and less harmful alternative.

Figure 11.30 **Many electric cars are currently under development. In fact, the idea of electric cars is not new. In the early days of the automobile, electric cars were more common than gasoline-powered cars. The production of electric cars peaked in 1912, but then completely stopped in the 1930s.**

Manufacturers and researchers have attempted to power electric cars with rechargeable batteries, such as modified lead-acid and nickel-cadmium batteries. However, rechargeable batteries run down fairly quickly. The distance driven before recharging a battery may be 250 km or less. The battery must then be recharged from an external electrical source. Recharging the lead-acid battery of an electric car takes several hours. Cars based on a version of the nickel-cadmium battery can be recharged in only fifteen minutes. However, recharging the batteries of an electric car is still inconvenient.

A new type of power supply for electric cars eliminates the need for recharging. A **fuel cell** is a battery that produces electricity while reactants are supplied continuously from an external source. Because reactants continuously flow into the cell, a fuel cell is also known as a *flow battery*. Unlike the fuel supply of a more conventional battery, the fuel supply in a fuel cell is unlimited. As in the combustion of gasoline in a conventional engine, the overall reaction in a fuel cell is the oxidation of a fuel by oxygen.

The space shuttle uses a fuel cell as a source of energy. This cell depends on the oxidation of hydrogen by oxygen to form water. The fuel cell operates under basic conditions, so it is sometimes referred to as an *alkaline fuel cell*. Figure 11.31, on the next page, shows the design of the cell. The half-reactions and the overall reaction are as follows.

Reduction (occurs at the cathode): $O_{2(g)} + 2H_2O_{(\ell)} + 4e^- \rightarrow 4OH^-_{(aq)}$

Oxidation (occurs at the anode): $H_{2(g)} + 2OH^-_{(aq)} \rightarrow 2H_2O_{(\ell)} + 2e^-$

Multiply the oxidation half-reaction by 2, add the two half-reactions, and simplify to obtain the overall cell reaction.

$$2H_{2(g)} + O_{2(g)} \rightarrow 2H_2O_{(\ell)}$$

Notice that the overall equation is the same as the equation for the combustion of hydrogen. The combustion of hydrogen is an exothermic reaction. In the fuel cell, this reaction produces energy in the form of electricity, rather than heat.

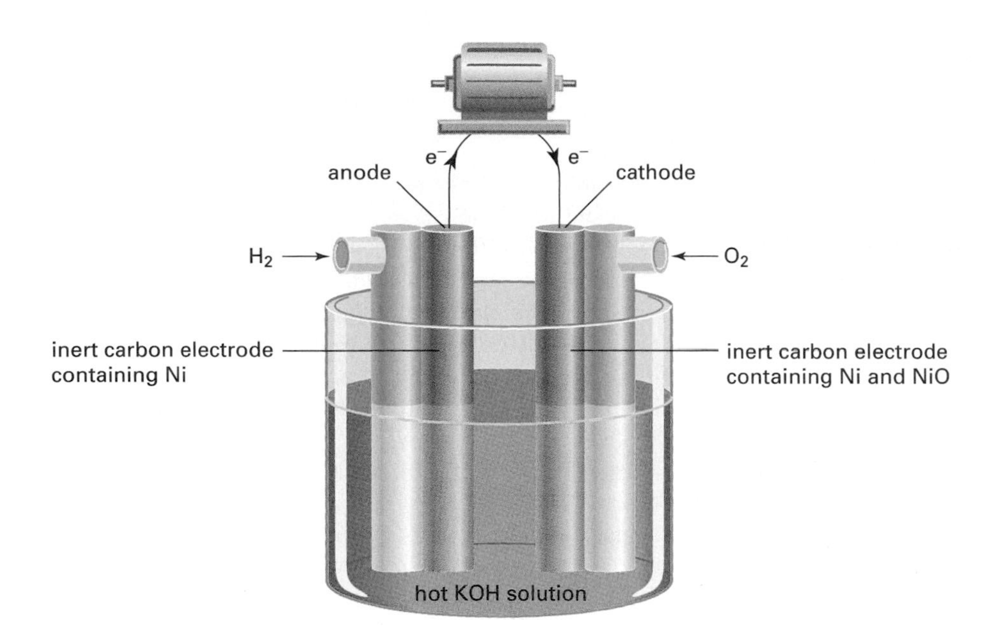

Figure 11.31 **In an alkaline fuel cell, the half-reactions do not include solid conductors of electrons. Therefore, the cell has two inert electrodes.**

The hydrogen fuel cell produces water vapour, which does not contribute to smog formation or to the greenhouse effect. This product makes the hydrogen fuel cell an attractive energy source for cars. Also, the hydrogen fuel cell is much more efficient than the internal combustion engine. A hydrogen fuel cell converts about 80% of the chemical energy of the fuel into the kinetic energy of the car.

There is a possible problem with the hydrogen fuel cell. The cell requires hydrogen fuel. Unfortunately, uncombined hydrogen is not found naturally on Earth. Most hydrogen is produced from hydrocarbon fuels, such as petroleum or methane. These manufacturing processes may contribute significantly to pollution problems. However, hydrogen can also be produced by the electrolysis of water. If a source such as solar energy or hydroelectricity is used to power the electrolysis, the overall quantity of pollution is low.

The following practice problems will allow you to test your understanding of fuel cells.

Practice Problems

29. Calculate E°_{cell} for a hydrogen fuel cell.

30. In one type of fuel cell, methane is oxidized by oxygen to form carbon dioxide and water.

(a) Write the equation for the overall cell reaction.

(b) Write the two half-reactions, assuming acidic conditions.

31. Reactions that occur in fuel cells can be thought of as being "flameless combustion reactions." Explain why.

32. If a hydrogen fuel cell produces an electric current of 0.600 A for 120 min, what mass of hydrogen is consumed by the cell?

Canadians in Chemistry

Dr. Viola Birss

As a science student, Viola Birss decided to focus on chemistry. "I have always had a concern for the environment," she says. "I am particularly interested in identifying new, non-polluting ways of converting, storing, and using energy." Dr. Birss's interest in non-polluting energy narrowed her field of interest to electrochemistry.

Today, Dr. Birss is a chemistry professor at the University of Calgary. Her research focuses on developing films to coat metal surfaces. Among other uses, these films can serve as protective barriers against corrosion, and as catalysts in fuel cells.

Magnesium alloys are very lightweight, and are being used in the aerospace industry. Because they are very reactive, these alloys need to be protected from corrosion. Dr. Birss holds a patent on a new approach to the electrochemical formation of protective oxide films on magnesium alloys. Dr. Birss also works on developing new catalysts for fuel cells, and studies the factors that lead to the breakdown of fuel cells.

After finishing an undergraduate degree at the University of Calgary, Dr. Birss went on to complete her doctoral degree as a Commonwealth Scholar at the University of Auckland in New Zealand. During her postdoctoral studies at the University of Ottawa, she worked with Dr. Brian Conway, a famous Canadian electrochemist. Dr. Conway's work in electrochemistry has led to progress in a range of electrochemical devices including fuel cells, advanced batteries, and electrolytic cells.

Dr. Birss takes pride in her team of undergraduate, graduate, and post-doctoral students. She works hard to provide a creative and inspiring environment for them. Together, they go on wilderness hiking and cross country ski trips. In addition to creating a sense of community within her team, Birss feels that sports and nature help to recharge her internal "battery." Dr. Birss comments, "These activities seem to provide me with mental and physical rest, so that my creativity and energy are catalyzed."

Water Treatment and the Chlor-Alkali Process

In many countries, water is not safe to drink. Untreated water is sometimes polluted with toxic chemicals. It may also carry numerous water-borne diseases, including typhoid fever, cholera, and dysentery. In Canada, the water that comes through your tap has been through an elaborate purification process. This process is designed to remove solid particles and toxic chemicals, and to reduce the number of bacteria to safe levels. Adding chlorine to water is the most common way to destroy bacteria.

You have already seen that chlorine gas can be made by the electrolysis of molten sodium chloride. In industry, some chlorine is produced in this way using the Downs cell described earlier. However, more chlorine is produced in Canada using a different method, called the **chlor-alkali process**. In this process, brine is electrolyzed in a cell like the one shown in Figure 11.32. *Brine* is a saturated solution of sodium chloride.

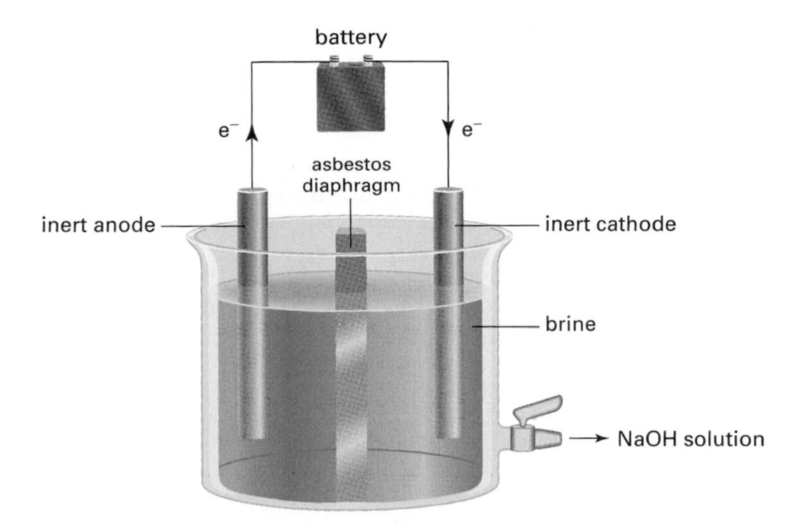

Figure 11.32 **The chlor-alkali cell in this diagram electrolyzes an aqueous solution of sodium chloride to produce chlorine gas, hydrogen gas, and aqueous sodium hydroxide. The asbestos diaphragm stops the chlorine gas produced at the anode from mixing with the hydrogen gas produced at the cathode. Sodium hydroxide solution is removed from the cell periodically, and fresh brine is added to the cell.**

The half-reactions and the overall cell reaction in the chlor-alkali process are as follows.

Oxidation: $2Cl^{-}_{(aq)} \rightarrow Cl_{2(g)} + 2e^{-}$

Reduction: $2H_2O_{(\ell)} + 2e^{-} \rightarrow H_{2(g)} + 2OH^{-}_{(aq)}$

Overall: $2Cl^{-}_{(aq)} + 2H_2O_{(\ell)} \rightarrow Cl_{2(g)} + H_{2(g)} + 2OH^{-}_{(aq)}$

Note that the sodium ion is a spectator ion, and does not take part in this reaction. However, it combines with OH^{-} to produce sodium hydroxide, as shown by the balanced chemical equation below.

$$2NaCl_{(aq)} + 2H_2O_{(\ell)} \rightarrow Cl_{2(g)} + H_{2(g)} + 2NaOH_{(aq)}$$

The products of the chlor-alkali process are all useful. Sodium hydroxide is used to make soaps and detergents. It is widely used as a base in many other industrial chemical reactions, as well. The hydrogen produced by the chlor-alkali process is used as a fuel. Chlorine has many uses besides water treatment. For example, chlorine is used as a bleach in the pulp and paper industry. Chlorine is also used in the manufacture of chlorinated organic compounds, such as the common plastic polyvinyl chloride (PVC).

The chlorination of water is usually carried out by adding chlorine gas, sodium hypochlorite, or calcium hypochlorite to the water in low concentrations. The active antibacterial agent in each case is hypochlorous acid, $HClO_{(aq)}$. For example, when chlorine gas is added to water, hypochlorous acid is formed by the following reaction.

$$Cl_{2(g)} + H_2O_{(\ell)} \rightarrow HClO_{(aq)} + HCl_{(aq)}$$

Some people object to the chlorination of water, and prefer to drink bottled spring water. There is controversy over the level of risk associated with chlorination, and over the possible benefits of spring water. For example, hypochlorous acid reacts with traces of organic materials in the water supply. These reactions can produce toxic substances, such as chloroform. Supporters of chlorination believe that these substances are present at very low, safe levels, but opponents of chlorination disagree. Complete the following practice problems to help you decide on your own opinion of chlorination.

Practice Problems

33. Show that the reaction of chlorine gas with water is a disproportionation reaction.

34. Would you predict the products of the chlor-alkali process to be hydrogen and chlorine? Explain.

35. Research and assess the most recent information you can find on the health and safety aspects of the chlorination of water. Are you in favour of chlorination, or opposed to it? Explain your answer.

36. Some municipalities use ozone gas rather than chlorine to kill bacteria in water. Research the advantages and disadvantages of using ozone in place of chlorine.

Electronic Learning Partner

Go to the Chemistry 12 Electronic Learning Partner for more information about aspects of material covered in this section of the chapter.

Section Summary

In this section, you learned about some important electrochemical processes. You had the opportunity to weigh some positive and negative effects of electrochemical technologies. The questions that follow in the section review and chapter review will encourage you to think further about the science of electrochemistry, and about its impact on society.

Section Review

1 MC Why does the use of road salt cause cars to rust faster than they otherwise would?

2 MC Aluminum is a more reactive metal than any of the metals present in steel. However, discarded steel cans disintegrate much more quickly than discarded aluminum cans when both are left open to the environment in the same location. Give an explanation.

3 K/U Explain why zinc acts as a sacrificial anode in contact with iron.

4 **(a)** C Identify two metals that do not corrode easily in the presence of oxygen and water. Explain why they do not corrode.

(b) C How are these metals useful? How do the uses of these metals depend on their resistance to corrosion?

5 I A silver utensil is said to *tarnish* when its surface corrodes to form a brown or black layer of silver sulfide. Research and describe a chemical procedure that can be used to remove this layer. Write balanced half-reactions and a chemical equation for the process.

6 I In a chlor-alkali cell, the current is very high. A typical current would be about 100 000 A. Calculate the mass of sodium hydroxide, in kilograms, that a cell using this current can produce in one minute.

7 MC Research the advances made in the development of fuel cells since this book was written. Describe how any new types of fuel cells operate. Evaluate their advantages and disadvantages, as compared to the internal combustion engine and other fuel cells.

CHAPTER 11 Review

Reflecting on Chapter 11

Summarize this chapter in the format of your choice. Here are a few ideas to use as guidelines:

- Represent one example of a galvanic cell, and one example of an electrolytic cell, using chemical equations, half-reactions, and diagrams.
- Calculate a standard cell potential using a table of standard reduction potentials.
- Compare primary and secondary batteries.
- Predict the products of the electrolysis of molten salts and aqueous solutions.
- Perform a sample stoichiometric calculation involving the quantity of electricity used in electrolytic processes.
- Describe some electrolytic processes involved in extracting and refining metals.
- Describe the process of corrosion. Explain some methods used to prevent it.
- Describe the design of fuel cells, and their potential use in automobiles.
- Describe the industrial production of chlorine, and its use in the purification of water.
- Give some examples of how electrochemistry affects the environment, human health, and safety.

Reviewing Key Terms

For each of the following terms, write a sentence that shows your understanding of its meaning.

electric current	electrochemistry
galvanic cell	voltaic cell
external circuit	electrodes
electrolytes	anode
cathode	salt bridge
inert electrode	electric potential
cell voltage	cell potential
dry cell	battery
primary battery	secondary battery
reduction potential	oxidation potential
electrolytic cell	electrolysis
overvoltage	electroplating
quantity of electricity	electric charge
Faraday's law	extraction
refining	corrosion
galvanizing	sacrificial anode
cathodic protection	chlor-alkali process
fuel cell	

Knowledge/Understanding

1. Explain the function of the following parts of an electrolytic cell.
   (a) electrodes
   (b) electrolyte
   (c) external voltage
2. In a galvanic cell, one half-cell has a cadmium electrode in a 1 mol/L solution of cadmium nitrate. The other half-cell has a magnesium electrode in a 1 mol/L solution of magnesium nitrate. Write the shorthand representation.
3. Write the oxidation half-reaction, the reduction half-reaction, and the overall cell reaction for the following galvanic cell.
   $Pt \mid NO_{(g)} \mid NO_3^-{}_{(aq)}, H^+{}_{(aq)} \parallel I^-{}_{(aq)} \mid I_{2(s)}, Pt$
4. What is the importance of the hydrogen electrode?
5. Lithium, sodium, beryllium, magnesium, calcium, and radium are all made industrially by the electrolysis of their molten chlorides. These salts are all soluble in water, but aqueous solutions are not used for the electrolytic process. Explain why.
6. Use the following two half-reactions to write balanced net ionic equations for one spontaneous reaction and one non-spontaneous reaction. State the standard cell potential for each reaction.
   $N_2O_{(g)} + 2H^+{}_{(aq)} + 2e^- \rightleftharpoons N_{2(g)} + H_2O_{(\ell)} \quad E^\circ = 1.770\ V$
   $CuI_{(s)} + e^- \rightleftharpoons Cu_{(s)} + I^-{}_{(aq)} \quad E^\circ = -0.185\ V$
7. Identify the oxidizing agent and the reducing agent in a lead-acid battery that is
   (a) discharging
   (b) recharging
8. Rank the following in order from most effective to least effective oxidizing agents under standard conditions.
   $Zn^{2+}{}_{(aq)}, Co^{3+}{}_{(aq)}, Br_{2(\ell)}, H^+{}_{(aq)}$
9. Rank the following in order from most effective to least effective reducing agents under standard conditions.
   $H_{2(g)}, Cl^-{}_{(aq)}, Al_{(s)}, Ag_{(s)}$
10. The ions $Fe^{2+}{}_{(aq)}$, $Ag^+{}_{(aq)}$, and $Cu^{2+}{}_{(aq)}$ are present in the half-cell that contains the cathode of an electrolytic cell. The concentration of each of these ions is 1 mol/L. If the external voltage is very slowly increased from zero, in what order will the three metals Fe, Ag, and Cu be plated onto the cathode? Explain your answer.

Inquiry

11. Write the half-reactions and calculate the standard cell potential for each reaction. Identify each reaction as spontaneous or non-spontaneous.
    (a) $Zn_{(s)} + Fe^{2+}{}_{(aq)} \rightarrow Zn^{2+}{}_{(aq)} + Fe_{(s)}$
    (b) $Cr_{(s)} + AlCl_{3(aq)} \rightarrow CrCl_{3(aq)} + Al_{(s)}$
    (c) $2AgNO_{3(aq)} + H_2O_{2(aq)} \rightarrow 2Ag_{(s)} + 2HNO_{3(aq)} + O_{2(g)}$
12. Calculate the mass of magnesium that can be plated onto the cathode by the electrolysis of molten magnesium chloride, using a current of 3.65 A for 55.0 min.
13. **(a)** Describe a method you could use to measure the standard cell potential of the following galvanic cell.
    $Sn \mid Sn^{2+}$ (1 mol/L) $\|$ Pb^{2+} (1 mol/L) $\mid Pb$
    (b) Why is this cell unlikely to find many practical uses?
14. The two half-cells in a galvanic cell consist of one iron electrode in a 1 mol/L iron(II) sulfate solution, and a silver electrode in a 1 mol/L silver nitrate solution.
    (a) Assume the cell is operating as a galvanic cell. State the cell potential, the oxidation half-reaction, the reduction half-reaction, and the overall cell reaction.
    (b) Repeat part (a), but this time assume that the cell is operating as an electrolytic cell.
    (c) For the galvanic cell in part (a), do the mass of the anode, the mass of the cathode, and the total mass of the two electrodes increase, decrease, or stay the same while the cell is operating?
    (d) Repeat part (c) for the electrolytic cell in part (b).
15. **(a)** Describe an experiment you could perform to determine the products from the electrolysis of aqueous zinc bromide. How would you identify the electrolysis products?
    (b) Zinc and bromine are the observed products from the electrolysis of aqueous zinc bromide solution under standard conditions. They are also the observed products from the electrolysis of molten zinc bromide. Explain why the first observation is surprising.
16. Use the half-cells shown in a table of standard reduction potentials. Could you build a battery with a potential of 8 V? If your answer is yes, give an example.
17. Suppose you produce a kilogram of sodium and a kilogram of aluminum by electrolysis. Compare your electricity costs for these two processes. Assume that electricity is used for electrolysis only, and not for heating.

Communication

18. Research the following information. Prepare a short presentation or booklet on the early history of electrochemistry.
    (a) the contributions of Galvani and Volta to the development of electrochemistry
    (b) how Humphry Davy and Michael Faraday explained the operation of galvanic and electrolytic cells. (Note that these scientists could not describe them in terms of electron transfers, because the electron was not discovered until 1897.)
19. How rapidly do you think that iron would corrode on the surface of the moon? Explain your answer.
20. Reactions that are the reverse of each other have standard cell potentials that are equal in size but opposite in sign. Explain why.
21. Use a labelled diagram to represent each of the following.
    (a) a galvanic cell in which the hydrogen electrode is the anode
    (b) a galvanic cell in which the hydrogen electrode is the cathode

Making Connections

22. A D-size dry cell flashlight battery is much bigger than a AAA-size dry cell calculator battery. However, both have cell potentials of 1.5 V. Do they supply the same quantity of electricity? Explain your answer.
23. **(a)** Would you use aluminum nails to attach an iron gutter to a house? Explain your answer.
    (b) Would you use iron nails to attach aluminum siding to a house? Explain your answer

24. Research the aluminum-air battery, and the sodium-sulfur battery. Both are rechargeable batteries that have been used to power electric cars. In each case, describe the design of the battery, the half-reactions that occur at the electrodes, and the overall cell reaction. Also, describe the advantages and disadvantages of using the battery as a power source for a car.

25. Explain why the recycling of aluminum is more economically viable than the recycling of many other metals.

26. Suppose you live in a small town with high unemployment. A company plans to build a smelter there to produce copper and nickel by roasting their sulfide ores and reducing the oxides formed. Would you be in favour of the plant being built, or opposed to it? Explain and justify your views.

27. Many metal objects are vulnerable to damage from corrosion. A famous example is the Statue of Liberty. Research the history of the effects of corrosion on the Statue of Liberty. Give a chemical explanation for the processes involved. Describe the steps taken to solve the problem and the chemical reasons for these steps.

28. **(a)** Estimate the number of used batteries you discard in a year. Survey the class to determine an average number. Now estimate the number of used batteries discarded by all the high school students in Ontario in a year.
(b) Prepare an action plan suggesting ways of decreasing the number of batteries discarded each year.

Answers to Practice Problems and Short Answers to Section Review Questions

Practice Problems: **1.(a)** oxidizing agent, Cu(II); reducing agent, Zn **(b)** same as previous **2.(a)** ox: $Sn_{(s)} \rightarrow Sn^{2+}{}_{(aq)} + 2e^-$, red: $Tl^+{}_{(aq)} + e^- \rightarrow Tl_{(s)}$, overall: $Sn_{(s)} + 2Tl^+{}_{(aq)} \rightarrow Sn^{2+}{}_{(aq)} + +2Tl_{(s)}$, tin anode, thallium cathode **(b)** ox: $Cd_{(s)} \rightarrow Cd^{2+}{}_{(aq)} + 2e^-$, red: $2H^+{}_{(aq)} + 2e^- \rightarrow H_{2(g)}$, overall: $Cd_{(s)} + 2H^+{}_{(aq)} \rightarrow Cd^{2+}{}_{(aq)} + H_{2(g)}$, cadmium anode, platinum cathode **3.(a)** ox: $2I^-{}_{(aq)} \rightarrow I_{2(s)} + 2e^-$, red: $MnO_4^-{}_{(aq)} + 8H^+{}_{(aq)} + 5e^- \rightarrow Mn^{2+}{}_{(aq)} + 4H_2O_{(\ell)}$, overall: $10I^-{}_{(aq)} + 2MnO_4^-{}_{(aq)} + 16H^+{}_{(aq)} \rightarrow 5I_{2(s)} + 2Mn^{2+}{}_{(aq)} + 8H_2O_{(\ell)}$ **(b)** oxidizing agent, $MnO_4{}^-$, reducing agent, I^- **(c)** the anode **4.** one electrolyte, no barrier **5.** ox: $2Br^-{}_{(aq)} \rightarrow Br_{2(\ell)} + 2e^-$, red: $Cl_{2(g)} + 2e^- \rightarrow 2Cl^-{}_{(aq)}$, $E^\circ_{cell} = 0.292$ V **6.** ox: $Cu^+{}_{(aq)} \rightarrow Cu^{2+}{}_{(aq)} + e^-$, red: $2H^+{}_{(aq)} + O_{2(g)} + 2e^- \rightarrow H_2O_{2(aq)}$, $E^\circ_{cell} = 0.542$ V **7.** ox: $Sn_{(s)} \rightarrow Sn^{2+}{}_{(aq)} + 2e^-$, red: $2H^+{}_{(aq)} + 2e^- \rightarrow H_{2(g)}$, $E^\circ_{cell} = 0.138$ V **8.** ox: $Cr_{(s)} \rightarrow Cr^{3+}{}_{(aq)} + 3e^-$, red: $AgCl_{(s)} + e^- \rightarrow Ag_{(s)} + Cl^-{}_{(aq)}$, $E^\circ_{cell} = 0.966$ V **9.(a)** $2Cl^- \rightarrow Cl_2 + 2e^-$ **(b)** $Ca^{2+} + 2e^- \rightarrow Ca$ **(c)** $CaCl_2 \rightarrow Ca + Cl_2$ **10.(a)** cathode: $Li^+ + e^- \rightarrow Li$ **(b)** anode: $2Br^- \rightarrow Br_2 + 2e^-$ **(c)** $2Li^+ + 2Br^- \rightarrow 2Li + Br_2$ **11.** direct current: reaction proceeds steadily in one direction **12.** reduction (gain of electrons) at cathode of electrolytic cell; electrons come from negative electrode (anode) of battery **13.** hydrogen and oxygen **14.** hydrogen, not calcium, produced at cathode **15.(a)** 0.146 V **(b)** Cd anode, Ni cathode **(c)** ox: $Cd \rightarrow Cd^{2+} + 2e^-$, red: $Ni^{2+} + 2e^- \rightarrow Ni$, overall: $Cd + Ni^{2+} \rightarrow Cd^{2+} + Ni$ **16.(a)** –0.146 V **(b)** Ni anode, Cd cathode **(c)** ox: $Ni \rightarrow Ni^{2+} + 2e^-$, red: $Cd^{2+} + 2e^- \rightarrow Cd$, overall: $Cd^{2+} + Ni \rightarrow Cd + Ni^{2+}$ **17.** No **18.(a)** spontaneous **(b)** non-spontaneous **(c)** spontaneous **19.** $2Cu^+{}_{(aq)} \rightarrow Cu^{2+}{}_{(aq)} + Cu_{(s)}$, $E^\circ_{cell} = 0.368$ V **20.(a)** non-spontaneous **(b)** spontaneous **21.** 2.97 g **22.** 8.90 min **23.** 1.85 A **24.(a)** increases by 1.09 g **(b)** You can use the stoichiometry of the equations. **25.(a)** 0.848 V **(b)** no; conditions are not standard **26.** aluminum is more easily oxidized than iron **27.** higher levels of salt (an electrolyte) and moisture **28.(a)** acidic conditions **(b)** promotes it **29.** 1.229 V **30.(a)** overall: $CH_{4(g)} + 2O_{2(g)} \rightarrow CO_{2(g)} + 2H_2O_{(\ell)}$ **(b)** ox: $CH_{4(g)} + 2H_2O_{(\ell)} \rightarrow CO_{2(g)} + 8H^+{}_{(aq)} + 8e^-$, red: $O_{2(g)} + 4H^+{}_{(aq)} + 4e^- \rightarrow 2H_2O_{(\ell)}$, **31.** same equation as combustion, but fuel does not burn **32.** 0.0451 g **33.** oxidation number of Cl increases from 0 to +1 in forming HClO, decreases from 0 to –1 in forming HCl; chlorine undergoes both oxidation and reduction **34.** hydrogen and oxygen, but conditions are far from standard

Section Review: 11.1: **1.** ox: manganese(IV) oxide, red: zinc **3.(a)** ox: $Zn_{(s)} + 2OH^-{}_{(aq)} \rightarrow ZnO_{(s)} + H_2O_{(\ell)} + 2e^-$, red: $Ag_2O_{(s)} + H_2O_{(\ell)} + 2e^- \rightarrow 2Ag_{(s)} + 2OH^-{}_{(aq)}$, overall: $Zn_{(s)} + Ag_2O_{(s)} \rightarrow ZnO_{(s)} + 2Ag_{(s)}$ **(b)** zinc anode, stainless steel cathode **4.** 3 V, two cells connected in series **5.** four dry cells (4×1.5 V = 6 V) **7.** it dissolves; inner container is zinc anode **8.** Fe anode, Ag cathode, ox: $Fe_{(s)} \rightarrow Fe^{2+}{}_{(aq)} + 2e^-$, red: $Ag^+{}_{(aq)} + e^- \rightarrow Ag_{(s)}$, overall: $Fe_{(s)} + 2Ag^+{}_{(aq)} \rightarrow Fe^{2+}{}_{(aq)} + 2Ag_{(s)}$ **11.2:** **1.(a)** 0.710 V **(b)** 5.797 V **(c)** 0.432 V **2.(a)** 0.599 V **(b)** 1.097 V **(c)** 0.994 V **3.** standard hydrogen electrode **4.** no; biggest difference in E°_{red} is less than 7 V **6.** 0.987 V **11.3:** **1.** copper, oxygen **2.(a)** copper(II) ions, hydrogen **(b)** hydrogen gas production requires overvoltage, so zinc forms at cathode **3.(a)** non-spontaneous **(b)** spontaneous **4.** ox: $Ni(OH)_{2(s)} + OH^-{}_{(aq)} \rightarrow NiO(OH)_{(s)} + H_2O_{(\ell)} + e^-$, red: $Cd(OH)_{2(s)} + 2e^- \rightarrow Cd_{(s)} + 2OH^-{}_{(aq)}$, overall: $Cd(OH)_{2(s)} + 2Ni(OH)_{2(s)} \rightarrow Cd_{(s)} + 2NiO(OH)_{(s)} + 2H_2O_{(\ell)}$ **5.** over 12 V **7.(a)** yes **11.4:** **1.(a)** cathode **(b)** 3.29×10^3 C **2.** 618 kg Na, 952 kg Cl_2 **3.** 2.08 t **4.** plentiful and inexpensive electricity **11.5:** **1.** salt acts as electrolyte **2.** aluminum protected by oxide layer **4.(a)** e.g., gold, platinum **6.** 2.49 kg

Design Your Own Investigation

SKILL FOCUS

- Initiating and planning
- Performing and recording
- Analyzing and interpreting
- Communicating results

Electroplating

Background

The most beautiful metals, such as silver and gold, can also be the most expensive. Unfortunately, pure gold is soft, and pure silver tarnishes quickly.

Electroplating is a common method used to produce beautiful and durable metal objects. Jewellery and other metal objects can be made cheaper, stronger, and more durable by coating a strong, cheaper metal with a thin layer of a more expensive metal. For example, cheaper jewellery can be made from copper or nickel, and then coated with silver. More expensive jewellery can be made from silver, and then coated with gold.

Industry also uses electroplating for many applications. Chromium, an extremely hard surface metal, is commonly electroplated over steel cores for heavy duty applications. Worn or damaged metal machine parts may be restored by re-plating the worn sections.

The bumper of this car is plated with chromium.

Pre-Lab Focus

In this investigation, you will design an electrolytic cell to plate zinc onto a metal object of your choice. You will repeat your procedure using three different currents, and then compare your final products. When designing your procedure, consider questions such as the following.

- How will you clean the pieces of metal you want to plate?
- What concentration of electrolyte will you use?
- What external voltage will you use?
- What three current values will you try?
- What will you use for the anode in your cell?
- What will you use for the cathode in your cell?
- What time limit will you set for the electroplating process?

Question

What conditions work best for electroplating zinc onto a metal object? Does changing the current affect the quality of the finished object?

Hypothesis

Predict whether the value of the current used to carry out the plating affects the quality of the finished object.

Materials

beaker or plastic egg carton
adjustable D.C. power supply (9 V max)
ammeter wires with alligator clips
metal objects to electroplate, such as copper wire, iron nails, nickel objects, paperclips, bronze objects
anode materials such as a zinc strip
zinc nitrate solution, provided by your teacher (0.5–1.0 mol)
steel wool
any other materials required for your procedure

Safety Precautions

- Wear gloves, safety goggles, and an apron while you carry out the investigation.
- If you use an acid or base to clean your electrodes, treat both the acid or base and the electrode with caution. Wash any spills on your skin with plenty of cool water. Inform your teacher.
- If you use acetone to clean your electrodes, make sure there are no open flames in the laboratory. Acetone is flammable.
- Use the power supply and electrical connections with care.
- Be sure that all materials are properly disposed of after your experiment.
- Do not use currents greater than 1.0 amp.
- Do not use a voltage greater than 9 V.

Procedure

1. Design a rubric, an assessment checklist, or some other means of assessing your experimental design and procedure.
2. Decide on a piece of metal or a metal object to be plated. You will need three identical metal objects. You may wish to use the copper wire to form a piece of jewellery to plate.
3. Design a procedure to electroplate zinc onto the metal object you have chosen. You will repeat your procedure three times, using three different values for the current, and three different metal objects.
4. Make a list of all the materials you will use. Check that all are available.
5. Do not start your procedure until you have shown it to your teacher for approval.
6. When you have finished your laboratory procedure, dispose of your materials as recommended by your teacher. Record your disposal methods in your report.

Assessment

After you complete this investigation:

- Compare your own procedure with procedures used by your classmates. In what ways did your procedure work better than others? How could you improve your procedure?
- Evaluate the quality of your resulting electroplated pieces.

7. Write up a full laboratory report of your procedure and results. Include:
   - a diagram of your experimental set-up
   - appropriate balanced redox equations
   - answers to the Analysis questions below

Analysis

1. What problems did you encounter when carrying out your investigation? How did you solve these problems?
2. Compare the three finished objects that you electroplated. What conditions worked best to electroplate zinc onto a metal object?
3. Were you satisfied with your procedure? What improvements would you make if you did the experiment again?
4. Could you build a galvanic cell using the same materials that you used in your procedure? If your answer is yes, explain how the galvanic cell would differ from the electrolytic cell that you made in this investigation.

Conclusion

5. Write a statement summarizing the results of your investigation.

Extension

In this investigation, you plated a coating of zinc onto metal objects. As you learned in Unit 5, Chapter 11, iron objects are sometimes coated with zinc, or galvanized, to protect the objects from corrosion. Research the processes used to galvanize an iron object. Compare the industrial processes to the process you carried out in this investigation.

COURSE CHALLENGE

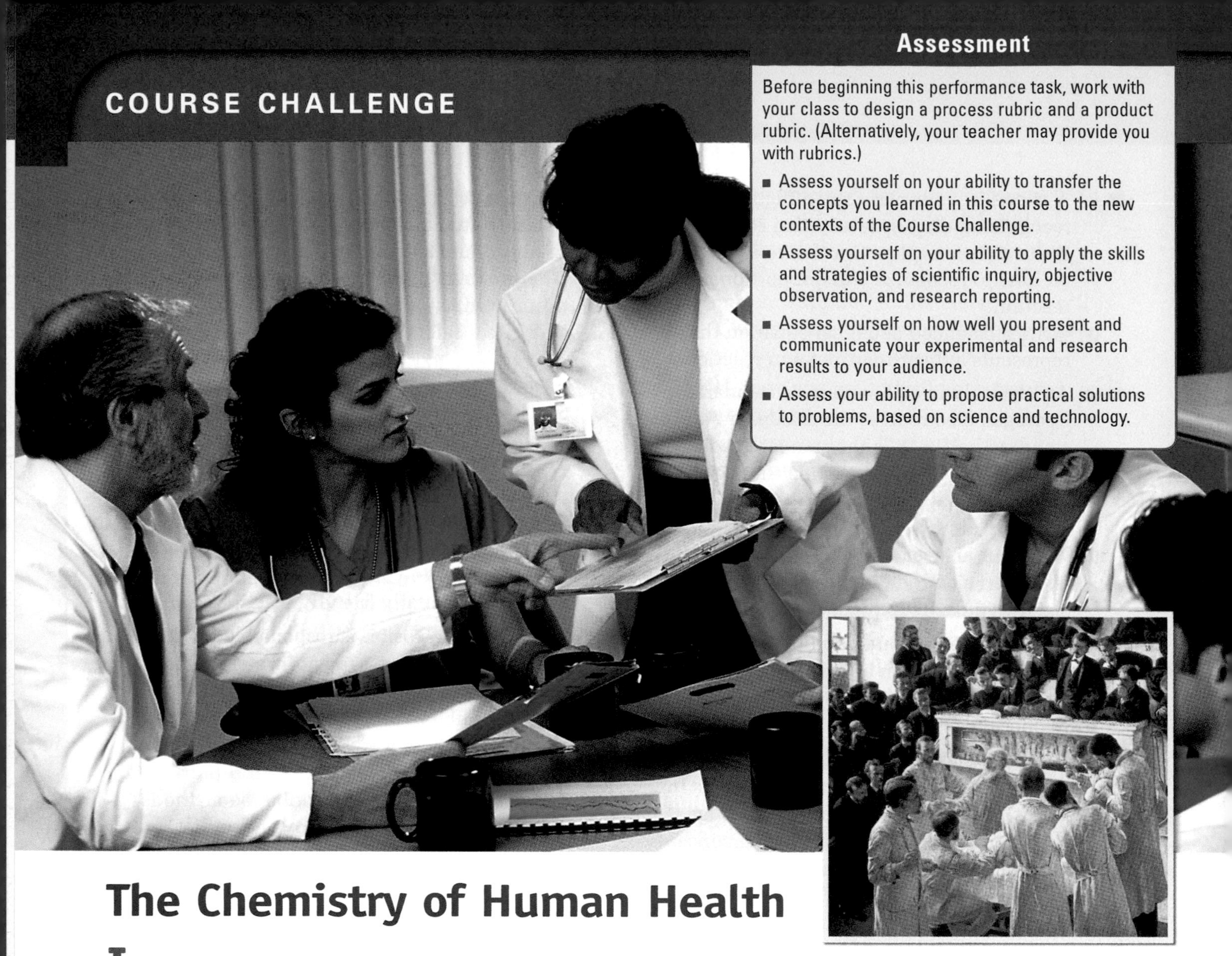

Assessment

Before beginning this performance task, work with your class to design a process rubric and a product rubric. (Alternatively, your teacher may provide you with rubrics.)

- Assess yourself on your ability to transfer the concepts you learned in this course to the new contexts of the Course Challenge.
- Assess yourself on your ability to apply the skills and strategies of scientific inquiry, objective observation, and research reporting.
- Assess yourself on how well you present and communicate your experimental and research results to your audience.
- Assess your ability to propose practical solutions to problems, based on science and technology.

The Chemistry of Human Health

It is tempting to think of science as a clearly defined set of disciplines, such as chemistry, physics, and biology. This is not wholly accurate, however. Scientific disciplines blend and overlap. Therefore, scientists need to know what is happening in relevant areas of different disciplines. For example, if you are studying physics, you have probably noticed that both chemists and physicists contributed to the development of the atomic theory. The physicists built on the discoveries of the chemists, and vice versa.

The health sciences are a group of disciplines that draw upon chemistry, physics, and biology. How does chemistry relate to the health sciences? Knowing about molecular structure and function allows biochemists to synthesize medicinal compounds on a large scale, and even to design new molecules to fight diseases and their symptoms. Understanding the chemical role of nutrients, vitamins, and minerals in the body helps dieticians develop healthy dietary guidelines. Knowing the properties of materials helps researchers develop new materials and devices—such as titanium joints, plastic heart valves, and artificial skin—to repair, maintain, and replace parts of the human body.

The developments just described, and many others, are possible because health scientists and chemists communicate by writing books, journal articles, and letters. They use the Internet to post results and hold discussions. They also communicate face-to-face by giving presentations and lectures, and by chatting over coffee. A scientific conference is the perfect place for such face-to-face interactions.

In this Course Challenge, you will investigate different facets of the intersection of chemistry with the health sciences. Then you will share your findings and learn from your peers in a conference about the chemistry of human health.

Part 1: The Citric Acid Cycle

Metabolism is a complex series of interdependent organic reactions that power the body by breaking down food molecules to provide energy. An important part of human metabolism is the breakdown of glucose to provide energy for the body. First, a metabolic process called *glycolysis* breaks down glucose into two molecules of pyruvate, $C_3H_4O_3$. The pyruvate molecules then react with a chemical species called *coenzyme A, CoA*, to form *acetyl CoA*. Acetyl CoA reacts with oxaloacetic acid to begin the metabolic cycle called the *citric acid cycle*, shown in the diagram below. The citric acid cycle is also called the Krebs cycle, after its discoverer, Hans Krebs.

Preview

Part 1 draws upon the skills and concepts you learned in Unit 1. You will use your understanding of organic molecules and their reactions to analyze a metabolic cycle.

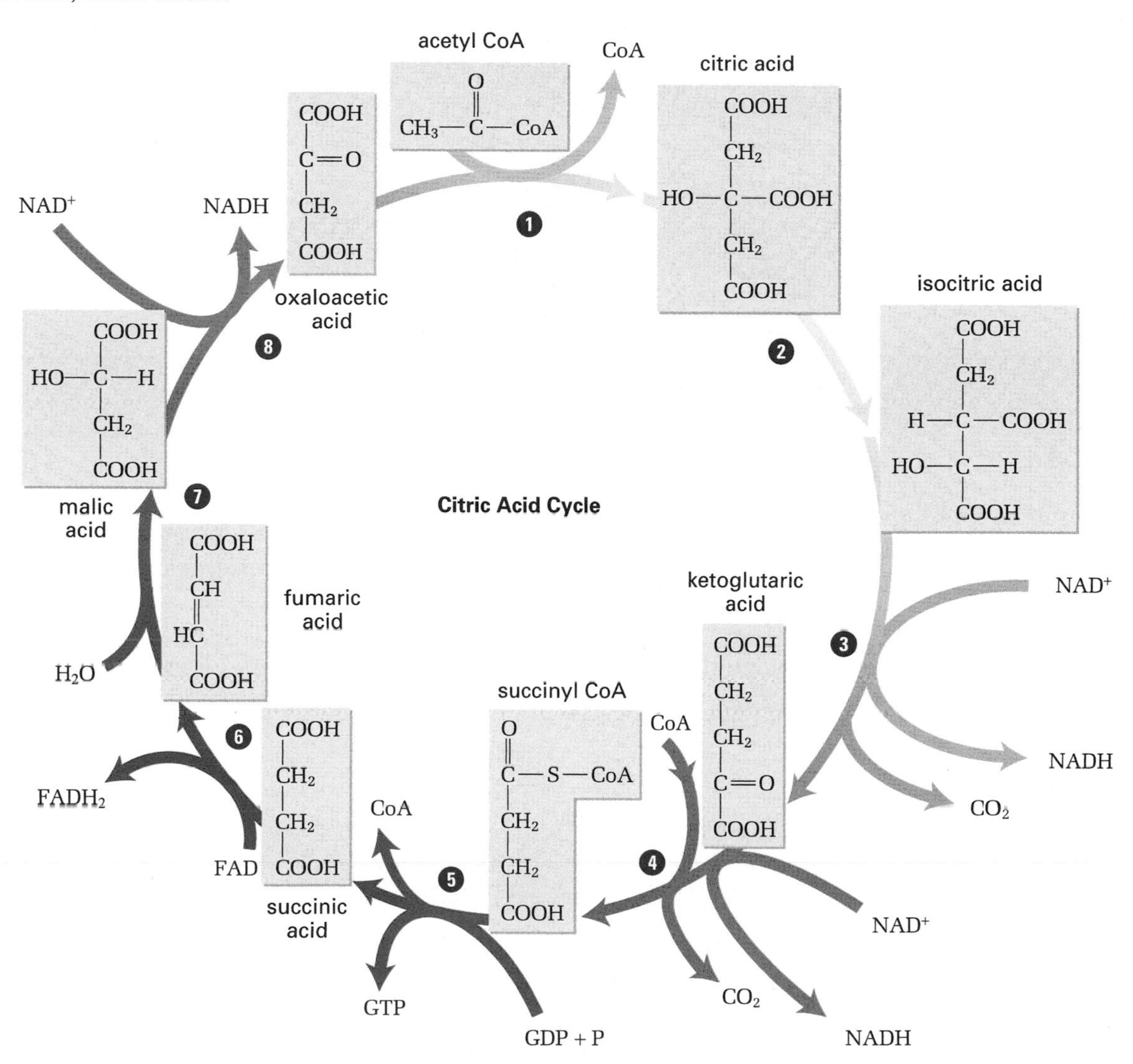

This diagram shows the citric acid cycle. For simplicity, the molecules in the citric acid cycle are represented here as acids. In the body, they are actually present as ions. For example, citric acid is present as citrate ions.

Challenge

Use the diagram of the citric acid cycle to answer the following questions.

1. K/U Examine the acids that are formed in each step of the citric acid cycle (do not include succinyl CoA). Redraw the acids. Then circle and identify the functional groups that are present on each acid.

PROBLEM TIP

You may have learned about NAD^+, GDP, and FAD in a biology course. For your purposes, assume that they represent chemical species used as electron acceptors, which are later used by the body to produce energy.

PROBLEM TIP

Recall, from your previous courses, that isomers are molecules that have the same formula but different arrangements of atoms. An isomerism occurs when a molecule reacts to form one of its isomers.

2. K/U Identify the steps in the citric acid cycle that fit the following descriptions.
   (a) steps that are oxidations
   (b) one step that combines addition, reduction, and condensation reactions
   (c) steps that are elimination reactions
   (d) one step that is an *isomerism*
   (e) one step that involves the addition of water
3. I Think about the mechanism of the reaction of acetyl CoA with oxaloacetate. Which carbons in the citric acid molecule do you think are likely the carbons that were added from acetyl CoA? Justify your answer.
4. K/U In the citric acid cycle, every step that is an oxidation reaction also produces energy, until the most oxidized form of carbon leaves the cycle as carbon dioxide. Which steps in the cycle produce energy?
5. C Create a large version of the citric acid cycle. You may want to create a large bulletin-board display or a hanging mobile, using three-dimensional molecular models. Whatever the format of your citric acid cycle, make sure that it is completely and accurately labelled.

Think and Research

1. MC Ketoacidosis is a dangerous condition that is characterized by the acidification of the blood and an acetone odour on the breath. The condition occurs when levels of oxaloacetic acid for the citric acid cycle are low. This leads to a buildup of acetyl CoA molecules, which the liver metabolizes to produce acidic ketone bodies. Since carbohydrates are the main source of oxaloacetic acid in the body, high-protein, low-carbohydrate diets have been linked to ketoacidosis.
   (a) Why do people follow these diets? Do you think the benefits of these diets justify the risks?
   (b) Research the connections between the citric acid cycle, ketoacidosis, and high-protein, low-carbohydrate diets. Present your findings as a web page or poster.

Part 2: Materials for Replacing Human Body Parts

Preview

Part 2 draws upon the skills and concepts you learned in Unit 2. You will use your understanding of molecular structure and properties to suggest appropriate materials for human prosthetics (artificial body parts).

Science has come a long way since wooden "pegs" were used to replace lost legs. Prosthetics are now made of lightweight, long-lasting materials. Engineers have developed replacement joints that function in much the same way as the original human joints. Understanding the properties of materials is key to developing prosthetics that function well with living human bodies, are relatively comfortable, and last a long time.

Challenge

Design a replacement joint or bone, or a prosthetic limb, using the following questions to guide you.

1. K/U Titanium is a metal that is often used in joint and bone replacement.

(a) Write the electron configuration of titanium.

(b) What are some properties of titanium?

(c) Why is titanium a suitable metal for replacing human joints and bones?

(d) A titanium alloy, rather than pure titanium, is often used as the material for replacement joints and bones. Suggest a reason why an alloy is used.

2 I Build a replacement joint or bone out of household materials. Create an information kit that explains why you chose the materials you did. For example, you will want to describe each material's reactivity, strength, and density, as well as other properties. Describe any problems you foresee with your replacement part. Will it react with fluids in the body? Will it wear out quickly? (Instead of building a replacement bone or joint, you may choose to create a prosthetic limb.)

3 I Conduct some tests on the materials you used for your replacement part to determine or confirm their physical properties. Relate the properties of each material to the type of bonding present. Obtain your teacher's permission before conducting your tests.

4 MC Use print and electronic resources to find out what kinds of materials are being used and developed for use as prosthetic limbs. Be sure to explain what properties make each material suitable for its use. How does the chemical bonding in each material relate to its properties? What are the risks and benefits associated with the materials?

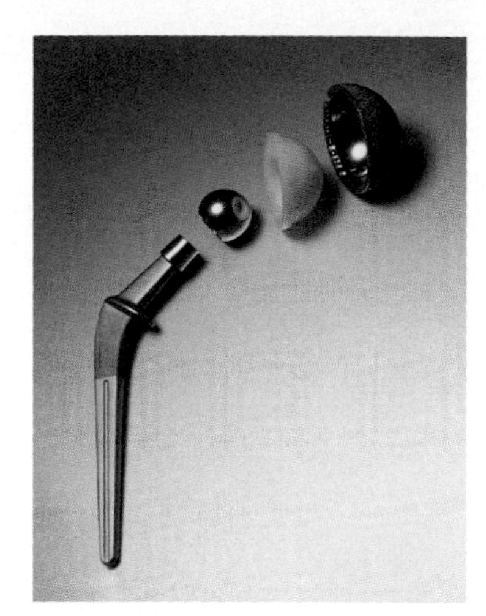

This photo shows the components of an artificial hip joint.

Technology LINK

Use print and electronic resources to study the historical use of prosthetic limbs. Why were certain materials chosen in the past? What problems did people have with materials that were used as prosthetics? How did they attempt to overcome these problems?

Think and Research

1 C Engineers use the macroscopic properties of matter when they design prosthetic limbs. Designing instruments for medical diagnosis, on the other hand, depends on knowing about the structure of matter at the atomic level. Use print and electronic resources to research the development, use, underlying theory, and function of the instruments that are used for medical diagnosis. Use your knowledge of atomic structure to explain how each instrument works. Choose one of the following areas of medical diagnosis: infrared spectroscopy, nuclear magnetic resonance spectroscopy, X-rays, nuclear medicine, optometrists' instruments, or magnetic resonance imaging. What contributions have Canadian scientists made to this area of medical diagnosis? Present your findings as a talk, supported by a slide show or overhead transparencies.

Part 3: The Enzyme Catalase

Enzymes control the rate of nearly every reaction in your body. They are absolutely essential for the efficient functioning of the chemical processes that take place continuously and keep the human body alive and healthy.

Preview

Part 3 draws upon the skills and concepts you learned in Unit 3. You will use your understanding of catalysis and energy to analyze a reaction that is catalyzed by an enzyme.

Challenge

The enzyme catalase is contained in human, animal, and plant tissues. Investigate catalase, and compare its action with the actions of several inorganic catalysts.

1 K/U Catalase is the enzyme that is responsible for the decomposition of hydrogen peroxide into water and oxygen gas.

(a) Write the balanced chemical equation for this reaction.

(b) What is the enthalpy change for this reaction? Write a thermochemical equation.

(c) How does catalase affect the enthalpy change of this reaction? Use a diagram to illustrate your answer.

(d) How does catalase affect the activation energy of this reaction? Use a diagram to illustrate your answer.

2 I Design an experiment that uses the technique of downward displacement to test the efficiency of enzymes and inorganic catalysts in the decomposition of 6% hydrogen peroxide, H_2O_2. (If 6% H_2O_2 is unavailable, use 3% H_2O_2). Test the catalytic ability of the following materials: manganese(IV) oxide, iron(III) chloride, potato, apple, banana, papaya, and pineapple, as well as uncooked meat, fish, and liver. Conduct several trials. For the manganese(IV) oxide and iron(III) chloride catalysts, determine the rates of reaction and graph the results. Do both catalyzed reactions have the same rate law? **CAUTION** 6% hydrogen peroxide is reactive and can burn your skin. Outline safety precautions in your procedure, and get your teacher's permission before proceeding.

3 I Perform the investigation a second time, but boil the foods for 10 min before adding the hydrogen peroxide. (Do not boil the inorganic catalysts or the hydrogen peroxide.) Get your teacher's approval before proceeding. Explain your results.

4 C Prepare a detailed report to share your findings.

Think and Research

1 MC Hydrogen peroxide is often used to sterilize wounds. The fizzing that occurs is caused by catalase in the blood acting on the hydrogen peroxide. Bacteria that do not have catalase to disable the hydrogen peroxide are killed by this chemical. Humans are protected by the presence of catalase. Not all bacteria are killed by hydrogen peroxide, however, because some bacteria do have catalase.

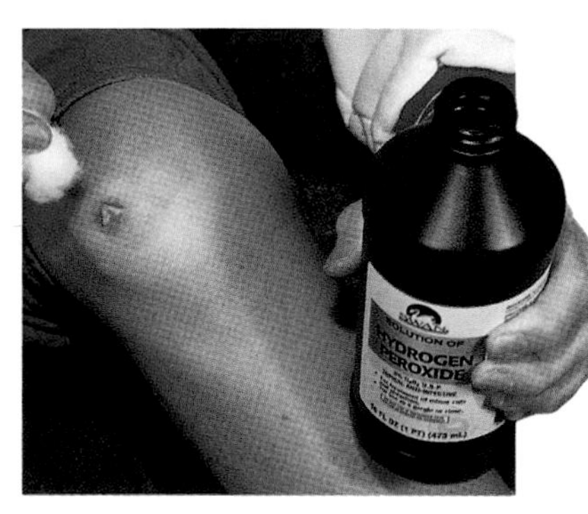

(a) Suggest a laboratory procedure that a researcher could use to classify species of bacteria as *catalase-positive* or *catalase-negative*.

(b) Use print and electronic resources to find three dangerous species of bacteria that are catalase-positive and three dangerous species that are catalase-negative. Propose safe ways to guard against both types.

Part 4: Blood and Le Châtelier's Principle

Preview

Part 4 draws upon the skills and concepts you learned in Unit 4. You will use your understanding of equilibrium to analyze the buffer system in blood serum.

In Unit 4, you learned that buffers resist changes in pH. Buffer systems are extremely important to human health, since many of the life-sustaining reactions in the body depend on a narrow pH range to proceed effectively.

The principal buffer system in blood serum is based on the equilibrium between carbonic acid, $H_2CO_{3(aq)}$, and the hydrogen carbonate ion, HCO_3^-. Carbonic acid is unstable, however. It is also in equilibrium with carbon dioxide. Therefore, a second equilibrium reaction is involved in the hydrogen carbonate buffer system in the blood: the reaction between carbon dioxide and water to produce carbonic acid, and its reverse. The two equilibrium reactions are summarized below.

$$CO_{2(aq)} + 2H_2O_{(\ell)} \rightleftharpoons H_2CO_{3(aq)} + H_2O_{(\ell)} \rightleftharpoons HCO_3^-{}_{(aq)} + H_3O^+{}_{(aq)}$$

Challenge

Refer to the description of the hydrogen carbonate buffer system in blood serum to answer questions 1 and 2. Then use questions 3 and 4 to design an investigation to find out how a buffer system resists changes in pH.

1. K/U One of the differences between a buffer system in the body and a simple buffer system in a lab is that the body can enhance the buffer's power to resist pH changes. For example, the lungs regulate the amount of carbon dioxide that enters and leaves the body. The kidneys help to regulate the pH of blood in several ways, such as increasing or decreasing excretion of hydronium ions, H_3O^+, in urine. Explain how the kidneys might respond to the following conditions.
   (a) The blood pH rises to 7.48.
   (b) The blood pH sinks to 7.33.

2. I Healthy blood has a pH of 7.4. Estimate the ratio of $[CO_2]$ to $[HCO_3^-]$ in the blood. Use the following assumptions and information to help you.
   - Assume that carbonic acid and hydrogen carbonate ions are the only contributors to blood pH.
   - Use $K_a = 4.3 \times 10^{-7}$ for H_2CO_3.
   - To express your answer in terms of CO_2, assume that all undissociated carbonic acid is present as dissolved CO_2.

3. I Because blood is buffered, it resists changes in pH. To model blood's resistance to changes in pH, design an investigation that compares the effects of adding an acid or a base to buffered and non-buffered systems. Your teacher will tell you what materials you may use. The following tips and suggestions will help you get started.
   - You can prepare a simple buffer system by mixing equal volumes of 0.1 mol/L acetic acid, CH_3COOH, and 0.1 mol/L sodium acetate, $NaCH_3COO$. Alternatively, you may wish to try more closely simulating the buffer system in blood. Use soda water, which contains dissolved CO_2, and sodium hydrogen carbonate (baking soda), $NaHCO_3$. Your teacher may suggest other buffer systems.
   - Consider carrying out your investigation on the microscale.
   - You will need to decide how to monitor pH changes. You may choose to use a pH meter or a universal indicator, for example.
   - Decide what type and concentration of acid and base you will add to your systems. Dilute solutions of strong acids and bases work well.
   - Include a step in which you determine the initial pH of your buffer solution and the unbuffered system.
   - Include all appropriate safety precautions.

   Obtain your teacher's approval for your procedure, and carry out your investigation.

4. C Prepare a detailed report to communicate your findings. Be sure to compare and contrast the system you used in your investigation with the carbonic acid/carbonate buffer system in the blood. How are the systems similar? How are they different?

Chemists sometimes simplify calculations that involve complex systems in order to make an estimate. They do this by making simplifying assumptions. You will try this technique for yourself in Challenge question 2.

CONCEPT CHECK

Panic attacks or illness can bring on hyperventilation—fast and heavy breathing.

- When hyperventilation occurs, large quantities of carbon dioxide are removed from the blood. What effect does hyperventilation have on the pH of the blood?
- Suppose that your friend is hyperventilating. You tell your friend to breathe from a paper bag. Explain why this technique helps.

Think and Research

1 MC *Respiratory acidosis* occurs when someone has difficulty breathing, as in an asthma attack, or when someone is in an area with insufficient ventilation. Symptoms of acidosis include fainting and depression of the nervous system.

(a) Based on the name of the condition, what do you think happens to the pH of the blood in respiratory acidosis?

(b) Explain why insufficient ventilation or breathing difficulties lead to respiratory acidosis. Refer to the carbonic acid/hydrogen carbonate buffer system in your answer.

(c) To help maintain the pH of the blood, the kidneys can increase or decrease reabsorption of HCO_3^- into the blood. They can also decrease or increase the excretion of H_3O^+ ions. How would the kidneys respond to respiratory acidosis? Explain your answer.

(d) Suggest ways that you could help someone who was suffering from respiratory acidosis as a result of having been in an area with insufficient ventilation.

Part 5: Vitamin C—An Important Antioxidant

Preview

Part 5 draws upon the skills and concepts you learned in Unit 5. You will use your understanding of redox reactions to design a titration for vitamin C, an antioxidant.

Oxygen and other oxidizing agents can react harmfully with your body. Vitamin C acts as an antioxidant. It is very easily oxidized, so it reacts with oxidizing agents, preventing them from reacting with other important molecules in the body. Vitamin C is relatively stable when it is oxidized. Therefore, it does not propagate a harmful series of oxidation reactions. The Chemistry Bulletin in Unit 5 explains the possible connection between oxidizing agents and aging.

One way to determine the vitamin C content of a sample is to titrate it with an iodine/iodide solution. The diagram below shows the reaction involved.

ascorbic acid (vitamin C) + I_3^- ⟶ dehydroascorbic acid + $2H^+ + 3I^-$

ascorbic acid (vitamin C) dehydroascorbic acid

Often, to make the endpoint of an iodine titration more obvious, an indicator solution that contains starch is added to the solution being titrated. Starch forms a deep blue complex with triiodide, I_3^-, but it is colourless with I^-. As long as there is unreacted vitamin C in solution, no triiodide ions will be present in solution. Therefore, the blue colour will appear only at the endpoint.

In the iodine solution, iodine, I_2, and iodide, I^-, ions are in equilibrium with triiodide, as shown below.

$$I_2 + I^- \rightleftharpoons I_3^-$$

Molecular iodine is a deep violet-red colour. Iodine ions are colourless. Thus, when an antioxidant reduces the iodine molecules in a solution, the iodine colour disappears completely.

Challenge

In this Challenge, you will design a titration to compare the vitamin C content of various samples. Your teacher will tell you what reactants are available for you to use in your titration.

1 I Iodine solutions may be used to titrate solutions that contain antioxidants, such as vitamin C.

(a) In storage, the concentration of iodine in solution decreases fairly quickly over time. Why do you think this happens?

(b) Because the iodine solution's concentration is not stable, it should be standardized frequently. To standardize an iodine solution, use it to titrate a solution that contains a known quantity of vitamin C. Then determine what volume of iodine solution is needed to react with the known quantity of vitamin C. Explain how you would standardize a solution of iodine using vitamin C tablets from a pharmacy.

(c) To standardize iodine using vitamin C tablets, you should use fresh tablets. Explain why.

2 I Design a titration that uses an iodine solution to determine and compare the concentration of vitamin C in a variety of fresh fruit juices. You may find it helpful to research appropriate procedures on the Internet. Include a method for standardizing the iodine solution. (See Challenge question 1.) Make sure that you list all the safety precautions you will use in your titration. Check your procedure with your teacher, and then carry out your titration.

3 C Prepare a detailed report to communicate your results. Include the half-reaction for iodine in your discussion of the titration.

4 K/U If you titrate orange juice that has been exposed to the air for a week, will the vitamin C concentration be different from the vitamin C concentration in fresh juice? If so, will it decrease or increase? Explain your prediction, in terms of redox reactions.

5 I If time permits, extend your investigation to compare the vitamin C content of fresh juice and juice that has been left exposed to air for varying periods of time. Ask your teacher to approve your procedure before carrying out this extension.

6 C A chemist adds a few drops of deep violet-red iodine solution to a vitamin C tablet. The iodine solution quickly becomes colourless. Then the chemist adds a solution that contains chlorine, Cl_2. The chemist observes that the violet-red colour of the iodine reappears. Explain the chemist's observations, in terms of redox reactions.

Think and Research

1 MC A reliable supply of fresh, nutritious food is essential to human health. To this end, the food industry has developed different types of food additives to keep food fresh longer. Numerous reactions are responsible for food spoilage. For example, a food may become stale and rancid due to the oxidation of the molecules in the food by oxygen in the air.

(a) Based on the information above, explain why vitamin C (ascorbic acid) is often added to foods, such as bread.

(b) Butylated hydroxyl toluene (BHT) is another compound that is added to food to prevent spoilage. Find out about BHT in a library or on the Internet. Draw its structure. How does BHT work to preserve food? What type of spoilage does it prevent? Are there risks or drawbacks to using BHT? How does adding BHT to products such as bread affect the cost of the product? Present your findings as a poster.

Web LINK

www.mcgrawhill.ca/links/chemistry12

Strong oxidizing agents create chemical species called *free radicals*. Researchers suggest that free radicals are responsible for reactions that lead to food spoilage, aging, and cancer. Antioxidants work by "deactivating" free radicals. Find out more about the chemical nature of free radicals and their reactions. Also find out how antioxidants interact with them. To begin your research, go to the web site above and click on **Web Links**.

Part 6: Final Presentation

Present your research and laboratory investigation reports for one or more sections of this Course Challenge at a conference called "The Chemistry of Human Health." Use various media to show your audience how chemical processes affect each aspect of human health that you studied. Make sure that you understand your topic well, and be prepared to answer questions about your presentations and displays. As a class, discuss the following questions before creating your presentations and displays.

- What purposes do professional conferences serve?
- How are conferences run?
- Who is your audience? For example, your audience might include health professionals, the general public, or science students. Will you invite other classes, your parents and guardians, and/or the public to your conference?
- What form should the presentations take? The conference could include some or all of the following:
  - information booths
  - posters, pamphlets, and handouts
  - models, laboratory reports, and computer simulations
  - formal presentations with slides or overhead transparencies
  - demonstrations of experiments
  - round-table discussions
  - formal debates
  - a conference web site
  - a schedule brochure with presentation summaries
  - a post-conference publication to summarize the information presented

Additional Topics

What other health-related challenges has chemistry helped scientists to understand or solve? You may wish to go beyond the challenges listed in Parts 1 to 5 of the Course Challenge and investigate other areas of human health. Some suggestions are listed below.

Unit 1

- Researchers can use computer programs to test the geometry of many molecular structures and determine their suitability for specific applications. For example, a specialized screen saver can check the shape of hypothetical molecules based on their suitability for cancer treatment. Anyone can upload the screen saver and use a computer's idle time to test molecules. Research this method, and write an article on the success of computer-assisted research.
- Pyruvate carboxylase (also called PC) is an enzyme that converts pyruvate to oxaloacetate (shown as oxaloacetic acid in the citric acid cycle diagram). Pyruvate carboxylase deficiency is a genetic disorder that is characterized by insufficient quantities of pyruvate carboxylate in the body. How do you think this disorder affects the citric acid cycle? Use print and electronic resources to research pyruvate carboxylase deficiency. Find out what its symptoms are, and how it affects the body at the molecular level. Also find out what percent of the population is affected, and how the deficiency can be relieved. Present your findings as an informative pamphlet. If possible, conduct an e-mail interview with an expert on the disorder.

- Use print and electronic resources to study compounds that have been produced as diet aids, such as aspartame and olestra. How do these compounds work? Are there drawbacks or risks involved in using these compounds? Choose one compound, and conduct a risk-benefit analysis.

Unit 2

- Watson and Crick's discovery of the structure of DNA revolutionized scientists' understanding of heredity and genetics. Find out how Watson and Crick showed that the structure of DNA is a double helix. How did Rosalind Franklin's work in X-ray crystallography play a role? Present your findings as an annotated, illustrated time line.
- It is relatively straightforward to extract DNA from animal matter. Research a procedure on the Internet. Obtain your teacher's approval, and then carry out the extraction. Use the concept of intermolecular forces to explain how the extraction works.

Unit 3

- Human disorders and conditions are sometimes caused by the lack of an enzyme. For example, lactose intolerance is caused by low concentrations of lactase, an enzyme that breaks down lactose (the sugar in milk). Find out about lactose intolerance or another condition that is caused by a missing enzyme. How are the conditions diagnosed? Have treatments been developed? How effective are the treatments? Who suffers from the condition? How harmful is it?
- Researchers have used a specially designed calorimeter, called a *human calorimeter*, to monitor the heat that is produced by metabolic reactions in humans. Research the human calorimeters that are used to study metabolic reactions. What are the design features of these calorimeters? What factors do researchers need to take into account? Why are the experiments with human calorimeters important? Design an investigation that incorporates the results of your research.

Unit 4

- In Part 4 of the Course Challenge, you looked at the carbonic acid/hydrogen carbonate buffer system in the blood. Another important buffer system in the body depends on the equilibrium between dihydrogen phosphate, $H_2PO_4^-$, and hydrogen phosphate, HPO_4^{2-}. Use print and electronic resources to learn where this buffer system is found in the body and how the body regulates concentrations of $H_2PO_4^-$ and $H_2PO_4^{2-}$.
- Acid indigestion, also known as "heartburn," is caused by the excess production of stomach acid. Eating a large meal may bring on acid indigestion. Commercial antacids contain basic salts that neutralize the excess acids. Design a titration experiment to test the effectiveness of various brands of antacids.

Unit 5

- Using only materials that are available in your school chemistry laboratory, design a pacemaker battery. Find out what voltage is appropriate for a pacemaker. Be sure to outline all design considerations. For example, how large will the battery need to be? How long will it last? Have your teacher approve your design. Then build your battery.
- Using print and electronic resources, research the work of Wilson Greatbatch, the inventor of the implantable pacemaker. What chemistry-related technological challenges did he face and overcome as he perfected his device?

Answers to Selected and Numerical Chapter and Unit Review Questions

UNIT 1

Chapter 1

15. **(a)** 2-bromopropane **(b)** cyclopentanol
(c) pentanoic acid **(d)** ethyl propanoate
(e) 5-ethyl-5,6-dimethyl-3-heptanone

16. **(a)** 3,5,7-trimethylcycloheptene
(b) 3-ethyl-2-hexanol
(c) 1-ethyl-4-methylbenzene
(d) 2-methylpentanoic acid
(e) N-ethyl-N-methyl-1-butanamine
(f) N-butylpropanamide

22. **(a)** either 2-methylhexanal or pentanal
(b) pentane
(c) 1,2-dimethylbenzene
(d) N-ethyl-N-methylpentanamide
(e) 2-methylpropanoic acid

23. 1-butanol, 2-butanol, 2-methyl-1-propanol, 1,1-dimethylethanol, 1-methoxypropane, 1-ethoxyethane, 2-methoxypropane

24. butanoic acid, 2-methylpropanoic acid, methyl propanoate, ethyl ethanoate, 1-propyl methanoate, 2-propyl methanoate (**Note:** Other isomers are possible that contain hydroxyl and carbonyl groups, or alkoxy and carbonyl groups, but these are beyond the scope of this course.)

Chapter 2

10. **(b)** 1-bromopropane, 2-bromopropane
(c) 2-bromopropane

13. **(a)** propene

18. **(a)** 2,3-dibromobutane **(b)** 1-bromopentane + water
(c) propanoic acid **(d)** propene + water
(e) 3-methyl-2-butanone **(f)** methyl propanoate + water
(g) pentane **(h)** ethanoic acid + ethanol

20. **(a)** ethene **(b)** 1-propanol
(c) bromine **(d)** 3-methyl-2-butanone
(e) propanoic acid + 1-butanol
(f) butanal **(g)** 2-ethyl-1-butanol, 2-ethylbutanal
(h) cyclobutanol **(i)** 1-butene

Unit 1 Review

25. ethanoic acid

27. 3-ethyl-4-methylhexanal, 3-ethyl-4-methylhexanoic acid

28. **(a)** butanoic acid **(b)** methanol
(c) methyl pentanoate **(d)** 2-butanone
(e) N-methylethanamine **(f)** 4,5-dimethylheptanal

30. **(a)** 3-ethyl-4-methyl-1-hexene
(b) methoxypropane
(c) 3-methyl-1-pentanamine
(d) 5-ethyl-3,7,9-trimethyl-2-decanol
(e) butyl propanoate

31. **(a)** 5-methyloctanoic acid
(b) N-ethyl-1-butanamine
(c) 6-ethyl-5-methyl-2-octyne
(d) butanamide
(e) 6-ethyl-7-methyl-4-decanone
(f) 2-ethyl-3-methylcyclohexanol
(g) 4,4,6,6-tetramethylheptanal
(h) 4-ethyl-3,5-dimethylcyclopentene
(i) N-ethyl-N-methylethanamide
(j) 3,6,7-trimethylcyclononanone

32. **(a)** 4-methyl-2-heptanone **(b)** 3,3-dimethylhexane
(c) 2,5-octanediol

36. **(d)** propyl methanoate + water
(e) 1-heptene + water

37. **(b)** decanoic acid + 1-propanol

38. **(a)** ethyne
(b) 1-chloro-2-methylpentane
(c) heptanoic acid + 1-butanol
(d) 2-methylpropanoic acid
(e) 3-ethyl-4,4-dimethyl-2-pentanol

UNIT 2

Chapter 3

7. $3p$ orbital in third energy level

8. S ground state: $1s^2 2s^2 2p^6 3s^2 3p^4$;
excited: $1s^2 2s^2 2p^6 3s^2 3p^3 4s^1$

9. **(a)** Na, smaller Z_{eff} **(b)** K, more energy levels
(c) Sr, more energy levels **(d)** Ga more energy levels
(e) O, smaller Z_{eff} **(f)** Br, more energy levels

10. U, *f* block, inner transition; Zr, *d* block, transition; Se, *p* block, main group; Dy, *f* block, inner transition; Kr, *p* block, main group; Rb, *s* block, main group; Re, *d* block, transition; Sr, *s* block, main group

11. Na_2O, K_2O, MgO, CaO, Al_2O_3, Ga_2O_3, SiO_2, GeO_2

12. **(a)** 18 **(b)** 15 **(c)** 4 **(d)** infinite

15. **(a)** Mg: $1s^2 2s^2 2p^6 3s^2$ **(b)** Cl: $1s^2 2s^2 2p^6 3s^2 3p^5$
(c) Mn: $1s^2 2s^2 2p^6 3s^2 3p^6 4s^2 3d^5$
(d) Y: $1s^2 2s^2 2p^6 3s^2 3p^6 4s^2 3d^{10} 4p^6 5s^2 4d^1$

18. $8s^2$, $n = 8$, $l = 0$

19. **(a)** Sr has greater n but K has a smaller Z_{eff}
(b) both have $4s^2$ valence energy level
(c) Ca has a greater n but also a greater IE_1.

21. **(a)** Rb **(b)** Ti **(c)** Mg **(d)** As

23. **(a)** Al: $1s^2 2s^2 2p^6 3s^2 3p^1$; Sc: $1s^2 2s^2 2p^6 3s^2 3p^6 4s^2 3d^1$
(b) 3+

Chapter 4

3. **(a)**
```
       ..
      :Br:
       |
 H — C — Br:
       |
      :Br:
       ..
```
(b) $[\,H—\ddot{S}:\,]^-$

6. **(a)** dipole-dipole and dispersion forces
   (b) ionic bonds
   (c) metallic bonds
   (d) dispersion forces

7. Cs, S, Kr, O_2

8. H_2O because of large *EN* difference between atoms and small size of O

9. $[:\ddot{O}=\ddot{O}:]^{2-} \longleftrightarrow [:\dot{O}\equiv\dot{O}:]^{2-} \longleftrightarrow [:\ddot{O}—\ddot{O}:]^{2-}$

13. C_3H_8 (dispersion forces), C_2H_5OH (dispersion forces and hydrogen bonding), K (metallic bonding), SiO_2 (network solid)

14. all tetrahedral; CCl_4 is non-polar; CH_3Cl and CH_2Cl_2 both polar with dipole-dipole forces.

16. **(a)** bent **(b)** octahedral
   (c) trigonal pyramidal

17. **(a)** Cl_2 **(b)** BrCl

19. linear with VSEPR notation AX_2E_3 (e.g., XeF_2) or square planar with notation AX_4E_2 (e.g., XeF_4).

20. non-polar

24. no; 7 pairs of electrons in the valence shell

26. SF_4: seesaw shape and polar; SiF_4: tetrahedral shape and non-polar

31. N_2H_4 hydrogen bonding, greater dispersion forces (18 electrons) higher boiling point; C_2H_4 lower dispersion forces (10 electrons)

Unit 2 Review

13. N: $1s^22s^22p^3$; P: $[Ne]3s^23p^3$; As: $[Ar]3d^{10}4s^24p^3$

14. Mg, Na, Ca, K

15. AX_2E or AX_2E_2

17. **(a)** Rb < K < Na < Li **(b)** Li < Be < B < C

29. titanium

30. group 14 (IVA)

31. LiBr

32. **(a)** As **(b)** P **(c)** Be

36. **(a)** N more electrongative **(b)** F more electronegative
   (c) Cl more electronegative

37. **(a)** Cl—Cl < Br—Cl < Cl—F
   (b) Si—Si < S—Cl < P—Cl < Si—Cl (except for non-polar Cl—Cl and Si—Si, left atom in each bond pair has partial positive charge, and right atom has partial negative charge)

38. trigonal pyramidal or T-shaped, but not trigonal planar

39. $[:\ddot{O}—H]^-\ [Ba]^{2+}\ [:\ddot{O}—H]^-$

40. dipole;
```
        :O:
         ‖
 :F̤ — P — F̤:
         |
        :F:
         ..
```

41. tetrahedral; bond dipoles for C—H shorter than for C—Cl

42. neither has expanded valence level;
```
 ⎡       ..       ⎤ −        ..
 ⎢      :O:       ⎥         :O:
 ⎢       |        ⎥          |
 ⎢ :O — Cl — O:   ⎥        :S — Cl:
 ⎢       |        ⎥          |
 ⎣      :O:       ⎦         :O:
         ..                  ..
```

43. $1s^22s^22p^63p^1$

44. **(a)** $n = 2,\ l = 0,\ m_l = 0,\ m_s = +\frac{1}{2}$
   (b) $n = 3,\ l = 1,\ m_l = 1,\ m_s = -\frac{1}{2}$

48. **(a)** shiny, malleable, conductor of heat and electricity, fairly high melting (or boiling) point
   (b) form cations by giving up electrons, form basic oxides when dissolved in water

49. **(b)** K (difference between $I.E._1$ and $I.E._2$); can infer the number of valence electrons because of the size of the I.E. difference)

UNIT 3

Chapter 5

8. **(a)** 60.0°C **(b)** 50.0°C

9. **(a)** −2225 kJ/mol

10. −14.5 kJ/g; −869 kJ/mol

11. +336.6 kJ

12. 466 kJ

13. 228 kJ

14. **(b)** −188 kJ/mol

15. 256 kJ

16. **(a)** −2657 kJ **(c)** 4.6×10^2 kJ
   (b) −1428.4 kJ/mol; −47.5 kJ/g

21. **(b)** 2 kJ

Chapter 6

7. **(a)** 4.00×10^{-4} mol/(L • s) **(b)** 3.40×10^{-2} g

8. **(a)** $k = 0.010\ s^{-1}$; $t_{1/2} = 69$ s

10. 14.2 s

11. **(b)** $k = 2.58 \times 10^{-28}$ L/(mol • s)

12. **(a)** rate = $k[A]^2[B]$ **(b)** third order
   (c) $k = 625\ L^2/(s \cdot mol^2)$

13. **(a)** rate = $k[Hg_2Cl_2][C_2O_4^{2-}]^2$
   (b) $k = 1.34 \times 10^{-3}\ L^2/(mol^2 \cdot s)$
   (c) rate = 1.20×10^{-6} mol/(L • s)

14. (a) 0.80 **(b)** 0.20

15. (a) 0.20 **(b)** 0.80

16. (a) −12.1 kJ/mol

Unit 3 Review

27. (a) 47 kJ/mol

29. 3.6 L

30. (b) 82.5 kJ

31. −747.6 kJ

38. (b) ~ − 6800 kJ/mol

UNIT 4

Chapter 7

7. decrease in temperature

8. left

10. $[H_2] = 5.6 \times 10^{-4}$ mol/L; $[HI] = 1.12 \times 10^{-3}$ mol/L

11. $[SO_3] = 0.13$ mol/L

12. 37.4 g ethyl propanoate

13. $[CO] = [H_2O] = 0.013$ mol/L; $[H_2] = [CO_2] = 0.037$ mol/L

14. $K_c = 2.5 \times 10^{-4}$

16. (a) $O_{2(g)} + 4HCl_{(g)} \rightarrow 2H_2O_{(g)} + 2Cl_{2(g)}$
(b) 1.7×10^{-2}

17. (a) $SO_{3(g)} + 6HF_{(g)} \rightarrow SF_{6(g)} + 3H_2O_{(g)}$

19. (a) $4NH_{3(g)} + 5O_{2(g)} \rightarrow 4NO_{(g)} + 6H_2O_{(g)}$
(b) 0.26

24. 1.1×10^{-17} mol/L; pollution sources

26. left; right; left; no effect; at constant volume, no effect

Chapter 8

2. (a) weak acid **(b)** weak acid
(c) weak base **(d)** weak base

3. (a) O^{2-} **(b)** NH_4^+
(c) H_2CO_3 **(d)** CO_3^{2-}

4. (a) F **(b)** T **(c)** T

5. 2.0 mol/L $HClO_4$ < 0.02 mol/L HCl < 0.20 mol/L CH_3COOH < 2.0 mol/L NaCl

6. (a) SO_4^{2-} **(b)** S^{2-}
(c) HPO_4^{2-} **(d)** CO_3^{2-}

7. (a) $HS^- = 1.0 \times 10^{-19}$; $HSO_3^- = 6.5 \times 10^{-8}$
(b) $HS^- + H_2O_{(\ell)} \rightleftharpoons H_2S_{(aq)} + OH^-_{(aq)}$;
$HSO_3^-{}_{(aq)} + H_2O_{(\ell)} \rightleftharpoons H_2SO_{3(aq)} + OH^-$
(c) K_b for $HS^- = 1.1 \times 10^{-7}$; K_b for $HSO_3^- = 7.7 \times 10^{-13}$
(d) HS^-

10. (a) number of O atoms bonded to central atom; number of H
(b) $HBrO_3$, H_2SO_3, $HClO_3$, $HClO_4$

11. 2

12. 0.300 mol/L

13. 2.52

14. (a) pH = 2.10; 0.79% dissociation
(b) pH = 2.60; 2.5% dissociation
(c) pH = 3.12; 7.6% dissociation

15. (a) pH = 4.36; 0.067% **(b)** OCl^-, 3.45×10^{-7}

16. pH = 9.10

17. pH = 8.07

18. $K_a = 1.1 \times 10^{-3}$

19. $HBrO < HBrO_2 < HBrO_3 < HBrO_4$; opposite order for conjugate bases

20. $H_2O < H_2S < CH_3COOH < HBrO_2$

22. 4.0×10^{-6}

23. (b) 2.6×10^{-10}

26. 2.5×10^{-6} mol/L; 12:1

Chapter 9

1. (a) NH_4Cl **(b)** $KHSO_4$ **(c)** $NaHPO_4$ **(d)** $Al(NO_3)_3$

7. (a) $(NH_4)_2SO_4$; NH_4Br; $Cu(NO_3)_2$
(b) KHS; LiCN; $Ca(NO_2)_2$

12. (a) $K_{sp} = [Cu^+][Br^-]$ **(b)** $K_{sp} = [Ca^{2+}][CrO_4^{2-}]$
(c) $K_{sp} = [Ni^{2+}][OH^-]^2$ **(d)** $K_{sp} = [Mg^{2+}]^3[PO_4^{3-}]^2$
(e) $K_{sp} = [Mg^{2+}][NH_4^+][PO_4^{3-}]$

15. $H_2PO_4^-{}_{(aq)} + H_2O_{(l)} \leftrightarrow H_3O^+_{(aq)} + HPO_4^{2-}{}_{(aq)}$

18. NaI since $[I^-]^2$

20. $[Pb^{2+}]:[Ba^{2+}] = 130:1$

21. 7.7×10^{-9} mol/L

22. 6.1×10^{-9} mol/L

25. (a) 19.44 mL **(b)** 3.29% **(c)** steps 1 and 4

26. AgCl, lower K_{sp}

27. $BaSO_4$

29. (a) $CH_3COOAg_{(s)} \rightleftharpoons Ag^+_{(aq)} + CH_3COO^-_{(aq)}$
(b) 23.3 g
(c) as Cu reacts, solid dissolves to replace the Ag^+

30. (a) 41.7 mL **(b)** 8.86
(c) phenolphthalein

31. 4.74

Unit 4 Review

15. 6.9×10^{-2}

17. 0.16

18. $\dfrac{[N_{2(g)}][H_{2(g)}]^3}{[NH_{3(g)}]^2}$

19. 6×10^3

20. (a) $HCO_3^-{}_{(aq)} + HPO_4^{2-}{}_{(aq)} \rightleftharpoons CO_3^{2-}{}_{(aq)} + H_2PO_4^-{}_{(aq)}$
(b) $A_1 = HCO_3^-{}_{(aq)}$ $B_1 = CO_3^{2-}{}_{(aq)}$ $A_2 = H_2PO_4^-{}_{(aq)}$ $B_2 = HPO_4^-{}_{(aq)}$
(c) reactants since K_a for $H_2PO_4^-{}_{(aq)}$ > K_a for $HCO_3^-{}_{(aq)}$

21. 2.4

22. 12.268

23. $[H_3O^+_{(aq)}] = 1.74 \times 10^{-3}$ mol/L, $[OH^-] = 5.75 \times 10^{-12}$ mol/L, pH = 2.76, pOH = 11.24

24. 0.0008

25. $2Ag^+_{(aq)} + CO_3^{2-}{}_{(aq)} \rightarrow Ag_2CO_{3(s)}$

26. 8.62×10^{-6} mol/L

28. 1.6×10^{-5} g

29. remove CO_2, increase temperature

30. (b) $[PCl_5] = 0.060$ mol/L. $[PCl_3] = 0.38$ mol/L, $[Cl_2] = 0.16$ mol/L

31. (a) 1.6×10^{-5} **(b)** 11.14

32. $NaOH_{(s)} \rightarrow Na^+_{(aq)} + OH^-_{(aq)}$; $NH_{3(aq)} + H_2O_{(\ell)} \rightarrow NH_4^+{}_{(aq)} + OH^-_{(aq)}$; $CH_3COO^-_{(aq)} + H_2O_{(\ell)} \rightarrow CH_3COOH_{(aq)} + OH^-_{(aq)}$; $CH_3COOH_{(\ell)} + H_2O_{(\ell)} \rightarrow CH_3COO^-_{(aq)} + H_3O^+_{(aq)}$

33. $K_{sp} = [Pb^{2+}][Cl^-]^2 = 1.6 \times 10^{-5}$

34. (a) decrease
(b) $Q_c > K_c$, $[NH_3]$ decreases

35. 0.091 mol

36. 12.3%

37. (c) $[XO_{2(g)}] = 0.00985$ mol/L; $[O_{2(g)}] = 0.992$ mol/L

38. 7.26×10^{-6}

39. 50 ppm = 1.4×10^{-3} mol/L; precipitation occurs

47. (b) 7.74

UNIT 5

Chapter 10

1. (a) $Zn_{(s)} + 2AgNO_{3(aq)} \rightarrow Zn(NO_3)_{2(aq)} + 2Ag_{(s)}$
(b) $3CoBr_{2(aq)} + 2Al_{(s)} \rightarrow 3Co_{(s)} + 2AlBr_{3(aq)}$
(c) $Cd_{(s)} + SnCl_{2(aq)} \rightarrow CdCl_{2(aq)} + Sn_{(s)}$

2. (a) $Zn_{(s)} + 2Ag^+_{(aq)} + 2NO_3^-{}_{(aq)} \rightarrow Zn^{2+}_{(aq)} + 2NO_3^-{}_{(aq)} + 2Ag_{(s)}$
$Zn_{(s)} + 2Ag^+_{(aq)} \rightarrow Zn^{2+}_{(aq)} + 2Ag_{(s)}$
(b) $3Co^{2+}_{(aq)} + 6Br^-_{(aq)} + 2Al_{(s)} \rightarrow 3Co_{(s)} + 2Al^{3+}_{(aq)} + 6Br^-_{(aq)}$
$3Co^{2+}_{(aq)} + 2Al_{(s)} \rightarrow 3Co_{(s)} + 2Al^{3+}_{(aq)}$
(c) $Cd_{(s)} + Sn^{2+}_{(aq)} + 2Cl^-_{(aq)} \rightarrow Cd^{2+}_{(aq)} + 2Cl^-_{(aq)} + Sn_{(s)}$
$Cd_{(s)} + Sn^{2+}_{(aq)} \rightarrow Cd^{2+}_{(aq)} + Sn_{(s)}$

3. (a) oxidizing agent: Ag^+; reducing agent: Zn
(b) oxidizing agent: Co^{2+}; reducing agent: Al
(c) oxidizing agent: Sn^{2+}; reducing agent: Cd

4. (a) oxidation half-reaction: $Zn \rightarrow Zn^{2+} + 2e^-$
reduction half-reaction: $Ag^+ + e^- \rightarrow Ag$
(b) oxidation half-reaction: $Al \rightarrow Al^{3+} + 3e^-$
reduction half-reaction: $Co^{2+} + 2e^- \rightarrow Co$
(c) oxidation half-reaction: $Cd \rightarrow Cd^{2+} + 2e^-$
reduction half-reaction: $Sn^{2+} + 2e^- \rightarrow Sn$

5. (a) metallic element
(b) non-metallic element
(c) non-metallic element
(d) metallic element

6. (a) Ba = +2; Cl = −1
(b) C = +4; S = −2
(c) Xe = +4; F = −1

7. (a) Ba +2; H −1
(b) Al +3; C −4
(c) K +1; C +2; N −3
(d) Li +1; N +3; O −2
(e) N −3; H +1; C +3; O −2
(f) S 0
(g) As +3; O −2
(h) V +5; O −2
(i) Xe +6; O −2; F −1
(j) S +2.5; O −2

8. ClO_2^-

9. (a) yes; oxidizing agent: O_2; reducing agent: C_6H_6
(b) no
(c) yes; oxidizing agent: I_2; reducing agent: H_2
(d) yes; oxidizing agent: $KMnO_4$; reducing agent: CuCl.

10. (a) yes; Cu undergoes oxidation; Ag^+ undergoes reduction.
(b) no
(c) yes; Mn^{2+} undergoes oxidation; BiO_3^- undergoes reduction.

11. (a) +5 in $V_2O_5, VO_2^+, VO_3^-, VO_4^{3-}, V_3O_9^{3-}$; +4 in VO_2, VO^{2+}
(b) no

12. (a) all three steps
(b) Step 1: oxidizing agent: O_2; reducing agent: NH_3
Step 2: oxidizing agent: O_2; reducing agent: NO
Step 3: oxidizing agent: NO_2; reducing agent: NO_2

14. (a) $I_2 + 2e^- \rightarrow 2I^-$
(b) $Pb \rightarrow Pb^{4+} + 4e^-$
(c) $AuCl_4^- + 3e^- \rightarrow Au + 4Cl^-$
(d) $C_2H_5OH + H_2O \rightarrow CH_3COOH + 4H^+ + 4e^-$
(e) $S_8 + 16H^+ + 16e^- \rightarrow 8H_2S$
(f) $AsO_2^- + 4OH^- \rightarrow AsO_4^{3-} + 2H_2O + 2e^-$

15. (a) $MnO_2 + 2Cl^- + 4H^+ \rightarrow Mn^{2+} + Cl_2 + 2H_2O$
(b) $2NO + 3Sn + 6H^+ \rightarrow 2NH_2OH + 3Sn^{2+}$
(c) $3Cd^{2+} + 2V^{2+} + 6H_2O \rightarrow 3Cd + 2VO_3^- + 12H^+$
(d) $2Cr + 6H_2O + 2OH^- \rightarrow 2Cr(OH)_4^- + 3H_2$
(e) $S_2O_3^{2-} + 2NiO_2 + H_2O + 2OH^- \rightarrow 2Ni(OH)_2 + 2SO_3^{2-}$
(f) $2Sn^{2+} + O_2 + 2H_2O \rightarrow 2Sn^{4+} + 4OH^-$

16. (a) $3SiCl_4 + 4Al \rightarrow 3Si + 4AlCl_3$
(b) $4PH_3 + 8O_2 \rightarrow P_4O_{10} + 6H_2O$
(c) $I_2O_5 + 5CO \rightarrow I_2 + 5CO_2$
(d) $2SO_3^{2-} + O_2 \rightarrow 2SO_4^{2-}$

17. (a) $2NO_2 + H_2O \rightarrow NO_2^- + NO_3^- + 2H^+$
(b) $Cl_2 + 2OH^- \rightarrow ClO^- + Cl^- + H_2O$

18. (a) $2Co^{3+} + Cd \rightarrow 2Co^{2+} + Cd^{2+}$
$2Co(NO_3)_{3(aq)} + Cd_{(s)} \rightarrow 2Co(NO_3)_{2(aq)} + Cd(NO_3)_{2(aq)}$
(b) $2Ag^+ + SO_2 + 2H_2O \rightarrow 2Ag + SO_4^{2-} + 4H^+$
$2AgNO_{3(aq)} + SO_{2(g)} + 2H_2O_{(\ell)} \rightarrow 2Ag_{(s)} + H_2SO_{4(aq)} + 2HNO_{3(aq)}$
(c) $Al + CrO_4^{2-} + 4H_2O \rightarrow Al(OH)_3 + Cr(OH)_3 + 2OH^-$
$Al_{(s)} + Na_2CrO_{4(aq)} + 4H_2O_{(\ell)} \rightarrow Al(OH)_{3(s)} + Cr(OH)_{3(s)} + 2NaOH_{(aq)}$

21. $O_2 + 2F_2 \rightarrow 2OF_2$

22. (a) $2P_4 + 12H_2O \rightarrow 5PH_3 + 3H_3PO_4$
(b) reduced

23. (a) $2Al + Fe_2O_3 \rightarrow 2Fe + Al_2O_3$
(b) yes; oxidizing agent: Fe_2O_3; reducing agent: Al

24. (a) $I_2 + 10HNO_3 \rightarrow 2HIO_3 + 10NO_2 + 4H_2O$
(b) 31.7 g

27. (a) $P_4 + 3H_2O + 3OH^- \rightarrow 3H_2PO_2^- + PH_3$
(c) 2.74 kg

32. (a) $C_2H_4 + H_2 \rightarrow C_2H_6$
$C_2H_5OH \rightarrow CH_3CHO + H_2$

33. (a) Na +1; Al +3; O −2; H +1; C +4
(b) $NaAl(OH)_2CO_3 + 4HCl \rightarrow NaCl + AlCl_3 + 3H_2O + CO_2$

34. (a) $2KClO_3 \rightarrow 2KCl + 3O_2$
$S + O_2 \rightarrow SO_2$
(b) oxidizing agent and reducing agent: $KClO_3$; oxidizing agent: O_2; reducing agent: S

35. $NH_4^+ + 2O_2 \rightarrow NO_3^- + H_2O + 2H^+$

36. (b) $CaC_2 + 2H_2O \rightarrow Ca(OH)_2 + C_2H_2$; not redox
(c) $2C_2H_2 + 5O_2 \rightarrow 4CO_2 + 2H_2O$; redox

Chapter 11

2. $Mg_{(s)} \mid Mg^{2+}{}_{(aq)}$ (1 mol/L) || $Cd^{2+}{}_{(aq)}$ (1 mol/L) | $Cd_{(s)}$

3. oxidation: $NO_{(g)} + 2H_2O_{(\ell)} \rightarrow NO_3^-{}_{(aq)} + 4H^+{}_{(aq)} + 3e^-$
reduction: $I_{2(s)} + 2e^- \rightarrow 2I^-{}_{(aq)}$
overall: $2NO_{(g)} + 4H_2O_{(\ell)} + 3I_{2(s)} \rightarrow 2NO_3^-{}_{(aq)} + 8H^+{}_{(aq)} + 6I^-{}_{(aq)}$

6. spontaneous: $N_2O_{(g)} + 2H^+{}_{(aq)} + 2Cu_{(s)} + 2I^-{}_{(aq)} \rightarrow N_{2(g)} + H_2O_{(\ell)} + 2CuI_{(s)}$ $E°_{cell} = 1.955$ V
non-spontaneous: $N_{2(g)} + H_2O_{(\ell)} + 2CuI_{(s)} \rightarrow N_2O_{(g)} + 2H^+{}_{(aq)} + 2Cu_{(s)} + 2I^-{}_{(aq)}$ $E°_{cell} = -1.955$V

7. (a) oxidizing agent: PbO_2; reducing agent: Pb
(b) oxidizing agent: $PbSO_4$; reducing agent: $PbSO_4$

8. $Co^{3+}{}_{(aq)}$, $Br_{2(aq)}$, $H^+{}_{(aq)}$, $Zn^{2+}{}_{(aq)}$

9. $Al_{(s)}$, $H_{2(g)}$, $Ag_{(s)}$, $Cl^-{}_{(aq)}$

11. (a) oxidation: $Zn_{(s)} \rightarrow Zn^{2+}{}_{(aq)} + 2e^-$
reduction: $Fe^{2+}{}_{(aq)} + 2e^- \rightarrow Fe_{(s)}$
cell potential: 0.315 V; spontaneous
(b) oxidation: $Cr_{(s)} \rightarrow Cr^{3+}{}_{(aq)} + 3e^-$
reduction: $Al^{3+}{}_{(aq)} + 3e^- \rightarrow Al_{(s)}$
cell potential: −0.918 V; non-spontaneous
(c) oxidation: $H_2O_{2(aq)} \rightarrow 2H^+{}_{(aq)} + O_{2(g)} + 2e^-$
reduction: $Ag^+_{(aq)} + e^- \rightarrow Ag_{(s)}$
cell potential: 0.105 V; spontaneous

12. 1.52 g

14. (a) $E°_{cell} = 1.247$ V
oxidation: $Fe_{(s)} \rightarrow Fe^{2+}{}_{(aq)} + 2e^-$
reduction: $Ag^+{}_{(aq)} + e^- \rightarrow Ag_{(s)}$
overall: $Fe_{(s)} + 2Ag^+{}_{(aq)} \rightarrow Fe^{2+}{}_{(aq)} + 2Ag_{(s)}$
(b) $E°_{cell} = -1.247$ V
oxidation: $Ag_{(s)} \rightarrow Ag^+{}_{(aq)} + e^-$
reduction: $Fe^{2+}{}_{(aq)} + 2e^- \rightarrow Fe_{(s)}$
overall: $Fe^{2+}{}_{(aq)} + 2Ag_{(s)} \rightarrow 2Ag^+{}_{(aq)} + Fe_{(s)}$

17. The cost for Al would be 2.56 times the cost for Na.

Unit 5 Review

14. (a) +3
(b) +5
(c) +4
(d) N −3; H +1; S +6; O −2

15. no

16. (a) $Ca(OH)_{2(aq)} + CO_{2(g)} \rightarrow CaCO_{3(s)} + H_2O_{(\ell)}$

17. (a) $Hg_2^{2+} + 2e^- \rightarrow 2Hg$
(b) $TiO_2 + 4H^+ + 2e^- \rightarrow Ti^{2+} + 2H_2O$
(c) $I_2 + 18OH^- \rightarrow 2H_3IO_6^{3-} + 6H_2O + 12e^-$

20. (a) $3Fe_{(s)} + 8HNO_{3(aq)} \rightarrow 3Fe(NO_3)_{2(aq)} + 2NO_{(g)} + 4H_2O_{(\ell)}$
(b) $3Fe(NO_3)_{2(aq)} + 4HNO_{3(aq)} \rightarrow 3Fe(NO_3)_{3(aq)} + NO_{(g)} + 2H_2O_{(\ell)}$
(c) $3Fe_{(s)} + 12HNO_{3(aq)} \rightarrow 3Fe(NO_3)_{3(aq)} + 3NO_{(g)} + 6H_2O_{(\ell)}$

21. (b) oxidation: $2I^- \rightarrow I_2 + 2e^-$
reduction: $Ag^+ + e^- \rightarrow Ag$
overall: $2I^- + 2Ag^+ \rightarrow I_2 + 2Ag$
(d) 0.264 V

22. (a) no
$Cl_2O_7 + H_2O \rightarrow 2HClO_4$
(b) yes; oxidizing agent: ClO_3^-; reducing agent: I_2
$3I_2 + 5ClO_3^- + 3H_2O \rightarrow 6IO_3^- + 5Cl^- + 6H^+$
(c) yes; oxidizing agent: Br_2; reducing agent: S_2^-
$S^{2-} + 4Br_2 + 8OH^- \rightarrow SO_4^{2-} + 8Br^- + 4H_2O$
(d) yes; oxidizing agent: HNO_3; reducing agent: H_2S
$2HNO_3 + 3H_2S \rightarrow 2NO + 3S + 4H_2O$

23. (a) −0.235 V; non-spontaneous
(b) 0.698 V; spontaneous
(c) 0.418 V; spontaneous

26. (a) $2MnO_4^- + 5C_2O_4^{2-} + 16H^+ \rightarrow 2Mn^{2+} + 10CO_2 + 8H_2O$
(b) 0.2249 mol/L

27. 20.73%

29. (a) 6
(b) 3

30. (a) 2.68×10^4 C
(b) 74.5 h

31. 1.16 g

32. 0.595 t

44. (a) $8Al + 3NH_4ClO_4 \rightarrow 4Al_2O_3 + 3NH_4Cl$
oxidizing agent: NH_4ClO_4; reducing agent: Al
(b) $2H_{2(\ell)} + O_{2(\ell)} \rightarrow 2H_2O_{(g)}$

Supplemental Practice Problems

UNIT 1

Chapter 1

1. Identify the type of organic compound in each diagram.

(a) 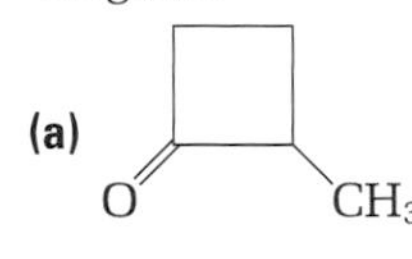

(b) $HC(=O)—CH_2—CH_3$

(c) 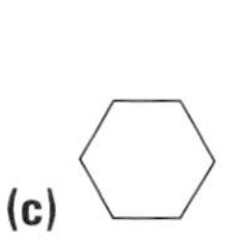

(d) $H_2C(—O—CH_3)—CH_3$

2. How are amines structurally different from amides?

3. The following compound has the IUPAC name 3-oxa-1,5-pentanediol.

$HO—H_2C—CH_2—O—CH_2—CH_2—OH$

(a) Based on its IUPAC name, is this compound classed as an ether or an alcohol?

(b) List the attractive forces between molecules of this compound, in order of strength.

(c) In which solvent, water or hexane, would this compound be more soluble?

4. Oxalic acid has the following structural formula.

$HO—C(=O)—C(=O)—OH$

This carboxylic acid is found in the leaves of rhubarb. Its IUPAC name is ethanedioic acid.

(a) Name the intermolecular forces that occur between molecules of oxalic acid.

(b) Is this acid soluble in water?

(c) Tartaric acid is a dicarboxylic acid that is used in baking powder. It has the following structural formula.

$HO—C(=O)—CH(OH)—CH(OH)—C(=O)—OH$

To name a hydroxyl group as a side-chain, use x-hydroxy- as a prefix, where x is the position number. Follow the example for naming oxalic acid, as well as the rules for naming the hydroxyl groups as side chains, to give the IUPAC name for tartaric acid.

5. Name each compound.

(a) $C_3H_7—C(=O)—O—C_5H_{11}$

(b) [skeletal structure: ethyl–O–CH(CH₃)–ethyl]

CH₃
O

(c) [cyclohexene ring with two Cl substituents]

Cl
Cl

6. Draw a condensed structural diagram for each molecule.

(a) 1,3-difluoro-4-methyl-1-pentyne

(b) *trans*-7,7-dimethyl-2-octene

(c) pentanamide

7. What is the IUPAC name for each molecule?

(a) $C_2H_5—CH_2—C(CH_3)_2—C(=O)—O—CH(CH_3)—CH_3$

(b) $H—N(C_2H_5)—CH(CH_3)—C_2H_5$

(c) $C_3H_7—C(=O)—C_3H_7$

8. How does a tertiary alcohol differ from a tertiary amine?

9. Draw and name all possible structural isomers of trichlorobenzene.

10. (a) Draw and name the carboxylic acid that is a structural isomer of ethyl pentanoate.

(b) Name the ether that is a structural isomer of 1-propanol.

(c) Name the ketone that is a structural isomer of propanal.

11. Consider the molecule below.

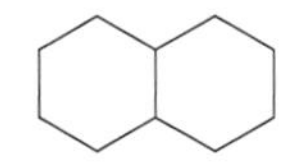

(a) Write the molecular formula of this molecule.

(b) Draw two structural isomers, one cyclic and one non-cyclic, for this molecule.

12. The following names are not correct. Identify the mistake in each name, and write the correct IUPAC name.

(a) methyl ethanoate, $C_2H_5COOCH_3$

(b) ethoxymethane

(c) 4-hexanol

(d) N-pentylethylethanamide
(e) 2-bromobenzene

Chapter 2

13. Name the product(s) formed in each addition reaction.
   (a) $CH_3-C\equiv C-CH_2-CH_3 + 1Br_2 \rightarrow$
   (b) 1-methylcyclopentene + $H_2 \rightarrow$
   (c) $C_5H_{11}-CH=CH_2 + H_2O \rightarrow$
   (d) 1,3-dimethylcyclobutene + HCl $\rightarrow$
14. Identify the alcohol that is oxidized to prepare each compound.
   (a) 2-pentanone
   (b) pentanal
15. Complete each substitution reaction.
   (a) 1-butanol + HBr $\rightarrow$
   (b) $CH_3-CH=CH_2 + Cl_2 \rightarrow$
   (c) benzene + $Br_2 \rightarrow$
   (d) 2-chloropentane + NaOH $\rightarrow$
16. Complete each elimination reaction.
   (a) $C_{16}H_{33}-CH_2-CH_2Br \rightarrow HBr + ?$
   (b) 3,3-dimethyl-2-butanol $\xrightarrow{H_2SO_4}$
17. Name the product that is formed in each oxidation.
   (a) 3-pentanol to a ketone
   (b) ethanal to a carboxylic acid
   (c) 2-ethyl-2-methyl-1-butanol to an aldehyde
   (d) cyclooctanol to a ketone
18. Name the product that is formed in each reduction.
   (a) propanoic acid to an alcohol
   (b) 3-hexanone to an alcohol
   (c) ethene to an alkane
   (d) formic acid to an alcohol
19. For each reaction, name the ester that forms and draw its condensed structural diagram.
   (a) methanol + heptanoic acid $\rightarrow$
   (b) octanoic acid + ethanol $\rightarrow$
   (c) propanoic acid + isopropyl alcohol $\rightarrow$
   (d) 2-butanol + pentanoic acid $\rightarrow$
20. Examine the reaction below, and answer the questions that follow.

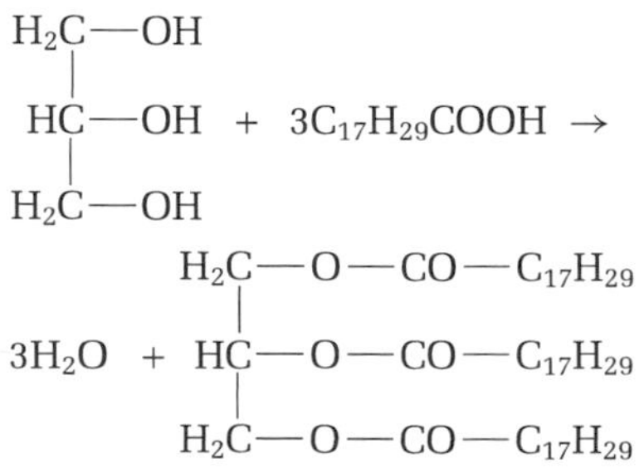

H_2C-OH
$HC-OH \; + \; 3C_{17}H_{29}COOH \rightarrow$
H_2C-OH

$H_2C-O-CO-C_{17}H_{29}$
$3H_2O \; + \; HC-O-CO-C_{17}H_{29}$
$H_2C-O-CO-C_{17}H_{29}$

   (a) State two names for the reaction as it proceeds from left to right.
   (b) If this reaction occurs in the reverse direction, what name is given to the reverse reaction?
   (c) Stearic acid, $C_{17}H_{35}COOH$, is a saturated carboxylic acid. What does this mean?
   (d) The carboxylic acid that is used in this reaction is called linolenic acid. Compare the formulas of linolenic acid and stearic acid, and suggest how the two acids differ in structure.
   (e) The ester that is made in this reaction is an oil. If stearic acid had been used instead, the product would have been a fat. Which of these two products has the lower melting point?
21. When PCBs burn, polychlorinated dibenzofurans (PCDFs) are among the products that may form. Examine the structure shown below, and circle and name the functional groups.

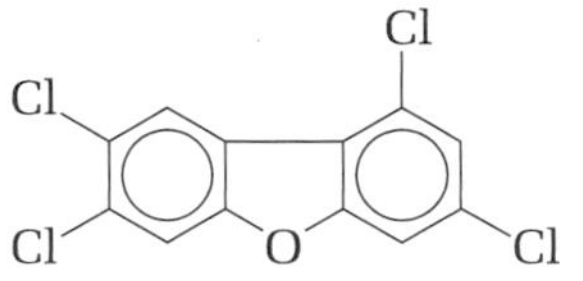

Cl Cl Cl O Cl

22. Name three types of molecules that are capable of forming polymers.
23. Name the alkene that is formed by the elimination of 2-methyl-2-propanol.
24. Polyurethane can be made by the reaction of a diol, such as $HO(CH_2)_4OH$ and toluene 2,4-diisocyanate, shown below.

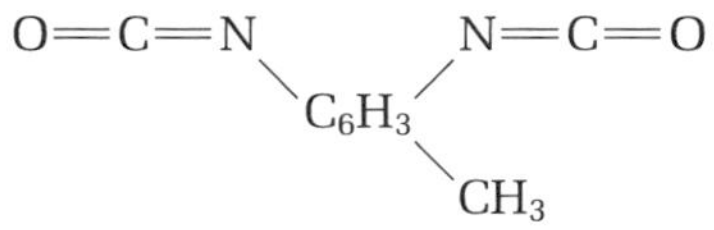

$O=C=N \quad N=C=O$
C_6H_3
CH_3

   Is this reaction going to be an addition or a condensation polymerization?
25. What two classes of monomers are used to make the polymer below?

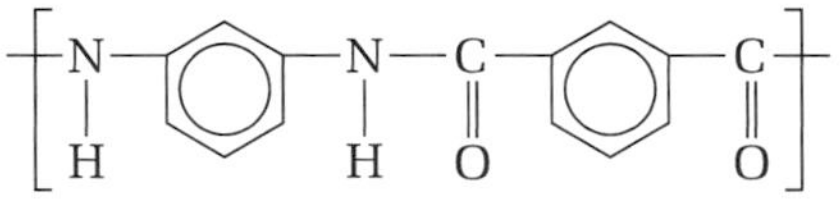

$[-N(H)-C_6H_4-N(H)-C(=O)-C_6H_4-C(=O)-]$

UNIT 2

Chapter 3

26. A sample of a comet has not yet been captured, but the elemental composition of a comet has been predicted. How do you think scientists can determine the composition of a distant celestial body, such as a comet?
27. Rank the elements Cs, Sr, and Rb in order of increasing IE_1. Then give a brief explanation for your order.
28. Compounds of Xe, but not He, have been synthesized. What properties of these two elements help to explain why?
29. (a) How many electrons must Sn lose to attain the electron configuration of Kr?
   (b) Why does this not happen?
   (c) The stable ions of Sn are Sn^{2+} and Sn^{4+}. Write the electron configuration for these two ions.
   (d) Suggest a reason why these ions are stable.
   (e) Which of these two ions has the larger radius?

(f) Sn is closer to the noble gas Xe than to the noble gas Kr. What charge would ions of Sn acquire if the electron configuration of Xe was attained?
(g) This negative ion of Sn does not exist. Suggest an explanation for this observation.

30. According to the order of filling atomic orbitals predicted by the aufbau principle, the 3*d* orbitals are filled after the 4*s* orbital. Explain why the 4*s* orbital is filled first, even though it has a higher pincipal quantum number than the 3*d* orbitals.

31. What type of atomic orbital is described by the following pairs of quantum numbers?
(a) $n = 3,\ l = 0$
(b) $n = 6,\ l = 3$
(c) $n = 3,\ l = 2$

32. For each grouping below, arrange the energy sub-levels in order from lowest energy to highest energy.
(a) 3*p*, 3*d*, 4*s*
(b) 4*f*, 6*s*, 6*d*
(c) 5*f*, 6*s*, 7*s*

33. A road flare contains a salt of strontium to give a red colour to the flame. Write the electron configuration of the strontium ion, with an electron in the least excited energy state.

34. In what ways are the 4d and 5d orbitals in an atom the same, and in what ways are they different?

35. The energy of an electron in its ground state in a hydrogen atom is -2.18×10^{-18} J.
(a) What is the significance of the negative sign given to this energy?
(b) How does the energy of the electron change if it is excited to the 2*s* orbital?
(c) Would the energy change be larger, smaller, or the same if the electron is excited to a 2*p* orbital, rather than the 2*s* orbital?

36. Select the largest ion in each grouping.
(a) Sr^{2+}, Ba^{2+}, Ca^{2+}
(b) N^{-}, N^{2-}, N^{3-}
(c) Se^{2-} Rb^{+}, Br^{-}
(d) N^{1-}, O^{1-}, F^{-}

37. (a) Arrange the elements As, Sb, and Sn in order of decreasing IE_1.
(b) Arrange the elements Ba, Mg, and Sr in order of increasing IE_2.
(c) Arrange the elements Br, Te, and I in order of increasing electronegativity.

Chapter 4

38. Use VSEPR theory to determine the shape of each molecule.
(a) SO_2Cl_2 (compound used to make some drugs and polymers)
(b) ClF_3 (oxidizing agent, mentioned in Chapter 10)
(c) XeO_2F_2 (rare stable compound of Xe)

39. Determine whether or not each molecule is polar.
(a) C_2HCl
(b) CF_3Cl
(c) CdI_2
(d) SnO_2

40. Sodium nitrite is used as a preservative in processed meats, such as bacon, ham, and sausage. The nitrite ion, NO_2^-, may enter the ground water in rural areas where manure is stored and be a serious health threat for infants. Draw a Lewis structure, and predict the shape of NO_2^-. Indicate any resonance structures.

41. Trisodium phosphate (TSP), Na_3PO_4, is used as a paint stripper and grease remover. Disodium hydrogen phosphate, Na_2HPO_4, can be found in some laxatives. Draw a Lewis structure for each molecule, and determine the shape of the phosphate ion, PO_4^{3-}, and the hydrogen phosphate ion, HPO_4^{2-}.

42. For a given atom, what is the maximum number of electrons that can have the following quantum numbers?
(a) $n = 3$
(b) $n = 2,\ l = 1,\ m_l = 0,\ m_s = +\frac{1}{2}$
(c) $n = 4,\ l = 2$

43. Comment on the validity of the following statement: "Dispersion forces are weak in comparison to other intermolecular forces."

44. Compare the molecules CH_3F and NH_2F with respect to their
(a) Lewis structures
(b) dispersion forces
(c) hydrogen bonding

45. From the information in the table, classify each compound as metallic, network, ionic, or molecular.

		Electrical conductivity	
Compound	Melting point (°C)	Solid	Liquid
A	848	no	yes
B	2300	no	no
C	1675	yes	yes
D	182	no	no

46. Explain why the melting points of sodium halides (NaCl, NaBr, and NaI) decrease as the molar masses increase, whereas the melting points of hydrogen halides (HCl, HBr, and HI) increase as the molar masses increase.

47. You have used an iodine solution to test for the presence of starch in many laboratory activities. Iodine has a very low solubility in water. When the iodide ion is present, however, the triiodide ion, $I_3^-{}_{(aq)}$, forms. Draw the Lewis structure for $I_3^-{}_{(aq)}$, and predict the shape of this ion.

48. Elements from different groups can form ions that have the same shape. Show that the ions $SbCl_4^+$ and $BeCl_4^{2-}$ have the same shape.

49. Lead(IV) oxide, PbO_2, is used as the cathode in a lead storage battery. Lead(IV) fluoride, PbF_4, is a strong oxidizing agent (Chapter 10). Determine if either of these lead compounds is a polar molecule.

50. Ozone, O_3, is an essential molecule in the ozone layer. It filters out some of the UV radiation that would otherwise destroy life on Earth. At ground level, ozone is a component in photochemical smog. Draw the Lewis structure for ozone, and indicate whether or not ozone is polar. Can ozone exist in resonance forms?

UNIT 3

Chapter 5

51. Using the compound ethanol, C_2H_5OH, write equations to distinguish between ΔH°_f and ΔH_{comb}.

52. Glass has a specific heat capacity of 0.84 J/g°C. A certain metal has a specific heat capacity of 0.500 J/g°C. A metal tray and a glass tray have the same mass. They are placed in an oven at 80°C.
- **(a)** After 1 h, how does the temperature of the two trays compare?
- **(b)** How does the quantity of heat absorbed by the two trays compare?
- **(c)** Does either tray feel hotter to the touch?

53. A 20.00 g sample of metal is warmed to 165°C in an oil bath. The sample is then transferred to a coffee-cup calorimeter that contains 125.0 g of water at 5.0°C. The final temperature of the water is 8.8°C.
- **(a)** Calculate the specific heat capacity of the metal.
- **(b)** What are three sources of experimental error that occur in this experiment?

54. ΔH°_f of $HI_{(g)}$ is + 25.9 kJ.
- **(a)** Write the equation that represents this reaction.
- **(b)** Which has more enthalpy, the elements $H_{2(g)}$ and $I_{2(g)}$, or the product $HI_{(g)}$?
- **(c)** What does the positive value of ΔH°_f indicate about the energy that is needed to break bonds in $H_{2(g)}$ and $I_{2(g)}$, compared with the energy released when H—I bonds form?

55. Use ΔH°_f data to determine the heat of reaction for each reaction.
- **(a)** $4NH_{3(g)} + 3O_{2(g)} \rightarrow 2N_{2(g)} + 6H_2O_{(\ell)}$
- **(b)** $2H_2O_{(g)} + CS_{2(g)} \rightarrow 2H_2S_{(g)} + CO_{2(g)}$
- **(c)** $4NO_{(g)} + 6H_2O_{(g)} \rightarrow 4NH_{3(g)} + 5O_{2(g)}$

56. A 120.0 g sample of water at 30.0°C is placed in the freezer compartment of a refrigerator. How much heat has the sample lost when it changes completely to ice at 0°C?

57. ΔH° is +106.9 kJ for the reaction $C_{(s)} + PbO_{(s)} \rightarrow Pb_{(s)} + CO_{(g)}$. How much heat is needed to convert 50.0 g of $PbO_{(s)}$ to $Pb_{(s)}$?

58. Use the enthalpies of combustion for the burning of $CO_{(g)}$, $H_{2(g)}$, and $C_{(s)}$ to determine ΔH° for the reaction $C_{(s)} + H_2O_{(g)} \rightarrow H_{2(g)} + CO_{(g)}$.

$CO_{(g)} + \frac{1}{2}O_{2(g)} \rightarrow CO_{2(g)}$ $\quad \Delta H^\circ_{comb} = -238$ kJ/mol

$H_{2(g)} + \frac{1}{2}O_{2(g)} \rightarrow H_2O_{(g)}$ $\quad \Delta H^\circ_{comb} = -241$ kJ/mol

$C_{(s)} + O_{2(g)} \rightarrow CO_{2(g)}$ $\quad \Delta H^\circ_{comb} = -393$ kJ/mol

59. ΔH_{comb} for ethene is −337 kJ/mol.
- **(a)** Write the balanced equation for the complete combustion of ethene in air. Include the heat term in your balanced equation.
- **(b)** The heat that is produced by burning 1.00 kg of ethene warms a quantity of water from 15.0°C to 85.0°C. What is the mass of the water if the heat transfer is 60.0% efficient?

60. Determine ΔH° for

$Ca^{2+}_{(aq)} + 2OH^-_{(aq)} + CO_{2(g)} \rightarrow CaCO_{3(s)} + H_2O_{(\ell)}$, given the following information.

$CaO_{(s)} + H_2O_{(\ell)} \rightarrow Ca(OH)_{2(s)}$ $\quad \Delta H^\circ = -65.2$ kJ/mol

$CaCO_{3(s)} \rightarrow CaO_{(s)} + CO_{2(g)}$ $\quad \Delta H^\circ = +178.1$ kJ

$Ca(OH)_{2(s)} \rightarrow Ca^{2+}_{(aq)} + 2OH^-_{(aq)}$ $\quad \Delta H^\circ = -16.2$ kJ

61. Calculate the enthalpy of formation of manganese(IV) oxide, based on the following information.

$4Al_{(s)} + 3MnO_{2(s)} \rightarrow 3Mn_{(s)} + 2Al_2O_{3(s)}$ $\quad \Delta H^\circ = -1790$ kJ

$2Al_{(s)} + \frac{3}{2}O_{2(g)} \rightarrow Al_2O_{3(s)}$ $\quad \Delta H^\circ_f = -1676$ kJ

62. ΔH_{comb} for toluene, $C_7H_{8(\ell)}$, is −3904 kJ/mol.
- **(a)** Write the equation for the complete combustion of toluene.
- **(b)** Use the combustion equation and ΔH°_f values to determine the enthalpy of formation of toluene.

63. An impure sample of zinc has a mass of 7.35 g. The sample reacts with 150.0 g of dilute hydrochloric acid solution inside a calorimeter. The calorimeter has a mass of 520.57 g and a specific heat capacity of 0.400 J/g°C. ΔH°_f for the following reaction is −153.9 kJ.

$Zn_{(s)} + 2HCl_{(aq)} \rightarrow ZnCl_{2(aq)} + H_{2(g)}$

When the reaction occurs, the temperature of the hydrochloric acid rises from 14.5°C to 29.7°C. What is the percentage purity of the sample? Assume that the specific heat capacity of the hydrochloric acid is 4.184 J/g°C. Also assume that all of the zinc in the impure sample reacts.

Chapter 6

64. Butane is burned in a lighter at a rate of 0.240 mol/min in the following reaction.

$C_4H_{10(g)} + \frac{13}{2}O_{2(g)} \rightarrow 4CO_{2(g)} + 5H_2O_{(g)}$
- **(a)** What is the rate at which $O_{2(g)}$ is consumed and $CO_{2(g)}$ is produced?
- **(b)** How long does 15.0 g of $C_4H_{10(g)}$ take to burn?

65. **(a)** The rate law equation for a reaction is rate $= k[A]^2[B]$. How does the rate change if [A] is halved and [B] is doubled?
(b) A reaction is zero order for one reactant and second order for another. What happens to the rate of the reaction if the concentrations of both reactants are doubled?

66. At 22°C, the rate constant is 1.02×10^7 L/mol • s for the following elementary reaction.
$O_{3(g)} + NO_{(g)} \rightarrow O_{2(g)} + NO_{2(g)}$
(a) What is the order of the reaction?
(b) Do you expect the reaction to proceed quickly or slowly at this temperature?
(c) What is the molecularity of the reaction?
(d) Based on the information that has been given, is it reasonable to predict that the rate law equation for this reaction is rate $= k[O_3][NO]$? Explain your answer.

67. Rate data are collected for the following reaction.
$2ClO_{2(aq)} + 2OH^-_{(aq)} \rightarrow ClO_3^-{}_{(aq)} + ClO_2^-{}_{(aq)} + H_2O_{(\ell)}$

$[ClO_{2(aq)}]$ (mol/L)	$[OH^-_{(aq)}]$ (mol/L)	Initial rate (mol/(L • s)
0.060	0.060	2.60×10^{-2}
0.12	0.060	1.04×10^{-1}
0.12	0.12	2.08×10^{-1}

(a) What is the order with respect to each reactant?
(b) What is the overall order of the reaction?
(c) What is the rate law equation for the reaction?
(d) Calculate the rate constant for the reaction.

68. The data below were collected for the following reaction.
$A_{(aq)} + 2B_{(aq)} \rightarrow 2C_{(aq)} + D_{(aq)}$

[A] (mol/L)	[B] (mol/L)	Initial rate (mol/(L • s)
0.050	0.040	6.2×10^{-3}
0.050	0.020	3.1×10^{-3}
0.10	0.080	1.2×10^{-2}

(a) Determine the rate law equation for the reaction.
(b) Determine the rate constant for the reaction.
(c) What is the overall order of the reaction?

69. At a temperature that is suitable for sucrose to be fermented, the concentration of a solution of sucrose changes from 0.20 mol/L to 0.10 mol/L in 12 h. After 24 h, the concentration is reduced to 0.050 mol/L.
(a) What is the order of the reaction?
(b) What is the half-life of the reaction?
(c) What is the rate constant for the reaction?

70. The following reaction is involved in the formation of photochemical smog.
$NO_{(g)} + O_{3(g)} \rightarrow O_{2(g)} + NO_{2(g)}$
The rate constant for the reaction is 1.02×10^7 L/(mol • s) at 22°C. The reaction is first order with respect to $O_{3(g)}$ and $NO_{(g)}$. Calculate the rate at which $NO_{2(g)}$ forms when $[O_{3(g)}] = 1.0 \times 10^{-8}$ mol/L and $[NO_{(g)}] = 3.5 \times 10^{-8}$ mol/L.

71. For reactions in the gaseous phase, a general guideline in chemistry is that the rate of reaction doubles for every 10°C rise in temperature.
(a) A 10°C rise in temperature does not double the kinetic energy of the molecules. Account for the increase in reaction rate.
(b) If the reaction rate does double for every 10°C rise in temperature, how does the rate at 40°C and at 80°C compare with the rate at 20°C?

72. A cube of zinc metal, measuring 1.0 cm on each side, reacts completely with dilute hydrochloric acid in 4 min. An identical cube of zinc is cut into four equal pieces. How long should it take for the pieces of zinc to react?

73. For the following reaction, a chemist determines by experiment that the rate law equation is rate $= k[H_{2(g)}][NO_{(g)}]^2$.
$2H_{2(g)} + 2NO_{(g)} \rightarrow N_{2(g)} + 2H_2O_{(g)}$
The chemist proposes three mechanisms for the reaction. The slow step in each mechanism is listed below. Which of these steps matches the rate law for the reaction? Explain your reasoning.
Slow step for mechanism I:
$2NO_{(g)} \rightarrow N_2O_{2(g)}$
Slow step for mechanism II:
$H_{2(g)} + NO_{(g)} \rightarrow H_2O_{(g)} + N_{(g)}$
Slow step for mechanism III:
$H_{2(g)} + 2NO_{(g)} \rightarrow H_2O_{(g)} + N_2O_{(g)}$

74. The mechanism for the decomposition of ethenal, CH_3CHO, in the presence of iodine vapour is written as follows:
Step 1 $CH_3CHO_{(g)} + I_{2(g)} \rightarrow CH_3I_{(g)} + HI_{(g)} + CO_{(g)}$ (slow)
Step 2 $CH_3I_{(g)} + HI_{(g)} \rightarrow CH_{4(g)} + I_{2(g)}$ (fast)
(a) Identify any intermediates in this reaction mechanism.
(b) Which step has the higher activation energy?
(c) What is the role of $I_{2(g)}$ in this reaction?
(d) What is the overall reaction?

75. A reaction proceeds in three elementary reactions. Sketch a potential energy diagram (using a reasonable scale) to show the following activation energies and enthalpies of reaction:
Step 1 $E_{a1} = 50$ kJ $\Delta H_1 = -30$ kJ
Step 2 $E_{a2} = 30$ kJ $\Delta H_2 = +20$ kJ
Step 3 $E_{a3} = 20$ kJ $\Delta H_3 = -40$ kJ
Mark the positions where activated complexes are expected. What is ΔH_{rxn} for the overall reaction?

UNIT 4

Chapter 7

76. Consider the equilibrium below.
$S_{8(s)} + 8O_{2(g)} \rightleftharpoons 8SO_{2(g)} + \text{heat}$
What is the effect on the concentration of each substance when the equilibrium is altered as follows.
(a) More $S_{8(s)}$ is added.
(b) The temperature is lowered.
(c) The volume of the container is increased.
(d) More SO_2 is injected into the system.

77. At 2000 K, the concentration of the components in the following equilibrium system are
$[CO_{2(g)}] = 0.30\text{mol/L}$, $[H_{2(g)}] = 0.20\text{mol/L}$, and $[H_2O_{(g)}] = [CO_{(g)}] = 0.55\text{mol/L}$.
$CO_{2(g)} + H_{2(g)} \rightleftharpoons H_2O_{(g)} + CO_{(g)}$
(a) What is the value of the equilibrium constant?
(b) When the temperature is lowered, 20.0% of the $CO_{(g)}$ is converted back to $CO_{2(g)}$. Calculate the equilibrium constant at the lower temperature.
(c) Rewrite the equilibrium equation, and indicate on which side of the equation the heat term should be placed.

78. After 5.00 g of SO_2Cl_2 are placed in a 2.00 L flask, the following equilibrium is established.
$SO_2Cl_{2(g)} \rightleftharpoons SO_{2(g)} + Cl_{2(g)}$
K_c for this equilibrium is 0.0410. Determine the mass of $SO_2Cl_{2(g)}$ that is present at equilibrium.

79. For the equilibrium below, $K_c = 6.00 \times 10^{-2}$.
$N_{2(g)} + 3H_{2(g)} \rightleftharpoons 2NH_{3(g)}$
Explain why this value of K_c does not apply when the equation is written as follows.
$\frac{1}{2}N_{2(g)} + \frac{3}{2}H_{2(g)} \rightleftharpoons NH_{3(g)}$

80. A 11.5 g sample of $I_{2(g)}$ is sealed in a 250 mL flask. An equilibrium, shown below, is established as this molecular form of I_2 dissociates into iodine atoms.
$I_{2(g)} \rightleftharpoons 2I_{(g)}$
K_c for the equilibrium is 3.80×10^{-5}. Calculate the equilibrium concentration of both forms of the iodine.

81. The greenhouse effect warms the surface of Earth. The increase of $CO_{2(g)}$ in the environment is one of the factors that contributes to this process. Many older buildings are made of limestone, $CaCO_{3(s)}$. Consider the following equilibrium.
$CaCO_{3(s)} \rightleftharpoons CaO_{(s)} + CO_{2(g)}\ \Delta H = +175\text{ kJ}$
(a) In theory, how can the greenhouse effect affect this equilibrium?
(b) Discuss the likelihood that this equilibrium can be altered by the greenhouse effect.

82. When iron rusts, is entropy increasing or decreasing?
$4Fe_{(s)} + 3O_{2(g)} \rightarrow 2Fe_2O_{3(s)}$
Why is the rusting of iron a spontaneous reaction at room temperature?

Chapter 8

83. A blood sample has a pH of 7.32. What is the hydronium ion concentration, $[H_3O^+]$? How can hydronium ions exist in this basic solution?

84. Hydrated metal ions act as weak acids by donating a proton. For example, Fe^{2+} becomes hydrated with six water molecules, $[Fe(OH_2)_6]^{2+}$. Write an equation to illustrate how the hydrated iron(II) ion acts as an acid in aqueous solution.

85. Imagine that you are asked to design a chemistry kit for young children to use. In one of the experiments, the properties of acids and bases are studied. What chemicals are appropriate for this experiment? What safety precautions would you include?

86. A toilet bowl cleaner contains $NaHSO_4$ as the active ingredient.
(a) Write an equation to show how this salt produces an acidic solution.
(b) What is the pH of a 0.203 mol/L solution of $NaHSO_4$?

87. A solution is made by combining 200.0 mL of 0.23 mol/L H_2SO_4, 600.0 mL of 0.16 mol/L KOH, and 200.0 mL of water. What is the pH of this solution?

88. K_a for CH_3COOH is 1.8×10^{-5}, and K_a for HNO_2 is 7.2×10^{-4}. When the conjugate bases of these two acids are compared, which has the larger K_b? What is the value of this larger K_b?

89. K_a for hydrazoic acid, HN_3, is 2.80×10^{-5}.
(a) Write the equation for the ionization of this acid in water.
(b) What is the K_b for the conjugate base of this acid?
(c) Compare the pH of 0.100 mol/L HN_3 with the pH of a sample of the same volume in which 0.600 g of sodium azide, NaN_3, has been dissolved.

90. A 250.0 mL sample of 0.100 mol/L $NaHCO_3$ contains 3.800 g of Na_2CO_3. What is the pH of the sample?

91. 10.0 mL of $NH_{3(aq)}$, with a concentration of 5.70×10^{-2} mol/L, is titrated to the endpoint with 2.85 mL of HBr solution. What is the concentration of the HBr solution?

92. Name the conjugate base and the conjugate acid of HPO_4^{2-}. What determines if HPO_4^{2-} will act as an acid or a base?

Chapter 9

93. K_{sp} for silver chromate, Ag_2CrO_4, is 1.12×10^{-12}. K_{sp} for silver chloride, AgCl, is 1.77×10^{-10}. Does this mean that the molar solubility of Ag_2CrO_4 is less than the molar solubility of AgCl? Use a calculation to illustrate your answer.

94. The solubility of a compound can be affected by the presence of a common ion.
(a) What does this statement mean?
(b) Compare the effect of dissloved Pb^{2+} to the effect of dissolved OH^- on the solubility of $Pb(OH)_2$.

95. What mass of NaOH is needed to start the precipitation of $Mg(OH)_2$ from 100 mL of a solution that contains 0.100 g of $MgCl_2$?

96. A warm saturated solution of $SrSO_4$, with a volume of 850 mL, is evaporated to dryness. The residue that remains has a mass of 0.160 g.
   (a) Use this information to calculate the K_{sp} for $SrSO_4$.
   (b) A reference book gives a K_{sp} value of 2.8×10^{-7}. What could account for the difference between the value in the reference book and the experimental value for this K_{sp}?

97. 0.800 L of 2.00×10^{-4} mol/L $Ba(NO_3)_2$ is added to 200.0 mL of 5.00×10^{-4} mol/L Li_2SO_4. Show, by calculation, whether or not a precipitate forms.

98. A radioactive form of barium, in the compound $BaSO_4$, is used to diagnose intestinal ailments. Why is a more soluble compound, such as $BaCl_2$, not used in this procedure?

99. A solution has a lead concentration, $[Pb^{2+}{}_{(aq)}]$, of 0.100 mol/L. 0.100 mol of NaI are added to 1.00 L of this solution. How much of the original $Pb^{2+}{}_{(aq)}$ precipitates?

100. In many parts of Ontario, the water is "hard" because of the presence of Ca^{2+} and Mg^{2+} ions.
   (a) Does hardness affect the F^- content of a fluoridated municipal water supply?
   (b) What is the maximum Ca^{2+} concentration that can be present if the F^- concentration is 80 ppm?

UNIT 5

Chapter 10

101. Identify the redox reactions.
   (a) $CCl_4 + HF \rightarrow CFCl_3 + HCl$
   (b) $Al_2O_3 + 3H_2SO_4 \rightarrow Al_2(SO_4)_3 + 3H_2O$
   (c) $CH_4 + 2O_2 \rightarrow CO_2 + 2H_2O$
   (d) $P_4 + 3OH^- + 3H_2O \rightarrow PH_3 + 3H_2PO_2^-$

102. Consider the following reaction.
   $S_8 + 8Na_2SO_3 \rightarrow 8Na_2S_2O_3$
   (a) Assign oxidation numbers to all the elements.
   (b) Identify the reactant that undergoes reduction.
   (c) Identify the reactant that is the reducing agent.

103. Find the oxidation number for each element.
   (a) sulfur in HS^-, S_8, SO_3^{2-}, $S_2O_3^{2-}$, and $S_4O_6^{2-}$
   (b) boron in $B_4O_7^{2-}$, BO_3^-, BO_2^-, B_2H_6, and B_2O_3

104. List the following oxides of nitrogen in order of decreasing oxidation number of nitrogen: NO_2, N_2O_5, NO, N_2O_3, N_2O, N_2O_4.

105. For the following redox reaction, indicate which statements, if any, are true.
   $C + D \rightarrow E + F$
   (a) If C is the oxidizing agent, then it loses electrons.
   (b) If D is the reducing agent, then it is reduced.
   (c) If C is the reducing agent, and if it is an element, then its oxidation number will increase.
   (d) If D is oxidized, then C must be a reducing agent.
   (e) If C is reduced, then D must lose electrons.

106. Consider the following reaction.
   $3SF_4 + 4BCl_3 \rightarrow 4BF_3 + 3SCl_2 + 3Cl_2$
   (a) Why is this reaction classified as a redox reaction?
   (b) What is the oxidizing agent in this reaction?

107. The metals $Ga_{(s)}$, $In_{(s)}$, $Mn_{(s)}$, and $Np_{(s)}$, and their salts, react as follows:
   $3Mn^{2+}{}_{(aq)} + 2Np_{(s)} \rightarrow 3Mn_{(s)} + 2Np^{3+}{}_{(aq)}$
   $In^{3+}{}_{(aq)} + Ga_{(s)} \rightarrow In_{(s)} + Ga^{3+}{}_{(aq)}$
   $Mn^{2+} + Ga_{(s)} \rightarrow$ no reaction
   Analyze this information, and list the reducing agents from the worst to the best.

108. The following redox reactions occur in basic solution. Balance the equations using the oxidation number method.
   (a) $Ti^{3+} + RuCl_5^{2-} \rightarrow Ru + TiO^{2+} + Cl^-$
   (b) $ClO_2 \rightarrow ClO_2^- + ClO_3^-$

109. The following redox reactions occur in basic solution. Use the half-reaction method to balance the equations.
   (a) $NO_2 \rightarrow NO_2^- + NO_3^-$
   (b) $CrO_4^- + HSnO_2^- \rightarrow CrO_2^- + HSnO_3^-$
   (c) $Al + NO_3^- \rightarrow Al(OH)_4^- + NH_3$

110. The following redox reactions occur in acidic solution. Balance the equations using the half-reaction method.
   (a) $ClO_3^- + I_2 \rightarrow IO_3^- + Cl^-$
   (b) $C_2H_4 + MnO_4^- \rightarrow CO_2 + Mn^{2+}$
   (c) $Cu + SO_4^{2-} \rightarrow Cu^{2+} + SO_2$

111. The following redox reactions occur in acidic solution. Use oxidation numbers to balance the equations.
   (a) $Se + NO_3^- \rightarrow SeO_2 + NO$
   (b) $Ag + Cr_2O_7^{2-} \rightarrow Ag^+ + Cr^{3+}$

112. Balance each equation for a redox reaction.
   (a) $P_4 + NO_3^- \rightarrow H_3PO_4 + NO$ (acidic)
   (b) $MnO_2 + NO_2^- \rightarrow NO_3^- + Mn^{2+}$ (acidic)
   (c) $TeO_3^{2-} + N_2O_4 \rightarrow Te + NO_3^-$ (basic)
   (d) $MnO_4^- + N_2H_4 \rightarrow MnO_2 + N_2$ (basic)
   (e) $S_2O_3^{2-} + OCl^- \rightarrow SO_4^{2-} + Cl^-$ (basic)
   (f) $Br_2 + SO_2 \rightarrow Br^- + SO_4^{2-}$ (acidic)
   (g) $PbO_2 + Cl^- \rightarrow PbCl_2 + Cl_2$ (acidic)
   (h) $MnO_4^- + H_2O_2 \rightarrow Mn^{2+} + O_2$ (acidic)

113. The following reaction takes place in acidic solution.
   $CH_3OH_{(aq)} + MnO_4^-{}_{(aq)} \rightarrow HCOOH_{(aq)} + Mn^{2+}{}_{(aq)}$
   (a) Balance the equation.
   (b) How many electrons are gained by the oxidizing agent when ten molecules of methanol are oxidized?
   (c) What volume of 0.150 mol/L MnO_4^- solution is needed to react completely with 20.0 g of methanol?

Chapter 11

114. Consider the galvanic cell represented as $Al|Al^{3+}||Ce^{4+}|Ce^{3+}|Pt$.

(a) What is the cathode?

(b) In which direction do electrons flow through the external circuit?

(c) What is the oxidizing agent?

(d) Will the aluminum electrode increase or decrease in mass as the cell operates?

115. (a) Which is a better oxidizing agent in acidic solution, MnO_4^- or $Cr_2O_7^{2-}$? From what information did you determine your answer?

(b) Suppose that you have the metals Ni, Cu, Fe, and Ag, as well as 1.0 mol/L aqueous solutions of the nitrates of these metals. Which metals should be paired in a galvanic cell to produce the highest standard cell potential? (Use the most common ion for each metal.)

116. E°_{cell} for the cell $No|No^{3+}||Cu^{2+}|Cu$ is 2.842 V. Use this information, as well as the standard reduction potential for the Cu^{2+}/Cu half-reaction given in tables, to calculate the standard reduction potential for the No/No^{3+} half-cell.

117. A galvanic cell is set up using tin in a 1.0 mol/L Sn^{2+} solution and iron in a 1.0 mol/L Fe^{2+} solution.

(a) Write the equation for the overall reaction that occurs in this cell.

(b) What is the standard cell potential?

(c) Which electrode is positive in this cell?

(d) What change in mass will occur at the anode when the cathode undergoes a change in mass of 1.50 g?

118. Predict the products that would be expected from the electrolysis of 1.0 mol/L NaI. Use the non-standard E values for water.

119. Given the half reactions below, determine if the thiosulfate ion, $S_2O_3^{2-}$, can exist in an acidic solution under standard conditions.

$S_2O_3^{2-} + H_2O \rightarrow 2SO_2 + 2H^+ + 4e^-$ $E^\circ = 0.40$ V

$2S + H_2O \rightarrow S_2O_3^{2-} + 6H^+ + 4e^-$ $E^\circ = -0.50$ V

120. Refer to the table of half-cell potentials to determine if MnO_2 can oxidize Br^- to Br_2 in acidic solution under standard conditions.

121. A galvanic cell contains 50.0 mL of 0.150 mol/L $CuSO_4$. If the Cu^{2+} ions are completely used up, what is the maximum quantity of electricity that the cell can generate?

122. Determine the standard cell potential for each redox reaction.

(a) $IO_3^- + 6ClO_2^- + 6H^+ \rightarrow I^- + 3H_2O + 6ClO_2$

(b) $2Cu^{2+} + Hg_2^{2+} \rightarrow 2Hg^{2+} + 2Cu^+$

(c) $Ba^{2+} + Pb \rightarrow Pb^{2+} + Ba$

(d) $Ni + I_2 \rightarrow Ni^{2+} + 2I^-$

123. A current of 3.0 A flows for 1.0 h during an electrolysis of copper(II) sulfate. What mass of copper is deposited?

124. To recover aluminum metal, Al_2O_3 is first converted to $AlCl_3$. Then an electrolysis of molten $AlCl_3$ is performed using inert carbon electrodes.

(a) Write the half-reaction that occurs at the anode and at the cathode.

(b) Can the standard reduction potentials be used to calculate the external voltage needed for this process? Explain your answer.

125. Summarize the differences between a hydrogen fuel cell and a dry cell.

Alphabetical List of Elements

Element	Symbol	Atomic Number	Element	Symbol	Atomic Number
Actinium	Ac	89	Neodymium	Nd	60
Aluminum	Al	13	Neon	Ne	10
Americium	Am	95	Neptunium	Np	93
Antimony	Sb	51	Nickel	Ni	28
Argon	Ar	18	Niobium	Nb	41
Arsenic	As	33	Nitrogen	N	7
Astatine	At	85	Nobelium	No	102
Barium	Ba	56	Osmium	Os	76
Berkelium	Bk	97	Oxygen	O	8
Beryllium	Be	4	Palladium	Pd	46
Bismuth	Bi	83	Phosphorus	P	15
Bohrium	Bh	107	Platinum	Pt	78
Boron	B	5	Plutonium	Pu	94
Bromine	Br	35	Polonium	Po	84
Cadmium	Cd	48	Potassium	K	19
Calcium	Ca	20	Praseodymium	Pr	59
Californium	Cf	98	Promethium	Pm	61
Carbon	C	6	Protactinium	Pa	91
Cerium	Ce	58	Radium	Ra	88
Cesium	Cs	55	Radon	Rn	86
Chlorine	Cl	17	Rhenium	Re	75
Chromium	Cr	24	Rhodium	Rh	45
Cobalt	Co	27	Rubidium	Rb	37
Copper	Cu	29	Ruthenium	Ru	44
Curium	Cm	96	Rutherfordium	Rf	104
Dubnium	Db	105	Samarium	Sm	62
Dysprosium	Dy	66	Scandium	Sc	21
Einsteinium	Es	99	Seaborgium	Sg	106
Erbium	Er	68	Selenium	Se	34
Europium	Eu	63	Silicon	Si	14
Fermium	Fm	100	Silver	Ag	47
Fluorine	F	9	Sodium	Na	11
Francium	Fr	87	Strontium	Sr	38
Gadolinium	Gd	64	Sulfur	S	16
Gallium	Ga	31	Tantalum	Ta	73
Germanium	Ge	32	Technetium	Tc	43
Gold	Au	79	Tellurium	Te	52
Hafnium	Hf	72	Terbium	Tb	65
Hassium	Hs	108	Thallium	Tl	81
Helium	He	2	Thorium	Th	90
Holmium	Ho	67	Thulium	Tm	69
Hydrogen	H	1	Tin	Sn	50
Indium	In	49	Titanium	Ti	22
Iodine	I	53	Tungsten	W	74
Iridium	Ir	77	Ununbium	Uub	112
Iron	Fe	26	Ununhexium	Uuh	116
Krypton	Kr	36	Ununnilium	Uun	110**
Lanthanum	La	57	Ununquadium	Uuq	114
Lawrencium	Lr	103	Unununium	Uuu	111
Lead	Pb	82	Uranium	U	92
Lithium	Li	3	Vanadium	V	23
Lutetium	Lu	71	Xenon	Xe	54
Magnesium	Mg	12	Ytterbium	Yb	70
Manganese	Mn	25	Yttrium	Y	39
Meitnerium	Mt	109	Zinc	Zn	30
Mendelevium	Md	101	Zirconium	Zr	40
Mercury	Hg	80			
Molybdenum	Mo	42			

**The names and symbols for elements 110 through 118 have not yet been chosen

Periodic Table of the Elements

6 12.01 2.5 +4 1086 +2 4765 4098 **C** carbon	
Atomic number	6
Average atomic mass*	12.01
Electronegativity	2.5
Common ion charge	+4
First ionization energy (kJ/mol)	1086
Other ion charges	+2
Melting point (K)	4765
Boiling point (K)	4098

- metals (main group)
- metals (transition)
- metals (inner transition)
- metalloids
- nonmetals

- Gases
- Liquids
- Synthetics

MAIN-GROUP ELEMENTS

TRANSITION ELEMENTS

Period	1 (IA)	2 (IIA)	3 (IIIB)	4 (IVB)	5 (VB)	6 (VIB)	7 (VIIB)	8 (VIIIB)	9 (VIIIB)
1	**1** 1.01 2.20 1+ 1312 1- 13.81 20.28 **H** hydrogen								
2	**3** 6.94 0.98 1+ 520 453.7 1615 **Li** lithium	**4** 9.01 1.57 2+ 899 1560 2744 **Be** beryllium							
3	**11** 22.99 0.93 1+ 496 371 1156 **Na** sodium	**12** 24.31 1.31 2+ 738 923.2 1363 **Mg** magnesium							
4	**19** 39.10 0.82 1+ 419 336.7 1032 **K** potassium	**20** 40.08 1.00 2+ 590 1115 1757 **Ca** calcium	**21** 44.96 1.36 3+ 631 1814 3109 **Sc** scandium	**22** 47.87 1.54 4+ 658 2+ 1941 3+ 3560 **Ti** titanium	**23** 50.94 1.63 5+ 650 2+ 2183 3+ 3680 4+ **V** vanadium	**24** 52.00 1.66 3+ 653 2+ 2180 6+ 2944 **Cr** chromium	**25** 54.94 1.55 4+ 717 2+ 1519 3+ 2334 7+ **Mn** manganese	**26** 55.85 1.83 3+ 759 2+ 1811 3134 **Fe** iron	**27** 58.93 1.88 2+ 760 3+ 1768 3200 **Co** cobalt
5	**37** 85.47 0.82 1+ 403 312.5 941.2 **Rb** rubidium	**38** 87.62 0.95 2+ 549 1050 1655 **Sr** strontium	**39** 88.91 1.22 3+ 616 1795 3618 **Y** yttrium	**40** 91.22 1.33 4+ 660 2128 4682 **Zr** zirconium	**41** 92.91 1.6 5+ 664 3+ 2750 5017 **Nb** niobium	**42** 95.94 2.16 6+ 685 2896 4912 **Mo** molybdenum	**43** (98) 2.10 7+ 702 4+ 2430 6+ 4538 **Tc** technetium	**44** 101.07 2.2 3+ 711 2607 4423 **Ru** ruthenium	**45** 102.91 2.28 3+ 720 2237 3968 **Rh** rhodium
6	**55** 132.91 0.79 1+ 376 301.7 944 **Cs** cesium	**56** 137.33 0.89 2+ 503 1000 2170 **Ba** barium	**57** 138.91 1.10 3+ 538 1191 3737 **La** lanthanum	**72** 178.49 1.3 4+ 642 2506 4876 **Hf** hafnium	**73** 180.95 1.5 5+ 761 3290 5731 **Ta** tantalum	**74** 183.84 1.7 6+ 770 3695 5828 **W** tungsten	**75** 186.21 1.9 4+ 760 6+ 3459 7+ 5869 **Re** rhenium	**76** 190.23 2.2 4+ 840 3+ 3306 5285 **Os** osmium	**77** 192.22 2.2 4+ 880 3+ 2719 4701 **Ir** iridium
7	**87** (223) 0.7 1+ ~375 300.2 – **Fr** francium	**88** (226) 0.9 2+ 509 973.2 – **Ra** radium	**89** (227) 1.1 3+ 499 1324 3471 **Ac** actinium	**104** (261) – 4+ – – – **Rf** rutherfordium	**105** (262) – – – – – **Db** dubnium	**106** (266) – – – – – **Sg** seaborgium	**107** (264) – – – – – **Bh** bohrium	**108** (269) – – – – – **Hs** hassium	**109** (268) – – – – – **Mt** meitnerium

INNER TRANSITION ELEMENTS

Period								
6	Lanthanoids	**58** 140.12 1.12 3+ 527 4+ 1071 3716 **Ce** cerium	**59** 140.91 1.13 3+ 523 1204 3793 **Pr** praseodymium	**60** 144.24 1.14 3+ 530 1294 3347 **Nd** neodymium	**61** (145) – 3+ 536 1315 3273 **Pm** promethium	**62** 150.36 1.17 3+ 543 2+ 1347 2067 **Sm** samarium	**63** 151.96 – 3+ 547 2+ 1095 1802 **Eu** europium	**64** 157.25 1.20 3+ 593 1586 3546 **Gd** gadolinium
7	Actinoids	**90** 232.04 1.3 4+ 587 2023 5061 **Th** thorium	**91** 231.04 1.5 5+ 568 4+ 1845 – **Pa** protactinium	**92** 238.03 1.7 6+ 584 3+ 1408 4+ 4404 5+ **U** uranium	**93** 237.05 1.3 5+ 597 3+ 917 4+ – 6+ **Np** neptunium	**94** (244) 1.3 4+ 585 3+ 913.2 5+ 3501 6+ **Pu** plutonium	**95** (243) – 3+ 578 4+ 1449 5+ 2284 6+ **Am** americium	**96** (247) – 3+ 581 1618 3373 **Cm** curium

*Average atomic mass data in brackets indicate atomic mass of most stable isotope of the element.

MAIN-GROUP ELEMENTS

10	11 (IB)	12 (IIB)	13 (IIIA)	14 (IVA)	15 (VA)	16 (VIA)	17 (VIIA)	18 (VIIIA)
								2 4.00 – – 2372 5.19 5.02 **He** helium
			5 10.81 2.04 – 800 2348 4273 **B** boron	**6** 12.01 2.55 – 1086 4765 4098 **C** carbon	**7** 14.01 3.04 3– 1402 63.15 77.36 **N** nitrogen	**8** 16.00 3.44 2– 1314 54.36 90.2 **O** oxygen	**9** 19.00 3.98 1– 1681 53.48 84.88 **F** fluorine	**10** 20.18 – – 2080 24.56 27.07 **Ne** neon
			13 26.98 1.61 – 577 933.5 2792 **Al** aluminum	**14** 28.09 1.90 – 786 1687 3538 **Si** silicon	**15** 30.97 2.19 – 1012 317.3 553.7 **P** phosphorus	**16** 32.07 2.58 2– 999 392.8 717.8 **S** sulfur	**17** 35.45 3.16 1– 1256 171.7 239.1 **Cl** chlorine	**18** 39.95 – – 1520 83.8 87.3 **Ar** argon
28 58.69 1.91 2+ 737 3+ 1728 3186 **Ni** nickel	**29** 63.55 1.90 2+ 745 1+ 1358 2835 **Cu** copper	**30** 65.39 1.65 2+ 906 692.7 1180 **Zn** zinc	**31** 69.72 1.81 3+ 579 302.9 2477 **Ga** gallium	**32** 72.61 2.01 – 761 1211 3106 **Ge** germanium	**33** 74.92 2.18 – 947 1090 876.2 **As** arsenic	**34** 78.96 2.55 2– 941 493.7 958.2 **Se** selenium	**35** 79.90 2.96 1– 1143 266 332 **Br** bromine	**36** 83.80 – – 1351 115.8 119.9 **Kr** krypton
46 106.42 2.20 2+ 805 3+ 1828 3236 **Pd** palladium	**47** 107.87 1.93 1+ 731 1235 2435 **Ag** silver	**48** 112.41 1.69 2+ 868 594.2 1040 **Cd** cadmium	**49** 114.82 1.78 3+ 558 429.8 3345 **In** indium	**50** 118.71 1.96 4+ 708 2+ 505 2876 **Sn** tin	**51** 121.76 2.05 – 834 903.8 1860 **Sb** antimony	**52** 127.60 2.1 – 869 722.7 1261 **Te** tellurium	**53** 126.90 2.66 1– 1009 386.9 457.4 **I** iodine	**54** 131.29 – – 1170 161.4 165 **Xe** xenon
78 195.08 2.2 4+ 870 2+ 2042 4098 **Pt** platinum	**79** 196.97 2.4 3+ 890 1+ 1337 3129 **Au** gold	**80** 200.59 1.9 2+ 1107 1+ 234.3 629.9 **Hg** mercury	**81** 204.38 1.8 1+ 589 3+ 577.2 1746 **Tl** thallium	**82** 207.20 1.8 2+ 715 4+ 600.6 2022 **Pb** lead	**83** 208.98 1.9 3+ 703 5+ 544.6 1837 **Bi** bismuth	**84** (209) 2.0 4+ 813 2+ 527.2 1235 **Po** polonium	**85** (210) 2.2 1– (926) 575 – **At** astatine	**86** (222) – – 1037 202.2 211.5 **Rn** radon
110 (271) – – – – – **Uun** ununnilium	**111** (272) – – – – – **Uuu** unununium	**112** (277) – – – – – **Uub** ununbium	**113**	**114** (285) – – – – – **Uuq** ununquadium	**115**	**116** (289) – – – – – **Uuh** ununhexium		

65 158.93 – 3+ 565 1629 3503 **Tb** terbium	**66** 162.50 1.22 3+ 572 1685 2840 **Dy** dysprosium	**67** 164.93 1.23 3+ 581 1747 2973 **Ho** holmium	**68** 167.26 1.24 3+ 589 1802 3141 **Er** erbium	**69** 168.93 1.25 3+ 597 1818 2223 **Tm** thulium	**70** 173.04 – 3+ 603 2+ 1092 1469 **Yb** ytterbium	**71** 174.97 1.0 3+ 524 1936 3675 **Lu** lutetium
97 (247) – 3+ 601 4+ 1323 – **Bk** berkelium	**98** (251) – 3+ 608 1173 – **Cf** californium	**99** (252) – 3+ 619 1133 – **Es** einsteinium	**100** (257) – 3+ 627 1800 – **Fm** fermium	**101** (258) – 3+ 635 2+ 1100 – **Md** mendelevium	**102** (259) – 3+ 642 2+ 1100 – **No** nobelium	**103** (262) – 3+ – 1900 – **Lr** lawrencium

Appendix D

Math and Chemistry

Significant Digits

All measurements involve uncertainty. One source of this uncertainty is the measuring device itself. Another source is your ability to perceive and interpret a reading. In fact, you cannot measure anything with complete certainty. The last (farthest right) digit in any measurement is always an estimate.

The digits that you record when you measure something are called *significant digits*. Significant digits include the digits that you are certain about, *and* a final, uncertain digit that you estimate. Follow the rules below to identify the number of significant digits in a measurement.

Rules for Determining Significant Digits

Rule 1 All non-zero numbers are significant.
- 7.886 has four significant digits.
- 19.4 has three significant digits.
- 527.266 992 has nine significant digits.

Rule 2 All zeros that are located between two non-zero numbers are significant.
- 408 has three significant digits.
- 25 074 has five significant digits.

Rule 3 Zeros that are located to the left of a measurement are *not* significant.
- 0.0907 has three significant digits: the 9, the third 0 to the right, and the 7.

Rule 4 Zeros that are located to the right of a measurement may or may not be significant.
- 22 700 may have three significant digits, if the measurement is approximate.
- 22 700 may have five significant digits, if the measurement is taken carefully.

When you take measurements and use them to calculate other quantities, you must be careful to keep track of which digits in your calculations and results are significant. Why? Your results should not imply more certainty than your measured quantities justify. This is especially important when you use a calculator. Calculators usually report results with far more digits than your data warrant. Always remember that calculators do not make decisions about certainty. You do. Follow the rules given below to report significant digits in a calculated answer.

Rules for Reporting Significant Digits in Calculations

Rule 1 Multiplying and Dividing
The value with the fewest number of significant digits, going into a calculation, determines the number of significant digits that you should report in your answer.

Rule 2 Adding and Subtracting
The value with the fewest number of decimal places, going into a calculation, determines the number of decimal places that you should report in your answer.

Rule 3 Rounding
To get the appropriate number of significant digits (rule 1) or decimal places (rule 2), you may need to round your answer.
- If your answer ends in a number that is greater than 5, increase the preceding digit by 1. For example, 2.346 can be rounded to 2.35.
- If your answer ends with a number that is less than 5, leave the preceding number unchanged. For example, 5.73 can be rounded to 5.7.
- If your answer ends with 5, increase the preceding number by 1 if it is odd. Leave the preceding number unchanged if it is even. For example, 18.35 can be rounded to 18.4, but 18.25 is rounded to 18.2.

Sample Problem

Using Significant Digits

Problem

Suppose that you measure the masses of four objects as 12.5 g, 145.67 g, 79.0 g, and 38.438 g. What is the total mass?

What Is Required?

You need to calculate the total mass of the objects.

What Is Given?

You know the mass of each object.

Plan Your Strategy

- Add the masses together, aligning them at the decimal point.
- Underline the estimated (farthest right) digit in each value. This is a technique you can use to help you keep track of the number of estimated digits in your final answer.
- In the question, two values have the fewest decimal places: 12.5 and 79.0. You need to round your answer so that it has only one decimal place.

Act on Your Strategy

$$
\begin{array}{r}
12.\underline{5} \\
145.6\underline{7} \\
79.\underline{0} \\
+\ 38.43\underline{8} \\
\hline
275.\underline{608}
\end{array}
$$

Total mass = 275.608 g
Therefore, the total mass of the objects is 275.6 g.

Check Your Solution

- Your answer is in grams. This is a unit of mass.
- Your answer has one decimal place. This is the same as the values in the question with the fewest decimal places.

Practice Problems

Significant Digits

1. Express each answer using the correct number of significant digits.
   a) 55.671 g + 45.78 g
   b) 1.9 mm + 0.62 mm
   c) 87.9478 L − 86.25 L
   d) 0.350 mL + 1.70 mL + 1.019 mL
   e) 5.841 cm × 6.03 cm
   f) $\frac{17.51\text{ g}}{2.2\text{ cm}^3}$

Scientific Notation

One mole of water, H_2O, contains 602 214 199 000 000 000 000 000 molecules. Each molecule has a mass of 0.000 000 000 000 000 000 000 029 9 g. As you can see, it would be very awkward to calculate the mass of one mole of water using these values. To simplify large numbers when reporting them and doing calculations, you can use scientific notation.

Step 1 Move the decimal point so that only one non-zero digit is in front of the decimal point. (Note that this number is now between 1.0 and 9.99999999.) Count the number of places that the decimal point moves to the left or to the right.

Step 2 Multiply the value by a power of 10. Use the number of places that the decimal point moved as the exponent for the power of 10. If the decimal point moved to the right, exponent is negative. If the decimal point moved to the left, the exponent is positive.

6.02 000 000 000 000 000 000 000.

23 21 18 15 12 9 6 3

6.02×10^{23}

Figure D.1 The decimal point moves to the left.

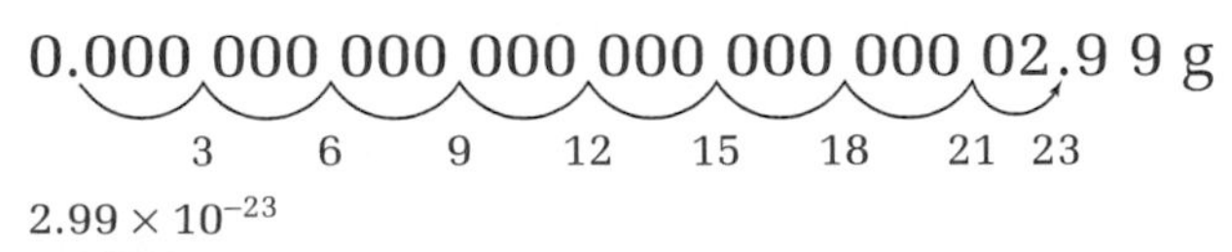

2.99×10^{-23}

Figure D.2 The decimal point moves to the right.

Figure D.3 shows you how to calculate the mass of one mole of water using a scientific calculator. When you enter an exponent on a scientific calculator, you do not have to enter (×10).

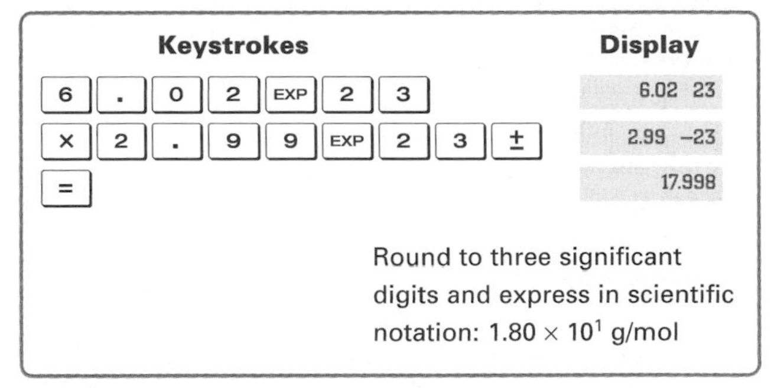

Keystrokes
Display
6 . 0 2 EXP 2 3
6.02 23
× 2 . 9 9 EXP 2 3 ±
2.99 −23
=
17.998
Round to three significant digits and express in scientific notation: 1.80 × 10¹ g/mol

Figure D.3 On some scientific calculators, the EXP key is labelled EE. Key in negative exponents by entering the exponent, then striking the ± key.

Rules for Scientific Notation

Rule 1 To multiply two numbers in scientific notation, add the exponents.

$(7.32 \times 10^{-3}) \times (8.91 \times 10^{-2})$
$= (7.32 \times 8.91) \times 10^{(-3 + -2)}$
$= 65.2212 \times 10^{-5}$
$\rightarrow 6.52 \times 10^{-4}$

Rule 2 To divide two numbers in scientific notation, subtract the exponents.

$(1.842 \times 10^{6}\text{ g}) \div (1.0787 \times 10^{2}\text{ g/mol})$
$= (1.842 \div 1.0787) \times 10^{(6-2)}$
$= 1.707611 \times 10^{4}\text{ g}$
$\rightarrow 1.708 \times 10^{4}\text{ g}$

Rule 3 To add or subtract numbers in scientific notation, first convert the numbers so they have the same exponent. Each number should have the same exponent as the number with the greatest power of 10. Once the numbers are all expressed to the same power of 10, the power of 10 is neither added nor subtracted in the calculation.

$(3.42 \times 10^{6}\text{ cm}) + (8.53 \times 10^{3}\text{ cm})$
$= (3.42 \times 10^{6}\text{ cm}) + (0.00853 \times 10^{6}\text{ cm})$
$= 3.42853 \times 10^{6}\text{ cm}$
$\rightarrow 3.43 \times 10^{6}\text{ cm}$

$(9.93 \times 10^{1}\text{ L}) - (7.86 \times 10^{-1}\text{ L})$
$= (9.93 \times 10^{1}\text{ L}) - (0.0786 \times 10^{1}\text{ L})$
$= 9.8514 \times 10^{1}\text{ L}$
$\rightarrow 9.85 \times 10^{1}\text{ L}$

Practice problems are given on the following page.

Practice Problems

Scientific Notation

1. Convert each value into correct scientific notation.
   (a) 0.000 934
   (b) 7 983 000 000
   (c) 0.000 000 000 820 57
   (d) 496×10^6
   (e) $0.000\,06 \times 10^1$
   (f) $309\,72 \times 10^{-8}$
2. Add, subtract, multiply, or divide. Round off your answer, and express it in scientific notation to the correct number of significant digits.
   (a) $(3.21 \times 10^{-3}) + (9.2 \times 10^2)$
   (b) $(8.1 \times 10^3) + (9.21 \times 10^2)$
   (c) $(1.010\,1 \times 10^1) - (4.823 \times 10^{-2})$
   (d) $(1.209 \times 10^6) \times (8.4 \times 10^7)$
   (e) $(4.89 \times 10^{-4}) \div (3.20 \times 10^{-2})$

Logarithms

Logarithms are a convenient method for communicating large and small numbers. The *logarithm*, or "log," of a number is the value of the exponent that 10 would have to be raised to, in order to equal this number. Every positive number has a logarithm. Numbers that are greater than 1 have a positive logarithm. Numbers that are between 0 and 1 have a negative logarithm. Table D1 gives some examples of the logarithm values of numbers.

Table D.1 Some Numbers and Their Logarithms

Number	Scientific notation	As a power of 10	Logarithm
1 000 000	1×10^6	10^6	6
7 895 900	7.8590×10^5	$10^{5.8954}$	5.8954
1	1×10^0	10^0	0
0.000 001	1×10^{-6}	10^{-6}	−6
0.004 276	4.276×10^{-3}	$10^{-2.3690}$	−2.3690

Logarithms are especially useful for expressing values that span a range of powers of 10. The Richter scale for earthquakes, the decibel scale for sound, and the pH scale for acids and bases all use logarithmic scales.

Logarithms and pH

The pH of an acid solution is defined as $-\log[H_3O^+]$. (The square brackets mean "concentration.") For example, suppose that the hydronium ion concentration in a solution is 0.0001 mol/L (10^{-4} mol/L). The pH is −log(0.0001). To calculate this, enter 0.0001 into your calculator. Then press the [LOG] key. Press the [±] key. The answer in the display is 4. Therefore, the pH of the solution is 4.

There are logarithms for all numbers, not just whole multiples of 10. What is the pH of a solution if $[H_3O^+]$ = 0.004 76 mol/L? Enter 0.00476. Press the [LOG] key and then the [±] key. The answer is 2.322. This result has three significant digits—the same number of significant digits as the concentration.

CONCEPT CHECK

For logarithmic values, only the digits to the right of the decimal point count as significant digits. The digit to the left of the decimal point fixes the location of the decimal point of the original value.

What if you want to find $[H_3O^+]$ from the pH? You would need to find 10^{-pH}. For example, what is $[H_3O^+]$ if the pH is 5.78? Enter 5.78, and press the [±] key. Then use the [10^x] function. The answer is $10^{-5.78}$. Therefore, $[H_3O^+]$ is 1.7×10^{-6} mol/L.

Remember that the pH scale is a negative log scale. Thus, a decrease in pH from pH 7 to pH 4 is an increase of 10^3, or 1000, in the acidity of a solution. An increase from pH 3 to pH 6 is a decrease of 10^3, or 1000, in acidity.

Practice Problems

Logarithms

1. Calculate the logarithm of each number. Note the trend in your answers.
   (a) 1 (b) 5 (c) 10 (d) 50 (e) 100 (f) 500 (g) 50 000 (h) 100 000
2. Calculate the antilogarithm of each number.
   (a) 0 (b) 1 (c) −1 (d) 2 (e) −2 (f) 3 (g) −3
3. (a) How are your answers for question 2, parts (b) and (c), related?
   (b) How are your answers for question 2, parts (d) and (e), related?
   (c) How are your answers for question 2, parts (f) and (g), related?
   (d) Calculate the antilogarithm of 3.5.
   (e) Calculate the antilogarithm of −3.5.
   (f) Take the reciprocal of your answer for part (d).
   (g) How are your answers for parts (e) and (f) related?
4. (a) Calculate log 76 and log 55.
   (b) Add your answers for part (a).
   (c) Find the antilogarithm of your answer for part (b).
   (d) Multiply 76 and 55.
   (e) How are your answers for parts (c) and (d) related?

The Unit Analysis Method of Problem Solving

The unit analysis method of problem solving is extremely versatile. You can use it to convert between units or to solve some formula problems. If you forget a formula during a test, you may still be able to solve the problem using unit analysis.

The unit analysis method involves analyzing the units and setting up conversion factors. You match and arrange the units so that they divide out to give the desired unit in the answer. Then you multiply and divide the numbers that correspond to the units.

Steps for Solving Problems Using Unit Analysis

Step 1 Determine which data you have and which conversion factors you need to use. (A conversion factor is usually a ratio of two numbers with units, such as 1000 g/1 kg. You multiply the given data by the conversion factor to get the desired units for the answer.) It is often convenient to use the following three categories to set up your solution: Have, Need, and Conversion factor.

Step 2 Arrange the data and conversion factors so that you can cross out the undesired units. Decide whether you need any additional conversion factors to get the desired units for the answer.

Step 3 Multiply all the numbers on the top of the ratio. Then multiply all the numbers on the bottom of the ratio. Divide the top result by the bottom result.

Step 4 Check that you have cancelled the units correctly. Also check that the answer seems reasonable, and that the significant digits are correct.

CONCEPT CHECK

Remember that counting numbers for exact quantities are considered to have infinite significant digits. For example, if you have 3 apples, the number is exact, and has an infinite number of significant digits. Conversion factors for unit analysis are a form of counting or record keeping. Therefore, you do not need to consider the number of significant digits in conversion factors, such as 1000 mL/1 L, when deciding on the number of significant digits in the answer.

Sample Problem
Active ASA

Problem

In the past, pharmacists measured the active ingredients in many medications in a unit called grains (gr). A grain is equal to 64.8 mg. If one headache tablet contains 5.0 gr of active acetylsalicylic acid (ASA), how many grams of ASA are in two tablets?

What Is Required?

You need to find the mass in grams of ASA in two tablets.

What Is Given?

There are 5.0 gr of ASA in one tablet. A conversion factor for grains to milligrams is given.

Plan Your Strategy

Multiply the given quantity by conversion factors until all the unwanted units cancel out and only the desired units remain.

Have	Need	Conversion factors
5.0 gr	? g	64.8 mg/1 gr and 1 g/1000 mg

Act on Your Strategy

$$\frac{5.0 \text{ gr}}{1 \text{ tablet}} \times \frac{64.8 \text{ mg}}{1 \text{ gr}} \times \frac{1 \text{ g}}{1000 \text{ mg}} \times 2 \text{ tablets}$$
$$= \frac{5.0 \times 64.8 \times 1 \times 2 \text{ g}}{1000}$$
$$= 0.648 \text{ g}$$
$$= 0.65 \text{ g}$$

There are 0.65 g of active ASA in two headache tablets.

Check Your Solution

There are two significant digits in the answer. This is the least number of significant digits in the given data.

Notice how conversion factors are multiplied until all the unwanted units are cancelled out, leaving only the desired unit in the answer.

The sample problem on the next page will show you how to solve a stoichiometric problem.

Sample Problem

Stoichiometry and Unit Analysis

Problem

What mass of oxygen, O_2, can be obtained by the decomposition of 5.0 g of potassium chlorate, $KClO_3$? The balanced equation is given below.

$$2KClO_3 \rightarrow 2KCl + 3O_2$$

What Is Required?

You need to calculate the mass of oxygen, in grams, that is produced by the decomposition of 5.0 g of potassium chlorate.

What Is Given?

You know the mass of potassium chlorate that decomposes.

$$\text{Mass} = 5.0 \text{ g}$$

From the balanced equation, you can obtain the molar ratio of the reactant and the product.

$$\frac{3 \text{ mol } O_2}{2 \text{ mol } KClO_3}$$

Plan Your Strategy

Calculate the molar masses of potassium chlorate and oxygen. Use the molar mass of potassium chlorate to find the number of moles in the sample.

Use the molar ratio to find the number of moles of oxygen produced. Use the molar mass of oxygen to convert this value to grams.

Act on Your Strategy

The molar mass of potassium chlorate is

$$\begin{aligned} 1 \times M_K &= 39.10 \\ 1 \times M_{Cl} &= 35.45 \\ 3 \times M_O &= 48.00 \\ \hline &122.55 \text{ g/mol} \end{aligned}$$

The molar mass of oxygen is

$$2 \times M_O = 32.00 \text{ g/mol}$$

Find the number of moles of potassium chlorate.

$$\begin{aligned} \text{mol } KClO_3 &= 5.0 \cancel{g} \times \left(\frac{1 \text{ mol}}{122.55 \cancel{g}\, KClO_3}\right) \\ &= 0.0408 \text{ mol} \end{aligned}$$

Find the number of moles of oxygen produced.

$$\frac{\text{mol } O_2}{0.0408 \text{ mol } KClO_3} = \frac{3 \text{ mol } O_2}{2 \text{ mol } KClO_3}$$

$$\begin{aligned} \text{mol } O_2 &= 0.0408 \text{ mol } KClO_3 \times \frac{3 \text{ mol } O_2}{2 \text{ mol } KClO_3} \\ &= 0.0612 \text{ mol} \end{aligned}$$

Convert this value to grams.

$$\begin{aligned} \text{mass } O_2 &= 0.0612 \text{ mol} \times \frac{32.00 \text{ g}}{1 \text{ mol } O_2} \\ &= 1.96 \text{ g} \\ &= 2.0 \text{ g} \end{aligned}$$

Therefore, 2.0 g of oxygen are produced by the decomposition of 5.0 g of potassium chlorate.
As you become more familiar with this type of question, you will be able to complete more than one step at once. Below, you can see how the conversion factors we used in each step above can be combined. Set these conversion ratios so that the units cancel out correctly.

$$\begin{aligned} \text{mass } O_2 &= 5.0 \text{ g } KClO_3 \times \left(\frac{1 \text{ mol}}{122.6 \text{ g } KClO_3}\right) \times \\ &\quad \left(\frac{3 \text{ mol } O_2}{2 \text{ mol } KClO_3}\right) \times \left(\frac{32.0 \text{ g}}{1 \text{ mol } O_2}\right) \\ &= 1.96 \text{ g} \\ &= 2.0 \text{ g} \end{aligned}$$

Check Your Solution

The oxygen makes up only part of the potassium chlorate. Thus, we would expect less than 5.0 g of oxygen, as was calculated.

The smallest number of significant digits in the question is two. Thus, the answer must also have two significant digits.

Practice Problems

Unit Analysis

Use the unit analysis method to solve each problem.

1. The molar mass of cupric chloride is 134.45 g/mol. What is the mass, in grams, of 8.19×10^{-3} mol of this compound?
2. To make a salt solution, 0.82 mol of $CaCl_2$ are dissolved in 1650 mL of water. What is the concentration, in g/L, of the solution?
3. The density of solid sulfur is 2.07 g/cm^3. What is the mass, in kg, of a 1.8 dm^3 sample?
4. How many grams of dissolved sodium bromide are in 689 mL of a 1.32 mol/L solution?

Chemistry Data Tables

Table E.1 Useful Math Relationships

$D = \frac{m}{V}$

$P = \frac{F}{A}$

$\pi = 3.1416$

Volume of sphere $V = \frac{4}{3}\pi r^3$

Volume of cylinder = $\pi r^2 h$

Table E.2 Fundamental Physical Constants (to six significant digits)

acceleration due to gravity (g)	9.806 65 m/s^2
Avogadro constant (N_a)	6.022 14 × 10^{23}/mol
charge on one mole of electrons (Faraday constant)	96 485.3 C/mol
mass of electron (m_s)	9.109 38 × 10^{-31} kg
mass of neutron (m_n)	1.674 93 × 10^{-27} kg
mass of proton (m_p)	1.672 62 × 10^{-27} kg
molar gas constant (R)	8.314 47 J/mol·K
molar volume of gas at STP	22.414 0 L/mol
speed of light in vacuo (c)	2.997 92 × 10^8 m/s
unified atomic mass (u)	1.660 54 × 10^{-27} kg

Table E.3 Common SI Prefixes

tera (T)	10^{12}
giga (G)	10^9
mega (M)	10^6
kilo (k)	10^3
deci (d)	10^{-1}
centi (c)	10^{-2}
milli (m)	10^{-3}
micro (μ)	10^{-6}
nano (n)	10^{-9}
pico (p)	10^{-12}

Table E.4 Conversion Factors

Quantity	Relationships between units
length	1 m = 10^{-3} km = 10^3 mm = 10^2 cm
	1 pm = 10^{-12} m
mass	1 kg = 10^3 g = 10^{-3} t
	1 u = 1.66 × 10^{-27} kg
temperature	0 K = –273.15°C
	T (K) = T (°C) + 273.15 T (°C) = T (K) – 273.15
	mp of H_2O = 273.15 K (0°C) bp of H_2O = 373.15 K (100°C)
volume	1 L = 1 dm^3 = 10^{-3} m^3 = 10^3 mL
	1 mL = 1 cm^3
pressure	101 325 Pa = 101.325 kPa = 760 mm Hg = 760 torr = 1 atm
density	1 kg/m^3 = 10^3 g/m^3 = 10^{-3} g/mL = 1 g/L
energy	1 J = 6.24 × 10^{18} eV

Table E.5 Alphabetical Listing of Common Polyatomic Ions

Most common ion		Common related ions	
acetate	CH_3COO^-		
ammonium	NH_4^+		
arsenate	AsO_4^{3-}	arsenite	AsO_3^{3-}
benzoate	$C_6H_5COO^-$		
borate	BO_3^{3-}	tetraborate	$B_4O_7^{2-}$
bromate	BrO_3^{3-}		
carbonate	CO_3^{2-}	bicarbonate (hydrogen carbonate)	HCO_3^-
chlorate	ClO_3^-	perchlorate chlorite hypochlorite	ClO_4^- ClO_2^- ClO^-
chromate	CrO_4^{2-}	dichromate	$Cr_2O_7^{2-}$
cyanide	CN^-	cyanate thiocyanate	OCN^- SCN^-
glutamate	$C_5H_8NO_4^-$		
hydroxide	OH^-	peroxide	O_2^{2-}
iodate	IO_3^-	iodide	I^-
nitrate	NO_3^-	nitrite	NO_2^-
oxalate	$OOCCOO^{2-}$		
permanganate	MnO_4^-		
phosphate	PO_4^{3-}	phosphite tripolyphosphate	PO_3^{3-} $P_3O_{10}^{5-}$
silicate	SiO_3^{2-}	orthosilicate	SiO_4^{4-}
stearate	$C_{17}H_{35}COO^-$		
sulfate	SO_4^{2-}	bisulfate (hydrogen sulfate) sulfite bisulfite (hydrogen sulfite) thiosulfate	HSO_4^- SO_3^{2-} HSO_3^- $S_2O_3^{2-}$
sulfide	S^{2-}	bisulfide (hydrogen sulfide)	HS^-

Table E.6 Summary of Naming Rules for Ions

Type of ion	Prefix or suffix	Example
Polyatomic Ions		
if the ion is the most common oxoanion	-ate	chlorate, ClO_3^-
if the ion has one O atom less than the most common oxoanion	-ite	chlorite, ClO_2^-
if the ion has two O atoms less than the most common oxoanion	hypo-___-ite	hypochlorite, ClO^-
if the ion has 1 O atom more than the most common oxoanion	per-___-ate	perchlorate, ClO_4^-
if the ion has 1 H atom added to the most common oxoanion	bi-	bicarbonate, HCO_3^-
if the ion has 1 O atom less and 1 S atom more than the most common oxoanion	thio-	thiosulphate, $S_2O_3^{2-}$
Metallic Ions		
if the ion has the higher possible charge	-ic	titanic, Ti^{4+}
if the ion has the lower possible charge	-ous	cuprous, Cu^+
Note: According to the Stock system, metallic ions are named using Roman numerals.	The Roman numeral shows the charge on the metal ion	titanium(IV) Ti^{4+} copper(I), Cu^+ manganese(VII), Mn^{7+}

Table E.7 Summary of Naming Rules for Acids

Modern name	Classical acid name	Example
aqueous hydrogen ___ide	hydro___ic acid	HCl, aqueous hydrogen chloride or hydrochloric acid
aqueous hydrogen ___ate	___ic acid	H_2CO_3, aqueous hydrogen carbonate or carbonic acid
aqueous hydrogen ___ite	___ous acid	HNO_2, aqueous hydrogen nitrite or nitrous acid

Table E.8 Standard Molar Enthalpies of Formation

Substance	ΔH°_f (kJ/mol)	Substance	ΔH°_f (kJ/mol)	Substance	ΔH°_f (kJ/mol)
$Al_2O_{3(s)}$	−1675.7	$HBr_{(g)}$	−36.3	$NH_{3(g)}$	−45.9
$CaCO_{3(s)}$	−1207.6	$HCl_{(g)}$	−92.3	$N_2H_{4(\ell)}$	+50.6
$CaCl_{2(s)}$	−795.4	$HF_{(g)}$	−273.3	$NH_4Cl_{(s)}$	−314.4
$Ca(OH)_{2(s)}$	−985.2	$HCN_{(g)}$	+135.1	$NH_4NO_{3(s)}$	−365.6
$CCl_{4(\ell)}$	−128.2	$H_2O_{(\ell)}$	−285.8	$NO_{(g)}$	+91.3
$CCl_{4(g)}$	−95.7	$H_2O_{(g)}$	−241.8	$NO_{2(g)}$	+33.2
$CHCl_{3(\ell)}$	−134.1	$H_2O_{2(\ell)}$	−187.8	$N_2O_{(g)}$	+81.6
$CH_{4(g)}$	−74.6	$HNO_{3(\ell)}$	−174.1	$N_2O_{4(g)}$	+11.1
$C_2H_{2(g)}$	+227.4	$H_3PO_{4(s)}$	−1284.4	$PH_{3(g)}$	+5.4
$C_2H_{4(g)}$	+52.4	$H_2S_{(g)}$	−20.6	$PCl_{3(g)}$	−287.0
$C_2H_{6(g)}$	−84.0	$H_2SO_{4(\ell)}$	−814.0	$P_4O_{6(s)}$	−2144.3
$C_3H_{8(g)}$	−103.8	$FeO_{(s)}$	−272.0	$P_4O_{10(s)}$	−2984.0
$C_6H_{6(\ell)}$	+49.1	$Fe_2O_{3(s)}$	−824.2	$KBr_{(s)}$	−393.8
$CH_3OH_{(\ell)}$	−239.2	$Fe_3O_{4(s)}$	−1118.4	$KCl_{(s)}$	−436.5
$C_2H_5OH_{(\ell)}$	−277.6	$FeCl_{2(s)}$	−341.8	$KClO_{3(s)}$	−397.7
$CH_3COOH_{(\ell)}$	−484.3	$FeCl_{3(s)}$	−399.5	$KOH_{(s)}$	−424.6
$CO_{(g)}$	−110.5	$FeS_{2(s)}$	−178.2	$Ag_2CO_{3(s)}$	−505.8
$CO_{2(g)}$	−393.5	$PbCl_{2(s)}$	−359.4	$AgCl_{(s)}$	−127.0
$COCl_{2(g)}$	−219.1	$MgCl_{2(s)}$	−641.3	$AgNO_{3(s)}$	−124.4
$CS_{2(\ell)}$	+89.0	$MgO_{(s)}$	−601.6	$Ag_2S_{(s)}$	−32.6
$CS_{2(g)}$	+116.7	$Mg(OH)_{2(s)}$	−924.5	$SF_{6(g)}$	−1220.5
$CrCl_{3(s)}$	−556.5	$HgS_{(s)}$	−58.2	$SO_{2(g)}$	−296.8
$Cu(NO_3)_{2(s)}$	−302.9	$NaCl_{(s)}$	−411.2	$SO_{3(g)}$	−395.7
$CuO_{(s)}$	−157.3	$NaOH_{(s)}$	−425.6	$SnCl_{2(s)}$	−325.1
$CuCl_{(s)}$	−137.2	$Na_2CO_{3(s)}$	−1130.7	$SnCl_{4(\ell)}$	−511.3
$CuCl_{2(s)}$	−220.1				

Note: The enthalpy of formation of an element in its standard state is defined as zero.

Table E.9 Ionization Constants for Acids

Acid	Formula	Conjugate base	K_a
acetic acid	CH_3COOH	CH_3COO^-	1.8×10^{-5}
benzoic acid	C_6H_5COOH	$C_6H_5COO^-$	6.3×10^{-5}
chlorous acid	$HClO_2$	ClO_2^-	1.1×10^{-2}
cyanic acid	$HOCN$	OCN^-	3.5×10^{-4}
formic acid	$HCHO_2$	CHO_2^-	1.8×10^{-4}
hydrobromic acid	HBr	Br^-	1.0×10^{9}
hydrochloric acid	HCl	Cl^-	1.3×10^{6}
hydrocyanic acid	HCN	CN^-	6.2×10^{-10}
hydrofluoric acid	HF	F^-	6.3×10^{-4}
hydrogen oxide	H_2O	OH^-	1.0×10^{-14}
hypobromous acid	$HOBr$	BrO^-	2.8×10^{-9}

continued...

Acid	Formula	Conjugate base	K_a
hypochlorous acid	$HClO$	ClO^-	4.0×10^{-8}
iodic acid	HIO_3	IO_3^-	1.7×10^{-1}
lactic acid	$CH_3CHOHCO_2H$	$CH_3CHOHCO_2^-$	1.4×10^{-4}
methanoic acid	$HCOOH$	$HCOO^-$	1.8×10^{-4}
nitric acid	HNO_3	NO_3^-	2.4×10^{1}
nitrous acid	HNO_2	NO_2^-	5.6×10^{-4}
phenol	C_6H_5OH	$C_6H_5O^-$	1.0×10^{-10}

Table E.10 Ionization Constants for Polyprotic Acids

Acid	Formula	Conjugate base	K_a
boric acid (aqueous hydrogen borate)	H_3BO_3 $H_2BO_3^-$	$H_2BO_3^-$ HBO_3^{2-}	5.4×10^{-10} $<1.0 \times 10^{-14}$
carbonic acid	H_2CO_3 HCO_3^-	HCO_3^- CO_3^{2-}	4.5×10^{-7} 4.7×10^{-11}
citric acid (aqueous hydrogen citrate)	$H_3C_6H_5O_7$ $H_2C_6H_5O_7^-$ $HC_6H_5O_7^{2-}$	$H_2C_6H_5O_7^-$ $HC_6H_5O_7^{2-}$ $C_6H_5O_7^{3-}$	7.4×10^{-4} 1.7×10^{-5} 4.0×10^{-7}
oxalic acid	$HOOCCOOH$ $HOOCCOO^-$	$HOOCCOO^-$ $OOCCOO^{2-}$	5.6×10^{-2} 1.5×10^{-4}
phosphoric acid (aqueous hydrogen phosphate)	H_3PO_4 $H_2PO_4^-$ HPO_4^{2-}	$H_2PO_4^-$ HPO_4^{2-} PO_4^{3-}	6.9×10^{-3} 6.2×10^{-8} 4.8×10^{-13}
hydrosulfuric acid	H_2S HS^-	HS^- S^{2-}	8.9×10^{-8} 1.0×10^{-19}
sulfuric acid	H_2SO_4 HSO_4^-	HSO_4^- SO_4^{2-}	1.0×10^{3} 1.0×10^{-2}
sulfurous acid	H_2SO_3 HSO_3^-	HSO_3^- SO_3^{2-}	1.4×10^{-2} 6.3×10^{-8}
tartaric acid	$H_2C_4H_4O_6$ $HC_4H_4O_6^-$	$HC_4H_4O_6^-$ $C_4H_4O_6^{2-}$	9.3×10^{-4} 4.3×10^{-5}

Table E.11 Ionization Constants for Nitrogen Bases

Base	Formula	Conjugate acid	K_b
1,2-diaminoethane (ethylenediamine)	$NH_2CH_2CH_2NH_2$	$NH_2CH_2CH_2NH_3^+$	8.4×10^{-5}
dimethylamine (N-methylmeth-anamine)	$(CH_3)_2NH$	$(CH_3)_2NH_2^+$	5.4×10^{-4}
ethanamine	$C_2H_5NH_2$	$C_2H_5NH_3^+$	4.5×10^{-4}
methanamine	CH_3NH_2	$CH_3NH_3^+$	4.6×10^{-4}
trimethylamine (N-N-dimethyl-methanamine)	$(CH_3)_3N$	$(CH_3)_3NH^+$	6.4×10^{-5}
ammonia	NH_3	NH_4^+	1.8×10^{-5}
hydrazine	N_2H_4	$N_2H_5^+$	1.3×10^{-6}
hydroxylamine	NH_2OH	NH_3OH^+	8.8×10^{-9}
pyridine	C_5H_5N	$C_5H_5NH^+$	1.7×10^{-9}
aniline	$C_6H_5NH_2$	$C_6H_5NH_3^+$	7.5×10^{-10}
urea	NH_2CONH_2	$NH_2CONH_3^+$	1.3×10^{-14}

Table E.12 Solubility Product Constants in Water at 25°C

Compound	K_{sp}
Bromates	
$AgBrO_3$	5.38×10^{-5}
$TlBrO_3$	1.10×10^{-4}
Bromides	
$AgBr$	5.35×10^{-13}
$CuBr$	6.27×10^{-9}
$PbBr_2$	6.60×10^{-6}
Carbonates	
Ag_2CO_3	8.46×10^{-12}
$BaCO_3$	2.58×10^{-9}
$CaCO_3$	3.36×10^{-9}
$MgCO_3$	6.82×10^{-6}
$PbCO_3$	7.40×10^{-14}
Chlorides	
$AgCl$	1.77×10^{-10}
$CuCl$	1.72×10^{-9}
Chromates	
Ag_2CrO_4	1.12×10^{-12}
$BaCrO_4$	1.12×10^{-10}
$PbCrO_4$	2.3×10^{-13}
Cyanides	
$AgCN$	5.97×10^{-17}
$CuCN$	3.47×10^{-20}
Fluorides	
BaF_2	1.84×10^{-7}
CdF_2	6.44×10^{-3}
CaF_2	3.45×10^{-11}
FeF_2	2.36×10^{-6}
Hydroxides	
$Be(OH)_2$	6.92×10^{-22}
$Cd(OH)_2$	7.2×10^{-15}
$Ca(OH)_2$	5.02×10^{-6}
$Co(OH)_2$	5.92×10^{-15}
$Eu(OH)_3$	9.38×10^{-27}
$Fe(OH)_2$	4.87×10^{-17}
$Fe(OH)_3$	2.79×10^{-39}
$Pb(OH)_2$	1.43×10^{-20}
$Mg(OH)_2$	5.61×10^{-12}
$Ni(OH)_2$	5.48×10^{-16}
$Sn(OH)_2$	5.45×10^{-27}
$Zn(OH)_2$	3×10^{-17}
Iodates	
$Ba(IO_3)_2$	4.01×10^{-9}
$Ca(IO_3)_2$	6.47×10^{-6}
$Sr(IO_3)_2$	1.14×10^{-7}
$Y(IO_3)_3$	1.12×10^{-10}
Iodides	
CuI	1.27×10^{-12}
PbI_2	9.8×10^{-9}
AgI	8.52×10^{-17}
Phosphates	
$AlPO_4$	9.84×10^{-21}
$Ca_3(PO_4)_2$	2.07×10^{-33}
$Co_3(PO_4)_2$	2.05×10^{-35}
$Cu_3(PO_4)_2$	1.40×10^{-37}
$Ni_3(PO_4)_2$	4.74×10^{-32}
Sulfates	
$BaSO_4$	1.08×10^{-10}
$CaSO_4$	4.93×10^{-5}
Hg_2SO_4	6.5×10^{-7}
Thiocyanates	
$CuSCN$	1.08×10^{-13}
$Pd(SCN)_2$	4.39×10^{-23}

Table E.13 Standard Reduction Potentials

Reduction half reaction	$E°$(V)
$F_{2(g)} + 2e^- \rightleftharpoons 2F^-_{(aq)}$	2.866
$Co^{3+}_{(aq)} + e^- \rightleftharpoons Co^{2+}_{(aq)}$	1.92
$H_2O_{2(aq)} + 2H^+_{(aq)} + 2e^- \rightleftharpoons 2H_2O_{(\ell)}$	1.776
$Ce^{4+}_{(aq)} + e^- \rightleftharpoons Ce^{3+}_{(aq)}$	1.72
$PbO_{2(s)} + 4H^+_{(aq)} + SO_4^{2-}{}_{(aq)} + 2e^- \rightleftharpoons PbSO_{4(s)} + H_2O_{(\ell)}$	1.691
$MnO_4^-{}_{(aq)} + 8H^+_{(aq)} + 5e^- \rightleftharpoons Mn^{2+}_{(aq)} + 4H_2O_{(\ell)}$	1.507
$Au^{3+}_{(aq)} + 3e^- \rightleftharpoons Au_{(s)}$	1.498
$PbO_{2(s)} + 4H^+_{(aq)} + 2e^- \rightleftharpoons Pb^{2+}_{(aq)} + 2H_2O_{(\ell)}$	1.455
$Cl_{2(g)} + 2e^- \rightleftharpoons 2Cl^-_{(aq)}$	1.358
$Cr_2O_7^{2-}{}_{(aq)} + 14H^+_{(aq)} + 6e^- \rightleftharpoons 2Cr^{3+}_{(aq)} + 7H_2O_{(\ell)}$	1.232
$O_{2(g)} + 4H^+_{(aq)} + 4e^- \rightleftharpoons 2H_2O_{(\ell)}$	1.229
$MnO_{2(s)} + 4H^+_{(aq)} + 2e^- \rightleftharpoons Mn^{2+}_{(aq)} + 2H_2O_{(\ell)}$	1.224
$IO_3^-{}_{(aq)} + 6H^+_{(aq)} + 6e^- \rightleftharpoons I^-_{(aq)} + 3H_2O_{(\ell)}$	1.085
$Br_{2(\ell)} + 2e^- \rightleftharpoons 2Br^-_{(aq)}$	1.066
$AuCl_4^-{}_{(aq)} + 3e^- \rightleftharpoons Au_{(s)} + 4Cl^-_{(aq)}$	1.002
$NO_3^-{}_{(aq)} + 4H^+_{(aq)} + 3e^- \rightleftharpoons NO_{(g)} + 2H_2O_{(\ell)}$	0.957
$2Hg^{2+}_{(aq)} + 2e^- \rightleftharpoons Hg_2^{2+}{}_{(aq)}$	0.920

continued...

Reduction half reaction	$E°$(V)
$Ag^+_{(aq)} + e^- \rightleftharpoons Ag_{(s)}$	0.800
$Hg_2^{2+}{}_{(aq)} + 2e^- \rightleftharpoons 2Hg_{(\ell)}$	0.797
$Fe^{3+}_{(aq)} + e^- \rightleftharpoons Fe^{2+}_{(aq)}$	0.771
$O_{2(g)} + 2H^+_{(aq)} + 2e^- \rightleftharpoons H_2O_{2(aq)}$	0.695
$I_{2(s)} + 2e^- \rightleftharpoons 2I^-_{(aq)}$	0.536
$Cu^+_{(aq)} + e^- \rightleftharpoons Cu_{(s)}$	0.521
$O_{2(g)} + 2H_2O_{(\ell)} + 4e^- \rightleftharpoons 4OH^-_{(aq)}$	0.401
$Cu^{2+}_{(aq)} + 2e^- \rightleftharpoons Cu_{(s)}$	0.342
$AgCl_{(s)} + e^- \rightleftharpoons Ag_{(s)} + Cl^-_{(aq)}$	0.222
$4H^+_{(aq)} + SO_4^{2-}{}_{(aq)} + 2e^- \rightleftharpoons H_2SO_{3(aq)} + H_2O_{(\ell)}$	0.172
$Cu^{2+}_{(aq)} + e^- \rightleftharpoons Cu^+_{(aq)}$	0.153
$2H^+_{(aq)} + 2e^- \rightleftharpoons H_{2(g)}$	0.000
$Fe^{3+}_{(aq)} + 3e^- \rightleftharpoons Fe_{(s)}$	−0.037
$Pb^{2+}_{(aq)} + 2e^- \rightleftharpoons Pb_{(s)}$	−0.126
$Sn^{2+}_{(aq)} + 2e^- \rightleftharpoons Sn_{(s)}$	−0.138
$Ni^{2+}_{(aq)} + 2e^- \rightleftharpoons Ni_{(s)}$	−0.257
$Cd^{2+}_{(aq)} + 2e^- \rightleftharpoons Cd_{(s)}$	−0.403
$Cr^{3+}_{(aq)} + e^- \rightleftharpoons Cr^{2+}_{(aq)}$	−0.407
$Fe^{2+}_{(aq)} + 2e^- \rightleftharpoons Fe_{(s)}$	−0.447
$Cr^{3+}_{(aq)} + 3e^- \rightleftharpoons Cr_{(s)}$	−0.744
$Zn^{2+}_{(aq)} + 2e^- \rightleftharpoons Zn_{(s)}$	−0.762
$2H_2O_{(\ell)} + 2e^- \rightleftharpoons H_{2(g)} + 2OH^-_{(aq)}$	−0.828
$Al^{3+}_{(aq)} + 3e^- \rightleftharpoons Al_{(s)}$	−1.662
$Mg^{2+}_{(aq)} + 2e^- \rightleftharpoons Mg_{(s)}$	−2.372
$La^{3+}_{(aq)} + 3e^- \rightleftharpoons La_{(s)}$	−2.379
$Na^+_{(aq)} + e^- \rightleftharpoons Na_{(s)}$	−2.711
$Ca^{2+}_{(aq)} + 2e^- \rightleftharpoons Ca_{(s)}$	−2.868
$Ba^{2+}_{(aq)} + 2e^- \rightleftharpoons Ba_{(s)}$	−2.912
$K^+_{(aq)} + e^- \rightleftharpoons K_{(s)}$	−2.931
$Li^+_{(aq)} + e^- \rightleftharpoons Li_{(s)}$	−3.040

Table E.14 Specific Heat Capacities of Various Substances

Substance	Specific heat capacity (J/g·°C at 25°C)
Element	
aluminum	0.900
carbon (graphite)	0.711
copper	0.385
gold	0.129
hydrogen	14.267
iron	0.444
Compound	
ammonia (liquid)	4.70
ethanol	2.46
water (solid)	2.01
water (liquid)	4.184
water (gas)	2.01
Other material	
air	1.02
concrete	0.88
glass	0.84
granite	0.79
wood	1.76

Table E.15 Average Bond Energies

Bond	Energy (kJ/mol)	Bond	Energy (kJ/mol)	Bond	Energy (kJ/mol)	Bond	Energy (kJ/mol)
Hydrogen		**Carbon**		**Nitrogen**		**Phosphorus and sulfur**	
H—H	436	C—C	347	N—N	160	P—P	210
H—C	338	C—N	305	N—O	201	P—S	444
H—N	339	C—O	358	N—F	272	P—F	490
H—O	460	C—F	552	N—Si	330	P—Cl	331
H—F	570	C—Si	305	N—P	209	P—Br	272
H—Si	299	C—P	264	N—S	464	P—I	184
H—P	297	C—S	259	N—Cl	200	S—S	266
H—S	344	C—Cl	397	N—Br	276	S—F	343
H—Cl	432	C—Br	280	N—I	159	S—Cl	277
H—Br	366	C—I	209			S—Br	218
H—I	298					S—I	170
H—Mg	126						
Oxygen		**Silicon**		**Halogens**		**Multiple bonds**	
O—O	204	Si—Si	226	F—Cl	256	C=C	607
O—F	222	Si—P	364	F—Br	280	C=N	615
O—Si	368	Si—S	226	F—I	272	C=O	745
O—P	351	Si—F	553	Cl—Br	217	N=N	418
O—S	265	Si—Cl	381	Cl—I	211	N=O	631
O—Cl	269	Si—Br	368	Br—I	179	O=O	498
O—Br	235	Si—I	293	F—F	159	C≡C	839
O—I	249	Si=O	640	Cl—Cl	243	C≡N	891
				Br—Br	193	C≡O	1077
				I—I	151	N≡N	945

continued...

Note: The values in this table represent average values for the dissociation of bonds between the pairs of atoms listed. The true values may vary for different molecules.

Table E.16 Average Bond Lengths

Bond	Length (pm)	Bond	Length (pm)	Bond	Length (pm)	Bond	Length (pm)
Hydrogen		**Carbon**		**Nitrogen**		**Phosphorus and sulfur**	
H—H	74	C—C	154	N—N	146	P—P	221
H—C	109	C—N	147	N—O	144	P—S	210
H—N	101	C—O	143	N—F	139	P—F	156
H—O	96	C—F	133	N—Si	172	P—Cl	204
H—F	92	C—Si	186	N—P	177	P—Br	222
H—Si	148	C—P	187	N—S	168	P—I	243
H—P	142	C—S	181	N—Cl	191	S—S	204
H—S	134	C—Cl	177	N—Br	214	S—F	158
H—Cl	127	C—Br	194	N—I	222	S—Cl	201
H—Br	141	C—I	213			S—Br	225
H—I	161					S—I	234
H—Mg	173						
Oxygen		**Silicon**		**Halogens**		**Multiple bonds**	
O—O	148	Si—Si	234	F—Cl	166	C=C	134
O—F	142	Si—P	227	F—Br	178	C=N	127
O—Si	161	Si—S	210	F—I	187	C=O	123
O—P	160	Si—F	156	Cl—Br	214	N=N	122
O—S	151	Si—Cl	204	Cl—I	243	N=O	120
O—Cl	164	Si—Br	216	Br—I	248	O=O	121
O—Br	172	Si—I	240	F—F	143	C≡C	121
O—I	194			Cl—Cl	199	C≡N	115
				Br—Br	228	C≡O	113
				I—I	266	N≡N	110

Note: The values in this table are average values. The length of a bond may be slightly different in different molecules, depending on the intramolecular forces within the molecules.

Titration Guidelines

Rinsing the Pipette

A pipette is used to measure and transfer a precise volume of liquid. You rinse a pipette with the solution whose volume you are measuring. This ensures that the solution will not be diluted or contaminated.

1. Pour a sample of standard solution into a clean, dry beaker.
2. Place the pipette tip in a beaker of distilled water. Squeeze the suction bulb. Maintain your grip while placing it over the stem of the pipette. Do not insert the stem into the bulb. (If your suction bulbs have valves, your teacher will show you how to use them.)
3. Relax your grip on the bulb to draw up a small volume of distilled water.
4. Remove the bulb, and discard the water by letting it drain out.
5. Rinse the pipette by drawing several millilitres of solution from the beaker into it. Rotate and rock the pipette to coat the inner surface with solution. Discard the rinse. Rinse the pipette twice in this way. It is now ready to fill with standard solution.

TITRATION TIP

Never use your mouth instead of a suction bulb to draw a liquid into a pipette. The liquid could be corrosive or poisonous. As well, you would contaminate the glass stem.

Filling the Pipette

6. Place the tip of the pipette below the surface of the solution.
7. Hold the suction bulb loosely on the end of the glass stem. Use the suction bulb to draw liquid up just past the etched volume mark. (See Figure F.1.)
8. As quickly and smoothly as you can, slide the bulb off and place your index finger over the end of the glass stem.

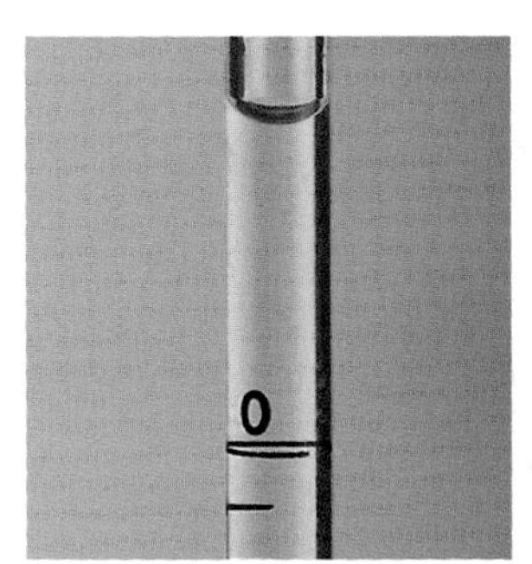

Figure F.1 Draw a bit more liquid than you need into the pipette. It is easier to reduce this volume than it is to add more solution to the pipette.

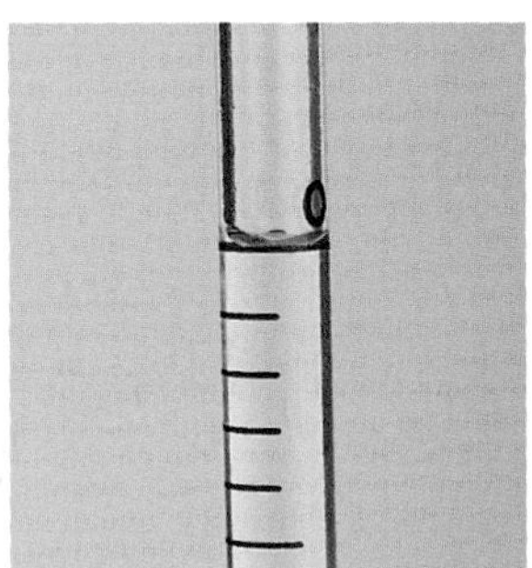

Figure F.2 The bottom of the meniscus must align exactly with the etched mark.

TITRATION TIP

Practice removing the bulb and replacing it with your index finger (or thumb). You need to be able to perform this action quickly and smoothly.

9. Gently roll your finger slightly away from end of the stem to let solution drain slowly out.
10. When the bottom of the meniscus aligns with the etched mark, as in Figure F.2, press your finger back over the end of the stem. This will prevent more solution from draining out.
11. Touch the tip of the pipette to the side of the beaker to remove any clinging drop. See Figure F.3. The measured volume inside the pipette is now ready to be transferred to an Erlenmeyer flask or a volumetric flask.

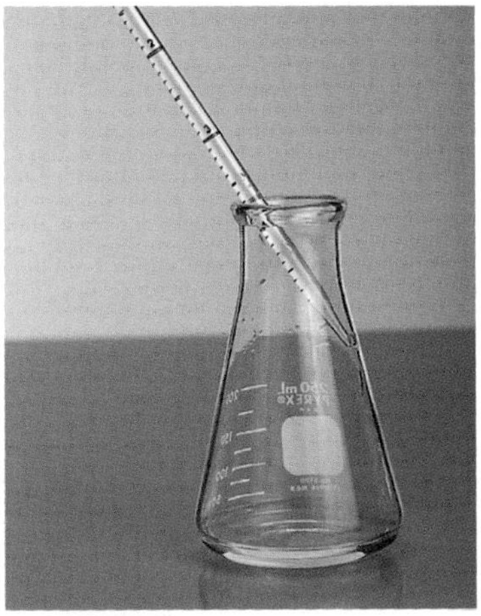

Figure F.3 You can prevent a "stubborn" drop from clinging to the pipette tip by touching the tip to the inside of the glass surface.

Transferring the Solution

12. Place the tip of the pipette against the inside glass wall of the flask. Let the solution drain slowly, by removing your finger from the stem.
13. After the solution drains, wait several seconds and then touch the tip to the inside wall of the flask to remove any drop on the end. **Note:** You may notice a small amount of liquid remaining in the tip. The pipette was calibrated to retain this amount. Do not try to remove it.

Adding the Indicator

14. Add two or three drops of indicator to the flask and its contents. Do not add too much indicator. Using more does not make the colour change easier to see. Also, indicators are usually weak acids. Too much can change the amount of base needed for neutralization. You are now ready to prepare the apparatus for the titration.

Rinsing the Burette

A burette is a graduated glass tube with a tap at one end. It is used to accurately measure the volume of liquid added during a titration experiment.

15. To rinse the burette, close the tap and add about 10 mL of distilled water from a wash bottle.
16. Tip the burette to one side, and roll it gently back and forth so that the water comes in contact with all inner surfaces.

TITRATION TIP

If you are right-handed, the tap should be on your right as you face the burette. Use your left hand to operate the tap. Use your right hand to swirl the liquid in the Erlenmeyer flask. If you are left-handed, reverse this arrangement.

17. Hold the burette over a sink. Open the tap, and let the water drain out. While you do this, check that the tap does not leak. Make sure that it turns smoothly and easily.
18. Rinse the burette with 5 mL to 10 mL of the solution that will be measured. Remember to open the tap to rinse the lower portion of the burette. Rinse the burette twice, discarding the liquid each time.

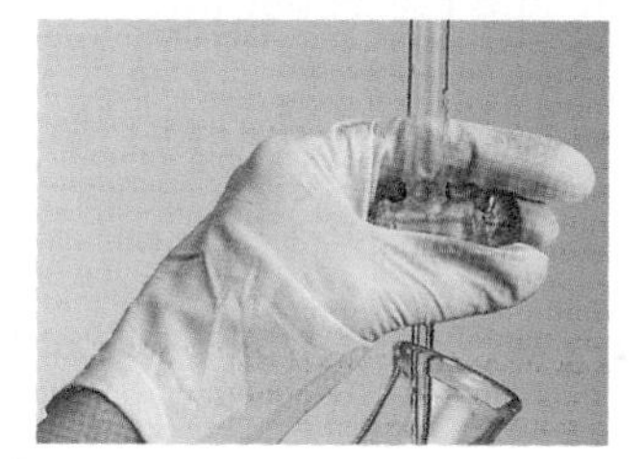

Filling the Burette

19. Assemble a retort stand and burette clamp to hold the burette. Place a funnel in the top of the burette.
20. With the tap closed, add solution until the liquid is above the zero mark. Remove the funnel. Carefully open the tap. Drain the liquid into a beaker until the bottom of the meniscus is at or below the zero mark.
21. Touch the tip of the burette against the beaker to remove any clinging drop. Check that the portion of the burette that is below the tap is filled with liquid and contains no air bubbles.
22. Record the initial burette reading in your notebook.
23. Replace the beaker with the Erlenmeyer flask that you prepared earlier. Place a sheet of white paper under the Erlenmeyer to help you see the indicator colour change that will occur near the endpoint.

TITRATION TIP

Near the endpoint, when you see the indicator change colour as liquid enters the flask from the burette, slow the addition of liquid. The endpoint can occur very quickly.

Reading the Burette

24. A meniscus reader is a small white card with a thick black line on it. Hold the card behind the burette, with the black line just under the meniscus, as in Figure F.4. Record the volume added from the burette to the nearest 0.01 mL.

TITRATION TIP

Observe the level of solution in the burette so that your eye is level with the bottom of the meniscus.

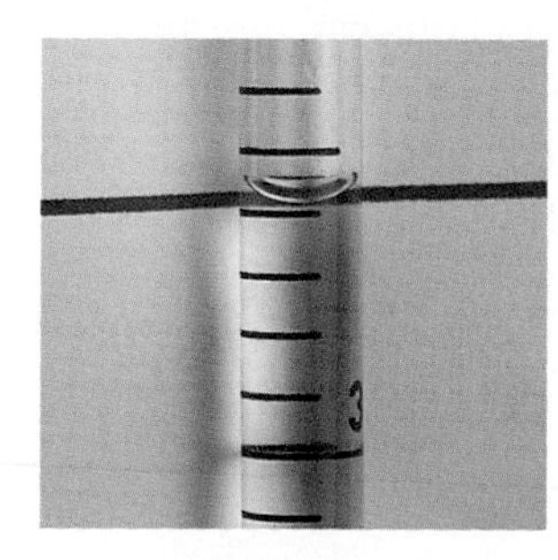

Figure F.4 A meniscus reader helps you read the volume of liquid more easily

Glossary

Section numbers are provided in parentheses, where applicable. Terms that appear in the Concepts and Skills Review are labelled (Review).

A

absorption spectrum: the spectrum that is produced when electrons of atoms absorb photons of certain wavelengths, causing the electrons to be excited from lower energy levels to higher energy levels (3.1)

accuracy: the closeness of a measurement to an accepted value

acid-base titration curve: the graph for an acid-base titration, in which the pH of an acid (or base) is plotted versus the volume of the base (or acid) added (8.4)

acid dissociation constant (K_a): the equilibrium constant for the ionization of a weak acid in water, giving the hydronium ion and the conjugate base of the acid (8.2)

activated complex: a highly unstable species that has partial bonds and is neither product nor reactant; shown at the transition state (6.3)

activation energy (E_a): the minimum energy that is required for a successful reaction between colliding molecules (6.3)

active site: the small portion of an enzyme that is involved in a catalyzed reaction (6.4)

addition polymerization: a reaction in which monomers with double bonds are joined together, through multiple addition reactions, to form a polymer (2.3)

addition reaction: an organic reaction in which atoms are added to a multiple carbon-carbon bond (2.1)

alcohol: an organic compound that contains the –OH functional group (1.3)

aldehyde: an organic compound that has a double-bonded oxygen and a hydrogen atom bonded to the first carbon of a carbon chain (1.4)

aliphatic hydrocarbon: an organic compound that contains carbon atoms bonded in chains and/or non-aromatic rings; alkanes, alkenes, and alkynes are aliphatic hydrocarbons (1.2)

alkane: a hydrocarbon that contains carbon-carbon single bonds (1.2)

alkene: a hydrocarbon that contains at least one carbon-carbon double bond (1.2)

alkoxy group: an –OR group, where O represents an oxygen atom and R represents an alkyl group (1.3)

alkyl group: an alkane branch that is obtained by removing one hydrogen atom from an alkane (1.2)

alkyl halide (haloalkane): an alkane in which one or more hydrogen atoms have been replaced with halogen atoms (1.3)

alkyne: a hydrocarbon that has at least one carbon-carbon triple bond (1.2)

allotropes: different forms of the same element that have different physical and chemical properties (4.3)

amide: an organic compound that has a carbon atom double-bonded to an oxygen atom and single-bonded to a nitrogen atom (1.4)

amine: an organic compound that has the functional group $–NR_2$, where N represents a nitrogen atom and R represents a hydrogen atom or an alkyl group (1.3)

amino acid: an organic compound that contains both the amine and carboxylic acid functional groups, connected by a central carbon atom; the monomer from which proteins are made (2.3)

amino group: the $–NH_2$ functional group (1.3)

amorphous solids: solids that have indistinct shapes, because their particle arrangements lack order (4.3)

amphoteric: the ability of a substance to act as a proton donor (an acid) in one reaction and a proton acceptor (a base) in a different reaction (8.1)

anion: a negatively charged ion

aromatic hydrocarbon: a hydrocarbon that contains one or more rings and has electrons shared in a particularly stable configuration; often drawn with alternating single and double bonds (1.2)

Arrhenius theory: (1887) the theory stating that acids and bases are defined in terms of their structure and the ions produced when they dissolve in water; defines an acid as a substance that produces hydrogen ions in water and a base as a substance that produces hydroxide ions in water (8.1)

asymmetrical alkene: an alkene that has different groups on either side of the double bond (2.2)

atom: the basic unit of an element, which still retains the element's properties (Review)

atomic mass number (mass number, A): the total number of protons and neutrons in the nucleus of an atom (Review)

atomic number (Z): the number of protons in the nucleus of one atom of a particular element (Review)

atomic symbol: a one-letter or two-letter abbreviation of the name of an element (Review)

atmosphere (atm): a unit of pressure; equal to 101.325 kPa or 760 mm Hg

aufbau principle: the principle behind an imaginary process of building up the electronic structure of the atoms, in order of atomic number (3.3)

average rate: the average change in concentration of a reactant or product over a given period of time (6.1)

Avogadro constant (N_A): the experimentally determined number of particles in 1 mol of a substance; the currently accepted value is $6.022\,141\,99 \times 10^{23}$ (Review)

B

bent shape: an angular molecular shape, e.g., around a central oxygen atom with two lone pairs that has formed two single bonds (1.1)

bimolecular elementary reaction: a reaction in which two particles collide and react (6.4)

binary compound: a compound that contains atoms of two elements (Review)

biochemistry: the study of the organic compounds and reactions that occur in living things (2.3)

bond dipole: a separation of negative and positive charge along the length of a bond (1.1)

bond energy: the energy required to break a bond (4.1)

Brønsted-Lowry theory: the theory recognizing acid-base reactions as a chemical equilibrium; defines an acid as a substance from which a proton can be removed and a base as a substance that can accept a proton (8.1)

buffer capacity: the amount of acid or base that can be added to a solution before significant change occurs to the pH of the solution (8.3)

buffer solution: a solution that resists changes in pH when moderate amounts of acid or base are added (8.3)

C

calorimeter: a device that is used to measure enthalpy changes for chemical and physical processes (5.2)

carbohydrate (saccharide): a biological molecule that, in its linear form, contains either an aldehyde group or a ketone group, along with two or more hydroxyl groups; when present in ring form, it contains an ether linkage and hydroxyl groups (2.3)

carbonyl group: the functional group of aldehydes and ketones, composed of a carbon atom double-bonded to an oxygen atom (1.4)

carboxyl group: the functional group of carboxylic acids, composed of a carbon atom double-bonded to an oxygen atom and single-bonded to an –OH group; written as –COOH (1.4)

carboxylic acid: an organic compound that contains the carboxyl functional group, composed of a carbon atom double-bonded to an oxygen atom and single-bonded to an –OH group (1.4)

catalyst: a substance that increases the rate of a reaction and is regenerated at the end of the reaction, so it can be re-used (6.1)

cation: a positively charged ion

Le Châtelier's principle: the principle stating that a dynamic equilibrium tends to respond so as to relieve the effect of any change in the conditions that affect the equilibrium (7.4)

chemical: traditionally, a substance that has been chemically produced in a laboratory; also defined as a substance that is composed of atoms (1.1)

chemical bond: the force that holds atoms together in the form of an ionic compound or molecule (4.1)

cis-isomer: a geometric isomer in which the two largest groups are on the same side of the double bond (Review)

cis-trans isomers: two compounds that are identical except for the arrangement of groups across a double bond; also see *geometric isomers, cis-isomer, trans-isomer* (Review)

coffee-cup calorimeter: a calorimeter that consists of two nested polystyrene cups (coffee cups) sitting in a 400 mL beaker (5.2)

collision theory: the theory stating that reacting particles (atoms, molecules, or ions) must collide with one another for a reaction to occur (6.3)

combustion reaction: the reaction of a substance with oxygen, producing oxides, heat, and light; burning (Review)

common ion effect: the effect on a system at equilibrium when a surplus of one of the reactant ions is added to the system; an application of Le Châtelier's principle (7.4)

complete combustion: combustion in which a hydrocarbon fuel is completely reacted in the presence of sufficient oxygen, producing only carbon dioxide, $CO_{2(g)}$, and water vapour, $H_2O_{(g)}$ (Review)

complete structural diagram: a symbolic representation of all the atoms in a molecule, showing how they are bonded (Review, 1.1)

compound: a pure substance that consists of two or more elements chemically combined in fixed proportions (Review)

concentration: an expression of the amount of solute that is present in a given volume of solution (Review)

condensation polymerization: a reaction in which monomers with two functional groups, usually one at each end, are joined together by the formation of ester bonds or amide bonds to produce a polymer (2.3)

condensation reaction: an organic reaction in which two molecules combine to form a larger molecule, producing a small molecule (such as water) as a second product (2.1)

condensed structural diagram: a symbolic representation of an organic compound, showing the atoms present and the bonds between carbon atoms (Review, 1.1)

conjugate acid-base pair: an acid and a base that differ by one proton (8.1)

constant-pressure calorimeter: a calorimeter (such as a coffee-cup calorimeter) that is open to the atmosphere (5.2)

coordinate covalent bond: a covalent bond in which one atom contributes both electrons to the shared electron pair bond (4.2)

covalent bond: a chemical bond in which electrons are shared between two atoms (1.1, 4.1)

crystalline solids: solids that have organized particle arrangements and, therefore, have distinct shapes (4.3)

cycloalkane: an alkane in the shape of a ring (1.2)

D

decomposition reaction: a chemical reaction in which a compound breaks down into elements or simpler compounds (Review)

derivatives: organic compounds that are structurally based on, or derived from, other organic compounds (1.4)

dilution: the addition of water or another appropriate solvent to a solution in order to reduce the concentration of solute

dipole: a separation of negative and positive charge by a short distance; called a bond dipole if the charge is separated along the length of a bond; called a dipole or molecular dipole if the charge is separated across a molecule (4.2)

dipole-dipole force: an intermolecular attraction between opposite charges of polar molecules (1.1, 4.3)

dipole-induced dipole force: an intermolecular force in which a polar molecule distorts the electron density of a non-polar molecule to produce a temporary dipole; similar to an ion-induced dipole force; non-polar gases (such as oxygen and nitrogen) dissolve, sparingly, in water because of dipole-induced dipole forces (4.3)

disaccharide: a biological molecule (carbohydrate) that is composed of two saccharide units; sucrose (table sugar) is a disaccharide (2.3)

dispersion force: a weak intermolecular attractive force that is present between all molecules, due to temporary dipoles (4.3)

disproportionation: a reaction in which a single substance undergoes both oxidation and reduction (10.1)

dissociation: the separation of a compound into oppositely charged ions when dissolved in water (Review)

DNA (2-deoxyribonucleic acid): essential genetic material that is found in the nucleus of every cell; a double helix with two polymeric strands, each strand composed of repeating units called nucleotides (2.3)

double bond: a covalent bond in which two pairs of electrons are shared between two atoms (1.1)

double displacement reaction: a chemical reaction in which the cations of two ionic compounds exchange places, resulting in the formation of two new compounds (Review)

E

electrolyte: a solute that conducts a current in an aqueous solution (11.1)

electromagnetic spectrum: the entire range of wavelengths of electromagnetic waves, including visible light waves

electron: a negatively charged particle with a mass of 0.000 545 u

electron affinity: the change in energy that accompanies the addition of an electron to an atom in the gaseous state

electron configuration: a shorthand notation that shows the number and arrangement of electrons in an atom's orbitals (3.3)

electronegativity: a relative measure of an atom's ability to attract shared electrons in a chemical bond (Review, 1.1)

element: a pure substance that cannot be separated chemically into any simpler substances (Review)

elementary reaction: a reaction that involves a single molecular event, such as a simple collision between atoms, molecules, or ions; can involve the formation of different molecules or ions; may involve a change in the energy or geometry of the starting molecules; cannot be broken down into simpler steps (6.4)

elimination reaction: a reaction in which atoms are removed from a compound and a multiple bond is formed (2.1)

emission spectrum: a characteristic spectrum of distinct, coloured lines that results when excited atoms emit light; can be observed through a spectroscope or a diffraction grating when a high voltage is passed through a gas in a glass tube (3.1)

empirical formula: a formula that is reduced to the smallest ratio of atoms present

endothermic reaction: a reaction that results in a net absorption of energy (5.1)

endpoint: the point in a reaction when an indicator changes colour; occurs over a range of about 2 pH units (8.4, 9.1)

enthalpy: the total kinetic energy and potential energy of a substance (5.1)

enthalpy change: the difference between the enthalpy of the reactants and the enthalpy of the products in a reaction (5.1)

entropy: a measure of the energy that spreads out during a process (7.2)

enzyme: an enormous protein molecule that acts as a biological catalyst (6.4)

equilibrium constant (K_{eq}): the ratio of the forward rate constant, k_f, divided by the reverse rate constant, k_r; also a ratio of products over reactants (7.3)

equilibrium (dynamic): an equilibrium in which opposing changes to a closed chemical system occur simultaneously at the same rate (7.1)

equivalence point: the point in a titration when just enough acid and base have been mixed so that a complete reaction occurs and there is no excess of either reactant (8.4, 9.1)

ester: an organic compound that contains a carbon atom double-bonded to an oxygen atom and single-bonded to an alkoxy group; the product of a reaction between a carboxylic acid and an alcohol (1.4)

esterification reaction: the reaction of a carboxylic acid with an alcohol to form an ester; a specific type of condensation reaction (2.2)

ether: an organic compound that has the general formula ROR; consists of two alkyl groups joined by an oxygen atom (1.3)

evidence of reaction: proof that a reaction has occurred; may be a change in colour, the formation of a new substance, or the evolution of heat

exclusion principle: the principle stating that a maximum of two electrons of opposite spin can occupy an orbital (3.3)

exothermic reaction: a reaction that results in a net release of energy (5.1)

expanded molecular formula: a symbolic representation that shows the arrangement of atoms in a molecule; an example is $CH_3CH_2CH_3$ for propane, C_3H_8 (Review)

expanded valence energy level: the valence energy level of a central atom, where the valence energy level has more than eight electrons (4.2)

F

fat: a lipid that contains a glycerol molecule bonded by ester linkages to three long-chain carboxylic acids (2.3)

favourable change: a change that has a tendency to occur naturally under certain conditions (7.2)

first-order reaction: a reaction with an overall reaction order of 1 (6.2)

formation reaction: a reaction in which a substance is formed from elements in their standard states (5.3)

fractional precipitation: a process by which ions are selectively precipitated from solution, leaving other ions still dissolved (9.3)

free-electron model: a model in which a metal is thought of as a densely packed core of metallic kernels (nuclei and inner electrons) within a delocalized region of shared, mobile valence electrons (4.1)

free energy: available energy; a measure of the useful work that can be obtained from a reaction (7.2)

functional group: a group of bonded atoms in an organic compound, which reacts in a characteristic way; if only one functional group is present, it determines the reactivity of the compound in which it occurs (1.2)

G

general formula: a formula that represents a family of simple organic compounds; written as R + functional group, where R represents an alkyl group (1.3)

geometric isomers: two compounds that are identical except for the arrangement of groups across a double bond; also see *cis-trans isomers, cis-isomer, trans-isomer* (Review)

global warming: a gradual increase in the average temperature of Earth's atmosphere (2.4)

greenhouse gas: a gas that prevents some of the heat produced by solar radiation from leaving the atmosphere (2.4)

ground state: the most stable energy state (3.2)

H

half-life ($t_{\frac{1}{2}}$): the time that is needed for the mass or concentration of a reactant to decrease by one-half of its initial value (6.2)

half-reaction: a balanced equation that shows the number of electrons involved in either oxidation or reduction (10.1)

half-reaction method: a method for balancing a redox reaction; the oxidation and reduction half-reactions are balanced separately, multiplied to have the same number of electrons, and then combined (10.3)

haloalkane (alkyl halide): an alkane in which one or more hydrogen atoms have been replaced with halogen atoms (1.3)

heat: the transfer of thermal energy (5.1)

heat capacity: the amount of energy that is needed to raise the temperature of an object 1°C or 1 K (5.2)

heat of solution (ΔH_{soln}): the enthalpy change that occurs when a solution forms from a solute and a solvent (5.1)

Hess's law of heat summation: the scientific law stating that the enthalpy change of a physical or chemical process depends only on the beginning conditions (reactants) and the end conditions (products) and is independent of the pathway of the reaction or the number of intermediate steps in the reaction (5.3)

heterogeneous catalyst: a catalyst that exists in a phase that is different from the phase of the catalyzed reaction (6.4)

heterogeneous equilibrium: an equilibrium in which reactants and products in the chemical system are in different phases; an example is a solution containing a crystal of solute (7.1)

heterogeneous mixture: a mixture in which the different components can be seen distinctly (Review)

homogeneous catalyst: a catalyst that exists in the same phase as the reactants (6.4)

homogeneous equilibrium: an equilibrium in which all the reactants and the products in the system are in the same phase (7.1)

homogeneous mixture: a mixture in which the different components are mixed so that they appear to be a single substance; a solution (Review)

Hund's rule: the rule stating that the lowest energy state for an atom has the maximum number of unpaired electrons allowed by the Pauli exclusion principle in a given energy sublevel. (3.3)

hydrocarbon: a molecular compound that contains only hydrogen and carbon atoms (1.2)

hydrogen bonding: the strong intermolecular attraction between molecules that contain a hydrogen atom bonded to an atom of a highly electronegative element, especially oxygen (1.3, 4.3)

hydrofluorocarbons (HFCs): organic compounds that are composed of hydrogen, carbon, and fluorine; can be used to replace chlorofluorocarbons (2.4)

hydrolysis reaction: a reaction in which a molecule is split in two by the addition of a water molecule (2.1)

hydronium ion: an aqueous hydrated proton, H_3O^+(8.1)

hydroxyl group: the functional group of the alcohol family of organic compounds; a non-basic –OH group (1.3)

I

ICE table: a problem-solving table that shows the initial conditions, the change, and the equilibrium conditions (7.3)

ideal gas equation: the equation that relates the pressure, volume, number of moles, and temperature of an ideal gas; $PV = nRT$ (Review)

incomplete combustion: combustion in which insufficient oxygen prevents a hydrocarbon fuel from reacting completely, resulting in a combination of products that may include carbon dioxide, $CO_{2(g)}$, water vapour, $H_2O_{(g)}$, carbon monoxide, $CO_{(g)}$, and solid carbon, $C_{(s)}$ (Review)

initial rates method: a method for measuring and comparing the initial rates of reactions (6.2)

instantaneous rate: the rate of a reaction at a particular time; can be determined by drawing a tangent to a concentration-time curve at the given time and finding the slope of the tangent (6.1)

intermolecular forces: forces that act *between* molecules or ions to influence the physical properties of compounds (1.3, 4.3)

intramolecular forces: forces that are exerted *within* a molecule or polyatomic ion; an example is covalent bonds (4.3)

ion: a positively or negatively charged particle that results from a neutral atom or group of atoms giving up or gaining electrons (Review)

ion-dipole force: the attraction between an ion and a polar molecule (4.3)

ionic bond: a bond between oppositely charged ions that arises from electron transfer; usually involves metal atoms and non-metal atoms (4.1)

ionic equation (total ionic equation): a form of chemical equation that shows dissociated ions of soluble ionic compounds (Review)

ion-induced dipole force: an intermolecular force in which an ion distorts the electron density of a non-polar molecule to produce a temporary dipole (4.3)

ion product (Q_{sp}): an expression that is identical to the solubility product constant, but its value is calculated using concentrations that are not necessarily those at equilibrium (9.3)

ion product constant for water (K_w): the product of the concentration of the hydronium ions and the hydroxide ions in a solution ($[H_3O^+][OH^-]$); always equal to 1.0×10^{-14} at 25°C (8.2)

isomers: compounds that have the same chemical formula but different molecular arrangements and properties (Review)

IUPAC: the acronym for *International Union of Pure and Applied Chemistry*, an organization that specifies rules for chemical names and symbols (1.2)

K

Kelvin scale: a temperature scale that begins at the theoretical point of absolute zero kinetic energy, or −273.15˚C; each unit (a kelvin) is equal to 1˚C

ketone: an organic compound that has an oxygen atom double-bonded to a carbon within a carbon chain (1.4)

L

lattice energy: the energy that is given off as an ionic crystal forms from the gaseous ions of its elements (4.1)

law of chemical equilibrium: the law stating that there is a constant ratio, at equilibrium, between the concentrations of products and reactants (7.3)

law of conservation of energy: the law stating that the total energy of the universe is constant (5.1)

Le Châtelier's principle: the principle stating that a dynamic equilibrium tends to respond so as to relieve the effect of any change in the conditions that affect the equilibrium (7.4)

Lewis structure: a symbolic representation of the arrangement of the valence electrons of an element or compound (1.1, 4.1)

limiting reactant: the reactant that is completely consumed during a chemical reaction, limiting the amount of product (Review)

linear shape: the molecular shape that occurs when a central atom is bonded to only two other atoms, when either zero or three lone pairs are present (1.1, 4.2)

line structural diagram: a graphical representation of the bonds between carbon atoms in an organic compound, usually omitting hydrogen atoms and carbon-hydrogen bonds (Review, 1.1)

lipid: a biological molecule that is not soluble in water but is soluble in a non-polar solvent, such as benzene or hexanes (2.3)

lone pairs: pairs of electrons in an atom's valence shell that are not involved in covalent bonding (1.1, 4.1)

M

magnetic quantum number (m_l): an integer that indicates the orientation of an orbital in the space around the nucleus; values range from $-l$ to $+l$, including 0 (3.2)

Markovnikov's rule: the rule stating that, in an addition reaction of two asymmetrical reactants, the halogen atom or –OH group is usually added to the *more substituted* carbon atom (the carbon atom that is bonded to the largest number of other carbon atoms); the hydrogen atom is added to the carbon atom that is bonded to the largest number of hydrogen atoms (2.2)

mass defect: the difference in mass between a nucleus and its nucleons (5.1)

mass number (atomic mass number): the number of protons plus the number of neutrons in the nucleus of an atom of a particular element (Review)

matter: anything that has mass and takes up space (Review)

metallic bond: the force of attraction between positively charged metal ions, and the pool of valence electrons that moves among them (4.1)

millimetre of mercury (mm Hg): a unit of pressure that is based on the height of a column of mercury in a barometer or manometer; equal to 1 torr

miscible: a term used to describe substances that are able to mix with each other in any proportion (1.3)

mixture: a physical combination of two or more kinds of matter, in which each component retains its own characteristics (Review)

mole (mol): the SI base unit for amount of substance; contains the same number of atoms, molecules, or formula units as exactly 12 g of carbon-12 (Review)

molecular compound: a non-conducting compound whose intramolecular bonds are not broken when the compound changes state

molecular formula: a chemical formula that gives the actual number of atoms of each element in a molecule or formula unit (Review)

molecular polarity: the distribution of charge on a molecule; if a molecule is polar, it has some molecular polarity (1.1)

molecularity: the number of reactant particles (molecules, atoms, or ions) that are involved in an elementary reaction (6.4)

mole ratio: a ratio that compares the number of moles of different substances in a balanced chemical equation (Review)

monomers: small molecules that are combined in long chains to produce very large molecules called polymers (2.3)

monosaccharide: the smallest molecule possible for a carbohydrate; composed of one saccharide unit (2.3)

Montréal Protocol: an international agreement that limits the global use of CFCs and other ozone-destroying chemicals (2.4)

N

natural: a substance that occurs in nature and is not artificial (1.1)

net ionic equation: a representation of a chemical reaction that shows only the ions and precipitate(s) involved in the chemical change (Review)

network solids: solids in which atoms are bonded covalently into continuous two-dimensional or three-dimensional arrays with a wide range of properties (4.3)

neutralization reaction: a double displacement reaction in which an acid and a base combine to form water and a salt

neutron: an uncharged subatomic particle in the nucleus of an atom; has a mass of 1 u

non-renewable resources: resources that will run out; once the supply is used up, there will be no more (5.4)

nuclear binding energy: the energy that is associated with the strong force holding a nucleus together; accounts for the mass defect (5.1)

nuclear fission: a reaction in which a heavy nucleus splits into lighter nuclei (5.1)

nuclear fusion: a reaction in which two smaller nuclei fuse to form a larger nucleus (5.1)

nuclear model of the atom: a model in which the electrons move around the nucleus; also called a *planetary model* (3.1)

nucleotide: the monomer from which DNA and RNA are composed; consists of three parts: a sugar, a phosphate group, and a base; a strand of DNA may be made of more than one million nucleotides (2.3)

nucleus: the central, heavy region of an atom that contains the protons and neutrons

nylon (polyamide): a condensation polymer that contains amide bonds (2.3)

O

octahedral shape: a molecular shape in which six atoms are bonded to a central atom at 90° angles (4.2)

octet rule: the rule stating that atoms bond in such a way as to attain eight electrons in their valence shells (Review, 4.1)

oil: a fat that is liquid at room temperature (2.3)

orbital diagram: a diagram that uses a box for each orbital in any given principal energy level (3.3)

orbital-shape quantum number (l): the second quantum number; refers to the energy sublevels within each principal energy level (3.2)

orbitals: solved wave functions that describe a region of probable location of electrons (3.2)

ore: a solid, naturally occurring compound, or mixture of compounds, from which a metal can be extracted (10.1)

organic chemistry: the study of carbon-based compounds (1.1)

organic compound: a molecular compound that is based on carbon; usually contains carbon-carbon and carbon-hydrogen bonds (1.1)

overall reaction order: the sum of the exponents in the rate law equation (6.2)

oxidation: defined as the loss of electrons (10.1); in organic chemistry, a reaction in which an organic compound is oxidized by forming more C—O bonds or fewer C—H bonds (2.1)

oxidation number method: a method for balancing redox equations; based on the fact that the total increase in the oxidation number(s) of the oxidized element(s) equals the total decrease in the oxidation number(s) of the reduced element(s) (10.4)

oxidation numbers: actual or hypothetical charges that are used to describe redox reactions with covalent reactants or products; assigned using a set of rules (10.2)

oxidation-reduction reaction (redox reaction): a reaction in which one reactant loses electrons (oxidation) and another reactant gains electrons (reduction) (10.1)

oxidizing agent: a reactant that oxidizes another reactant and gains electrons itself (is reduced) in the process (10.1)

oxoacid: an acid that contains oxygen atoms (8.1)

P

parent alkane: an alkane that contains the same number of carbons and the same basic structure as a more complex organic compound; a concept used primarily for nomenclature (1.3)

periodic table: a system for organizing the elements, by atomic number, into groups (columns) and periods (rows) (Review)

pH: the logarithm of the hydronium ion concentration (8.2)

photons: a quantum of electromagnetic energy that has particle-like properties (3.1)

plastic: a synthetic polymer that can be heated and moulded into specific shapes and forms (2.3)

pOH: the logarithm of the hydroxide ion concentration (8.2)

polar: having an uneven distribution of charge (1.1)

polar covalent bond: a bond in which electrons are unequally shared between two atoms (Review, 1.1, 4.1)

polyamide (nylon): a condensation polymer that contains amide bonds (2.3)

polyatomic ion: an ion, usually negatively charged, that contains more than one atom

polyester: a condensation polymer that contains ester bonds (2.3)

polymer: a large long-chain molecule with repeating units; made by linking many small molecules called monomers (2.3)

polymerization reaction: a reaction in which monomers are joined into a long chain called a polymer (2.1)

polyprotic acid: an acid that has more than one hydrogen atom that dissociates in water (8.1)

polysaccharide: a carbohydrate that consists of ten or more saccharide units (2.3)

potential energy diagram: a diagram that charts the potential energy of a reaction against the progress of the reaction; can represent the increase in potential energy during a chemical reaction (6.3)

precipitate: an insoluble solid that is formed by a chemical reaction between two soluble compounds

precision: the closeness of a measurement to other measurements of the same object or phenomenon

primary alcohol: an alcohol that has the –OH group bonded to a carbon atom, which is bonded to one carbon atom and two hydrogen atoms (1.3)

primary amide: an amide that has only hydrogen atoms, and no alkyl groups, attached to the nitrogen atom (1.4)

primary amine: an amine that has one alkyl group and two hydrogen atoms attached to the nitrogen atom (1.3)

principal quantum number (*n*): a positive whole number (integer) that indicates the energy level and relative size of an atomic orbital (3.2)

product: a new substance that is formed in a chemical reaction (Review)

protein: a natural polymer that is present in plants and animals; composed of monomers called amino acids (2.3)

proton: a positive subatomic particle that has an approximate mass of 1 u; exists in the nucleus of the atom

pure substance: a material that consists of only one type of particle; a substance that has a definite chemical composition (Review)

Q

qualitative analysis: the branch of analytical chemistry that involves identifying elements, compounds, and ions in samples of unknown or uncertain composition (9.3)

quantitative analysis: the branch of analytical chemistry that involves determining *how much* of a compound, element, or ion is in a sample (9.3)

quantum: the discrete quantity of energy that an atom can absorb or emit (3.1)

quantum mechanical model of the atom: a model that describes atoms as having certain allowed quantities of energy because of the wave-like properties of their electrons (3.2)

R

rate constant (*k*): a proportionality constant; different for each reaction at any given temperature (6.2)

rate-determining step: the slowest reaction step; determines the rate of the reaction (6.4)

rate law equation: an equation that describes the rate of a reaction by the concentration of the reactants raised to an exponent, e.g. Rate = $k[A]^m[B]^n$ (6.2)

rate of reaction: the rate at which a reaction occurs; measured in terms of reactant used or product formed per unit time (6.1)

reactant: a starting substance in a chemical reaction (Review)

reaction intermediates: molecules, atoms, or ions that appear in the elementary reactions but not in the overall reaction (6.4)

reaction mechanism: a series of steps that make up an overall reaction (6.4)

reaction quotient (*Q*): an expression that is identical to the equilibrium constant expression but is calculated using concentration values that are not necessarily those at equilibrium (7.4)

redox reaction (oxidation-reduction reaction): a reaction in which one reactant loses electrons (oxidation) and another reactant gains electrons (reduction) (10.1)

reducing agent: a reactant that reduces another reactant and loses electrons (is oxidized) in the process (10.1)

reduction: defined as the gain of electrons (10.1); in organic chemistry, a reaction in which an organic compound is reduced by forming more C — H bonds or fewer C — O bonds (2.1)

renewable resources: resources that exist in infinite supply (5.4)

resonance structures: two or more Lewis structures that show the same relative position of atoms but different positions of electron pairs (4.2)

RNA (ribonucleic acid): a nucleic acid that is present in the body's cells; works closely with DNA to produce proteins in the body (2.3)

S

salt: any ionic compound that is formed in a neutralization reaction from the anion of an acid and the cation of a base

saturated: containing only single bonds (1.1)

secondary alcohol: an alcohol that has the –OH group bonded to a carbon atom, which is bonded to two other carbon atoms and one hydrogen atom (1.3)

secondary amide: an amide that has one alkyl group attached to the nitrogen atom (1.4)

secondary amine: an amine that has two alkyl groups and one hydrogen atom attached to the nitrogen atom (1.3)

second law of thermodynamics: the law stating that the total entropy of the universe is increasing (7.2)

second order reaction: a reaction in which the overall reaction order is 2 (6.2)

SI: the international system of measurement units, including units such as the metre, the kilogram, and the mole; from the French *Système internationale d'unités* (Review)

single bond: a covalent bond in which one pair of electrons is shared between two atoms (1.1)

single displacement reaction: a chemical reaction in which one element in a compound is replaced (displaced) by another element (Review)

solubility: the amount of solute that dissolves in a given quantity of solvent at a specific temperature

solubility product constant (K_{sp}): the product of the concentrations of the ions of an ionic compound in aqueous solution at 25°C (9.2)

solute: a substance that is dissolved in a solution (Review)

solvent: a substance that has other substances (solutes) dissolved in it

specific heat capacity: a measure of the amount of heat that is needed to raise the temperature of 1 g of a substance 1°C or 1 K (5.2)

spectator ions: ions that are present in a solution but do not participate in the chemical reaction taking place (Review)

spin quantum number (m_s): the quantum number that specifies the direction in which the electron is spinning with values of $+\frac{1}{2}$ and $-\frac{1}{2}$ (3.3)

standard enthalpy of formation: the quantity of energy that is absorbed or released when 1 mol of a compound is formed directly from its elements in their standard states (5.3)

standard enthalpy of reaction: the enthalpy change associated with a reaction that occurs at SATP (5.1)

starch: a glucose polysaccharide that is used by plants to store energy; humans can digest starch (2.3)

stoichiometry: the study of the mass-mole-number relationships in chemical reactions and formulas (Review)

structural diagram: a two-dimensional representation of the structure of a compound; can be a complete diagram, a condensed diagram, or a line diagram (Review)

structural isomers: two compounds that contain the same number of each atom but in a different arrangement; bonds would have to be broken to rearrange one isomer into another (Review)

substitution reaction: a reaction in which a hydrogen atom or a functional group is replaced by a different functional group (2.1)

substrate: the reactant molecule (6.4)

surroundings: everything in the universe outside the system (5.1)

symmetrical alkene: an alkene that has identical groups on either side of the double bond (2.2)

synthesis reaction: a chemical reaction in which two or more reactants combine to produce a single, different substance (Review)

system: the part of the universe that is being studied and observed (5.1)

T

temperature: a measure of the average kinetic energy of the particles that make up a substance or system (5.1)

termolecular: a term that describes a reaction involving three particles colliding (6.4)

tertiary alcohol: an alcohol that has the –OH group bonded to a carbon atom, which is bonded to three other carbon atoms (1.3)

tertiary amide: an amide that has two alkyl groups attached to the nitrogen atom (1.4)

tertiary amine: an amine that has three alkyl groups attached to the nitrogen atom (1.3)

tetrahedral shape: the most stable shape for a compound that contains four atoms bonded to a central atom with no lone pairs; the atoms are positioned at the four corners of an imaginary tetrahedron, and the angles between the bonds are approximately 109.5° (1.1, 4.2)

thermochemical equation: a balanced chemical equation that indicates the amount of energy absorbed or released by the reaction it represents (5.1)

thermochemistry: the study of the energy that is involved in chemical reactions (5.1)

thermodynamics: the study of energy and energy transfer (5.1)

torr: a unit of pressure; equal to 1 mm of mercury in the column of a barometer or manometer

total ionic equation: a form of chemical equation that shows dissociated ions of soluble ionic compounds (Review)

trans-isomer: a geometric isomer in which the two largest groups are on different sides of the double bond (Review)

transition state: the top of the E_a barrier (the "hill") in a potential energy diagram (6.3)

transition state theory: the theory explaining what happens when molecules collide in a reaction (6.3)

trigonal bipyramidal: a molecular shape in which five atoms surround a central atom; three atoms are on a horizontal plane at 120° angles to each other, and two more atoms point straight up and straight down from the plane (4.2)

trigonal planar: a molecular shape in which three bonding groups surround a central atom; the three bonded atoms are all in the same plane as the central atom, at the corners of an invisible triangle (1.1, 4.2)

triple bond: a covalent bond in which three pairs of electrons are shared between two atoms to create a strong, inflexible bond (1.1)

U

unimolecular elementary reaction: a reaction in which one molecule or ion reacts (6.4)

unsaturated: having one or more double or triple bonds (1.2)

V

valence electrons: electrons that occupy the outer occupied energy level of an atom in its ground state (Review)

valence shell: the outermost shell of electrons

W

wax: a biological molecule that is an ester of a long-chain alcohol or a long-chain carboxylic acid (2.3)

Z

zero sum rule: the rule stating that, for chemical formulas of neutral compounds involving ions, the sum of the positive charges and the negative charges must equal zero (Review)

Index

The page numbers in **boldface** type indicate the pages where the terms are defined. Terms that occur in Sample Problems (*SP*), Investigations (*inv*), ExpressLabs (*EL*), and Thoughtlabs (*TL*) are also indicated.

1,1,2,2-tetrabromopropane, 68
1,1-difluoroethane, 101
1,1-dimethyl ethanol, 102
1,2,2,2-tetrafluoroethane, 101
1,2-dibromobutene, 58
1,2-dibromopropene, 68
1,2-dichlorobutane, 66
1,2-dimethylbenzene. *See* Ortho-xylene
1,3-butadiene, *6EL*
1,3-dibromo-4-methylcyclohexane, *28SP*
1,3-dimethylbenzene meta-dimethylbenzene. *See* Meta-xylene
1,4-dimethylbenzene para-dimethylbenzene. *See* Para-xylene
1,6-diaminohexane, 83
1,6-hexamethylenediamine, 205
1-bromo-1-butene, 68
1-bromobutane, 66
1-bromoethanol, 7
1-butene, *6EL*, 66
1-butyne, 68
1-ethoxypropane, *30SP*
1-ethylcyclopentane, *15SP*
1-methoxypropane, 29
1-propanol, boiling point, 37
10-deacetylbaccatin III, 56

2,2-dimethylpropanal, *36SP*
2,4,6-trinitromethylbenzene (*Also* TNT), 18
2-bromo-1-butene, 68
2-bromopropane, *72SP*
2-butanamide, 58
2-butanone, *72SP*
2-butene, 66
2-butyne, *6EL*
2-chloro-2-methylpentane, *67SP*
2-deoxyribose, 93
2-hexyne, 12
2-methoxy-2-methyl propane, 102
2-methylbutanal, *36SP*
2-methyl-pentene, *67SP*
2-methylpropane, *15SP*
2-pentanone, *36SP*
2-propanol, 21
  addition reaction product, 66
  addition reaction, 66
  boiling point, 25
  elimination reactions, 59
  predicting reaction, *72SP*
  reduction product, 61, *77SP*
  substitution reaction, 58

3-chloro-2-methylpentane, *67SP*
3-ethyl-2,2-dimethyl-3-heptene, *15SP*
3-ethyl-2-methylhexane, *16SP*
3-methyl-1-hexanol, *26SP*
3-methyl-2-butanone, *36SP*
3-methylbutanal, *36SP*
3-methylpentanoic acid, 39
3-pentanone, *36SP*

Absorption spectrum, **128**
Acetaldehyde. *See* Ethanal
Acetaminophen, 48, 98
Acetic acid, 39, 421
  acid dissociation constant, 393, 394–*395inv*, 396
  buffers, 410
  dissociation reaction, 380–381
  properties, 377
Acetyl CoA, structure, 565
Acetyl salicylic acid, 377
Acid,
  Arrhenius definition, 378, **380**
  Brønsted-Lowry definition, **380**
  ionization constants, 597
  naming rules, 596
  properties, 377
  strength, 383–384
Acid-base indicator, 412, **425**
Acid-base titration curves, **412**–413, 425
Acid dissociation constant (*Also* Acid ionization constant) (K_a), **393**
Acidic hydrolysis, 76
Acidic solution
  balancing half reactions, 484
  balancing redox equation, 489–*490SP*
  ion concentration, 385–*386SP*
  oxidation number method, 497
  salts in, 421
Acrylonitrile, 83
Activated complex, 292
Activation energy (E_a), **290**
  catalysts' effect, 302
Active site, enzyme, 304
Addition polymerization, 82
Addition reaction, **57**
  alkenes, 57, 66
  alkynes, 57, 68
  identifying, *62SP*
  polymers, 82
Adhesive forces, *196TL*
Adrenaline, 33
Adenine, 93
Adipic acid, 83
Adsorption, 69
Airy, George, 362

Alcohol, **25**–26
  condensation reaction, 61
  elimination reactions, 59, 70
  functional group, 22
  oxidation, 60, 71, *74inv*
  physical properties, 27
  predicting reaction, *72SP*
  substitution reaction, 58, 70
Aldehyde, **35**, *36SP*
  functional group, 22
  oxidation, 60, 71, 75
  physical properties, 37
  reduction, 60, 61, 75
Aldehyde group, carbohydrate component, 90
Aliphatic hydrocarbon, **12**
Alizarin, 425
Alkaline cell battery, 513
Alkaline fuel cell. *See* Fuel cell
Alkane, **12**, *15SP*, *16SP*
  functional group, 22
Alkene, **12**, 14, *15SP*
  addition reaction, 57, 66
  functional group, 22
  reaction, 65–67
  reduction, 60, 61
Alkoxy group, 29
Alkyl group, **14**
  general formula, 21
Alkyl halide (haloalkane), **28**
  elimination reactions, 59
  substitution reactions, 58
Alkyne, **12**, 14
  addition reaction, 57, 68
  functional group, 22
  reactions 65, 68
  reduction, 60
Allotropes, **197**
Aluminum, 234, 449
Amide, **46**–48
  functional group, 22
Amide bonds, 83
Amine, **31**–33
  amides with, 46
  functional group, 22
Amino acid, 90
Amino group
  amino acids component, 90
Ammonia, 421
  bond angles, 181
  combustion, 495
  dissociation, 381
  in amides, 46
  pH at equivalence, *426–428SP*
  production, *355SP*, 357, 367–369
  properties, 377
  substitution reaction, 58

Ammonia nitrate, 514
Ammonium chloride
  electrolysis, *539–540SP*
  formation, 379
  salt bridge, 507
Ammonium cyanate, 5
Ammonium ion,
  Lewis structure, *175–176SP*
Amorphous solids, 196, **204**–205
Ampere (A), 538
Ampère, André, 538
Amphoteric, **381**
Amylopectin, 91
Amylose, 91
Anesthesiology, 371
Angular molecular shape, 182
Aniline, 33, 377, 404
Anode, **506**
Antioxidants, 469
Aqueous solutions, 528, 529–*531SP*
Aromatic chemicals, 17
Aromatic compounds, 18–19
  reactions of, 70
  substitution reactions, 58
Aromatic hydrocarbon, **12**
Arrhenius theory, 378–379
Ascorbic acid, structure, 570
Asymmetrical alkene, 66
Atmosphere, 456
Atomic model
  Bohr's, 126–129
  Dalton's, 119–120
  quantum mechanical, 132–133
  Rutherford's, 120–122
  Thomson's, 120
Atomic radius
  ionization energy and, 154
  periodic trends, 152–153
Atomic solids, 197
  properties of, 201
Atomic spectra, 122–123, *124–125inv*
Atoms in Molecules (AIM) theory, 186
Aufbau principle, **142**
Aurora borealis, 118
Avogadro constant (N_a), 539

Bader, Richard, 186
Balancing equations
  oxidation number method, 495–497
Ball-and-stick-model, 7
Base, 377– 378, **380**
  strength, 383–384
Base dissociation constant (K_b), 404, *405–406SP*
Bases, nucleotides, 92–93
Basic hydrolysis, 76
Basic solutions
  balancing redox equations, *496–497SP*
  balancing half reactions, 485–486
  ion concentration, *385–386SP*
  salts in, 421
Battery, **512**
Bauxite, 544
Becquerel, Henri, 120
Bent shape, 7
Benzene, 18–19
  addition reaction, 70
  standard molar enthalpy of formation, 250–251
  substitution reaction, 58
Benzoic acid, 39
Bernstein, Dr. Richard, 200
Beta-damascone, 37
Beta-ionone, 37
Bimolecular elementary reaction, **297**–298
Binary acids, 383
Biochemistry, **88**
Biodegradable plastics, 89
Biofuel, 110
Biological catalysts, 304
Biosphere, 456
Birss, Dr. Viola, 552
Blood, 371, 568
  pH 411
  phosphate ions, 418
Blue Bottle™, 69
Bohr, Niels, 126
Bohr's atomic model, 126–129
Boiling point
  alcohols, 27
  aldehydes and ketones, 37
  amides, 48
  amines, 33
  carbon atoms, 24
  carboxylic acids, 40
  esters, 46
  ethers, 31
  metals, 171
  molecular polarity, 23
  organic compounds, 41
Body-centred cubic structure, 199
Bond angles, 179
Bond dipole, **8**, 187
Bond energy, **168**
Bonding pair-bonding pair repulsion, 181
Bond strength, periodic trends of binary acids, 383
Bonding
  solids, 196
Bonding pair
  bond representation, 163
  covalent bond, 167
  VSEPR theory, 179
Borkenstein, Robert, 491
Borosilicate glass, 204
Bosch, Carl, 368–369
Breathalyzer test, 491
Bromine
  addition reaction, 58, 68, 70
  test for alkanes, 68
Bromobenzene, 70
Bromochloromethane, 7
Bromocresol green, 412–413
Bromoethane, 58, 59
Brønsted, Johannes, 380
Brønsted-Lowry theory, **380**
Buckminster Fuller, R., 198
Buckminsterfullerene, 198
  structure, 197
Buckyballs. *See* Fullerenes
Buffers, **409**–411
  common ion effect, 440
Butane, *6EL*, 12
  CFC substitute, 101
Butanoic acid, 39, 44
Button cell battery, 513–514
Butyl group, 14
Butyne, addition reaction, 58

Cadavarine, 33
Cadmium precipitate, 449
Calcium oxalate, 418
Calorimeter, **236**
Carbohydrate (saccharide), **90**–91
Carbon, 5
  allotropic forms, 197
  average bond energy, 168
  electron configuration, 144–145
  nuclear binding energy, 229
Carbon-based network solids, 197
Carbon-carbon bonds, 5
Carbon-hydrogen bonds, 5
Carbon dioxide, 8, 9
  citric acid cycle, 565
  formation, 243
Carbon tetrachloride, 8, 9
Carbonyl group, **35**
Carboxyl group
  amino acids component, 90
Carboxylic acid, **39**–41
  aldehyde oxidation product, 75
  amides with, 46
  condensation reaction, 61
  esters with, 44
  derivatives, 42–*43inv*, 44–48
  functional group, 22
  hydrolysis reaction product, 61
  oxidation product, 71
  physical properties, 40
  reactions, 76
  reduction, 60
Catalase, 565
Catalysts, 276, **302**–304, *305TL*, 306–*307inv*, 314, 364
  equilibrium effects, 364
  fuel cells, 522
  Haber process, 369
Cathode, **506**
  electroplating, *542–543inv*
Cathodic protection, **549**

Cell Potential (*Also* Electric potential), **509**
measuring, *510–511SP*
Cell voltage. *See* Electric potential
Cellular respiration, 91
Cellulose, 88, 91
Celsius degrees (°C), 222
Central atoms, expanded valence level, 177
Cesium chloride, 199
Chemical, **5**
Chemical bond, **163**
energy in, 223
Chemical vapour deposition, 198
Chlor-alkali process, 552–553
Chloric acid, structure, 384
Chlorination, 553
Chlorine
addition reaction, 66
Lewis structure, 474
oxidation number, 474
Chloroform, 9
Chlorofluorocarbons (CFCs), 101
Chlorous acid, structure, 384
Chromium, 558
Chromomorphic groups, 89
Cinnamaldehyde, 37
Citric acid, 39, 565
Citric acid cycle, 565
Closed system, 456
Coffee-cup calorimeter, **236**, *238SP*, *240–241inv*
Cold packs, 228
Collision theory, **289**–290
Cohesive forces, *196TL*
Common ion effect, 436–437
buffers 440
Le Châtelier's principle, 363
solubility, 437–*438SP*
Complete structural diagram, 6
Concentration-time curve, 269
Condensation
polymerization, **83**
Condensation reaction, **61**
polymers, 82
Condensed electron configuration, 145
Condensed structural diagram, 6
Conductivity, metals, 170
Conjugate acid-base pair, **380**
identifying, *381–382SP*
dissociation constant relationship, 407, *408–409SP*
Constant-pressure calorimeter. *See* Coffee-cup calorimeter
Continuous spectrum, 123
Conversion factors, 595
Co-ordinate covalent bond, **175**
Copper, refining, 544
Copper(I) solution, 468
Copper(II) sulfate solution
cell conversion, 528
cell electrolyte, 505
electroplating, *542–543inv*
redox reaction, 465–466, 478, 486
Corrosion, **546**
prevention, 548–549
Coulomb (C), **538**
Coulomb, Charles, 538
Covalent bond, **167**, 195
diamond, 198
in organic molecules, 5
predicting, 168
Covalent network solid, properties of, 201
Crowfoot Hodgkin, Dorothy, 199
Crystal lattice, 199
Crystalline solids, 196–199
Curie, Pierre, 120
Curie, Marie, 120
Cyanide ion, 475
Cyclic compound, 18
Cycloalkane, **12**
Cyclopentane, 12
Cytosine, 93

d block elements, 148–149
d orbital, 137
Dacron™, 84
Dalton, John, 118
Dalton's atomic model, 119–120
Damascenone, 37
Daniell cell, 505–508
conversion to electrolytic cell, 528
standard cell potentials, 517–518
Daniell, John Frederic, 505
Davy, Sir Humphry, 526
de Broglie, Louis, 131
Degradable plastics, **88**–89
Dehydration reaction, **70**
Dehydroascorbic acid, 570
Deuterium, 231
Dichlorodiphenyltrichloro-ethane (DDT), 102
Diamond, 212
covalent bonds, 198
enthalpy of formation, 250–251
structure, 197
Diethyl ether
enthalpies, 228
See also Ethoxyethane
Dinitrogen peroxide, reaction rate determination, 270, *271SP*, 273, 281–282, 285
Dinitrogen tetroxide, 335
Dioxin, 88
Dipole-dipole force, **22**, 190–191, 195
Dipole-induced dipole force, 191, 195
Disaccharide, 91
Dispersion force, 22, 24, **192**, 195
graphite, 198
Disposable batteries, 512–514
Disproportionation reaction, **468**
Dissolution, 228
DNA (2-deoxyribonucleic acid), 92–94
hydrogen bonds, 193
Downs cell, 525
Double bond, 5, 167
Dry cell, **512**
Ductility, metals, 170

Earth, 456
Ecosystems, 362
Effective nuclear charge (Z_{eff}), **153**
Einstein, Albert, 126, 229
Electric cars, 550
Electric charge, **538**
Electric current, **505**, 538
Electric potential (*E*), **509**
Electrochemical cell, 525
Electrochemistry, **505**
Electrodes, **506**
Electrolysis, **524**–527
aqueous solutions, 528, *529–531SP*
calculating mass of product, *539–540SP*
potassium iodide, 532–*533inv*
Electrolytes, **506**
Electrolytic cell, **524**
electroplating, *542–543inv*
refining copper, 544
Electromagnetic spectrum
properties, 123
Electromotive force (emf), 509
Electron
mass, 120
probability density maps, 133
spin, 140
Electron affinity, **156**–157
Electron density maps, 186
Electron configuration, **142**–147
Electron sharing, 167
Electronegativity (*EN*), **8**
bonding character determination, 168
ionic bonding, 165
oxidation numbers, 473
periodic trend of binary acids, 383
values, 474
Electronegativity difference (Δ*EN*)
bond character determination, 168–169
Electron-group arrangement, **179**, 182–183
Electroplating, **528**, *542–543inv*, 558
Elementary reactions, 297–298
Elimination reaction, 59, 70
identifying, *62SP*
Emission spectrum, **127**
Enamel, 548
Endothermic reaction, **223**–224, 328, 363
potential energy diagram, 292
End-point, **425**

Energy
atom ground state, 133
atomic levels, 128, 131
bonding systems, 163
changes, 232
chemical bonds, 223
efficiency, 256–257
Einstein's equation, 229
fat storage, 94
non-renewable, 258
processes, 231–232
quanta, 126
renewable, 258
sources in Canada, 256
Sun, 231
technologies, 257
Enthalpy (*H*), **222**
favourable changes, 328
Gibbs free energy, 331
temperature, 329
Enthalpy change, (ΔH), 223–229
algebraic determination, 244–245
Hess's law, *245–246SP*
calculating, 252–253
of a neutralization reaction, *240–241inv*
transition state theory, 291
Enthalpy of combustion (ΔH_{comb}), **223**
magnesium, *248–249inv*
Enthalpy of condensation (ΔH_{cond}), **227**
Enthalpy of formation, ***253**–254SP*
Enthalpy of freezing (ΔH_{fre}), **227**
Enthalpy of fusion, **227**
Enthalpy of melting (ΔH_{melt}), **227**
Enthalpy of reaction (ΔH_{rxn}), **223**, 236, *237–238SP*
Enthalpy of solution (ΔH_{soln}), **228**
Enthalpy of vaporization (ΔH_{vap}), **227**
Entropy, **329**
favourable changes, 329–331
Gibbs free energy, 331
Enzymes, 304, 308, 314
Equilibrium, **323**
catalyst, 364
changing conditions, 365
common ion effect, 363
conditions that apply, 326–327
free energy, 331
heterogeneous, 326
homogeneous, 326
modelling, *325EL*
perturbing, *358–361inv*
pressure changes, 364
temperature change effect, 363
volume change, 364
Equilibrium concentrations
approximation method, *350–352SP*
calculating, 339, *344–345SP*
Equilibrium constant (K_{eq}), **335**
calculating, *337–338SP*, *340–343SP*
qualitative interpretation, 348
reaction quotient, 354
small values of, 350
temperature, 337
Equilibrium expression
quadratic equation solution, *345–347SP*
writing, *336SP*
Equilibrium shift, *355SP*
Equivalence point, **412**, 425
Ester, 44–46
bonds in condensation polymerization, 83
esterification product, 76
functional group, 22
hydrolysis reaction, 61, 76
Esterification reaction, **76**
Ethanal, 35, 60
Ethanamide, 46
Ethane, 303
Ethanoic acid
esterification, *77SP*
See also Acetic acid
Ethanol, 21, 44, 76
elimination reaction, 71
enthalpies, 228
esterification, *77SP*
fuel catalyst, 102
oxidation, 60
substitution reaction, 58, 70
Ethene
addition polymerization, 82, 83
addition reaction, 58
elimination reactions product, 59, 71
symmetry, 66
Ether, **29**–31
Ethoxyethane, 29
See also Diethyl ether
Ethyl alcohol, 25
See also Ethanol
Ethyl butanoate, 44, 76
Ethyl ethanoate, *45SP*, *77SP*
Ethyl group, 14
Ethylene glycol, 25, 100
Ethylene hydrogenation, 303
Exothermic reaction, **223**–224, 328, 332, 363
potential energy diagram, 292
Expanded valence energy level, 177–178
Explosives, 514
External circuit, **505**
Extraction, **544**

f block elements, 148–149
Face-centred cubic structure, 199
Faraday (F), 539
Faraday, Michael, 506, 538–539
Faraday's law, **541**, *542–543inv*
Fat, **94**
Fatty acids, 94
Fatty tissues, 95
Favourable change, **328**–329
signs of ΔH and ΔS, 332
Filipovic, Dusanka, 69
First ionization energy, 154
First-order reactions, **279**
half-life, 285, *286–287SP*
Flint glass, 204
Flow battery. *See* Fuel cell
Fluorine,
electron affinity, 156
electron configuration, 144–145
hydrogen bonded, 193
Fluoroapatite, 423
Ford, Henry, 102
Formaldehyde. *See* Methanal
Formation reactions, **250**
Fractional precipitation, **448**
Franklin, Rosalind, 199
Free-electron model, **170**
Free radical, 469
Freezing point, 227
Freon®, 101
Fructose, structure, 90
Fuel cell, **550**
Fulhame, Elizabeth, 314
Fullerenes, 198
Fumaric acid, structure, 565
Functional group, **12**
carbonyl groups, 35–41
single bonds, 21–33
Fusion, phase change, 227

Galvani, Luigi, 506
Galvanic cell, **505**–508
measuring cell potentials, *510–511SP*
rusting, 547–548
Galvanizing, **548**
Gas
collision theory, 290
entropy, 330
phase change with heat, 227
Geiger, Hans, 121
Geosphere, 456
Gibbs free energy, 331–332, 430
Gibbs, Josiah Willard, 331
Gillespie, Ronald, 178
Glass, 204, 234
Glucose, structure, 90
Glycerol, lipid, 94
Glyceryl trioleate, 94
Glycogen, 91
Gold, 116
Goulenchyn, Dr. Karen, 129
Grain alcohol, 25
See also Ethanol
Graphite, 212
enthalpy of formation, 250–251
dispersion forces in, 198
structure, 197

Greenhouse gas, 101
Ground state, 133
Guanine, 93
Guillet, Dr. James, 89
Guldberg, Cato, 334–335

Haber, Fritz, 367–369
Haber process, *355SP*, 367–369
Haber-Bosch process, 369
Half-life ($t_{1/2}$) of a reaction, **285**
Half-reaction, **467**
  balancing, 483–485
Half-reaction method, 486–489
Haloalkane. *See* Alkyl halide
Halozite™, 69
Heat, **222**, 226–228
Heat capacity (*C*), **235**
Heat transfer
  measuring, 236
  specific heat capacity, 235
Heisenberg, Werner, 132
Heisenberg's uncertainty principle, 132
Helium atom,
  emission spectrum, 131
  electron configuration, 144–145
  quantum numbers, 140
Henri, Victor, 308
Hess's law of heat summation, **243**, *248–249inv*
Heterogeneous catalyst, **303**
Heterogeneous equilibrium, **326**
Homeostasis, 362
Homogeneous catalyst, **303**
Homogeneous equilibrium, **326**
Hund's rule, 143–144
Hydrobromic acid
  addition reaction product, 70
  addition reaction, 58, 66, 68
  elimination reaction product, 59
  substitution reaction product, 58
Hydrocarbon, **12**
  naming, 14
  drawing, 16
Hydrochloric acid, *67SP*
Hydrofluoric acid, 376
Hydrofluorocarbons (HFCs), 101
Hydrogen atom,
  absorption spectrum, 128
  electron probability density map, 133
  electron configuration, 144–145
  emission spectrum, 131
  line spectrum, 123
  orbital energy, 139
  spectrum and Bohr's atomic model, 126–127
  quantum numbers, 140
Hydrogen bonding, 22, 192–193, 195
  alcohols, 27
  aldehydes and ketones, 37
  amides, 48
  amines, 33
  boiling point effect, 23
  carboxylic acids, 40
  ethers, 31
  esters, 46
  solubility in water effect, 23
Hydrogen chloride gas, 377
Hydrogen fuel cell, 552
Hydrogenation, 303
Hydrohalic acids, 383
Hydroiodic acid, 58
Hydrolysis reaction, **61**, 421
  carbohydrates, 91
  esters, 61, 76
  fatty acids, 94
Hydronium ion, 379
  concentration, 388, *389SP*, 392
  molecular shape, *184–185SP*
  polyprotic acids, 402
Hydrosphere, 456
Hydroxide ion
  concentration, 389, 392
  collision geometry, 293
Hydroxyapatite, 422
Hydroxyl group, 21
  carboxylic acids, 41
  carbohydrate component, 90
Hypochlorous acid, 384

ICE table, 339, 344, *345SP*, *346SP*, *351SP*
Induced fit-model, enzyme, 304
Inert electrodes, **508**
Inert gases, 364
Infrared spectroscopy, 38
Initial rates method, 280
Inner transition elements, 149
Integrated rate law, 285
Intermolecular forces, 22, **190**
  comparing, *24TL*
  physical properties, 22–24
International Union of Pure and Applied Chemistry (IUPAC), 13
Intramolecular forces, **190**
Iodoethane, 58
Iodine titration, 570
Ion, naming rules, 596
Ion product constant (Q_{sp}), **443**–444
Ion product constant of water (K_w), **388**
Ion-dipole forces, **191**, 195
Ionic bonding, **165**–166, 195
  lattice energy, 166
  predicting, 168
Ionic crystal, **199**
Ion-induced dipole force, **191**, 195
Ionic liquids, 203
Ionic solids, 166, 201
Ionization constants, 597
Ionization energy, **153**–155
Iron, galvanizing, 548
Iron(III) nitrate, 339
Iron(III) oxide, 547
Iron(III) thiocyanate ion, 339, *340–343SP*
Isobutane, 101
Iso-butyl group, 14
Isocitric acid, structure, 565
Isomers, **19**
Isopropanol (Isopropyl alcohol), 25
  See also 2-propanol
Isopropyl group, 14
Isostacy, 362
Isotropic materials, 204

Joules (J), 222

Kamerlingh Onnes, Heike, 206
Karwa, Dr. Laila Rafik, 371
Kelvin (K), 222
Ketoglutaric acid, 565
Ketone, **35**
  functional group, 22, 90
  oxidation, 71, 75
  physical properties, 37
  reduction, 60, 61, 75
Kettering, Charles, 102
KEVLAR®, 205–206
Kinetic energy
  collision theory, 290
  heat definition, 222
  transition state theory, 291
Krebs cycle. *See* Citric acid cycle
Krebs, Hans, 565
Kwolek, Stephanie, 205–206

Lattice energy, **166**
Law of chemical equilibrium, **334**
Law of conservation of energy, **221**
Law of mass action, 335
Lead-acid battery, 535
Lead(II) chromate, common ion effect, 436, *437–438SP*
Leaded fuels, 102
Le Châtelier, Henri, 357
Le Châtelier's principle, 356–**357**, *358–361inv*, 418
  applying, *365–366SP*
  blood, 568
  buffers, 410
  gas equilibrium, 439
  Haber process, 367–369
  non-chemistry applications, 362
  water dissociation, 388
Leclanché cell. *See* Dry cell
Leclanché, Georges, 512
Le Roy radius, 200
Le Roy-Bernstein theory, 200
Lenz, Heinrich, 362
Lenz's law, 362
Leroy, Dr. R.J., 200

Lewis structure
molecules, 173–176
oxidation numbers from, 473–475
representing chemical bonds, 163
Light, 126
Linear electron-group
arrangement, 179, 182
Linear shape, 7
Linear molecular shape, 182
Lipid, **94**–95
Liquid
entropy, 330
phase change with heat, 227
Liquidmetal™ alloy, 205
Lithium
electron configuration, 144–145
orbital energy, 139
quantum numbers, 142
Lithium hydroxide, 488
Lock-and-key model, 304
Logarithm, 592
London forces. *See* Dispersion force
London, Fritz, 192
Lone pair, 163
covalent bonds, 167
VSEPR theory, 179
resonance structures, 176
Lone pair-bonding pair repulsion, 181
Lonsdale, Kathleen Yardley, 19
Lowry, Thomas, 380

MacInnes, Dr. Joseph, 439
Magnetic quantum number (m_l), **135**
Magnesium
alloys, 552
enthalpy of combustion, *248–249inv*
Magnesium fluoride
crystal structure, 165
lattice energy, 166
Lewis structure, 165
orbital diagram, 165
Magnetite, 476
Main group elements, 147, 148
atomic radii, 152–153
electron affinity, 157
Malic acid, structure, 565
Malleability, metals, 170
Markovnikov's rule, **66**, *67SP*, 68
Marsden, Ernest, 121
Mass defect, **229**
Matter waves, 131–132
Maxwell-Boltzmann distribution, 290
Melting points, 227
alcohols, 27
aldehydes and ketones, 37
amides, 48
amines, 33
carboxylic acids, 40
esters, 46
ethers, 31
metals, 171
organic compounds, 41
Menten, Dr. Maud L., 308
Menthol, 97, 99
Metabolism, 565
Metallic bonding, 170, 195
Metal activity series, 470
Metals, 170
extraction, 544
properties, 170–171, 201
reactivity and periodic trends, 155
stress, 170
Meta-xylene, 19
Methanal (formaldehyde), 35
Lewis structure, *174–175SP*
oxidation, *77SP*
Methanamide, 46
Methanamine, 76
Methane
bond angles, 181
energy efficiency, 257
enthalpies, 228
enthalpy changes from
combustion, 252
greenhouse gas, 260
hydrates, 260
Lewis structure, 5
Methanoic acid, 39, *77SP*
See also Formic acid
Methanol (Also Methyl
alcohol), 25, 293
Methyl benzene, 18
Methyl bromide, 293
Methyl group, 14
Methyl red, 412–413, 425
Michaelis, Leonor, 308
Michaelis-Menten equation, 308
Molar absorption coefficient, 200
Molar solubility, calculating
from K_{sp}, *435–436SP*
Molecular geometry, 181
Molecular polarity, **8**
boiling point effect, 23
molecular shape
relationship, 187–188
predicting, 8–9, *10SP*
solubility effect, 10
Molecular shape, 7, 8, 179,
182–183, *184SP*
modelling, *180EL*
molecular polarity, 187–188
Molecular solids, 197
properties of, 201
Molecularity, **297**
Molecule, **173**
Lewis structures, 173–175
polarity, 178
shapes, 178
Monomers, 81
Monoprotic acids, **384**
Monosaccharide, **91**
Montréal Protocol, 101
Mueller, Paul, 102
Multi-electron atoms,
orbital energy, 139

N-methylpropanamide, 46
acid hydrolysis, 76
Nagaoka, Hantaro, 122
Nanotubes, 198
structure, 197
superconductors, 207
Natural essential oils, 17
Natural gas, 257
Haber process, 369
Natural polymers, 88
Natural substance, **5**
Net ionic equation, 487
Network solids, **197**
Neutral solutions, salts in, 420–421
Niacin, 393
Nickel-cadmium (nicad) cell, 536
Nickel tetracarbonyl, 150
Nitrates, 322
Nitrobenzene, 58
Nitrogen, 364
average bond energy, 168
cycle, 322
electron configuration, 144–145
hydrogen bonded, 193
in amines, 31
Nitrogen dioxide, 334
Nitroglycerin, 218, 514
Nitrous acid, 58
Nobel, Alfred B., 218
Noble gases, 364
Non-polar molecule, 9, 12, 187
dispersion forces, 192
Non-renewable energy, 258
Non-spontaneous change.
See Unfavourable change
Noronha, Jennifer, 259
Nuclear binding energy, **229**
Nuclear fission, **230**
Nuclear fusion, **231**
energy source, 261
Nuclear medicine, 129
Nuclear model of the atom.
See Rutherford's atomic model
Nuclear safety, 259
Nucleotide, 92–94
Nyholm, Ronald, 178
Nylon (polyamide), 83, 84

Octrahedral electron-group
arrangement, 179, 183
Octahedral molecular shape, 183
Octane numbers, 102
Octyl ethanoate, 46
Oil, 94

One-electron systems, 129
Orbital diagram, 143
Orbitals, 132–137
covalent bonds, 167
filling guidelines, 143
relative energies of, 142
Orbital-shape quantum number, **134**–135
Organic chemistry, **4**
Organic compounds, **5**, 78
boiling points of 41
general formula, 21
general rules for naming, 13
melting points of, 41
organic shapes of, 7
reactions, 56
risk-benefit analysis, 97–98
predicting products of, *77SP*
three-dimensional structural diagrams, 7
Ortho-xylene, 19
Overvoltage, **527**
Oxaloacetic acid, 565
Oxidants, 469
Oxidation, 465–**466**, 479
alcohols, 60, 71, *74inv*
aldehydes, 60, 75
identifying, *62SP*
ketones, 75
organic chemistry definition, **60**
Oxidation half-reaction, **506**
Oxidation number method, **495**–497
Oxidation numbers, 473–475
assigning, *477SP*
redox reactions, 478
rules for finding, 476
Oxidation potential, **518**
Oxidation-reduction reaction. *See* Redox reactions
Oxidizing agent, 60, **466**
Oxoacids, 383
Oxygen
corrosion agent, 546
electron configuration, 144–145
enthalpy of formation, 250–251
hydrogen bonded, 193
ionization energy, 154–155
Ozone, 101
enthalpy of formation, 250–251

p block elements, 147, 149
ionic bonding, 165
p orbital, 137
Palladium catalyst, 303
Para-xylene, 19
Parent alkane, **25**
Partition coefficient, 371
Pauli, Wolfgang, 140
Pauli's exclusion principle, **140**
Pentoic acid, 76
Perchloric acid, 384
Periodic table, 118
arrangement, *151TL*
atomic radius, 152–153
electron affinity, 156–157
electron configurations, 147–148
electronegativity values, 474
ionization energy trends, 153–155
pH, 390
blood, 411
buffers, 410
calculation at equivalence, 425, *426–428SP*
common ion effect, *440–441SP*
logarithm, 592
scale, 341
Phase changes, 227
Phenolphthalein, 412–413
Phenylethene, 18
Phosphate group, 92–93
Phosphate ions, blood, 418
Phospholipids, 95
Phosphoric acid, pH calculation, *400–402SP*
Phosphorous pentachloride, expanded valence level, 177
Photochromic lenses, 320
Photodegradable plastics, 89
Photons, 126
Physical constants, 595
Planck, Max, 126
Plastic, 81, 89
Plastic sulfur, 204
Platinum catalyst, 303
pOH, 390
Polar covalent bond, 8, **169**
Polar molecule, 8, 187
Polarity of functional group
amines, 33
alcohols, 27
aldehydes and ketones, 37
amides, 48
carboxylic acids, 40
esters, 46
ethers, 31
Polyacrylonitrile, uses, 83
Polyamide (nylon), 83
Polyatomic ion, 596
Lewis structures, 173–175
Polyester, 83
Polyethene, 82, 83
Polyethylene terephthalate (PET), 81
Polymer, **81**–88
Polymeric sulfur. *See* Plastic sulfur
Polymerization reaction, classifying, *84SP*
Polyprotic acids, 384
hydronium ion concentration, 402
ionization constants, 597
pH calculation, *400–402SP*
Polysaccharide, **91**
Polystyrene, uses, 83
Polyvinylchloride (PVC, vinyl), 83, 88
Positron emission tomography (PET), 129
Potassium chloride, 483
Potassium iodide, electrolysis, *532–533inv*
Potassium permanganate, oxidizing agent, *74inv*
Potassium thiocyanate, 339
Potential energy, 509
Potential energy diagram, **291**–293, *293–294SP*
catalyst effect on, 302
transition state theory and, 291–292
Potential ladder diagram, 517–518, *531SP*
Pratt, John, 362
Precipitation, predicting, *445SP, 446–447SP*
Pressure, effecting equilibrium, 363
Primary alcohol, 25, *26SP*
aldehyde reduction product, 75
oxidation, 71
Primary amide, 47
Primary amine, 31, 33
Primary battery, **512**
Principle quantum number (*n*), **134**–135
See also Quantum numbers
Product, 270–271
Propanal, boiling point, 37
Propanamide, 46
Propane, 101, *24TL*
Propanoic acid, 76, *397–380SP*
Propanol, *24TL*
Propanone, 61, *77SP*
Propene, 12, *72SP*
asymmetry, 66
elimination reaction product, 59
Propyl group, 14
Propyne, 68
Protein, **90**
Pyrex™. *See* Borosilicate glass
Pyridine, 404, *406SP*
Pyruvate carboxylase, 572

Qualitative analysis, **449**, *450TL*
Quantitative analysis, **449**
Quantum energy, **126**
Quantum mechanical model, **132**–133
Quantum mechanics, 132
Quantum number, 127
See also Principle quantum numbers
determining, *135–136SP, 149–150SP*
orbitals, 133–136
Pauli's exclusion principle, 140–142
periodic table arrangement, *151TL*
Quartz, structure, 198
Quigg, Jeff, 17
Quinine, 33

pH calculation, *404–405SP*

Radioactive particles, 120
Rate constant (k), **278**–279
determining, 282, 299–300
Rate law
determination, 299, *306–307inv*
elementary reactions, 298
Rate law equation, **278**
calculation, *282–283SP*
Reactant, 270–271
Reaction intermediates, 297
Reaction mechanism, **297**
evaluating, *300–301SP*
Reaction quotient (Q_c), **354**
Reaction rate, **267**
average, **268**, *270TL*
calculating, *271SP*
colour changes, 273
conductivity changes, 273
experiments, 281
factors affecting, 276
instantaneous, **268**, *270TL*
mass changes, 272
pH changes, 272
pressure changes, 273
reactant concentrations, 278–279
studying, *274–275SP*
temperature dependence, 295
volume changes, 273
Reactions
enthalpy changes in, 223–224
half-life, 285
first-order, 279
initial rates method, 280
rate determining step, 299–300
second order, 279
spontaneity, *534SP*
temperature effect, *349SP*
Rechargeable battery, **512**, 535–536
Redox reactions (oxidation-reduction reaction), **466**, *492–493inv*
half-reaction method, 486–488
identifying, *479–480SP*
oxidation numbers applied, 478
Reducing agent, 60, **466**
Reduction, 465–**466**, 479
hydrocarbons, 60 , 61, 75
identifying, *62SP*
organic chemistry definition, **60**
Reduction half-reaction, **506**
Reduction potential, **516**
Reference values, *522TL*
Refining, **544**
Reflux, 295
Renewable energy, 258
Resonance structures, **176**–177
Ribose, 93
RNA (ribonucleic acid), 92–93
Rose ketones, 37
Rubbing alcohol, 25
See also 2-propanol
Rutherford, Ernest, 120
Rutherford's atomic model, 120–123
Rutherford's gold-foil experiment, 121
Rusting, 547–549

s block elements, 147, 149
ionic bonding, 165
s orbital, 137
Saccharide. *See* Carbohydrates
Sacrificial anode, **548**–549
Salt solutions, pH, *429EL*
Salt bridge, **507**
Salts
acid-base properties, 419, 422
neutral solutions, 420
predicting acidity/basicity, *423–424SP*
Saturated carbon bonds, 5
Schrödinger, Erwin, 132
Scientific notation, 591
Sec-butyl group, 14
Secondary alcohol, 25
ketone reduction product, 75
oxidation, 71
Secondary amide, 47
Secondary amine, 31, 33
naming *32SP*
Secondary battery, **512**
Second law of thermodynamics, 330
Second-order reactions, **279**–280
Seesaw molecular shape, 182
Semi-permeable membrane, 505
Significant digits, 590
SI prefixes, 595
Silica, 198
Silver chloride
photochromic lenses, 320
precipitation, *445SP*
Silver halides, 448
Silver mirror test, 65
Silver nitrate test, *445SP*
Silver precipitate, 449
Simple cubic crystal, 199
Single bond carbon bonds, 5
Single displacement reactions, *470–471inv*
Soda-lime glass, 204
Sodium acetate, 324
conjugate acid-base dissociation constants, *408–409SP*
Sodium butanoate, 76
Sodium chloride, 422, *423–424SP*
crystal cubic structure, 199
electrolysis, 524–525
enthalpies, 228
Sodium fluoride, 422–423
Sodium hydroxide, 377
Solids
bonding, 196
entropy, 330
phase change with heat, 227
properties of, 201
Solubility
general guidelines, 444
molecular polarity, 10
Solubility in water
alcohols, 27
aldehydes and ketones, 37
amides, 48
amines, 33
carboxylic acids, 40
esters, 46
ethers, 31
hydrogen bonds, 23
hydrogen bonds in polar covalent compounds, 194
molecular polarity, 23
strong bases, 384
Solubility product
constant (K_{sp}), **431**, 598
determining, *432–433SP*, *434inv*
ion product constant relationship, 443–444
Specific heat capacity (c), **234**–235
Spectrophotometer, 273
Spin quantum number (m_s), **140**
Spontaneous change. *See* Favourable change
Square planar molecular shape, 183
Square pyramidal molecular shape, 183
Stainless steel, 548
Standard cell potential, *519–521SP*
Standard enthalpy of reaction (ΔH°_{rxn}), **223**
Standard half-cell potentials, 516–518
Standard hydrogen electrode, 517
Standard molar enthalpy of formation (ΔH°_f), **250**–251, 597
Standard reduction potentials, 598
Starch, 91
Steady state systems, 362
Stoichiometry
thermochemical equations, 224, *225SP*
unit analysis method, *594SP*
Stress, metals, 170
Strong acids, 383, 421
Strong base, 421
Styrene, 83
See also Phenylethene
Substitution reaction, **58**
alcohols, 58, 70
alkyl halides, 58
aromatic compounds, 58
identifying, *62SP*
Substrate, enzyme, 304
Succinic acid, structure, 565
Succinyl CoA, structure, 565
Sucrose, 90
Sugar, nucleotides, 92–93

Sulfur dioxide, 176–177
Sulfuryl chloride, *286–287SP*
Superconductors, **206**–207
Surface area, collision theory, 289
Surroundings, **221**
Symmetrical alkene, 66
Synthetic polymers, 82
System, **221**

T-shaped molecular shape, 183
TAXOL™, 56
Teflon™, *84SP*
Temperature change, **222**
  collision theory, 290–291
  enthalpy, 329
  equilibrium constant, 337
  equilibrium effect, 363
  favourable changes, 329
  Gibbs free energy, 331
  reaction extent, *349SP*
  reaction rates, 276, 295
Terephthalic acid, 205
Termolecular elementary reaction, 298
Tert-butyl group, 14
Tertiary alcohol, 25, 71
Tertiary amide, 47
Tertiary amines, 31, 33
Tetra-ethyl lead, 102
Tetrahedral
  electron-group arrangement, 179, 182
  molecular shape, 7, 182
Thermochemical equation, **223**–224, *225SP*
  adding algebraically, 244–245
Thermochemistry, **221**
Thermodynamics, **221**
Thomson, Joseph John, 120
Thomson's atomic model, 120
Three-dimensional structural diagrams of organic molecules, 7
Thymol blue, 425
Thymine, 93
Tin, cathodic protection, 549
Toluene. *See* Methylbenzene
Transition elements, 148–149
atomic radii, 152
Transition state, **292**
Transition state theory, **291**–292
Trigonal bipyramidal
  electron-group arrangement, 179, 182
  molecular shape, 182
Trigonal planar
  electron-group arrangement, 179, 182
  molecular shape, 7, 182
Trigonal pyramidal molecular shape, 182
Triple bonds
  carbon bonds, 5
  covalent molecules, 167
Tritium, nuclear fusion, 231
Tungsten, as a catalyst, 280

Unfavourable change, 328
Unimolecular elementary reaction, **298**
Unit analysis method, 593, *594SP*
Unit cell, 199
Universe, energy of, 221
Unleaded fuels, 102
Unsaturated carbon bonds, 5
Uracil, 93
Uranium, 230–231
Urea, 5, 48, 314

Valence electrons, 148
Valence-Shell Electron-Pair Repulsion (VSEPR) theory, **178**–179
  predicting molecular shape, *184–185SP*
van der Waals, Johannes, 190
Vectors, 8
Villamagna, Fortunato, 514
Vinegar, 39
Vinyl chloride, 83, 88
Vitamins, 95
Volume changes
  equilibrium effect, 364
Volt (V), 509
Voltaic cell. *See* Galvanic cell

Waage, Peter, 334–335
Warm packs, 228
Water, 9, 23
  amphoteric character, 381
  bond angles, 181
  dissociation, 388
  electrolysis, 526–527
  enthalpies, 228
  hydrogen bonds, 194
  hydrolysis reaction, 61
  ion product constant (K_w), 388
  measuring heat of reaction, 235
  physical changes with heat, 226–227
  properties, 194
  solubility constants, 598
  treatment, 552–553
Wax, 94
Weak acid
  acid dissociation constant, 396, *397–380SP*
  percent dissociation, 397
  pH calculation, *398–399SP*
  salts of, 421–422
Weak base, salts of, 421–422
Wohler, Friedrich, 5, 314
Wood alcohol, 25. *See also* Methanol

X-ray diffraction, 199

Zeolites, 69
Zinc, galvanizing, 548
Zinc chloride, catalyst, 303
Zinc sulfate
  cell conversion, 528
  cell electrolyte, 505

Photo Credits

iv (top right), © Wally Eberhart/Visuals Unlimited, Inc.; **v** (top right), © Daryl Benson/Masterfile; **v** (bottom right), Artbase Inc.; **vi** (centre left), © Leonard Rue III/Visuals Unlimited, Inc.; **vii** (top right), © 1997 Brownie Harris/The Stock Market/Firstlight.ca; **xii** (top left), Paul McCormick/Image Bank; **xii** (centre right), Artbase Inc.; **xii** (center left), Minnesota Historical Society/CORBIS/MAGMA; **xii** (bottom right), Malcolm Hanes/Bruce Coleman Inc.; **xiii** (top left), Artbase Inc.; **xiii** (centre right), Jose L. Pelaez/The Stock Market/Firstlight.ca; **4** (bottom right), Stephen Saks/Photo Researchers Inc.; **9** (top right), From Chemistry: The Molecular Nature of Matter and Change, © 2000, The McGraw-Hill Companies, Inc.; **9** (centre right), From Chemistry: The Molecular Nature of Matter and Change, © 2000, The McGraw-Hill Companies, Inc.; **13** (top), From Chemistry 11, © 2001, McGraw-Hill Ryerson, a subsidiary of The McGraw-Hill Companies; **2–3** (centre), Paul McCormick/Image Bank; **35** (bottom right), Artbase Inc.; **37** (bottom centre), Dick Keen/Visuals Unlimited; **37** (centre right), Paul Eekhoff/Masterfile; **39** (top right), © Wally Eberhart/Visuals Unlimited; **46** (centre left), Artbase Inc.; **56** (bottom centre), Visuals Unlimited; **57** (centre left), Artbase Inc.; **65** (centre), © E.R. Degginger/Color-Pic, Inc.; **68** (centre left), © Chris Sorensen; **69** (top left), Blue-Zone Technologies; **82** (top centre), Lynn Goldsmith/CORBIS/MAGMA; **82** (top right), Jacqui Hurst/CORBIS/MAGMA; **89** (top right), Herb Segars/Animals Animals; **91** (top left), © Gunther/Explorer/Photo Researchers Inc.; **91** (bottom left), From Chemistry: The Molecular Nature of Matter and Change, © 2000, The McGraw-Hill Companies, Inc.; **92** (top centre), Photo Courtesy Vince Satira; **92** (top right), © Science Pictures Ltd/Science Photo Library/Photo Researchers Inc.; **92** (centre), Artbase Inc.; **92** (centre right), © Studio/Science Photo Library/Photo Researchers Inc.; **94** (top left), From Chemistry: The Molecular Nature of Matter and Change, © 2000, The McGraw-Hill Companies, Inc.; **97** (top centre), © Jerome Wexler/Visuals Unlimited; **110** (top right), Photo Researchers Inc.; **116–117** (centre), Artbase Inc.; **118** (bottom left), © Daryl Benson/Masterfile; **119** (bottom right), From Chemistry 11, © 2001, McGraw-Hill Ryerson, a subsidiary of The McGraw-Hill Companies; **121** (centre), From Chemistry: The Molecular Nature of Matter and Change. © 2000, The McGraw-Hill Companies, Inc.; **123** (top right), From Chemistry: The Molecular Nature of Matter and Change. © 2000, The McGraw-Hill Companies, Inc.; **123** (centre), From Chemistry: The Molecular Nature of Matter and Change. © 2000, The McGraw-Hill Companies, Inc.; **123** (centre right), From Chemistry: The Molecular Nature of Matter and Change. © 2000, The McGraw-Hill Companies, Inc.; **123** (bottom), © John Talbot; **125** (bottom), © John Talbot; **128** (top), From Chemistry 11, © 2001 McGraw-Hill Ryerson Limited, a subsidiary of The McGraw-Hill Companies; **128** (bottom), From Chemistry: The Molecular Nature of Matter and Change. © 2000, The McGraw-Hill Companies, Inc.; **129** (centre left), Lawrence Berkeley National Laboratory/PhotoDisc; **131** (centre), From Chemistry: The Molecular Nature of Matter and Change. © 2000, The McGraw-Hill Companies, Inc.; **132** (top left), From Chemistry: The Molecular Nature of Matter and Change. © 2000, The McGraw-Hill Companies, Inc.; **133** (centre), From Chemistry 11, © 2001, McGraw-Hill Ryerson, a subsidiary of The McGraw-Hill Companies; **137** (bottom), From Physics 12. © 2002, McGraw-Hill Ryerson, a subsidiary of The McGraw-Hill Companies; **145** (top), From Chemistry: The Molecular Nature of Matter and Change. © 2000, The McGraw-Hill Companies, Inc.; **148** (top), From Chemistry: The Molecular Nature of Matter and Change. © 2000, The McGraw-Hill Companies, Inc.; **148** (centre left), From Chemistry: The Molecular Nature of Matter and Change. © 2000, The McGraw-Hill Companies, Inc.; **152** (bottom), From Chemistry: The Molecular Nature of Matter and Change. © 2000, The McGraw-Hill Companies, Inc.; **154** (centre), From Chemistry: The Molecular Nature of Matter and Change. © 2000, The McGraw-Hill Companies, Inc.; **155** (bottom left), From Chemistry: The Molecular Nature of Matter and Change, © 2000, The McGraw-Hill Companies, Inc.; **155** (bottom), From Chemistry: The Molecular Nature of Matter and Change, © 2000, The McGraw-Hill Companies, Inc.; **155** (bottom right), From Chemistry: The Molecular Nature of Matter and Change, © 2000, The McGraw-Hill Companies, Inc.; **156** (centre), From Chemistry: The Molecular Nature of Matter and Change, © 2000, The McGraw-Hill Companies, Inc.; **167** (top right), From Chemistry: The Molecular Nature of Matter and Change. © 2000, The McGraw-Hill Companies, Inc.; **168** (bottom left), From Chemistry: The Molecular Nature of Matter and Change. © 2000, The McGraw-Hill Companies, Inc.; **169** (top), Stephen Frisch; **169** (bottom), From Chemistry: The Molecular Nature of Matter and Change. © 2000, The McGraw-Hill Companies, Inc.; **170** (top left), From Chemistry: The Molecular Nature of Matter and Change. © 2000, The McGraw-Hill Companies, Inc.; **170** (bottom left), From Chemistry: The Molecular Nature of Matter and Change. © 2000, The McGraw-Hill Companies, Inc.; **172** (top), From Chemistry: The Molecular Nature of Matter and Change. © 2000, The McGraw-Hill Companies, Inc.; **179** (bottom), From Chemistry: The Molecular Nature of Matter and Change. © 2000, The McGraw-Hill Companies, Inc.; **186** (bottom left), Photo Courtesy of Dr. Richard Bader; **186** (centre left), Photo Courtesy of Dr. Richard Bader; **190** (bottom), From Chemistry: The Molecular Nature of Matter and Change. © 2000, The McGraw-Hill Companies, Inc.; **191** (centre), From Chemistry: The Molecular Nature of Matter and Change. © 2000, The McGraw-Hill Companies, Inc.; **192** (centre), From Chemistry: The Molecular Nature of Matter and Change. © 2000, The McGraw-Hill Companies, Inc.; **193** (centre left), From Chemistry: The Molecular Nature of Matter and Change. © 2000, The McGraw-Hill Companies, Inc.; **193** (centre right), From Chemistry: The Molecular Nature of Matter and Change. © 2000, The McGraw-Hill Companies, Inc.; **194** (bottom), From Chemistry: The Molecular Nature of Matter and Change. © 2000, The McGraw-Hill Companies, Inc.; **195** (top), From Chemistry: The Molecular Nature of Matter and Change, © 2000, The McGraw-Hill Companies, Inc.; **195** (top), From Chemistry: The Molecular Nature of Matter and Change, © 2000, The McGraw-Hill Companies, Inc.; **195** (centre), From Chemistry: The Molecular Nature of Matter and Change, © 2000, The McGraw-Hill Companies, Inc.; **195** (centre), From Chemistry: The Molecular Nature of Matter and Change, © 2000, The McGraw-Hill Companies, Inc.; **195** (centre), From Chemistry: The Molecular Nature of Matter and Change, © 2000, The McGraw-Hill Companies, Inc.; **195** (centre), From Chemistry: The Molecular Nature of Matter and Change, © 2000, The McGraw-Hill Companies, Inc.; **195** (bottom), From Chemistry: The Molecular Nature of Matter and Change, © 2000, The McGraw-Hill Companies, Inc.; **195** (bottom), From Chemistry: The Molecular Nature of Matter and Change, © 2000, The McGraw-Hill Companies, Inc.; **195** (bottom), From Chemistry: The Molecular Nature of Matter and Change, © 2000, The McGraw-Hill Companies, Inc.; **197** (bottom left), From Chemistry 11, © 2001, McGraw-Hill Ryerson, a subsidiary of The McGraw-Hill Companies; **197** (bottom centre), From Chemistry 11, © 2001, McGraw-Hill Ryerson, a subsidiary of The McGraw-Hill Companies; **197** (bottom right), From Chemistry 11, © 2001, McGraw-Hill Ryerson, a subsidiary of The McGraw-Hill Companies; **199** (top right), From Chemistry: The Molecular Nature of Matter and Change. © 2000, The McGraw-Hill Companies, Inc.; **199** (centre right), © Rosalind Franklin/Science Photo Library/Photo Researchers Inc.; **199** (bottom right), Hulton Archive/STONE; **200** (top right), Photo courtesy the Photo Imaging Department at the University of Waterloo;

206 (bottom right), AT&T Bell Labs/Science Photo Library/Photo Researchers Inc.; **212** (bottom left), David Parker/Science Photo Library/Photo Researchers Inc.; **216** (bottom right), From Chemistry: The Molecular Nature of Matter and Change, © 2000, The McGraw-Hill Companies, Inc.; **218–219** (centre), Minnesota Historical Society/CORBIS/MAGMA; **220** (bottom right), Artbase Inc.; **222** (centre), From Physics 11. © 2001, McGraw-Hill Ryerson Limited, a subsidiary of The McGraw-Hill Companies; **230** (top left), From Chemistry: The Molecular Nature of Matter and Change, © 2000, The McGraw-Hill Companies, Inc.; **231** (bottom left), From Chemistry: The Molecular Nature of Matter and Change, © 2000, The McGraw-Hill Companies, Inc.; **234** (centre left), © Trevor Bonderud/Firstlight.ca; **243** (bottom right), From Chemistry 11, © 2001, McGraw-Hill Ryerson Limited, a subsidiary of The McGraw-Hill Companies; **251** (top centre), © Thom Lang/The Stock Market/Firstlight.ca; **251** (top left), Artbase Inc.; **257** (bottom left), Artbase Inc.; **259** (top left), Courtesy of OPG; **260** (bottom right), Tom Pantages; **266** (bottom right), VCL-TCL/Masterfile; **267** (bottom right), Stu Forster/Allsport; **273** (centre left), Courtesy of Thermo Spectronic, Rochester, NY A Thermo Electron Business; **285** (centre), From Chemistry: The Molecular Nature of Matter and Change, © 2000, The McGraw-Hill Companies, Inc.; **289** (bottom right), Paul A. Souders/CORBIS/MAGMA; **290** (top left), From Chemistry: The Molecular Nature of Matter and Change, © 2000, The McGraw-Hill Companies, Inc.; **291** (top), From Chemistry: The Molecular Nature of Matter and Change, © 2000, The McGraw-Hill Companies, Inc.; **293** (top), From Chemistry: The Molecular Nature of Matter and Change, © 2000, The McGraw-Hill Companies, Inc.; **297** (bottom left), Mike Clemmer/The Stock Market/Firstlight.ca; **302** (centre), From Chemistry: The Molecular Nature of Matter and Change, © 2000, The McGraw-Hill Companies, Inc.; **303** (bottom left), From Chemistry: The Molecular Nature of Matter and Change, © 2000, The McGraw-Hill Companies, Inc.; **304** (centre), From Chemistry: The Molecular Nature of Matter and Change, © 2000, The McGraw-Hill Companies, Inc.; **308** (top left), With Permission of The Canadian Medical Hall of Fame and artist Irma Coucill; **314** (top left), Cover of Mrs. Fulhame's Book from, http://bip.cnrs-mrs.fr/bip10/jbiosci.htm; **314** (centre right), From Chemistry 11, © 2001, McGraw-Hill Ryerson, a subsidiary of The McGraw-Hill Companies; **320–321** (centre), Malcolm Hanes/Bruce Coleman Inc.; **322** (bottom), © Marcos Welsh/Firstlight.ca; **323** (bottom left), © Pat Ivy/Ivy Images; **329** (centre), From Chemistry: The Molecular Nature of Matter and Change, © 2000, The McGraw-Hill Companies, Inc.; **330** (top left), From Chemistry 11, © 2001 McGraw-Hill Ryerson Limited, a subsidiary of The McGraw-Hill Companies; **330** (top centre), From Chemistry 11, © 2001 McGraw-Hill Ryerson Limited, a subsidiary of The McGraw-Hill Companies; **330** (top right), From Chemistry 11, © 2001 McGraw-Hill Ryerson Limited, a subsidiary of The McGraw-Hill Companies; **334** (centre left), From Chemistry: The Molecular Nature of Matter and Change, © 2000, The McGraw-Hill Companies, Inc.; **334** (centre), From Chemistry: The Molecular Nature of Matter and Change, © 2000, The McGraw-Hill Companies, Inc.; **334** (centre), From Chemistry: The Molecular Nature of Matter and Change, © 2000, The McGraw-Hill Companies, Inc.; **334** (centre right), From Chemistry: The Molecular Nature of Matter and Change, © 2000, The McGraw-Hill Companies, Inc.; **354** (bottom), From Chemistry: The Molecular Nature of Matter and Change, © 2000, The McGraw-Hill Companies, Inc.; **357** (centre right), Science Photo Library/Photo Researchers Inc.; **362** (top left), © Leonard Rue III/Visuals Unlimited Inc.; **369** (top), From Chemistry: The Molecular Nature of Matter and Change, © 2000, The McGraw-Hill Companies, Inc.; **371** (top left), Adrian Holmes Photography Ltd.; **376** (bottom centre), Carl & Anne Purcell/CORBIS/MAGMA; **379** (centre left), E.F. Smith Collection/University of Pennsylvania Library; **391** (centre left), From Chemistry 11, © 2001, McGraw-Hill Ryerson, a subsidiary of The McGraw-Hill Companies; **391** (centre), From Chemistry 11, © 2001 McGraw-Hill Ryerson Limited, a subsidiary of The McGraw-Hill Companies; **410** (bottom left), From Chemistry: The Molecular Nature of Matter and Change, © 2000, The McGraw-Hill Companies, Inc.; **412** (bottom left), From Chemistry: The Molecular Nature of Matter and Change, © 2000, The McGraw-Hill Companies, Inc.; **413** (centre), From Chemistry: The Molecular Nature of Matter and Change, © 2000, The McGraw-Hill Companies, Inc.; **413** (centre), From Chemistry: The Molecular Nature of Matter and Change, © 2000, The McGraw-Hill Companies, Inc.; **418** (bottom right), © Stephen J. Krasemann/Photo Researchers, Inc.; **425** (centre), From Chemistry: The Molecular Nature of Matter and Change, © 2000, The McGraw-Hill Companies, Inc.; **430** (bottom left), CNRI/Science Photo Library/Photo Researchers Inc.; **437** (top left), Stephen Frisch; **437** (top centre), Stephen Frisch; **439** (centre left), AP/Wide World Photos; **448** (bottom right), Jeff Greenberg/Photo Edit; **456** (centre left), Photo Courtesy NASA; **456** (top right), Mangelsen/Firstlight.ca; **462–463** (centre), Artbase Inc. **464** (bottom centre), Artbase Inc.; **465** (bottom left), Stephen Frisch; **465** (bottom centre), Stephen Frisch; **469** (top right), Josh Mitchell/Index Stock Imagery/Picture Quest; **473** (top left), © Jeff Daly/Stock Boston; **474** (centre right), From Chemistry 11, © 2001, McGraw-Hill Ryerson, a subsidiary of The McGraw-Hill Companies; **482** (centre right), © 1997 Brownie Harris/The Stock Market/Firstlight.ca; **483** (centre left), Stephen Frisch; **483** (centre), Stephen Frisch; **483** (centre right), From Chemistry: The Molecular Nature of Matter and Change, © 2000, The McGraw-Hill Companies, Inc.; **488** (centre left), Stephen Frisch; **488** (centre), Stephen Frisch; 488 (centre right), Stephen Frisch; **491** (bottom right), © Phil Martin/Photo Edit; **506** (bottom left), From Chemistry Concepts and Applications © The McGraw-Hill Companies Inc.; **512** (bottom centre), StudiOhio; **513** (top left), From Chemistry: The Molecular Nature of Matter and Change, © 2000, The McGraw-Hill Companies, Inc.; **513** (centre right), The McGraw-Hill Companies, Inc/ Photographer Pat Watson; **513** (bottom right), The Bakken Library and Museum Minneapolis; **514** (top centre), From Chemistry: The Molecular Nature of Matter and Change, © 2000, The McGraw-Hill Companies, Inc.; **514** (centre left), Photo courtesy of Fortunato Villamagna; **517** (top left), From Chang's Chemistry. © 1998, The McGraw-Hill Companies Inc, 6th edition; **524** (bottom centre), From Chemistry Concepts and Applications © The McGraw-Hill Companies Inc.; **525** (centre), From Chemistry: The Molecular Nature of Matter and Change, © 2000, The McGraw-Hill Companies, Inc.; **526** (bottom centre), Stephen Frisch; **535** (centre right), © Charles D. Winters/ Photo Researchers Inc.; **535** (bottom left), From Chemistry Concepts and Applications © The McGraw-Hill Companies Inc.; **536** (top left), Myrleen Ferguson Cate/Photo Edit; **536** (bottom left), From Chemistry Concepts and Applications © The McGraw-Hill Companies Inc.; **538** (top right), Hulton/Archive by Getty Images; **544** (top left), Artbase Inc.; **544** (bottom right), © Chris Sharp/Photo Researchers Inc.; **546** (top right), Bill Lisenby/CORBIS/MAGMA; **546** (bottom right), AP Wide World; **547** (top left), From Chang's Chemistry. © 1998, The McGraw-Hill Companies Inc, 6th edition; **549** (top right), RC Hall Photography; **550** (centre), Artbase Inc.; **551** (top left), From Chang's Chemistry. © 1998, The McGraw-Hill Companies Inc, 6th edition; **552** (top left), Chipperfield Photography/ Photographed By Jennifer Chipperfield; **553** (top left), From Chang's Chemistry. © 1998, The McGraw-Hill Companies Inc, 6th edition; **558** (bottom left), Ken Stepnell/Bruce Coleman; **564** (top), Jose L. Pelaez/The Stock Market/Firstlight.ca; **564** (centre right), Christel Gerstenberg/CORBIS/MAGMA; **567** (top right), Artbase Inc.; **568** (bottom left), Steve Callahan/Visuals Unlimited

Periodic Table of the Elements

Key (carbon): 6 = Atomic number; 12.01 = Average atomic mass*; 2.5 = Electronegativity; +4 = Common ion charge; 1086 = First ionization energy (kJ/mol); +2 = Other ion charges; 4765 = Melting point (K); 4098 = Boiling point (K); C carbon

Legend: Gases; Liquids; Synthetics; metals (main group); metals (transition); metals (inner transition); metalloids; nonmetals

Cell format: atomic number, symbol, name, average atomic mass; electronegativity; first ionization energy; melting point; boiling point; ion charges

MAIN-GROUP ELEMENTS | TRANSITION ELEMENTS | MAIN-GROUP ELEMENTS

Period	1 (IA)	2 (IIA)	3 (IIIB)	4 (IVB)	5 (VB)	6 (VIB)	7 (VIIB)	8	9 (VIIIB)	10	11 (IB)	12 (IIB)	13 (IIIA)	14 (IVA)	15 (VA)	16 (VIA)	17 (VIIA)	18 (VIIIA)
1	1 H hydrogen 1.01; 2.20; 1312; 13.81; 20.28; 1+, 1-																	2 He helium 4.00; –; 2372; 5.19; 5.02
2	3 Li lithium 6.94; 0.98; 520; 453.7; 1615; 1+	4 Be beryllium 9.01; 1.57; 899; 1560; 2744; 2+											5 B boron 10.81; 2.04; 800; 2348; 4273	6 C carbon 12.01; 2.55; 1086; 4765; 4098	7 N nitrogen 14.01; 3.04; 1402; 63.15; 77.36; 3-	8 O oxygen 16.00; 3.44; 1314; 54.36; 90.2; 2-	9 F fluorine 19.00; 3.98; 1681; 53.48; 84.88; 1-	10 Ne neon 20.18; 2080; 24.56; 27.07
3	11 Na sodium 22.99; 0.93; 496; 371; 1156; 1+	12 Mg magnesium 24.31; 1.31; 738; 923.2; 1363; 2+											13 Al aluminum 26.98; 1.61; 577; 933.5; 2792	14 Si silicon 28.09; 1.90; 786; 1687; 3538	15 P phosphorus 30.97; 2.19; 1012; 317.3; 553.7	16 S sulfur 32.07; 2.58; 999; 392.8; 717.8; 2-	17 Cl chlorine 35.45; 3.16; 1256; 171.7; 239.1; 1-	18 Ar argon 39.95; 1520; 83.8; 87.3
4	19 K potassium 39.10; 0.82; 419; 336.7; 1032; 1+	20 Ca calcium 40.08; 1.00; 590; 1115; 1757; 2+	21 Sc scandium 44.96; 1.36; 631; 1814; 3109; 3+	22 Ti titanium 47.87; 1.54; 658; 1941; 3560; 4+, 3+	23 V vanadium 50.94; 1.63; 650; 2183; 3680; 5+, 2+, 3+, 4+	24 Cr chromium 52.00; 1.66; 653; 2180; 2944; 3+, 2+, 6+	25 Mn manganese 54.94; 1.55; 717; 1519; 2334; 4+, 2+, 3+, 7+	26 Fe iron 55.85; 1.83; 759; 1811; 3134; 3+, 2+	27 Co cobalt 58.93; 1.88; 760; 1768; 3200; 2+, 3+	28 Ni nickel 58.69; 1.91; 737; 1728; 3186; 2+, 3+	29 Cu copper 63.55; 1.90; 745; 1358; 2835; 2+, 1+	30 Zn zinc 65.39; 1.65; 906; 692.7; 1180; 2+	31 Ga gallium 69.72; 1.81; 579; 302.9; 2477; 3+	32 Ge germanium 72.61; 2.01; 761; 1211; 3106	33 As arsenic 74.92; 2.18; 947; 1090; 876.2	34 Se selenium 78.96; 2.55; 941; 493.7; 958.2; 2-	35 Br bromine 79.90; 2.96; 1143; 266; 332; 1-	36 Kr krypton 83.80; 1351; 115.8; 119.9
5	37 Rb rubidium 85.47; 0.82; 403; 312.5; 941.2; 1+	38 Sr strontium 87.62; 0.95; 549; 1050; 1655; 2+	39 Y yttrium 88.91; 1.22; 616; 1795; 3618; 3+	40 Zr zirconium 91.22; 1.33; 660; 2128; 4682; 4+	41 Nb niobium 92.91; 1.6; 664; 2750; 5017; 5+, 3+	42 Mo molybdenum 95.94; 2.16; 685; 2896; 4912; 6+	43 Tc technetium (98); 2.10; 702; 2430; 4538; 7+, 4+, 6+	44 Ru ruthenium 101.07; 2.2; 711; 2607; 4423; 3+	45 Rh rhodium 102.91; 2.28; 720; 2237; 3968; 3+	46 Pd palladium 106.42; 2.20; 805; 1828; 3236; 2+, 3+	47 Ag silver 107.87; 1.93; 731; 1235; 2435; 1+	48 Cd cadmium 112.41; 1.69; 868; 594.2; 1040; 2+	49 In indium 114.82; 1.78; 558; 429.8; 3345; 3+	50 Sn tin 118.71; 1.96; 708; 505; 2875; 4+, 2+	51 Sb antimony 121.76; 2.05; 834; 903.8; 1860	52 Te tellurium 127.60; 2.1; 869; 722.7; 1261	53 I iodine 126.90; 2.66; 1009; 386.9; 457.4; 1-	54 Xe xenon 131.29; 1170; 161.4; 165
6	55 Cs cesium 132.91; 0.79; 376; 301.7; 944; 1+	56 Ba barium 137.33; 0.89; 503; 1000; 2170; 2+	57 La lanthanum 138.91; 1.10; 538; 1191; 3737; 3+	72 Hf hafnium 178.49; 1.3; 642; 2506; 4876; 4+	73 Ta tantalum 180.95; 1.5; 761; 3290; 5731; 5+	74 W tungsten 183.84; 1.7; 770; 3695; 5828; 6+	75 Re rhenium 186.21; 1.9; 760; 3459; 5869; 4+, 6+, 7+	76 Os osmium 190.23; 2.2; 840; 3306; 5285; 4+, 3+	77 Ir iridium 192.22; 2.2; 880; 2719; 4701; 4+, 3+	78 Pt platinum 195.08; 2.2; 870; 2042; 4098; 4+, 2+	79 Au gold 196.97; 2.4; 890; 1337; 3129; 3+, 1+	80 Hg mercury 200.59; 1.9; 1107; 234.3; 629.9; 2+, 1+	81 Tl thallium 204.38; 1.8; 589; 577.2; 1746; 1+, 3+	82 Pb lead 207.20; 1.8; 715; 600.6; 2022; 2+, 4+	83 Bi bismuth 208.98; 1.9; 703; 544.6; 1837; 3+, 5+	84 Po polonium (209); 2.0; 813; 527.2; 1235; 4+, 2+	85 At astatine (210); 2.2; (926); 575; 1-	86 Rn radon (222); 1037; 202.2; 211.5
7	87 Fr francium (223); 0.7; ~375; 300.2; 1+	88 Ra radium (226); 0.9; 509; 973.2; 2+	89 Ac actinium (227); 1.1; 499; 1324; 3471; 3+	104 Rf rutherfordium (261); 4+	105 Db dubnium (262)	106 Sg seaborgium (266)	107 Bh bohrium (264)	108 Hs hassium (269)	109 Mt meitnerium (268)	110 Uun ununnilium (271)	111 Uuu unununium (272)	112 Uub ununbium (277)	113	114 Uuq ununquadium (285)	115	116 Uuh ununhexium (289)		

INNER TRANSITION ELEMENTS

Period															
6	Lanthanoids	58 Ce cerium 140.12; 1.12; 527; 1071; 3716; 3+, 4+	59 Pr praseodymium 140.91; 1.13; 523; 1204; 3793; 3+	60 Nd neodymium 144.24; 1.14; 530; 1294; 3347; 3+	61 Pm promethium (145); 536; 1315; 3273; 3+	62 Sm samarium 150.36; 1.17; 543; 1347; 2067; 3+, 2+	63 Eu europium 151.96; 547; 1095; 1802; 3+, 2+	64 Gd gadolinium 157.25; 1.20; 593; 1586; 3546; 3+	65 Tb terbium 158.93; 565; 1629; 3503; 3+	66 Dy dysprosium 162.50; 1.22; 572; 1685; 2840; 3+	67 Ho holmium 164.93; 1.23; 581; 1747; 2973; 3+	68 Er erbium 167.26; 1.24; 589; 1802; 3141; 3+	69 Tm thulium 168.93; 1.25; 597; 1818; 2223; 3+	70 Yb ytterbium 173.04; 603; 1092; 1469; 3+, 2+	71 Lu lutetium 174.97; 1.0; 524; 1936; 3675; 3+
7	Actinoids	90 Th thorium 232.04; 1.3; 587; 2023; 5061; 4+	91 Pa protactinium 231.04; 1.5; 568; 1845; 5+, 4+	92 U uranium 238.03; 1.7; 584; 1408; 4404; 6+, 3+, 4+, 5+	93 Np neptunium 237.05; 1.3; 597; 917; 5+, 3+, 4+, 6+	94 Pu plutonium (244); 1.3; 585; 913.2; 3501; 4+, 3+, 5+, 6+	95 Am americium (243); 578; 1449; 2284; 3+, 4+, 5+, 6+	96 Cm curium (247); 581; 1618; 3373; 3+	97 Bk berkelium (247); 601; 1323; 3+, 4+	98 Cf californium (251); 608; 1173; 3+	99 Es einsteinium (252); 619; 1133; 3+	100 Fm fermium (257); 627; 1800; 3+	101 Md mendelevium (258); 635; 1100; 3+, 2+	102 No nobelium (259); 642; 1100; 3+, 2+	103 Lr lawrencium (262); 1900; 3+

*Average atomic mass data in brackets indicate atomic mass of most stable isotope of the element.

periods

(s)

(d) but is one less than # at side

(p)

(f) but is two less than # at side.